MW01626787

Searching for ORDER IN the COMPLEXITY of Evolving Worlds ·

ACKNOWLEDGMENTS

The SFI Press is supported by William H. Miller and the Miller Omega Program.

◊ ◊ ◊

To produce a mighty book, you must choose a mighty theme. No great and enduring volume can ever be written on the flea, though many there be who have tried it.

—HERMAN MELVILLE
Moby-Dick (1851)

These four volumes are a product of collective intelligence. They have come into existence through the coordinated insights of a global network of complexity scientists. We thank every one of them for their insights and efforts.

We thank our generous Board of Trustees, research foundations, and federal agencies for their support of science, reason, and debate.

We thank our colleagues Kate Joyce, Tim Taylor, Renée Tursi, Ellis Wylie, Katherine Mast, and Bronwynn Woodsworth for reading, commenting, adding to, and improving on the project.

We dedicate these four volumes to the friends and colleagues we have lost during their making: Phil Anderson, Dan Dennett, Herb Gintis, James Hartle, Erica Jen, Richard Lewontin, Dan Lynch, Robert May, Cormac McCarthy, David Padwa, James Pelkey, William Sick, Chuck Stevens, and Douglas White.

David C. Krakauer
Laura Egley Taylor
Sienna Latham
Zato Hebbert

FOUNDATIONAL PAPERS IN COMPLEXITY SCIENCE

Volume One

1922–1962

DAVID C. KRAKAUER

editor

1399 Hyde Park Road
Santa Fe, New Mexico 87501

Foundational Papers in Complexity Science, Vol. 1
ISBN (PAPERBACK): 978-1-947864-56-6
Library of Congress Control Number: 2024938011

The SFI Press is generously supported by
the Miller Omega Program.

WE NATURALLY THINK that we can more easily reach the center of things than embrace their circumference. The visible bulk of the world visibly exceeds us, but as we exceed little things, we think ourselves more capable of possessing them. Yet we need no less capacity to attain the nothing than the whole.

—BLAISE PASCAL
Pensées, "Man's Disproportion" (1885)

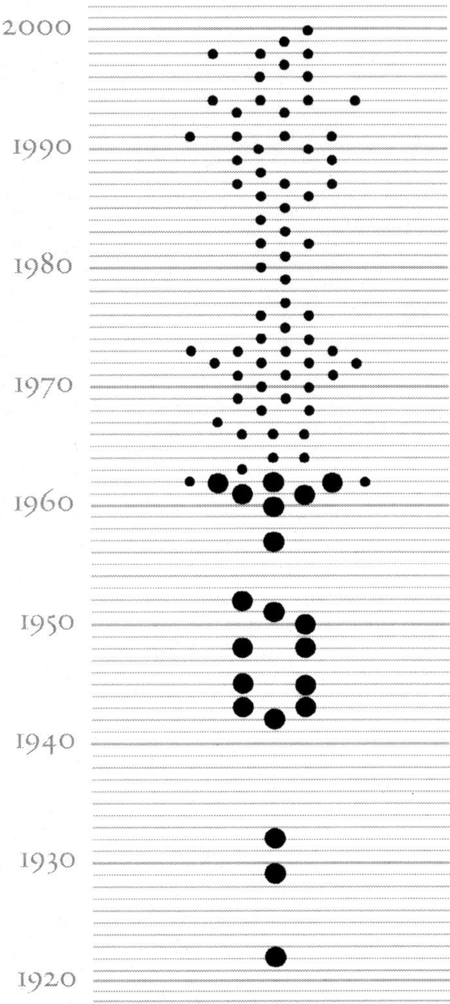

THE PAPERS

Large dots represent papers included in this volume; small dots are the remaining papers in the collection. Dots are positioned from top to bottom according to the year the paper was published. Volume I begins with Lotka (1922) and ends with Holland (1962).

TABLE OF CONTENTS

— Introduction —

David C. Krakauer

— Volume One —

Listed chronologically, with the introduction to each paper followed by the (☛) annotated paper

— FOUNDATIONAL PAPERS: Volume Two —

Available for purchase. Visit www.sfipress.org *to learn more.*

21: *Simon Levin*
☛ H. A. Simon, "The Architecture of Complexity" (1962)

22: *Erica Jen*
☛ S. Ulam, "On Some Mathematical Problems Connected with Patterns of Growth in Figures" (1962)

23: *J. Doyne Farmer*
☛ E. N. Lorenz,"Deterministic Nonperiodic Flow" (1963)

24: *Cristopher Moore*
☛ A. Cobham, "The Intrinsic Computational Difficulty of Functions" (1964)

25: *Paul M.B. Vitányi*
☛ R. J. Solomonoff, "A Formal Theory of Inductive Inference, Part 1" (1964)

26: *Simon DeDeo*
☛ G. J. Chaitin, "On the Length of Programs for Computing Finite Binary Sequences" (1966)

27: *Ricard Solé*
☛ D. M. Raup, "Geometric Analysis of Shell Coiling; General Problems" (1966)

28: *Neil Gershenfeld*
☛ J. von Neumann, "Theory of Self-Reproducing Automata" (1966)

29: *Geoffrey B. West*
☛ B. B. Mandelbrot, "How Long is the Coast of Britain? Statistical Self-Similarity and Fractional Dimension" (1967)

30: *Carl Bergstrom and Michael Lachmann*
☛ M. Kimura, "Evolutionary Rate at the Molecular Level" (1968)

31: *Simon DeDeo*
☛ A. N. Kolmogorov, "Three Approaches to the Quantitative Definition of Information" (1968)

32: *Dan Schrag*
☛ M. I. Budyko, "The Effect of Solar Radiation Variations on the Climate of the Earth" (1969)

33: *Sanjay Jain*
☛ S. Kauffman, "Metabolic Stability and Epigenesis in Randomly Constructed Genetic Nets" (1969)

— FOUNDATIONAL PAPERS: Volume Three —

— FOUNDATIONAL PAPERS: **Volume Four** —

Available for purchase. Visit www.sfipress.org *to learn more.*

67: *Scott Page*

☛ W. B. Arthur, "Competing Technologies, Increasing Returns, and Lock-In by Historical Events" (1989)

68: *Rob de Boer*

☛ A. S. Perelson, "Immune Network Theory" (1989)

69: *Sidney Redner*

☛ J. D. Farmer, "A Rosetta Stone for Connectionism" (1990)

70: *Jessica C. Flack*

☛ J. A. Wheeler, "Information, Physics, Quantum: The Search for Links" (1990)

71: *Vijay Balasubramanian*

☛ W. Bialek, F. Rieke, R. R. de Ruyter van Steveninck, and D. Warland, "Reading a Neural Code" (1991)

72: *Richard Bookstaber*

☛ J. H. Holland and J. H. Miller, "Artificial Adaptive Agents in Economic Theory" (1991)

73: *W. Brian Arthur*

☛ K. Lindgren, "Evolutionary Phenomena in Simple Dynamics" (1991)

74: *Suresh Naidu*

☛ H. A. Simon, "Organizations and Markets" (1991)

75: *Mirta Galesic*

☛ J. S. Lansing and J. M. Kremer, "Emergent Properties of Balinese Water Temple Networks: Coadaptation on a Rugged Fitness Landscape" (1993)

76: *David Ackley*

☛ M. Mitchell, P. T. Hraber, and J. P. Crutchfield, "Revisiting the Edge of Chaos: Evolving Cellular Automata to Perform Computations" (1993)

77: *Willemien Kets*

☛ W. B. Arthur, "Inductive Reasoning and Bounded Rationality" (1994)

78: *Peter Sloot*

☛ J. P. Crutchfield, "The Calculi of Emergence: Computation, Dynamics, and Induction " (1994)

79: *Anil Somayaji*

☛ S. Forrest, A.S. Perelson, L. Allen, and R. Cherukuri, "Self–Nonself Discrimination in a Computer" (1994)

80: *Evandro Ferrada*

☛ P. Schuster, W. Fontana, P.F. Stadler, and I.L. Hofacker, "From Sequences to Shapes and Back: A Case Study in RNA Secondary Structures" (1994)

81: *Miguel Fuentes*

☛ M. Gell-Mann and S. Lloyd, "Information Measures, Effective Complexity, and Total Information" (1996)

82: *David C. Krakauer*

☛ F.J. Odling-Smee, K.N. Laland, and M.W. Feldman, "Niche Construction" (1996)

83: *Pablo Marquet*

☛ G.B. West, J.H. Brown, and B.J. Enquist, "A General Model for the Origin of Allometric Scaling Laws in Biology" (1997)

84: *Nihat Ay*

☛ S. Amari, "Natural Gradient Works Efficiently in Learning" (1998)

85: *Rajiv Sethi*

☛ S. Bowles, "Endogenous Preferences: The Cultural Consequences of Markets and Other Economic Institutions " (1998)

86: *Michelle Girvan*

☛ D. Watts and S. Strogatz, "Collective Dynamics of 'Small-World' Networks" (1998)

87: *Chris Kempes*

☛ R.B. Laughlin, D. Pines, J. Schmalian, B.P. Stojkovic, and P. Wolynes, "The Middle Way" (1999)

88: *Marty Anderies*

☛ E. Ostrom, "Collective Action and the Evolution of Social Norms" (2000)

VOLUME 1 CONTRIBUTORS

Robert L. Axtell *George Mason University; Santa Fe Institute*

Luís M.A. Bettencourt *University of Chicago; Santa Fe Institute*

Samuel Bowles *Santa Fe Institute*

Daniel Dennett *Tufts University; Santa Fe Institute; New College of the Humanities, London*

Walter Fontana *Harvard University*

Sergey Gavrilets *University of Tennessee*

John Geanakoplos *Yale University; Santa Fe Institute*

Christopher Hillar *Redwood Center for Theoretical Neuroscience*

Dawn E. Holmes *University of California Santa Barbara*

Erica Jen *Santa Fe Institute*

Stuart Kauffman *University of Pennsylvania*

David C. Krakauer *Santa Fe Institute*

Simon Levin *Princeton University*

Seth Lloyd *Massachusetts Institute of Technology*

John H. Miller *Carnegie Mellon University; Santa Fe Institute*

Melanie Mitchell *Santa Fe Institute*

Bruno Olshausen *Redwood Center for Theoretical Neuroscience, University of California, Berkeley; Santa Fe Institute*

Karen Page *University College London*

Andrew Pickering *University of Exeter*

Maxim Raginsky *University of Illinois, Urbana-Champaign*

Cosma Rohilla Shalizi *Carnegie Mellon University; Santa Fe Institute*

Susanne Still *University of Hawai'i*

David H. Wolpert *Santa Fe Institute*

HOW TO CITE

When citing *Foundational Papers of Complexity Science* **in full**, please use the following approach:

BIBLIOGRAPHY:

Foundational Papers in Complexity Science. Edited by David C. Krakauer. 4 volumes. Santa Fe, NM: SFI Press, 2024.

LATEX:

```
@book{Krakauer_FP_2024,
       editor = {Krakauer, David C.},
       title = {Foundational Papers in Complexity Science},
       year = {2024},
       publisher = {SFI Press},
       location = {Santa Fe, NM}}
```

When citing a particular **volume** of *Foundational Papers in Complexity Science*, please use the following approach (*volume 1 serves as an example here*):

BIBLIOGRAPHY:

Foundational Papers in Complexity Science. Edited by David C. Krakauer. Volume 1. Santa Fe, NM: SFI Press, 2024.

LATEX:

```
@book{Krakauer_FP_1_2024,
       editor = {Krakauer, David C.},
       title = {Foundational Papers in Complexity Science},
       volume = {1},
       year = {2024},
       publisher = {SFI Press},
       location = {Santa Fe, NM}}
```

When citing a specific **chapter** from this project, please treat the introduction and the annotated paper as a single unit, as follows:

BIBLIOGRAPHY:

Bettencourt, Luís M.A. "Maximum Power as a Physical Principle of Evolution." In *Foundational Papers in Complexity Science.* Edited by David C. Krakauer. Volume 1. Santa Fe, NM: SFI Press, 2024, 1–15.

LATEX:

```
@incollection{Bettencourt_Lotka_2024,
      author = {Bettencourt, Luís M. A.},
      title = {Maximum Power as a Physical Principle of Evolution},
      year = {2024},
      editor = {Krakauer, David C.},
      booktitle = {Foundational Papers in Complexity Science},
      volume = {1},
      publisher = {SFI Press},
      location = {Santa Fe, NM}
      pages = {1–15}}
```

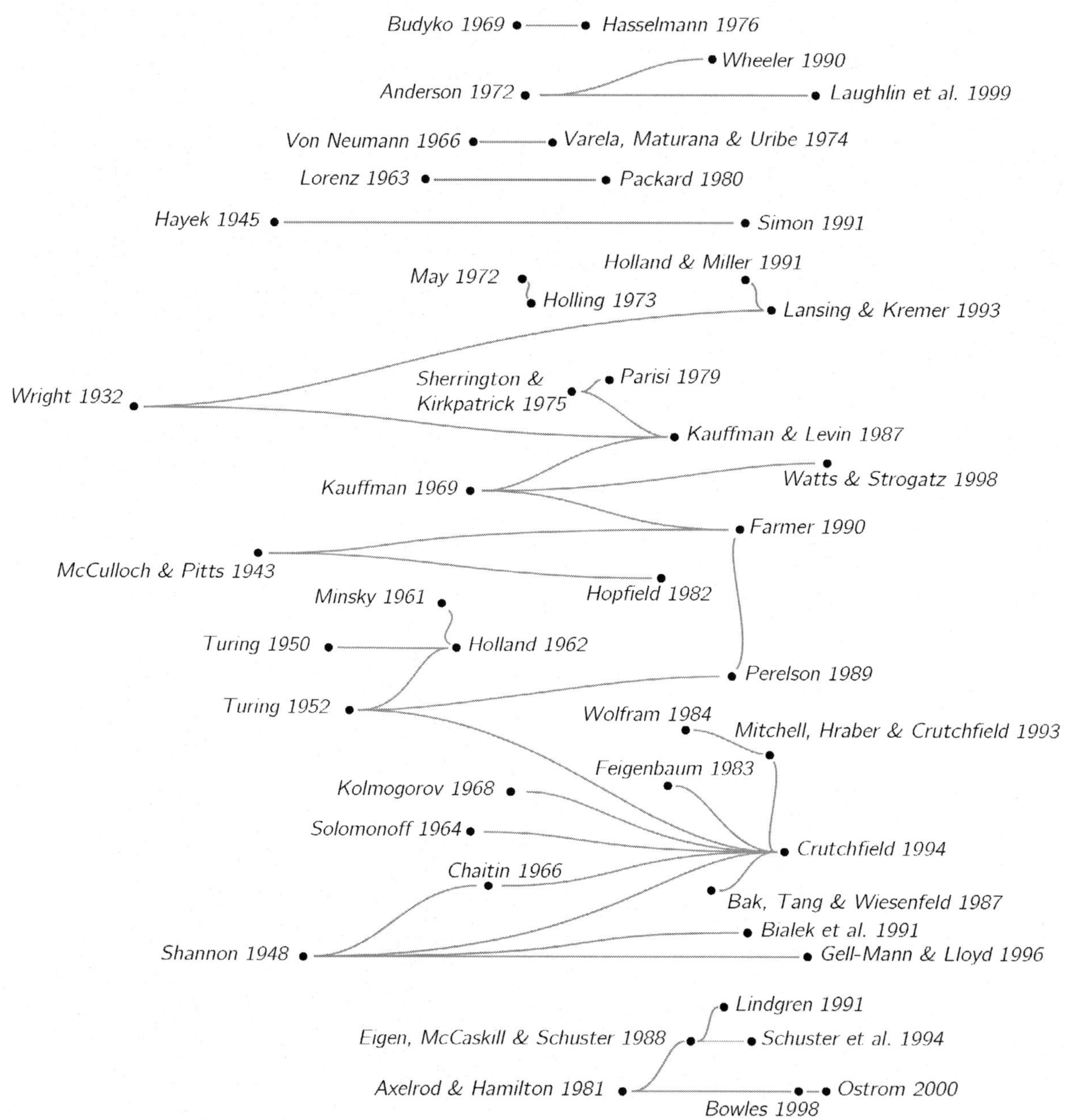

FOUNDATIONAL PAPERS: CITATION CONNECTIVITY

This visualization illustrates citation connections between foundational papers. Only papers that cite or are cited by other foundational papers are included. The papers are arranged left to right according to publication date. The gray connection lines signify that the later paper cites the earlier. *Examples*: Hasselmann 1976 cites Budyko 1969. Both Wheeler 1989 and Laughlin et al. 1999 reference Anderson 1972.

RATIONALE

1. *Foundational Papers in Complexity Science* is a project to discern the unity of an evolving inquiry.
2. After polling members of the extended network of Santa Fe Institute complexity researchers, we selected 88 papers spanning just under a century that chart the formation of the field.
3. These papers—some classics, others cultish—collectively investigate the principles governing open, out-of-equilibrium systems that are self-organizing or selected, in the natural and cultural world.
4. The papers are ordered chronologically to establish patterns of influence and an emerging consensus.
5. Each paper has a unique introduction by a complexity researcher placing its ideas in historical context, and highlighting its perdurable contributions, and the new ideas that it has spawned.
6. Each paper is annotated by a researcher to underscore points where critical insights are made.
7. The year 2000 was established as the cut-off year for the *Foundational Papers*, recognizing that not enough time has elapsed to label subsequent work foundational.
8. These papers are being made available as a print-only four-volume set. The decision in favor of print is based on the prohibitive cost of licensing these papers as online materials.
9. A great deal of effort has gone into making these four volumes as beautiful and practical as possible in order to engage the senses and the mind.
10. *Searching for Order in the Complexity of Evolving Worlds*

—DAVID C. KRAKAUER AND THE SFI PRESS TEAM

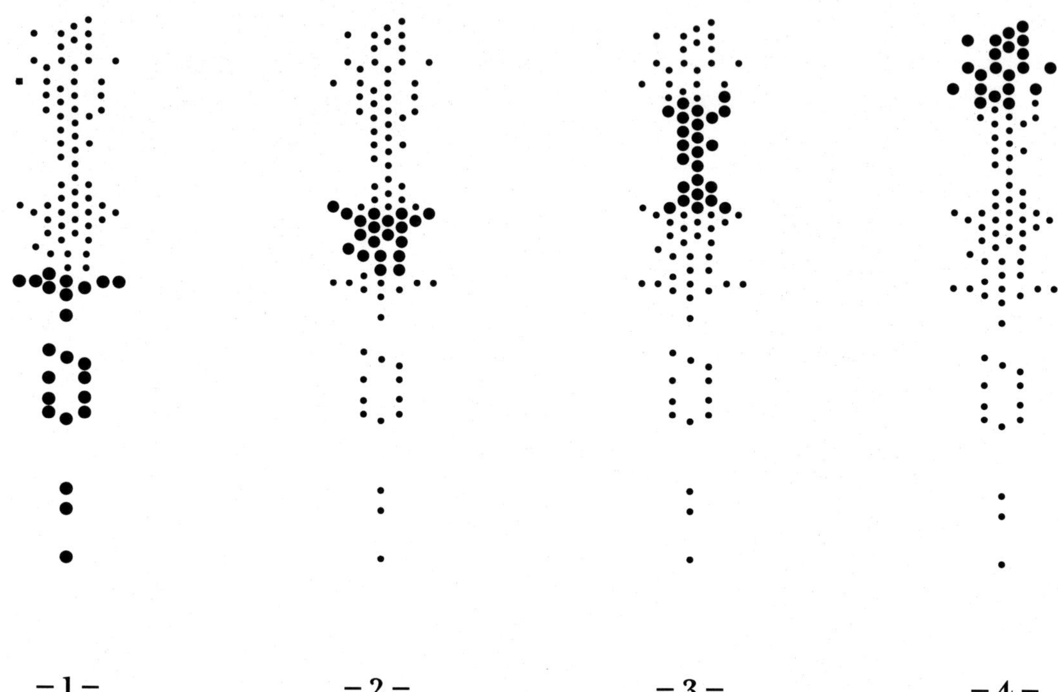

THE FOUR VOLUMES AND A GRAPHICAL REPRESENTATION OF THEIR CONTENT.

THE COMPLEX WORLD: An Introduction to the Foundations of Complexity Science

David C. Krakauer, Santa Fe Institute

Two World Systems

The scientific and social implications of differences between **(A)** closed, reversible, symmetry-dominated, and predictable classical domains and **(B)** open, self-organizing, dissipative, uncertain, and adaptive domains are the subject of this book.[1]

[1]**OPEN** implies positioned within a chemical or electrochemical gradient; **SELF-ORGANIZING** refers to rules amplifying fluctuations into coordinated space-time patterns; **DISSIPATIVE** refers to dynamics that are irreversible and that break time-reversal symmetry; **UNCERTAIN** describes an open-ended space of possible states not enumerable in initial conditions; and **ADAPTIVE** describes the capture of information by a system from the environment that promotes persistence and increases multiplicity.

At one limit—**A**—are fundamental regularities described using minimal assumptions, simple rule system, and few initial conditions. At the other—**B**—emergent regularities constructed from contingent histories, described with coarse-grained rules and nested boundary conditions.

Between **A** and **B** there is an uneven spectrum, shaped like a dumbbell with equilibrium structures at one end, non-equilibrium forms of self-organization in the middle, and fully adaptive self-synthesizing organizations with long evolutionary histories at the other.

Analyzing the connections between the simple **A** and the complex **B** requires much more than a powerful measurement device. Here there exist differences that can only be resolved by principles, models, and theories. Interestingly, the more powerful the device—the more finely grained the measurement—the less easily **B** can be distinguished from **A**. Hence reductionism in the units of analysis not only *fails to explain complexity; it fails to detect it*.

It is not possible to describe differences between **A** and **B** in terms of the fundamental laws of physics and chemistry. Both obey these laws. There is no new physics in a replicating virus not already found in a crystallizing mineral. Indeed, viruses exploit properties of physics far-from-equilibrium to self-assemble within the cell. Human brains are no more or less concentrations of particles than cannon balls: both depend on covalent chemistry, and both respect the law of gravity. Yet meaningful theories of brains bear no resemblance to fundamental theory in physics.

And the complex world defies many of our best normative intentions. By working along the **A–B** spectrum, engineers have built ingenious electromagnetic communications networks that coordinate to span the globe. But when connecting diverse communities of different basic beliefs and values, the challenging dynamics of **B** often lead to outcomes described as unstable and iniquitous.

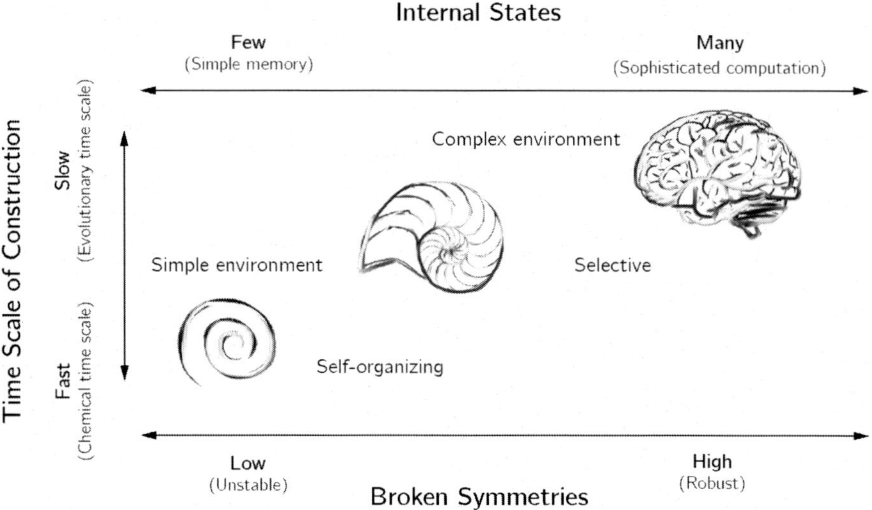

Figure 1. The complex domain: spanning rapidly established self-organizing patterns with few degrees of freedom and slowly evolving lineages with robust information storage. All complex systems emerge first from spontaneous broken symmetries. Simple emergent patterns remain sensitive to changes in boundary conditions. Evolved lineages accumulate a considerable number of metabolically protected "frozen accidents"—intergenerational information-bearing degrees of freedom. These are used to encode adaptive schema (computations) thereby achieving a high degree of autonomy from environmental fluctuations.

Prequel

In 1632 Galileo Galilei wrote of two world systems: the Ptolemaic (geocentric) and the Copernican (heliocentric) (Galilei 2001). Through a series of dialogues, one gradually comes to understand the magnitude of the disruption to metaphysics and empirical knowledge the replacement of Ptolemy will entail. Galileo, played by his alter ego Salviati, nevertheless perseveres in the face of opposition from Simplicio (Galileo 1632, 371):

> SIMPLICIO. *The first and greatest difficulty is the repugnance and incompatibility between being at the center and being distant from it. For if the terrestrial globe must move in a year around the circumference of a circle—that is, around the zodiac—it is impossible for it at the same time to be in the center of the zodiac. But the earth is at that center, as is proved in many ways by Aristotle, Ptolemy, and others.*
>
> SALVIATI. *Very well argued. There can be no doubt that anyone who wants to have the Earth move along the circumference of a circle must first prove that it is not at the center of that circle. The next thing is for us to see whether the earth is or is not at that center around which I say it turns, and in which you say it is situated. And prior to this, it is necessary that we declare ourselves as to whether or not you and I have the same concept of this center. Therefore tell me what and where this center is that you mean.*

Often painful, albeit amusing, Salviati faces extraordinary resistance. If Salviati is correct, it is not merely a matter of redrawing a few celestial atlases. The implication is the abandonment of scholastic authority (Heilbron 2010). It is not easy to relinquish a paradigm.

A similar argument could be made for two contemporary world systems: **A** of symmetry and law; and **B** of information and adaptation. For a wide range of phenomena, it has not always seemed obvious which world system applies, and various attempts have been made to squeeze biology, medicine, economics, sociology, and political economy into **A**. Only in recent decades have these migrated towards a more natural home in **B**.

Reports from Two Worlds

Based on profound ideas from condensed matter, in conjunction with advances in the design of semiconductor fabrication plants, engineers are capable of building and controlling three nanometer (3nm) integrated circuits. This exploits the evolution of complementary metal–oxide–semiconductors (CMOS), roughly following Moore's Law, leading to miniaturized transistor gate lengths of 3nm (around 250 million transistors per square nanometer) and below. Miniaturization has been accompanied by energy savings (lower supply voltages) and faster switching speeds (Jin 2023).

Viruses span a comparable range of scales from 20 to 200nm in diameter. This range accommodates genomes from several thousand to millions of base pairs (Holland 1998). The coronaviruses range from 50nm to 140nm (Holmes 1999). COVID-19 is about 100nm, the size of a transistor gate from 2000. COVID-19 encodes around fifty distinct proteins, each of which performs a rather specific function in the completion of the virus life cycle. The logic of gene regulation in COVID-19 does not come close to the diversity of operations that can be performed by a modern, vastly smaller, *integrated circuit* (**IC**).

How might we explain the discrepancy of control and understanding we possess over these two machines of comparable scale? It is not a question of numbers of components, or their complications and interactions, total functional repertoire, inorganic chemistry or biochemistry, or energetics. Something else is going on that makes us masters of miniature artifacts and victims of miniature organisms.

The key to understanding a virus is that it has evolved in order to achieve the unrelenting objective of replication in a variety of host cells. The virus is an agent with an evolution-ordained function. All practical differences between minerals, machines, and microbes follows from this fact of natural history.

A virus, a cell, an organism, an ecosystem, and a society all evolve—both organically and culturally—and exploit the out-of-equilibrium affordances of active matter. Because they evolve, they use sources of metabolic free energy efficiently to adapt. Through adaptation, the basic rules describing their properties and behavior are constantly changing, and doing so in response to stochastic processes and parameters in the environment. Unlike any purely physico-chemical system, the logic of their operation changes, and in such a way as to encode and anticipate salient features of the volatile world around them. There are laws of physics and there are rules of life—and rules are meant to be broken.

A virus is not at all like a modern integrated circuit; it is, rather, like an old integrated circuit in an organic computer (cell and organism), surrounded by a large team of programmers, all of whom respond to signals in the world (natural selection) so as to maximize viral growth rates. And the same basic logic can be applied to many complex systems.

The Second World System

Over the last century countless researchers have sought to capture the essence of complexity in ideas as wide-ranging as "self-organizing systems," "voluntary activity," "cybernetic control," "goal-seeking," "self-reproducing," "representational," "schematic," "autopoietic," "cognitive," and "information gathering and utilizing system" (IGUS).

See table 2 on pages 38–39.

Each of these ideas, and many more like them, attempts to identify the key character of mechanisms that promote persistent information-rich couplings, or long histories of strategic interaction. All slowly build upon simple out-of-equilibrium structures observed in nonlinear regimes.

Complexity science is also concerned with how life can be different from physics and chemistry but dependent on them: how to move **A** into **B** (**A→B**). One might just as well say how society comes to be different from brains and minds but remains dependent on them. The complex domain encompasses the world of self-organizing and evolving agents at all scales, from organisms to whole societies, and, increasingly, software and machines.

Complexity science is one of the most radical new scientific paradigms of the twentieth century. It includes paradoxes and challenges as puzzling as those in quantum mechanics and general relativity. Many of these puzzles are connected to the idea of emergence, including the origin of life from small molecule abiotic chemistry (the dynamics of **A→B**); how organisms coordinate into functional collectives; how "free will" (or the illusion thereof) might emerge from the biochemistry of brains; and how societies and their laws emerge from collectives of semi-autonomous agents.

Beyond these fundamental scientific difficulties there are enormous practical challenges. The application of complexity scholarship to global commons problems, including disease, climate, conflict, and political economy, is likely to be an essential component in any effort to ensure the prosperity and survival of life on Earth (Levin 1999; West 2018).

On the Origin of Paradigms

Where do new fields or disciplines come from? How do we establish when a series of inventions and discoveries warrants an entirely new paradigm, or whether existing ones might be modified to accommodate novelty? For example, at what limit of observation was it determined that physics was insufficient and that we needed to maintain the separate disciplines of chemistry and biology? Or why are there English, Italian, and other language departments and not just one monolithic linguistics department? How do we determine the variety and resolution of our epistemological commitments?

We should like to know what contributed to the origin and introgression of complexity science over the last century. And what new ideas, or new combinations of ideas, beyond those of physics, chemistry, economics, and evolution needed to be introduced and thereafter reconciled? Three ideas from philosophy help to flesh out the idea of a paradigm: language games, hermeneutic circles, and disciplinary matrices.

The philosopher Ludwig Wittgenstein suggested that these kinds of questions might be answered through the identification of new rule systems, or *language games*, as he called them (Ahmed 2010). A new language game is made up from a set of rules that are substantially different from, perhaps even mutually unintelligible to, those that came before. Different games have different features, and, as a practical matter, it is not always easy to determine by what sequence of behaviors we even identify the transition to a new game.

> *Compare chess with noughts and crosses. Or is there always winning and losing, or competition between players? Think of patience. In ball games there is winning and losing; but when a child throws his ball at the wall and catches it again, this feature has disappeared. Look at the parts played by skill and luck; and at the difference between skill in chess and skill in tennis. Think now of games like ring-a-ring-a-roses; here is the element of amusement, but how many other characteristic features have disappeared! And we can go through the many, many other groups of games in the same way; can see how similarities crop up and disappear* (Wittgenstein 1953, §66).

Scientific research is not merely a process of fitting new data into old models or into constellations of models comprising larger theories. New data often break the old models and theories. One cannot use the rules of chess to analyze a game of Go, nor can we use particle physics to understand genetics. Wittgenstein thought of games as the supreme metaphor for the specificity and context-dependence of formal thought: new possibilities need to be explained with new rules. And, we might add, in what "language" are these new rules written—natural language, mathematics, computer code?

In 1900 Wilhelm Dilthey described the experience of understanding a new phenomenon through a strict context-dependence in terms of a *hermeneutic circle* (Palmer 1969):

> *It is at this point that the central difficulty of all exegetical practice makes itself felt. The whole of a work is to be understood from the individual words and their connections with each other, and yet the full comprehension of the individual part already presupposes comprehension of the whole* (Dilthey and Jameson 1972, 243).

Dilthey was in effect pointing out a "chicken-and-egg" problem for understanding: parts are required to comprehend the whole but the whole is needed to make sense of the parts. This seems to be particularly apropos of complexity, which, by some definitions, is the science associated with the maxim "the whole is greater than the sum of its parts."

Thomas Kuhn (2012) expanded on Dilthey's hermeneutics and Wittgenstein's language games to explain how scientific ideas evolve through the persistence and overthrow of *paradigms* constituted by a *disciplinary matrix* describing how ideas relate to each other. The matrix describes how ideas are mutually dependent, and how some might be reduced into elementary facts and others aggregated into synthetic propositions. It is a defining feature of paradigms (like language games and hermeneutic circles) that they are incommensurable, or discordant, with one another.

> *For present purposes I suggest "disciplinary matrix": "disciplinary" because it refers to the common possession of the practitioners of a particular discipline; "matrix" because it is composed of ordered elements of various sorts, each requiring further specification.* (Kuhn 2012, 181)

> *Though the strength of group commitment varies, with non-trivial consequences, along the spectrum from heuristic to onto-logical models, all models have similar functions. Among other things they supply the group with preferred or permissible analogies and metaphors. By doing so they help to determine what will be accepted as an explanation and as a puzzle- solution; conversely, they assist in the determination of the*

> *roster of unsolved puzzles and in the evaluation of the importance of each* (Kuhn 2012, 183).

Disciplinary matrices might be likened to a mechanical diagram of an automobile illustrating how each part is connected, assembled, and made interoperable with other parts. Quantum mechanics is a paradigm, plate tectonics is a paradigm, organic chemistry is a paradigm, and neoclassical economics is a paradigm. None of these fields is defined by a single anomalous experiment, model, or idea, but by a matrix of compatible elements. When an incompatible observation or model is thrown into the mix, it threatens to demolish the matrix. This more often than not leads to the rejection of the novel element, or what the philosopher of science Gunther Stent called *prematurity*.

> "A discovery is premature if its implications cannot be connected by a series of simple logical steps to contemporary canonical [or generally accepted] knowledge."
>
> —GUNTHER STENT
> *Prematurity in Scientific Discovery: On Resistance and Neglect.*

Whether an idea is accepted or rejected depends to a large extent on how central it is to keeping a paradigm connected. A car might function without its heater, but it is worthless without an engine block. Adding a large battery to a combustion engine does not make sense.

And yet multiple paradigms can coexist as long as each provides instrumental value in its respective domain. We launch satellites with classical mechanics, navigate by satellite using relativistic mechanics, and exploit understanding of quantum mechanics to build semiconductor-based photovoltaics powering the satellite. Three paradigms encased in one celestial artifact. Incommensurability need not imply incompatibility. This is one of the explanations for pluralism in science: different paradigms play different roles and their practical value often outweighs their incommensurability.

Periods of "normal" science consist in modifying or adding to existing paradigms—adding rows and columns to the matrix. Periods of "revolutionary" science consist in rewiring concepts and forming novel communities, through this process generating, in Kuhn's language, a "plurality of worlds."

> "In science, as in the playing card experiment, novelty emerges only with difficulty, manifested by resistance, against a background provided by expectation."
>
> —THOMAS S. KUHN
> *The Structure of Scientific Revolutions*

From the perspective of rule systems and paradigms, disciplines like geology and anthropology are not simply two different views onto reality using the same underlying universal rationality. They are describing two different, albeit overlapping, sets of empirical principles, including different tools for investigating emergent worlds. Chemistry is not just physics-at-scale, but a set of mechanisms that requires distinct and extra-physical principles and models to be usable. Similarly, biology is not just chemistry-at-scale, and so forth through the spatial and temporal hierarchy.

The history of complexity science represents a revolutionary transformation in our way of understanding the world that forged four areas of research into a new way of seeing the world: a new world system. We might abbreviate these as: evolution, entropy, dynamics, and computation. In the process of connecting these areas, principles from each had to be reconciled, modified, and extended. This resulted in a greater understanding of the concenter of complex systems—*the theory of far-from-equilibrium, purposeful machines.*

Constructing the Complexity Paradigm

New paradigms can come into existence through a number of parallel processes, including the discovery or proposal of radically new phenomena (e.g., genetic dominance, quantum entanglement, localization of function in the brain), from the development of new methods and models for studying phenomena (e.g., population genetics, Bell's theorem, linguistics, connectionism), and from the unification of existing fields through the invention of laws or principles that reconfigure the boundaries of understanding (e.g. evolution, game theory)—rewiring the disciplinary matrix (see Kuhn 2000; Callebaut and Pinxten 2012; Brad Wray 2021; and Guerra, Capitelli, and Longo 2012 for review of this area).

Take molecular biology. Rosalind Franklin's use of X-ray crystallography to produce Photo 51 provided somewhat cryptic information about the structure of DNA (Pederson 2020). James Watson and Francis Crick used Franklin's insights as the basis for their inference of a double-helical molecule of inheritance. The structure of the molecule led more or less directly to a new mechanism of genetic replication. This mechanism could be made compatible with both meiosis and mitosis during cell division, and, more circuitously, it could provide a justification for the "central dogma" of molecular biology (the now-refuted one-directional flow of information from DNA to RNA to proteins) (Shapiro 2009). Hence a new technique revealed a new structure that in turn supported a novel suite of functions and thereupon a rather profound mechanistic refutation of Lamarckian inheritance (the genetic transmission of acquired characters).

Teleology refers to a goal-oriented explanation, interpreting structures or dynamics in terms of their functions. In biological contexts, teleology is a consequence of natural selection, and in cultural contexts a result of learning and planning. Typically teleology implies internal states supporting conditional branching—pursuing different behaviors as a result of contingent history and context.

The same logic could be unfolded for numerous fields, including Max Planck's study of black-body radiation, Vera Rubin's observations of galaxy rotation to infer the existence of dark matter, and Carl Woese's use of sequencing and phylogenetic techniques to identify a third branch of life using ribosomal RNAs.

For complexity science, the new phenomena relate principally to machines, organisms, and collectives with purpose: function or *teleology*. The new methods investigate the laws and rules that govern these machines, including their efficiency, predictability, control, and coordination. And unification has taken the form of frameworks that balance and connect the ideas of evolution, entropy-production, nonlinear dynamics, and computation. These concepts are typically expressed in terms of processes of origination, optimization, stability, robustness, resilience, energy dissipation, regulatory mechanics, informational constraints, and computational costs and limits. These elements of the complexity matrix are familiar to most researchers in the field despite working on very different problems.

One of several public-facing "manifestos" for this new integrated science, an effort to describe this new matrix, was Norbert Wiener's 1948 *Cybernetics: Or Control and Communication in the Animal and the Machine.* Written in eight chapters, starting with classical mechanics, and proceeding through statistical mechanics, communication, control theory, computation, and ending with models of society. Another was Claude Shannon and Warren Weaver's 1949 *The Mathematical Theory of Communication,* written in two long sections, the first laying out the mathematics of information theory as applied to machines, and the second, limits and possible extensions into a theory of semantics and behavior in living systems.

Complexity science seeks to understand an adaptive ontology—infiltrated by far-from-equilibrium, nonlinear, dissipative mechanisms, found in both nonliving and living systems. In the Kuhnian sense, the complexity paradigm is incommensurable with many that came before and yet remains compatible with them. In order to address the limitations of existing ideas, models, and frameworks, new effective laws have needed to be discovered. These account for the emergent properties that Philip Anderson, in his paradigm-supporting paper "More Is Different" (1972) sought to describe. New effective theories might not be fully computable, and many will be "inelegant" by the standards of fundamental theory. One could describe a bacterium like *Escherichia coli* as a point mass, but it would be pointless. Much that is of interest beyond the physico-chemistry related to the historical logic of adaptive mechanisms.

There has been need for new methods, system-level descriptions, and new technical insights to grapple with broken symmetry, non-ergodicity, dissipative dynamics, and their sequels. These are the crucial epistemological and auxiliary considerations attendant on building up the disciplinary matrix of complexity.

Reducing complexity science exclusively to methods diminishes its paradigmatic status. These have been techniques for studying out-of-equilibrium patterns or new quantitative techniques for finding statistical order in large datasets. At its worst, this tendency equates the many principles of complexity with the search for a universal metric that might apply to all observables: the complexity of a brain, societies, and technologies, all on one single axis of quantification. This is a case of techniques substituting for theories. It would be like trying to measure the degree of "chemistry" in all chemical reactions or the quotient of "anthropology" in a society.

There are schools of thought in which everything is mathematics, everything is physics, everything is poetry, and everything is religion. These kinds of framing of reality are lacking in richness, subtlety, modesty, and value. The last thing we want is for everything to be complexity. An important step towards this restraint is not to confuse the application of a method with a field of inquiry. The insights of Dilthey, Wittgenstein, Kuhn, and Anderson all help us to understand why this is the case. Complexity science should help us to understand why a plurality of paradigms is not only of utility but inevitable.

PREFOUNDATION

The development of complexity science was strongly influenced by the emerging industries, records of scientific exploration, and new theories produced in the age of steam and machines. If modern physics and chemistry have their roots in the scientific revolution of the seventeenth century, complexity science has its roots in the Industrial Revolution and its aftermath in the nineteenth century. This was the age of working matter: manufactured machines and evolved organisms. In the nineteenth century parallels were consciously sought that might connect manufactured technology with the mechanics of life, and entirely new vocabularies needed to be invented to describe this interface. Tom Peters (1996, 356), writing on the topic of machines and manufacturing in the nineteenth century, explains how these ideas were connected through technology:

> *The machine, war, and life are concepts that we can trace posteriori in older building processes. But their conscious application to construction dates to that period in which machine-making, war, and biology themselves evolved from arts to technologies.*

Charles Babbage (1832), a notable inventor and analyst of technology, who would exert a lasting influence on Karl Marx's theories of industry, wrote optimistically on the substitution of organic traits with mechanical prosthetics:

> *The hand of man is now too slow for the demands of his curiosity, but the power of steam comes to his assistance.* (Babbage 1832, 218)

The machine was self-evidently a purely physical system, entirely constructed, and yet described in a language different from physics. The requirements of mechanical work engendered a new vocabulary of tools and equipment emphasizing concepts particular to these artifacts. These include the ideas of cost, production, construction, efficiency, control, utility, durability, life-span, value, and market. Latently, each of these concepts rooted in technology would provide the seeds for the development of entirely new natural scientific and mathematical theories and models.

Milestones in the Prefoundations of Complexity

Searching for prequels to complexity science in the nineteenth century—the prefoundational period—it is most evident in four research areas. These include ideas relating to entropy by Sadi Carnot, Rudolf Clausius, James Clerk Maxwell, Josiah Willard Gibbs, and Ludwig Boltzmann; theories of evolution and adaptation in the work of Charles Darwin, Alfred Russel

Wallace, and Gregor Mendel; the mathematics of nonlinear dynamics and control in the work of Maxwell and Henri Poincaré; and the concept and machinery of difference and analytical engines and logic in Charles Babbage, Ada Lovelace, and George Boole.

All fields are porous, and several key figures from the eighteenth century—most notably Gottfried Leibniz (monadology and mechanical calculators), Alexander von Humboldt (biogeography and ecology), Thomas Malthus (population growth and collapse), and Adam Smith (invisible hand and division of labor)—might justly be teleported into the nineteenth century for purposes of extending the primary contributors to the new science; but the formal, or mathematical, exposition of key ideas is very much of the nineteenth century.

From a contemporary, disciplinary perspective, we might group these four areas into the separate histories of physics, chemistry, engineering, biology, mathematics, and computation. It is not clear that this is what these pioneers would have wanted. After all, Darwin and Wallace thought of selection in terms of dynamical regulation and physical laws, Babbage and Boole thought about natural law and mind in terms of calculation and dynamics, Gibbs thought of statistical physics in terms of probability theory, Boltzmann thought of evolution in terms of statistical physics, and along with Poincaré, interpreted mathematics in light of evolution. Nevertheless, complexity science proper is a twentieth-century program in the unification of natural and engineering sciences bearing directly on out-of-equilibrium and living phenomena, including a variety of extensions into society, technology, and history.

Four pillars—*entropy* (**E**), *evolution* (**V**), *dynamics* (**D**), and *computation* (**C**)—had been established in the nineteenth century, and much twentieth-century effort consists in integrating these four areas of research: (1) the principles of order and disorder in relation to the production of mechanical work; (2) the mathematics of the evolutionary and adaptive process; (3) the regulation, control and prediction of non-linear dynamical systems; and (4) the informational and computational requirements of purposeful or intelligent behavior.

The task has been not only to integrate these four, but, where possible, to discover general principles that might subsume them. These principles include understanding, stability, order, predictability, adaptability, function, aging, and collapse in far-from-equilibrium systems and "teleonomic" matter.

Table 1. A chronology of some of the major figures and milestones from the prefoundational period. This table is far from comprehensive yet represents a fairly uniform sample of ideas of incontestable importance in seeking to understand the roots of complexity. This is a necessary, albeit insufficient, nineteenth-century proto-complexity reading list.

Year	*Authors*	*Code Letters*	*Core Complexity Connections*	*Key Ideas in Paper*
1824	Sadi Carnot	E	Thermodynamics; statistical mechanics	The efficiency of heat engines; the Carnot cycle
1837	Charles Babbage	C	Computation; cognition	Analytical engine; mechanical general-purpose computer
1841	Charles Babbage	C	Computation; cognition	The Ninth Bridgewater Treatise: on mathematical induction; low probability events; and miracles (simulation theory)
1843	Ada Lovelace Luigi Menabrea	C	Computation; cognition	Analytical engine suggests a generalized science of "operations" including mechanical creativity (in music)
1847	George Boole	C	Computation; cognition	Theory of probability; algebraic methods of logic
1854	George Boole	C	Computation; cognition	The fundamental operations of mind as described using a probabilistic and universal logic; thought as a natural law
1858	Charles Darwin Alfred Wallace	V	Evolution; adaptation	The principle of natural selection; selection as analogous to a centrifugal governor
1859	Charles Darwin	V	Evolution; adaptation	On the origin of species; selection as analogous to a physical force (gravity)
1865/6	Gregor Mendel	V	Evolution; adaptation	The inheritance/transmission and discrete segregation of heritable factors
1868	J. Clerk Maxwell	E	Dynamical systems; stability and chaos	The stability of integral feedback control; negative feedback
1871	J. Clerk Maxwell	E	Thermodynamics; statistical mechanics	The mechanical basis of statistical phenomena; "Maxwell's demon"
1872	Ludwig Boltzmann	E	Thermodynamics; statistical mechanics	The increase in entropy of a system towards the Maxwell–Boltzmann distribution (H-theorem)

Table continues on next page.

Year	*Authors*	*Code Letters*	*Core Complexity Connections*	*Key Ideas in Paper*
1875	J. Willard Gibbs	E	Thermodynamics; statistical mechanics	The purely probabilistic interpretation of the second law
1879	Rudolf Clausius	E	Thermodynamics; statistical mechanics	The mechanical theory of heat; the theorem of equivalence of transformations
1890	Henri Poincaré	D	Dynamical systems; stability and chaos	The three-body problem; transverse homoclinic orbits; Poincaré maps

In the following sections, I briefly enumerate the contributions of each of these researchers with an emphasis on ideas that have proven to be of significance in twentieth-century complexity science.

CARNOT, BOLTZMANN, MAXWELL, CLAUSIUS, AND GIBBS: THE EFFICIENCY OF MACHINES AND THE ARROW OF TIME

1824 ◊ Sadi Carnot, an engineer, concerned himself with the problem of how to maximize the amount of mechanical work performed during the transfer of heat from a boiler to a condenser in a steam engine. The problem was highly practical but his insights enormously general. Carnot intuited much that would come later with the development of the entropy concept. Carnot verbally introduced the idea of a reversible thermodynamic cycle which we now think about in terms of two isothermal processes (constant temperature) and two adiabatic processes (constant heat or entropy). Clapeyron translated Carnot's work into mathematics and visualized the cycle using the Watt Indicator diagram. Carnot understood that only the differences in temperature between the two reservoirs mattered, and not the medium (gas, fluid, etc.) performing the work. Carnot showed that the key to maximizing the efficiency of any work cycle is to minimize the loss of heat (in modern terms strive to keep the entropy of the system constant over the cycle). The Carnot cycle is an ideal engine and can only be approximated in practice. The discrepancy between the ideal cycle and physical reality provides an instrumental definition of the second law of thermodynamics.

1871 ◊ James Clerk Maxwell in his *The Mechanical Theory of Heat* set out to establish the status of thermodynamics as a physical theory. In practice this meant "investigating the relations between the thermal and mechanical properties of substances" or exploring the epistemological limits of grounding thermodynamics in classical mechanics. In order to accomplish this Maxwell adopted his method of "physical analogies" whereby all hypotheses of assumed collective molecular geometry and motion were grounded in the constraints of Newton's laws and the conservation of energy. By pursuing an hypothesis involving many spherical particles engaged in perfectly elastic collisions, Maxwell was able to deduce the stationary distribution of particle velocities (the distribution left unchanged by further collisions). Both Clausius and Boltzmann had sought to use this kind of reasoning to explain the law-like status of Carnot's observation that heat cannot pass from a colder body to a warmer without some other change occurring. They were unsuccessful. Maxwell explained why this was so through his thought experiment with a small mechanism (at the same scale as the particles) able to distinguish each particle and guide and control their actions (never violating the conservation of energy) so as to violate the second law: allowing heat to pass from a colder (low-energy) state to a warmer (high-energy) state. The importance of what Lord Kelvin called "Maxwell's demon" was to show that thermodynamics had to be built on statistical foundations, to be a highly probable process, and to give up its ambition at inter-theoretic deduction from more fundamental principles. In this regard, Maxwell's work foreshadowed one of the core ideas behind emergence, as described in Philip Anderson's paper "More Is Different" a century later in 1972. It also foresaw the possibility and potential challenges of building mechanisms to capture "thermal information" in nonequilibrium computing devices.

1872 ◊ Boltzmann was placed in a quandary through the insights of Maxwell. He had made it his objective to reconcile time-reversible mechanics of point masses with thermodynamics without introducing microscopic stochasticity. On the other hand, Boltzmann needed to start with probability in order to derive his expression for entropy increase: the *H-theorem*. Boltzmann developed his ergodic conjecture as an interface between these levels of descriptions: leave particles bouncing around for long enough and through "molecular chaos" they will even out in space. Now replace the microscopic description with a distributional (probabilistic) description and operate at the probabilistic level. Note that classical mechanics is abandoned in this move towards statistics. Whereas Maxwell used the "demon" to demonstrate the intrinsically statistical nature of the second law, Boltzmann outlawed demons by replacing microscopic observables with continuous density functions. Boltzmann was thereby free to write down the dynamics of a gas through a time-dependent density of particles in an appropriate phase space (positions

The **H-Theorem** was developed by Ludwig Boltzmann to describe the evolving energy distribution of molecules. Over the course of numerous elastic collisions with associated transfers of energy, the speed distribution of molecules converges on a stationary distribution—the Maxwell-Boltzmann distribution—which minimizes the value of *H* (increases the entropy).

and velocities), and show that H (the distance to the Maxwell distribution) always decreases in time, or the entropy (-H) increases in time. Thus Boltzmann sought to derive a time-asymmetric irreversible process from an underlying time-symmetric reversible mechanics. Regardless of the degree of success of his project, the profound implication of Boltzmann's work was to explain the apparent asymmetry of the past and the future and thus lay the foundations for some of the most important concepts in complex systems: causality, inefficiency, robustness, resilience, aging, and death.

1875 ◊ Josiah Willard Gibbs did not share the Maxwell–Clausius–Boltzmann preoccupation with Newtonian fundamentals. Reviewing his ideas in the 1902 monograph *Elementary Principles in Statistical Mechanics,* Gibbs jumped straight into probabilities:

"*For some purposes, however, it is desirable to take a broader view of the subject. We may imagine a great number of systems of the same nature, but differing in the configurations and velocities which they have at a given instant, and differing not merely infinitesimally, but it may be so as to embrace every conceivable combination of configuration and velocities. And here we may set the problem, not to follow a particular system through its succession of configurations, but to determine how the whole number of systems will be distributed among the various conceivable configurations and velocities at any required time, when the distribution has been given for some one time.... Such inquiries have been called by Maxwell statistical.*" (Gibbs 1902, vii-viii)

Freed in some sense from traditional "physics," Gibbs could also free himself from the coils of historical thermodynamics, and in so doing move considerably beyond it. Throughout the 1870s, Gibbs introduced a series of new formalisms and ideas, including the vector calculus, the "thermodynamic equation of a fluid," the "surface of absolute stability," the "Gibbs paradox," and "Gibbs free energy" (Gibbs 1873, 1875; Wilson and Gibbs 1901). In many ways Gibbs is the first post-emergence thinker—having no need to pursue reductionist physics in order to discover a new fundamental physics, expressed through a purely effective (non-fundamental) theory.

1879 ◊ The career of Clausius spans the contributions of Carnot, Maxwell, Boltzmann, and Gibbs. The name entropy was provided by Clausius in 1865 as well as many of its practical implications. Clausius concluded his entropy paper with the famous couplet: "The energy of the universe is constant. The entropy of the universe tends to a maximum." Herein lies one of the many fractures dividing the nineteenth century from the twentieth century: we might call it the transition from simple symmetry to complex broken symmetry. *The Mechanical Theory of Heat* from 1871 represents a mature form of mechanical thermodynamics with a strong emphasis on work and might be read as the culmination of the engineering concerns of Carnot. It is full of engines, heat baths, and tubes. One idea that Clausius championed, and that has faded from discussion, is his idea of "disgregation," a concept describing the geometric dispersion of molecules in the generation of mechanical work. Clausius saw entropy as a sum of a heat term and a "disgregation" term, and felt that this term captured something more fundamental about the hidden mechanics of molecules than their energy.

BABBAGE, LOVELACE, AND BOOLE: CALCULATORS, COMPUTERS, AND THE LOGIC OF MIND

1837–1851 ◊ Charles Babbage, having made a name for himself with the publication in 1824 of *On the Economy of Machinery and Manufacture*, becoming the Lucasian Chair of Mathematics at Cambridge in 1828, and co-founder of the Royal Astronomical Society, begins his contribution to computation with plans for a *Difference Engine* (**DE**). A device to be used in the automatic calculation of the Nautical Almanac, comprising tables of data bearing on the determination of longitude while at sea. After studying Joseph Marie Jacquard's 1805 punch-card mechanism for weaving arbitrary designs when attached to looms, Babbage conceived of his dual store–mill mechanism for a general-purpose computer, the *Analytical Engine* (**AE**). Ada Lovelace, having translated Luigi Menabrea's lucid description of Babbage's AE from a series of 1840 lectures in Turin, wrote, "We may say most aptly that the Analytical Engine *weaves algebraical patterns* just as the Jacquard-loom weaves flowers and leaves. Here, it seems to us, resides much more of originality than the Difference Engine can be fairly entitled to claim" (Taylor 1843, 696). Lovelace was likely the first to conceive of the art of computer programming, describing an algorithm for computing the Bernoulli numbers, as well as speculating on the computational generation of musical and art works; a computational perspective going far beyond look-up tables and ephemerides. Babbage's sweeping vision of computation led him to write *The Ninth Bridgewater Treatise*, an attempt to demonstrate that "there exists no such fatal collision between the words of Scripture and the facts of nature." The monograph is the first thorough working out of what we now call *simulation theory*. It presents a computational analog to Maxwell's demon in the form of programs that conditionally branch to produce rare events—miracles—based on purely mechanical principles. The profundity of Babbage's work remained in dormancy until the 1940s and 1950s when they were rediscovered by Howard Aiken and Alan Turing.

1847–1854 ◊ George Boole's professional career represents an approach to mathematics and logic distinct from those of his contemporaries. As Professor of Mathematics at Queen's College in Cork from 1849, for Boole, mathematical research represented a means of investigating "the fundamental laws of those operations of the mind, by which reasoning is performed." This was the stated aim of his 1854 monograph, *An Investigation of the Laws of Thought*; in it he constructed a framework in which "processes of symbolical reasoning are independent of the conditions of their interpretation." Boole's approach to thought closely resembles Claude Shannon's later approach to both circuit design and information—both sought to make progress by eliminating semantic content from automatic procedure, and constructing these procedures independently from machinery. Boole's earlier work, *The Mathematical Analysis of Logic* (1847), starts by making the same observation: "Thus the abstractions of the modern Analysis, not less than the ostensive diagrams of the ancient Geometry, have encouraged the notion, that Mathematics are essentially, as well as actually, the Science of Magnitude." Boole's technique was to introduce a new algebra (related to Leibniz's algebra of concepts), which we now describe in terms of a binary-valued Boolean logic, or an algebra of sets. In concluding his work, Boole alluded in embryo to the topic of mathematical intuition in connection to the power of "unconscious" thought—a topic in torpor for a century until the publication of *The Psychology of Invention in the Mathematical Field* by Jacques Hadamard in 1945. Boole wrote, "It is not contended that it is necessary for us to acquaint ourselves with those laws in order to think coherently, or, in the ordinary sense of the terms, to reason well. Men draw inferences without any consciousness of those elements upon which the entire procedure depends" (Boole 1854, 422–423).

Logic and the unconscious. Hadamard's interest in the role of the unconscious in analytical thought can be traced back to the writings of Arthur Schopenhauer in his essay, "Transcendent Speculation on the Apparent Deliberateness in the Fate of the Individual" (1851), further developed by Eduard von Hartmann in "Philosophy of the Unconscious: Speculative Results According to the Induction Method of the Physical Sciences" (1869) and applied to mathematics by Henri Poincaré in his book *Science and Method* (1908) in a chapter on mathematical Creation.

DARWIN, WALLACE AND MENDEL: EVOLUTION, ADAPTATION, AND GENETIC TRANSMISSION

1858–1859 ◊ Charles Darwin and Alfred Russel Wallace introduced the essential elements of their theory of natural selection, competition, divergence, speciation, and sexual selection in 1858. The theory placed material laws at the center of biology as they had been in physics and chemistry for over a century. Natural history was thereby reconciled with a universal mechanism. Darwin and Wallace's view of selection is as a powerful distributed machine for making discriminations: "Now suppose there were a being who did not judge by mere external appearances, but who could study the whole internal organization, who was never capricious, and should go on selecting for one object during millions of generations" (Darwin and Wallace 1858, 51). The selective environment in this account is not reducible to a simple phenomenological variable—the Malthusian parameter—but is conceived of as a high-dimensional filter capable of detecting countless internal degrees of freedom. Darwin described this environment in terms of "an entangled bank, clothed with many plants of many kinds . . . elaborately constructed forms, so different from each other, and dependent upon each other in so complex a manner, have all been produced by laws acting around us" (Darwin 1859, 489). Darwin was describing his own "demon" in analogy to Maxwell's Newtonian demon, capable of seeing into the "black box" of the organism. The way that selection achieves this X-ray-like capability is in terms that what we would now think of as the *ergodic hypothesis*, whereby countless generations—time—allow that the space of microstates ("the whole internal organization") is explored and evaluated. The mutation-selection process is an algorithmic procedure for generating and propagating neutral families (ensembles) of approximate solutions to adaptive challenges in high-dimensional fitness landscapes, a perspective subsequently codified in modern approaches to genetic algorithms and genetic programming.

1865–1866 ◊ Gregor Mendel's paper of 1866 provided the "genetic" basis for heritability that enabled the Darwin–Wallace mechanism to function (Mendel 1866). Darwin's own theory of pangenesis proved to be incorrect. Mendel's mechanism was very popular for espousing an idea resembling an atomic theory of variation. The implications of Mendel's theory are captured by a simple recurrence equation whose fixed point has come to be known as the Hardy–Weinberg equilibrium. It is a useful null expectation under very strict conditions of invariance. Mendelian genetics is, however, highly non-ergodic, or saltationist, and explores a very small volume of sequence space. In the early twentieth century, Raphael Weldon suggested a more realistic theory, incorporating multiple internal and external causal factors, allowing for approximately continuous variation in organisms. Weldon died before his book, *Theory of Inheritance*, was published. Structurally similar modifications were made later by such figures as C.H. Waddington and James Mark Baldwin, key contributors to complexity science (Radick 2023).

MAXWELL AND POINCARÉ: THE STABILITY OF MACHINES AND INSTABILITY OF SOLAR SYSTEMS

1868. ◊ James Maxwell's paper "On Governors" represents the first rigorous analysis of feedback control, the founding paper of what later would be called cybernetics, and a pioneering application of dynamical systems to modeling machines—as it happens, the same machines that inspired Sadi Carnot to invent the field of thermodynamics: steam engines. Maxwell's interest was piqued by the challenge of maintaining a constant speed of a steam engine in the face of uneven surfaces and supplies of fuel. Maxwell used differential equations to describe two forms of feedback mechanism controlling the delivery of fuel to an engine: moderators and regulators. He was able to show, by analysis of the characteristic equation of these systems, that differential control (widely adopted) was unstable, whereas integral control was stable. Maxwell's contribution was ignored for over eighty years until it was rediscovered by Norbert Wiener in 1948. Maxwell's work introduced several concepts that lie at the heart of the study of complex systems: (1) an elementary form of agency whereby a purely physical system pursues a target by minimizing an error function; (2) an emphasis on stability which undergirds ideas of robustness, error-correction, and resilience; and (3) an insight into the role of a system memory (integral control) able to store a cumulative history of error in order to regulate a system reliably.

1890. ◊ The question of stability also motivated Henri Poincaré's analysis of the three-body problem. If we assume three gravitationally interacting bodies, all subject to Newton's laws of motion and gravitation, of arbitrary mass and initial condition, can we write down a solution for the three orbits into the past and into the future? The answer discovered by Poincaré remains the answer today: no. The equations of motion are nonintegrable and no general closed-form solution exists. However, in studying this system and simplified subproblems, Poincaré was the first to conceive of the idea of chaos, recognizing the three-body problem's extreme sensitivity to initial conditions—related to problems of diverging trajectories—compromising the ability to mathematically integrate out future orbits. The larger philosophical implications of Poincaré's findings relate to prediction and, by extension, free will. As Poincaré wrote himself in 1903: "If we knew exactly the laws of nature and the situation of the universe at the initial moment, we could predict exactly the situation of that same universe at a succeeding moment . . . it may happen that small differences in the initial conditions produce very great ones in the final phenomena. A small error in the former will produce an enormous error in the latter. Prediction becomes impossible . . ." (Poincaré 1903, 68). The development of nonlinear dynamics subsequent to Poincaré, including the crucial contributions of Edward Lorenz in the 1960s and Robert May and Mitchell Feigenbaum in the 1970s, established the generality of chaos. And the very high dimensionality of most complex systems, including the diversity of interactions among their parts, describes worlds where the elegant mathematics of celestial mechanics hits a wall. Perhaps descends into an analytical singularity?

CONNECTIONS AND SOCIAL NETWORKS

One of the appealing features of nineteenth-century science is the way that its modest scale promoted familiarity outside of the vehicle of the journal system. Many of the major contributors in table 1 (See page 15.) knew each other personally, if not in person, then at a short remove, through universities, institutions, and scholarly societies. Hence ideas and influence flowed as an advance wave through social and professional networks before publications became generally available.

Charles Babbage was one of three founders of the Cambridge Analytical Society (along with William Herschel and George Peacock) bent on a mission to bring the European emphasis on algebraic abstraction and analysis to a Britain hitherto dominated by physical intuition and synthetic geometric reasoning. Babbage (1864), like many other students of mathematics at the time, had hoped that Cambridge would open his mind to the full range of research on the Continent but often found its materials rather provincial:

> *Thus it happened that when I went to Cambridge I could work out such questions as the very moderate amount of mathematics which I then possessed admitted, with equal facility, in the dots of Newton, the d's of Leibnitz, or the dashes of Lagrange. I had, however, met with many difficulties, and looked forward with intense delight to the certainty of having them all removed on my arrival at Cambridge. [. . .] I had heard of the great work of Lacroix, on the "Differential and Integral Calculus," which I longed to possess . . . .* (Babbage 1864, 26)

CHARLES BABBAGE. PASSAGES FROM THE LIFE OF A PHILOSOPHER

Through the formation and meetings of the Analytical Society, its members would exert a slow but sustained pressure, largely through the Royal Society in London, on British mathematical thought and notational conventions. An obvious example is the adoption of the differential form of Maxwell's equations (1865), which makes use of the very convenient Laplace operator to express his theory of electromagnetism. And Maxwell continued to use the differential form in his more popular books, including *An Elementary*

Treatise on Electricity (1881). Similarly, over the course of his career, George Boole would feature the differential notation for the calculus, which he used analogically to introduce his theory of elective symbols and functions, which one might broadly interpret as binary values and Boolean operators.

Charles Babbage was also a notorious raconteur, bon vivant, and London socialite, throwing regular weekend parties at 1 Dorset Street. Guests included Ada Lovelace, Michael Faraday, Charles Lyell, Charles Dickens, Felix Mendelssohn, Charles Darwin, and literally hundreds more (Snyder 2011). These events featured both scientific discussions and free-ranging polemics:

> *I used to call pretty often on Babbage & regularly attended his famous evening parties. He was always worth listening to, but he was a disappointed & discontented man; & his expression was often or generally morose. I do not believe that he was half as sullen as he pretended to be. One day he told me that he had invented a plan by which all fires could be effectively stopped, but added, — "I shan't publish it — damn them all, let all their houses be burnt." They all were the inhabitants of London* (Charles Darwin 1876, 133).

Researchers in this period would also meet at Exhibitions. Boole and Babbage met at the 1862 International Exhibition in London, where Babbage displayed diagrams and parts of his unbuilt Analytical Engine:

> *My dear Sir, It is a source of regret to me that I was quite unable to avail myself of your kind invitation to call upon you on my return from Cambridge to London . . . Meanwhile, I shall endeavor to acquaint myself with Menabrea's paper and the principle of the Jacquard loom. But I cannot allow this opportunity of writing to you to pass without thanking you very warmly for the kind explanations you gave me of the working of the Difference Engine, and without saying that it was a pleasure and an honour to me to meet you* (Boole 1862).

Boole's books on the calculus of logic and thought were written in 1847 and 1854 and could not therefore be inspired by meetings with Babbage. Boole's critical predecessors, according to his own writing, were Aristotle and Augustus De Morgan, neither of whom in their logical publications were much concerned with physical reality. Nevertheless, *An Investigation of the Laws of Thought* from 1854 did purport to be a treatise on the natural intellect and not merely an exposition on a new mathematical logic. Boole does not at any point mention physics, chemistry, or natural history, or any of the scientist practitioners he would have met in society or their opinions on his ideas. The only mention of empiricism, in an age dominated by experiments, comes in a general allusion to experimental evidence in support of scientific knowledge:

> *Thus the necessity of an experimental basis for all positive knowledge, viewed in connexion with the existence and the peculiar character of that system of mental laws, and principles, and operations, to which attention has been directed, tends to throw light upon some important questions by which the world of speculative thought is still in a great measure divided. How, from the particular facts which experience presents, do we arrive at the general propositions of science?* (Boole 1854, 402)

It is interesting to note Alfred Russel Wallace's comments on mathematics. They provide a clue to why Boole ignored underlying physiological and behavioral mechanisms and dispositions:

> *We have to ask, therefore, what relation the successive stages of improvement of the mathematical faculty had to the life or death of its possessors; to the struggles of tribe with tribe, or nation with nation; or to the ultimate survival of one race and the extinction of another. If it cannot possibly have had any such effects, then it cannot have been produced by natural selection . . .*
>
> [...]
>
> *We conclude, then, that the present gigantic development of the mathematical faculty is wholly unexplained by the theory of natural selection, and must be due to some altogether distinct cause* (Wallace 1889, 466–467).

Wallace concludes his pro-Darwinian monograph with an anti-Darwinian appeal to extra-physical capability:

> *These three distinct stages of progress from the inorganic world of matter and motion up to man, point clearly to an unseen universe—to a world of spirit, to which the world of matter is altogether subordinate. To this spiritual world we may refer the marvelously complex forces which we know as gravitation, cohesion, chemical force, radiant force, and electricity . . .* (Wallace 1889, 476).

There seems to be little doubt that Wallace's intentions in the final section of his book mirror the earlier opinions of Boole, particularly those Boole expresses in the final chapter of his own book, *An Investigation of the Laws of Thought:*

> *If the mind, in its capacity of formal reasoning, obeys, whether consciously or unconsciously, mathematical laws, it claims through its other capacities of sentiment and action, through its perceptions of beauty and*

> *of moral fitness, through its deep springs of emotion and affection, to hold relation to a different order of things.* (Boole 1854, 444)

The nominal alliance in thought between Wallace and Boole is in stark contrast to the intellectual connection established between Darwin and Boltzmann on similar topics. Boltzmann, an ardent Darwinian and unrepentant materialist, found arguments like those of Boole and Wallace esoteric and muddleheaded.

> *Thus it may happen to the mathematician that he, always occupied with his equations and dazed by their internal perfection, takes their mutual relationships for what truly exists, and that he turns away from the real world* (Boltzmann [1890] 1973, 24).

For Boltzmann, mathematics was merely one useful representational system, and not a source of ultimate truth. Certain areas of mathematics have been retained due to their success at explaining the world not because they express an unfathomable platonic or spiritual reality. Boltzmann expressed what we might now think of as an evolutionary epistemology—one where ideas, in the struggle for explanatory preeminence, are either retained or discarded. Boltzmann went so far as to describe the nineteenth century as the century of Darwin and ruefully thought of himself as the Darwin of physics.

Unlike Wallace, and like Darwin, Boltzmann had no qualms about positing continuities in all traits between humans and non-humans:

> *The analogy between the sensations of man with those of the highest animals is so perfect that we absolutely must ascribe objective existence to the latter sensations* (Boltzmann 1897, 30).

And Boltzmann had little patience for the immutability and perfection of logic as described by Boole:

> *In the history of science there are many cases where theorems were either proved or refuted through evidence which was thought to correspond to laws of thinking, while now we are convinced of its futility* (Boltzmann [1899] 1973, 38).

Darwin's influential writings on the intellectual faculties and their selective value made the case for the evolutionary benefit of the intellect in his book *The Descent of Man:*

> *These faculties are variable; and we have every reason to believe that the variations tend to be inherited. Therefore, if they were formerly of high importance to primeval man and to his ape-like progenitors, they would have been perfected or advanced through natural selection. Of the high importance of the intellectual faculties there can be no doubt, for man*

> *mainly owes to them his preëminent position in the world. We can see, that, in the rudest state of society, the individuals who were the most sagacious, who invented and used the best weapons or traps, and who were best able to defend themselves, would rear the greatest number of offspring* (Darwin 1871, 153).

Henri Poincaré, much like Ludwig Boltzmann, was an admirer of Charles Darwin and was evidently influenced by his style of naturalistic reasoning. In his philosophical essays, Poincaré made arguments that were rather similar to those of Boltzmann, explaining the appeal of mathematical theories in material terms: their origin in empirical observation, their ability to encode the regular structure of observation, and through correspondence with reality demonstrate significant utility.

> *Whence comes this concordance? Is it merely that things which seem to us beautiful are those which are best adapted to our intelligence, and that consequently they are at the same time the tools that intelligence knows best how to handle? Or is it due rather to evolution and natural selection?* (Poincaré [1903] 2017, 23)

COMPLEX TIME

French scholars tended to resist Darwin's works for either nationalistic reasons (favoring Jean-Baptiste Lamarck and George Cuvier) or on grounds of its irreligiousity (Farley 1974). Germany was far more receptive having been primed for transmutation by both Immanuel Kant and Johann Wolfgang von Goethe and championed by Ernst Haeckel (Richards 2013). Most prominent among Darwin's more formidable critics were two of the founders of statistical mechanics in Britain, James Clerk Maxwell and William Thomson (Lord Kelvin) (Burchfield 2009). Their criticisms are of particular interest because they represent an attempt to establish a direct connection between physical and biological theories: an effort to establish a mechanistic incompatibility between physical law and two of Darwin's assumptions: (1) the time required for natural selection to have evolved complex forms of life; and (2) the mechanism of inheritance required for selection to operate. Unlike Wallace's criticisms, which might be ascribed to a lack of imagination, Maxwell and Thomson's criticisms were full of scientific ingenuity, rigor, and represent a generative clash of paradigms that continued into the twentieth century as complexity continued its materialization.

There is no procedure in Darwin's published work for estimating the time required for a given increment in evolution to take place. Darwin was writing before dendrochronology had become a calibrated technique for chronometry (interestingly, Charles Babbage used dendrochronology to estimate the age of trees in peat bogs in *The Ninth Bridgewater Treatise* in 1938). There was no radio-carbon dating (1940s–) and no techniques of phylogenetic inference (1950s–). There was evidence from the fossil record and techniques of stratigraphic superposition. Darwin had discussed the imperfections of the fossil record in chapter 10 of *The Origin of Species* and concluded that it could not be relied upon for any kind of accurate estimates. Lyell in the *Principles of Geology* summarized the logic of stratigraphy in terms of the relationships between age and vertical position, unconformities, erosion, weathering, and uniformitarianism. By means of which Lyell had estimated the earth's age to the order of a hundred million years old.

Starting in the 1860s, Lord Kelvin made a series of calculations in which he estimated the age of the earth based on the radiative life-span of the sun, heating of the earth by tidal friction, and geochronological calculations. Kelvin's geochronology assumed the nebular hypothesis of planet formation: start with a molten ball of uniform temperature and subject it to cooling. In 1863 Kelvin's estimates were in accordance with those of Lyell to the order of a hundred million years. By the 1890s, these estimates had been reduced to twenty million based in part on an estimate of the sun's age at twenty million.

Despite the lack of a theory of evolutionary time, Darwin considered these estimates damaging to his position, going so far as to write to one correspondent, "I am greatly troubled at the short duration of the world according to Sir W. Thompson, for I require for my theoretical views a very long period *before* the Cambrian formation." (F. Darwin and Seward 1903, 164)

Darwin recruited his son George (Fellow of Trinity College Cambridge) to his cause, asking for his help in reanalyzing Lord Kelvin's estimates. These calculations demonstrated that tidal friction would have considerably slowed the cooling of the earth. When used in combination with the observations of John Perry on the correct estimation of geothermal gradients at the earth surface based on interior convection (rather than Kelvin's diffusion), and buttressed by the discovery of a molten core, the estimate of the age of the earth was in the billions. Charles Darwin crowed in a letter to George on October 29, 1878, "Hurrah for the bowels of the earth & their viscosity & for the moon & for all the Heavenly bodies & for my son George." (Mason 1994, 119) We now estimate the age of the earth at approximately four and half billion years.

James Clerk Maxwell had a rather different objection to Darwin's theories. Maxwell saw flaws in Darwin's theory of inheritance, the so-called theory of pangenesis. Darwin's idea was that every cell in the body shed smaller undifferentiated cells (gemmules) into the bloodstream. Gemmules migrated into the germ line, constituting a memory of a generation, thence to be transmitted through reproduction to offspring. Each offspring blends the gemmules of its parents, thereby supporting the adaptive correlations within a lineage. We now know this theory to be incorrect based on techniques of genetics and cell biology.

Maxwell's objections were based on reasoning derived from his kinetic theory of heat (Maxwell 1871, 42):

> *Some of the exponents of this theory of heredity have attempted to elude the difficulty of placing a whole world of wonders within a body so small and so devoid of visible structure as a germ, by using the phrase structureless germs. Now, one material system can differ from another only in the configuration and motion which it has at a given instant. To explain differences of function and development of a germ without assuming differences of structure is therefore to admit that the properties of a germ are not those of a purely material system.*

Maxwell's observations were not as quantitative as Lord Kelvin's but proved to be more lasting. It was not until the publication of Erwin Schrödinger's *What Is Life? The Physical Aspect of the Living Cell* in 1944 that the basis of "differences in structure" was ascribed to an aperiodic crystal. Like Maxwell before him, Schrödinger was thought to reconcile the stability of inheritance with the known facts of statistical mechanics. It is fair to say that to this day the tension between biological form and function and physical theory remains an extremely generative connection.

The creators of statistical mechanics, evolutionary theory, formal logic, dynamics, and computation were members of a scholarly community connected by a dense web of institutions and societies. Through both publications and meetings they were influenced by shared cultural values and by each other's research results. Many of their ideas originated in the effort to understand the properties of organisms and machines. All of them struggled to explain the origin of stable, efficient, predictable, and practical technologies—both designed and evolved. It would take the twentieth century to commence their integration. And accompanying this synthesis, a reduced emphasis on physics and an increasing dependence on computational principles, which disclosed the information schemas supporting complex reality.

On the Origin of Design

One challenge for science at this time was that many machine-age concepts lived an anthropomorphic existence: mechanical function was used to make sense of the domains of purposeful traits and behaviors in society. And by the metaphysical norms of the period their origin was presumed to lie squarely within the purview of God's design extending subcreatively into human lives. This is a perspective much at odds with contemporary thinking (paradigmatically so) and is often forgotten in our ahistorical endeavors to construct an unbroken narrative linking sequential epochs in scientific belief.

The challenge was therefore to naturalize mechanics, generalize its insights into mechanical work and natural history, and do so respecting the postulates of natural theology. Much of this effort was accomplished through ideas circulating after the publication of a hugely ambitious scholarly project.

There are few nineteenth-century publications more ambitious in drawing parallels between the designed and the natural by pursuing evidence for the "teleological thesis" than the eight-plus-one *Bridgewater Treatises*. The first eight volumes were published between 1833 and 1836 and now stand as rather inspired tracts on mechanics suspended in an incongruous distillate of apologetics (Topham 2022).

The Ninth Bridgewater Treatise was an unauthorized contribution to the series published by Charles Babbage in 1837, proposing a new theory of "miracles" achieved by reconciling natural law, contingency, and computation. The first treatise was published in two volumes by Thomas Chalmers, establishing the logic and tone of the whole series: building a rhetorical connection between scientific principles—including parsimony—the new mechanical machines, and design.

> *Now it is a commonly received, and has indeed been raised into a sort of universal maxim, that the highest property of wisdom is to achieve the most desirable end, or the greatest amount of good, by the fewest possible means, or by the simplest machinery. When this test is applied to the laws of nature—then we esteem it, as enhancing the manifestation of intelligence, that one single law, as gravitation, should, as from a central and commanding eminence, subordinate to itself a whole host of most important phenomena; or that from one great and parent property, so vast a family of beneficial consequences should spring. And when the same test is applied to the dispositions, whether of nature or art—then it enhances the manifestation of wisdom, when some great end is brought about with a less complex or cumbersome instrumentality, as often takes place in the simplification of machines, when, by the device of some ingenious ligament or wheel, the apparatus is made equally, perhaps more effective, whilst less unwieldy . . .* (Chalmers 1833, 49–50).

The most far-reaching and complete vision of design in nature, adduced through analogy with machines, was expressed in the third Bridgewater Treatise by William Whewell. Whewell describes mechanisms at many scales, from the global atmospheric system to the motion and stability of the solar system:

> *The contemplation of the atmosphere as a machine which answers all these purposes, is well suited to impress upon us the strongest conviction of the most refined, far-seeing, and far-ruling contrivance. It seems impossible to suppose that these various properties were so bestowed and so combined, any otherwise than by a beneficent and intelligent Being, able and willing to diffuse organization, life, health, and enjoyment through all parts of the visible world . . .* (Whewell 1833, 127–128).

> *We know at present very little indeed of the construction of this machine. Its* existence *is, perhaps, satisfactorily made out; in order that we may not interrupt the progress of our argument, we shall refer to other works for the reasonings which appear to lead to this conclusion. But whether heat, electricity, galvanism, magnetism, be fluids; or effects or modifications of fluids; and whether such fluids or* ethers *be the same with the luminiferous ether, or with each other; are questions of which all or most appear to be at present undecided, and it would be presumptuous and premature here to take one side or the other* (Whewell 1833, 139–140).

The Bridgewater Treatises would largely be of historical interest, and not a core concern for the origins of complexity, if they had not proven to be of such importance to its prefoundational architects, including Charles Darwin, Alfred Russel Wallace, James Clerk Maxwell, Robert Boole, and Charles Babbage. *The Bridgewater Treatises* were demonstrably of direct influence on science, but one should not underestimate the order they lent to the zeitgeist of natural science.

The larger narrative of *The Origin of Species*, Darwin's so-called "one long argument," was a naturalistic account of the origin of novelty or species, and was strongly influenced by Darwin's repeat readings of Whewell and his natural theology. This influence extended to a quote from Whewell as the opening epigraph of *The Origin of Species*:

> *"But with regard to the material world, we can at least go so far as this—we can perceive that events are brought about not by insulated interpositions of divine power, exerted in each particular case, but by the establishment of general laws" William Whewell,* Astronomy and General Physics Considered with Reference to Natural Theology (Darwin 1859, epigraph).

Darwin used Whewell's system as a pretext to invoke Newton's gravitational law as analogy and widely acclaimed, authoritative prequel, for the generality of natural selection presented in the third edition of *The Origin of Species*.

> *It has been said that I speak of natural selection as an active power or Deity; but who objects to an author speaking of the attraction of gravity*

> *as ruling the movements of the planets? Everyone knows what is meant and is implied by such metaphorical expressions; and they are almost necessary for brevity. So again it is difficult to avoid personifying the word Nature; but I mean by nature, only the aggregate action and product of many natural laws, and by laws the sequence of events as ascertained by us.* (Darwin 1861, 85)

Darwin chose the gravity metaphor as a mechanical metaphor for selection in 1859, but had agreed to Wallace's far more fitting suggestion of Watt's centrifugal governor in the first selection paper with Alfred Russel Wallace in 1858:

> *The action of this principle is exactly like that of the centrifugal governor of the steam engine, which checks and corrects any irregularities almost before they become evident; and in like manner no unbalanced deficiency in the animal kingdom can ever reach any conspicuous magnitude, because it would make itself felt at the very first step, by rendering existence difficult and extinction almost sure soon to follow.* (Darwin and Wallace 1858, 62)

Darwin had switched out an engineered mechanism of contingent generality for a law of universal application. By the dictates of natural theology, neither metaphor questioned the ultimate origin of design. But the choice of gravity pushed causality back into a primordial and parsimonious beginning, and suggests a far more expansive vision for selection than a practical regulator engine speed.

It was a decade after the publication of Darwin and Wallace's paper, in 1868, that James Clerk Maxwell explained the logical basis of Darwin and Wallace's intuition for the balancing of the natural world, demonstrating under what conditions regulation or moderation would stabilize or destabilize a dynamical system. In so doing Maxwell invented the field of control theory, and established one of the foundational ingredients of complexity science—the principle of feedback control—the kernel of the selection process.

Maxwell himself was an avid reader of Whewell and, like Darwin before him, sought a means of reconciling physics with machines and theology. In one of the more extraordinary convolutions of the entwining of incompatible ideas, Maxwell speculated on the uniformity of the fundamental particles of nature in relation to the production-line in an article from 1873:

> *We are thus assured that molecules of the same nature as those of our hydrogen exist in those distant regions, or at least did exist when the light by which we see them was emitted. . . .* (Maxwell 1873, 375).

> *Each molecule, therefore, throughout the universe, bears impressed on it the stamp of a metric system as distinctly as does the metre of the Archives at Paris, or the double royal cubit of the Temple of Karnac . . .*
>
> [. . .]
>
> *None of the processes of Nature, since the time when Nature began, have produced the slightest difference in the properties of any molecule. We are therefore unable to ascribe either the existence of the molecules or the identity of their properties to the operation of any of the causes which we call natural . . . On the other hand, the exact equality of each molecule to all the others of the same kind gives it, as Sir John Herschel has well said, the essential character of a manufactured article, and precludes the idea of its being eternal and self-existent.* (Maxwell 1873, 376)

Maxwell's observation, by the standards of our own science, strikes us as utterly surreal.

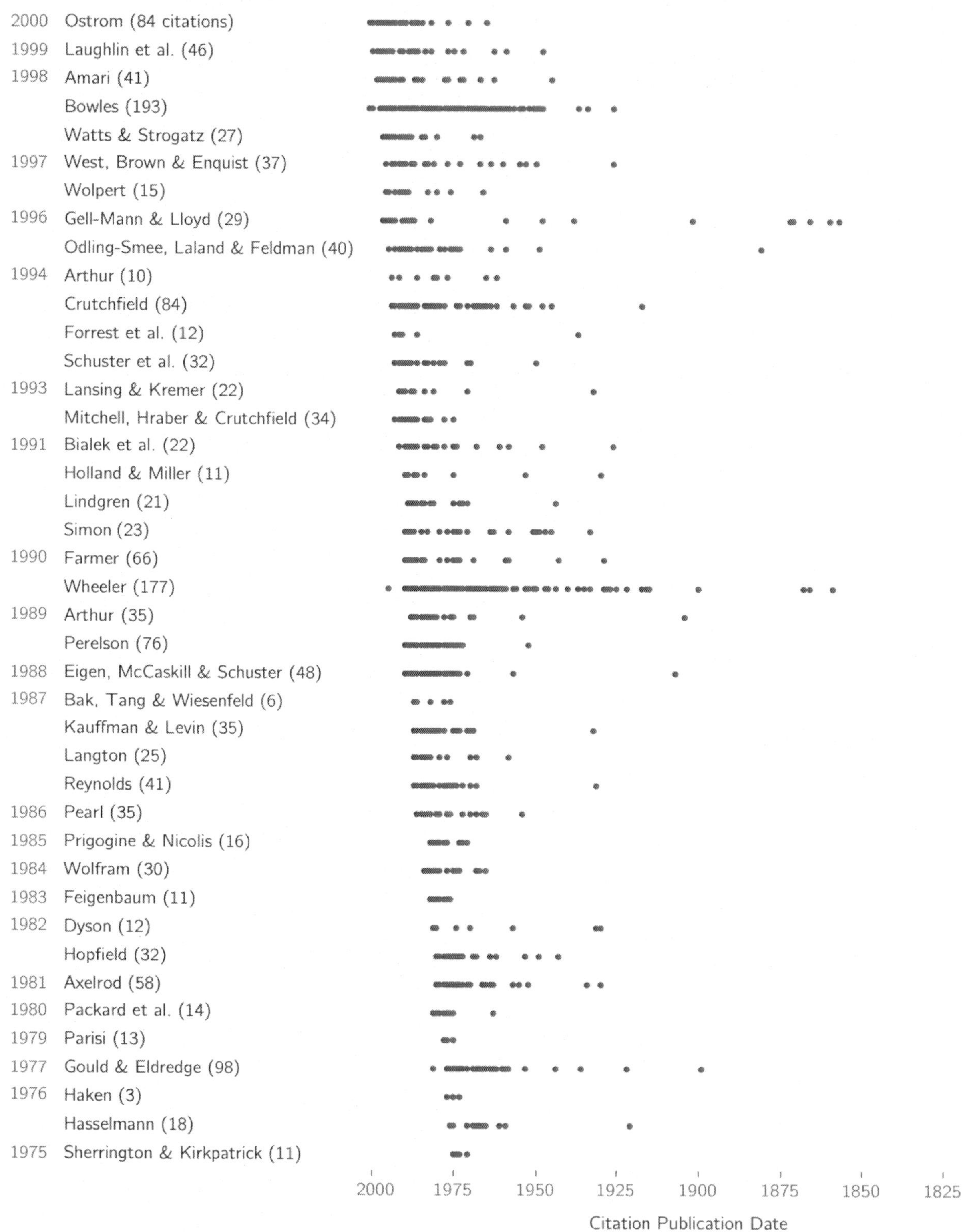

Figure 2. This visualization represents the date spans of works cited by the foundational papers that were published between 1800 and 2000 (only polymathic Wheeler 1989 went back further, to the 1600s). Each citation appears as a dot that corresponds to year of publication on the x-axis. The number of citations is indicated in parentheses. Foundational papers that did

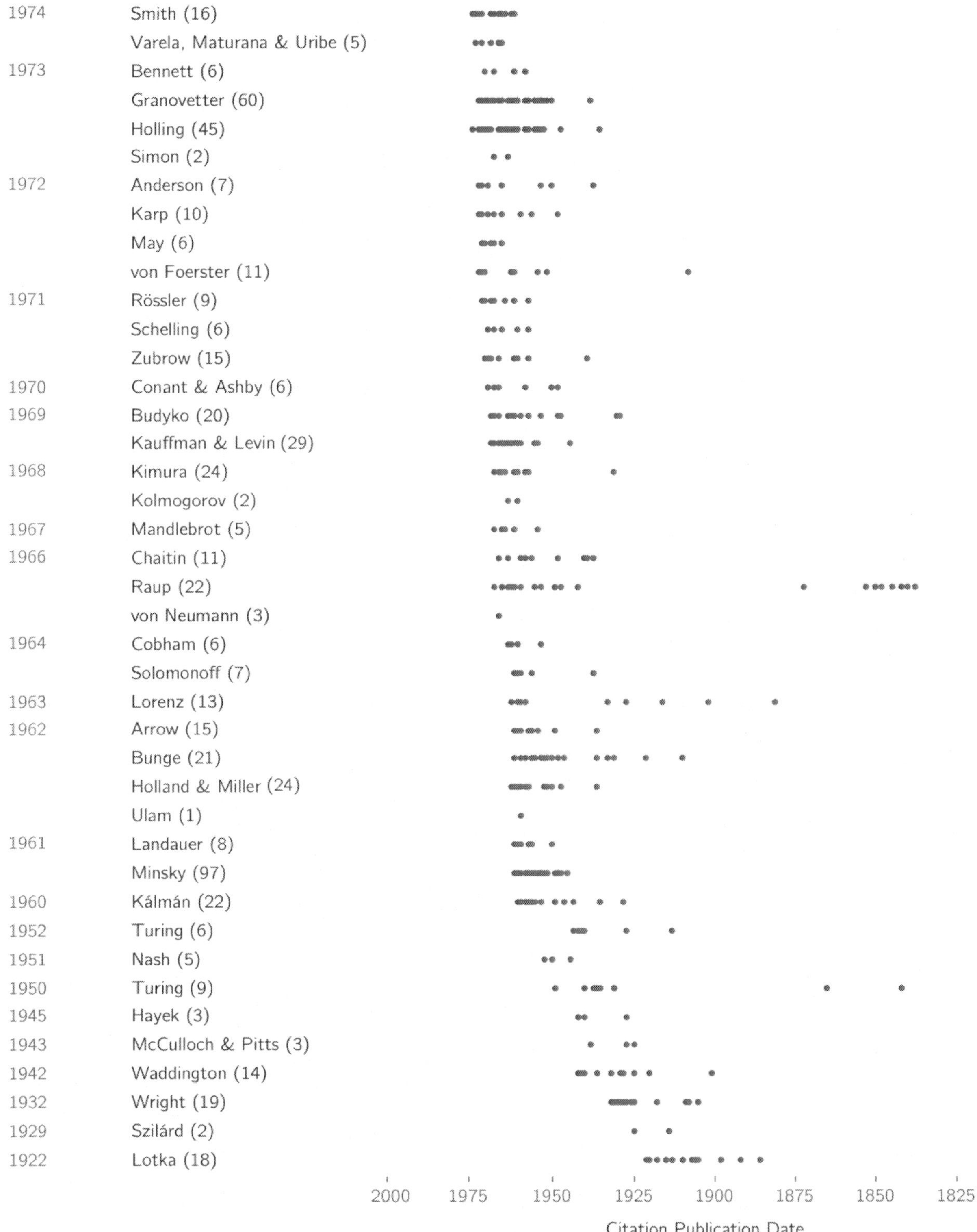

not reference other works are excluded from this chart. *Examples*: Forrest *et al.* 1994 cites only a handful of recent publications, but Raup 1966 was significantly influenced by both work published in the twenty-five years prior and earlier publications from the first half of the nineteenth century.

DEFINITIONAL PURSUITS

Outline of an Ontology

This section presents a chronology and distillation of papers and books that have elucidated and provided definitions for both the purview and principles of complexity science. These are all drawn from a *superset* of the *foundational papers*. They reflect in profound ways the four pillars of complexity established in the nineteenth century. In almost every case we can chart the two or more fields that they are seeking to integrate.

Definitional projects are not synonymous with foundational contributions; hence many important foundational papers do not appear in this section. Illustrative exclusions include those by Claude Shannon on communication and Alan Turing on computation. Both are incontestably foundational for complexity, yet neither sought to articulate the character of the complex domain. Both theoretical computer science and information theory provide important ideas and methods in complexity science while not being synonymous with it. After all, ideas from information theory and computer science are just as important to engineering classical machines and to fundamental physics as they are to complexity science. The ideas of Shannon and Turing, when supplemented by considerations from nonlinear dynamics, thermodynamics, or evolution, have informed more explicit definitions of complexity provided by Warren Weaver, Herbert Simon, John Holland, Stephen Wolfram, and others.

It is also often the case that new ideas from seemingly unrelated fields—those without connections to questions of complexity—prove to be important. The Brouwer fixed-point theorem, conceived as a means of characterizing the topology of Euclidean spaces, was the central analytical technique deployed by John Nash in developing his competitive game solution concept—the Nash equilibrium. The method is agnostic toward complexity and is a valuable pragmatic approach from applied mathematics.

Many recent efforts make use of power laws (Brown and West 2000), perturbation theory (Simmonds and Mann, Jr. 1986), and networks (Newman 2010). All introduce powerful frameworks and methods from physics and the social sciences to study complex systems. But these need not themselves elucidate principles of complexity. Typically they are important ingredients in more inclusive attempts at coming to terms with robust and sometimes near-universal patterns of organization. These include the use of scaling theory to analyze evolved metabolic networks and the use of network statistics to make sense of the structure of neural circuits or social structures.

Complex phenomena are not to be confused with methods; they are what the methods discover: we might describe these as the *syntax* (rules revealed by analysis) and *semantics* (functions revealed through analysis) ascribed to the results of the analysis. Thus a perturbation theory can be applied to derive a scaling relationship between mass and spin in particle physics or area and species diversity in ecology; it is only the latter by virtue of the relationships arising through the constrained optimization achieved via evolutionary history (near-countless frozen accidents fixed by selection) that benefits from the label complex.

The primary purpose of this section is to demonstrate two things: (1) the emphasis placed on revealing a new domain of inquiry, or an ontology, that does not fit naturally into any one existing disciplinary focus and thereby requires combining new ideas (many from the prefoundational period); and (2) the historical accumulation, or ratcheting-up, of principles required to adequately address this domain through an historical timeline. The more recent definitions of complexity build on principles provided in prior publications—not always explicitly. Complexity is an evolving field accumulating concepts; it is not a stationary pursuit, no more than biology, archaeology, or any other vibrant area of inquiry.

A Chronology of Definitions

The ongoing challenge for all of these definitional projects has been to arrive at concepts that are inclusive of physical processes but sufficiently extended to contend with extra-physical or adaptive phenomena (e.g., life and intelligence). It is very natural that some researchers place greater emphasis on the recruitment of physical laws with key modification (typically those coming from nonequilibrium physics), whereas others introduce entirely new concepts that they seek to connect—inter-theoretically—with established physical processes (those coming from biology, computer science, and the social sciences). Nonetheless, the overall impression is one of surprising coherence. In most definitions there is a primary thesis of stable nonsymmetric hierarchies (in both space and time) supporting transmissible, adaptive functions.

What we now think of as complexity science appears to have become organized into a loosely bound set of related principles (systems) governing the adaptive world between the 1940s and early 1990s. The earliest publications sought to differentiate fledgling complexity science (cybernetics, general systems theory, living systems theory, etc.) from physics. These consisted largely in definitions emphasizing out-of-equilibrium behavior, dissipative or irreversible dynamics, and self-regulating and self-reproducing systems. Later definitions place greater emphasis on integrating nonequilibrium

Table 2. The core definitional ingredients of complexity science slowly begin to coalesce in the mid-twentieth-century. Through a series of highly influential books and papers, the complex domain comes into focus. Through time, an implicational scale (a series of entailments) is built up, culminating in an expansive definition for a phenomenal world of self-organizing and selected adaptive patterns.

Reference/Definition	*Year*	*Ontology/Domain*	*Stated Field*
ROSENBLUETH, WIENER, AND BIGELOW: Dynamical systems with active or purposeful behavior; directed towards the attainment of a goal; "voluntary activity" supported by feedback and prediction (extrapolative)	1943	Biology, engineering	"Philosophy of science"
HAYEK: Systems planning through local/decentralized knowledge of "time and place": markets and the price system; spontaneous sources of order (catallaxy; 1978)	1945	Resource allocation in the economy and society	"Economics"
WEAVER: Systems of organized complexity; sizeable numbers of factors which are interrelated into an organic (functional) whole	1948	Biology; society; economy	"Complexity science"
WIENER: Systems controlling functional information through feedback to achieve homeostasis; stability and goal-seeking	1948	Biology; society; engineering; culture	"Cybernetics"
VON BERTALANFFY: Open systems; dynamic equilibrium; efficient metabolism; principle of equifinality; thermodynamics of irreversibility (building on Prigogine 1947)	1950	Biology, chemical engineering	"General systems theory"
MILLER: Tripartite concrete, conceptual, and abstract systems; structured subsystems; living processes; relationships between growth, decay, and termination	1955	Biology; organizations; nations	"Living systems theory"
SIMON: Systems whose whole is more than the sum of its parts; hierarchy; evolution; functional decomposition; self-reproducible	1962	Society; biology; physics; chemistry	"Behavioral science," "complexity science"
HOLLAND: Adaptive systems; adaptation as the generation of programs that address environmental distributions of problems; feedback operating through differential activation	1962	Biology; computation; logic	"Complexity science"
CONANT AND ASHBY: Regulatory systems; feedback control; prediction, inference; and complexity of controller matching that of environment	1970	Engineering; neuroscience	"Systems science," "control theory"
ANDERSON: "More is Different": complex systems dominated by broken symmetries; reductionism does not imply constructionism: each level requires a different conceptual structure	1972	Many-body theory; chemistry; biology; culture	"Complexity science"

Reference/Definition	*Year*	*Ontology/Domain*	*Stated Field*
VON FOERSTER: A theory of "observers": "description-invariant subjective worlds"; representational systems; recursive and self-referring systems	1972	Living systems	"Epistemology of living things"
SIMON: Near-decomposable systems; evolvable hierarchies; alphabet-based; systems of languages and programs	1973	Biochemistry; biology; engineering; society	"Complex systems"
VARELA, MATURANA, AND URIBE: Unitary and autopoietic organizations; recursive networks of unitary processes supporting mechanisms of self-production; non-determinant elements; autonomous systems	1974	Living systems	"Autopoiesis," "complex systems"
HAKEN: Self-organized structures emergent from chaos; structures induced by fluxes of energy and matter; biological function combining dissipative and non-dissipative structures; the "slaving principle"	1977	Statistical physics; chemistry; biology; society	"Synergetics"
PRIGOGINE AND ALLEN: Dissipative dynamical systems; symmetry breaking; far from equilibrium; long-range correlations; selection and algorithmic complexity	1982	Dynamical systems; statistical physics; pattern formation	"Complexity"
WOLFRAM: Discrete spatially extended dynamical systems with nonlinear cooperative effects (rule systems); spatial computation; complexity and non-computability of rule systems	1984	Dynamical systems; emergence; pattern formation; computation	"Complex systems"
KAUFFMAN: Systems at the edge of chaos; multi-peaked fitness landscapes; replication as autocatalysis; evolvability as selective meta-dynamics	1990	Dynamical systems; evolution; gene regulation; development	"Complex systems"
FARMER: Connectionism: networks of interacting agents with time-varying—adaptive—connection strengths	1990	Neural networks; classifier systems; immune networks; autocatalytic networks	"Complex systems"
GELL-MANN AND HARTLE: Information gathering and utilizing systems (IGUS); the intersection of computational devices with physical reality to create quasiclassical realms	1994	Society; biology; physics; chemistry	"Complex systems"

dynamics into functional capabilities, including adaptation, exploration, robustness-related mechanisms, and computation.

These publications also emphasize the broad compass of complexity principles—that they could not be defined in any restricted relation to existing disciplinary boundaries. In this respect, the emerging paradigm not only introduced new principles, but is fundamentally iconoclastic with respect to existing scholarly institutions structured by disciplines. Researchers such as Norbert Wiener and Warren Weaver made this a very explicit part of their articulation of a new science.

Complexity science increasingly pursues scholarship structured by shared principles, forms, and functions. Through this new approach, domains that diverge when conditioned on discipline (e.g., economics and ecology or evolution and archaeology), came to be seen as variations on a common theme, or case studies, for similar patterns manifesting in different structures and at different space and time scales. A common feature of these contributions is that they simply do not see the disciplinary boundary as particularly meaningful. This is in contrast to more conventional interdisciplinary work that seeks to make progress by combining ideas from previous areas to solve a new problem, but maintain disciplinary distinctions as a justification for their novelty.

Extracting Principles

From the definitional table (tab. 2) we can extract from each study, in chronological order, a central principle and consider the implications of their combinations and conjunctions. After all, classificatory definition is never satisfying, and even a real definition that alludes to contingent material properties will be constrained by a field of inquiry and can often feel restrictive. A more useful approach is to provide an explication of shared areas of interest in order to arrive at a more exact understanding of the richness of the complex domain.

1940s

ROSENBLUETH, WIENER & BIGELOW: ***Behavior, Purpose, and Teleology***

YEAR	NEW CONCEPTS	INHERITED CONCEPTS
1943	T	

A founding paper in cybernetics in which behaviorism (reinforced input–output system) and purpose (**T** = *teleology*) are realized through feedback in a dynamical system.

Cybernetics integrates earlier work from engineering, dynamical systems, and psychology. The paper contains the seeds for many ideas in contemporary cognitive science, including policies that minimize prediction errors, and extrapolation through active inference. Complexity is treated as a self-regulating dynamical system.

HAYEK: ***The Use of Knowledge in Society***

YEAR	NEW CONCEPTS	INHERITED CONCEPTS
1945	M	T

Hayek builds on economic insights from Adam Smith on the role of the "invisible hand" to construct systems of tacit knowledge for allocating scarce resources through *markets* (**M**). Hayek was one of the earliest proponents of complexity as collective knowledge generation through decentralized processes mediated by market-based coordination mechanisms. Hayek's ideas share with cybernetics an interest in regulation towards an efficient goal but its logic comes from evolution, psychology, and political economy.

WEAVER: ***Science and Complexity***

YEAR	NEW CONCEPTS	INHERITED CONCEPTS
1948	OC	T, M

Warren Weaver was the first to use the term complexity (or *organized complexity,* **OC**) as a means of identifying a domain of inquiry spanning biology, markets, societies, and polities. A colleague of Norbert Wiener's when both worked on electrical anti-aircraft systems—a professional connection that informed Weaver's understanding of purposeful, or teleological, dynamics. Weaver's ideas on **OC** are defined in distinction to both simplicity, understood as classical dynamics, and *disorganized complexity*, derived from a close reading of the work of Willard Gibbs and the statistical-mechanics literature. Weaver's understanding of science is influenced by Hayek's market-based reasoning concerning the division of labor required to make progress on complex problems.

WIENER: ***Cybernetics and Control***

YEAR	NEW CONCEPTS	INHERITED CONCEPTS
1948	IT	T, M, OC

A monumental effort to establish cybernetics at "the boundary regions of science" by integrating ideas from information theory, control theory, physiology, cognition, computer science, and social science. Cybernetics seeks to analyze a broad range of human functions based on analogies to "mechanico-electrical systems." The central organizing principle is an expansive conception of *information transmission* (**IT**) in machines, brains, and societies. For Wiener, complexity and cybernetics are largely synonymous.

1950s

VON BERTALANFFY: ***Open Systems in Physics and Biology***

YEAR	NEW CONCEPTS	INHERITED CONCEPTS
1950	OS	OC

An approach to complexity stressing the importance of the physics of *open systems* (**OS**) as these manifest in all life-like phenomena. A synoptic investigation of time-independent steady states that do not correspond to a maximum of entropy and demonstrate an independence from initial conditions (equifinality). A project to connect classical dynamics and thermodynamics to augmented models suitable for life. Best seen as an effort to establish the necessary but not sufficient conditions for complexity of a kind considered by Wiener, Hayek, and Weaver.

MILLER: ***General Theory of Behavioral Science***

YEAR	NEW CONCEPTS	INHERITED CONCEPTS
1948	LST	T, IT, OS

A descriptive framework in which life is presented as an hierarchically structured organization of increasingly open systems. The lowest levels are described by equilibrium physical principles and subsequent inclusive levels by far-from-equilibrium physics and organic reactions. The highest levels are characterized by functional mutual dependencies among evolved structures characteristic of life. *Living systems theory* (**LST**) asserts that organisms are built outwards from a simple core to a complex periphery. The core is populated by chemical elements, or basic organisms, and the periphery by *larger organisms*, including cells. An effort to defy the *continuum hypothesis* of the disciplines that tend to dichotomize ontology and advocate for a complex spectrum of life.

1960s

SIMON: ***The Architecture of Complexity***

YEAR	NEW CONCEPTS	INHERITED CONCEPTS
1962	ND	T, M, OC, IT, OS

Simon proposes after Weaver a second explicit definition for complexity science. Simon extends organized complexity into a focus on the value of nested spacetime structures. Complex structures are characterized as *near-decomposable* (**ND**) hierarchies. These resemble Miller's description of living systems, with a focus on robust assembly, the importance of a separation of time scales, information-processing layers, and greater comprehensibility through modularity.

HOLLAND: ***Adaptive Systems***

YEAR	NEW CONCEPTS	INHERITED CONCEPTS
1962	AS	T, OC, IT

Generalizing the Darwin–Fisher conception of an *adaptive system* (**AS**) into a broad range of logical processes described by automata theory. Formally, each adaptive agent is a generative procedure running on a universal computer to solve problems posed by the environment. Evolution is conceptualized as the interface of a universal generation procedure outputting populations of programs that solve contingent problems. A complex system defined as an abstract form of life characterized by information describing an adaptive program.

1970s

CONANT AND ASHBY: ***Good Regulators***

YEAR	NEW CONCEPTS	INHERITED CONCEPTS
1970	GR	T, OC, IT, AS

Complex systems as models of their environment, or *good regulators* (**GR**). Building on ideas from control theory, information theory, and psychology in order to open up the black box (behaviorism) of adaptive feedback. Hypothesizing that all complex systems need to have "agentic" features—an internal model—when dealing with nontrivial environmental dynamics. The brain is described as the prototypical complex system that models the world around it in order to promote control and thereby survival.

ANDERSON: ***More Is Different***

YEAR	NEW CONCEPTS	INHERITED CONCEPTS
1972	E	OC, IT, OS, AS

Introduces the idea that, for systems at any reasonable mesoscopic scale, symmetry-breaking abrogates the power of the fundamental laws of physics. This implies that effective theories are not merely useful approximations of reality but strictly necessary to model collective phenomena—*emergence* (**E**). Emergence is therefore a signature of complexity and a profound justification for a pluralism of theories and of understanding. An inescapable refutation of ontological reductionism based on many-body physics. Complex systems as increasingly complicated histories of broken symmetries.

VON FOERSTER: ***Epistemology of Life***

YEAR	NEW CONCEPTS	INHERITED CONCEPTS
1972	SB	OC, IT, OS, AS

Introduces a conceptual framework for exploring a recursive extension of Conant and Ashby through the idea of *self-observation* (**SB**). How to pursue an objective or description-invariant account for the subjective world; or an epistemology true to the living world. Building on ideas from logic and computation rooted in the ideas of Gödel and Turing, psychology derived from Piaget, and the linguistic philosophy of Wittgenstein. Complex systems as self-aware, reflexive, or, in the language of Douglas Hofstadter, "strange-loops."

SIMON: *Organization of Complexity*

YEAR	NEW CONCEPTS	INHERITED CONCEPTS
1973	SF	OC, IT, OS, ND, AS, E

Building on his own ideas of near-decomposability and hierarchy from 1962, and extending these to emergent hierarchies of programs that are *sealed-off* (**SF**) from lower levels. This is an emergent approach to computation that uses the idea of SF to justify a hierarchy of *effective theories* within a single system. Introduces an informal theory of types through level-dependent alphabets and languages. Complex systems as nested information-processing systems or hierarchies of formal languages.

VARELA, MATURANA, AND URIBE: *Autopoietic Organization*

YEAR	NEW CONCEPTS	INHERITED CONCEPTS
1974	AU	OC, IT, OS, ND, AS, E

Combining ideas of emergence, self-modeling, adaptation, universal construction, and computation to formalize an idea of *autopoiesis*. An autopoietic system is a dynamical network of interacting components capable of self-synthesis, self-reproduction, and autonomy. Autonomy is a degree of independence from the environment, resembling contemporary ideas of the *Markov blanket* in Bayesian networks. A complex system as an autonomous open system that transcends its materials to synthesize a robust replicate.

HAKEN: *Synergetic Self-Organization*

YEAR	NEW CONCEPTS	INHERITED CONCEPTS
1977	SO	OC, OS, AS, E

The application of ideas from the theory of phase transitions and bifurcation theory to explore a potentially universal class of synergetic *self-organizing* (**SO**) system. In practice, focusing on a variety of nonequilibrium collective dynamics under the control of a relatively small number of internal or external control parameters. Self-organization is an important ingredient of complex systems. It is difficult to distinguish between the spontaneous emergence of ordered states and those resulting from prolonged evolutionary histories.

1980s

PRIGOGINE AND ALLEN: *Dissipative Systems*

YEAR	NEW CONCEPTS	INHERITED CONCEPTS
1982	DS	OC, OS, E, SO

The importance of the arrow of time in contributing to far-from-equilibrium organization or *dissipative structures* (**DS**). A cognate to synergetics, with a strong focus on simple physico-chemical processes that span both biotic and abiotic phenomena. An emphasis on the dynamics of irreversible processes giving rise to an arrow of time. Complexity emerges through self-organization understood in terms of long-range order and symmetry-breaking sensitively coupled to boundary conditions. Complexity theory as a theory of patterning, not a theory of function.

WOLFRAM: *Rule Systems*

YEAR	NEW CONCEPTS	INHERITED CONCEPTS
1984	RS	T, OC, IT, E, SO

A science of self-organizing pattern formation based on discrete, spatially extended dynamical systems. An analysis of both the rules governing these dynamics and different classes of spacetime patterns with varying degrees of computational power. Complexity as an emergent form of spatial computation achieved through collective dynamics of simple *rule systems* (**RS**).

1990s

FARMER: *Adaptive Connectionism*

YEAR	NEW CONCEPTS	INHERITED CONCEPTS
1990	AC	TT, OC, IT, AS, E, SO

Complex systems as networks of adaptive agents that achieve a separation of time scales between slowly evolving connectivity and quickly changing coupling strength. The identification of common structures and functions in autocatalytic chemistry, nervous systems, immune systems, and ecosystems suggestive of a universal class of *agentic connectionist* (**AC**) system.

KAUFFMAN: *Edge of Chaos*

YEAR	NEW CONCEPTS	INHERITED CONCEPTS
1993	EC	T, OC, IT, OS, AS, E, SO

A project to integrate selection and self-organization by building on structural correspondences between physical energy landscapes and biological fitness landscapes. A common feature of ordered states in nonequilibrium physical systems and functional states in adaptive systems is a tuning to the

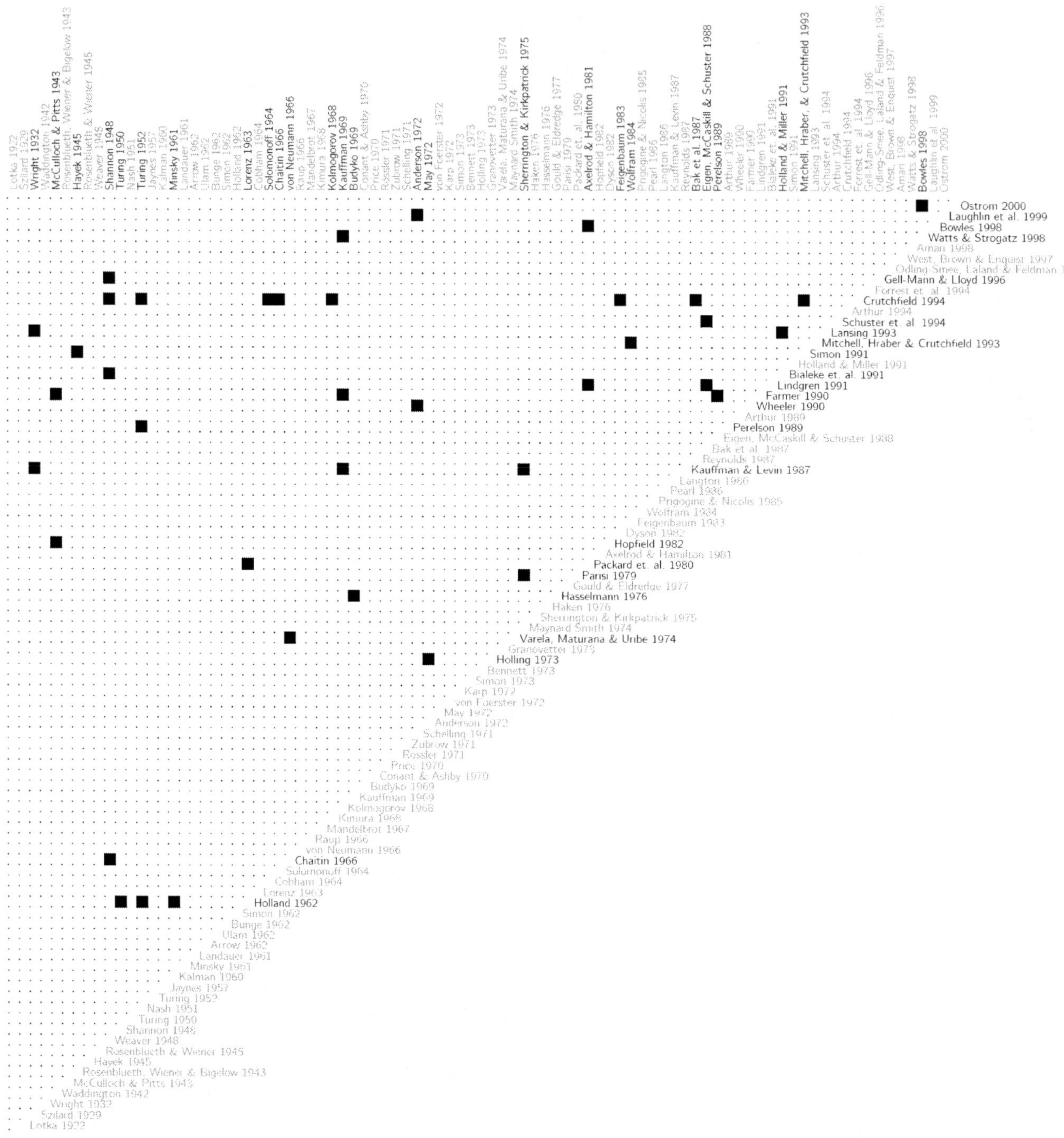

Figure 3. This chart illustrates citation connections between all 88 foundational papers. The black squares represent a connection between the referencing paper (on the slanted axis) and cited papers (on the top axis). *Examples:* Holland 1962 (along the slanted axis) references three papers: Turing 1950, Turing 1952, and Minsky 1961 (all on the top axis). Ostrom 2000 references Bowles 1998.

edge of chaos (**EC**). Complex systems as a broad class of phenomena that exist at the edge of chaos in order to more effectively catalyze reactions, form stable patterns, store memories, and ensure continued adaptability.

GELL-MANN AND HARTLE: ***Information Gathering and Utilizing Systems***

YEAR	NEW CONCEPTS	INHERITED CONCEPTS
1994	IGUS	T, OC, AS, GR

Complex systems as *information gathering and utilizing systems* (**IGUS**). Any evolved or engineered system capable of adaptive and predictive information processing. The IGUS establishes a principled basis for the definition of what were once thought to be physics concepts: past, present, and future. Through the lens of the IGUS these three temporalities require complexity-based definitions. These are in the form of agent-centric computations embedded in Minkowski space whereby each agent experiences a unique world line.

Explicating Complexity

This sparse chronology serves to demonstrate both the specificity and the generality of definitions proposed for complex phenomena. Each definition might conveniently be placed along a spectrum in order to explicate complexity in relation to a continuum. Toward one end are those approaches exploring the reach of an extended physical theory of *open systems* (**OS**) dominated by *dissipative structures* (**DS**) and *self-organization* (**SO**), all displaying simple forms of *sealed-off* (**SF**) *emergence* (**E**). And toward the other end, systems whose *teleonomic* (**T**) features need to be accounted for in terms of *adaptive systems* (**AS**) producing *organized complexity* (**OC**) in a variety of *autopoietic* (**AU**) organizations capable of both modeling the environment through *good regulators* (**GR**), which in some cases achieve *self-observation* (**SB**).

These definitions suggest that the distinction sometimes made between *complex systems* (**CS**) (non-evolved) and *complex adaptive systems* (**CAS**) (evolved), is too dichotomized. Very few systems of interest outside of idealized mathematical models or computer simulations conform to either of these two strict limits. It is more straightforward to describe all these systems as *complex systems*, recognizing that any constructed systems—evolved or engineered—will need to exploit far-from-equilibrium processes—and will fall somewhere in the interval described by a history of selected broken symmetries. *Complexity science is from this perspective a family of models and theories, all of which seek to explain the emergence of far from equilibrium structures, which vary in their adaptive capacity and length of evolutionary histories.* Taking our lead from both James Clerk Maxwell and Carl Linnaeus, we might describe the full complexity science spectrum as the genus (the domain of inquiry) and each point on the continuum as the species (particular processes required by the explanandum).

ENGINEERING EMERGENCE

There have been efforts to discover non-living phenomena at the CS limit: to find systems that self-organize into nontrivial ordered states spontaneously without an evolutionary history. Consider one of the favorite reactions adduced as a candidate: the *Belousov–Zhabotinsky* (**BZ**) reaction (Miyazaki 2013). The BZ reaction comprises around eighteen distinct chemical reactions organized into two autocatalytic cycles that are able to produce nonlinear chemical oscillations. This is an open system, showing a simple form of organized complexity, based on dissipative reactions, manifesting a simple form of hierarchy in the form of self-similarity—fractality. At present, the BZ reaction has only been observed in the laboratory when engineered by purposeful chemists and maintained under very exacting conditions. Candidate BZ reactions have been hypothesized to explain the growth patterns of evolved life-forms including for amoebae and slime molds. The BZ is in other words a complex system made by complex adaptive systems. And the same argument applies to Turing patterns, and to a rather broad range of non-equilibrium structures not yet observed unengineered in the abiotic universe.

A far more compelling candidate for a limit-case complex system is a spin-glass (Stein and Newman 2013).[2] A spin-glass is a disordered magnetic material. The magnetic moments of a spin-glass freeze/quench into orientations that are arbitrary or random as a result of a mixture of positive and negative interactions among atoms. Unlike simple materials, spin-glasses show significant levels of frustration (unsolvable constraints) generating complicated energy landscapes with many metastable states. When moved out of equilibrium (e.g., when placed in magnetic fields for some duration) they have been shown to possess aging and memory properties. These are all features of engineered and evolved systems. This is why spin-glass models (typically built on the Sherrington–Kirkpatrick mean field limit of the Edwards–Anderson model) provide a very powerful framework for studying a range of complex systems.

[2]Stein and Newman's monograph *Spin Glasses and Complexity* (2013) builds up from the physics of simple magnets to spin glasses and their applications. It is notable for its engagement with the history of ideas of complexity, including the limitations of physical models and frameworks applied to the complex domain.

Daniel Stein and Charles Newman have addressed the complexity status of spin-glasses directly. As regards the use of *spin-glass models* (**SGM**) as *mathematical tools*: they are useful for analyzing phenomena as diverse as computational complexity, protein-folding, content-addressable memory, and social dynamics. SGM, much like empirical networks abstracted into matrices or biological growth abstracted into systems of differential equations, have proven to be an important tool in the analysis of complex systems. As regards *spin-glasses* (**SG**) as *physical systems* (e.g,. dilute magnetic alloys), Stein and Newman propose that:

> *spin glasses deserve at least the rubric of quasi-complex system. Whether truly complex or not, they have provided mathematical descriptions of important aspects of complexity*... (Stein and Newman 2013, 237–238)

Any rich field has at its boundary an escarpment not a wall. The SG forces us to reckon with mechanisms that we might wish to position closer or further from the center of any inquiry into complexity.

How to Think about Post-Evolutionary Physics

Applications of both equilibrium and non-equilibrium physics principles to complex systems often consist in the applications of variational principles to existing adaptive structures. The physics of adaptive matter is the search for parsimonious principles that might coexist with, or be placed above, unparsimonious mechanisms. A few examples worth analyzing in a little more detail include theories of development and morphology and theories of metabolic scaling. All of these pursue a minimal approach to effective theories, whose screened or sealed off constituent parts, are contingent forms of adaptive matter. In this respect, these are all kinds of post-evolutionary physics: theories of emergent properties that place evolution in a fundamental position relative to the variational principle—an inversion of the usual sense of fundamental.

D'ARCY WENTWORTH THOMPSON ON GROWTH AND FORM

There are foundational works inimical to adaptive complexity in a philosophical sense: ideas like those in D'Arcy Wentworth Thompson's 1917 book *On Growth and Form*. This is very much a self-styled work of biophysics, seeking to reduce complexity by focusing on structural phenomena dominated by physical laws while deprecating questions of adaptation and function. The importance of *On Growth and Form* was to demonstrate that a significant degree of functional variation ascribed to selection could be derived from variational principles. It is in this way crucial to our understanding of complexity despite its ambition to eliminate it.

> *How far, even then, mathematics will* suffice *to describe, and physics to explain, the fabric of the body no man can foresee. It may be that all the laws of energy, and all the properties of matter, and all the chemistry of all the colloids are as powerless to explain the body as they are impotent to comprehend the soul. For my part, I think it is not so.* (Thompson 1917, 8)

There are several notable chapters where Thompson challenges the selective account of a biological structure. In "A Note Upon Torsion," Thompson explains the twining of a stem about its own axis as a:

> *temporary adhesion or "clinging" between it and the growing stem which twines around it; and a system of forces is thus set up, producing a "couple," just as it was in the case of the ram's or antelope's horn through direct adhesion of the bony core to the surrounding sheath. The twist is*

> *the direct result of this couple, and it disappears when the support is so smooth that no such force comes to be exerted.* (Thompson 1917, 626)

Darwin explains the same phenomenon thus:

> *The stem probably gains rigidity by being twisted (on the same principle that a much twisted rope is stiffer than a slackly twisted one), and is thus indirectly benefited so as to be able to pass over inequalities in its spiral ascent, and to carry its own weight when allowed to revolve freely.* (Darwin 1891, 10)

There is no conflict between these accounts. We resolve each of them into proximate and ultimate explanations. Thompson constructed a nice physical rule that in Darwin's estimation achieves a selective reward. Despite the compatibility of these accounts, debates in ink like the one between Thompson and Darwin continue to this day, typically because the variational argument fails to recognize that it has been made possible by a very long evolutionary history. That is to say, elegant forms of self-organizing are given license through rather messy genetic processes.

NICOLAS RASHEVSKY ON TOPOLOGY

Following in Thompson's footsteps, Nicolas Rashevsky proposed a biophysics based on mathematical and fundamental physical principles including the principles of diffusion, drag forces described using hydrodynamics, and topological theorems.

The abstract of Rashevsky's paper on "The Geometrization of Biology" (1956) is worth reproducing in full:

> *The twentieth century has witnessed a geometrization of physics, that is, a reduction of the basic concepts of physics to geometric concepts. The topological approach to biology, recently proposed and to some extent developed by the author, is a small step in the direction of geometrization of biology, but is unable to achieve the main purpose of such a geometrization of biology, namely, the reduction to geometric concepts of such purely biological concepts as ingestion, digestion, assimilation, etc. To achieve this purpose we must find geometric structures or spaces, in which different geometric properties stand to each other in the same formal logical relation, as the different concepts of biology stand to each other. If this were possible, then a set of geometric theorems could be "translated" by an appropriate "glossary" into a set of biological laws* (Rashevsky 1956, 31).

It goes without saying that Rashevsky's contributions to current understanding of ingestion and digestion are negligible. This is not because topological considerations are without merit. The problem is the topological principles were not adequately constrained by the evolved physiology of life

Table 3. A small sample of books that set out to provide a synoptic introduction to principles bearing on complexity science or its adjacent domains. These are organized by publication date spanning the last four decades.

Author(s)	*Title*	*Year*	*Abbreviated Definition*
DOUGLAS R. HOFSTADTER	*Gödel, Escher, Bach: An Eternal Golden Braid*	1979	Living systems as pattern-forming mechanisms for encoding the world and themselves through recursion ("strange loops")
MANFRED EIGEN AND RUTHILD WINKLER	*Laws of the Game: How the Principles of Nature Govern Chance*	1982	Systems capable of exploiting intrinsic sources of randomness to produce emergent forms of combinatorial organization
ILYA PRIGOGINE AND ISABELLE STENGERS	*Order out of Chaos: Man's New Dialogue with Nature*	1984	Complex systems as intrinsically irreversible mechanisms exerting force
JAMES GLEICK	*Chaos: Making a New Science*	1989	How simple dynamical rules, often deterministic, can generate complicated and often unpredictable patterns in nature and culture
M. MITCHELL WALDROP	*Complexity: Science at the Edge of Order and Chaos*	1989	Self-organizing systems of interacting strategic agents
MURRAY GELL-MANN	*The Quark and the Jaguar: Adventures in the Simple and the Complex*	1994	Contrasting a universe dominated by symmetry with life, defined in terms of frozen accidents describing effective information
JOHN H. HOLLAND	*Hidden Order: How Adaptation Builds Complexity*	1995	Nested models of agents and meta- agents built from seven putatively universal properties and mechanisms
ROGER HIGHFIELD AND PETER COVENEY	*Frontiers of Complexity: The Search for Order in a Chaotic World*	1996	Emergence of functional states of order from collective dynamics—mechanisms moving in the opposite direction to reductionism
HAROLD J. MOROWITZ	*The Emergence of Everything: How the World Became Complex*	2002	A survey of 28 candidates of instances of emergence from the origin of the universe to technology and urbanization
STEPHEN WOLFRAM	*A New Kind of Science*	2002	The complex world as an instantiation of many simple programs that can be understood in correspondence (equivalence) to formal languages executed on a digital computer
JOHN GRIBBIN	*Deep Simplicity: Bringing Order to Chaos and Complexity*	2005	Rules of interaction and nonlinear dynamics underlying far-from-equilibrium systems, including biological life
MELANIE MITCHELL	*Complexity: A Guided Tour*	2009	Networks of agents capable of information processing and collective adaptation through learning and evolution

Table continues on next page.

Author(s)	*Title*	*Year*	*Abbreviated Definition*
NEIL JOHNSON	*Simply Complexity: A Clear Guide to Complexity Theory*	2009	The study of phenomena emergent from a collection of interacting objects including in physics and biology
SCOTT E. PAGE	*Diversity and Complexity*	2010	Emergent outcomes of diverse rule-following adaptive agents interacting in networks
MARK NEWMAN	*Networks: An Introduction*	2010	Network structure as a defining feature of engineered and evolved phenomena
JOHN H. HOLLAND	*Complexity: A Very Short Introduction*	2014	The emergent properties of adaptive populations of agents
JOHN H. MILLER	*A Crude Look at the Whole: The Science of Complex Systems in Business, Life, and Society*	2016	Mechanisms of construction utilizing noisy and nested feedback loops connecting distributed adaptive agents
GEOFFREY B. WEST	*Scale: The Universal Laws of Life, Growth, and Death in Organisms, Cities, and Companies*	2018	The thermodynamics of living systems and their macroscopic consequences—species abundance, life spans, and urban scaling
STEFAN THURNER, RUDOLF HANEL, AND PETER KLIMEK	*Introduction to the Theory of Complex Systems*	2018	The science of generalized matter interacting through algorithmic rule systems
JAMES LADYMAN AND KAROLINE WIESNER	*What Is a Complex System?*	2020	Open systems with collective dynamics classified by mechanisms of emergence: including their numerosity, feedback, and diversity
HENRIK JELDTOFT JENSEN	*Complexity Science: The Study of Emergence*	2023	All those sciences providing principles for explaining emergent phenomena (including purely physical and biological transformations)

forms. If Rashevsky had understood his own post-evolutionary physics, constraining models by common anatomy and physiology, one suspects his ideas would have proved far more influential.

GEOFFREY WEST AND JAMES BROWN ON ALLOMETRY

At the opposite end of the spectrum is metabolic scaling theory. This is a theory that makes selection fundamental and then pursues surprisingly universal consequences. The challenge is to explain the widespread observation that macroscopic features of organisms and ecosystems scale as powers of one-quarter.

Geoffrey West and James Brown (2005) explain this regularity as the outcome of natural selection pushing up against the bounds of energetic constraints:

> *We have proposed a set of principles based on the observation that almost all life is sustained by hierarchical branching networks, which we assume have invariant terminal units, are space-filling and are optimized by the process of natural selection. We show how these general constraints explain quarter power scaling and lead to a quantitative, predictive theory that captures many of the essential features of diverse biological systems.* (West and Brown 2005, 1575)

The metabolic scaling theory is a sophisticated demonstration of the full complexity spectrum. The theory rests on a teleology of adaptation, a functional hierarchy of circulation, dissipative dynamics of resource allocation, and self-organization during the formation of networks. And the theory relies on emergent properties of fractal networks, which permits the derivation of a biological effective theory, captured in terms of the physical dimensions of space. Through evolution, one of the most fundamental features of the universe—spatial dimensionality—comes to play a central role in adaptation. One might say that scaling theory shows us how evolution has discovered physics.

Synoptic Surveys

Complexity science has been well served by books and monographs that both review and analyze the field. Despite the different scholarly provenances of authors, there is substantial overlap in interests and topics in many books.

A clue to the unity of complexity science is how many of the books explaining complexity science have very similar chapter headings. Table 3 is a representative list organized chronologically.

Nearly all of the books discuss chaos, cellular automata, social insects, nervous systems, immune systems, brains, and markets. And all of the authors make a form of emergence the central organizing concept. This agreement is achieved despite differential emphasis on the role of noise, adaptation, and determinism in the emergence of physical phases, dynamical patterns, or biological functions.

There is a slight chronological trend in which principles and ontology are emphasized in the earlier books and methods and models in the later books. Presumably this reflects some combination of the maturation of the field and a growing pressure from society to demonstrate the practical utility of complexity science.

EMERGENCE AND EFFECTIVE THEORIES

The prefoundational pillars, entropy, evolution, dynamics, and computation, all contend with questions related to the interface of local mechanisms with collective global properties. Maxwell, Clausius, Boltzmann, and Gibbs sought to map fundamental physical laws onto statistical regularities, or adduce justifications for a purely statistical theory of collectives. Darwin, Wallace, and Mendel explored the nature of persistent variation in populations, and how competition after sufficient time, produces an average macroscopic fit to the environment. Maxwell and Poincaré calculated how pairwise dynamical interactions might lead to long term stability or chaos at a system level. And Babbage, Lovelace, and Boole investigated how simple machines, implementing local logical operations, might support computational functions and even underpin patterns of thought.

We can with justification describe all four pillars as considerations of problems of emergence: the emergence of variables of state, the emergence of organisms and species, the emergence of fixed points and attractors, and the emergence of mind from machine. This is of course a very informal use of the word emergence, but nevertheless captures its essential concern: how to characterize and think about the origin of aggregate or "macroscopic" states of matter with apparently causal properties that are not physically, and "microscopically," fundamental.

The name given to theories that are not causally fundamental is *effective theories*. Effective theories discover and "label" large groupings ("coarse-graining") of fundamental or non-fundamental materials and properties that maintain coherence through time. These coarse-grainings are the collective, coordinated dimensions of matter and provide indirect evidence for emergence. More inclusive effective theories support a variety of models that exploit coupling among different effective theories achieving *inter-theoretic compatibility*. For example, the framework of population genetics connects two effective theories: (1) discrete, low-dimensional, genetic transmission (Mendelian genetics), and mutation-selection dynamics (Fisher–Wright model).

Consider for example how theories of condensed matter provide insights into the crystalline nature of metals. Engineers with limited knowledge of condensed-matter physics, including the condensed-matter physics of metallic bonds, or how material properties emerge from certain configurations of structured iron atoms and alloys, can still effectively shape metals into an engine block, and do so using abstract design principles.

At a far higher level of abstraction, we might be interested in the relative consistency and stability of psychologically and sociologically grounded affiliations within a polity: left or right leaning. These labels, and others like them, can serve as predictive factors for a wide range of seemingly independent behavioral and economic preferences, ignoring input from neural correlates and reported cognitive bias.

As long as variables assigned to each dimension maintain sufficient physical or organizational integrity through time, an effective theory has a chance of succeeding. In the final analysis, effective theories describe reduced dimensions along which coordinated aggregates of fundamental matter transform. The more expansive problem of emergence is to explain both the origin and stability of these dimensions, and how these dimensions might influence subsequent emergence. Effective dimensions can be both many-to-one outputs of processes (e.g., atoms to metals; many neurons to one mind) and one-to-many inputs outcomes (e.g., educational attainment to income, mobility, and health).

Factor, Compression and Basis

Any model or theory confronted with non-fundamental observables needs to justify its choice of variables. More often than not, a weak form of justification consists in borrowing from the outlines of reality itself: clouds, cells, tissues, trees, species, cities, etc. This can of course be highly problematic since it relies on the stability of perception and description to isolate effective dimensions. More principled approaches abound. These include factor analysis, compression, and establishing the correct basis. None of these generate in themselves "effective theories," but they do provide useful pointers to where theories might apply.

FACTORS

In data analysis a very convenient approach is factor analysis. The basic principle is to relate a large number of observables, $(O_1, O_2, ..., O_n)$, to a smaller number of inferred factors, $(F_1, F_2, ..., F_m)$. If the observables are related linearly, then an obvious determination to make is how each hidden factor contributes (loadings) to each of the data points. If only the variance in the data needs to be fitted, *principal component analysis* (**PCA**) is a preferred method, where the right singular vectors from a *singular value decomposition* (**SVD**) are the effective dimensions of interest.

PCA provides some rather nice insights into a very limited, linear approach to data reduction and indirectly to emergence. It also illustrates the challenge of assigning the right labels to the inputs and outputs of emergence claims.

If every observable requires its own factor (they are independent variables), then the whole is equal to the sum of its parts. If the number of factors is small relative to the number of observables, then there is evidence of correlated behavior. These correlations shine a light on candidate effective dimensions.

Effective dimensions. The orthogonal dimensions of either a dataset or independent variables required by an effective theory.

There is nothing in SVD–PCA ensuring that the factors are internal to a system. If the observables were measurements of some species' features, including abundance, metabolic rate, life span, and mobility rate, a perfectly credible candidate factor for every one is temperature. We might say that life span and mobility are the emergent outputs from the input temperature. This is nothing like water emerging from many H_2O molecules.

We also know that mass is a credible factor for these observables. Building an emergence explanation on mass is more familiar since it is a bulk property of the system itself. The way that mass influences each of these variables is provided by metabolic scaling theory: mass governs the efficiency of energy flows through fractal circulatory systems. In these situations the observables are emergent outputs from the input dimension mass.

COMPRESSION

Techniques of compression involve transforming data structures of a given size into smaller structures with a minimal loss of information. As with PCA, compression exploits the elimination of redundancies in data and compression does not discover effective theories but suggests where they might be useful. Shannon–Fano codes and Huffman codes deploy codeword lengths in inverse proportion to word frequencies in order to achieve a lossless compression of a source file. Lempel–Ziv–Welch achieves lossless compression by building up a string-to-code lookup table as it receives a string, systematically replacing longer strings with shorter codewords.

The benefits of all forms of compression typically relate to the minimization of a resource requirement—space or time. This is somewhat analogous to replacing astronomical ephemerides with Newton's laws of motion, except in the case of a true law, the compression is asymptotically infinite as the dataset size increases.

BASIS

A different means of achieving compression is through a principled choice of *basis*. A basis is a set of vectors in a vector space that can encode every vector in the space. This is achieved using linear combinations from the basis.

The choice of basis tends to be domain specific: the spherical basis is useful in quantum mechanics; radial basis functions are commonly used to approximate partial differential equations; the Mexican Hat Wavelet is used to encode visual images. Some vector spaces are infinite or real-valued and involve infinite linear combinations. These include the periodic trigonometric functions combined into a Fourier series. These provide an orthonormal basis for the vector space of all real or complex valued functions.

Developing an effective theory is very closely related to the choice of basis; in fact it is typically dependent on the right choice of basis. The development of natural gradient optimization in neural networks was based on the field of information geometry (connecting probability theory to geometry by placing distributions on statistical manifolds).

A natural metric on the manifold is the *Fisher metric*, a Riemannian metric, typically defined in terms of the exponential family of probability distributions. These distributions provide the basis for most of the results in the application of information geometry in natural science. The same could be said for the use of the Gaussian basis set in computational chemistry. These bases are commonly used to model the molecular orbitals (wave functions) of electrons.

Every effective theory has a basis and the basis provides either an explicit definition of dimension (e.g., Fourier basis) or some approximation thereof. Most commonly accepted examples of emergence, at least in relation to phase transitions, are expressed in terms of a basis. For example, Landau's theory of second-order phase transitions deploys a linear combination of the *k-q* function (quasimomentum and quasicoordinate space). Without the correct choice of basis, it is not at all clear whether an emergent phenomenon would be recognized.

More Is Different

For many scientists, the defining paper on emergence is Philip Anderson's "More Is Different" (1972). Its many profound ideas are lost when it is invoked to support a rather overworked sense of emergence, as understood by the Aristotelian aphorism "the whole is greater than the sum of its parts." It is in many ways a rather unfortunate adage. In a literal sense it has to be wrong, and in a conceptual sense it fails to convey why effective theories are important. Literally, a whole made from parts implies a reduction in number ($N \rightarrow 1$). Technically, an effective theory is always simpler, or more parsimonious, than the microscopic processes that it approximates. In both colloquial and technical usage, it would be more accurate to say, "the whole is less than the sum of its parts," in contrast to the expectation that it be "equal to the sum of its parts." This is what Newton had in mind when he wrote, "We are to admit no more causes of natural things than such as are both true and sufficient to explain their appearances" (Newton 1934, 398). What Aristotle and most successors have had in mind is better captured by the phrase, "different than the sum of its parts." It is this difference that Anderson explored in "More Is Different": the fundamental origin of epistemological irreducibility arising from scaling.

A single molecule of ammonia (NH_3) flips between two states—up and down—at 30 billion times a second. One consequence is that the stationary distribution of ammonia is a mixture of two mutually invertible pyramids. The barrier between two configurations of ammonia implies, as Anderson writes, that "the state of the system, if it is to be stationary, must always have the same symmetry as the laws of motion which govern it" (Anderson 1972, 394). When one considers larger molecules like phosphine (PH_3), the inversion rate is at least an order of magnitude slower. Phosphorus trifloride (PF_3) is yet more massive and an order of magnitude slower than PH_3. When one reaches biologically active molecules, at the scale of even the simplest carbohydrates, parity symmetry breaks down. The stationary distribution of molecules is dominated by initial conditions rather than fundamental laws of motion. The extra parameterization required in the initial conditions "counts" the broken symmetries, which stand as the foundation stones of emergence. One of the neglected consequences of accumulating broken symmetries is that the most fundamental theories will be the least parsimonious.

Anderson extends these insights in "Some General Thoughts about Broken Symmetry" (1994). Landau writes that "Every transition from a crystal to a liquid or to a crystal of a different symmetry is associated with the disappearance or appearance of some elements of symmetry" (Landau 1969, 25). For Anderson, the order parameter of a state of matter that changes through a phase transition should be thought of as a measure of the degree of broken symmetry, which in Landau theory is typically the degree of increased order. Since the order parameter is derived from an understanding of the symmetrical laws of motion, we get to see very precisely the degree to which the fundamental laws and emergent structures deviate from one another. This is why phase transitions are often mentioned in the same sentences as emergence: not because they are synonymous with emergence but because they sometimes provide evidence of symmetry breaking. Not always. The transition from a liquid to a gas is symmetry preserving.

In "More Is Different," Anderson presents a table with two columns, X and Y. Under X he includes a list of sciences that obey the laws of the sciences in column Y. For example, X (= chemistry) obeys Y (= many-body physics), and X (= cell biology) obeys Y (= molecular biology). The point is that "obey" and "determine" are not synonyms. We might say that the differences between obey and determine are statements about the number of symmetries that have been broken. An operational definition of ontological reductionism is the limiting case where the symmetries of the laws of physics determine the empirical distribution of observables. Under reductionism, the meaning of obey and determine converge on identity.

In the study of complexity, effective theories that are determinate for quantities of interest bear little resemblance to the physics that they obey. At a certain point the symmetries become so peripheral to a macroscopic behavior that even the idea of broken symmetry loses its moorings, such that, "At some point we have to stop talking about decreasing symmetry and start calling it increasing complication. Thus, with increasing complication at each stage, we go on up the hierarchy of the sciences" (Anderson 1972, 396).

Mesoscopic Protectorates

Robert Laughlin, David Pines, and colleagues discuss a very compelling extension of the ideas of Anderson, particularly ideas of mesoscale organization, in "The Middle Way" (2000). Building up beyond broken symmetries, they generalize Anderson's analysis of ground states into *protected states*. These are states stable against small perturbations of the underlying equations of motion. Hence, "superfluidity, ferromagnetism, metallic conduction, hydrodynamics, and so forth are 'protected' properties of matter—generic behavior that is reliably the same one system to the next, regardless of details" (Laughlin et al. 2000, 32). Protectorates are the dimensions of effective theories. In the case of phosphorus trifloride, protection against tunneling is provided by properties of the chemical potential and the molecular wave-function. In the case of a large macromolecule like DNA, protection is achieved through a very complicated, evolved suite of error-correction mechanisms imposed on the primary information-transmitting structures. Emergence at living scales is perhaps best thought of in terms of mechanisms of protection, or robustness, against large perturbations enabling "effective" equations of motion "screening off" subjacent foundations.

The Republic of Effective Theories

In the natural and the social sciences, as well as in mathematics, nearly every theory is an effective theory: thermodynamics, condensed-matter physics, the theory of chemical reactions, population genetics, neural networks, neoclassical economics, game theory; theoretical computer science, set theory, and structural engineering. The domain of fundamental theory is highly restrictive and of little use either in theory-building or in practical applications. One might say to a first approximation that every model and theory deals with non-fundamental interactions and emergent properties.

Given this state of affairs, it is a little surprising that ideas of emergence seem so enigmatic. Part of the problem, as always, is the desire to find a single and simple model for a concept that describes a large family of processes.

The emergence of a liquid from a solid and the emergence of a mind from a brain have little in common materially. What they do have in common is a contingent set of effective theories (different ones) relating parts to wholes: in the first case, single molecules to incompressible fluids, and in the second, populations of cells to behaviors and perhaps even propositions of thought.

Practitioners are interested in the rich details of mechanisms, interactions, scaling, and limits. These details necessarily lie within a specialist domain, be it condensed matter, cognitive science, or psychology.

The value of framing dissimilar mechanisms as comparably emergent features derives from family resemblances. These are the structural and logical analogies observed across material systems. Through analogy we discover applications of frameworks and theories of great power far from their fields of origination. This includes the rather modest, albeit far-reaching, calculation of solutions to many-body problems (e.g., the use of the Ising model in physics, neuroscience, and social science), to the profound discovery of universal properties of collectives that transcend material differences (i.e., the universality classes).

Philosophies

The philosophy of emergence is largely concerned with the problem of relating fundamental theories to macroscopic features described through a suitable effective theory. This work has produced a number of fascinating concepts, including (Corradini and O'Connor 2010; Gibb 2019):

☛ *property dualism*—emergent states are produced by physical states but can not be reduced to them.

☛ *coinstantiation*—mental events perfectly correlated with physical events can be treated as causal.

☛ *hierarchies*—nested effective theories, or degrees of coarse-graining.

☛ *transphysical laws*—or "emergent" laws describing persistent relationships between physical properties and secondary properties: from sensory experiences to metaphysical concepts.

☛ *multiple realizability*—many physical processes mapping onto a single aggregate process.

☛ *reductive unity*—a putative physical level underpinning all emergent phenomena.

☛ *conceptual entailment failure*—emergent phenomena that require different concepts to those used to explain parts.

Some of these concepts can be related to simple symmetry breaking during phase transitions—hierarchies, multiple realizability, conceptual entailment failure. Others can only be restrictively applied to function and minds—transphysical laws and property dualism.

Some philosophies are motivated by the puzzle that David Chalmers (2006) calls *strong emergence* or supervenience: how can effective theories be causal when there is only fundamental causality? There are areas of particle physics populated by those Daniel Dennett describes as "greedy reductionists" (Dennett 1995, 82), only satisfied by explanations that miniaturize matter into particles and fields, achieving *reductive unity*. In the Appendix to Emergence, I discuss a possible resolution to this problem.

For most scholars, causality is almost always "effective causality," which makes it no less powerful. The steps in a mathematical proof follow from axioms in accordance with strict rules of deduction—that is, causal rules. The fact that these deductions are not described in the language of quarks and gluons has no bearing on the correctness of a proof.

The primary challenges of emergence are not concerned with relating the effective to the fundamental. This connection was even short-circuited in physics with Gibbs's revolutionary effective theory of thermodynamics; the most powerful and practical theory of thermodynamics to date. It is an approach to statistical physics that was supported by Maxwell's demonstration that intertheoretic reduction (from probability distributions to classical trajectories) is almost always impossible (logically and computationally). Greater attention is paid to inter-theoretic compatibility: Gibbs does not contradict Newton but can not be derived from him. Life does not contradict quantum mechanics but can not be derived from it. The pursuit of emergent descriptions is more often concerned with compatibility.

A confusion also arises from the way the word reductionism can be used as an antonym of emergence. It should be stated forcefully that there can be no theories of emergence without reductionism. Reductionism is used in two senses: reductionism as reducing all phenomena to a minimal set of elementary particles, waves, fields, symmetries and so forth (ontological); and reductionism as reducing a model and theory to an empirically justifiable minimal set of assumptions, transformations and interactions (epistemological). The first kind of reduction is only practiced by particle physicists and other fields that envy their style of demolition. The second kind of reduction, described as parsimony or compression, is pursued to varying degrees by all scholars, from cell biologists to historians, for the simple reason that there needs to be a basis for selecting a finite set of causal factors from an infinite set.

In figure 4, four of the contributors to thermodynamics and statistical mechanics are placed in an epistemological–ontological quadrant. Only Maxwell's theory can be classified as a fundamental theory operating at the

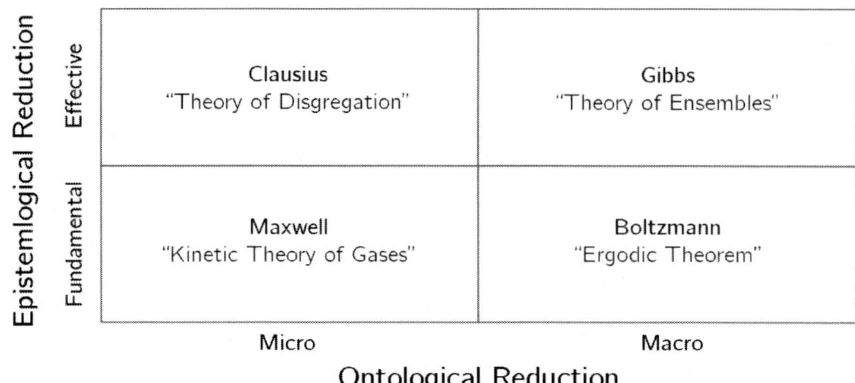

Figure 4: A quadrant of degrees of reduction. On one axis, ontological or spatial reduction, and on the other, epistemological or descriptive reduction. Even within statistical mechanics and thermodynamics, there are very different perspectives on what constitutes a fundamental and effective theory.

microscopic level, where fundamental causality applies. Boltzmann's ergodic hypothesis is macroscopic and he desired that it be proved on a fundamental basis. Clausius's theory of disgregation is described as microscopic but had no basis in mechanics; it has the peculiar status of being an effective theory at the microscopic scale. The Gibbs theory of ensembles is purely statistical and operates at a macroscopic level deploying effective theories. Practicing physicists would not hesitate to call the Gibbs theory causal—fully understanding that it rests on the unproven, perhaps unprovable, limit of a fundamentally causal microscopic mechanics.

An Emergence Lattice

With these ideas in mind, and in order to establish analogical correspondences across different domains of emergence, this section introduces an *emergence lattice*. It is a simple schematic upon which different fields and disciplines can affix phenomena and theories. The purpose of the diagram is to establish the nature of the transformative processes that are being suggested as emergent and at what level in a hierarchy they apply.

The elements of the diagram can be enumerated:

- *Collections* (y-axis): the aggregation of individual units into collections of units
- *Transitions* (x-axis): the transformation of units or their collectives through suitable local or global mechanisms.
- Independent elements appear in lists surrounded by curly brackets: $\{x,y\}$.

- Elements conjoined (∪) into stable units through transitions are surrounded by parentheses: (x,y).
- Collections indicate parts or units aggregated (∈) into multiplicities of parts or units: $n\{x,y\}$, $n(x,y)$.
- Entailment (↔) and failures of entailment (↮) indicate intertheoretic reduction and non-reduction.
- The preservation (~) of molecular identity and abundance in different "phases": $n(xy)$ or $n(xy)'$.
- Different effective theories are indexed by their position on the lattice: ET$\{i,j\}$.

The first objective of the lattice is to clarify the very different senses of emergence implied by transitions and collections. In each domain of inquiry these map onto distinct mechanisms. *Transitions* (**T**) are typically local and pairwise. *Collections* (**C**) are typically global and involve many-body interactions. Secondly, whereas mechanisms of transitions are more often highly domain-specific, mechanisms of collections tend to be more generic (e.g., universality as captured through the renormalization group). Thirdly, different positions on the lattice do not all equate to different effective theories. Intertheoretic reduction, or entailment, can reduce the need for a new effective theory—no new theory is required to describe $n\{x,y\}$ not already contained within $\{x,y\}$.

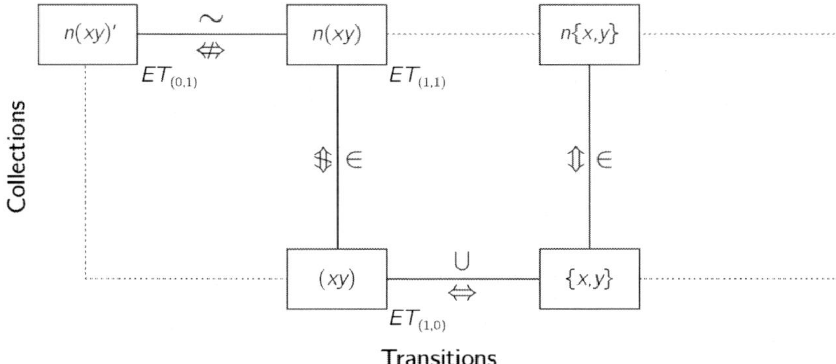

Figure 5: The emergence lattice is intended to keep track of emergence claims that are "horizontal"—Transitions derived from local or global interactions; and those that are "vertical"—Collections that move between individual and aggregate descriptions. Both possess system-specific and generic properties. Transitions are often described in terms of symmetry-breaking, Collections, in terms of coarse-graining and renormalization. Effective theory is abbreviated as *ET* in this and subsequent diagrams.

A Thermodynamic Lattice

The relationships between hydrogen and oxygen, individually and collectively, at different temperatures and pressures, provides a convenient example for the use of the emergence lattice. Enumerating each term individually:

- *Collections.* Molecular (H_2) and (O_2) exist as diatomic gases until sufficient energy initiates an exothermic reaction.
- *Transition.* The transition from molecular hydrogen and oxygen to a single water molecule involves the formation of three sigma bonds. These bonds are described using molecular orbital theory, which details constructive interference of the wave function of each atomic orbital.
- ($\leftrightarrow$). The structure of (H_2O) is entailed by knowledge of (H_2) and (O_2).
- ($\in$). The polarity of water molecules allows them to form hydrogen bonds in bulk, thereby aggregating into characteristic tetrahedral configurations in water.
- ($\not\leftrightarrow$). The many properties of water, including ionic diffusivity, solvation, heat capacity, etc, cannot simply be derived from the properties of (H_2O) molecules.
- ($\sim$). At different temperatures and pressures, water undergoes phase transitions to solid and gas. The same Collections can exhibit different symmetries.
- (ET). Each phase has its own effective theory: Kinetic theory of gasses, Navier–Stokes equations for incompressible liquids, theory of crystal lattices for solids.

Emergence as a Transition (e.g., solid to liquid) is not at all like emergence as a collection (molecule to water). When we speak of emergence in the context of the formation of water, we are using ambiguous shorthand for the theory of inelastic collisions (T); the theory of chemical bonds (T); the theory of phase transitions (T); the renormalization group (C); and collective currents (C). Each one of these theories relates to a large class of distinct mechanisms (of varying degrees of "fundamentalness") often highly dependent on the matter involved.

An Evolutionary Lattice

A far more complicated case to be analyzed with the emergence lattice involves an evolving lineage. Entailment fails nearly everywhere, and effective theories proliferate. There are no canonical theories for transitions, and collectives involve so many heterogeneous interactions that it is impossible to predict detailed properties of the aggregates. In addition, each configuration is able to evolve an internal *schematic* encoding of its environment (genome,

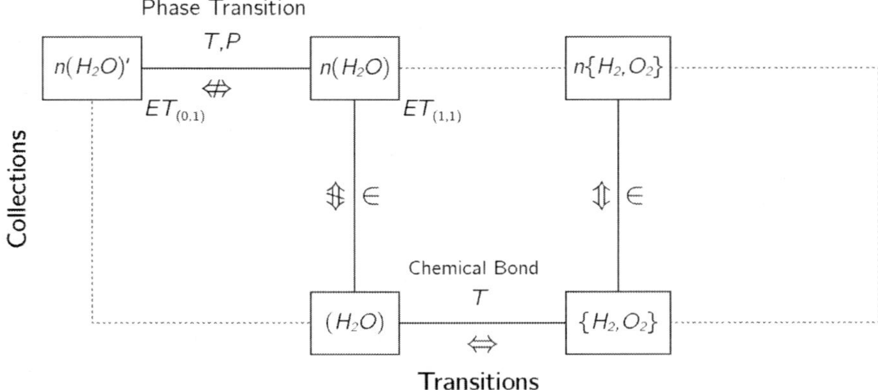

Figure 6: The Emergence Lattice of Thermodynamics. Temperature is abbreviated as T and pressure is abbreviated as P in this diagram.

brain, etc.)—constituting an endogenous effective theory. In the following list, terms are placed together as they appear grouped on the lattice:

- *Collections.* (∪), (↮), (ET). Individual lineages form into an ecosystem defined by energy and mass transfers.
- *Transitions.* (↮), (ET). Lineages can fuse through symbiosis, forming a stable cellular unit.
- *Transitions.* (↮), (ET). Cells transition through genotypic or phenotypic modification.
- *Collection*, (∈), (↮), (ET). Cells cooperate into multicellular organisms with significant cellular division of labor.
- Evolution, (ET). Lineages of organisms do not transition directly one into the other. Transitions rely on subjacent transitions, followed by collections at the cellular level and below, in order to produce variation for competing populations at the multicellular organismal level. This selection process relies upon extraordinarily high-dimensional inputs from the environment (treated below as emergence in knowledge environment).
- (*S*). Unique to evolving lineages are internal schema (a simple internal effective theory of the world) required to persist and survive. These schemas encode sensory, memory, information processing, and behavioral capabilities. They are not constant over time.

The evolutionary example almost breaks the lattice. There are so many candidates for emergence and so few established mechanistic theories for how these might come about. Emergence refers to the origin of cells with cellular-properties, tissues and organisms with unique functions, the introduction of novelties through time over the course of evolution, and most importantly, an endogenous form of effective theory, a schema. The immediate consequence of a schema is that it blurs the boundary between the epistemological and the ontological. Every organism is both the object of a theory and a theorizer about the environment in which it lives. Human epistemology is the cognitive and cultural aggrandizement of this schematic feature of complex systems.

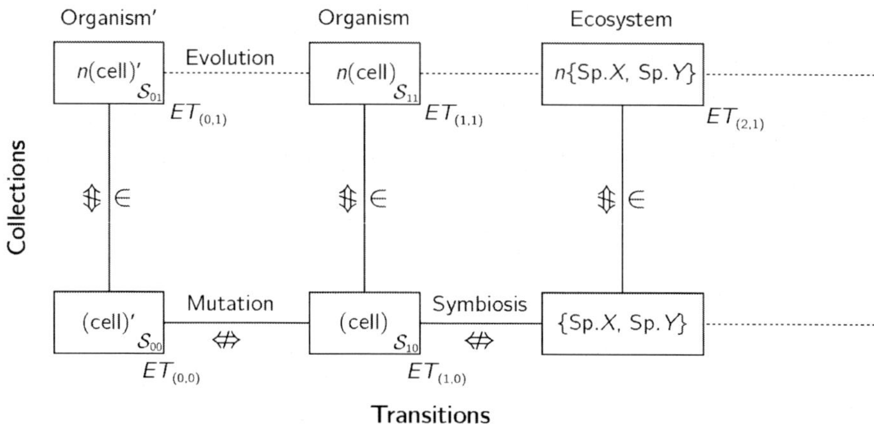

Figure 7: The Emergence Lattice of Evolution. Schema is abbreviated as S in this diagram.

Knowledge, Transitions, and Collections

Directly comparing the thermodynamic lattice and the evolutionary lattice obscures one of the most important considerations of complexity science: the disparity between a small number of global control parameters in a purely physical setting and a vast number of local selection pressures in the adaptive setting.

Darwin described the environment as an almost infinitely discerning "observer" of biotic variability—"an entangled bank." (Darwin 1859, 489) In order to facilitate a transition between any two lineages or species, an unknown, very large number of internal degrees of freedom must be modified; these form the basis of each genome, agent, or organismal schema. This is not a context where the theory of phase transitions has much to offer about biological emergence.

Let's call the typical physics theory "knowledge second" or **K2**, and the complex situation "knowledge first" or **K1**. A **K2** system produces knowledge from very limited inputs (knowledge from limited data). In physical and chemical theory, these are the laws of physics, initial conditions, and suitable control parameters. Using these minimal assumptions we are able to provide models and explanations for a wide range of observations from quantum mechanics to condensed matter and celestial mechanics. Eugene Wigner (1964, 995) describes physics in the following terms: "The elements of the behavior which are not specified by the laws of nature are called initial conditions. These, then, together with the laws of nature, specify the behavior as far as it can be specified at all: if a further specification were possible, this specification would be considered as an added initial condition."

A **K1** system produces knowledge from a vast number of inputs (knowledge from nearly unlimited data). In biological theory these are described as the environmental selection pressures acting on lineages of organisms; in psychology the childhood environment of assimilation; and in anthropology the facts of culture. Unlike the initial conditions and simple macroscopic variables, including temperature and pressure, selective inputs are able to fine-tune through feedback high-resolution states of everything from DNA molecules to the form and function of tissues, and the "circuits" of the brain. These are the schema (*S*) of complex matter.

Ideas related to emergence, or how scale affects phenomenology, sit very naturally in **K2**. This is because the objective of **K2** theories is to demonstrate how parsimonious mechanisms applied to large collections of identical, or similar, elements produce outputs that require much scientific knowledge to explain. As described previously, the canonical example is how the collective properties of vast numbers of identical H_2O molecules produce a new phase of matter (liquid), by changing a single global input (e.g., temperature). This produces a state of order that is explained in terms of new forms of knowledge—effective theories—including the ideas of compressibility, boiling point, viscosity, solvency, and solubility.

When it comes to **K1** systems, the status of emergence is more ambiguous. This is because the ratio of inputs to knowledge outputs is large. A child learning a second language in a classroom is learning a preexisting knowledge structure. In fact, the language that any one individual holds in their head is self-evidently a tiny fraction of the total "dictionary" language available. It is also a small sample of the language to which any one individual has been exposed over the course of their education.

Many effective theories in complex systems are of type **K1** and are therefore theories stressing the environmental inputs: natural selection, linguistics, ecology, archaeology, economics, etc. It is for this reason that biologists and social scientists use the words adaptation, learning, and evolution far more frequently than they use the word emergence.

An important challenge for complexity science is not only to explain how a small theory can account for considerable variability across a diverse range of physical phenomena (**K2,** e.g. Newton's theory of gravity), but how a small algorithmic theory (**K1**, natural selection, reinforcement learning, etc.) explains how vast numbers of input parameters produce a diverse range of adaptive phenomena—the difference between Ockham's razor, which applies to **K2** and meta–Ockham's razor (**K1**).(Krakauer 2023).

The **K1–K2** distinction lies at the heart of the emergentist attack on selection from biophysics and self-organization, and the more recent attack on intelligence in Large Language Models. Starting with D'Arcy Wentworth Thompson in his magnum opus, *On Growth and Form*, and continuing through the works of Alan Turing, Nicolas Rashevsky, Hermann Haken, Ilya Prigogine, John Conway, and Stuart Kauffman, there has been a concerted project to transform **K2** systems into **K1** systems. This has the character of substituting out a large number of local selection pressures or learning parameters (input contingent parameters) with a far smaller number of global control parameters.

INFORMATION, COMPUTATION, AND COGNITION

The development of complexity science is intertwined with the development of information theory, computing, computer science and engineering as they relate to information processing—cognition (Pylyshyn 1984; Piccinini and Scarantino 2011).

This is partly instrumental. The absence of closed-form solutions for many nonlinear dynamical systems, the use of stochastic simulations, the explicit use of computational concepts in fundamental scientific theories, and the purposeful or teleonomic behavior of agents induced an important mutual dependence between new complexity-related concepts and new information-processing frameworks. These include the consideration of computing resources and a variety of algorithms and heuristics. It is also partly fundamental—the complex domain as one where general information-processing principles apply at many scales.

In this section there is focus on a computationally oriented selection of the *Foundational Papers*: these describe a variety of proto-computational, computational, and cognitive systems (computing system, algorithms, and concepts).

Computational concepts began to be used in the 1920s and increased in use dramatically into the 1940s and 1950s, as physical computers became less scarce and computational principles became more important (Ceruzzi 2003). These principles include ideas relating to thermodynamics, calculation, filtering, sorting, controllers, Turing machines, transistor-like switches, and digital memories. By the 1960s there is scarcely a paper that does not rely on either computational hardware and software, and more importantly, place computational (information-processing) principles on an equal footing with those of physics, chemistry, and biology (Zenil 2013). This section is not about the "use of computers" in complexity research, or the extraction/discovery of patterns in complex systems through simulation, but a consideration of the diversity of computational principles and their sequelae as they *inhere* in complex systems.

Table 4 is organized along similar lines to David Marr's 1982 framework for levels of information-processing system (Marr 2010). The "Computing System" column describes the "category" of logical-machine upon which processing is assumed to take place; physical (as implied in the original Marr scheme) or abstract (e.g., cellular automata on a digital computer). The "Algorithm/Rule" describes the local operations acting on states of the machine. "Computational Concepts and Functions" are the desired outputs, their properties, and the principles of the computation. "Research Hardware" (when relevant and where available) describes the particular devices used by researchers in production of the publication when essential for the findings.

Table 4. A chronology of foundational papers exploring the interface of complexity and computation. Each contribution is described in terms of (1) the basic computing system or logical machines on which (2) algorithms or rule systems operate in order to (3) produce a desired function or output.

Foundational Paper & Year	*Computing System*	*Algorithm/Rule*	*Computational Concepts & Functions*	*Research Hardware*
Szilárd (1929)	Physical systems and observing device	Measure/filter	"Intelligent" selecting mechanism with a memory	N/A
McCulloch and Pitts (1943)	Physical neurons	Integrate and fire	Recurrent propositional calculus; "emergent parallel computation"	N/A
Rosenblueth, Wiener, and Bigelow (1943)	N/A	Negative feedback	Servomechanisms: passive and predictive; (extrapolative) input–output behavior	N/A
Shannon (1948)	Telegraphy; vocoder; SIGSALY system	Discrete transducer	Optimal encoding and decoding of messages; information; error correction	N/A
Turing (1950)	Universal machines	Store; execute; control	Universal machine; imitation game; artificial intelligence	N/A
Kálmán (1960)	N/A	Control grammarian; linear quadratic regulator	Controllability and observability	N/A
Landauer (1961)	Magnetic films; cryotron; flip-flop circuits	Moving particle in bistable potential	Irreversible switches; minimum energy bounds; entropy production; entropy reduction	N/A
Minsky (1961)	Turing machines; digital computers	Search; learning; planning; induction; pattern-matching	Artificial intelligence; problem-solving	N/A
Holland (1962)	Iterative circuit computer	Generators; generation trees	Logic of adaptation; universal generations procedures; populations of programs; adaptive computation	N/A
Ulam (1962)	Cellular automata	Recursion relations; discrete dynamics	Pattern formation; growth and morphogenesis	IBM 704 36 bit; vacuum, tube
Lorenz (1963)	N/A	Hydrodynamics: forced dissipative flow	Deterministic chaos; divergent trajectories; long-range unpredictability	Royal McBee LPG-20; 31 bit; vacuum, tube

Foundational Paper & Year	*Computing System*	*Algorithm/Rule*	*Computational Concepts & Functions*	*Research Hardware*
Solomonoff (1964)	Turing machines	Formal grammars; stochastic processes; Bayes' theorem	Universal induction; algorithmic information theory	N/A
Cobham (1965)	N/A	Addition and multiplication	Computational complexity; polynomial versus exponential time	N/A
von Neumann (1966)	Cellular automata	"Copy" and "paste"	Self-replications ("autoreproduction"); controller or "constructor" operations	N/A
Chaitin (1966)	Turing machines	Bounded-transfer Turing machine	Effective computability; randomness and incompressibility	N/A
Kolmogorov (1968)	Theory of recursive functions	Asymptotically optimal programs	Kolmogorov complexity; computability and randomness	N/A
Conant (1970)	Morphisms of Turing machines	Regulation by error-control	Theory of model-based regulation	N/A
Schelling (1971)	Cellular automata; agent-based model	Discrete dynamical updating in Moore neighborhood	Spontaneous order; emergent sorting/segregation	"Tabletop" experiments
Karp (1972)	Turing machines	Reduction; language; recognition problems	NP-complete property; combinatorial complexity	N/A
von Foerster (1972)	Recursive relations	"Computations" on representations and their relations	Self-reference; theory of the observer	N/A
Bennett (1973)	Turing machine	Logically reversible Turing machine	Thermodynamical reversibility; operating close to thermodynamic equilibrium; efficient computation	N/A
Varela, Maturana, and Uribe (1974)	Cellular automata	Catalysis and linkage	Autopoiesis; self-production; self-distinction	N/A
Hopfield (1982)	Recurrent neural networks	Hebbian learning	Content-addressable memory; generalization; error-correction	N/A

Table continues on next page.

Foundational Paper & Year	*Computing System*	*Algorithm/Rule*	*Computational Concepts & Functions*	*Research Hardware*
Wolfram (1984)	Cellular automata	Finite state transitions in 1D	CA classes (1–4); dynamical and computational classes; limits to prediction; computational irreducibility	N/A
Langton (1986)	Cellular automata	Finite state transitions in 2D; sparsity control parameter λ	Equivalence of operators and operands; emergent phase-diagrams; molecular logic as propagating periodic structure	Apollo Corporation DN600; Dual 6800
Wheeler (1989)	Physical universe and observing device	Measurement of bits; self-referential deductive systems	It from bit: "Communication employed to establish meaning"; most elementary physics (quantum) involve observer-participant in a Shannon channel	N/A
Holland and Miller (1991)	Agent-based model	Genetic algorithms; classifier systems	Artificial adaptive agents (AAA); emergent economic phenomena	N/A
Mitchell, Hraber, and Crutchfield (1993)	Cellular automata	Genetic algorithms	Emergent computational capability; evolutionary "programming" of parallel computers	N/A
Arthur (1994)	Agent-based model	Genetic algorithms; temporally fulfilled expectations	Co-evolutionary induction	N/A
Crutchfield (1994)	Bernoulli Turing machine	Epsilon machine reconstruction	Causal states; statistical complexity; entropy rates; finitary stochastic hierarchy	N/A
Forrest, Perelson, Allen, and Cherukuri (1994)	Digital computer; SPARC OS; DOS	Negative selection algorithm	Computational immune systems; self–nonself in computer security	N/A
Gell-Mann and Lloyd (1996)	Turing machines	Algorithmic information content	Effective complexity; total information	N/A

Conceptual Material

There is something wanting in the usual description of computation as the "execution sequences of halting Turing machines (or their equivalents)," just as there is limited insight derived from defining cognition in terms of "acquiring knowledge and understanding through thought, experience, and the senses." Both definitions clearly refer to a vast subterranean knowledge base. And yet anything we might care to describe or model in nature requires that we operationalize Turing machines, understanding, and thought.

In an important clarification for both computation and cognition, Edsger Dijkstra distinguished computation—the trajectory of a dynamical system—from an algorithm—the logical specification of a procedure guiding the computation. Algorithms are like recipes and computation is like cooking. The programmer, by applying an algorithm, structures "what is happening where in a useful way."

A decade later David Marr made a similar distinction when describing the visual system. Marr thought of computation in terms of processing of sensory inputs, algorithms as the specifications for how this processing should proceed, and the implementation layer described in terms of neurons, their properties, and their connections.

Both Marr and Dijkstra proposed that a science of computation or cognition should not be confounded with physical material and artifact but understood in terms of *conceptual materials*. This suggestion shuffles around the subject-object boundaries established in the physical sciences where matter moves and minds cogitate. A fairly representative take on the role of traditional physical theory is described by Roger Penrose (2004, 1,024):

> *Nevertheless, I should make it absolutely clear that the apparent lack of objectivity is not the fault of Nature herself. There is an objective physical world out there, and physicists correctly regard it as their job to find out its nature and to understand its behaviour.*

Which is as it should be. But when a natural structure is a purposeful mechanism, encoding both history and policy, and adapting to its own environment, then a *Dijkstra–Marr decomposition* (**DMD**) might be a better fit. It is rather revealing to look at a number of the Foundational Papers through the lens of the DMD. These illustrate how divergent studies in complexity can be from the standard model of a material or physical reality (ontology) cleanly separated from the logical methods of their analysis (epistemology).

In the following sections, the insights of the contributors are recruited to explain key ideas in the development of computation and cognition applied to complexity.

1920–1930

DEMONS

SZILÁRD (1929)

The first study to take a DMD-like approach did so in order to resolve the challenge to the second law made by James Clerk Maxwell—the problem of "Maxwell's demon" as described by Lord Kelvin (Thomson 1874). Leo Szilárd in his paper "On the Decrease of Entropy in a Thermodynamic System by the Intervention of Intelligent Beings" (1929) opted to put the ghost back into the machine. The demon's intelligence consists in measuring and remembering microscopic states of a stochastic process. The curious thing about the Szilárd approach, building on Maxwell, is that it performs computational thought experiments at a fundamental scale. Susanne Still notes how "Szilárd's work . . . plays a role in the recent developments in far-from-equilibrium (also called "stochastic") thermodynamics, in situations in which the protocol applied to a thermodynamic system is dependent on measurements of (some aspects of) the system's state. . . ."

1930–1950

NEURONS

MCCULLOCH AND PITTS (1943)

Warren McCulloch and Walter Pitts (1943) present the first successful effort to connect a physical entity—the neuron—to both algorithms (logical circuits) and functions (logical propositions). A brain is every bit as physical as a galaxy, but is insufficiently understood by any physical effective theory. The brain needs to be understood by a qualitatively different class of effective theories, those we describe as cognitive or computational. Bruno Olshausen and Christopher Hillar dwell on one of the more surprising contributions of this work: "Among the most enduring contributions of McCulloch and Pitts's paper are their hand-drawn diagrams of neural circuits hypothesized to perform the fundamental operations that would constitute a system of logical calculus. Today we readily recognize these diagrams as the logic circuits—for example, and-gates, or-gates, and not-gates—that lie at the core of every digital computer. However such circuits were not at all obvious at the time, and in fact had not yet been conceived." How delightful, and so at odds with much utilitarian thinking, that a project to connect the soft tissue of the brain with the lofty philosophy of the *Principia Mathematica* led to the logic circuit.

FEEDBACK

ROSENBLUETH, WIENER, AND BIGELOW (1943)

Arturo Rosenblueth, Norbert Wiener, and Julian Bigelow (1943) "erased the difference between machines, animals, and indeed, humans," as Andrew Pickering describes their contribution. The strategy they pursued was to abstract an object from its environment and analyze the relation of its outputs to its input in terms of an error minimization (function). And this was to be achieved through negative feedback (algorithm). No longer a computational description of a natural system per McCulloch and Pitts, but a project to engineer the ultimate tracking device. The cybernetic movement was an impulse that over time morphed into the current interest in agency. And it is perhaps surprising that, as Pickering sees it, "Cybernetics is much discussed these days in the arts and humanities, and less so in engineering, but echoes of 'Behavior, Purpose and Teleology' are heard all over the disciplinary map."

INFORMATION

SHANNON (1948)

Working at Bell Labs, Claude Shannon wrestled with the practical computational challenge of transmitting a message reliably across space. This requires an algorithmic means of encoding a message with sufficient redundancy to be able to error-correct. The surprising thing about information theory is how widely applicable it has become, and how little this seems to depend on the underlying physical system. Seth Lloyd suggests that Information theory has in part been successful because "[t]he mathematical techniques that Shannon introduces are powerful but subtle. One of the triumphs of 'A Mathematical Theory of Communication' is that it uses multiple examples to present these techniques in a manner accessible to a broad audience." There is perhaps a lesson to be learned from Shannon's felicitous presentation of demanding ideas, which no doubt contributes to the consensus view that, "information and information processing lie at the heart of the sciences of complexity."

1950–1960

INTELLIGENCE

TURING (1950)

By the time Alan Turing wrote his paper, "Computing Machinery and Intelligence" (1950), he had jettisoned all consideration of the physics and chemistry of matter in favor of a purely functionalist approach. It is like Marr without the implementation layer (one might say he reads like Marr without a brain). Rather than design an experiment to probe the laws governing matter, Turing invented a new kind of experiment—"The Imitation Game"—to determine whether a computer is intelligent. It is hard to overstate how radical this move was: away from the methodology of particle accelerators, towards the analyst's couch. The many and various implications of this new and distinctly positivist verification principle are explored by Daniel Dennett through "what I will call anthroponormative thinking about thinking: our way of thinking, the human way of thinking, is the only kind of thinking worth considering."

1960–1970

STATES

KÁLMÁN (1960)

Several of the logical extensions of cybernetics to include multi-input–output, non-stationary processes, and internal states, were made by Rudolf Kálmán in the 1960s by means of the concepts of controllability and observability. Maxim

Raginsky argues that "Kálmán made it possible to speak in a unified way to the interplay between structure, organization, and function of open dynamical systems." Kálmán augments the feedback narrative of Wiener and colleagues by integrating in ideas from nonlinear dynamics and thermodynamics.

BOUNDS

LANDAUER (1961)

In 1961 Rolf Landauer reintroduced physics to computation. Reestablishing the connection that had begun with Szilárd and that Turing, following a purely logical approach, had severed. In the creation of the field of the "thermodynamics of computation," Landauer demonstrated how any elementary computation implied a minimum energetic costs (kT ln2), now described as the minimal entropy flow. As David Wolpert observes, in order to arrive at an understanding of the physics of computation, a "revolution in physics" was required, and this in turn has provided some of the theories to "fully and formally analyze the thermodynamics of what we now call complex systems," Thus a computational approach along the lines of DMD loops back down to change the way physicists conceive of reality.

MINDS

MINSKY (1961)

In the same year, 1961, Marvin Minsky pursued the functionalists' approach to its limits: replacing algorithms with approximate heuristics and, like Turing, paying no heed to physics, chemistry, or biology. For Minsky, intelligence—which he thought of as generalized problem solving—was largely a matter of solving problems through "incomplete analysis." As Melanie Mitchell explains, Minsky's deliberate disavowal of the work of McCulloch and Pitts was one of its failings, and many today are "proposing that true machine intelligence will require deep neural networks to be integrated with . . . symbolic processing."

ADAPTATION

HOLLAND (1962)

Despite the obvious evolutionary foundations of complex systems, it was not until John Holland's paper in 1962 that a serious attempt was made to integrate population genetics with the theory of computation. Rather like McCulloch and Pitts seeking to reconcile mind-level-logic with a cell-level-calculus, Holland wanted to re-describe adaptation as a logical procedure: adaptation as a large distributed natural algorithm. Holland's approach, notes John Miller, was to imagine "a machine that is able to generate any possible program that could be run on Turing's universal computer. Adaptation is viewed as a process that modifies such generation process in response to feedback from the environment . . ."

PATTERNS

ULAM (1962)

It is natural to relate computation and cognition to questions of intelligence. One might just as effectively consider computation as an alternative framework for deriving the rules governing complex reality more generally. This was the perspective articulated in 1962 by Stanisław Ulam. Rather than "solve problems" à la Turing and Minsky, Ulam sought to generate patterns. Erica Jen explains that "The objective was to determine global characteristics such that limiting density of cells and to describe the dynamics . . . of the coherent structures generated in the spatiotemporal evolutions." By inventing cellular automata and exploring their behavior through exquisite visualizations, Ulam established one of complexity science's credos—unexpected forms of order arising through the collective iteration of simple, discrete rules.

CHAOS

LORENZ (1963)

There are few words as closely allied to complexity as chaos. Whereas complexity encompasses a very large number of phenomena, chaos is more restrictive in both derivation and application. And yet chaos strikes at the very heart of all mechanisms, exposing the ultimate limitations on prediction, and thereby erodes one of the dominant criteria used to evaluate the correctness of a theory. Chaos was discovered through computation, but, more profoundly, it reveals the limitations of any analog computational system. Given that the natural world is analog, and only approximates digital behavior, the reach of chaos is large. Doyne Farmer writes that "without chaos, the world would be extraordinarily boring: life and thought would be impossible without it."

DIFFICULTY

COBHAM (1965)

The concept of time is a fascinating conundrum in physics, but it is an existential challenge for complex systems. Questions relating to estimated life span, developmental rate, transient states of order, evolutionary fixation, and solution time, are all apposite for understanding a complex system. The problem of understanding the time-scaling of algorithms (how solution time scales with problem size) has emerged as one of the more rigorous definitions for computational complexity. Alan Cobham's paper, "The Intrinsic Computational Difficulty of Functions" (1965), provides an insight into what is meant when we state that a solution will take a long time to calculate independently of the physical computer upon which it is realized. Cris Moore writes that "Cobham was among the first to argue that problems have an intrinsic complexity—an objective fact about their structure, not a subjective one about whether or not we have found a clever algorithm." Cobham provided a principled insight into those problems for which the choice of physics is merely a constant of proportionality.

INDUCTION

SOLOMONOFF (1964)

A question related to intrinsic difficulty is how much contingent information from a process is required in order to identify its generative model—the problem of induction. Evolution through natural selection samples a small number of environments in order to produce adaptations of general value;

learning involves experiencing a small training set in order to establish rules for unfamiliar contexts. In both evolution and learning, traits and knowledge are acquired that enable subsequent induction. Ray Solomonoff tackled these problems in his "Formal Theory of Inductive Inference" (1964). As Paul Vitányi describes it, "Solomonoff's ultimate goal in AI, as pursued in many subsequent papers, was a general system for machine learning. The goal is not so much to acquire knowledge itself but rather to determine how learning is performed by machines." Solomonoff's solution was to build a hierarchy of functions in which new problems access lower levels from earlier inputs. Unlike Turing, who presented a deductive framework with a one-dimensional architecture, Solomonoff arrives at a modular architecture for induction comparable to those described by Warren Weaver and Herbert Simon as hallmarks of complex systems.

COMPRESSION

CHAITIN (1966)

Whereas Solomonoff sought to induce general rules from short sequences, Gregory Chaitin sought to find the shortest programs to encode deductive systems. It seems somewhat intuitive that the length of a program to solve a problem should be related to the difficulty of the problem, following a logic along the same lines as Cobham's investigations. If one were to build a Turing machine, would there be a minimal way to do so, and how would this scale with the challenge? Simon DeDeo describes Gregory Chaitin's paper, "On the Length of Programs For Computing Finite Binary Sequences" (1966), as one that "twists back and forth between the mind-numbingly tedious pretense that a Turing machine is something one might actually want to construct in a materially efficient fashion (e.g., the endless design specifications of section 1.3, and the excursion of section 2, which feel like a user manual for a science-fictional device), and the mathematically mysterious idea that some sequences of binary digits might be more difficult—as a matter of engineering—to calculate than others." Chaitin's idea of complexity can be mapped in approximate language to genomes, neural schema, and machine weight vectors. To the extent that these have been minimized, they tell us something about the complexity of the problems that they are solving.

REPLICATION

VON NEUMANN (1966)

There is no evidence that Dijkstra and Marr ever met. But one might be forgiven for believing that they met virtually—in the imagination of John von Neumann. In an effort to understand the elaborate architecture of the brain, and of life itself, von Neumann chased down the quantitative implications of a number of computational metaphors and models. Von Neumann demonstrated considerable wisdom when he wrote "Now, none of this can get out of the realm of vague statement until one has defined the concept of complication correctly. And one cannot define the concept of complication correctly until one has seen in greater detail some critical examples, that is, some of the constructs which exhibit the critical and paradoxical properties of complication." Despite this assertion, von Neumann left a great deal in his writings ambiguous, but this barely diminished their influence. As Neil Gershenfeld describes the research: "Unusually, given von Neumann's mathematical rigor, he introduced a key concept of 'complication' without defining it beyond an ability 'to do things.' He doesn't quite mean complexity; he's getting at the minimum requirements for a system to reproduce and adapt. His description of complexity anticipates important results to come, including the work he presented in lectures at Caltech in 1952 on computing reliably with unreliable devices and the recognition of the physical limits of computing."

COMPLEXITY

KOLMOGOROV 1968

Much as Chaitin builds on Turing to explain complexity, Andrey Kolmogorov arrives at similar conclusions by building on Shannon and the theory of partially recursive functions. Simon DeDeo explains the idea in terms of "'[s]imple' objects, ones with little algorithmic information, [which] have succinct programs that generate them; the Mandelbrot set may look complicated, but only a short program is needed to cover a wall with its patterns. 'Complex,' 'information-bearing' objects, meanwhile, have only long programs, full of exception cases and details." Kolmogorov moves in the direction of restoring to communication what Shannon very deliberately left out—semantics and behavior. Whereas information is rather like reporting only the caloric content of a recipe, Kolmogorov complexity tells us something about its ingredients. This is a computational step closer to what we mean to convey by "complex."

1970–1980

REGULATION

CONANT AND ASHBY (1970)

Roger Conant and W. Ross Ashby pick up the threads of the cybernetic–control narrative from Arturo Rosenblueth, Norbert Wiener, Julian Bigelow, and Rudolf Kálmán. Rather than move down in DMD into the implementation, they move several levels up into algorithmic abstraction. Like Kálmán, their interests lie in opening the black box of behaviorism favored by Rosenblueth et al. Somewhere in this box are echoes of Solomonoff, Chaitin, and Kolmogorov. The challenge of regulation must relate in some fundamental way to the complexity of the regulated. James Crutchfield writes, "Perhaps stating the opposite helps: if the regulator knew nothing about the system—nothing about its accessible configurations and states—it could do no effective control. The regulator could perturb the overall system, but those changes would not be coordinated with the system's moment-by-moment condition. And so the regulator would help as much as it hindered reaching the goal." Through the regulatory lens, complex reality is like a hall of mirrors composed of a near-endless series of reflections.

SEGREGATION

SCHELLING (1971)

How do these several computational ruminations bear on the complex systems in which we swim? Thomas Schelling took it upon himself to answer this question in "Dynamic Models of Segregation" (1971). Schelling is in the lineage of McCulloch, Pitts, and Ulam searching for simple rules that might reproduce the outlines of complicated patterns; he channels the parsimonious philosophy of Chaitin and Kolmogorov, for whom the complexity of a model might have something to say about the complexity of the social phenomenon. It is also an attempt to place Hayek's theories into a computational language. Peyton Young considers that "Schelling's 1971 article represents a major departure from this way of thinking. Indeed, it represents one of the clearest expressions of the notion of spontaneous order and how it can be applied to a contemporary matter of practical significance, namely, racial segregation." Through an *agent-based model* (**ABM**), Schelling is able to connect ideas related to collective dynamics with simple heuristics to derive an emergent pattern of enormous social consequence.

REDUCTION

KARP (1972)

There is little sense to the question, "How intrinsically difficult is it to synthesize ethanol from ethylene and water?" We have a perfectly adequate theory of chemical reactions and the competence of a chemist should play no part in the correct answer. But there are problems where such difficult questions are natural, and these relate to calculating solutions to combinatorial problems. Some of the hardest of these problems are those that take an exponentially long time to discover but only polynomial time to verify: NP-hard. These problems might sound rather exotic, but in 1972 Richard Karp showed just how widespread NP-hard problems really are. As Cris Moore describes the situation, "We can then build simulations on top of simulations, creating a family tree of NP-complete problems. . . . This family tree has since grown to include hundreds of problems in algebra, automata, machine learning, statistical physics, calculus, and virtually any other system in which complexity can live." Once again, it is not the materials that pose the problem but the internal structure of a system—a property that yields to an algorithmic and computational description.

SUBJECTIVITY

VON FOERSTER (1972)

One of the consequences of the cybernetic movement was to alert the world to the ubiquity of feedback. Kálmán and Conant and Ashby presented some of these implications in terms of control and regulation. It was Heinz von Foerster (1972) who extrapolated these ideas to the limit where the regulator and the regulated coalesce into self awareness: "Abandon all hope, ye who enter here." Manfred Laubichler summarizes the dilemma: "Von Foerster's starting point is as simple as it is profound. Insofar as the observer is now a crucial part of any scientific account, we need a description of the observer to be part of any science. And, as the only observers we know are living beings (so far limited to our planet), we need a biologically sound theory of these observers and their ability to make these observations and express them in a coherent form." The information-processing implications of this situation remain an outstanding challenge for complexity, and are only likely to get worse, as we continue to replicate ourselves imperfectly in machine-learning memory.

REVERSIBILITY

BENNETT (1973)

Szilárd vanquished perpetual motion machines of the second kind. Landauer calculated the minimal costs of running a computing machine. Charles Bennett demonstrated that if a computation was reversible these costs might be driven to zero. Jon Machta explains how "Bennett refuted the belief that computing requires logical irreversibility and heat dissipation by showing explicitly that it is possible to design a general purpose logically reversible computer. . . What Bennett showed by construction is that any computation that can be carried out by an irreversible one-tape Turing machine can also be carried out by a three-tape Turing machine that is logically reversible." Bennett's extraordinary insight was to suggest a way in which an algorithmic account of computational work might trump the physical laws mandated by mechanical work. This has not yet happened.

AUTOPOIESIS

MATURANA, VARELA AND URIBE (1974)

Investigations into the origin of life are traditionally conducted in laboratories. Prebiotic conditions are replicated and small molecules are provided. Elementary forms of biosynthesis are construed as plausible models for the transition from an abiotic to a biotic planet. Humberto Maturana, Francisco Varela and Ricardo Uribe started somewhere else: with the computational principles that these chemistries were aiming to manifest. Randall Beer writes, "an autopoietic system is one organized as a network of processes that have the dual properties of self-production and self-distinction. In order to concretely illustrate the idea of autopoiesis and with the hope of spurring the development of formal tools for its analysis, Varela, Maturana, and Uribe developed and simulated a minimal computer model of the central ideas. The model takes the form of an artificial chemistry playing out upon a rectangular grid." The final sentence is an echo of Ulam and von Neumann and the prequel to a growing class of models that seek to discover the minimal set of rules inherent in life within the reasonable confines of an expanded Go board.

1980–1990

MEMORY

HOPFIELD (1982)

The logical properties of small neural circuits had been established by McCulloch and Pitts. The brain is made of billions of neurons and this begs the question of what new principles of computation might operate at larger scales. Drawing on insights from the collective dynamics of condensed matter, John Hopfield sought computational correlates corresponding to coordinated macroscopic states. David Sherrington explains the connections between computational neuroscience, dynamical systems, information theory, and statistical mechanics: "He also employed a very simple formulation for information coding in the synaptic connections, philosophically encompassing Hebb's observations on synaptic evolution, and with asynchronous dynamics. Hopfield noted that, for symmetric synapses, these assumptions led to a Lyapunov minimization dynamics resulting in associative cooperative dynamics toward retrieval of patterns coded in Hebbian synapses. This opened the door to Gibbsian statistical physics analysis and simulation, of a type that had recently been employed for spin-glasses." To this day, a tension remains between the intuitive appeal of simple circuits whose causal flows can be described, and the emergent properties of large networks, whose coordinated states perform computational work.

SPACE

WOLFRAM (1984)

Neither small Boolean circuits nor large collective networks elucidate the role that spatial patterns might play in computation. Building on Ulam's and von Neumann's interest in cellular automata, Stephen Wolfram initiated a project to connect these ideas directly to the logical concerns of Turing, Solomonoff, Chaitin, and Kolmogorov. What emergent forms of computation might we expect from compact rule-systems placed into discrete spaces? Wolfram suggested a taxonomy of pattern-formation corresponding to distinct dynamical and computational complexity classes. According to Hector Zenil, "The study of the limit behavior of small computer programs can provide an early indication of their computational capabilities without having to prove universality, which is known to be extremely difficult for general cases. The proofs of von Neumann, Konrad Zuse, or Conway for universality took decades to develop. It has been shown that pervasive computational universality, as defined by Wolfram's complexity classes, is likely to be very common and is suggestive of spatial computation in nature."

SIMULATION

LANGTON (1986)

The research of von Neumann, Holland, Varela, Maturana, and Uribe established connections between principles of computation and principles of life: algorithms that achieve replication, error-correction, and adaptation. Chris Langton pursued the DMD down into biochemistry, seeking a dynamical logic in which computational substrates could mirror those of chemistry. Langton was more interested in the dynamics of computation than the sequential recipe defined by algorithms. Sara Walker describes the profound functionalist implications of Langton's work: "If all the functions of life can be simulated in something like a cellular automaton, we can come to understand life purely by its logical and informational properties. We don't need to build life to understand it, we only need to simulate it." The debate over the adequacy of simulation is ongoing, ultimately turning on the question of whether the most causally significant factors live in the materials or the conceptual materials.

1990–2000

BITS

WHEELER (1990)

It is always exciting to encounter an apostate. And when they are leading congregationalists it is even better. John Archibald Wheeler was by any measure a considerable force in theoretical physics: an important contributor to general relativity and the theory of gravity, to fundamental theories of quantum mechanics, and to quantum information. Wheeler's apostasy was to take a coarse-grained theory of information and claim it is more fundamental than the microscopic laws of physics. Wheeler called this inversion "it from bit." Jessica Flack captures Wheeler's heresy: "Wheeler's most intoxicating conclusion is that information precedes matter, rather than the other way around, as is typically the case in grand unified theories. Wheeler's story turns not on objects like strings but on information-theoretic entities—decoders or registrants—his so-called observer-participants or choice-makers, as I call them" The extraordinary implications of this thesis is that physics should be viewed as a computation acting on information: complexity is foundational and physics an epiphenomenon.

AGENTS

HOLLAND AND MILLER (1991)

James Clerk Maxwell was astounded by the identical nature of every electron. His hypothesis to explain this uniformity was to posit a divine factory tuned to infinite tolerance. Physical theory takes homogeneity at scale as an input in order to develop powerful collective theories. Life is otherwise—fully heterogeneous and deeply history-dependent. This is not more in evidence than in societies, and yet canonical economics models are often in thrall to the uniform simplicities of the physical universe. John Holland and John Miller pursued a computational approach to economics that fully encompassed human diversity, developing *agent-based models* (**ABMs**). As Rick Bookstaber writes, "It seems self-evident that people are different from one another. We each have different ways of looking at the world, different approaches for making decisions. People also change. We are tempered by our individual experience and by changes in our environment, so how we make decisions today might not be the way we do so tomorrow. Such is the world in which we reside. It is also a world in which the artificial adaptive agents described by Holland and Miller reside." The reluctance to adopt ABMs is a complicated issue, in part a question of sensitivity analysis, but perhaps to a greater degree a persistent preference for mathematical formalisms in closed form, and an instinctual distrust of the computationally undecidable.

λ

MITCHELL, HRABER, AND CRUTCHFIELD (1993)

One of the greatest scientific insights into the natural world has been the realization that different forms of matter can be different phases of the same matter: water vapor, liquid water, and ice. The selection of phase depends on the value of a suitable control parameter (e.g., temperature or pressure). Using *cellular automata* (**CA**), Christopher Langton sought out a means of finding a control parameter (λ) describing the phases of computational patterns in terms of Wolfram's four complexity classes. But he went one step further: just as physical matter can be tuned to the vicinity of a phase transition, Langton suggested that computational forms of life would live at the "edge of chaos"—the boundaries of the complexity classes. Melanie Mitchell, Peter Hraber, and James Crutchfield discovered that, rather than tune to a critical value, evolved CAs tend towards values associated with the performance of a particular task; they tend to avoid edges. Dave Ackley summarizes the controversy: "For me, this particular story is about a shift from theoretical physics to empirical computer science. The idea that persists throughout is that physics and life are fundamentally linked, somehow, via computation and programming. The change is that later work focuses less on global properties to be derived or formally proven, and more on concrete examples to be implemented and studied. In this story, statistical physics as a formal framework gives way to 'emergent physics' observed in empirical results."

IMMUNITY

FORREST *ET AL* (1994)

The immune system is for many the canonical complex system: autonomous, decentralized, adaptive, coordinated, feedback regulated, and communicative. If nature can be rationalized as computational, what about rethinking computation in terms of the complexity of the immune system—or at least some algorithmic abstractions of immunity? This is what Stephanie Forrest and Alan Perelson did in their "Self–Nonself Discrimination in a Computer" (1994). Anil Somayaji describes the three influential contributions of this work: "The first was its demonstration that computational models of the immune system can be used to create novel algorithms. . . . The second key influence was its presentation of the negative selection algorithm. . . . The third key contribution of this paper was the idea of self versus non-self as applied to computer systems, particularly computer security." Forrest and Perelson's paper is a pioneering example of the use of biological models for technology development. Until recently, this has been rather limited. With the growth of brain-inspired neural networks and their surprising and astronomical economic impact, research along these synthetic lines is likely to exponentiate.

ε

CRUTCHFIELD (1994)

Since Kálmán, Conant and Ashby, and von Foerster, there has been a concerted effort to derive some index of the computational complexity of an agent. The approach of Chaitin and Kolmogorov assumed that the problem is known and needs to be parsimoniously described; Solomonoff came closer with his theory of inductive inference but, by basing his ideas on Turing machines, came up against the problem of uncomputability. In "The Calculi of Emergence: Computation, Dynamics, and Induction" (1994), James Crutchfield sought to derive a measure for an observed time series by constructing a minimal stochastic generator for its statistics—the (ε)-machine. Peter Sloot, Rick Quax, and Mile Glu describe Crutchfield's approach: "This agent is tasked with describing the observed process as efficiently as possible, i.e., by constructing an internal model which uses a minimal number of states to predict the observed future of the process. The overarching idea is that the extent to which the process is deemed 'complex' is defined by the agent's autonomous deliberation. More specifically, it is defined through the evolution of the size of the minimal model that the agent can formulate." Through this Conant–Ashby-styled lens, an agent induces a model from the world and then deduces its minimum—it thereby achieves an inferential cycle of adaptivity.

EL FAROL

ARTHUR (1994)

Game theory is a powerful formalism for studying strategic decisions at equilibrium. Over the past couple of decades, a number of formalisms have been developed to deal with the computational and cognitive challenge attendant on finding these equilibria; these include behavioral economics, Bayesian games, and algorithmic game theory. Brian Arthur's "Inductive Reasoning and Bounded Rationality" (1994) was one of the first to sacrifice the analytic for the synthetic, and do so using ABMs. Willemien Kets describes the problem as follows: "Suppose going to a bar is only pleasant if not too many people show up. How do you decide whether to go? This 'bar problem' was inspired by the popular El Farol bar in Santa Fe. But, of course, this toy problem is an exemplar of a much more general class of problems: How do people divide a scarce resource among themselves? . . . If the system is not in equilibrium, then at least one of the players could gain by changing her action. For example, if few people are going to the bar one night, then at least one player would be better off if she attended rather than staying home. And even if people face this problem repeatedly (say, every Thursday night) it is far from clear the system would converge to an equilibrium." Whereas the analytic approach divorces ontology and epistemology, an ABM builds learning and memory into the theory. It is like a mathematical morphism where we use a computational model of the mind to study the mind in the wild.

REGULARITY

GELL-MANN AND LLOYD (1996)

Theoreticians are nearly all minimalists. Given a choice between a Watteau or LeWitt painting, it would have to be the latter's bands, lines, and cubes over the former's wigs, gardens, and picnics. And yet the natural world can only be compressed so far, and at a certain limit, features begin to be lost that are constitutive of phenomena. Deriving the information-theoretic limit that is "as simple as possible but no simpler" is what Murray Gell-Mann and Seth Lloyd (1996) described as "effective complexity." Miguel Fuentes writes, "The effective complexity could be defined in colloquial language as the length of the compressed description of its regularities." In other words, not of its randomness—the bane of information-theoretic measures of complexity. But of course also of theorizing: "The sum of the effective complexity and entropy is called total information. . . . it is all the information required to describe an entity (or phenomenon). The importance of having this form of description lies, from my point of view, in the fact that it is very close to the task of critical knowledge taking, providing a way to evaluate or compare possible descriptions of the same phenomenon." Since William of Ockham onwards, careful thought has been synonymous with frugal hypothesis, and effective complexity puts more bits on the bone of the idea.

GEOMETRY

AMARI (1998)

There is a natural relationship established between learning and geometry through gradient descent. The question is, which gradient? The Euclidean gradient is the most familiar but it might not be the "natural" choice. Shun'ichi Amari showed that the natural gradient depends on the Riemannian metric tensor of the parameter space. In other words, the landscape should tell us how to move across it. As Nihat Ay writes, "The main idea of the natural gradient method is that a particular parameterization serves as a coordinate system for learning and has no special meaning. One can parameterize the actual search space in many equivalent ways. The geometry that we should use for optimization of the search is the one that is naturally defined on that space . . . respecting the geometry of the search space as outlined in this introduction greatly improves previously proposed gradient learning algorithms." An outcome of Amari's search for the natural gradient was the development of the field of information geometry—the study of statistical manifolds—spaces where every coordinate is an hypothesis. This brings us full circle to where theories of physical spaces (parameters) can be related to inferential landscapes (hypotheses): materials and conceptual materials are integrated within a suitable variational framework.

CONCLUDING UNSCIENTIFIC POSTSCRIPT TO COMPLEX FRAGMENTS

Complexity science has been evolving for the last two centuries, starting with science inspired by machines—manufactured and evolved—and slowly fusing into a new epistemology.

Complexity science is perhaps the first modern science to transcend disciplines. The phenomena that it investigates, and the models and theories it deploys, move fluidly across fields.

This fluidity reflects an underlying ontological watercourse best described by self-organizing and selective principles operating on broken symmetries, and not by the time and space symmetries that yield to parsimonious frameworks of least action.

Information, energy, and computation provide a principled way to talk about randomness and order and define a continuum upon which to place both natural processes and cultural creations.

The adaptive sciences—evolution, learning, control theory, and computation—provided ideas to make sense of this reality. And, over the last several decades, these have recombined to generate many of the most novel ideas in complexity science.

The growing computational–cognitive perspective on complexity shows how abstract ideas need to be rooted in physical reality (thermodynamics of computation and cognition) and how physical reality might emerge from information ("it from bit").

With the origins of life, multicellular organisms, and systems of knowledge, physics and chemistry have become auxiliary scientific frameworks in support of a variety of mechanisms of function (teleology).

The integrated nature of complexity science aligns with the connected nature of the modern world. Complexity science will be essential to all future projects that aim to escape terminal planetary decline.

Concluding Unscientific Postscript to Philosophical Fragments was Søren Kierkegaard's 1846 pseudonymous defense of a subjective approach to knowledge. In it he writes that "The subjective existing thinker is aware of the dialectic of communication." This is an apposite framing for the spirit of this final section."

APPENDIX TO EMERGENCE

Supervenience, Programs, and Compilers

This section draws on emergence and computation to explore the idea of strong emergence in a causally consistent fashion. The world of apparently top-down causal engineered systems is presented as a useful analogy.

Supervenient phenomena that one should like to understand all connect *low-information sources* to *high-information targets*. This of course requires that the target augment information from the source in a predictable way. A few candidates include how a bird listening to a sound (information source) is able to reproduce that sound with its syrinx (information target); how reading a musical score (information source) can through performance produce structured sound waves perceived as musical melodies (information target); how learning calculus (information source) can through industry put a rocket into orbit (information target); and how vast corpora of data (information source) enable a large language model to pass a Turing test (information target). In each of these examples, low-dimensional inputs—frequencies or symbols derived from a finite alphabet—influence the form and function of very high-dimensional coordinated physical matter at a fundamental level.

The concepts that we shall use to explore supervenience are derived from computer science: programming languages (Sammet 1969; Davis, Sigal, and Weyuker 1994) and compilers (Wulf 1981; Su and Yan 2011). Programming languages consist of human-interpretable symbols and grammars capable of encoding an algorithm. Compilers take languages as inputs (low-information source) and expand these into machine-executable operations (high-information target). Starting with a thought that is expressed in a formal language—an algorithmic effective theory—a compiler extracts and maps tokenized elements of the language onto registers in a computer. These ultimately entrain the flow of fundamental particles, electrons, through semiconductors. Computers, programming languages, and compilers provide the constituent exemplars required to make sense of supervenience—they offer an engineered and highly practical proof of principle of the flow from low to high information. The key requirement for this kind of "downward causation" is that the initial whole needs to be far less than the sum of its final parts: *anti-emergence mechanisms*. The preoccupation with coarse-graining and information channels (both of which imply information loss to various degrees) has obscured the reality of programmed information expansion in engineered and living systems.

LANGUAGES

Programming languages are often described in terms of three major characteristics: (1) expressivity; (2) syntax; and (3) semantics. Expressivity captures the human-language interface with the computer-language in terms of naturalness, effort-requirements, and range of thought that can be translated into language. Syntax refers to the grammar of a language. Grammar is used in the sense of fixed rules describing interpretable sequences of tokens and operations, including the power of these rules (*Chomsky hierarchy*). The semantics of a language describe its meaning in terms of the global function of a program (denotational semantics) or the local meaning of individual operations within the program (operational semantics). *Domain-specific languages* (**DSL**) impose restrictions on expressivity in order to align language syntax and semantics with the structure of their domain (Mathematica for symbolic manipulation, R for statistics, HTML for web pages, etc.)[3]

Chomsky hierarchy. A containment hierarchy of formal grammars—admissible strings produced by concatenating letters from an alphabet within a rule system—nested according to the complexity of their languages.

[3] The most expressive computer language might be assumed to be a programming language that resembles spoken language, in which case programming would involve speaking a thought out loud. Spoken thoughts have the disadvantage of hiding various assumptions from the programmer. Making all assumptions explicit, understanding how results follow from these, and how ideas might be combined all provide important justifications for the choice of programming language: functional languages (composition of functions without regard for state and execution flow—highly abstract and efficient), procedural languages (execution of modular algorithms realized through concatenated subroutines—robust and generalizable), and imperative languages (direct control of execution flow and state changes—compact and problem-specific).

COMPILERS

The role of a compiler is to turn a high-level "source" language into register-level and memory-level "object" languages (targets). These objects are then optimized for a given hardware to produce machine-language operations. Compilers transform symbolic thoughts into instructions for the control of matter. Interpreters perform the same basic operations according to a different real-time model.

Compilation proceeds in distinct phases: (1) lexical analysis; (2) syntax analysis; (3) global optimization; (4) code generation; and (5) local optimization. Phases (1–3) can be thought of as the front-end of the computation process since they are at furthest remove from the details of the hardware. Phases (3–5) are back-end and highly machine-dependent.

☛ *Front-End Logic.* Lexical analysis discovers the boundaries of words in order to build up a symbol table and establish data types. Syntax analysis inputs sequences of tokens and constructs syntax trees corresponding to sequence of operations in target machines. Trees can be optimized to eliminate redundancies and improve the efficiency of control loops. Front-end logic is associated with distinct computational machines. Lexical analysis makes use of finite state machines and regular expressions. Syntax analysis implements a context-free grammar using a deterministic push-down machine.

☛ *Back-End Machine.* Tokens are converted into sequences of instructions that are expressed in a suitable assembly language, including operations loading into individual registers, into a global accumulator, or saving to memory. Tokens are scanned multiple times to ensure that conditional jumps are appropriately coded. The tokens are collected in a look-up table that assigns each a correct machine address.

☛ *Back-End Matter.* Each processor chip includes micro-code, typically in the form of a ROM chip, that reads the opcode or machine code connecting to control lines that send read/write signals to transistors.

THE LOGICAL CONTROL OF MATTER

Supervenience in a complex system closely resembles compilation. For example, cells possess dedicated receptors that "read" and error-correct ligands binding at the cell surface. These activate a multiplicity of signaling pathways that culminate in differential gene expression in a context-sensitive way. Different gene regulatory networks (back-end) make use of the same upstream signaling pathways to translate extracellular inputs into "tokenized" activators of gene expression (front-end). The same logic can be extended to sensory activation and perception at the level of the whole nervous system: electromagnetic waves focused by the lens induce electrochemical activation in the retina, and signals propagate through the optic nerve and tracts, until they eventually converge and expand throughout visual cortex.

Computer languages are a means of expressing symbolic effective theories. Through compilation these languages are transformed from minimal sequences of signals into maximal states of material activation. Coordinated and collective modes of causal interaction at the material level are derived—top-down—from effectively causal algorithms.

The situation is of course materially different in biology and culture, at least as regards the logical sequence of compilation-like transformations, from inputs to tokens to registers. And the origin of these *organic generalized compilers* is of course a complicated matter that needs to be described through an organic or cultural evolutionary process. An *organic-interpreter-model* (line by line tokenization and parsing) seems like a better fit to epigenetic modification and short-term memory.

The supposed paradox of low-dimensional inputs entraining high-dimensional outputs is resolved by compilers and interpreters. And top-down causality in the complex world arises from evolved pathways of information expansion—"many its from few bits."

BIBLIOGRAPHY

Ahmed, A. 2010. *Wittgenstein's Philosophical Investigations: A Reader's Guide.* London, UK: Continuum.

Amari, S.-I. 1998. "Natural Gradient Works Efficiently in Learning." *Neural Computation* 10 (2): 251–276. https://doi.org/10.1162/089976698300017746.

Anderson, P. W. 1972. "More Is Different." *Science* 177 (4047): 393–396. https://doi.org/10.1126/science.177.4047.393.

Arthur, W. B. 1994. "Inductive Reasoning and Bounded Rationality." *American Economic Review* 84 (2): 406–411.

Babbage, C. 1832. *On the Economy of Machinery and Manufactures.* London, UK: Charles Knight.

———. 1837. *The Ninth Bridgewater Treatise: A Fragment.* London, UK: John Murray.

———. 1864. *Passages from the Life of a Philosopher.* London, UK: Longman.

———. 1982. "On the Mathematical Powers of the Calculating Engine." In *The Origins of Digital Computers*, 19–54. Berlin, Germany: Springer.

Bennett, C. H. 1973. "Logical Reversibility of Computation." *IBM Journal of Research and Development* 17 (6): 525–532. https://doi.org/10.1147/rd.176.0525.

Boltzmann, L. 1973. *The Boltzmann Equation: Theory and Applications.* Edited by E. G. D. Cohen and W. Thirring. New York, NY: Springer-Verlag Wien. https://doi.org/10.1007/978-3-7091-8336-6.

———. 1974. "On the Question of the Objective Existence of Processes in Inanimate Nature." In *Theoretical Physics and Philosophical Problems*, 57–76. Dordrecht, Netherlands: Springer.

———. 2012. *Theoretical Physics and Philosophical Problems: Selected Writings.* Berlin, Germany: Springer Science & Business Media.

Boole, G. 1847. *The Mathematical Analysis of Logic.* Cambridge, UK: Philosophical Library.

———. 1854. *An Investigation of the Laws of Thought on which are Founded the Mathematical Theories of Logic and Probabilities.* London, UK: Walton & Maberly.

———. 1859. *A Treatise on Differential Equations.* Cambridge, UK: Macmillan.

Boole, G. *Letter to Charles Babbage (October 15, 1862).* Add. MS 37198, no. 414. British Library, London, UK.

Bowler, P. J. 2005. "Revisiting the Eclipse of Darwinism." *Journal of the History of Biology* 38 (1): 19–32. https://doi.org/10.1007/s10739-004-6507-0.

Brush, S. G. 2003. *The Kinetic Theory Of Gases: An Anthology of Classic Papers with Historical Commentary.* Edited by N. S. Hall. River Edge, NJ: World Scientific. https://doi.org/10.1142/p281.

Burchfield, J. D. 2009. *Lord Kelvin and the Age of the Earth*. Chicago, IL: University of Chicago Press.

Callebaut, W., and R. Pinxten. 2012. *Evolutionary Epistemology: A Multiparadigm Program*. Dordrecht, Netherlands: Springer Science & Business Media.

Carnot, S. 1824. *Reflections on the Motive Power of Fire, and on Machines Fitted to Develop That Power.* Paris, France: Chez Bachelier.

Ceruzzi, P. E. 2003. *A History of Modern Computing, Second Edition.* Cambridge, MA: MIT Press.

Chaitin, G. J. 1966. "On the Length of Programs for Computing Finite Binary Sequences." *Journal of the ACM* 13 (4): 547–569. https://doi.org/10.1145/321356.321363.

Chalmers, D. J. 2006. "Strong and Weak Emergence." In *The Re-Emergence of Emergence*, edited by P. Clayton and P. Davies, 244–256. Oxford, UK: Oxford University Press.

Chalmers, T. 1853. *On the Power, Wisdom, and Goodness of God: As Manifested in the Adaptation of External Nature, to the Moral and Intellectual Constitution of Man*. London, UK: Bohn.

Clausius, R. 1879. *The Mechanical Theory of Heat*. London, UK: Macmillan.

Cobham, A. 1965. "The Intrinsic Computational Difficulty of Functions." In *Logic, Methodology and Philosophy of Science: Proceedings of the 1964 International Congress (Studies in Logic and the Foundations of Mathematics)*, edited by Y. Bar-Hillel, 24–30. Amsterdam, Netherlands: North- Holland Publishing.

Conant, R. C., and W. R. Ashby. 1970. "Every Good Regulator of a System Must be a Model of that System." *International Journal of Systems Science* 1 (2): 89–97. https://doi.org/10.1080/00207727008920220.

Corradini, A., and T. O'Connor. 2010. *Emergence in Science and Philosophy.* Hoboken, NJ: Taylor & Francis.

Crutchfield, J. P. 1994. "The Calculi of Emergence: Computation, Dynamics and Induction." *Physica D* 75 (1): 11–54. https://doi.org/10.1016/0167-2789(94)90273-9.

Darwin, C. 1859. *On the Origin of Species.* London, UK: John Murray.

——. 1861. *On the Origin of Species.* 3rd ed. London, UK: John Murray.

——. 1871. *The Descent of Man, and Selection in Relation to Sex.* Vol. 1. New York, NY: Appleton.

——. 1876. *The Movements and Habits of Climbing Plants*. 2nd ed. New York, NY: Appleton.

——. [1876]1958. *The Autobiography of Charles Darwin: 1809-1882.* Edited by N. Barlow. Vol. 29. New York, NY: Collins.

Darwin, C., and A. Wallace. 1858. "On the Tendency of Species to form Varieties; and on the Perpetuation of Varieties and Species by Natural Means of

Selection." *Zoological Journal of the Linnean Society* 3 (9): 45–62. https://doi.org/10.1111/j.1096-3642.1858.tb02500.x.

Darwin, F., and A. C. Seward. 1903. *More Letters of Charles Darwin*. Vol. 2. New York, NY: Appleton. Darwin Correspondence Project. 1869. Letter no. 6585. https://www.darwinproject.ac.uk/letter/?docId=letters/DCP-LETT-6585.xml.

——. 1878. *Letter no. 11729*. https://www.darwinproject.ac.uk/letter/?docId=letters/DCP- LETT- 11729.xml.

Davis, M., R. Sigal, and E. J. Weyuker. 1994. *Computability, Complexity, and Languages: Fundamentals of Theoretical Computer Science*. Amsterdam, Netherlands: Elsevier.

Dennett, D. 1995. *Darwin's Dangerous Idea: Evolution and the Meanings of Life*. New York, NY: Simon & Schuster.

Dilthey, W., and F. Jameson. 1972. "The Rise of Hermeneutics." *New Literary History* 3 (2): 229–244. https://doi.org/10.2307/468313.

Eigen, M., and R. Winkler. 1993. *Laws of the Game: How the Principles of Nature Govern Chance*. Princeton, NJ: Princeton University Press.

Farley, J. 1974. "The Initial Reactions of French Biologists to Darwin's 'Origin of Species'." *Journal of the History of Biology* 7 (2): 275–300.

Farmer, J. D. 1990. "A Rosetta Stone for Connectionism." *Physica D* 42 (1): 153–187. https://doi.org/10.1016/0167-2789(90)90072-W.

Feynman, R. P., R. B. Leighton, and M. Sands. 1963. *The Feynman Lectures on Physics*. Reading, MA: Addison-Wesley.

Forrest, S., A. S. Perelson, L. Allen, and R. Cherukuri. 1994. "Self–Nonself Discrimination in a Computer." In *Proceedings of 1994 IEEE Computer Society Symposium on Research in Security and Privacy*, 202–212. IEEE. https://doi.org/10.1109/RISP.1994.296580.

Galilei, G. 2001. *Dialogue Concerning the Two Chief World Systems, Ptolemaic and Copernican*. Edited by S. J. Gould. Translated by S. Drake. New York, NY: Modern Library.

Gell-Mann, M. 1994. *The Quark and the Jaguar: Adventures in the Simple and the Complex*. New York, NY: W. H. Freeman and Company.

Gell-Mann, M., and J. B. Hartle. 1994. *Equivalent Sets of Histories and Multiple Quasiclassical Realms*. arXiv: gr-qc/9404013 [gr-qc].

Gibb, S., R. F. Hendry, and T. Lancaster. 2019. *The Routledge Handbook of Emergence*. Abingdon, UK: Routledge.

Gibbs, J. W. 1875. "On the Equilibrium of Heterogeneous Substances." *Transactions of the Connecticut Academy of Arts and Sciences* 3:108–248.

——. 1902. *Elementary Principles in Statistical Mechanics*. New York, NY: Charles Scribner's Sons.

Gleick, J. 2008. *Chaos: Making a New Science*. New York, NY: Penguin.

Gribbin, J. 2004. *Deep Simplicity: Bringing Order to Chaos and Complexity.* New York, NY: Random House.

Guerra, C., M. Capitelli, and S. Longo. 2012. "The Role of Paradigms in Science: A Historical Perspective." In *Paradigms in Theory Construction*, edited by L. L'Abate, 19–30. New York, NY: Springer New York.

Hadamard, J. 1945. *The Psychology of Invention in the Mathematical Field.* New York, NY: Princeton University Press.

Haken, H. 1977. *Synergetics*. Berlin: Springer-Verlag.

Heilbron, J. L. 2010. *Galileo*. Oxford, UK: Oxford University Press.

Highfield, R., and P. Coveney. 1995. *Frontiers of Complexity.* New York, NY: Fawcett Columbine.

Hofstadter, D. R. 1979. *Gödel, Escher, Bach: An Eternal Golden Braid.* New York, NY: Basic Books.

Holland, J., and E. Domingo. 1998. "Origin and Evolution of Viruses." *Virus Genes* 16 (1): 13–21.

Holland, J. H. 1962. "Outline for a Logical Theory of Adaptive Systems." *Journal of the ACM* (New York, NY) 9 (3): 297–314. https://doi.org/10.1145/321127.321128.

——. 1995. *Hidden Order: How Adaptation Builds Complexity.* Cambridge, MA: Perseus Books.

——. 2014. *Complexity: A Very Short Introduction.* Oxford, UK: Oxford University Press.

Holland, J. H., and J. H. Miller. 1991. "Artificial Adaptive Agents in Economic Theory." *The American Economic Review* 81 (2): 365–370.

Holmes, K. V. 1999. "Coronaviruses (Coronaviridae)." *Encyclopedia of Virology* (San Diego, CA), 291.

Hopfield, J. J. 1982. "Neural Networks and Physical Systems with Emergent Collective Computational Abilities." *Proceedings of the National Academy of Sciences* 79 (8): 2554–2558. https : / / doi . org / 10.1073/pnas.79.8.2554.

Jensen, H. J. 2022. *Complexity Science: The Study of Emergence.* Cambridge, UK: Cambridge University Press.

Jin, H. 2023. "The History, Current Applications and Future of Integrated Circuit." *Highlights in Science, Engineering and Technology*, https://doi.org/10.54097/hset.v31i.5146.

Johnson, N. 2009. *Simply Complexity: A Clear Guide to Complexity Theory.* Oxford, UK: Oneworld.

Kalman, R. E. 1960. "Contributions to the Theory of Optimal Control." *Boletín de la Sociedad Matemática Mexicana* 5:102–119.

Karp, R. M. 1972. "Reducibility among Combinatorial Problems." In *Complexity of Computer Computations*, edited by R. E. Miller and J. W. Thatcher, 85–103. New York, NY: Plenum Press.

Kauffman, S. A. 1993. *The Origins of Order: Self-organization and Selection in Evolution.* New York, NY: Oxford University Press. https://doi.org/10.1093/oso/9780195079517.001.0001.

Kierkegaard, S. 1992. *Concluding Unscientific Postscript to Philosophical Fragments, A Mimical-Pathetic-Dialectical Compilation an Existential Contribution.* Translated by H. V. Wong and E. H. Wong. Vol. I. Princeton, NJ: Princeton University Press.

Kolmogorov, A. N. 1968. "Three Approaches to the Quantitative Definition of Information." *International Journal of Computer Mathematics* 2 (1–4): 157–168.

Krakauer, D. 2023. "Unifying Complexity Science and Machine Learning." *Frontiers in Complex Systems* 1. https://doi.org/10.3389/fcpxs.2023.1235202.

Kuhn, T. S. 2000. *The Road Since Structure: Philosophical Essays, 1970–1993, with an Autobiographical Interview.* Chicago, IL: University of Chicago Press.

——. [1962] 2012. *The Structure of Scientific Revolutions.* Chicago, IL: University of Chicago Press.

Ladyman, J., and K. Wiesner. 2020. *What Is a Complex System?* New Haven, CT: Yale University Press.

Landau, L. 2008. "On the Theory of Phase Transitions." Originally published in Zh. Eksp. Teor. Fiz. 7, pp. 19–32 (1937), *Ukrainian Journal of Physics* 53:25–35.

Landauer, R. 1961. "Irreversibility and Heat Generation in the Computing Process." *IBM Journal of Research and Development* 5 (3): 183–191. https://doi.org/10.1147/rd.53.0183.

Langton, C. G. 1986. "Studying Artificial Life with Cellular Automata." *Physica D* 22 (1): 120–149. https://doi.org/10.1016/0167-2789(86)90237-X.

Laughlin, R. B., D. Pines, J. Schmalian, B. P. Stojković, and P. Wolynes. 2000. "The Middle Way." *Proceedings of the National Academy of Sciences* 97 (1): 32–37. https://doi.org/10.1073/pnas.97.1.32.

Levin, S. 1999. *Fragile Dominion: Complexity and the Commons.* Reading, MA: Basic Books.

Lorenz, E. N. 1963. "Deterministic Nonperiodic Flow." *Journal of the Atmospheric Sciences* 20 (2): 130–141. https://doi.org/10.1175/1520-0469(1963)020<0130:DNF>2.0.CO;2.

Lyell, C. 1842. *Principles of Geology.* Boston, MA: Hilliard, Gray & Company.

Marr, D. 2010. *Vision: A Computational Investigation into the Human Representation and Processing of Visual Information.* Cambridge, MA: MIT Press. https://doi.org/10.7551/mitpress/9780262514620.001.0001.

Mason, R. 1994. *Cambridge Minds.* Cambridge, UK: Cambridge University Press.

Maxwell, J. C. 1868. "I. On Governors." P*roceedings of the Royal Society of London* 16:270–283. https://doi. org/10.1098/rspl.1867.0055.

——. 1871. *Theory of Heat.* London, UK: Longmans, Green and Company.

——. 1872. "Molecules." In *The Scientific Papers of James Clerk Maxwell,* edited by W. D. Niven, vol. 2. Cambridge, UK: Cambridge University Press.

——. 1881. *An Elementary Treatise on Electricity.* Oxford, UK: Clarendon.

——. 1990. *The Scientific Letters and Papers of James Clerk Maxwell: Volume 2, 1862-1873.* Cambridge, UK: CUP Archive.

McCulloch, W. S., and W. Pitts. 1943. "A Logical Calculus of the Ideas Immanent in Nervous Activity." *Bulletin of Mathematical Biology* 5 (4): 115–133. https://doi.org/10.1007/BF02478259.

Menabrea, L. F., and A. Lovelace. 1843. *Sketch of the Analytical Engine Invented by Charles Babbage, Esq.* London, UK: Taylor & Francis.

Mendel, G. [1866]1925. *Experiments in Plant Hybridization.* Cambridge, MA: Harvard University Press.

Miller, J. G. 1955. "Toward a General Theory for the Behavioral Sciences." *American Psychologist* 10 (9): 513–531.

Miller, J. H. 2015. *A Crude Look at the Whole: The Science of Complex Systems in Business, Life, and Society.* New York, NY: Basic Books.

Minsky, M. 1961. "Steps Toward Artificial Intelligence." *Proceedings of the IRE* 49 (1): 8–30. https://doi. org/10.1109/JRPROC.1961.287775.

Mitchell, M. 2009. *Complexity: A Guided Tour.* Oxford, UK: Oxford University Press.

Mitchell, M., P. Hraber, and J. P. Crutchfield. 1993. "Revisiting the Edge of Chaos: Evolving Cellular Automata to Perform Computations." *Complex Systems* 7:89–130.

Miyazaki, J. 2013. *Pattern Formations and Oscillatory Phenomena: 2. Belousov–Zhabotinsky Reaction.* Amsterdam, Netherlands: Elsevier Inc.

Morowitz, H. J. 2002. *The Emergence of Everything: How the World Became Complex.* New York, NY: Oxford University Press.

Newton, I. 1934. *Sir Isaac Newton's Mathematical Principles of Natural Philosophy and His System of the World.* Translated by A. Motte. Berkeley, CA: University of California Press.

Page, S. E. 2010. *Diversity and Complexity.* New Jersey, NJ: Princeton University Press.

Palmer, R. E. 1969. *Hermeneutics: Interpretation Theory in Schleiermacher, Dilthey, Heidegger, and Gadamer.* Evanston, IL: Northwestern University Press.

Pederson, T. 2020. "The Double Helix: "Photo 51" Revisited." *The FASEB Journal* 34 (2): 1923–1927. https://doi.org/10.1096/fj.202000119.

Penrose, R. 2004. *The Road to Reality: A Complete Guide to the Laws of the Universe.* New York, NY: Knopf, Borzoi.

Peters, T. F. 1996. *Building the Nineteenth Century.* Cambridge, MA: MIT Press.

Piccinini, G., and A. Scarantino. 2011. "Information Processing, Computation, and Cognition." *Journal of Biological Physics* 37 (1): 1–38. https://doi.org/10.1007/s10867-010-9195-3.

Poincaré, H. 2017. The Three-Body Problem and the Equations of Dynamics. Translated by B. D. Popp. Cham, Switzerland: Springer International Publishing. https://doi.org/10.1007/978-3-319-52899-1.

——. [1903]2017. Science and Method. Translated by F. Maitland. New York, NY: Dover.

Prigogine, I., and P. M. Allen. 1982. "The Challenge of Complexity." In *Self-Organization and Dissipative Structures,* edited by W. C. Schieve and P. M. Allen, 1–39. Austin, TX: University of Texas Press.

Prigogine, I., and I. Stengers. 2018. *Order Out of Chaos: Man's New Dialogue with Nature.* London, UK: Verso Books.

Pylyshyn, Z. W. 1984. *Computation and Cognition: Toward a Foundation for Cognitive Science.* Cambridge, MA: MIT Press.

Radick, G. 2023. *Disputed Inheritance: The Battle over Mendel and the Future of Biology.* Chicago, IL: University of Chicago Press.

Rashevsky, N. 1956. "The Geometrization of Biology." *Bulletin of Mathematical Biophysics* 18 (1): 31–56. https://doi.org/10.1007/BF02477842.

Richards, R. J. 2013. "The German Reception of Darwin's Theory, 1860–1945." In *The Cambridge Encylopedia of Darwin and Evolutionary Thought,* edited by M. Ruse, 235–242. Cambridge, UK: Cambridge University Press.

Rosenblueth, A., N. Wiener, and J. Bigelow. 1943. "Behavior, Purpose and Teleology." *Philosophy of Science* 10 (1): 18–24.

Sammet, J. E. 1969. *Programming Languages: History and Fundamentals.* Englewood Cliffs, NJ: Prentice- Hall.

Schelling, T. C. 1971. "Dynamic Models of Segregation." *Journal of Mathematical Sociology* 1 (2): 143–86. https://doi.org/10.1080/0022250X.1971.9989794.

Schrödinger, E. 1944. *What is Life? The Physical Aspect of the Living Cell.* Cambridge, UK: Cambridge University Press.

Shannon, C. E., and W. Weaver. 1949. *The Mathematical Theory of Communication.* Urbana, IL: University of Illinois Press.

Shapiro, J. A. 2009. "Letting Escherichia coli Teach Me About Genome Engineering." *Genetics* 183 (4): 1205– 1214. https://doi.org/10.1534/genetics.109.110007.

Simon, H. A. 1962. "The Architecture of Complexity." *Proceedings of the American Philosophical Society* 106 (6): 467–482.

——. 1977. "The Organization of Complex Systems." In *Models of Discovery: And Other Topics in the Methods of Science*, edited by H. A. Simon, 245–261. Springer Netherlands.

Snyder, L. 2011. *The Philosophical Breakfast Club: Four Remarkable Friends*. New York, NY: Broadway Books.

Solomonoff, R. J. 1964. "A Formal Theory of Inductive Inference. Part I." *Information and Control* 7 (1): 1–22. https://doi.org/10.1016/S0019-9958(64)90223-2.

Stein, D. L., and C. M. Newman. 2013. *Spin Glasses and Complexity.* Princeton, UK: Princeton University Press.

Stent, G. 2002. *Prematurity in Scientific Discovery: On Resistance and Neglect.* Edited by E. B. Hook. Berkeley, CA: University of California Press.

Su, Y., and S. Y. Yan. 2011. *Principles of Compilers.* Berlin, Germany: Springer.

Szilárd, L. [1929]1964. "On the Decrease of Entropy in a Thermodynamic System by The Intervention of Intelligent Beings." *Behavorial Science* 9 (4): 301–310. https://doi.org/10.1002/bs.3830090402.

Taylor, R., ed. 1843. *Scientific Memoirs, Selected from the Transactions of Foreign Academies of Science and Learned Societies.* Translated by A. Lovelace. "Translator's Notes to Mr. Menabrea's Memoir." London, UK: Longman.

Thompson, D. W. 1917. *On Growth and Form*. Cambridge, UK: Cambridge University Press.

Thomson, W. 1874. "Kinetic Theory of the Dissipation of Energy." *Nature* 9:441–444. https://doi.org/10.1038/009441c0.

Thurner, S., R. Hanel, and P. Klimek. 2018. Introduction to the Theory of Complex Systems. Oxford, UK: Oxford University Press.

Topham, J. R. 2022. *Reading the Book of Nature: How Eight Best Sellers Reconnected Christianity and the Sciences on the Eve of the Victorian Age.* Chicago, IL: University of Chicago Press.

Turing, A. M. [1950]2012. "Computing Machinery and Intelligence." In *The Essential Turing: the Ideas That Gave Birth to the Computer* Age, 433–464. Oxford, UK: Oxford University Press.

Ulam, S. 1962. "On Some Mathematical Problems Connected with Patterns of Growth in Figures." In *Mathematical Problems in the Biological Sciences,* edited by R. Bellman, 215–224. American Mathematical Society. https://doi.org/10.1090/psapm/014/9947.

Varela, F. G., H. R. Maturana, and R. Uribe. 1974. "Autopoiesis: The Organization of Living Systems, its Characterization and a Model." *Biosystems* 5 (4): 187–196. https://doi.org/10.1016/0303-2647(74)90031-8.

von Bertalanffy, L. 1950. "The Theory of Open Systems in Physics and Biology." *Science* 111 (2872): 23–29.

von Foerster, H. 2003. "Notes on an Epistemology for Living Things." In *Understanding Understanding: Essays on Cybernetics and Cognition*, 247–259. New York, NY: Springer New York.

von Hayek, F. A. 1945. "The Use of Knowledge in Society." *American Economic Review* 35:519–530.

von Neumann, J. 1966. *Theory of Self-Reproducing Automata.* Edited by A.W. Burks. Urbana, IL: University of Illinois Press. https://doi.org/10.5555/1102024.

Waldrop, M. M. 1992. *Complexity: The Emerging Science at the Edge of Order and Chaos.* New York, NY: Simon & Schuster.

Wallace, A. R. 1889. *Darwinism: An Exposition of the Theory of Natural Selection with Someof Its Applications.* 2nd ed. London, UK: Macmillan.

Weaver, W. 1948. "Science and Complexity." *American Scientist* 36 (4): 536–544.

West, G. B. 2018. Scale: *The Universal Laws of Life, Growth, and Death in Organisms, Cities, and Companies.* New York, NY: Penguin.

West, G. B., and J. H. Brown. 2005. "The Origin of Allometric Scaling Laws in Biology from Genomes to Ecosystems: Towards a Quantitative Unifying Theory of Biological Structure and Organization." *The Journal of Experimental Biology* 208 (9): 1575–1592. https://doi.org/10.1242/jeb.01589.

Wheeler, J. A. 1990. "Information, Physics, Quantum: The Search for Links." In *Complexity, Entropy, and the Physics of Information,* edited by W. H. Zurek, 3–28. Reading, MA: Addison–Wesley.

Whewell, W. 1833. *Astronomy and General Physics Considered with Reference to Natural Theology.* London, UK: Pickering.

Wiener, N. 1948. *Cybernetics: or Control and Communication in the Animal and the Machine.* New York, NY: John Wiley and Sons.

Wigner, E. 1964. "Events, Laws of Nature, and Invariance Principles." *Science* 145 (3636): 995–999. https://doi.org/10.1126/science.145.3636.995.

Wittgenstein, L. [1953]2009. *Philosophical Investigations.* 4th ed. Edited by P. M. S. Hacker and J. Schulte. Translated by G. E. M. Anscombe. Malden, WI: Wiley-Blackwell.

Wolfram, S. 1984a. "Cellular Automata as Models of Complexity." *Nature* 311 (5985): 419–424. https ://doi.org/10.1038/311419a0.

——. 1984b. "Universality and Complexity in Cellular Automata." *Physica D* 10 (1): 1–35. https://doi. org/10.1016/0167-2789(84)90245-8.

——. 2018. A New Kind of Science. Champaign, IL: Wolfram Media.

Wray, B. K. 2021. *Interpreting Kuhn: Critical Essays.* Cambridge, UK: Cambridge University Press.

Wulf, W. A. 1981. "Compilers and Computer Architecture." *Computer* 14 (07): 41–47. https://doi.org/10.1109/C-M.1981.220527.

Zenil, H. 2013. *A Computable Universe: Understanding and Exploring Nature as Computation.* River Edge, NJ: World Scientific.

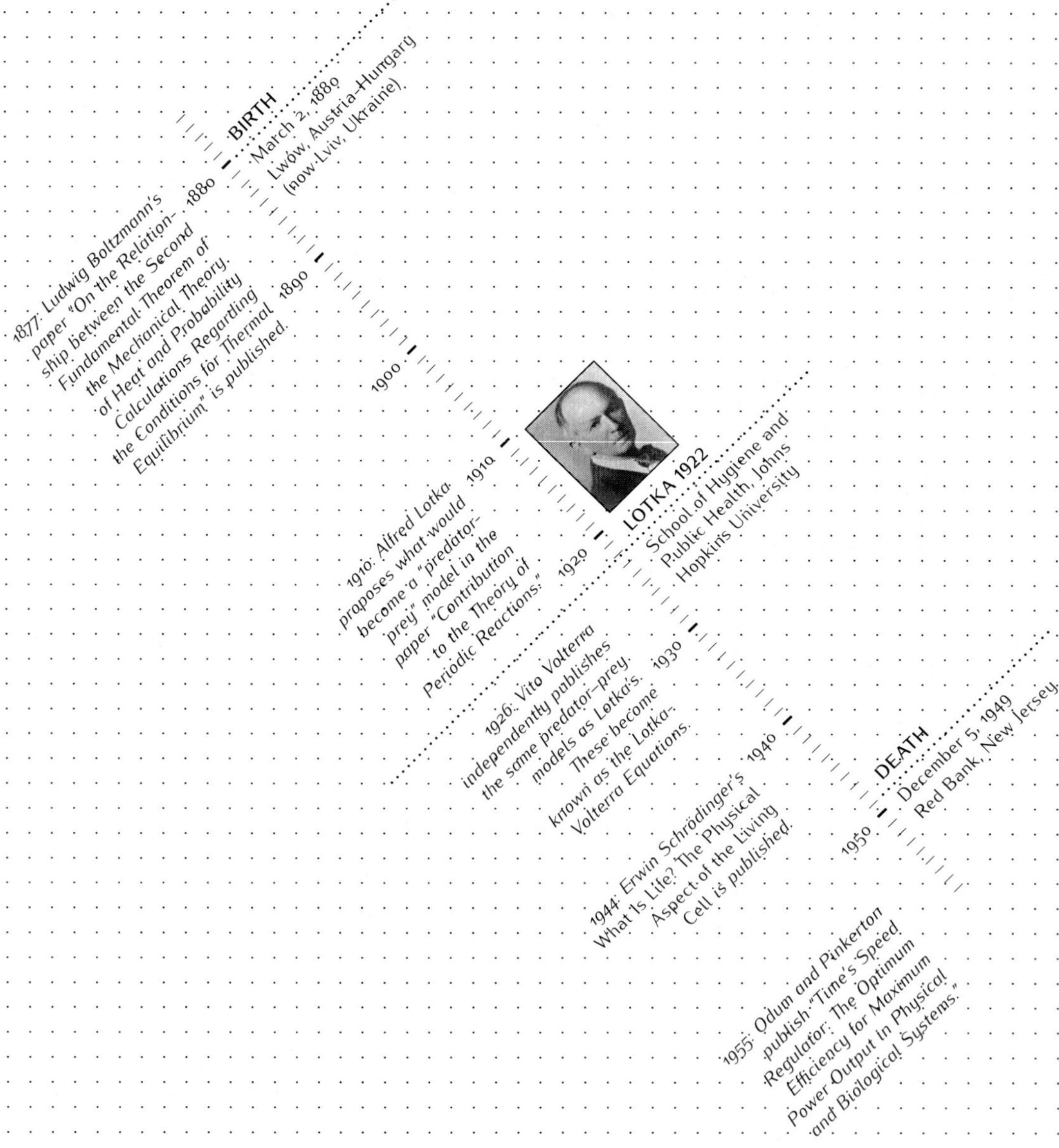

ALFRED J. LOTKA

[1]

MAXIMUM POWER AS A PHYSICAL PRINCIPLE OF EVOLUTION

Luís M. A. Bettencourt, University of Chicago and Santa Fe Institute

A. J. Lotka, "Contribution to the Energetics of Evolution," *Proceedings of the National Academy of Sciences* 8 (6), 147–151 (1922).

The reconciliation of evolution—the theory underlying all of biology and society—and physical theory, which explains energy and matter and underlies most current technology and engineering, remains one of the most important problems in science. A synthesis of these very different theories is essential for creating a general, predictive theory of complex systems (Mitchell 2009; Thurner, Hanel, and Klimek 2018). Some of the deepest unanswered questions in science, such as the origins of life or the sustainability of future human societies, also hinge on a theory that can articulate the physical and biological worlds and produce extrapolations to entirely new situations (Elmqvist *et al.* 2021).

An important step in this direction was taken a century ago by Alfred Lotka (1922a) with a very simple deductive observation that was then, and it is now, as powerful as it is intriguing. His insight, following closely on developments in statistical physics and specifically the work of Boltzmann and the second law of thermodynamics, is simple and direct. It first recognizes that 1) any living organism requires "available energy," which must be harvested from its environment, to live and thrive. It follows that 2) organisms that can harvest more energy from their environment relative to others will do better in the sense of sustaining larger numbers of their kind and more mass. This putative advantage is then interpreted by Lotka as the physical instantiation of natural selection (Lotka 1922a, 1922b), equating evolutionary fitness to the ability to obtain energy inflows from the environment. In Lotka's own words, this principle "may be expressed by saying that natural selection tends to make this energy flux a maximum, so far as compatible with the constraints to which the system is subject" (Lotka 1922a). This idea must be true because it follows from the laws of physics: This is what it makes it powerful.

By contrast, what makes it intriguing is that it is not a complete idea (Lotka 1922b). In the language of logic, we can say that energy influx is a *necessary* condition for life but it is not *sufficient* to generate a theory of biology. Some of the ways in which it is not sufficient were already clear to Lotka and briefly discussed in his paper (Lotka 1922a). Others, however, have become better understood and formally developed since. Their discussion brings us to several important developments in evolutionary theory and statistical physics in subsequent decades, and to critical questions for complex systems still open today. I will briefly outline these here, taking Lotka's arguments as a starting point.

Let us start with what Lotka already realized was incomplete or unspecified in enunciating his principle. First, it is important to specify what Boltzmann meant by "available energy." In modern language, we refer to this quantity as *free energy* (Kardar 2007), which is the form of energy (at a given ambient temperature) that can be used to perform work such as running a machine or keeping an organism's internal order (homeostasis). All free energy is subject to degradation—by living and non-living processes—into heat, a form of energy that can no longer be used once its temperature is that of the surrounding environment. The expression for free energy, F, is $F = E - TS$, where E is the total energy under consideration, T is the ambient temperature, and S the entropy of the substance. Heat, at the ambient temperature, has $E_{heat} = TS_{heat}$, so that $F_{heat} = 0$; this is also the maximum entropy (most disordered) state for a substance, $S_{heat} \geq S$. This means that the kind of energy living organisms must obtain from their environment is "organized" (low entropy). In the case of herbivores or predator–prey interactions, sunlight or another organism's tissue are the most common examples, though this fact is not explicit in Lotka's discussion. In his famous "What is Life?" lectures in 1944, Erwin Schrödinger (1992) referred to the kind of flow needed by living organisms as "negative entropy," rather than an energy flux. This pivot—from talking about energy to talking about "negative entropy" (information; see Cover and Thomas 2005)—presages many of the difficulties with Lotka's proposal.

A subtler set of questions arises about what time spans and organizational scales we may expect Lotka's principle to apply to. When we refer to energy per unit time (i.e., power), should we expect to maximize it instantaneously, or over some biologically relevant timescale such as a lifetime, or even across generations? Should it be a property of single cells? Organisms? Or ecosystems?

In his own language, Lotka refers to such questions in terms of the "constraints" that may apply to living organisms and ecosystems (Lotka 1922a). A few examples illustrate the problem: Perhaps surprisingly, plants are rather inefficient at converting sunlight into usable energy. They convert only a small percentage of incoming sunlight into chemical energy via photosynthesis (about 1%; see Matthews 2023). While there are constraints to these biochemical processes, this rate pales in comparison to bioengineered plants and engineered photovoltaic cells (currently > 20%). Why have photosynthetic organisms, which have existed for over three billion years (Blankenship 2010), not developed much higher free energy capture efficiencies? Among other reasons, plants are limited by the availability of water, carbon dioxide and other nutrients. They also tend to be limited by available physical space (at least on land): Trees can only grow so tall before they become structurally unstable or too prone to falling. Thus, a plant that would use more available energy to become taller or bigger may be checked by other environmental constraints; in such circumstances their ability to capture more solar energy may not be their main optimization goal. Other examples come from the metabolism of multicellular organisms (such as ourselves), which is known to be mass dependent (a regularity known as Kleiber's law; Brown *et al.* 2004), in such a way that larger organisms use less energy per unit time and unit mass—less power (West, Brown, and Enquist 1999). While this varies across the life course and between major phyla—with mammals, for example, consuming two to five times more energy per unit time and mass than reptiles (Hulbert and Else 1981)—such patterns of development seem contrary to Lotka's principle. A final example from multicellular organisms deals with cancer (Aktipis 2021; Curtius, Wright, and Graham 2018), which in evolutionary terms can be interpreted as a lineage of cells going "rogue" and consuming more resources (power), which reduces the organism's integrity and fitness. This maximum power drive is checked by many biological mechanisms, from genetically encoded programmed cell death (apoptosis) to germ line separation (not allowing cancers to reproduce outside the organism), to organismal termination by death.

In each of these cases, we may attempt to salvage Lotka's principle by arguing that it may still apply at longer times or on different scales of organization, where particular biological constraints are effectively lifted. So, for example, trees may capture as much sunlight as possible given nutrient availability and long-term survival and reproduction. Multicellular organisms may slow down their power consumption with age and mass as a tradeoff with their smaller and faster offspring, which operate at higher power per unit mass, and so on. And even ageing and death—at face value, violations of a maximum power principle—may be a consequence of longer time scale tradeoffs, as developed in the context of life history theory (Stearns 1992). In this sense, maximum power principles may apply macroscopically, on longer time scales, and across lineages made up of individual organisms of different ages, subjected to certain other environmental constraints. Alas, much of the clarity and simplicity of Lotka's physical principle is lost when these considerations become essential. What was physics became biology and what was an energy principle applying in the moment became the consideration of multiple environmental factors over an extended time frame.

Besides these insufficiencies, it is remarkable that Lotka did not mention Darwin even once and does not really cite much of the biological literature in his paper. This separation has detracted from his work's impact in biology (Kingsland 2015). Moreover, the time at which Lotka is writing was pregnant with possibility; it became pivotal in biology for the clarification and quantification of evolutionary dynamics, a set of developments that were necessary for the eventual acceptance of evolution as the underlying general theory of biology. The body of work developing at this time became known as the "modern synthesis" (Bowler 2009): It combines ideas of Mendelian genetics with the general principles of variation, heritability and selection introduced by Darwin, now expressed mathematically in the rigorous form of population dynamics. Lotka was one of the early contributors to population dynamics (the famous model of predator–prey dynamics bears his name), but the leap into evolutionary dynamics was made by others.

The ideas and methods of the modern synthesis became, in due time, the basis for our understanding of evolution as a theory of population dynamics and information (Fisher 1999; Krakauer *et al.* 2020). These ideas anticipated, and were in turn reinforced by, the discovery of the structure of DNA (Watson and Crick 1953), its connection to genetic coding and translation into proteins, and the birth of modern genetics and biotechnology. As a result, when biology emerged as the main focus of scientific developments at the turn of the twenty-first century, energy principles had been pushed to the background and hardly ever considered in the context of evolutionary dynamics by biologists.

Instead, the more formal focus on theoretical principles guiding evolution shifted towards ideas of genetics, information, and computation that developed during and immediately after WWII and have come to define the other main strand of discovery in science and technology today. These ideas were also critical for the birth of modern complexity science some fifty years ago (Mitchell 2009; Gleick 1988). They express a renewed interest in systems dynamics emerging from engineering, information theory and organizational theory and the advent of new computational tools allowing us to explore their consequences. An interesting example are "genetic algorithms" first proposed by John Holland (1992, 1973), as a means to search for solutions to given problems in complex systems using computational methods that operate on a population of symbolic strings (analogous to genomes; see Mitchell 2001). These strings are selected for maximizing the objective function and mutated and recombined towards producing better and better solutions. These dynamics attempt to mimic the models of evolution that emerged out of the modern synthesis. In them, the principle of maximum power does not play much of a role at all.

But the idea introduced by Lotka has lingered elsewhere and inspired other bodies of work. It became part of the background of ideas of macroecology (broad patterns in ecology, mostly to do with energetics) in terms of a "metabolic theory of ecology" (Brown *et al.* 2004). Lotka's discussion of human agriculture anticipated emerging theories in the 1970s of energy use in human societies. Specifically, they influenced the development of frameworks for conceptualizing sustainable development, which are now central to how to manage and transcend climate change. In this context, H. T. Odum (2007) recast Lotka's ideas as the "Principle of Maximum Power" and applied it to a number of different systems, starting with ecosystems, as Lotka had proposed, but also to human societies, including cities and urban systems (Odum and Pinkerton 1955). Central to these developments, Odum picked up Lotka's "paths of energy flux" to develop the concept of energy hierarchies. These hierarchies describe the large-scale structure of free energy flow in ecosystems and settlement systems in human societies (Odum 2007; Bettencourt 2021). Such concepts remain at this point still relatively underdeveloped, but the idea of organized energy flows through scales of organization, their differential uses, and transformations (Elmqvist *et al.* 2021) as we forego fossil fuels constitutes an inspiring set of principles that will no doubt become important in the decades ahead. Key to these ideas of energy flow is the adoption of a very macroscopic and general perspective, where most particulars fade away.

As we know, biology is in part physics. Complex systems are, at their most fundamental level, physical systems. But biology is also more than physics and "more is different" (Anderson 1972). Alfred Lotka might have glimpsed all this when he bravely proposed the idea of equating maximal energy flows to natural selection (Lotka 1922a, 1922b). It is a principle that remains central to complex systems, as much as it remains also incomplete. Learning how to complete it, in the simplest possible way that makes it relevant to biology and society, is one good way to summarize the main challenge to formulating a general theory of complex systems capable of explaining life from its very beginnings (Morowitz and Smith 2007) at the edge of physics, to its ever more complex future forms (Elmqvist *et al.* 2021; Bettencourt 2021).

REFERENCES

Aktipis, A. 2021. "How Evolution Helps Us Understand Cancer and Control It." This article was originally published with the title "Malignant Cheaters" in *Scientific American* 324, 1, 62-67 (January 2021), doi:10.1038/scientificamerican0121-62, *Scientific American,* https://www.scientificamerican.com/article/how-evolution-helps-us-understand-cancer-and-control-it/.

Anderson, P. W. 1972. "More Is Different." *Science* 177 (4047): 393–396. https://doi.org/10.1126/science.177.4047.393.

Bettencourt, L. M. A. 2021. *Introduction to Urban Science: Evidence and Theory of Cities as Complex Systems.* Cambridge, MA: MIT Press.

Blankenship, R. E. 2010. "Early Evolution of Photosynthesis." *Plant Physiology* 154 (2): 434–438. https://doi.org/10.1104/pp.110.161687.

Bowler, P. J. 2009. *Evolution: The History of an Idea.* Berkeley, CA: University of California Press.

Brown, J. H., J. F. Gillooly, A. P. Allen, V. M. Savage, and G. B. West. 2004. "Toward a Metabolic Theory of Ecology." *Ecology* 85:1771–1789. https://doi.org/10.1890/03-9000.

Cover, T. M., and J. A. Thomas. 2005. "Information Theory and Statistics." In *Elements of Information Theory,* 347–408. New York, NY: John Wiley and Sons.

Curtius, K., N. A. Wright, and T. A. Graham. 2018. "An Evolutionary Perspective on Field Cancerization." *National Review of Cancer* 18:19–32. https://doi.org/10.1038/nrc.2017.102.

Elmqvist, T., E. Andersson, T. McPhearson, X. Bai, L. Bettencourt, E. Brondizio, J. Colding, *et al.* 2021. "Urbanization in and for the Anthropocene." *npj Urban Sustainability* 1 (6). https://doi.org/10.1038/s42949-021-00018-w.

Fisher, R. A. 1999. *The Genetical Theory of Natural Selection.* Edited with a foreword and notes by J. H. Bennett. Oxford, UK: Oxford University Press.

Gleick, J. 1988. *Chaos: Making a New Science.* New York, NY: Penguin Books.

Holland, J. H. 1973. "Genetic Algorithms and the Optimal Allocation of Trials." *SIAM Journal on Computing* 2:88–105. https://doi.org/10.1137/0202009.

———. 1992. "Genetic Algorithms." *Scientific American* 267:66–73. https://doi.org/10.1038/scientificamerican0792-66.

Hulbert, A. J., and P. L. Else. 1981. "Comparison of the 'Mammal Machine' and the 'Reptile Machine': Energy Use and Thyroid Activity." *American Journal of Physiology* 241 (5): R350–356. https://doi.org/10.1152/ajpregu.1981.241.5.R350.

Kardar, M. 2007. *Statistical Physics of Particles.* Cambridge, UK: Cambridge University Press.

Kingsland, S. 2015. "Alfred J. Lotka and the Origins of Theoretical Population Ecology." *Proceedings of the National Academy of Sciences* 112 (31): 9493–9495. https://doi.org/10.1073/pnas.1512317112.

Krakauer, D., N. Bertschinger, E. Olbrich, J. C. Flack, and N. Ay. 2020. "The Information Theory of Individuality." *Theory in Biosciences* 139:209–223. https://doi.org/10.1007/s12064-020-00313-7.

Lotka, A. J. 1922a. "Contribution to the Energetics of Evolution*." *Proceedings of the National Academy of Sciences* 8 (6): 147–151. https://doi.org/10.1073/pnas.8.6.147.

Lotka, A. J. 1922b. "Natural Selection as a Physical Principle*." *Proceedings of the National Academy of Sciences* 8 (6): 151–154. https://doi.org/10.1073/pnas.8.6.151.

Matthews, M. L. 2023. "Engineering Photosynthesis, Nature's Carbon Capture Machine." *PLoS Biology* 21 (7): e3002183. https://doi.org/10.1371/journal.pbio.3002183.

Mitchell, M. 2001. *An Introduction to Genetic Algorithms.* Cambridge, MA: MIT Press. https://doi.org/10.7551/mitpress/3927.001.0001.

———. 2009. *Complexity: A Guided Tour.* Oxford, UK: Oxford University Press.

Morowitz, H., and E. Smith. 2007. "Energy Flow and the Organization of Life." *Complexity* 13 (1): 51–59. https://doi.org/10.1002/cplx.20191.

Odum, H. T. 2007. *Environment, Power, and Society for the Twenty-First Century: The Hierarchy of Energy.* New York, NY: Columbia University Press.

Odum, H. Y., and R. C. Pinkerton. 1955. "Time's Speed Regulator: The Optimum Efficiency for Maximum Power Output in Physical and Biological Systems." *American Scientist* 43 (2): 331–343. https://www.jstor.org/stable/27826618.

Schrödinger, E. 1992. *What is Life? The Physical Aspect of the Living Cell.* Cambridge, UK: Cambridge University Press.

Stearns, S. C. 1992. *The Evolution of Life Histories.* Oxford, UK: Oxford University Press.

Thurner, S., R. A. Hanel, and P. Klimek. 2018. *Introduction to the Theory of Complex Systems.* Oxford, UK: Oxford University Press.

Watson, J. D., and F. H. Crick. 1953. "Molecular Structure of Nucleic Acids: A Structure for Deoxyribose Nucleic Acid." *Nature* 171:737–738. https://doi.org/10.1038/171737a0.

West, G. B., J. H. Brown, and B. J. Enquist. 1999. "The Fourth Dimension of Life: Fractal Geometry and Allometric Scaling of Organisms." *Science* 284 (5420): 1677–1689. https://doi.org/10.1126/science.284.5420.1677.

CONTRIBUTION TO THE ENERGETICS OF EVOLUTION

Alfred J. Lotka, Johns Hopkins University

It has been pointed out by Boltzmann[1] that the fundamental object of contention in the life-struggle, in the evolution of the organic world, is available energy.[2] In accord with this observation is the principle[3] that, in the struggle for existence, the advantage must go to those organisms whose energy-capturing devices are most efficient[4] in directing available energy into channels favorable to the preservation of the species.

The first effect of natural selection thus operating upon competing species will be to give relative preponderance (in number or mass) to those most efficient in guiding available energy in the manner indicated. Primarily the *path* of the energy flux through the system will be affected.

But the species possessing superior energy-capturing and directing devices may accomplish something more than merely to divert to its own advantage energy for which others are competing with it. If sources are presented, capable of supplying available energy in excess of that actually being tapped by the entire system of living organisms, then an opportunity is furnished for suitably constituted organisms to enlarge the total energy flux[5] through the system. Whenever such organisms arise, natural selection will operate to preserve and increase them. The result, in this case, is not a mere diversion of the energy flux through the

[1] *Der sweite Hauptsats der mechanischen Wärmetheorie*, 1886 (Gerold, Vienna), p. 21; *Populäre Schriften*, No. 3, Leipsic, 1905; Nernst, *Theoretische Chemie*, 1913, p. 819; Burns and Paton, *Biophysics*, 1921, p. 8, H. F. Osborn, *The Origin and Evolution of Life*, 1918, p. XV.

[2] Compare also Sir Oliver Lodge, *Life and Matter*, 1906, pp. 139, 140.

[3] Lotka, A. J., *Ann. Naturphil*, 1910, pp. 67, 68; *Proc. Nat. Acad., Sci.*, 1921, pp. 194, 195.

[4] Lotka, A. J., *Proc. Washington Acad.*, 5, 1915, pp. 360, 397.

[5] The term *energy flux* is here used to denote the available energy absorbed by and dissipated within the system per unit of time.

Boltzmann had understood the second law of thermodynamics and the necessity of sources of free energy for any device to perform work. Here Lotka extends this fundamental concept to living organisms, following from statistical physics. It is the main thesis of this paper. As we will see, it glosses over the fact that organisms are often limited by other resources and constraints, so that available free energy is necessary but not sufficient as a principle.

The idea Lotkta introduces here of the path of energy flux—i.e., how free energy is transformed successively and ultimately dissipated into heat—will turn out to be a very generative idea to organize complex systems.

Here, Lotka formulates the principle of maximum power as an ecological hypothesis. However, plants, for example, absorb only a fraction of available solar energy presumably because of other constraints in terms of space and nutrients.

This idea justifies thinking of evolution as maximizing energy intake and dissipation over time, which became known as the "maximum power principle."

Here, Lotka suggests that more power translates into more standing living mass. This neglects considerations of efficiency and metabolism by which it became later known that larger (and more massive) biological systems use less energy per unit mass. In general, biological and ecological metabolism is characterized by nonlinear relationships between mass and energy use per unit time (power), such as in Kleiber's law.

system of organic nature along a new path, but an increase of the total flux through that system.

Again, so long as sources exist, capable of supplying matter, of a character suitable for the composition of living organisms, in excess of that actually embodied in the system of organic nature, so long is opportunity furnished for suitably constituted organisms to enlarge the total mass of the system of organic nature. Whenever such organisms arise, natural selection will operate to preserve and increase them, provided always that there is presented a residue of untapped available energy. The result will be to increase the total mass of the system, and, with this total mass, also the total energy flux through the system, since, other things equal, this energy flux is proportional to the mass of the system.

Where a limit, either constant or slowly changing,[6] is imposed upon the total mass available for the operation of life processes, the available energy per unit of time (available power) placed at the disposal of the organisms, for application to their life tasks and contests, may be capable of increase by increasing the rate of turnover of the organic matter through the life cycle. So, for example, under present conditions,[7] the United States produce annually a crop of primary and secondary food amounting to about 1.37×10^{14} kilogramcalories per annum, enough to support a population of about 105 million persons (equivalent to about 88 million adults) at the present rate of food consumption (4, 270 kilogram-calories per adult per day). Suppose, as a simple, though rather extreme illustration, that man found means of doubling the rate of growth of crops, and of growing two crops a year instead of one. Then, without changing the average crop actually standing on the fields, the land would be capable of supporting double the present population. If this population were attained, the energy flux through the system composed of the human population and the organisms upon which it is dependent for food, would also be doubled. This result would be

[6]As, for example, if the total mass of the system is capable of accretion, but only at a limited velocity, in which case the phenomenon of a moving equilibrium may present itself. Compare Lotka, A. J., *Proc. Nat. Acad. Sci.*, 7, 1921, p. 168.

[7]Pearl, R., *The Nation's Food*, 1920, pp. 218, 203, 258, 80, 245; *Amer. J. Hygiene*, 1, 1921, p. 598.

attained, not by doubling the mass of the system (for the matter locked up in crops, etc., at a given moment would be, on an average, unchanged) but by increasing the velocity of circulation of mass through the life cycle in the system. Once more it is evident that, whenever a group of[8] organisms arises which is so constituted as to increase the rate of circulation of matter through the system in the manner exemplified, natural selection will operate to preserve and increase such a group, provided always that there is presented a residue of untapped available energy, and, where circumstances require it, also a residue of mass suitable for the composition of living matter.

The speeding up of energy use per unit time, given the same mass, is common in human systems like agriculture as well as cities and urban systems.

To recapitulate: In every instance considered, natural selection will so operate as to increase the total mass of the organic system, to increase the rate of circulation of matter through the system, and to increase the total energy flux through the system, so long as there is presented an unutilized residue of matter and available energy.

This is the main statement of the paper, the principle of maximum power, equated to natural selection. Note the last sentence emphasizing "constraints."

This may be expressed by saying that *natural selection* tends to make the energy flux through the system a maximum, so far as compatible with the constraints to which the system is subject.

It is not lawful to infer immediately that *evolution* tends thus to make this energy flux a maximum. For in evolution two kinds of influences are at work: selecting influences, and generating influences. The former select, the latter furnish the material for selection. If the material furnished for selection is strictly limited, as in the case of a simple chemical reaction,[9] which gives rise to a finite number of products, the range of operation of the selective influences is equally limited.

This refers to mutation and other diversity-generating mechanisms.

In the case of organic evolution the situation is very different. We have no reason to suppose that there is any finite limit to the number of possible types of organisms. In the present state of our knowledge, or rather our ignorance, regarding the generating influences that furnish material for natural selection, for organic evolution, an element of

[8]Owing to the fact that in existing organisms the anabiotic and catabiotic functions are very largely segregated in different types (plants and animals), evolution will here operate upon systems or groups of at least two species, one species of autotrophic ana bions, and one of heterotrophic catabions.

[9]Compare Lotka, A. J., *Amer. J. Sci.*, 24, 1907, pp. 204, 216.

uncertainty enters here. It appears, however, at least *a priori* probable that, among the certainly very large (if not infinite) variety of types presented for selection, sooner or later those will occur which give the opportunity for selection to operate in the direction indicated, namely so as to increase the total mass of the system, the rate of circulation of mass through the system, and the total energy flux through the system. If this condition is satisfied, the law of selection becomes also the law of evolution:

☞ Lotka uses a device from statistical physics called ergodicity that posits that a system can explore all possible states, given constraints; in this situation energy becomes the principal determinant of statistically observed states. This principle does not apply to biology in a straightforward way, as Darwin had realized, since historical contingencies expressed of descent with modifications play a critical role in the dynamics. This is because of the truly gigantic number of states in a combinatorial (e.g., genetic) system. This issue exposes some of Lotka's blind spots in conceptualizing biology in terms of information.

Evolution, in these circumstances, proceeds in such direction as to make the total energy flux through the system a maximum compatible with the constraints.

We have thus derived, upon a deductive basis, at least a preliminary answer to a question proposed by the writer in a previous publication.[10] It was there pointed out that the influence of man, as the most successful species in the competitive struggle, seems to have been to accelerate the circulation of matter through the life cycle, both by "enlarging the wheel," and by causing it to "spin faster." The question was raised whether, in this, man has been unconsciously fulfilling a law of nature, according to which some physical quantity in the system tends toward a maximum. This is now made to appear probable; and it is found that the physical quantity in question is of the dimensions of power, or energy per unit time, as was hinted by the writer on an earlier occasion.[11]

It may be remarked that the principle of maximum energy flux here set forth bears a certain outward resemblance to a principle enunciated by Ostwald:[12] "Of all possible energy transformations, that

---

[10] Lotka, A. J., *Proc. Nat. Acad. Sci.*, 7, 1921, p. 172.

[11] Idem., *Ann. Naturphil.*, 1910, p. 70. It is there suggested that the continuous energy transformations associated with the maintenance of a steady state would probably be found to play the dominant rôle, while any "latent heat" effect associated with a change in the distribution of matter among the several species composing the system, would probably play a subordinate rôle; in contrast with the condition of affairs familiar in ordinary physico-chemical systems. This is an obvious inference from the observation that the several species of organisms are distinguished much more by structural differences than by differences in chemical composition.

[12] Ostwald, W., *Lehrbuch der allgemeinen Chemie*, 1892, vol. 2, p. 37; Siebel, J. E. *Compend of Mechanical Refrigeration*, 1915, p. 88. For a discussion of the validity and limitations of Ostwald's principle see Helm G., *Die Energetik*, 1898, pp. 248; Neumann, C., *Leipziger Berichte*, 1892, p. 184.

one takes place, which brings about the maximum transformation in a given time." This principle of Ostwald's, however, is based on entirely different grounds from those here brought forward. It is not of general applicability, and in particular, its application to systems of the kind here considered does not appear warranted.

Addendum. Since the paragraphs above were penned, the writer has received from the booksellers a copy of Professor J. Johnstone's book, "The Mechanism of Life" (1921), in which (pp. 217–221) that author touches on matters closely related to those here discussed. Professor Johnstone draws, however, a somewhat different conclusion, namely that "In living processes the increase of entropy is retarded."[13] He points out that this is true, primarily, of plants; but that among animals[14] also natural selection must work toward the weeding out of unnecessary and wasteful activities, and thus toward the conserving of free energy, or, what amounts to the same thing, toward retarding energy dissipation.

This issue deals with the time scale at which energy is stored and dissipated by biological systems. Clearly no maximum power principle can apply instantaneously for a growing organism, at least not for the total energy involved.

This is perhaps not wholly convincing, for the first effect of the advent of animal organisms in a world peopled with a purely vegetable population, would certainly seem to be an acceleration of the process of dissipation. It appears, therefore, that at certain stages in the evolution of the system, at the least, life must have tended to increase rather than decrease dissipation. And even if animals ultimately evolve in the direction of decreased dissipative effect, they still remain essentially a dissipative type, as compared with plants, and, to make Professor Johnstone's argument conclusive, it would seem necessary to show, not merely that the animal organism evolves in that direction, but that the system of coupled transformers, plant and animal, as a whole has so evolved.

Both authors could be right here. What is at stake are different scales in time and in population size. Lotka invokes the fact that a maximum power principle should apply over large scales, as is typical of any statistical law.

Professor Johnstone's conclusion is, however, not essentially incompatible with the one developed in this paper. Where the supply of available energy is limited, the advantage will go to that organism which is most efficient, most economical, in applying to preservative uses such energy as it captures. Where the energy supply is capable of

[13] That living organisms may be capable of retarding the energy flux through the system of nature was suggested by the present writer in *Ann. Naturphil.*, 1910, p. 60.

[14] Johnstone, J., *The Mechanism of Life*, 1921, p. 220.

expansion, efficiency or economy, though still an advantage, is only one way of meeting the situation, and, so long as there remains an unutilized margin of available energy, sooner or later the battle, presumably, will be between two groups or species equally efficient, equally economical, but the one more apt than the other in tapping previously unutilized sources of available energy.

There is here a problem that will call for further investigation. In particular, it remains to be established just what is the significance of the phrase "compatible with the constraints" which, in the presentation here given, modifies the maximum principle enunciated. The present communication is intended rather as preliminary than as an attempt to say the last word on the subject. More detailed discussion is planned for another occasion.

The importance of other constraints, such as available space, nutrients, predation, and many more, ended up emphasizing issues of information in the definition of biological fitness in the last century. This makes consideration of energy flux certainly necessary but not sufficient for developing a theory of living complex systems.

Lotka seems to anticipate many of the considerations that would make biology and ecology mostly concerned with information, expressed primarily via genes but also as behavior and organization. Nevertheless, Lotka's principle endures as a necessary basis for understanding evolution, awaiting a more complete reckoning between the dynamics of energy and information not only in biology but also in human societies.

REFERENCES

Boltzmann, L. 1905. *Populare Schriften.* 3. Leipsic.

Burns, D., and D. N. Paton. 1921. *An Introduction to Biophysics.*

Clausius, R. 1886. *Der Zweite Hauptsatz der mechanischen Wärmetheorie.* Vienna: Gerold.

Helm, G. 1898. *Die Energetik.*

Johnstone, J. 1921. *The Mechanism of Life.*

Lodge, Sir Oliver. 1906. *Life and Matter.*

Lotka, A. J. 1907. "Studies on the Mode of Growth of Material Aggregates." *American Journal of Science* 24:199–216.

———. 1910. "Die Evolution vom Standpunkte der Physik." In *Annalen der Naturalphilosophie,* edited by W. Ostwald.

———. 1915. "Efficiency as a factor in organic evolution. I and II." *Proceedings of the Washington Academy of Sciences* 5 (10, 11): 360–368, 397–403.

———. 1921a. "Note on Moving Equilibra." *Proceedings of the National Academy of Sciences* 7 (6): 168–172.

———. 1921b. "Note on the Economic Conversion Factors of Energy." *Proceedings of the National Academy of Sciences* 7 (7): 192–197.

Nernst, W. 1913. *Theoretische Chemie.*

Neumann, C. 1892. *Leipziger Berichte.*

Osborn, H. F. 1918. *The Origin and Evolution of Life.*

Ostwald, W. 1892. *Lehrbuch der Allgemeinen Chemie.* Vol. 2. Leipzig: Verlag von Wilhelm Engelmann.

Pearl, R. 1920, 218, 203, 258, 80, 245.

———. 1921. "The Vitality of the Peoples of America." *American Journal of Hygiene* 1:592–674.

Siebel, J. E. 1915. *Compend of Mechanical Refrigeration.*

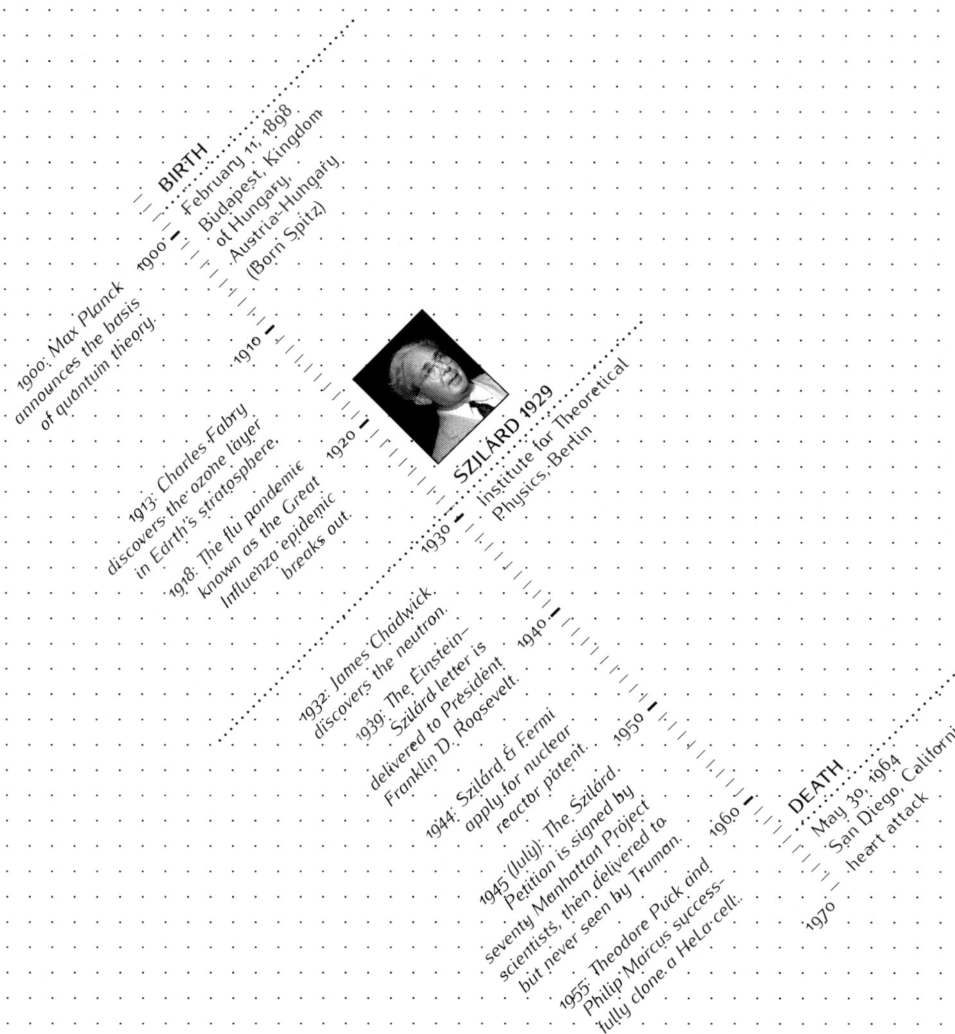

LEÓ SZILÁRD (born Spitz)

[2]

TO WORK IS TO THINK

Susanne Still, University of Hawai'i

Léo Szilárd made important contributions to the foundations of statistical thermodynamics and the physics of information with work he completed in 1922. This appeared in print much later (Szilárd 1925, 1929), and was not translated into English until 1972 and 1964, respectively.[1] The second paper (Szilárd 1929), reprinted in this chapter, provides a physical basis for quantifying information. This groundbreaking work was not immediately fully absorbed into the developing literature on information theory, cybernetics, and physics of computation; its recognition came with some delay. The paper has had a recent revival, recognized as the founding paper on information engines.

L. Szilárd, "On the Decrease of Entropy in a Thermodynamic System by the Intervention of Intelligent Beings" (trans. A. Rapoport and M. Knoller), *Behavioral Science* 9 (4), 301–310 (1964). Originally published in *Zeitschrift für Physik* (1929).

At the age of twenty-one, Szilárd fled Hungary and continued his engineering studies in Berlin, where he arrived in January 1920.[2] Berlin was an epicenter of modern physics at the time. Soon Szilárd switched from engineering to physics, and in 1921 he asked Albert Einstein to teach a course on statistical mechanics. In the audience were several of Szilárd's friends, including (from Budapest, like Szilárd) John von Neumann, Eugene Wigner, and Dennis Gábor. Over winter break 1921/1922, Szilárd worked out how to extend phenomenological thermodynamics to fluctuations of thermodynamic quantities, bridging Clausius's approach with that of Boltzmann and Einstein.

Einstein was impressed with Szilárd's work, which was accepted in 1922 as Szilárd's dissertation at the University of Berlin.[3] Within a few months, Szilárd had worked out the contents of the paper reprinted in

[1] Szilárd's dissertation (Szilárd 1925) was translated in Szilárd (1972). The translation of the paper reprinted in this chapter (Szilárd 1929) appeared as an homage to Szilárd after his death (Szilárd 1964), not in a physics journal, but rather in *Behavioral Science*. The German Szilárd and his contemporaries wrote has a somewhat different structure compared to modern scientific English, which poses an intrinsic challenge for translation.

[2] My main reference for information relating to Szilárd's life is Lanouette (2013).

[3] Gutachten uber die Dissertation von Léo Szilárd, Berlin 10.7.1922. Archived in *The UC San Diego Collection of Leo Szilárd's papers.*

this chapter as a logical continuation of his dissertation work. This would eventually form the core of his *Habilitation*,[4] granting the right to teach at the University of Berlin based on work establishing a scholar's ability to do independent research.

Szilárd's results were thus known as early as 1922 to a significant part of the physics community, including the great physicists Szilárd had studied with, such as Einstein, Max von Laue and Max Planck, and Szilárd's friends who would go on to become great physicists themselves. But years passed before Szilárd's work was published in *Zeitschrift für Physik*, and later translated into English, which might explain why there does not seem to have been a direct influence on the early work in information theory emerging during the 1920s, mostly in engineering circles centered around research at American telecommunication companies.[5] Von Neumann was influenced directly by Szilárd's ideas and included references in his 1932 book on quantum mechanics.[6] However, the influence of Szilárd's work did not seem to penetrate fully. In 1948 neither Claude Shannon's information theory nor Norbert Wiener's cybernetics carefully discussed Szilárd's work.[7]

Both cybernetics and information theory became popular. Connections to Szilárd's work were pointed out by von Neumann and Gábor, and Szilárd had conversations with Léon Brillouin[8] in the early 1950s, but otherwise he did not seem to have been too involved in discussions about information theory, cybernetics, and the physics of information. In the 1930s, political circumstances and his scientific interests had taken him in a very different direction.

---

[4]Gutachten zum Habilitationsgesuch des Dr. L. Szilárd's, Berlin 11.12.1926. Archived in *The UC San Diego Collection of Leo Szilárd's papers*.

[5]See Nyquist (1924) and Hartley (1928).

[6]Chapter V.2 in von Neumann (1932): *Anmerkung* (note) 184 cites Szilárd (1925), *Anmerkung* (note) 194 cites Szilárd (1929).

[7]According to Lanouette (2013, 468, ref. 58), bearing in mind Szilárd's ideas, von Neumann urged Shannon to use the term "entropy." Von Neumann (1932) wasn't translated into English until (1955), and I do not see any mention of either von Neumann or Szilárd in the place where entropy is introduced in Shannon (1948). Likewise, Wiener, despite having had interactions with von Neumann, does not discuss Szilárd's work in the chapter on Maxwell's demon in his classic book on cybernetics (Wiener 1948, Chapter II).

[8]See, e.g., Brillouin (1949, 1951), von Neumann (1949), and Gábor (1961)

For roughly a decade after his groundbreaking 1922 work, Szilárd remained mostly in Berlin. He became von Laue's assistant, and later a *Privatdozent* (lecturer). He co-taught classes on new developments in theoretical physics with von Neumann and Erwin Schrödinger, and on atomic physics with Lise Meitner. Szilárd was affected by the horrifying situation in Germany, emigrating first to England and later to the USA. He helped scholars in exile, and in 1933 his life took a difficult, in my opinion even tragic, turn when he conceived of the nuclear chain reaction, apparently before anyone else, and henceforth dealt with the consequences of this insight. Much has been written about Szilárd's involvement in atomic energy, the creation of the nuclear bomb, and how he eventually tried to convince people not to use atomic bombs.

One may ponder if and how things might have been different, had Szilárd's 1922 work become more broadly known and been translated earlier. However, by the late 1940s, the importance of Szilárd's contributions was certainly recognized within physics, e.g.:[9] "Central to v. Neumann's more detailed discussion is Szilard's idea of a phenomenological thermodynamics which does not neglect fluctuations and Brownian movement. This idea, developed in two admirable papers," (Szilárd 1925, 1929), "stands rather isolated apart from the flow of modern physical ideas; but I am inclined to regard it as one of the greatest achievements of modern theoretical physics, and believe that we are still very far from being able to evaluate all its consequences. The present discussion of these issues is very incomplete, and I refer the reader to Szilard's own writings for more careful exposition."

In Szilárd's own words: "My thesis showed that the Second Law of Thermodynamics covers not only the mean values, as was up to then believed, but also determines the general form of the law that governs the fluctuations of the values." The 1929 paper, reprinted here, "established the connection between entropy and information which forms part of present day information theory" (Szilárd 1972, 3).

To this day, Szilárd's 1929 paper continues to inspire. Recent interest in Szilárd's work is sparked by experimental realizations[10] of the thought

---

[9] Jordan (1949, 275)

[10] Some of the first experiments were Toyabe *et al.* (2010) and Bérut *et al.* (2012).

experiment Szilárd used to illustrate his results (known as "Szilárd's engine"), which have led to a sharp increase in work on information engines. Szilárd's ideas play a role in recent developments in far-from-equilibrium (also called "stochastic") thermodynamics, in situations in which the protocol applied to a thermodynamic system is dependent on measurements of the system's state.[11] Szilárd's approach can be *generalized* to provide a physical basis for a modern understanding of information processing, decision making under uncertainty, adaptation and learning.[12]

Szilárd's 1929 paper addresses a problem Maxwell introduced in a thought experiment over half a century prior.[13] A *"very observant and neat-fingered being"* (later called "Maxwell's demon") could sort fast from slow molecules, without doing any work, by conveniently timing the opening and closing of an idealized door in a wall separating two compartments of a gas-filled container. Starting initially with the same temperature in both compartments, after the demon's interventions they would have different temperatures, in apparent violation of the second law of thermodynamics. Maxwell's thought experiment thus makes it look as if one could build a perpetual motion machine, simply by using intelligent interventions.

Szilárd's new insight was that the important feature of the intelligent intervention lies in the demon's ability to adjust a control parameter in a way that contains memory of aspects of the thermodynamic system's state. The "demon" plays the role of an observer who decides which intervention to apply, based on an observation of the system at an earlier time. Consequently, Szilárd points out that Maxwell's demon can be replaced by machinery that has the same effect. The important thing is not that the demon is a sentient being, but rather that the demon can effect a memory-carrying coupling between a measurable parameter of the thermodynamic system and a control parameter. Szilárd focuses on "an unusually simple type" of such coupling, which he uses to define *measurement* as the process

[11]This is a large active research area, and providing a complete list of references is far beyond the scope of this introduction. But citing only a few papers seems unfair, as it would reflect my personal preference. Therefore, I refer the interested reader to the literature.

[12]For my detailed view on this, see Still (2014, 2020), further developed in Still and Daimer (2022) and Daimer and Still (2023a, 2023b).

[13]First in a letter to Tait on December 11, 1867 (see Knott 1911, 213), later in his textbook *Theory of Heat* (Maxwell 1871).

of coupling (and later de-coupling) a stochastically fluctuating variable x at time t to a parameter y, in such a way that the value of y after time t is determined by the value of x at time t, and y remains stable over some time interval, while x continues to fluctuate.

Szilárd's approach takes phenomenological thermodynamics as given, the second law is central. Szilárd argues that if the so-defined measurement process[14] would not necessitate an increase in entropy, then the second law could be violated, because a cyclic process could be built such that the entropy of the environment would decrease without compensation. Therefore, the second law dictates that such a measurement process must produce entropy, at least as much as the amount of entropy reduction it enables. From this requirement Szilárd calculates a bound on the entropy production accompanying the measurement process. That is his main result, displayed in his equation (1).

Szilárd focuses on the case in which y can have two outcomes, that is, x can be classified into two different types: if x is in some interval $X = (x_1, x_2)$, then y is set to value 1 (and we could say x is of type 1), otherwise y is set to -1 (x is of type 2). The probability of finding x in X is denoted by w_1, and the probability of finding it outside of X by w_2. The average entropy production going along with the process of setting y to either of its possible values is denoted by S_i, where $i = 1, 2$ (I write terms indexed with subscripts, because it reduces visual clutter, and because Szilárd's arguments generalize to $i = 1, \ldots, n$). The total average entropy production encountered in the process of setting y to the appropriate measurement outcome, averaged over all possible outcomes, is then $\bar{S} = \sum_i w_i S_i$, and normalization dictates $\sum_i w_i = 1$.

Szilárd uses the construct of his figure 1 to argue that, equipped with the measurement outcome, a cycle can be built resulting in an average entropy decrease in the environment of $\bar{s} = k \sum_i w_i \log(w_i)$ (a negative

[14] Perhaps it makes sense not to focus too much on the *word* "measurement." Szilárd's argument holds for the physical process he described. He called that process measurement, but whether or not this was the best nomenclature does not change the results. The essence here is that at the end of the process, the outcome of a random variable has a stable value, determined by the value of a fluctuating variable at an earlier time.

number)[15] per molecule (his eq. (10)). His argument starts by invoking a container filled with molecules, which all have an observable, fluctuating, parameter x. There are two semipermeable membranes, each of which lets pass only one of the possible types of molecules (what Szilárd calls "modifications in x"). Using these semipermeable walls as pistons, one can separate a large number of particles, mixed in some ratio $w_1 : w_2$, without doing work on the system. Afterwards, each container has only molecules of one of the two types, while the $w_1 : w_2$ ratio in the combined volume has remained the same. If the equilibrium distribution is now restored reversibly in each container, holding volume and temperature fixed, and letting the x parameter change, then, Szilárd argues, the entropy of the system increases, while the entropy of the environment decreases by the same amount, which is, on average, $\bar{s} = k \sum_i w_i \log(w_i)$ per molecule.

To have a cyclic process, one would have to bring the gas back to its original volume, and this would require work, because the gas is now no longer unmixed. But, Szilárd points out, if for each molecule, the value of x at the time when it passed the semipermeable wall[16] was recorded by adjusting parameter y accordingly, then one could exchange the semipermeable walls for two others, one of which would let only molecules with $y = 1$ pass, and the other only molecules with $y = -1$. With those new semipermeable walls, one could bring the gas back to its original state without doing work, but one would have used a memory of which type of "modification in x" each molecule was earlier.

During this cycle the total average entropy production is given by the reduction $\bar{s}$ plus the average entropy produced during the measurement process, $\bar{S}$. According to the second law of thermodynamics, this sum must be non-negative. Thus, the second law directly implies the requirement $\bar{S} \geq -k \sum_i w_i \log(w_i)$, equivalently (see his eq. (12))

$$\bar{s} + \bar{S} = \sum_i w_i \left(S_i + k \log(w_i)\right) \geq 0\,. \qquad (1)$$

[15] Szilárd uses log for the natural logarithm. I will do the same throughout this introduction.

[16] Szilárd actually considers a synchronous time of measurement, but that requires that the "modification in x" be assumed stable over some time interval. The asynchronous alternative I'm suggesting would not require this assumption.

The left hand side of this inequality depends on the w_i, and Szilárd points out that the inequality must be true for *all* possible values of w_i, including those that *minimize* the l.h.s. (subject to the constraint imposed by normalization). He shows that at the minimum the following condition has to be fulfilled between the S_i and the logarithm of the inverse probability[17] (see his eq. (15)):

$$S_i = -k \log(w_i) + k\lambda, \quad \forall i \text{ , with } \lambda = const. \tag{2}$$

The constant can be determined from normalization, $\lambda = -\log\left(\sum_i e^{-\frac{S_i}{k}}\right)$, and since the total average dissipation, $\bar{s} + \bar{S}$, evaluated at the minimum, is given by $k\lambda$, and has to be non-negative due to the second law, λ thus also has to be non-negative, whereby Szilárd's main result follows (his eq. (20)):

$$\sum_i e^{-\frac{S_i}{k}} \leq 1\,. \tag{3}$$

Equation (2) is a relationship between, on the one hand, the average physical entropy produced during a process in which the value of a random variable is set to reflect (in a stable way) a particular outcome of another random variable (which is here the type of "modification in x," not the exact value of x), and thereby eliminates *all* uncertainty about the outcome at the time of recording, and, on the other hand, the negative log probability of finding the outcome, given all *a priori* knowledge. Average *entropy production* is proportional to average *uncertainty reduction* (which, as Shannon (1948) pointed out, serves as a measure of information, if one accepts a few basic axioms). As long as the average entropy production associated with this physical measurement process, averaged over all possible outcomes, is at least $\bar{S} = -k\sum_i w_i \log(w_i)$, the second law is not violated.

Szilárd's first example, discussed in his paper *before* the main result is derived, has inspired what is often referred to as "Szilárd's engine": a one-particle gas in a container separated into two volumes by a divider which can act as a piston. Knowledge of which side is empty enables the observer

[17] The term $-\log(w_i)$ is the negative "log probability" of finding x to be of type i, given everything we know about the physics of the setup. In information theory, this is called "self-information" (Cover 1999), or "surprisal" (Tribus 1961), as it measures how informative, or how surprising, it is to discover the outcome of the random variable.

to extract work from the surrounding heat bath, with the help of a piston, by isothermal volume expansion into the empty side.

Szilárd points out that his main result applies here. The average entropy produced when y is set to one of its values,[18] say $y = 1$, which has the magnitude $\bar{S}_1$, could be small, but at the expense of $\bar{S}_2$ having to be larger, as dictated by Szilárd's inequality: $e^{-\frac{\bar{S}_1}{k}} + e^{-\frac{\bar{S}_2}{k}} \leq 1$. He points out the important fact that $\bar{S}_1 = \bar{S}_2 = k \log(2)$ ensures that the second law holds. Szilárd presents this insight in form of an Ansatz[19] and then shows, by explicitly calculating $\bar{s}$ for the example, that the second law is guaranteed not to be violated.[20]

It is easily noted that for two outcomes ($n = 2$), the distribution $w_1 = w_2 = 1/2$ results in the largest possible entropy reduction, namely $\bar{s} = -k \log(2)$. This tells us that the entropy reduction achievable from recording *one bit* of information[21] is *exactly* compensated by the entropy production of Szilárd's Ansatz, where $\bar{S} = k \log(2)$. One bit is, of course, the maximum amount of information a *binary* random variable can capture. Thus, if any procedure that sets the value of a binary variable was universally accompanied by an average entropy production of $k \log(2)$, then the second law would certainly not be violated.

After a measurement is recorded, the variable y has a definite outcome, but $y = 1$ gets recorded with probability w_1, and $y = -1$ with probability w_2. Knowing which specific value y is set to, changes the probability

[18] In his first example, Szilárd denotes the average entropy produced when y is set to 1, or -1, respectively, by $\bar{S}_1$ and $\bar{S}_2$. In the second part of his paper, he uses S_1 and S_2 for seemingly the same quantity (unless I am missing a subtle distinction). Nonetheless, I am trying to stick to Szilárd's notation to ease readability.

[19] We can see why this Ansatz makes sense by considering, for example, that Szilárd's inequality can be written as $e^{-\frac{S}{k}}(1 + e^{-\frac{\delta S}{k}}) \leq 1$, where $S := \min(S_1, S_2)$, and $\delta S := |S_1 - S_2|$. Note that $(1 + e^{-\frac{\delta S}{k}}) \leq 2$. For any fixed value $S = \hat{S}$, Szilárd's inequality has to hold for all δS. The maximum of the l.h.s. occurs at $\delta S = 0 \Leftrightarrow S_1 = S_2$. Therefore, we have the requirement $2e^{-\frac{\hat{S}}{k}} \leq 1 \Leftrightarrow \hat{S} \geq k \log(2)$, for $\hat{S} = S_1 = S_2$. The choice $\bar{S}_1 = \bar{S}_2 = k \log(2)$ thus ensures that the second law is not violated.

[20] A statement regarding the generality of $k \log(2)$ (in the binary case) occurs in a very difficult-to-translate sentence in Szilárd (1929), but the phrase appears to have gotten lost in the English translation.

[21] The term *bit* (attributed to John Tukey) became popular from Shannon (1948); obviously, this term did not exist in 1922.

distribution from $p(y=1)=1/2$ (for $w_1=w_2$) to either $p(y=1)=1$, or $p(y=1)=0$, depending on which value was chosen. This requires a reduction in the accessible phase space of the information bearing physical degree of freedom, which explains that the entropy production has to be at least $k\log(2)$. The argument can be made explicit with various models for a bistable memory, as, for example, in Landauer (1961) where a connection was drawn to computation and logical irreversibility.[22]

The example of the one-particle gas in a container serves to illustrate Szilárd's main insight: the second law of thermodynamics directly implies that, if there exists a method by which memory of an observation can be used to *reduce* entropy by a certain average amount in a cyclic process, then any physical process that records the memory (what Szilárd calls a *measurement* process) must, on average, *produce* at least as much entropy.

Personally, I prefer to think about it in terms of work potential. In Szilárd's engine, inserting the partition, plus recording an observation that reflects which side of the container the particle resides in, results in the same work potential as a simple isothermal compression of the gas would have resulted in. The first of these two alternative processes is a clever preparation that apparently can be done without having to do work on the system,[23] together with what Szilárd defined as a measurement process. The second process is a regular state preparation that obviously requires work. According to Szilárd's argument, these two processes are *thermodynamically equivalent*, in the sense that their minimal entropy production is the same. This equivalence is dictated by the second law.

Szilárd's engine is an idealization, a thought experiment, providing a physical mechanism to convert *information to work*, and vice versa, i.e., an *information engine*. The clever process, which we could interpret as a most basic "computation," or a very primitive form of "thinking," allows the observer to use knowledge to extract work from a heat bath, but does itself also require work, some of which has to be dissipated.

---

[22] Landauer (1961) does not give a detailed discussion of the connection to Szilárd (1929).

[23] Not to forget that this is an idealization.

Importantly, Szilárd's bound holds regardless of the specific details of the physical process implementing such a most basic "computation." The general insight Szilárd provided is that the second law of thermodynamics implies that one could *never* find a way to get around the corresponding minimum entropy production, no matter how clever an information processing scheme one develops. In other words, if we agree that the second law holds, which all evidence suggests, then we must *admit* that we won't be able to outfox it.

To close, let me take the liberty of using the word "work" in two of its meanings. We have learned from Szilárd's paper that thinking must require work. Conversely, for Léo Szilárd, work meant thinking![24] Society would do well to create and maintain, at least in academia, conditions enabling that.

Acknowledgements

I thank Dorian Daimer, Robert Shaw, Thomas Redford, and Xerxes Tata for discussions and detailed comments on this manuscript.

REFERENCES

Bérut, A., A. Arakelyan, A. Petrosyan, S. Ciliberto, R. Dillenschneider, and E. Lutz. 2012. "Experimental Verification of Landauer's Principle Linking Information and Thermodynamics." *Nature* 483 (7388): 187–189. https://doi.org/10.1038/nature10872.

Brillouin, L. 1949. "Life, Thermodynamics, and Cybernetics." *American Scientist* 37 (4): 554–568.

———. 1951. "Maxwell's Demon Cannot Operate: Information and Entropy. I." *Journal of Applied Physics* 22 (3): 334–337. https://doi.org/10.1063/1.1699951.

Cover, T. M. 1999. *Elements of Information Theory.* New York, NY: John Wiley & Sons.

Daimer, D., and S. Still. 2023a. "The Physical Observer in a Szilárd Engine with Uncertainty." *arXiv preprint arXiv:2309.10580,* https://doi.org/10.48550/arXiv.2309.10580.

———. 2023b. "Thermodynamically Rational Decision Making under Uncertainty." *arXiv preprint arXiv:2309.10476,* https://doi.org/10.48550/arXiv.2309.10476.

Gábor, D. 1961. "Light and Information." In *Progress in Optics,* 1:109–153. Amsterdam, Netherlands: Elsevier. https://doi.org/10.1016/S0079-6638(08)70609-7.

[24] This was extracted from Pyke and Medawar (2000). It appears consistent with everything else I have read about Szilárd.

Hartley, R. V. L. 1928. "Transmission of Information." *Bell System Technical Journal* 7 (3): 535–563. https://doi.org/10.1002/j.1538-7305.1928.tb01236.x.

Jordan, P. 1949. "On the Process of Measurement in Quantum Mechanics." *Philosophy of Science* 16 (4): 269–278. https://doi.org/10.1086/287049.

Knott, C. G., ed. 1911. *Life and Scientific Work of Peter Guthrie Tait.* Includes Maxwell's Letter to P. G. Tait, 11 December 1867. London, UK: Cambridge University Press.

Landauer, R. 1961. "Irreversibility and Heat Generation in the Computing Process." *IBM Journal of Research and Development* 5 (3): 183–191. https://doi.org/10.1147/rd.53.0183.

Lanouette, W. 2013. *Genius in the Shadows: A Biography of Leo Szilárd, the Man behind the Bomb.* New York, NY: Skyhorse.

Maxwell, J. C. 1871. *Theory of Heat.* London, UK: Longmans, Green & Co.

Nyquist, H. 1924. "Certain Factors Affecting Telegraph Speed." *Transactions of the American Institute of Electrical Engineers* 43:412–422. https://doi.org/10.1109/T-AIEE.1924.5060996.

Pyke, D., and J. Medawar. 2000. "The Pest from Budapest." August 12, 2000, *The Guardian,* https://www.theguardian.com/books/2000/aug/12/politics.history.

Shannon, C. E. 1948. "A Mathematical Theory of Communication." *The Bell System Technical Journal* 27 (3): 379–423. https://doi.org/10.1002/j.1538-7305.1948.tb01338.x.

Still, S. 2014. "Lossy is Lazy." In *Proceedings of the Seventh Workshop on Information-Theoretic Methods in Science and Engineering,* 17–21. Helsinki, Finland: University of Helsinki. http://hdl.handle.net/10138/153411.

———. 2020. "Thermodynamic Cost and Benefit of Memory." *Physical Review Letters* 124 (5–7): 050601. https://doi.org/10.1103/PhysRevLett.124.050601.

Still, S., and D. Daimer. 2022. "Partially Observable Szilárd Engines." *New Journal of Physics* 24 (7): 073031. https://doi.org/10.1088/1367-2630/ac6b30.

Szilárd, L. 1925. "Über die Ausdehnung der phänomenologischen Thermodynamik auf die Schwankungserscheinungen." *Zeitschrift für Physik* 32 (1): 753–788. https://doi.org/10.1007/BF01331713.

———. 1929. "Über die Entropieverminderung in einem thermodynamischen System bei Eingriffen intelligenter Wesen." *Zeitschrift für Physik* 53 (11-12): 840–856. https://doi.org/10.1007/BF01341281.

———. 1964. "On the Decrease of Entropy in a Thermodynamic System by the Intervention of Intelligent Beings." *Behavioral Science* 9 (4): 301–310. https://doi.org/10.1002/bs.3830090402.

———. 1972. "On the Extension of Phenomenological Thermodynamics to Fluctuation Phenomena." In *Collected Works of Leo Szilárd,* 70–102. Cambridge, MA: MIT Press.

The UC San Diego Collection of Leo Szilárd's papers. https://library.ucsd.edu/dc/object/bb3073224j/_1.pdf.

Toyabe, S., T. Sagawa, M. Ueda, E. Muneyuki, and M. Sano. 2010. "Experimental Demonstration of Information-to-Energy Conversion and Validation of the Generalized Jarzynski Equality." *Nature Physics* 6 (12): 988–992. https://doi.org/10.1038/nphys1821.

Tribus, M. 1961. "Information Theory as the Basis for Thermostatics and Thermodynamics." *Journal of Applied Mechanics* 28 (1): 1–8. https://doi.org/10.1115/1.3640461.

von Neumann, J. 1932. *Mathematische Grundlagen der Quantenmechanik.* Berlin, Germany: Springer-Verlag.

———. 1949. "Cybernetics or Control and Communication in the Animal and the Machine, by Norbert Wiener." *Physics Today* 2 (5): 33–34. https://doi.org/10.1063/1.3066516.

———. 1955. *Mathematical Foundations of Quantum Mechanics.* Translated by R. T. Beyer. Princeton, NJ: Princeton University Press.

Wiener, N. 1948. *Cybernetics: Or, Control and Communication in the Animal and the Machine.* Cambridge, MA: MIT Press.

ON THE DECREASE OF ENTROPY IN A THERMODYNAMIC SYSTEM BY THE INTERVENTION OF INTELLIGENT BEINGS°

Leo Szilárd, University of Berlin

Abstract

The objective of the investigation is to find the conditions which apparently allow the construction of a perpetual-motion machine of the second kind, if one permits an intelligent being to intervene in a thermodynamic system. When such beings make measurements, they make the system behave in a manner distinctly different from the way a mechanical system behaves when left to itself. We show that it is a sort of a memory faculty, manifested by a system where measurements occur, that might cause a permanent decrease of entropy and thus a violation of the Second Law of Thermodynamics, were it not for the fact that the measurements themselves are necessarily accompanied by a production of entropy. At first we calculate this production of entropy quite generally from the postulate that full compensation is made in the sense of the Second Law (Equation [1]). Second, by using an inanimate device able to make measurements—however under continual entropy production—we shall calculate the resulting quantity of entropy. We find that it is exactly as great as is necessary for full compensation. The actual production of entropy in connection with the measurement, therefore, need not be greater than Equation (1) requires.

There is an objection, already historical, against the universal validity of the Second Law of Thermodynamics, which indeed looks rather ominous. The objection is embodied in the notion of Maxwell's demon, who in a different form appears even nowadays again and again; perhaps not unreasonably, inasmuch as behind the precisely formulated question quantitative connections seem to be hidden which to date have not been clarified. The objection in its original formulation concerns a demon who catches the fast molecules and lets the slow ones pass. To be sure, the objection can be met with the reply that man cannot

°Translated by Anatol Rapoport and Mechthilde Knoller from the original article "Über die Entropieverminderung in einem thermodynamischen System bei Eingriffen intelligenter Wesen." *Zeitschrift für Physik*, 1929, 53, 840–856.

in principle foresee the value of a thermally fluctuating parameter. However, one cannot deny that we can very well measure the value of such a fluctuating parameter and therefore could certainly gain energy at the expense of heat by arranging our intervention according to the results of the measurements. Presently, of course, we do not know whether we commit an error by not including the intervening man into the system and by disregarding his biological phenomena.

Apart from this unresolved matter, it is known today that in a system left to itself no "perpetuum mobile" (perpetual motion machine) of the second kind (more exactly, no "automatic machine of continual finite work-yield which uses heat at the lowest temperature") can operate in spite of the fluctuation phenomena. A perpetuum mobile would have to be a machine which in the long run could lift a weight at the expense of the heat content of a reservoir. In other words, if we want to use the fluctuation phenomena in order to gain energy at the expense of heat, we are in the same position as playing a game of chance, in which we may win certain amounts now and then, although the expectation value of the winnings is zero or negative. The same applies to a system where the intervention from outside is performed strictly periodically, say by periodically moving machines. We consider this as established (Szilárd 1925) and intend here only to consider the difficulties that occur when intelligent beings intervene in a system. We shall try to discover the quantitative relations having to do with this intervention.

Smoluchowski (1914, p. 89) writes: "As far as we know today, there is no automatic, permanently effective perpetual motion machine, in spite of the molecular fluctuations, but such a device might, perhaps, function regularly if it were appropriately operated by intelligent beings....

"A perpetual motion machine therefore is possible if—according to the general method of physics—we view the experimenting man as a sort of *deus ex machina*, one who is continuously and exactly informed of the existing state of nature and who is able to start or interrupt the macroscopic course of nature at any moment without expenditure of work. Therefore he would definitely not have to possess the ability to catch single molecules like Maxwell's demon, although he would definitely be different from real living beings in possessing the above abilities. In eliciting any physical effect by action of the sensory as well as the motor nervous systems a degradation of energy is always involved, quite apart from the fact that the very existence of a nervous system is dependent on continual dissipation of energy.

"Whether—considering these circumstances—real living beings could continually or at least regularly produce energy at the expense of heat of the lowest temperature appears very doubtful, even though our ignorance of the biological phenomena does not allow a definite answer. However, the latter questions lead beyond the scope of physics in the strict sense. . . "

The Smoluchowski (1914) quote ends here. The 1964 English translation accidentally ended it after the first paragraph.

It appears that the ignorance of the biological phenomena need not prevent us from understanding that which seems to us to be the essential thing. We may be sure that intelligent living beings—insofar as we are dealing with their intervention in a thermodynamic system—can be replaced by non-living devices whose "biological phenomena" one could follow and determine whether in fact a compensation of the entropy decrease takes place as a result of the intervention by such a device in a system.

In the first place, we wish to learn what circumstance conditions the decrease of entropy which takes place when intelligent living beings intervene in a thermodynamic system. We shall see that this depends on a certain type of coupling between different parameters of the system. We shall consider an unusually simple type of these ominous couplings.[1] For brevity we shall talk about a "measurement" if we succeed in coupling the value of a parameter y (for instance the position co-ordinate of a pointer of a measuring instrument) at one moment with the simultaneous value of a fluctuating parameter x of the system, in such a way that, from the value y, we can draw conclusions about the value that x had at the moment of the "measurement." Then let x and y be uncoupled after the measurement, so that x can change, while y retains its value for some time. Such measurements are not harmless interventions. A system in which such measurements occur

Szilárd defines measurement here.

[1] The author evidently uses the word "ominous" in the sense that the possibility of realizing the proposed arrangement threatens the validity of the Second Law.—*Translator*

I will expand on this in note 1 on the next page.

shows a sort of memory faculty, in the sense that one can recognize by the state parameter y what value another state parameter x had at an earlier moment, and we shall see that simply because of such a memory the Second Law would be violated, if the measurement could take place without compensation. We shall realize that the Second Law is not threatened as much by this entropy decrease as one would think, as soon as we see that the entropy decrease resulting from the intervention would be compensated completely in any event if the execution of such a measurement were, for instance, always accompanied by production of $k \log 2$ units of entropy. In that case it will be possible to find a more general entropy law, which applies universally to all measurements. Finally we shall consider a very simple (of course, not living) device, that is able to make measurements continually and whose "biological phenomena" we can easily follow. By direct calculation, one finds in fact a continual entropy production of the magnitude required by the above-mentioned more general entropy law derived from the validity of the Second Law.

"In that case" should have been translated as "Then" (see note 2, next page). Alternative translation: "A more general entropy law will then be derived, which applies universally to all measurements."

The first example, which we are going to consider more closely as a typical one, is the following. A standing hollow cylinder, closed at both ends, can be separated into two possibly unequal sections of volumes V_1 and V_2 respectively by inserting a partition from the side at an arbitrarily fixed height. This partition forms a piston that can be moved up and down in the cylinder. An infinitely large heat reservoir of a given temperature T insures that any gas present in the cylinder undergoes isothermal expansion as the piston moves. This gas shall consist of a single molecule which, as long as the piston is not inserted into the cylinder, tumbles about in the whole cylinder by virtue of its thermal motion.

Here Szilárd introduces the thought experiment that has become known as Szilárd's engine.

Imagine, specifically, a man who at a given time inserts the piston into the cylinder and somehow notes whether the molecule is caught in the upper or lower part of the cylinder, that is, in volume V_1 or V_2. If he should find that the former is the case, then he would move the piston slowly downward until it reaches the bottom of the cylinder. During this slow movement of the piston the molecule stays, of course, above the piston. However, it is no longer constrained to the upper part of the cylinder but bounces many times against the piston which is already moving in the lower part of the cylinder. In this way the molecule does a certain amount of work on the piston. This is the work that corresponds to the isothermal expansion of an ideal gas—consisting of one single molecule—from volume V_1 to the volume $V_1 + V_2$. After some time, when the piston has reached the bottom of the container, the molecule has again the full volume $V_1 + V_2$ to move about in, and the piston is then removed. The procedure can be repeated as many times as desired. The man moves the piston up or down depending on whether the piston up or down depending on whether the molecule is trapped in the upper or lower half of the piston. In more detail, this motion may be caused by a weight, that is to be raised, through a mechanism that transmits the force from the piston to the weight, in such a way that the latter is always displaced upwards. In this way the potential energy of the weight certainly increases constantly. (The transmission of force to the weight is best arranged so that the force exerted by the weight on the piston at any position of the latter equals the average pressure of the gas.) It is clear that in this manner energy is constantly gained at the expense of heat, insofar as the biological phenomena of the intervening man are ignored in the calculation.

In order to understand the essence of the man's effect on the system, one best imagines that the movement of the piston is performed mechanically and that the man's activity consists only in determining the altitude of the molecule and in pushing a lever (which steers the piston) to the right or left, depending on whether the molecule's height requires a down- or upward movement. This means that the intervention of the human being consists only in the coupling of two position co-ordinates, namely a co-ordinate x, which determines the altitude of the molecule, with another co-ordinate y, which determines the position of the lever and therefore also whether an upward or downward motion is imparted to the piston. It is best to imagine the mass of the piston as large and its speed sufficiently great, so that the thermal agitation of the piston at the temperature in question can be neglected.

(cont. from previous page)

Note 1: Expanding on Szilárd's argument, note that to *utilize* detailed knowledge of (aspects of) the microscopic state of a thermodynamic system, the "demon" has to choose some control parameter (or several) appropriately. The detailed knowledge comes through an observation of the value of some parameter of the system at a certain time. The appropriate adjustment of the control parameter(s) must reflect that knowledge in order to utilize it. The control protocol thus has to be correlated with the observation, which occurred at a time before the adjustment happens. In that sense, the control protocol *must* contain memory of the observation, for it to be possible for the "demon" to use the knowledge.

Szilárd argues that the observer can be replaced by machinery to clarify the role the observer plays in their interaction with the system.

Note 2: Szilárd gives the outline of his contributions. The phrase "*Es wird dann gelingen . . .*" ("It will then be possible . . .") uses "*dann*" ("then") in the temporal meaning, in my opinion.

In the typical example presented here, we wish to distinguish two periods, namely:

1. The period of *measurement* when the piston has just been inserted in the middle of the cylinder and the molecule is trapped either in the upper or lower part; so that if we choose the origin of co-ordinates appropriately, the x-co-ordinate of the molecule is restricted to either the interval $x > 0$ or $x < 0$;

2. The period of *utilization of the measurement,* "the period of decrease of entropy," during which the piston is moving up or down. During this period the x-co-ordinate of the molecule is certainly not restricted to the original interval $x > 0$ or $x < 0$. Rather, if the molecule was in the upper half of the cylinder during the period of measurement, i.e., when $x > 0$, the molecule must bounce on the downward-moving piston in the lower part of the cylinder, if it is to transmit energy to the piston; that is, the co-ordinate x has to enter the interval $x < 0$. The lever, on the contrary, retains during the whole period its position toward the right, corresponding to downward motion. If the position of the lever toward the right is designated by $y = 1$ (and correspondingly the position toward the left by $y = -1$) we see that during the period of measurement, the position $x > 0$ corresponds to $y = 1$; but afterwards $y = 1$ stays on, even though x passes into the other interval $x < 0$. We see that in the utilization of the measurement the coupling of the two parameters x and y disappears.

Here Szilárd makes another comment regarding his notion of measurement.

We shall say, quite generally, that a parameter y "measures" a parameter x (which varies according to a probability law), if the value of y is directed by the value of parameter x at a given moment. A measurement procedure underlies the entropy decrease effected by the intervention of intelligent beings.

One may reasonably assume that a measurement procedure is fundamentally associated with a certain definite average entropy production, and that this restores concordance with the Second Law. The amount of entropy generated by the measurement may, of course, always be greater than this fundamental amount, but not smaller. To put it precisely: we have to distinguish here between two entropy values. One of them, $\bar{S}_1$, is produced when during the measurement y assumes the value 1, and the other, $\bar{S}_2$, when y assumes the value -1. We cannot expect to get general information about $\bar{S}_1$ or $\bar{S}_2$ separately, but we shall see that *if* the amount of entropy produced by the "measurement" is to compensate the entropy decrease affected by utilization, the relation must always hold good.

$$e^{-\bar{S}_1/k} + e^{-\bar{S}_2/k} \leqq 1 \tag{1}$$

☞ Statement of main result, which is derived later (see eqs. 9–20).

One sees from this formula that one can make one of the values, for instance $\bar{S}_1$, as small as one wishes, but then the other value $\bar{S}_2$ becomes correspondingly greater. Furthermore, one can notice that the magnitude of the interval under consideration is of no consequence. One can also easily understand that it cannot be otherwise.

Conversely, as long as the entropies $\bar{S}_1$ and $\bar{S}_2$, produced by the measurements, satisfy the inequality (1), we can be sure that the expected decrease of entropy caused by the later utilization of the measurement will be fully compensated.

Before we proceed with the proof of inequality (1), let us see in the light of the above mechanical example, how all this fits together. For the entropies $\bar{S}_1$ and $\bar{S}_2$ produced by the measurements, we make the following Ansatz:

$$\bar{S}_1 = \bar{S}_2 = k \log 2 \tag{2}$$

This ansatz satisfies inequality (1) and the mean value of the quantity of entropy produced by a measurement is (of course in this special case independent of the frequencies w_1, w_2 of the two events):

$$\bar{S} = k \log 2 \tag{3}$$

In this example one achieves a decrease of entropy by the isothermal expansion:[2]

$$\begin{aligned} -\bar{s}_1 &= -k \log \frac{V_1}{V_1 + V_2}; \\ -\bar{s}_2 &= -k \log \frac{V_2}{V_1 + V_2}, \end{aligned} \tag{4}$$

[2] The entropy generated is denoted by $\bar{s}_1, \bar{s}_2$.

depending on whether the molecule was found in volume V_1 or V_2 when the piston was inserted. (The decrease of entropy equals the ratio of the quantity of heat taken from the heat reservoir during the isothermal expansion, to the temperature of the heat reservoir in question). Since in the above case the frequencies w_1, w_2 are in the ratio of the volumes V_1, V_2, the mean value of the entropy generated is (a negative number):

$$\bar{s} = w_1 \cdot (+\bar{s}_1) + w_2 \cdot (+\bar{s}_2) = \frac{V_1}{V_1+V_2} k \log \frac{V_1}{V_1+V_2} + \frac{V_2}{V_1+V_2} k \log \frac{V_1}{V_1+V_2} \tag{5}$$

As one can see, we have, indeed

$$\frac{V_1}{V_1+V_2} k \log \frac{V_1}{V_1+V_2} + \frac{V_2}{V_1+V_2} \cdot k \log \frac{V_2}{V_1+V_2} + k \log 2 \geqq 0 \tag{6}$$

and therefore:

$$\bar{S} + \bar{s} \geqq 0. \tag{7}$$

In the special case considered, we would actually have a full compensation for the decrease of entropy achieved by the utilization of the measurement.

☞ The argumentation used for the derivation of the main result begins here.

We shall not examine more special cases, but instead try to clarify the matter by a general argument, and to derive formula (1). We shall therefore imagine the whole system—in which the co-ordinate x, exposed to some kind of thermal fluctuations, can be measured by the parameter y in the way just explained—as a multitude of particles, all enclosed in one box. Every one of these particles can move freely, so that they may be considered as the molecules of an ideal gas, which, because of thermal agitation, wander about in the common box independently of each other and exert a certain pressure on the walls of the box—the pressure being determined by the temperature. We shall now consider two of these molecules as chemically different and, in principle, separable by semipermeable walls, if the co-ordinate x for one molecule is in a preassigned interval while the corresponding co-ordinate of the other molecule falls outside that interval. We also shall look upon them as chemically different, if they differ only in that the y coordinate is $+1$ for one and -1 for the other.

We should like to give the box in which the "molecules" are stored the form of a hollow cylinder containing four pistons. Pistons A and A' are fixed while the other two are movable, so that the distance BB' always equals the distance AA', as is indicated in Figure 1 by the two brackets. A', the bottom, and B, the cover of the container, are impermeable for all "molecules," while A and B' are semipermeable; namely, A is permeable only for those "molecules" for which the parameter x is in the preassigned interval, i.e., (x_1, x_2) , B' is only permeable for the rest.

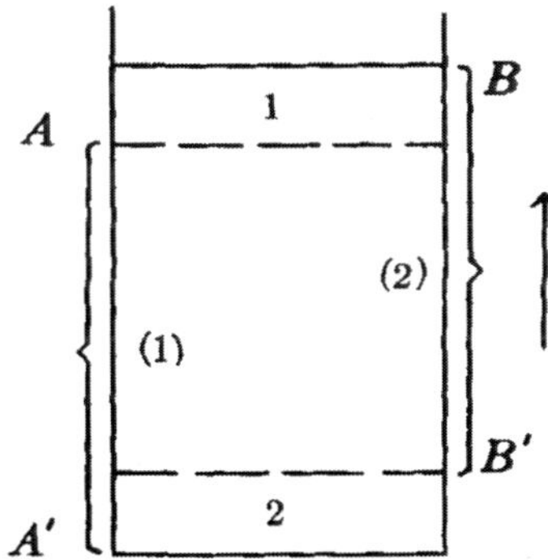

Figure 1.

In the beginning the piston B is at A and therefore B' at A', and all "molecules" are in the space between. A certain fraction of the molecules have their co-ordinate x in the preassigned interval. We shall designate by w_1 the probability that this is the case for a randomly selected molecule and by w_2 the probability that x is outside the interval. Then $w_1 + w_2 = 1$.

Let the distribution of the parameter y be over the values $+1$ and -1 in any proportion but in any event independent of the x-values. We imagine an intervention by an intelligent being, who imparts to y the value 1 for all "molecules" whose x at that moment is in the selected interval. Otherwise the value -1 is assigned. If then, because of thermal fluctuation, for any "molecule," the parameter x should come out of the preassigned interval or, as we also may put it, if the "molecule" suffers a monomolecular chemical reaction with regard to x (by which it is transformed from a species that can pass the semipermeable piston A into a species for which the piston is impermeable), then the parameter y

retains its value 1 for the time being, so that the "molecule," because of the value of the parameter y, "remembers" during the whole following process that x originally was in the preassigned interval. We shall see immediately what part this memory may play. After the intervention just discussed, we move the piston, so that we separate the two kinds of molecules without doing work. This results in two containers, of which the first contains only the one modification and the second only the other. Each modification now occupies the same volume as the mixture did previously. In one of these containers, if considered by itself, there is now no equilibrium with regard to the two "modifications in x." Of course the ratio of the two modifications has remained w_1:w_2. If we allow this equilibrium to be achieved in both containers independently and at constant volume and temperature, then the entropy of the system certainly has increased. For the total heat release is 0, since the ratio of the two "modifications in x", w_1:w_2 does not change. If we accomplish the equilibrium distribution in both containers in a reversible fashion then the entropy of the rest of the world will decrease by the same amount. Therefore the entropy increases by a negative value, and, the value of the entropy increase per molecule is exactly:

Two types of "modifications in x": Type 1: $x \in (x_1, x_2)$ occurs with probability w_1 Type 2: $x \notin (x_1, x_2)$ occurs with probability w_2.

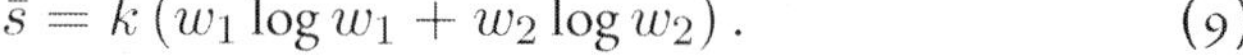

$$\bar{s} = k\left(w_1 \log w_1 + w_2 \log w_2\right). \tag{9}$$

(The entropy constants that we must assign to the two "modifications in x" do not occur here explicitly, as the process leaves the total number of molecules belonging to the one or the other species unchanged.)

Now of course we cannot bring the two gases back to the original volume without expenditure of work by simply moving the piston back, as there are now in the container—which is bounded by the pistons BB'—also molecules whose x-co-ordinate lies outside of the preassigned interval and for which the piston A is not permeable any longer. Thus one can see that the calculated decrease of entropy (Equation [9]) does not mean a contradiction of the Second Law. *As long as we do not use the fact that the molecules in the container BB', by virtue of their coordinate y, "remember" that the x-co-ordinate for the molecules of this container originally was in the preassigned interval, full compensation exists for the calculated decrease of entropy*, by virtue of the fact that the partial pressures in the two containers are smaller than in the original mixture.

But now we can use the fact that all molecules in the container BB' have the y-co-ordinate 1, *and in the other accordingly* -1, *to bring all molecules back again to the original volume.* To accomplish this we only need to replace the semipermeable wall A by a wall A^*, which is semipermeable not with regard to x but with regard to y, namely so that it is permeable for the molecules with the y-co-ordinate 1 and impermeable for the others. Correspondingly we replace B' by a piston B'^*, which is impermeable for the molecules with $y = -1$ and permeable for the others. Then both containers can be put into each other again without expenditure of energy. The distribution of the y-co-ordinate with regard to 1 and -1 now has become statistically independent of the x-values and besides we are able to re-establish the original distribution over 1 and -1. Thus we would have gone through a complete cycle. The only change that we have to register is the resulting decrease of entropy given by (9):

$$\bar{s} = k\,(w_1 \log w_1 + w_2 \log w_2)\,. \tag{10}$$

If we do not wish to admit that the Second Law has been violated, we must conclude *that the intervention which establishes the coupling between y and x, the measurement of x by y, must be accompanied by a production of entropy.* If a definite way of achieving this coupling is adopted and if the quantity of entropy that is inevitably produced is designated by S_1 and S_2, where S_1 stands for the mean increase in entropy that occurs when y acquires the value 1, and accordingly S_2 for the increase that occurs when y acquires the value -1, we arrive at the equation:

$$w_1 S_1 + w_2 S_2 = \bar{S} \tag{11}$$

In order for the Second Law to remain in force, this quantity of entropy must be greater than the decrease of entropy $\bar{s}$, which according to (9) is produced by the utilization of the measurement. Therefore the following inequality must be valid:

$$\begin{gathered} \bar{S} + \bar{s} \geqq 0 \\ w_1 S_1 + w_2 S_2 \\ + k\,(w_1 \log w_1 + w_2 \log w_2) \geqq 0 \end{gathered} \tag{12}$$

This equation must be valid for any values of w_1 and w_2,[3] and of course the constraint $w_2 + w_2 = 1$ cannot be violated. We ask, in particular, for which w_1 and w_2 and given S-values the expression becomes a minimum. For the two minimizing values w_1 and w_2 the inequality (12) must still be valid. Under the above constraint, the minimum occurs when the following equation holds:

Minimizing $\sum_i w_i\,(S_i + k\log(w_i))$ over w_i, subject to the constraint $\sum_i w_i = 1$. Using Lagrange parameter μ, we take the derivative of $\sum_i w_i\,(S_i + k\log(w_i) + \mu)$, with respect to w_i. At the minimum, $S_i + k\log(w_i) + 1 + \mu = 0 \Leftrightarrow \frac{S_i}{k} + \log(w_i) = \lambda$ (with $\lambda = -\frac{\mu+1}{k}$).

$$\frac{S_1}{k} + \log w_1 = \frac{S_2}{k} + \log w_2 \tag{13}$$

But then:

$$e^{-\overline{S}_1/k} + e^{-\overline{S}_2/k} \leqq 1. \tag{14}$$

This is easily seen if one introduces the notation

$$\frac{S_1}{k} + \log w_1 = \frac{S_2}{k} + \log w_2 = \lambda; \tag{15}$$

then:

$$w_1 = e^{\lambda} \cdot e^{-S_1/k}; \quad w_2 = e^{\lambda} \cdot e^{-S_2/k}. \tag{16}$$

If one substitutes these values into the inequality (12) one gets:

$$\lambda e^{\lambda}\left(e^{-S_1/k} + e^{-S_2/k}\right) \geqq 0. \tag{17}$$

Therefore the following also holds:

$$\lambda \geqq 0. \tag{18}$$

If one puts the values w_1 and w_2 from (16) into the equation $w_1 + w_2 = 1$, one gets

$$e^{-S_1/k} + e^{-S_2/k} = e^{-\lambda}. \tag{19}$$

And because $\lambda \geqq 0$, the following holds:

$$e^{-S_1/k} + e^{-S_2/k} \leqq 1. \tag{20}$$

This equation must be universally valid, if thermodynamics is not to be violated.

As long as we allow intelligent beings to perform the intervention, a direct test is not possible. But we can try to describe simple nonliving

[3] The increase in entropy can depend only on the types of measurement and their results but not on how many systems of one or the other type were present.

devices that effect such coupling, and see if indeed entropy is generated and in what quantity. Having already recognized that the only important factor is a certain characteristic type of coupling, a "measurement," we need not construct any complicated models which imitate the intervention of living beings in detail. We can be satisfied with the construction of this particular type of coupling which is accompanied by memory.

In our next example, the position co-ordinate of an oscillating pointer is "measured" by the energy content of a body K. The pointer is supposed to connect, in a purely mechanical way, the body K—by whose energy content the position of the pointer is to be measured—by heat conduction with one of two intermediate pieces, A or B. The body is connected with A as long as the coordinate—which determines the position of the pointer—falls into a certain preassigned, but otherwise arbitrarily large or small interval a, and otherwise if the co-ordinate is in the interval b, with B. Up to a certain moment, namely the moment of the "measurement," both intermediate pieces will be thermally connected with a heat reservoir at temperature T_0. At this moment the insertion A will be cooled reversibly to the temperature T_A, e.g., by a periodically functioning mechanical device. That is, after successive contacts with heat reservoirs of intermediate temperatures, A will be brought into contact with a heat reservoir of the temperature T_A. At the same time the insertion B will be heated in the same way to temperature T_B. Then the intermediate pieces will again be isolated from the corresponding heat reservoirs.

Max von Laue questioned in his "Gutachten zum Habilitationsgesuch des Dr. L. Szilárd's" whether this last example was the best choice. So do I, but for different reasons. I don't feel that going into them would add value to the discussion.

We assume that the position of the pointer changes so slowly that all the operations that we have sketched take place while the position of the pointer remains unchanged. If the position co-ordinate of the pointer fell in the preassigned interval, then the body was connected with the insertion A during the above-mentioned operation, and consequently is now cooled to temperature T_A.

In the opposite case, the body is now heated to temperature T_B. Its energy content becomes—according to the position of the pointer at the time of "measurement"—small at temperature T_A or great at temperature T_B and will retain its value, even if the pointer eventually leaves the preassigned interval or enters into it. After some time, while the pointer is still oscillating, one can no longer draw any definite conclusion from the energy content of the body K with regard to the momentary position of the pointer but one can draw a definite conclusion with regard to the position of the pointer at the time of the measurement. Then the measurement is completed.

After the measurement has been accomplished, the above-mentioned periodically functioning mechanical device should connect the thermally isolated insertions A and B with the heat reservoir T_0. This has the purpose of bringing the body K—which is now also connected with one of the two intermediate pieces—back into its original state. The direct connection of the intermediate pieces and hence of the body K—which has been either cooled to T_A or heated to T_B—to the reservoir T_0 consequently causes an increase of entropy. This cannot possibly be avoided, because it would make no sense to heat the insertion A reversibly to the temperature T_0 by successive contacts with the reservoirs of intermediate temperatures and to cool B in the same manner. After the measurement we do not know with which of the two insertions the body K is in contact at that moment; nor do we know whether it had been in connection with T_A or T_B in the end. Therefore neither do we know whether we should use intermediate temperatures between T_A and T_0 or between T_0 and T_B.

The mean value of the quantity of entropy S_1 and S_2, per measurement, can be calculated, if the heat capacity as a function of the temperature $\bar{u}(T)$ is known for the body K, since the entropy can be calculated from the heat capacity. We have, of course, neglected the heat capacities of the intermediate pieces. If the position co-ordinate of the pointer was in the preassigned interval at the time of the "measurement," and accordingly the body in connection with insertion A, then the entropy conveyed to the heat reservoirs during successive cooling was

$$\int_{T_A}^{T_0} \frac{1}{T}\frac{d\bar{u}}{dT}. \tag{21}$$

However, following this, the entropy withdrawn from the reservoir T_0 by direct contact with it was

$$\frac{\bar{u}\left(T_0\right)-\bar{u}\left(T_A\right)}{T_0}. \tag{22}$$

All in all the entropy was increased by the amount

$$S_A=\frac{\bar{u}\left(T_A\right)-\bar{u}\left(T_0\right)}{T_0}+\int_{T_A}^{T_0} \frac{1}{T}\frac{d\bar{u}}{dT}dT. \tag{23}$$

Analogously, the entropy will increase by the following amount, if the body was in contact with the intermediate piece B at the time of the "measurement":

$$S_B = \frac{\bar{u}\left(T_B\right) - \bar{u}\left(T_0\right)}{T_0} + \int_{T_B}^{T_0} \frac{1}{T}\frac{d\bar{u}}{dT}dT. \tag{24}$$

We shall now evaluate these expressions for the very simple case, where the body which we use has only two energy states, a lower and a higher state. If such a body is in thermal contact with a heat reservoir at any temperature T, the probability that it is in the lower or upper state is given by respectively:

$$\left.\begin{aligned} p(T) &= \tfrac{1}{1+ge^{-u/kT}} \\ q(T) &= \tfrac{ge^{-u/kT}}{1+ge^{-u/kT}} \end{aligned}\right\} \tag{25}$$

Here u stands for the difference of energy of the two states and g for the statistical weight. We can set the energy of the lower state equal to zero without loss of generality. Therefore:[4]

$$\left.\begin{aligned} S_A &= q\left(T_A\right) k \log \tfrac{q(T_A)p(T_0)}{q(T_0)p(T_A)} \\ &\quad + k \log \tfrac{p(T_A)}{p(T_0)} \\ S_B &= p\left(T_B\right) k \log \tfrac{q(T_0)p(T_B)}{q(T_B)p(T_0)} \\ &\quad + k \log \tfrac{q(T_B)}{q(T_0)} \end{aligned}\right\}. \tag{26}$$

Here q and p are the functions of T given by equation (25), which are here to be taken for the arguments T_0, T_A, or T_B.

If (as is necessitated by the above concept of a "measurement") we wish to draw a dependable conclusion from the energy content of the body K as to the position co-ordinate of the pointer, we have to see to it that the body surely gets into the lower energy state when it gets into contact with T_B. In other words:

$$\begin{aligned} p\left(T_A\right) &= 1, q\left(T_A\right) = 0; \\ p\left(T_B\right) &= 0, q\left(T_B\right) = 1. \end{aligned} \tag{27}$$

---

[4]See the Appendix.

This of course cannot be achieved, but may be arbitrarily approximated by allowing T_A to approach absolute zero and the statistical weight g to approach infinity. (In this limiting process, T_0 is also changed, in such a way that $p(T_0)$ and $q(T_0)$ remain constant.) The equation (26) then becomes:

$$S_A = -k \log p(T_0); \qquad S_B = -k \log q(T_0) \tag{28}$$

and if we form the expression $e^{-S_A/k} + e^{-s_B/k}$, we find:

$$e^{-S_A/k} + e^{-S_B/k} = 1. \tag{29}$$

Our foregoing considerations have thus just realized the smallest permissible limiting care. The use of semipermeable walls according to Figure 1 allows a complete utilization of the measurement: inequality (1) certainly cannot be sharpened.

As we have seen in this example, a simple inanimate device can achieve the same essential result as would be achieved by the intervention of intelligent beings. We have examined the "biological phenomena" of a nonliving device and have seen that it generates exactly that quantity of entropy which is required by thermodynamics.

APPENDIX

In the case considered, when the frequency of the two states depends on the temperature according to the equations:

$$p(T) = \frac{1}{1 + ge^{-u/kT}}; q(T) = \frac{ge^{-u/kT}}{1 + ge^{-u/kT}} \tag{30}$$

and the mean energy of the body is given by:

$$\bar{u}(T) = uq(T) = \frac{uge^{-u/kT}}{1 + ge^{-u/kT}}, \tag{31}$$

the following identity is valid:

$$\frac{1}{T}\frac{d\bar{u}}{dT} = \frac{d}{dT}\left\{\frac{\bar{u}(T)}{T} + k \log\left(1 + e^{-u/kT}\right)\right\}. \tag{32}$$

Therefore we can also write the equation:

$$B_A = \frac{\bar{u}(T_A) - \bar{u}(T_0)}{T_0} + \int_{T_A}^{T_0} \frac{1}{T}\frac{d\bar{u}}{dT} dT \tag{33}$$

as

$$S_A = \frac{\bar{u}(T_A) - \bar{u}(T_0)}{T_0} + \left\{ \frac{\bar{u}(T)}{T} + k \log \left(1 + g e^{-ukT}\right) \right\}_{T_A}^{T_0}, \tag{34}$$

and by substituting the limits we obtain:

$$S_A = \bar{u}(T_A) \left(\frac{1}{T_0} - \frac{1}{T_A} \right) + k \log \frac{1 + g e^{-u/kT_0}}{1 + g e^{-u/kT_A}}. \tag{35}$$

If we write the latter equation according to (25):

$$1 + g e^{-u/kT} = \frac{1}{p(T)} \tag{36}$$

for T_A and T_0, then we obtain:

$$S_A = \bar{u}(T_A) \left(\frac{1}{T_0} - \frac{1}{T_A} \right) + k \log \frac{p(T_A)}{p(T_0)} \tag{37}$$

and if we then write according to (31):

$$\bar{u}(T_A) = uq(T_A) \tag{38}$$

we obtain:

$$S_A = q(T_A) \left(\frac{u}{T_0} - \frac{u}{T_A} \right) + k \log \frac{p(T_A)}{p(T_0)}. \tag{39}$$

If we finally write according to (25):

$$\frac{u}{T} = -k \log \frac{q(T)}{gp(T)} \tag{40}$$

for T_A and T_0, then we obtain:

$$S_A = q(T_A) k \log \frac{p(T_0)}{q(T_0)} \frac{q(T_A)}{p(T_A)} + k \log \frac{p(T_A)}{p(T_0)}. \tag{41}$$

We obtain the corresponding equation for S_B, if we replace the index A with B. Then we obtain:

$$S_B = q(T_B) k \log \frac{p(T_0)}{q(T_0)} \frac{q((T_B)}{p((T_B)} + k \log \frac{p(T_B)}{p(T_0)}. \tag{42}$$

Formula (41) is identical with (26), given, for S_A, in the text.

We can bring the formula for S_B into a somewhat different form, if we write:

$$q(T_B) = 1 - p(T_B) \tag{43}$$

expand and collect terms, then we get

$$S_B = p(T_B)\, k \log \frac{q(T_0)\, p(T_B)}{p(T_0)\, q(T_B)} + k \log \frac{q(T_B)}{q(T_0)}. \tag{44}$$

This is the formula given in the text for S_B.

REFERENCES

Smoluchowski, F. 1914. *Vorträge über die kinetische Theorie der Materie u. Elektrizitat.* Leipzig.

Szilárd, L. 1925. "Über die Ausdehnung der phänomenologischen Thermodynamik auf die Schwankungserscheinungen." *Zeitschrift für Physik* 32:753–788.

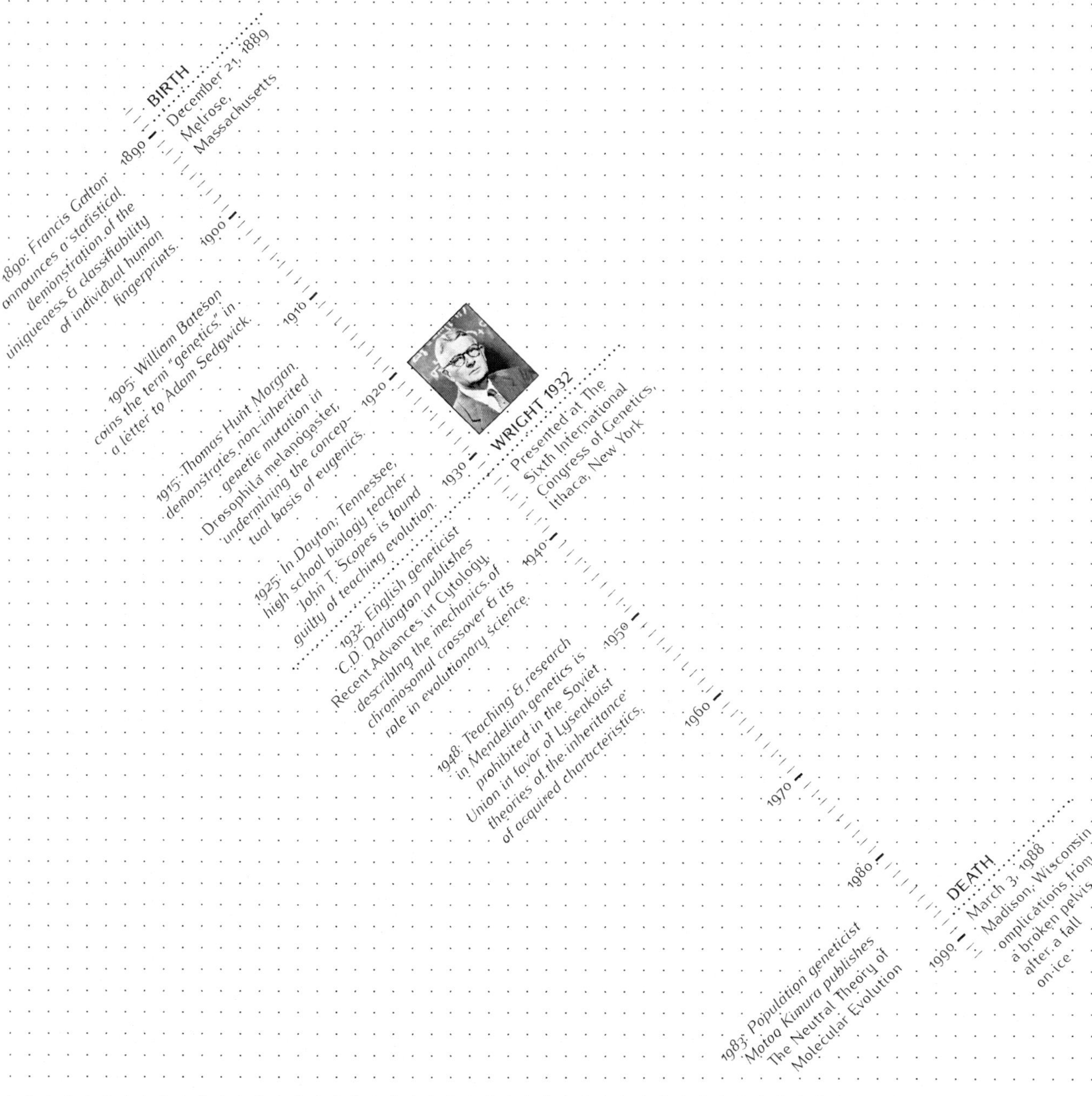

SEWALL GREEN WRIGHT

[3]

NAVIGATING THE LANDSCAPE OF COMPLEXITY: SEWALL WRIGHT'S SEMINAL CONTRIBUTIONS AND THEIR FAR-REACHING IMPACT

Sergey Gavrilets, University of Tennessee

Sewall Wright's landmark 1932 paper highlighted the intricate nature of biological evolution and laid the groundwork for population genetics. Wright, along with Ronald Fisher and J. B. S. Haldane, is widely recognized as a founder of population genetics. These authors' theoretical work paved the way for the modern evolutionary synthesis of the 1930–1940s (Provine 1971). Wright's paper, presented at the 1932 International Congress of Genetics, aimed to explain his earlier mathematical theory (Wright 1931) in a clear, nontechnical manner (Provine 1986; Wright 1988). This paper made several significant contributions to our understanding of evolutionary processes.

S. Wright, "The Roles of Mutation, Inbreeding, Crossbreeding, and Selection in Evolution," in *Proceedings of the Sixth International Congress of Genetics, Ithaca, NY, 1932, Volume I: Transactions and Genetics*, ed. D. F. Jones, Menasha, WI: Brooklyn Botanic Garden, 356–366 (1932).

Wright employed simple mathematical models to illustrate how mutation, selection, migration, and random genetic drift interact to drive population evolution. His work led to a greater appreciation of random genetic drift, formerly known as the "Sewall Wright effect," and the role of population subdivision. Wright also emphasized the significance of inbreeding in small populations, which can increase homozygosity and reduce genetic variation. His models underscored the advantages of crossbreeding, or mating between different populations or species, as a way to introduce genetic variation and potentially enhance fitness. Wright proposed that crossbreeding could help populations counteract inbreeding depression and genetic drift. His ideas about the prevalence of nonadaptive genetic differentiation in natural populations were later expanded by Motoo Kimura (1983) in his *Neutral Theory of Molecular Evolution*, forming the basis for modern statistical analysis methods of genetic variation (Griffiths *et al.* 2015).

Wright's paper also introduced two contentious ideas: the shifting balance theory and (inter)group selection. A common result in population genetics theory is that selection increases the mean fitness of the population. Due to epistasis and pleiotropy, the mean fitness is expected to have multiple local "peaks" separated by "adaptive valleys" (Wright 1931; 1988, see below). Natural selection guides a population to a nearby peak. The question then arises: how can a population cross an adaptive valley to reach a higher peak? Wright proposed that this could occur in a population divided into many partially isolated groups. In Wright's model, a new adaptive gene combination first establishes itself in a single subpopulation through selection and random drift, then spreads throughout the entire population due to differential migration and competition between subpopulations/groups. However, the theoretical plausibility of the shifting balance theory has been questioned (Gavrilets 1996; Wade and Goodnight 1998). The role and power of group selection in biology also remain controversial (Wilson and Sober 1994; Rubenstein and Alcock 2019). Nevertheless, the significance of group selection in human evolution, based on cultural rather than genetic variation, is now widely recognized (Soltis, Boyd, and Richerson 1995; Henrich 2004, 2020).

In the realm of complexity science, Wright's paper is most renowned for its concepts of the genotype space and fitness landscapes, which he referred to as the "field of gene combinations" and "levels of adaptive values" or "surfaces of selective value," respectively. Assuming there are L loci with two alleles each, each haploid genotype can be defined by a binary sequence of length L. Wright depicted the genotype space (the set of all possible genotypes) as a hypercube, where genotypes differing by a single locus are connected by an edge (fig. 1 in his paper). He emphasized that with the number of loci in the thousands, the number of possible genotypes exceeds "the total number of electrons and protons in the whole visible universe." This suggests that stochasticity in mutations, recombination, and reproduction plays a significant role in population evolution.

Wright introduced two versions of fitness landscapes, which he used interchangeably, causing some confusion (Provine 1986; Gavrilets 2004). The first version, simpler and more fundamental, is essentially a fitness function assigning a fitness value to each genotype, which can be specified by a gene sequence. This concept can be extended to continuous traits rather than discrete genes and to mating pairs, with fitness being a measure of a mating pair's fertility or the probability of mating between a specific male–female pair (Gavrilets 2004). The second version represents the population's average fitness as a function of variables specifying the population state, such as genotype frequencies (Ewens 1979). This version is practical only in simple cases due to the high dimensionality L^N of the space of all possible population states compared to the genotype state's dimensionality L. An example is a one-locus–two-allele randomly mating diploid population with discrete generations, which can be completely specified by a single variable: an allele frequency (fig. 2 in Wright's paper). The second version of fitness landscapes assumes that selection always increases mean fitness, leading the population up the gradient of the average fitness landscape, as predicted by Fisher's fundamental theorem of natural delection (Fisher 1930; Crow 2002). However, there are situations where fitness can decrease, cycle, or even change chaotically, such as in the presence of linkage disequilibrium or frequency-dependent selection (Frank and Slatkin 1992; Gavrilets and Hastings 1998). Wright used this version of fitness landscapes to illustrate the effects of different evolutionary forces in figures 3–4 of his paper.

Beyond these mathematical constructs, Wright introduced a metaphor of fitness landscapes, a one- or two-dimensional visualization of certain features of multidimensional fitness landscapes (Wright 1988; Gavrilets 2004). Given the enormous dimensionality of real fitness landscapes, any low-dimensional visualization is a simplification that emphasizes only those specific features of fitness landscapes deemed most important while neglecting many others. Wright envisioned typical fitness landscapes as rugged terrains with multiple isolated high-fitness peaks separated by low-fitness valleys (fig. 1a). Interactions due to pleiotropy and epistasis make multiple fitness peaks inevitable, representing alternative survival solutions. Peaks distant in genotype space might correspond to different species, while peak clusters could represent species within the same genera. Valleys represent low-fitness gene combinations, including deleterious genes or incompatible genes.

The expectation is that natural selection pushes populations towards these fitness peaks, making adaptive evolution akin to "hill climbing." However, once a population reaches a local peak, any movement away is restricted by selection, potentially trapping the population on an intermediate fitness peak while higher ones exist. This raises the question of how fitness can be further increased and how new species form. Solutions lie in factors like random genetic drift that occasionally push the population across a valley, and temporal changes in the fitness landscape, causing a continuous "uphill chase" of a moving fitness peak—as in the "Red Queen" scenario (van Valen 1973). Wright's metaphor of rugged landscapes simplifies the understanding of multidimensional fitness landscapes and evolutionary dynamics, emphasizing the existence of different survival solutions and illustrating the complex path between them.

The strength of Wright's metaphor lies in its translation of intricate concepts—multidimensional fitness landscapes and evolutionary dynamics—into the more familiar terms of geographical landscapes. This perspective of rugged landscapes accentuates the existence of diverse survival solutions, represented by alternative fitness peaks. Just as one cannot traverse from one peak to another in a three-dimensional geographical landscape without descending into a valley, the rugged fitness landscapes metaphor instills the notion that a similar principle applies in the realm of multidimensional fitness landscapes.

Contrary to Wright's perspective, Fisher proposed that as the dimensions in a fitness landscape increase, local peaks in lower dimensions tend to morph into saddle points in higher dimensions (Provine 1986). Under this paradigm, Fisher suggested that natural selection alone could navigate the population to the absolute peak, eliminating the need for genetic drift or other influences. Fisher's perspective suggests a singular peak fitness landscape, grounded in the belief that there exists one superior gene combination, which can be discovered purely through selection, negating the need for other factors like genetic drift (fig. 1b). However, Fisher's critique may not be warranted: the peaks that become saddle points due to increased dimensionality in the genotype space are significantly outnumbered by the new local peaks that arise from the same process (Kauffman and Levin 1987). This implies that a typical fitness landscape is dotted with local peaks, making the singular pursuit of the global peak through selection a generally unattainable task.

Kimura's neutral theory of evolution offers a different perspective (Kimura 1983). In this theory, fitness landscapes are flat: they have large regions of equal fitness, meaning that different genotypes can have the same fitness (fig. 1c). Evolution on these landscapes is primarily driven by genetic drift rather than natural selection.

Wright's, Fisher's, and Kimura's classical landscapes were challenged in the late 1990s when mathematical analyses of fitness landscapes led to the realization that the dimensionality of the landscape can drastically affect its topology. In high-dimensional spaces, fitness landscapes are likely to be "holey" (Gavrilets 1997, 2004), with many connected pathways of equivalent or nearly equivalent fitness ("nearly neutral networks" Reidys, Stadler, and Schuster 1997; Reidys and Stadler 2001) spreading across the whole genotype space and leading to the neighborhoods of different fitness peaks. The representation of a holey fitness landscape can be simplified as illustrated in figure 1d, where all genotypes within a nearly neutral network are labeled with a fitness of 1, while all other genotypes are labeled with a fitness of 0. The existence of nearly neutral networks allows for a substantial genetic divergence and speciation without any need to cross any deep fitness valleys (Gavrilets 2004). Holey landscapes, with their numerous alternative paths, can also make populations more robust to environmental changes. If the environment changes, populations can potentially move along these networks of near-equivalent fitness to new peaks that are better suited to the new conditions. The concept of "holey" fitness landscapes also suggests that randomness and contingencies play a more pronounced role in biological evolution.

Wright's ideas on fitness landscapes were not confined within biology's boundaries; they crossed disciplinary borders, influencing other fields grappling with complex systems. Optimization problems in computer science adopted the concept of fitness landscapes, where potential solutions traverse the landscape, guided by algorithms that mirror biological processes (Goldberg 1989). Similarly, the concept permeated physics and material science, economics, and social science, influencing the study of complex materials, strategic business interactions, and cultural evolution, respectively (Levinthal 1997; Gavetti and Levinthal 2000;

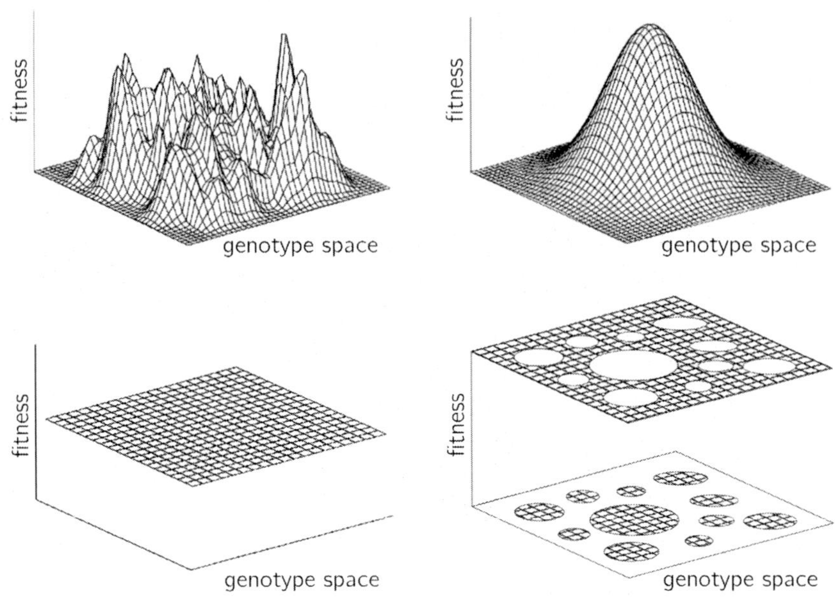

Figure 1. Canonical fitness landscapes. (a) Rugged landscape. (b) Single-peak landscape. (c) Flat landscape. (d) Holey landscape.

Rivkin 2000; Wales 2003; Henrich 2004; Aleti, Moser, and Grunske 2017; Giannoccaro, Nair, and Choi 2018; Brunswicker, Priego, and Almirall 2019; Khraisha 2020).

Not only has Wright's concept of the fitness landscape been a vital tool for thinking about the process of evolution—it has also inspired a vast body of theoretical and empirical research, leading to a deeper understanding of how populations navigate the complex terrain of genetic, and cultural, variation and selection.

REFERENCES

Aleti, A., I. Moser, and L. Grunske. 2017. "Analyzing the Fitness Landscape of Search-Based Software Testing Problems." *Automated Software Engineering* 24:603–662. https://doi.org/10.1007/s10515-016-0197-7.

Brunswicker, S., L. P. Priego, and E. Almirall. 2019. "Transparency in Policy Making: A Complexity View." *Government Information Quarterly* 36 (3): 571–591. https://doi.org/10.1016/j.giq.2019.05.005.

Crow, J. F. 2002. "Here's to Fisher, Additive Genetic Variance, and the Fundamental Theorem of Natural Selection." *Evolution* 56 (7): 1313–1316. https://doi.org/10.1111/j.0014-3820.2002.tb01445.x.

Ewens, W. J. 1979. *Mathematical Population Genetics.* Berlin, Germany: Springer–Verlag.

Fisher, R. A. 1930. *The Genetical Theory of Natural Selection.* Oxford, UK: Oxford University Press.

Frank, S. A., and M. Slatkin. 1992. "Fisher's Fundamental Theorem of Natural Selection." *Trends in Ecology and Evolution* 7 (3): 92–95. https://doi.org/https://doi.org/10.1016/0169-5347(92)90248-A.

Gavetti, G., and D. Levinthal. 2000. "Looking Forward and Looking Backard: Cognitive and Experiential Search." *Administrative Science Quarterly* 45 (1): 113–137. https://doi.org/10.2307/266698.

Gavrilets, S. 1996. "On Phase Three of the Shifting-Balance Theory." *Evolution* 50 (3): 1034–1041. https://doi.org/10.1111/j.1558-5646.1996.tb02344.x.

———. 1997. "Evolution and Speciation on Holey Adaptive Landscapes." *Trends in Ecology and Evolution* 12 (8): 307–312. https://doi.org/10.1016/S0169-5347(97)01098-7.

———. 2004. *Fitness Landscapes and the Origin of Species.* Princeton, NJ: Princeton University Press.

Gavrilets, S., and A. Hastings. 1998. "Intermittency and Transient Chaos from Simple Frequency-Dependent Selection." *Proceedings of the Royal Society London B* 261 (1361): 233–238. https://doi.org/10.1098/rspb.1995.0142.

Giannoccaro, I., A. Nair, and T. Choi. 2018. "The Impact of Control and Complexity on Supply Network Performance: An Empirically Informed Investigation Using NK Simulation Analysis." *Decision Sciences* 49 (4): 625–659. https://doi.org/10.1111/deci.12293.

Goldberg, D. E. 1989. *Genetic Algorithms in Search, Optimization, and Machine Learning.* Boston, MA: Addison-Wesley Professional.

Griffiths, A. J. F, S. R. Wessler, S. B. Carroll, and J. Doebley. 2015. *An Introduction to Genetic Analysis.* New York, NY: W. H. Freeman.

Henrich, J. 2004. "Cultural Group Selection, Coevolutionary Process and Large-Scale Cooperation." *Journal of Economic Behavior & Organization* 53 (1): 3–35. https://doi.org/10.1016/S0167-2681(03)00094-5.

———. 2020. *The WEIRDest People in the World: How the West Became Psychologically Peculiar and Particularly Prospective.* New York, NY: Farrar, Straus, and Giroux.

Kauffman, S. A., and S. Levin. 1987. "Towards a General Theory of Adaptive Walks on Rugged Landscapes." *Journal of Theoretical Biology* 128 (1): 11–45. https://doi.org/10.1016/S0022-5193(87)80029-2.

Khraisha, T. 2020. "Complex Economic Problems and Fitness Landscapes: Assessment and Methodological Perspectives." *Structural Change and Economic Dynamics* 52:390–407. https://doi.org/10.1016/j.strueco.2019.01.002.

Kimura, M. 1983. *The Neutral Theory of Molecular Evolution.* New York, NY: Cambridge University Press.

Levinthal, D. A. 1997. "Adaptation on Rugged Landscapes." *Management Science* 43:934–950. https://doi.org/10.1287/mnsc.43.7.934.

Provine, W. B. 1971. *The Origins of Theoretical Population Genetics.* Chicago, IL: University of Chicago Press.

Provine, W. B. 1986. *Sewall Wright and Evolutionary Biology.* Chicago, IL: University of Chicago Press.

Reidys, C. M., and P. F. Stadler. 2001. "Neutrality in Fitness Landscapes." *Applied Mathematics and Computation* 117 (2–3): 321–350. https://doi.org/10.1016/S0096-3003(99)00166-6.

Reidys, C. M., P. F. Stadler, and P. Schuster. 1997. "Generic Properties of Combinatory Maps: Neutral Networks of RNA Secondary Structures." *Bulletin of Mathematical Biology* 59 (2): 339–397. https://doi.org/10.1007/BF02462007.

Rivkin, J. W. 2000. "Imitation of Complex Strategies." *Management Science* 46 (6): 824–844. https://doi.org/10.1287/mnsc.46.6.824.11940.

Rubenstein, D. R., and J. Alcock. 2019. *Animal Behavior.* Sunderland, MA: Oxford University Press.

Soltis, J., R. Boyd, and P. J. Richerson. 1995. "Can Group-Functional Behaviours Evolve by Cultural Group Selection? An Empirical Test." *Current Anthropology* 36 (3): 473–494.

van Valen, L. 1973. "A New Evolutionary Law." *Evolutionary Theory* 1 (1): 1–30.

Wade, M. J., and C. J. Goodnight. 1998. "The Theories of Fisher and Wright in the Context of Metapopulations: When Nature Does Many Small Experiments." *Evolution* 52 (6): 1537–1553. https://doi.org/10.2307/2411328.

Wales, D. J. 2003. *Energy Landscapes: Applications to Clusters, Biomolecules, and Glasses.* Cambridge, UK: Cambridge University Press.

Wilson, D. S., and E. Sober. 1994. "Reintroducing Group Selection to the Human Behavioral Sciences." *Behavioral and Brain Sciences* 17 (4): 585–654. https://doi.org/10.1017/S0140525X00036104.

Wright, S. 1931. "Evolution in Mendelian Populations." *Genetics* 16 (2): 97–159. https://doi.org/10.1093/genetics/16.2.97.

———. 1988. "Surfaces of Selective Value Revisited." *American Naturalist* 131 (1): 115–123. https://doi.org/10.1086/284777.

THE ROLES OF INBREEDING, CROSSBREEDING, AND SELECTION IN EVOLUTION

S. Wright, University of Chicago

The enormous importance of biparental reproduction as a factor in evolution was brought out a good many years ago by East. The observed properties of gene mutation-fortuitous in origin, infrequent in occurrence and deleterious when not negligible in effect-seem about as unfavorable as possible for an evolutionary process. Under biparental reproduction, however, a limited number of mutations which are not too injurious to be carried by the species furnish an almost infinite field of possible variations through which the species may work its way under natural selection.

Estimates of the total number of genes in the cells of higher organisms range from 1000 up. Some 400 loci have been reported as having mutated in Drosophila during a laboratory experience which is certainly very limited compared with the history of the species in nature. Presumably, allelomorphs of all type genes are present at all times in any reasonably numerous species. Judging from the frequency of multiple allelomorphs in those organisms which have been studied most, it is reasonably certain that many different allelomorphs of each gene are in existence at all times. With 10 allelomorphs in each of 1000 loci, the number of possible combinations is 10^{1000} which is a very large number. It has been estimated that the total number of electrons and protons in the whole visible universe is much less than 10^{100}.

This is believed to be the first quantitative discussion of genetic complexity in biological literature.

However, not all of this field is easily available in an interbreeding population. Suppose that each type gene is manifested in 99 percent of the individuals, and that most of the remaining 1 percent have the most favorable of the other allelomorphs, which in general means one with only a slight differential effect. The average individual will show the

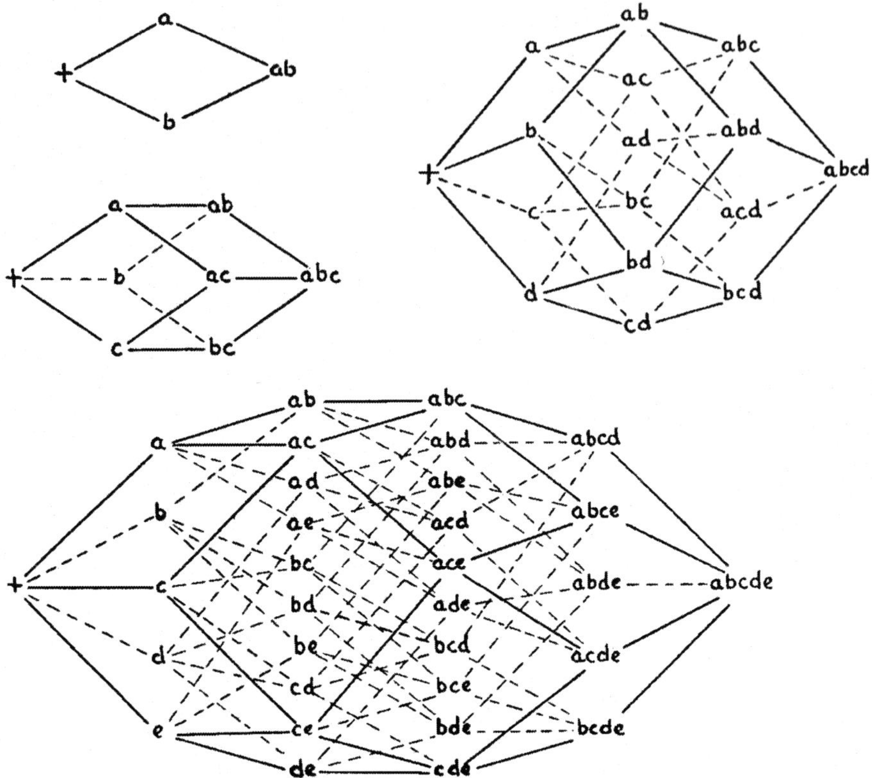

Figure 1. The combinations of from 2 to 5 paired allelomorphs.

effects of 1 percent of the 1000, or 10 deviations from the type, and since this average has a standard deviation of $\sqrt{10}$ only a small proportion will exhibit more than 20 deviations from type where 1000 are possible. The population is thus confined to an infinitesimal portion of the field of possible gene combinations, yet this portion includes some 10^{40} homozygous combinations, on the above extremely conservative basis, enough so that there is no reasonable chance that any two individuals have exactly the same genetic constitution in a species of millions of millions of individuals persisting over millions of generations. There is no difficulty in accounting for the probable genetic uniqueness of each individual human being or other organism which is the product of biparental reproduction.

If the entire field of possible gene combinations be graded with respect to adaptive value under a particular set of conditions, what would be its nature? Figure 1 shows the combinations in the cases of 2

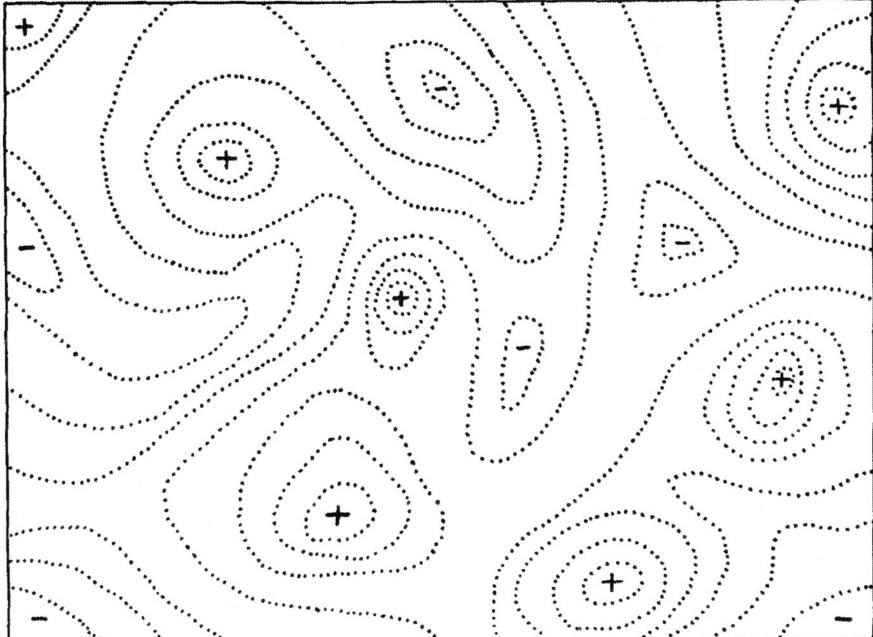

Figure 2. Diagrammatic representation of the field of gene combinations in two dimensions instead of many thousands. Dotted lines represent contours with respect to adaptiveness.

In this figure and in the rest of the discussion, Wright employs fitness landscapes—an approach he introduced within evolutionary biology—that represent the average fitness of the population as a function of allele frequencies.

to 5 paired allelomorphs. In the last case, each of the 32 homozygous combinations is at one remove from 5 others, at two removes from 10, etc. It would require 5 dimensions to represent these relations symmetrically; a sixth dimension is needed to represent level of adaptive value. The 32 combinations here compare with 10^{1000} in a species with 1000 loci each represented by 10 allelomorphs, and the 5 dimensions required for adequate representation compare with 9000. The two dimensions of figure 2 are a very inadequate representation of such a field. The contour lines are intended to represent the scale of adaptive value.

One possibility is that a particular combination gives maximum adaptation and that the adaptiveness of the other combinations falls off more or less regularly according to the number of removes. A species whose individuals are clustered about some combination other than the highest would move up the steepest gradient toward the peak, having reached which it would remain unchanged except for the rare occurrence of new favorable mutations.

In this version of fitness landscapes, fitness is assigned to each particular genotype.

The idea of a fitness peak was first introduced here.

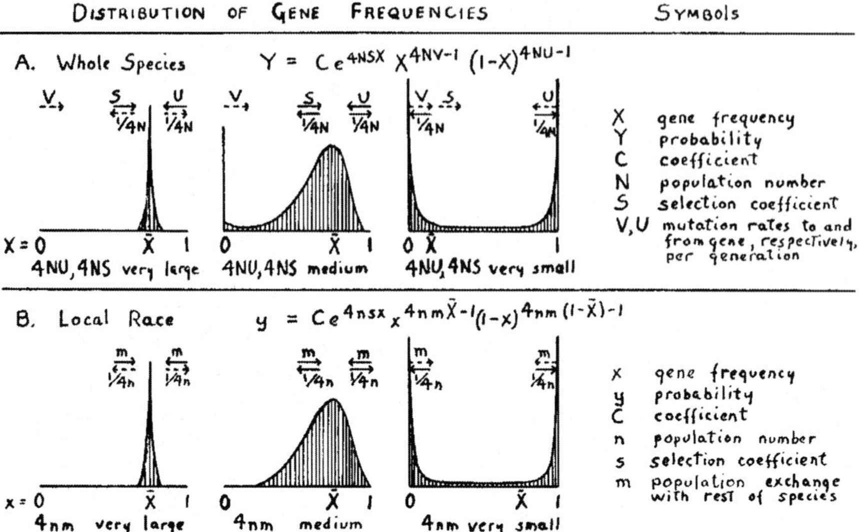

Figure 3. Random variability of a gene frequency under various specified conditions.

But even in the two factor case (figure 1) it is possible that there may be two peaks, and the chance that this may be the case greatly increases with each additional locus. With something like 10^{1000} possibilities (figure 2) it may be taken as certain that there will be an enormous number of widely separated harmonious combinations. The chance that a random combination is as adaptive as those characteristic of the species may be as low as 10^{-100} and still leave room for 10^{800} separate peaks, each surrounded by 10^{100} more or less similar combinations. In a rugged field of this character, selection will easily carry the species to the nearest peak, but there may be innumerable other peaks which are higher but which are separated by "valleys." The problem of evolution as I see it is that of a mechanism by which the species may continually find its way from lower to higher peaks in such a field. In order that this may occur, there must be some trial and error mechanism on a grand scale by which the species may explore the region surrounding the small portion of the field which it occupies. To evolve, the species must not be under strict control of natural selection. Is there such a trial and error mechanism?

Wright clearly describes the metaphor of "rugged fitness landscapes" with multiple peaks and valleys.

At this point let us consider briefly the situation with respect to a single locus. In each graph in figure 3 the abscissas represent a scale of gene frequency, 0 percent of the type genes to the left, 100 percent to the right. The elementary evolutionary process is, of course, change of gene frequency, a practically continuous process. Owing to the symmetry of the Mendelian mechanism, any gene frequency tends to remain constant in the absence of disturbing factors. If the type gene mutates at a certain rate, its frequency tends to move to the left, but at a continually decreasing rate. The type gene would ultimately be lost from the population if there were no opposing factor. But the type gene is in general favored by selection. Under selection, its frequency tends to move to the right. The rate is greatest at some point near the middle of the range. At a certain gene frequency the opposing pressures are equal and opposite, and at this point there is consequently equilibrium. There are other mechanisms of equilibrium among evolutionary factors which need not be discussed here. Note that we have here a theory of the stability of species in spite of continuing mutation pressure, a continuing field of variability so extensive that no two individuals are ever genetically the same, and continuing selection.

If the population is not indefinitely large, another factor must be taken into account: the effects of accidents of sampling among those that survive and become parents in each generation and among the germ cells of these, in other words, the effects of inbreeding. Gene frequency in a given generation is in general a little different one way or the other from that in the preceding, merely by chance. In time, gene frequency may wander a long way from the position of equilibrium, although the farther it wanders the greater the pressure toward return. The result is a frequency distribution within which gene frequency moves at random. There is considerable spread even with very slight inbreeding and the form of distribution becomes U-shaped with close inbreeding. The rate of movement of gene frequency is very slow in the former case but is rapid in the latter (among unfixed genes). In this case, however, the tendency toward complete fixation of genes, practically irrespective of selection, leads in the end to extinction.

In a local race, subject to a small amount of crossbreeding with the rest of the species (figure 3, lower half), the tendency toward random fixation is balanced by immigration pressure instead of by mutation and selection. In a small sufficiently isolated group all gene frequencies can drift irregularly back and forth about their mean values at a rapid rate, in

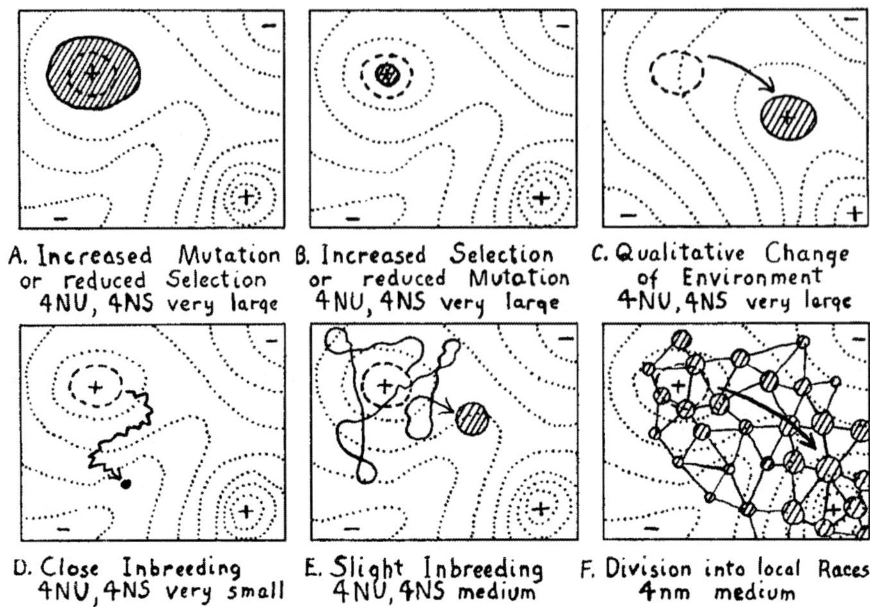

Figure 4. Field of gene combinations occupied by a population within the general field of possible combinations. Type of history under specified conditions indicated by relation to initial field (heavy broken contour) and arrow.

terms of geologic time, without reaching fixation and giving the effects of close inbreeding. The resultant differentiation of races is of course increased by any local differences in the conditions of selection.

Let us return to the field of gene combinations (figure 4). In an indefinitely large but freely interbreeding species living under constant conditions, each gene will reach ultimately a certain equilibrium. The species will occupy a certain field of variation about a peak in our diagram (heavy broken contour in upper left of each figure). The field occupied remains constant although no two individuals are ever identical. Under the above conditions further evolution can occur only by the appearance of wholly new (instead of recurrent) mutations, and ones which happen to be favorable from the first. Such mutations would change the character of the field itself, increasing the elevation of the peak occupied by the species. Evolutionary progress through this mechanism is excessively slow since the chance of occurrence of such mutations is very small and, after occurrence, the time required for attainment of sufficient frequency to be subject to selection to an appreciable extent is enormous.

The general rate of mutation may conceivably increase for some reason. For example, certain authors have suggested an increased incidence of cosmic rays in this connection. The effect (figure 4A) will be as a rule a spreading of the field occupied by the species until a new equilibrium is reached. There will be an average lowering of the adaptive level of the species. On the other hand, there will be a speeding up of the process discussed above, elevation of the peak itself through appearance of novel favorable mutations. Another possibility of evolutionary advance is that the spreading of the field occupied may go so far as to include another and higher peak, in which case the species will move over and occupy the region about this. These mechanisms do not appear adequate to explain evolution to an important extent.

The effects of reduced mutation rate (figure 4B) are of course the opposite: a rise in average level, but reduced variability, less chance of novel favorable mutation, and less chance of capture of a neighboring peak.

The effect of increased severity of selection (also 4B) is, of course, to increase the average level of adaptation until a new equilibrium is reached. But again this is at the expense of the field of variation of the species and reduces the chance of capture of another adaptive peak. The only basis for continuing advance is the appearance of novel favorable mutations which are relatively rapidly utilized in this case. But at best the rate is extremely slow even in terms of geologic time, judging from the observed rates of mutation.

Relaxation of selection has of course the opposite effects and thus effects somewhat like those of increased mutation rate (figure 4A).

The environment, living and non-living, of any species is actually in continual change. In terms of our diagram this means that certain of the high places are gradually being depressed and certain of the low places are becoming higher (figure 4C). A species occupying a small field under influence of severe selection is likely to be left in a pit and become extinct, the victim of extreme specialization to conditions which have ceased, but if under sufficiently moderate selection to occupy a wide field, it will merely be kept continually on the move. Here we undoubtedly have an important evolutionary process and one which has been generally recognized. It consists largely of change without

Wright implies here that fitness landscapes may also change continuously. Therefore in some cases a metaphor of "fitness seascapes" can be more appropriate than that of "fitness landscapes."

advance in adaptation. The mechanism is, however, one which shuffles the species about in the general field. Since the species will be shuffled out of low peaks more easily than high ones, it should gradually find its way to the higher general regions of the field as a whole.

Figure 4D illustrates the effect of reduction in size of population below a certain relation to the rate of mutation and severity of selection. There is fixation of one or another allelomorph in nearly every locus, largely irrespective of the direction favored by selection. The species moves down from its peak in an erratic fashion and comes to occupy a much smaller field. In other words there is the deterioration and homogeneity of a closely inbred population. After equilibrium has been reached in variability, movement becomes excessively slow, and, such as there is, is nonadaptive. The end can only be extinction. Extreme inbreeding is not a factor which is likely to give evolutionary advance.

With an intermediate relation between size of population and mutation rate, gene frequencies drift at random without reaching the complete fixation of close inbreeding (figure 4E). The species moves down from the extreme peak but continually wanders in the vicinity. There is some chance that it may encounter a gradient leading to another peak and shift its allegiance to this. Since it will escape relatively easily from low peaks as compared with high ones, there is here a trial and error mechanism by which in time the species may work its way to the highest peaks in the general field. The rate of progress, however, is extremely slow since change of gene frequency is of the order of the reciprocal of the effective population size and this reciprocal must be of the order of the mutation rate in order to meet the conditions for this case.

These processes are usually divided into three stages. The first stage is stochastic: a new adapted combination of genes is formed and increases in numbers above a certain threshold. The second stage, when the new adapted combination of genes becomes dominant in a local population, is driven by selection. During the third stage, the new combination of genes takes over the whole population as a result of differential migration and between-deme competition.

Finally (figure 4F), let us consider the case of a large species which is subdivided into many small local races, each breeding largely within itself but occasionally crossbreeding. The field of gene combinations occupied by each of these local races shifts continually in a nonadaptive fashion (except in so far as there are local differences in the conditions of selection). The rate of movement may be enormously greater than in the preceding case since the condition for such movement is that the reciprocal of the population number be of the order of the proportion of crossbreeding instead of the mutation rate. With many local races, each spreading over a considerable field and moving relatively rapidly in the more general field about the controlling peak, the chances are good that one at least will come under the influence of another peak. If a higher peak, this race will expand in numbers and by crossbreeding with the others will pull the whole species toward the new position. The average adaptiveness of the species thus advances under intergroup selection, an enormously more effective process than intragroup selection. The conclusion is that subdivision of a species into local races provides the most effective mechanism for trial and error in the field of gene combinations.

Wright's ideas of intergroup selection were later developed into theories of genetic group selection and cultural group selection.

It need scarcely be pointed out that with such a mechanism complete isolation of a portion of a species should result relatively rapidly in specific differentiation, and one that is not necessarily adaptive. The effective intergroup competition leading to adaptive advance may be between species rather than races. Such isolation is doubtless usually geographic in character at the outset but may be clinched by the development of hybrid sterility. The usual difference of the chromosome complements of related species puts the importance of chromosome aberration as an evolutionary process beyond question, but, as I see it, this importance is not in the character differences which they bring (slight in balanced types), but rather in leading to the sterility of hybrids and thus making permanent the isolation of two groups.

How far do the observations of actual species and their subdivisions conform to this picture? This is naturally too large a subject for more than a few suggestions.

That evolution involves nonadaptive differentiation to a large extent at the subspecies and even the species level is indicated by the kinds of differences by which such groups are actually distinguished by systematists. It is only at the subfamily and family levels that clear-cut adaptive differences become the rule (Robson, Japot). The principal evolutionary mechanism in the origin of species must thus be an essentially nonadaptive one.

Wright highlights the importance of neutral (nonadaptive) variation in biological evolution, an idea that was further developed by Motoo Kimura (1983) and others.

That natural species often are subdivided into numerous local races is indicated by many studies. The case of the human species is most familiar. Aside from the familiar racial differences recent studies indicate a distribution of frequencies relative to an apparently nonadaptive series of allelomorphs, that determining blood groups, of just the sort discussed above. I scarcely need to labor the point that changes in the average of mankind in the historic period have come about more by expansion of some types and decrease and absorption of others than by uniform evolutionary advance. During the recent period, no doubt, the phases of intergroup competition and crossbreeding have tended to overbalance the process of local differentiation, but it is probable that in the hundreds of thousands of years of prehistory, human evolution was determined by a balance between these factors.

Subdivision into numerous local races whose differences are largely nonadaptive has been recorded in other organisms wherever a sufficiently detailed study has been made. Among the land snails of the Hawaiian Islands, Gulick (sixty years ago) found that each mountain valley, often each grove of trees, had its own characteristic type, differing from others in "nonutilitarian" respects. Gulick attributed this differentiation to inbreeding. More recently Crampton has found a similar situation in the land snails of Tahiti and has followed over a period of years evolutionary changes which seem to be of the type here discussed. I may also refer to the studies of fishes by David Starr Jordan, garter snakes by Ruthven, bird lice by Kellogg, deer mice by Osgood, and gall wasps by Kinsey as others which indicate the role of local isolation as a differentiating factor. Many other cases are discussed by Osborn and especially by Rensch in recent summaries. Many of these authors insist on the nonadaptive character of most of the differences among local races. Others attribute all differences to the environment, but this seems to be more an expression of faith than a view based on tangible evidence.

An even more minute local differentiation has been revealed when the methods of statistical analysis have been applied. Schmidt demonstrated the existence of persistent mean differences at each collecting station in certain species of marine fish of the fjords of Denmark, and these differences were not related in any close way to the environment. That the differences were in part genetic was demonstrated in the laboratory. David Thompson has found a correlation between water distance and degree of differentiation within certain fresh water species of fish of the streams of Illinois. Sumner's extensive studies of subspecies of Peromyscus (deer mice) reveal genetic differentiations, often apparently nonadaptive, among local populations and demonstrate the genetic heterogeneity of each such group. The modern breeds of livestock have come from selection among the products of local inbreeding and of crossbreeding between these, followed by renewed inbreeding, rather than from mass selection of species. The recent studies of the geographical distribution of particular genes in livestock and cultivated plants by Serebrovsky, Philiptschenko and others are especially instructive with respect to the composition of such species.

The paleontologists present a picture which has been interpreted by some as irreconcilable with the Mendelian mechanism, but this seems to be due more to a failure to appreciate statistical consequences of this mechanism than to anything in the data. The horse has been the standard example of an orthogenetic evolutionary sequence preserved for us with an abundance of material. Yet Matthew's interpretation as one in which evolution has proceeded by extensive differentiation of local races, intergroup selection, and crossbreeding is as close as possible to that required under the Mendelian theory.

Summing up: I have attempted to form a judgment as to the conditions for evolution based on the statistical consequences of Mendelian heredity. The most general conclusion is that evolution depends on a certain balance among its factors. There must be gene mutation, but an excessive rate gives an array of freaks, not evolution; there must be selection, but too severe a process destroys the field of variability, and thus the basis for further advance; prevalence of local inbreeding within a species has extremely important evolutionary consequences, but too close inbreeding leads merely to extinction. A certain amount of crossbreeding is favorable but not too much. In this dependence on balance the species is like a living organism. At all levels of organization life depends on the maintenance of a certain balance among its factors.

More specifically, under biparental reproduction a very low rate of mutation balanced by moderate selection is enough to maintain a practically infinite field of possible gene combinations within the species. The field actually occupied is relatively small though sufficiently extensive that no two individuals have the same genetic constitution.

The course of evolution through the general field is not controlled by direction of mutation and not directly by selection, except as conditions change, but by a trial and error mechanism consisting of a largely nonadaptive differentiation of local races (due to inbreeding balanced by occasional crossbreeding) and a determination of long time trend by intergroup selection. The splitting of species depends on the effects of more complete isolation, often made permanent by the accumulation of chromosome aberrations, usually of the balanced type. Studies of natural species indicate that the conditions for such an evolutionary process are often present.

REFERENCES

Crampton, H. E. 1925. "Contemporaneous Organic Differentiation in the Species of Partula Living in Moorea, Society Islands." *The American Naturalist* 59 (660): 5–35.

East, E. M. 1918. "The Role of Reproduction in Evolution." *The American Naturalist* 52 (618/619): 273–289.

Gulick, J. T. 1905. *Evolution, Racial and Habitudinal.* Washington, DC: Carnegie Institution of Washington.

Japot, A. P. 1908. "The Status of the Species and the Genus." *The American Naturalist* 66 (705): 346–364.

Jordan, D. S. 1908. "The Law of Germinate Species." *The American Naturalist* 42 (494): 73–80.

Kellogg, V. L. 1908. *Darwinism Today.* New York, NY: Henry Holt / Co.

Kinsey, A. C. 1930. *The Gall Wasp Genus Cynips: A Study in the Origin of Species.* Vol. 16. Indiana University Studies. Bloomington, IN: Indiana University.

Matthew, W. D. 1926. "The Evolution of the Horse. A Record and Its Interpretation." *The Quarterly Review of Biology* 1 (2): 139–185.

Osborn, H. F. 1927. "The Origin of Species. V. Speciation and Mutation." *The American Naturalist* 61 (672): 5–42.

Osgood, W. H. 1909. "Revision of the Mice of the Genus *Peromyscus*," North American Fauna, 28:1–285.

Philiptschenko, J. 1927. *Variabilität und Variation.* Berlin, Germany: Gebrüder Borntraeger.

Rensch, B. 1929. *Das Prinzip geographischer Rassenkreise und das Problem der Artbildung.* Berlin, Germany: Gebrüder Borntraeger.

Robson, G. C. 1928. *The Species Problem: An Introduction to the Study of Evolutionary Divergence in Natural Populations.* Edinburgh and London, UK: Oliver / Boyd.

Ruthven, A. G. 1908. "Variation and Genetic Relationships of the Garter-Snakes." *Bulletin of the United States National Museum* 61:1–301.

Schmidt, J. 1918. "Racial Studies in Fishes. Statistical Investigations with *Zoarces viviparus*." *Journal of Genetics* 7 (2): 105–118.

Serebrovsky, A. S. 1929. "Beitrag zur geographischen Genetic des Haushuhns in Sowjet-Rußland." *Archiv für Geflügelkunde* 3:161–169.

Sumner, F. B. 1932. "Genetic, Distributional, and Evolutionary Studies of the Subspecies of Deer Mice." *Bibliography of Genetics* 9:1–106.

Thompson, D. H. 1931. "Variation in Fishes as a Function of Distance." *Transactions of the Illinois State Academy of Science* 23:276–281.

Wright, S. 1931. "Evolution in Mendelian Populations." *Genetics* 16:97–159.

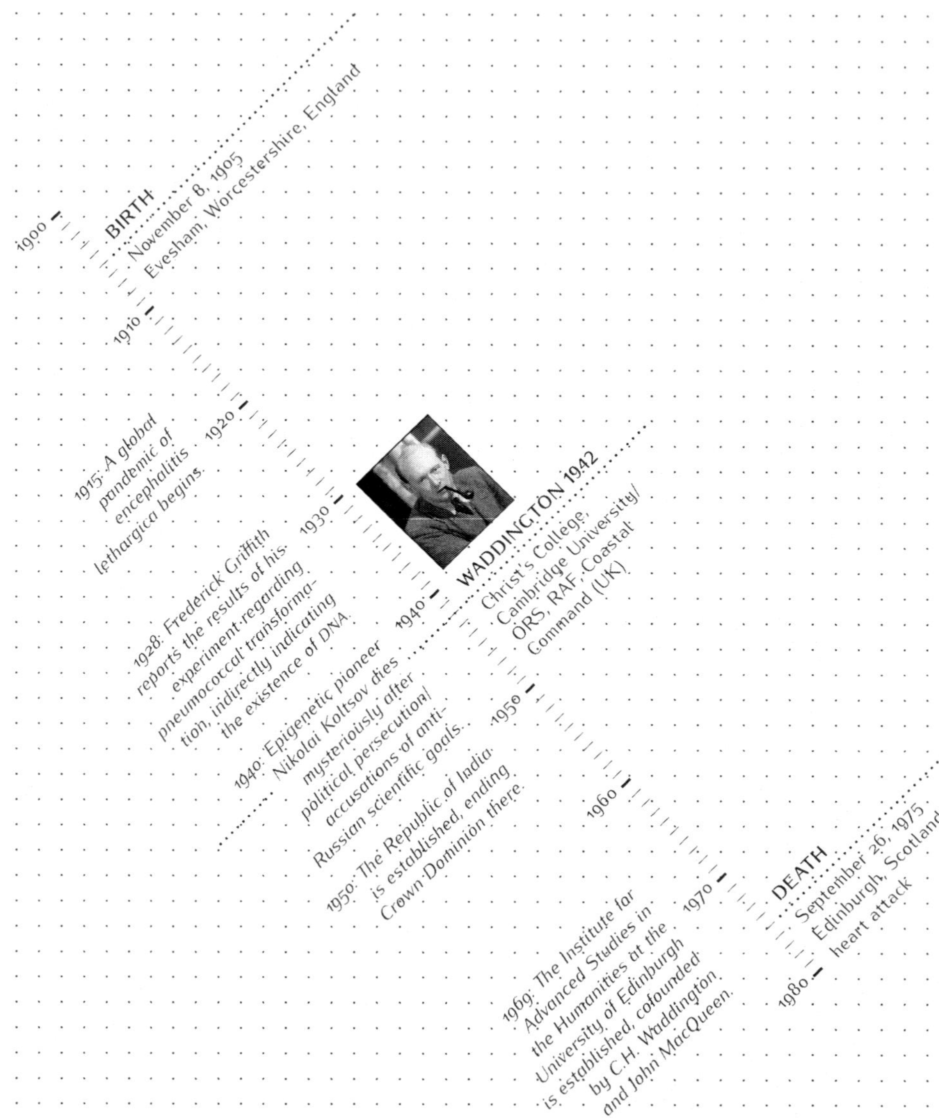

CONRAD HAL WADDINGTON

[4]

THE SHADOW OF A MECHANISM

Walter Fontana, Harvard University

I stumbled across Conrad H. Waddington's 1942 paper a few years after my PhD while unwittingly retracing some of his thinking in the context of a narrow but illustrative toy-model of development, understood as a mapping from genotype to phenotype(s) (see Ancel and Fontana 2000; Fontana 2002). Digesting his paper and following its thread through his other writings made it clear that I hadn't said much that Waddington didn't already say fifty years prior. His prescience and subtle precision of thought at a time when far less was known than today are stunning. This is my brief take on it.

C. H. Waddington, "Canalization of Development and the Inheritance of Acquired Characters," *Nature* 150, 563–565 (1942).

The Darwinian framework of evolution through heritable variation and natural selection is the foundation for explaining adaptation, the functional integration of living systems and their environment. Consider, for example, the evolution of different shapes of bird beaks, such as long and pointed or short and stout, adapted to tapping distinct sources of food like fruits, seeds, or insects. A short and thick beak is not a useful tool for picking seeds from a cactus. As a consequence, there is a tendency for evolution to remodel the beak in a search process based on mutation and selection as posited by the Darwinian framework. Varying the timing of the action of "calmodulin," "bone morphogen protein 4," and other molecular players makes mechanistically intelligible how a blind evolutionary process can find a suitable beak (Wu *et al.* 2004; Abzhanov *et al.* 2004; Mallarino *et al.* 2011). Waddington did not know about any of these, but not knowing exactly which levers mutation and selection can play with is not an impediment to being, in principle, satisfied with the conceptual adequacy of the Darwinian foundation. Today, given the impressive tapestry of empirical evidence grounding Darwin's framework in mechanism, there is nothing implausible about the evolution of strong yet hollow bones in birds, even though we do not know in every detail how it happened.

The conceptual sufficiency of the Darwinian search process derives from a mechanism that does not prejudge what the search comes up with. It is a matter of theory to figure out what is contingent and what is necessary about evolutionary outcomes. The dearth of theory in regard does not affect the plausibility that any theory, once built, would be situated within the general framework. Human cognition, attuned to functional relations, might predict a long and pointed beak as a solution to the cactus problem, but this does not make it implausible that a search devoid of cognition comes up with that same solution. There is no final cause, only local guidance from gradients on a fitness landscape whose structure may be complex, shifting in time, and ill-understood. Darwin proposed a paradigm, not a theory.

Contrast this with situations that involve another kind of adaptation, one in which an individual organism adapts *during its lifetime* to environmental circumstances. Such real-time adjustment is a consequence of the varying degrees of plasticity that living systems are capable of. Today, we know that this flexibility arises from manifold interconnected processes of regulation in response to physical and chemical cues within and outside the organism. A so-adapted state is not hereditary, since it requires a persistent interaction with the environment to come about. Waddington notes that many cases exist in which evolution appears to have genetically "hardwired" an organismic state that was previously attainable only through experience, as it were. The problem now is that we are asked to believe in the blindness and rapidity of a search process arriving at the same solution that was successfully tried out in advance through real-time adaptation. This hinges on too much coincidence. It seems more plausible to assume that the suitability of an organismic reaction to the environment orchestrates the necessary genetic changes that make it heritable. Such an assumption, however, would shatter the Darwinian framework into Lamarckian pieces.

Waddington mentions by way of example the callosities of ostriches. These patches of thickened skin are situated on the sternum and near the tail, both of which rub against the ground when the ostrich is in a crouching position. Callouses like these arise plastically from continued friction, and since they are useful for the ostrich, it might be advantageous for them to be genetically hardwired. Indeed, these callosities have become hereditary and form already in the embryo who has never been subject to friction on hard ground. Many cases of this sort exist in nature and many can be generated through breeding in the laboratory, as Waddington (1959) himself did. In sum, the Darwinian paradigm is plausible when its outcomes are *not directly* inducible by the environment and appears at risk when they are, because it raises the suspicion that "use and disuse" are *directly* complicit in causing genetic change rather than being jury of random trials.

This apparent Lamarckian phenomenon was debated by several evolutionists—among them J. M. Baldwin (1896), L. Morgan (1896) and H. F. Osborn (1896)—around 1896, forty-six years before Waddington's paper. It is instructive to follow Waddington's explanation by way of contrast to Baldwin's, which was laid out in a review by G. G. Simpson (1953). Simpson provided such a clear exposition of Baldwin's explanation that it henceforth became known as the Simpson–Baldwin effect, ironically despite Simpson suggesting that Baldwin's explanation is no explanation at all. Simpson summarizes Baldwin's reasoning as follows:

1. Individual organisms interact with the environment in such a way as systematically to produce in them non-hereditary adaptations that are advantageous to the individuals having them.

2. Mutations ("genetic factors") producing similar traits as in 1 occur in the population.

3. These mutations are favored by natural selection (as they don't require the machinery for dynamically generating the adaptation) and tend to spread in the population.

The net result is that adaptation, originally individual and nonhereditary, has become hereditary. The crux, according to Simpson, is the *absence* of any logical connection between 1 and 2, for if there is one it could only be Lamarckian.

To fix what we are talking about, consider a simple hypothetical example at the molecular scale. Suppose an animal finds better food at high altitude. In response to lower oxygen levels the animal adapts by producing 2,3-diphosphoglycerate (DPG), which is a molecule that binds hemoglobin, altering its conformation, thereby lowering its oxygen affinity and causing it to unload oxygen more thoroughly. This non-primary but advantageous behavior of hemoglobin, induced by an environmental condition via DPG, permits the population to linger in a hostile environment that provides a benefit. Following Baldwin's argument, eventually a random mutation occurs as a result of which hemoglobin achieves on its own what it previously required DPG for. Because of the continued exposure to the hostile environment, selection kicks in and spreads the mutant to fixation in the population. This process hardwires genetically a previously acquired characteristic (of hemoglobin) in a fashion wholly compatible with the Darwinian framework. Baldwin and everyone else at the time emphasized the independence of 2 (the origination of appropriate mutations) from 1 (plasticity). In particular, the random mutation would have eventually occurred regardless of plasticity and exposure to high altitude—it just would not have been selected; the only contribution from plasticity is that it permits continued exposure to a specific selection pressure. This, however, makes Baldwin's solution nearly vacuous, since we are being asked again to suspend disbelief that somehow the right mutation (or, more likely, several required mutations) show up whose effect happens to be precisely the previously plastically induced property.

Simpson could be read as dismissing the significance of the whole phenomenon alongside Baldwin's explanation, whereas Waddington must have felt that the phenomenon of a plastic adaptation becoming heritable tells us something important about evolution. Waddington (1953) thus wrote a reply to Simpson's paper, in which he objects that Simpson has overlooked a possible logical connection between 1 and 2 that is neither Lamarckian nor Baldwinian: Genotypes with "the ability to produce an adaptive phenotype would [. . .] encourage the appearance of genetically controlled variants mimicking the adaptive type." This is classic Waddingtonian and needs some decryption. "The ability to produce an adaptive phenotype" refers to plasticity and "the appearance of genetically controlled variants" refers to mutations. The operative word here is "encourage," as in "making more probable." Back to the hemoglobin story.

A protein, such as hemoglobin, is a sequence of amino acids folding into a native three-dimensional structure by virtue of a complex network of interactions between the side chains. This structure conveys chemical and biological function. Given a protein, we can think of a conformation as a point on a landscape that assigns to every three-dimensional conformation a free energy. This landscape is high-dimensional, in indication of the many possible physical displacements that lead from a given conformation to neighboring ones—an Alpine terrain but with high-dimensional versions of passes, plateaus, saddle points, troughs, peaks, ridges, and valleys. The folding process traces a path guided by energy gradients on that landscape and ends up in a particular trough: the native configuration. It is easy to imagine the existence of nearby troughs, albeit separated by barriers. Our sequence could fold into one of the nearby troughs, but to do so would require a kick, such as an interaction with DPG. In the case of DPG, the interaction occurs between its negatively charged phosphate groups and specific positively charged lysine and histidine residues in hemoglobin. Thus, alternative shapes reveal themselves in response to exogenous triggers. However, chemical interactions, such as hydrogen bonds or electrostatic bonds, are completely *fungible*: nothing rides on the binding partner being specifically DPG; it could be any other molecule as long as it provides chemical groups with electrostatic charges in the proper positions. In fact, it could be a chemical group within hemoglobin itself, such as a suitable amino acid residue! It is, therefore, highly *likely* that some amino acid substitution can coax the network of extant interactions to switch into the conformation they were already capable of adopting in the presence of DPG. Conversely, it seems highly *unlikely* for such a substitution to achieve the same outcome in the context of an interaction network that cannot switch into the advantageous conformation even in the presence of DPG. This is Waddington's explanation. It obviates the need to suspend disbelief in the magical appearance of the right mutation(s) that would hardwire a plastically attained adaptation, because these mutations are bound to be readily accessible due to the mechanism enabling plasticity in the first place. The proverbial monkey being tasked with writing a meaningful sentence is vastly more likely to do so if given a typewriter than a pen. Waddington's insight could not be more different than Baldwin's.

Waddington develops his explanation in a different, necessarily far less mechanistic language and against the more complex background of developmental processes. The stylized hemoglobin example provided us with discrete (coarse-grained) conformational states, so we could speak of a "switch" at the level of a single molecule by virtue of its interaction with DPG. For Waddington's argument to work, development of an organism must have likewise acquired some quasi-discrete structure that allows one to speak of developmental trajectories that are meaningfully distinguishable from one another in the face of noise or perturbation. Waddington (1940, p.91) uses the term "epigenetic landscape" for what mathematicians call the *phase space* of a dynamical system. The landscape of conformations in the hemoglobin example plays the role of the epigenetic landscape. Waddington spends most of his paper arguing that natural selection has sculpted the epigenetic landscape into identifiable valleys separated by barriers. This structure creates developmental trajectories and enables them to reliably reach defined endpoints. He refers to such a phase space as "canalized" and the action of selection leading to this structure as "canalizing." (See Wagner, Booth, and Bagheri-Chaichian 1997; Ancel and Fontana 2000; Siegal and Bergman 2002 for computational models that have contributed to elucidating the circumstances under which canalization occurs.) Waddington dwells on canalization, because it allows him to cast the effects of mutations in semi-discrete terms such as the switching of states by crossing or shifting thresholds and sculpting the epigenetic landscape (Waddington 1957) to internalize external signals.

While all this is sensible, I felt that the quote from his reply to Simpson was the closest he came to succinctly stating a principle. *Waddington is asserting what even today is all too often forgotten: that a mutation at the genetic level may be random, but its consequences at the phenotypic level are not, because a mechanism necessarily biases in specific ways the effects of its modification.* This is true in particular for mechanisms with a prior evolutionary history, since they are canalized. A mechanism that implements plasticity by enabling input signals to switch among a repertoire of behaviors is vastly more likely to be modified by a genetic mutation in such a way as to make any of these behaviors independent of the trigger signal. I refer to these preferred directions of change as the "shadow" of a mechanism.

As a final thought I would like to draw attention to an unspoken assumption. Waddington's theory implicitly posits an alignment of sorts between possibilities whose realization is dependent on agents external to the organism, such as signals, and possibilities whose realization is entirely dependent on agents internal to the organism, such as genes. In the hemoglobin story, for example, this alignment is built into the physics of folding (where the "genes" are the amino acids along the sequence). As if by revenge, contemplating the limits of the alignment between plastic and genetic realizability leads us back to Baldwin. Baldwin was an evolutionary psychologist and his paradigmatic case of plasticity was *learning*. Even if the capacity to learn is entirely genetically determined, what we learn is certainly not. Depending on the specific mechanism, learning can open up vast ranges of plasticity. It stands to reason that many ranges of plastic possibility are *not* aligned, and perhaps even impossible to align, with genetic realizability. If someone has learned how to solve Rubik's cube in a few seconds, it does not follow that mutations are readily at hand whose effect is to hardwire this skill genetically. Some mechanisms may cast no shadow. It is tempting to speculate that the elaboration of learning mechanisms eventually opened the floodgates to the infinitely plastic and genetically non-alignable, morphing biology into something alien to the Darwinian framework that gave rise to it.

Acknowledgments

I'm grateful to Aviv Bergman, Marc Kirschner, and Gunter Wagner for helpful comments.

REFERENCES

Abzhanov, A., M. Protas, B. R. Grant, P. R. Grant, and C. J. Tabin. 2004. "*Bmp4* and Morphological Variation of Beaks in Darwin's Finches." *Science* 305 (5689): 1462–1465. https://doi.org/10.1126/science.1098095.

Ancel, L., and W. Fontana. 2000. "Plasticity, Evolvability, and Modularity in RNA." *Journal of Experimental Zoology*, https://doi.org/10.1002/1097-010X(20001015)288:3<242::AID-JEZ5>3.0.CO;2-O.

Baldwin, J. M. 1896. "A New Factor in Evolution." *The American Naturalist* 30 (354): 441–451. https://doi.org/10.1086/276408.

Fontana, W. 2002. "Modelling 'Evo–Devo' with RNA." *BioEssays* 24 (12): 1164–1177. https://doi.org/10.1002/bies.10190.

Mallarino, R., P. R. Grant, B. R. Grant, A. Herrel, W. P. Kuo, and A. Abzhanov. 2011. "Two Developmental Modules Establish 3D Beak-Shape Variation in Darwin's Finches." *Proceedings of the National Academy of Sciences* 108 (10): 4057–4062. https://doi.org/10.1073/pnas.1011480108.

Morgan, C. L. 1896. *Habit and Instinct.* London, UK: Arnold.

Osborn, H. F. 1896. "Ontogenic and Phylogenic Variation." *Science* 4 (100): 786–789. https://doi.org/10.1126/science.4.100.786.

Siegal, M. L., and A. Bergman. 2002. "Waddington's Canalization Revisited: Developmental Stability and Evolution." *Proceedings of the National Academy of Sciences* 99 (16): 10528–10532. https://doi.org/10.1073/pnas.102303999.

Simpson, G. G. 1953. "The Baldwin Effect." *Evolution* 7 (2): 110–117. https://doi.org/10.1111/j.1558-5646.1953.tb00069.x.

Waddington, C. H. 1940. *Organisers & Genes.* 182. Cambridge, UK: Cambridge University Press.

———. 1942. "Canalization of Development and the Inheritance of Acquired Characters." *Nature* 150 (3811): 563–565. https://doi.org/10.1038/150563a0.

———. 1953. "The 'Baldwin Effect,' 'Genetic Assimilation' and 'Homeostasis'." *Evolution* 7 (4): 386–387. https://doi.org/10.1111/j.1558-5646.1953.tb00099.x.

———. 1957. *The Strategy of the Genes.* London, UK: George Allen & Unwin.

———. 1959. "Canalization of Development and Genetic Assimilation of Acquired Characters." *Nature* 183 (4676): 1654–1655. https://doi.org/10.1038/1831654a0.

Wagner, G. P., G. Booth, and H. Bagheri-Chaichian. 1997. "A Population Genetic Theory of Canalization." *Evolution* 51 (2): 329–347. https://doi.org/10.1111/j.1558-5646.1997.tb02420.x.

Wu, P., T.-X. Jiang, S. Suksaweang, R. B. Widelitz, and C.-M. Chuong. 2004. "Molecular Shaping of the Beak." *Science* 305 (5689): 1465–1466. https://doi.org/10.1126/science.1098109.

CANALIZATION OF DEVELOPMENT AND THE INHERITANCE OF ACQUIRED CHARACTERS

C. H. Waddington, Cambridge University

The battle, which raged for so long between the theories of evolution supported by geneticists on one hand and by naturalists on the other, has in recent years gone strongly in favour of the former. Few biologists now doubt that genetical investigation has revealed at any rate the most important categories of hereditary variation; and the classical 'naturalist' theory—the inheritance of acquired characters—has been very generally relegated to the background because, in the forms in which it has been put forward, it has required a type of hereditary variation for the existence of which there was no adequate evidence. The long popularity of the theory was based, not on any positive evidence for it, but on its usefulness in accounting for some of the most striking of the results of evolution. Naturalists cannot fail to be continually and deeply impressed by the adaptation of an organism to its surroundings and of the parts of the organism to each other. These adaptive characters are inherited and some explanation of this must be provided. If we are deprived of the hypothesis of the inheritance of the effects of use and disuse, we seem thrown back on an exclusive reliance on the natural selection of merely chance mutations. It is doubtful, however, whether even the most statistically minded geneticists are entirely satisfied that nothing more is involved than the sorting out of random mutations by the natural selective filter. It is the purpose of this short communication to suggest that recent views on the nature of the developmental process make it easier to understand how the genotypes of evolving organisms can respond to the environment in a more co-ordinated fashion.

It will be convenient to have in mind an actual example of the kind of difficulties in evolutionary theory with which we wish to deal.

We may quote from C. and Richards (1936): "A single case will make the difficulty clear. Duerden (1920) has shown that the sternal, alar, etc., callosities of the ostrich, which are undoubtedly related to the crouching position of the bird, appear in the embryo. The case is analogous to the thickening of the soles of the feet of the human embryo attributed by Darwin (1901) 'to the inherited effects of pressure.' As Detlefsen (1925) points out, this would have to be explained on selectionist grounds by the assumption that it was of advantage to have the callosities, as it were, preformed at the place at which they are required in the adult. But it is a large assumption that variations would arise at this place and nowhere else."

In this case we have an adaptive character (the callosities) of a kind which it is known can be provoked by an environmental stimulus during a single lifetime (since skin very generally becomes calloused by continued friction) but which is in this case certainly inherited. The standard hypotheses which come in question are the two considered by Robson and Richards: the Lamarckian explanation in terms of the inheritance of the effects of use, which they cannot bring themselves to support at all strongly, and the 'selectionist' explanation, which, in the form in which they understand it, leaves entirely out of account the fact that callosities may be produced by an environmental stimulus and postulates the occurrence of a gene with the required developmental effect. A third possible type of explanation is to suppose that in earlier members of the evolutionary chain, the callosities were formed as responses to external friction, but that during the course of evolution the environmental stimulus has been superseded by an internal genetical factor. It is an explanation of this kind which will be advanced here.

The first step in the argument is one which will scarcely be denied but is perhaps often overlooked. The capacity to respond to an external stimulus by some developmental reaction, such as the formation of a callosity, must itself be under genetic control. There is little doubt, though no positive evidence in this particular case so far as I know, that individual ostriches differ genetically in the responsiveness of their skin to friction and pressure. If we suppose, then, that in the early ostrich ancestors callosities were formed by direct response to external pressure, there would be a natural selection among the birds for a genotype which gave an optimum response.

The next point to be put forward is the one which is, perhaps, new in such discussions, and which therefore requires the most careful scrutiny. It is best considered as one general thesis and one particular application of it.

The main thesis is that developmental reactions, *as they occur in organisms submitted to natural selection*, are in general canalized. That is to say, they are adjusted so as to bring about one definite end-result regardless of minor variations in conditions during the course of the reaction.

The evidence for this comes from two sides, the embryological and the genetical. In embryology we have abundant evidence of canalization on two scales. On the small scale of single tissues, one may direct attention to the obvious but not unimportant fact that animals are built up of sharply defined different tissues and not of masses of material which shade off gradually into one another. Similarly, from the experimental point of view, it is usual to find that, while it may be possible to steer a mass of developing tissue into one of a number of possible paths, it is difficult to persuade it to differentiate into something intermediate between two of the normal possibilities. Passing from the scale of tissues to that of organs, it is not too much to claim it as a general rule that there is some stage in every life-history (though it may be an extremely early and short stage) when minor variations in morphology become 'regulated' or regenerated; and that is, again, a tendency to produce the standard end-product. Of course neither of these types of canalization is absolute. Morphological regulation may fail if the abnormalities are too great or occur too late in development; and intermediate types of tissue can occasionally be found, particularly in pathological conditions.

The limitations on canalization which are important for our present purposes can better be seen when the problem is viewed from the other, genetical, side. The canalization, or perhaps it would be better to call it the buffering, of the genotype is evidenced most clearly by constancy of the wild type. It is a very general observation to which little attention has been directed (but see Huxley 1942; Plunkett 1932;

Ford 1940) that the wild type of an organism, that is to say, the form which occurs in Nature under the influence of natural selection, is much less variable in appearance than the majority of the mutant races. In Drosophila the phenomenon is extremely obvious; there is scarcely a mutant which is comparable in constancy with the wild type, and there are very large numbers whose variability, either in the frequency with which the gene becomes expressed at all or in the grade of expression, is so great that it presents a considerable technical difficulty. Yet the wild type is equally amazingly constant. If wild animals of almost any species are collected, they will usually be found 'as like as peas in a pod.' Variation there is, of course, but of an altogether lesser order than that between the different individuals of a mutant type.

The constancy of the wild type must be taken as evidence of the buffering of the genotype against minor variations not only in the environment in which the animals developed but also in its genetic make-up. That is to say, the genotype can, as it were, absorb a certain amount of its own variation without exhibiting any alteration in development. Considerable stress has been laid in recent years on certain aspects of this buffering. Fisher (1928) and many authors following him have discussed 'the evolution of dominance,' by which the genotype comes to be able to produce the standard developmental effects even when certain genes have been replaced by others of less efficiency. Again, Stern (1929) and Muller (1932) directed attention to the phenomenon of 'dosage compensation,' by which it comes about that a single dose of a sex-linked gene in the heterogametic sex has the same developmental effect as a double dose in the homogametic. These two processes are part of the larger phenomenon which we have called the canalization of development. This also includes other, at first sight unrelated, features of the genotypic control of development. For example, attention has been directed (Waddington 1940a) to genes which cause certain regions of developing tissue to take an abnormal choice out of a range of alternative possible paths; K. and de Winton (1941) have recently spoken of such genes as 'switch genes.' Finally, Goldschmidt has shown that environmental stimuli may, by switching development into a path which is usually only followed under the influence of some particular gene, produce what he has called a 'phenocopy' of a previously known mutant type.

There seems, then, to be a considerable amount of evidence from a number of sides that development is canalized in the naturally selected animal. At the same time, it is clear that this canalization is not a necessary characteristic of all organic development, since it breaks down in mutants, which may be extremely variable, and in pathological conditions, when abnormal types of tissue may be produced. It seems, then, that the canalization is a feature of the system which is built up by natural selection; and it is not difficult to see its advantages, since it ensures the production of the normal, that is, optimal, type in the face of the unavoidable hazards of existence.

The particular application of this general thesis which we require in connexion with 'the inheritance of acquired characters' is that a similar canalization will occur when natural selection favours some characteristic in the development of which the environment plays an important part. It is first necessary to point out the ways in which the environment can influence the developmental system. If we conceptually rigidify such a system into a definite formal scheme, we can think of it as a set of alternative canalized paths; and the environment can act either as a switch, or as a factor involved in the system of mutually interacting processes to which the buffering of the paths is due. This is, of course, too dead and formal a scheme to be a true picture of development as it actually occurs. In so far as it is always to some extent, but not entirely, a matter of convenience what we decide to call a complete organ, so far will it be a matter of convenience what we consider to be different alternative paths; and the question of whether a given influence is thought of as a switch mechanism or a modification of a path will depend on how we choose our alternatives. There are some cases, however, in which the alternatives are very clearly defined. Thus it is commonly assumed that the evolution of sexuality passed through a stage in which, as in Bonellia, the environment acted as a switch between two well-defined alternatives; later, genetic factors arose which superseded the environmental determination by an internal one.

More commonly, however, the original environmental effect will be to produce a modification of an already existent developmental path. Thus in the case of the ostrich ancestors, the formation of callosities following environmental stimulation is a response by a developmental system which is normally present in vertebrates. This system must, in all species, be subject to natural selection; outside certain limits, too great or too low a reactivity of the skin would be manifestly disadvantageous. If we suppose that the callosities, when they were first evolved, were dependent on the environmental stimulus, then the evolution appears as a readjustment of the reactivity of the skin to such a degree that a just sufficient thickening is produced with the normally occurring stimulus.

There would appear to be two possible ways in which such a development might be organized. It might on one hand remain uncanalized, the formation of the thickening in each individual depending on the reception of the adequate stimulus, to which the response remained strictly proportional. If this possibility was realized, the well-known difficulty of accounting for the hereditary fixation of the character remains unimpaired. The alternative is that the development does become canalized, to a greater or lesser extent. In that case, the magnitude of the response would not be proportional to that of the stimulus; there would be a threshold of stimulus, above which the optimum (that is, naturally selected) response would be formed. In so far as the response became canalized, the environment would be acting as a switch.

Systems of either type can be built up by natural selection, and one can point to examples of them in animals at the present day. The reaction of the patterns on Lepidopteran wings (for example, in Ephestia, Kühn 1936) to temperature during the sensitive period scarcely seems to involve thresholds, while the metamorphosis of the axolotl, for example, clearly does. In general, it seems likely that the optimum response to the environment will involve both some degree of proportionality and some restriction of this by canalization. The most favourable mixture of the two tendencies will presumably differ for different characters. It is easy to see why a much sharper distinction between alternatives is generally evolved in connexion with sex differences than with the degree of muscular development, for example; but even the former is to some extent modifiable by extreme and specialized environmental disturbances (heavy and early hormone treatment), and even the latter has some degree of genetic determination.

The canalization of an environmentally induced character is accounted for if it is an advantage for the adult animal to have some optimum degree of development of the character irrespective of the exact extent of stimulus which it has met in its early life; if, for example, it is an advantage to the young ostrich going out into the hard world to have adequate callosities even if it were reared in a particularly soft and cosy nest. Now in so far as the development of the character becomes canalized, the action of the external stimulus is reduced to that of a switch mechanism, simply in order that the optimum response shall be regularly produced. But switch mechanisms may notoriously be set off by any of a number of factors. The choice between the alternative developmental pathways open to gastrula ectoderm, for example, may be made by the normal evocator or by a number of other things (the mode of action of which may be through the release of the normal evocator (cf. Waddington 1940b), but which remain different to the normal evocator nevertheless). Again, we know many instances in which several different genes, by switching development into the same path, produce similar effects; and attention has already been directed to the 'phenocopying' of a gene by a suitable environmental stimulus. Thus once a developmental response to an environmental stimulus has become canalized, it should not be too difficult to switch development into that track by mechanisms other than the original external stimulus, for example, by the internal mechanism of a genetic factor; and, as the canalization will only have been built up by natural selection if there is an advantage in the regular production of the optimum response, there will be a selective value in such a supersession of the environment by the even more regularly acting gene. Such a gene must always act before the normal time at which the environmental stimulus was applied, otherwise its work would already be done for it, and it could have no appreciable selective advantage.

Summarizing, then, we may say that the occurrence of an adaptive response to an environmental stimulus depends on the selection of a suitable genetically controlled reactivity in the organism. If it is an advantage, as it usually seems to be for developmental mechanisms, that the response should attain an optimum value more or less independently of the intensity of stimulus received by a particular animal, then the reactivity will become canalized, again under the influence of natural selection. Once the developmental path has been canalized, it is to be expected that many different agents, including a number of mutations available in the germplasm of the species, will be able to switch development into it; and the same considerations which render the canalization advantageous will favour the supersession of the environmental stimulus by a genetic one. By such a series of steps, then, it is possible that an adaptive response can be fixed without waiting for the occurrence of a mutation which, in the original genetic background, mimics the response well enough to enjoy a selective advantage.

REFERENCES

C., Robson G., and O. W. Richards. 1936. *The Variation of Animals in Nature.* London, UK: Longmans, Green / Co.

Darwin, C. 1901. *The Descent of Man, and Selection in Relation to Sex.* London, UK: John Murray.

Detlefsen, J. A. 1925. "The Inheritance of Acquired Characters." *Physiological Reviews* 5 (2): 244–278.

Duerden, J. E. 1920. "The Inheritance of the Callosities in the Ostrich." *The American Naturalist* 54:289.

Fisher, R. A. 1928. "The Possible Modification of the Response of the Wild Type to Recurrent Mutations." *The American Naturalist* 62 (679): 115–126.

Ford, E. B. 1940. "Genetic Research in the Lepidoptera." *Annals of Human Genetics* 10 (1): 227–252.

Huxley, J. H. 1942. *Evolution: The Modern Synthesis.* 74. London, UK: Allen & Unwin.

K., Mather, and D. de Winton. 1941. "Adaptation and Counter-Adaptation of the Breeding System in Primula." *Annals of Botany* 5 (2): 297–311.

Kühn, A. 1936. "Versuche über die Wirkungsweise der Erbanlagen." *Naturwissenschaften* 24:1–10.

Muller, H. J. 1932. "Further Studies on the Nature and Causes of Gene Mutations." In *Proceedings of the Sixth International Congress of Genetics,* edited by D. F. Jones, 1:213–255. Menasha, WI: Brooklyn Botanic Garden.

Plunkett, C. C. 1932. "Temperature as a Tool in Research in Phenogenetics." In *Proceedings of the Sixth International Congress of Genetics,* edited by D. F. Jones, 2:158–160. Menasha, WI: Brooklyn Botanic Garden.

Stern, C. 1929. "Über die additive Wirkung multipler Allele." *Biologisches Zentralblatt* 49:261–290.

Waddington, C. H. 1940a. "Genes as Evocators in Development." *Growth (Supplement),* 37–44.

———. 1940b. *Organisers and Genes.* Cambridge, UK: Cambridge University Press.

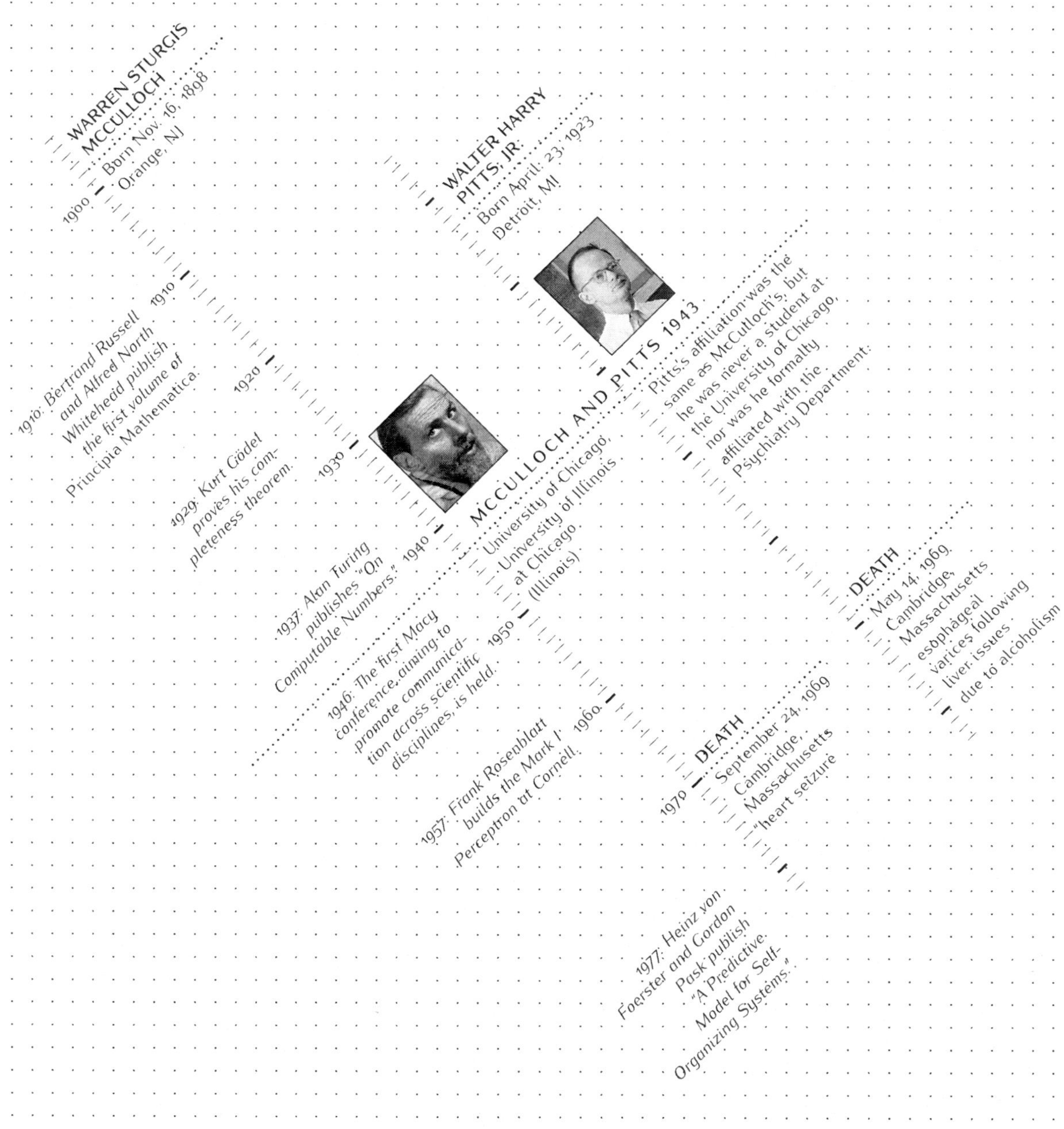

MCCULLOCH & PITTS

[5]

THE GENESIS OF MODERN COMPUTING AND AI

Bruno Olshausen, University of California, Berkeley; Redwood Center for Theoretical Neuroscience; and Santa Fe Institute, and Christopher Hillar, Redwood Center for Theoretical Neuroscience

W.S. McCulloch and W. Pitts, "A Logical Calculus of the Ideas Immanent in Nervous Activity," *Bulletin of Mathematical Biophysics* 5, 115–133 (1943).

Digital computers and neural networks would seem to be profoundly different systems for computation: one represents information with strings of 1s and 0s, the other as a set of analog values. One manipulates these representations through a cascade of logical operations coordinated by a central clock and processing unit, while the other transforms representations via matrix-vector multiplications and thresholding operations that run in loosely coordinated, parallel streams. The first served as the workhorse of the computing industry for more than half a century, while the second has only recently seen widespread deployment as the new computing paradigm powering artificial intelligence. Given such stark differences—both in their computational architectures and the technologies they have enabled—it may come as a surprise to learn that both architectures share a common origin in this landmark paper.

Published in 1943 amidst a world in upheaval, Warren McCulloch and Walter Pitts's paper presented a highly original and thought-provoking hypothesis: that human mental abilities, especially our capacity for logical thought, stem directly from neuronal circuits in the brain *that themselves perform logical operations.* While others such as Turing had previously speculated about mental processes as a kind of computation, no one had attempted to develop a computational theory of mind rooted in the biophysical substrates of the brain—that is, neurons. Their paper thus marks an important turning point in our intellectual approach to studying and understanding the brain. It is the product of an interdisciplinary collaboration born from a coming together of minds at the University of Chicago, where a revolutionary new movement in "mathematical biology" was taking place under the guidance of Nicolas Rashevsky. The premise of their approach was that mathematical models could be used to gain new insight into biological processes in much the same way that they had enabled powerful theoretical advances in physics (Abraham 2002).

It was in Rashevsky's group that McCulloch and Pitts first met. Though McCulloch was trained as a neurologist, he was driven by the desire to understand the connection between neural processes in the brain and human thought. He had experimented with Boolean algebra, but lacked the formal background to fully develop a mathematical theory. Enter Walter Pitts, a self-educated wunderkind who had fully mastered Whitehead and Russell's *Principia Mathematica* at the age of twelve and had begun working as a member of Rashevky's group in his late teens, not having obtained any formal degree. Pitts had been working on a theory of neuron networks, but had not yet realized its potential as a computational theory of the brain. Soon after meeting McCulloch around the end of 1941, the two began working closely together and became good friends, remaining so until their deaths in 1969 (Piccinini 2004).

Among the most enduring contributions of McCulloch and Pitts's paper are their hand-drawn diagrams of neural circuits hypothesized to perform the fundamental operations that would constitute a system of logical calculus. Today we readily recognize these diagrams as the logic circuits—for example, and-gates, or-gates, and not-gates—that lie at the core of every digital computer. However such circuits were not at all obvious at the time, and in fact had not yet been conceived. Interestingly, these inventions came about by contemplating how the illusory sensation of heat, when preceded by touching cold, could be explained from neuronal mechanisms. John von Neumann subsequently took inspiration from these drawings as he was pondering the design of the EDVAC computer, and his final report (von Neumann 1945)—considered highly influential to the first generation of computer engineers (Godfrey 1993)—contained just two citations, both to McCulloch and Pitts (1943).

McCulloch and Pitts considered two types of networks, those without loops (feedforward networks) and those containing loops (recurrent networks). The first type are logical circuits supported on a directed acyclic graph, and their achievement is a procedure to compute any boolean function using them. As for nets with circles, they would be later clarified by Stephen Kleene into what is now known as the class of finite automata. More generally, by adding scanners, tape, and motors, McCulloch and Pitts claimed that it was easily shown that their nets were computationally equivalent to Turing machines. Their logical conclusion was that an organism's output was Turing computable if its brain circuitry consisted of their proposed networks. While Alonzo Church arrived at the same result on the level of behavior, McCulloch and Pitts argued this from the vantage point of brain computation, which was remarkably novel for the time.

For those of us working in computational neuroscience today, McCulloch and Pitts's paper is widely recognized as the first serious attempt to model computational processes in the brain. Nearly two decades later, Frank Rosenblatt took their work a step further by incorporating the idea of *learning*, an important aspect they had largely ignored. In particular, he showed how the synaptic weights in a neural network could be automatically adjusted from fed-back error signals in order to perform specific tasks (Rosenblatt 1958, 1962). For this reason the Perceptron model has largely subsumed the McCulloch–Pitts model and today forms the core of the deep-learning algorithms powering artificial intelligence.

It is remarkable that one paper could have such far-reaching, transformational impact. It is a testament to the power of the synthetic approach in biology: just as the Wright brothers' pondering of bird flight forty years earlier led them not only to hypothesize, but to build and test their designs in a flying machine (McCullough 2015), McCulloch and Pitts's pondering of the internal workings of the brain led them to create a system of computation based on logic circuits that engineers could build and test to great effect. In other words, the very act of *thinking deeply about biology's solutions* focuses the mind on a different set of problems and a way of thinking that transcends the current technological paradigm and stimulates the development of new ideas and innovation.

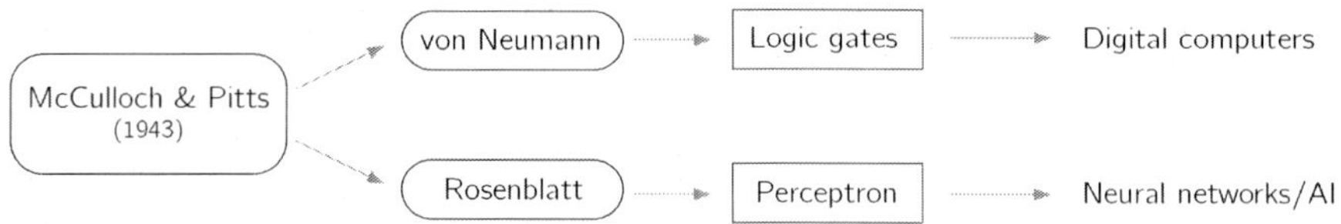

Figure 1. The dual legacies of McCulloch and Pitts.

A final note about Walter Pitts: He was widely recognized as a genius not only by McCulloch and von Neumann but also Norbert Wiener and other great minds of the time. Shortly after completing this landmark paper, Pitts went on to write another highly original and creative work, "How We Recognize Universals," which sought to understand the neural mechanisms by which transformations on patterns such as shift or scaling could be implemented by neural circuits in the brain (Pitts and McCulloch 1947). His solution—a kind of dynamic routing circuit for remapping from one reference frame to another—presaged and inspired similar proposals made many decades later by other investigators (Hinton 1981; Olshausen, Anderson, and Van Essen 1993; Arathorn 2002). For his PhD thesis at MIT, under the guidance and encouragement of Wiener, Pitts was working on probabilistic three-dimensional neural networks. Tragically, this brilliant trajectory was brought to an end after a falling out between McCulloch and Wiener. Pitts fell into depression, burned his thesis, and ultimately died from alcoholism (Gefter 2015).

REFERENCES

Abraham, T. H. 2002. "(Physio)Logical Circuits: The Intellectual Origins of the McCulloch–Pitts Neural Network." *Journal of the History of the Behavioral Sciences* 38 (1): 3–25. https://doi.org/10.1002/jhbs.1094.

Arathorn, D. W. 2002. *Map-Seeking Circuits in Visual Cognition: A Computational Mechanism for Biological and Machine Vision.* Redwood City, CA: Stanford University Press.

Gefter, A. 2015. "The Man Who Tried to Redeem the World with Logic." January 29, 2015, *Nautilus*, https://nautil.us/the-man-who-tried-to-redeem-the-world-with-logic-235253/.

Godfrey, M. D. 1993. "Introduction to "First Draft of a Report on the EDVAC" by John von Neumann." *IEEE Annals of the History of Computing* 15 (4): 27–75. https://doi.org/10.1109/85.238389.

Hinton, G. F. 1981. "A Parallel Computation that Assigns Canonical Object-Based Frames of Reference." In *IJCAI '81: Proceedings of the 7th International Joint Conference on Artificial Intelligence, Vancouver, BC, Canada,* 683–685. San Francisco, CA: Morgan Kaufmann Publishers Inc.

McCullough, D. 2015. *The Wright Brothers.* New York, NY: Simon and Schuster.

Olshausen, B. A., C. H. Anderson, and D. C. Van Essen. 1993. "A Neurobiological Model of Visual Attention and Invariant Pattern Recognition Based on Dynamic Routing of Information." *Journal of Neuroscience* 13 (11): 4700–4719. https://doi.org/10.1523/JNEUROSCI.13-11-04700.1993.

Piccinini, G. 2004. "The First Computational Theory of Mind and Brain: A Close Look at McCulloch and Pitts's 'Logical Calculus of Ideas Immanent in Nervous Activity'." *Synthese* 141:175–215. https://doi.org/10.1023/B:SYNT.0000043018.52445.3e.

Pitts, W., and W. S. McCulloch. 1947. "How We Know Universals; The Perception of Auditory and Visual Forms." *The Bulletin of Mathematical Biophysics* 9:127–147. https://doi.org/https://doi.org/10.1007/BF02478291.

Rosenblatt, F. 1958. "The Perceptron: A Probabilistic Model for Information Storage and Organization in the Brain." *Psychological Review* 65 (6): 386–408. https://doi.org/10.1037/h0042519.

———. 1962. *Principles of Neurodynamics: Perceptrons and the Theory of Brain Mechanisms.* Vol. 55. Washington, DC: Spartan Books.

von Neumann, J. 1945. *First Draft of a Report on the EDVAC.* Technical report. Contract no. W-670-ORD-4926 between the United States Army Ordnance Department and the University of Pennsylvania. Moore School of Electrical Engineering, University of Pennsylvania.

A LOGICAL CALCULUS OF THE IDEAS IMMANENT IN NERVOUS ACTIVITY

Warren S. McCulloch, University of Illinois
and Walter Pitts, University of Chicago

Abstract

Because of the "all-or-none" character of nervous activity, neural events and the relations among them can be treated by means of propositional logic. It is found that the behavior of every net can be described in these terms, with the addition of more complicated logical means for nets containing circles; and that for any logical expression satisfying certain conditions, one can find a net behaving in the fashion it describes. It is shown that many particular choices among possible neurophysiological assumptions are equivalent, in the sense that for every net behaving under one assumption, there exists another net which behaves under the other and gives the same results, although perhaps not in the same time. Various applications of the calculus are discussed.

I. Introduction

Theoretical neurophysiology rests on certain cardinal assumptions. The nervous system is a net of neurons, each having a soma and an axon. Their adjunctions, or synapses, are always between the axon of one neuron and the soma of another. At any instant a neuron has some threshold, which excitation must exceed to initiate an impulse. This, except for the fact and the time of its occurrence, is determined by the neuron, not by the excitation. From the point of excitation the impulse is propagated to all parts of the neuron. The velocity along the axon varies directly with its diameter, from less than one meter per second in thin axons, which are usually short, to more than 150 meters per second in thick axons, which are usually long. The time for axonal conduction is consequently of little importance in determining the time of arrival of impulses at points unequally remote from the same source. Excitation across synapses occurs predominantly from axonal terminations to somata. It is still a moot point whether this depends

Though many of their assumptions still hold, it is certainly not the case that axonal conduction time can be ignored.

upon irreciprocity of individual synapses or merely upon prevalent anatomical configurations. To suppose the latter requires no hypothesis *ad hoc* and explains known exceptions, but any assumption as to cause is compatible with the calculus to come. No case is known in which excitation through a single synapse has elicited a nervous impulse in any neuron, whereas any neuron may be excited by impulses arriving at a sufficient number of neighboring synapses within the period of latent addition, which lasts less than one quarter of a millisecond. Observed temporal summation of impulses at greater intervals is impossible for single neurons and empirically depends upon structural properties of the net. Between the arrival of impulses upon a neuron and its own propagated impulse there is a synaptic delay of more than half a millisecond. During the first part of the nervous impulse the neuron is absolutely refractory to any stimulation. Thereafter its excitability returns rapidly, in some cases reaching a value above normal from which it sinks again to a subnormal value, whence it returns slowly to normal. Frequent activity augments this subnormality. Such specificity as is possessed by nervous impulses depends solely upon their time and place and not on any other specificity of nervous energies. Of late only inhibition has been seriously adduced to contravene this thesis. Inhibition is the termination or prevention of the activity of one group of neurons by concurrent or antecedent activity of a second group. Until recently this could be explained on the supposition that previous activity of neurons of the second group might so raise the thresholds of internuncial neurons that they could no longer be excited by neurons of the first group, whereas the impulses of the first group must sum with the impulses of these internuncials to excite the now inhibited neurons. Today, some inhibitions have been shown to consume less than one millisecond. This excludes internuncials and requires synapses through which impulses inhibit that neuron which is being stimulated by impulses through other synapses. As yet experiment has not shown whether the refractoriness is relative or absolute. We will assume the latter and demonstrate that the difference is immaterial to our argument. Either variety of refractoriness can be accounted for in either of two ways. The "inhibitory synapse" may be of such a kind as to produce a substance which raises the threshold of the neuron, or it may be so placed that the local disturbance produced by its excitation opposes the alteration induced by the otherwise excitatory synapses. Inasmuch as position is already known to have such effects in the case of electrical stimulation, the first hypothesis is to be excluded unless and until it be substantiated, for the second involves no new hypothesis. We have, then, two explanations of inhibition based on the same general premises, differing only in the assumed nervous nets and, consequently, in the time required for inhibition. Hereafter we shall refer to such nervous nets as *equivalent in the extended sense*. Since we are concerned with properties of nets which are invariant under equivalence, we may make the physical assumptions which are most convenient for the calculus.

A key assumption of their theory was that neurons were equivalent to propositions, and thus a level of functional equivalence could be established among neural circuits even with quite different biophysical mechanisms.

Many years ago one of us, by considerations impertinent to this argument, was led to conceive of the response of any neuron as factually equivalent to a proposition which proposed its adequate stimulus. He therefore attempted to record the behavior of complicated nets in the notation of the symbolic logic of propositions. The "all-or-none" law of nervous activity is sufficient to insure that the activity of any neuron may be represented as a proposition. Physiological relations existing among nervous activities correspond, of course, to relations among the propositions; and the utility of the representation depends upon the identity of these relations with those of the logic of propositions. To each reaction of any neuron there is a corresponding assertion of a simple proposition. This, in turn, implies either some other simple proposition or the disjunction or the conjunction, with or without negation, of similar propositions, according to the configuration of the synapses upon and the threshold of the neuron in question. Two difficulties appeared. The first concerns facilitation and extinction, in which antecedent activity temporarily alters responsiveness to subsequent stimulation of one and the same part of the net. The second concerns learning, in which activities concurrent at some previous time have altered the net permanently, so that a stimulus which would previously have been inadequate is now adequate. But for nets undergoing both alterations, we can substitute equivalent fictitious nets composed of neurons whose connections and thresholds are unaltered. But one point must be made clear: neither of us conceives the formal equivalence to be a factual explanation. *Per contra!*—we regard facilitation and extinction as dependent upon continuous changes in threshold related to electrical and chemical variables, such as after-potentials and ionic concentrations; and learning as an enduring change which can survive sleep, anaesthesia, convulsions and coma. The importance of the formal equivalence lies in this: that the alterations actually underlying facilitation, extinction and learning in no way affect the conclusions which follow from the formal treatment of the activity of nervous nets, and the relations of the corresponding propositions remain those of the logic of propositions.

The nervous system contains many circular paths, whose activity so regenerates the excitation of any participant neuron that reference to time past becomes indefinite, although it still implies that afferent activity has realized one of a certain class of configurations over time. Precise specification of these implications by means of recursive functions, and determination of those that can be embodied in the activity of nervous nets, completes the theory.

II. The Theory: Nets Without Circles

We shall make the following physical assumptions for our calculus.

1. The activity of the neuron is an "all-or-none" process.

2. A certain fixed number of synapses must be excited within the period of latent addition in order to excite a neuron at any time, and this number is independent of previous activity and position on the neuron.

3. The only significant delay within the nervous system is synaptic delay.

4. The activity of any inhibitory synapse absolutely prevents excitation of the neuron at that time.

5. The structure of the net does not change with time.

Again, these assumptions, though perhaps reasonable in 1943, are not consistent with our modern understanding of the brain.

To present the theory, the most appropriate symbolism is that of Language II of R. Carnap (1938), augmented with various notations drawn from B. Russell and A.N. Whitehead (1925), including the *Principia* conventions for dots. Typographical necessity, however, will compel us to use the upright 'E' for the existential operator instead of the inverted, and an arrow ("$\rightarrow$") for implication instead of the horseshoe. We shall also use the Carnap syntactical notations, but print them in boldface rather than German type; and we shall introduce a functor S, whose value for a property P is the property which holds of a number when P holds of its predecessor; it is defined by '$S(P)(t) \,.\equiv .\, P(Kx) \,.\, t = x'$'; the brackets around its argument will often be omitted, in which case this is understood to be the nearest predicate-expression [$\boldsymbol{Pr}$] on the right. Moreover, we shall write $S^2\boldsymbol{Pr}$ for $S(S(\boldsymbol{Pr}))$, etc.

The neurons of a given net $\mathcal{N}$ may be assigned designations 'c_1', 'c_2', . . . , 'c_n'. This done, we shall denote the property of a number, that a neuron c_i fires at a time which is that number of synaptic delays from the origin of time, by 'N' with the numeral i as subscript, so that $N_i(t)$ asserts that c_i fires at the time t. N_i is called the *action* of c_i. We shall sometimes regard the subscripted numeral of 'N' as if it belonged to the object-language, and were in a place for a functoral argument, so that it might be replaced by a number-variable [z] and quantified; this enables us to abbreviate long but finite disjunctions and conjunctions by the use of an operator. We shall employ this locution quite generally for sequences of $\boldsymbol{Pr}$; it may be secured formally by an obvious disjunctive definition. The predicates 'N_t', 'N_2', . . . , comprise the syntactical class '$\boldsymbol{N}$'.

Let us define the *peripheral afferents of* $\mathcal{N}$ as the neurons of $\mathcal{N}$ with no axons synapsing upon them. Let $\boldsymbol{N}_1, \ldots, \boldsymbol{N}_p$, denote the actions of such neurons and $\boldsymbol{N}_{p+1}, \boldsymbol{N}_{p+2}, \ldots, \boldsymbol{N}_n$, those of the rest. Then a *solution of* $\mathcal{N}$ will be a class of sentences of the form $\boldsymbol{S}_i : \boldsymbol{N}_{p+1}(\boldsymbol{z}_1) \,.\equiv .\, \boldsymbol{Pr}_i(\boldsymbol{N}_1, \boldsymbol{N}_2, \ldots, \boldsymbol{N}_p, \boldsymbol{z}_1)$, where $\boldsymbol{Pr}_i$ contains no free variable save $\boldsymbol{z}_1$ and no descriptive symbols save the $\boldsymbol{N}$ in the argument [$\boldsymbol{Arg}$], and possibly some constant sentences [$\boldsymbol{sa}$]; and such that each $\boldsymbol{S}_i$ is true of $\mathcal{N}$. Conversely, given a $\boldsymbol{Pr}_1({}^1p_1^1, {}^1p_2^1, \ldots, {}^1p_p^1, \boldsymbol{z}_1, \boldsymbol{s})$, containing no free variable save those in its $\boldsymbol{Arg}$, we shall say that it is *realizable in the narrow sense* if there exists a net $\mathcal{N}$ and a series of $\boldsymbol{N}_i$ in it such that $\boldsymbol{N}_1(\boldsymbol{z}_1) \,.\equiv .\, \boldsymbol{PR}_1(\boldsymbol{N}_1, \boldsymbol{N}_2, \ldots, \boldsymbol{z}_1, \boldsymbol{sa}_1)$ is true of it, where $\boldsymbol{sa}_1$ has the form

$N(0)$. We shall call it *realizable in the extended sense*, or simply *realizable*, if for some $nS^n(\boldsymbol{Pr}_1)(\boldsymbol{p}_1, \ldots, \boldsymbol{p}_p, z_1, s)$ is realizable in the above sense. c_{pi} is here the realizing neuron. We shall say of two laws of nervous excitation which are such that every $\boldsymbol{S}$ which is realizable in either sense upon one supposition is also realizable, perhaps by a different net, upon the other, that they are equivalent assumptions, in that sense.

The following theorems about realizability all refer to the extended sense. In some cases, sharper theorems about narrow realizability can be obtained; but in addition to greater complication in statement this were of little practical value, since our present neurophysiological knowledge determines the law of excitation only to extended equivalence, and the more precise theorems differ according to which possible assumption we make. Our less precise theorems, however, are invariant under equivalence, and are still sufficient for all purposes in which the exact time for impulses to pass through the whole net is not crucial.

Our central problems may now be stated exactly: first, to find an effective method of obtaining a set of computable $\boldsymbol{S}$ constituting a solution of any given net; and second, to characterize the class of realizable $\boldsymbol{S}$ in an effective fashion. Materially stated, the problems are to calculate the behavior of any net, and to find a net which will behave in a specified way, when such a net exists.

A net will be called *cyclic* if it contains a circle, i.e. if there exists a chain $c_i, c_{i+1}, \ldots$ of neurons on it, each member of the chain synapsing upon the next, with the same beginning and end. If a set of its neurons $c_i, c_2, \ldots, c_p$ is such that its removal from $\mathcal{N}$ leaves it without circles, and no smaller class of neurons has this property, the set is called a *cyclic* set, and its cardinality is the *order* of $\mathcal{N}$. In an important sense, as we shall see, the order of a net is an index of the complexity of its behaviour. In particular, nets of zero order have especially simple properties; we shall discuss them first.

Let us define a *temporal propositional expression* (a *TPE*), designating a *temporal propositional function* (*TPF*), by the following recursion.

1. A ${}^1\boldsymbol{p}^1[z_1]$ is a *TPE*, where $\boldsymbol{p}_1$ is a predicate-variable.

2. If $\boldsymbol{S}_1$ and $\boldsymbol{S}_2$ are *TPE* containing the same free individual variable, so are SS_1, $\boldsymbol{S}_1 \mathrm{v} \boldsymbol{S}_2$, $\boldsymbol{S}_1 . \boldsymbol{S}_2$, and $\boldsymbol{S}_i . \sim . \boldsymbol{S}_2$.

3. Nothing else is a *TPE*.

Theorem I. *Every net of order* 0 *can be solved in terms of temporal propositional expressions.*

Let c_i be any neuron of $\mathcal{N}$ with a threshold $\theta_i > 0$, and let $c_{i1}, c_{i2}, \ldots, c_{ip}$ have respectively $n_{i1}, n_{i2}, \ldots, n_{ip}$ excitatory synapses upon it. Let $c_{j1}, c_{j2}, \ldots, c_{jq}$ have inhibitory synapses upon it. Let κ_i be the set of the subclasses of $\{n_{i1}, n_{i2}, \ldots, n_{ip}\}$ such that the sum of their members exceeds θ_i. We shall then be able to write, in accordance with the assumptions mentioned above,

$$N_i(z_1). \equiv . S\left\{\prod_{m=1}^{q} \sim N_{jm}(z_1). \sum_{\alpha\varepsilon\kappa_i} \prod_{s\varepsilon\alpha} N_{is}(z_1)\right\} \qquad (1)$$

where the '$\sum$' and '$\prod$' are syntactical symbols for disjunctions and conjunctions which are finite in each case. Since an expression of this form can be written for each c_i which is not a peripheral afferent, we can, by substituting the corresponding expression in (1) for each N_{jm} or N_{is} whose neuron is not a peripheral afferent, and repeating the process on the result, ultimately come to an expression for N_i in terms solely of peripherally afferent N, since $\mathcal{N}$ is without circles. Moreover, this expression will be a *TPE*, since obviously (1) is; and it follows immediately from the definition that the result of substituting a *TPE* for a constituent $p(z)$ in a *TPE* is also one.

Theorem II. *Every* TPE *is realizable by a net of order zero.*

The functor S obviously commutes with disjunction, conjunction, and negation. It is obvious that the result of substituting any S_i, realizable in the narrow sense (i.n.s.), for the $p(z)$ in a realizable expression S_1 is itself realizable i.n.s.; one constructs the realizing net by replacing the peripheral afferents in the net for S_1 by the realizing neurons in the nets for the S_i. The one neuron net realizes $p_1(z_1)$ i.n.s., and figure 1-a shows a net that realizes $Sp_1(z_1)$ and hence SS_2, i.n.s., if S_2 can be realized i.n.s. Now if S_2 and S_3 are realizable then $S^m S_2$ and $S^n S_3$ are realizable i.n.s., for suitable m and n. Hence so are $S^{m+n} S_2$ and $S^{m+n} S_3$. Now the nets of figures 1b, c and d respectively realize $S(p_1(z_1) \vee p_2(z_1))$, $S(p_1(z_1) . p_2(z_1))$, and $S(p_1(z_1) . \sim p_2(z_1))$ i.n.s. Hence $S^{m+n+1}(S_1 \vee S_2)$, $S^{m+n+1}(S_1 . S_2)$, and $S^{m+n+1}(S_1 . \sim S_2)$ are realizable i.n.s. Therefore $S_1 \mathbf{v} S_2 S_1 . S_2 S_1 . \sim S_2$ are realizable if S_1 and S_2 are. By complete induction, all *TPE* are realizable. In this way all nets may be regarded as built out of the fundamental elements of figures 1a, b, c, d, precisely as the temporal propositional expressions are generated out of the operations of precession, disjunction, conjunction, and conjoined negation. In particular, corresponding to any description of state, or distribution of the values *true* and *false* for the actions of all the neurons of a net save that which makes them all false, a single neuron is constructible whose firing is a necessary and sufficient condition for the validity of that description. Moreover, there is always an indefinite number of topologically different nets realizing any *TPE*.

Theorem III. *Let there be given a complex sentence* S_1 *built up in any manner out of elementary sentences of the form* $\mathbf{p}(\mathbf{z}_1 - \mathbf{zz})$ *where* **zz** *is any numeral, by any of the propositional connections: negation, disjunction, conjunction, implication, and equivalence. Then* S_1 *is a* TPE *and only if it is false when its constituent* $\mathbf{p}(\mathbf{z}_1 - \mathbf{zz})$ *are all assumed false—i.e., replaced by false sentences—or that the last line in its truth-table contains an* '**F**',*—or there is no term in its Hilbert disjunctive normal form composed exclusively of negated terms.*

These latter three conditions are of course equivalent (Hilbert and Ackermann 1927). We see by induction that the first of them is necessary, since $p(z_1 - zz)$ becomes false when it is replaced by a false sentence, and $S_1 \mathbf{v} S_2, S_1 . S_2$ and $S_1 . \sim S_2$ are all false if both their constituents are. We see that the last condition is sufficient by remarking that a disjunction is a *TPE* when its constituents are, and that any term

$$S_1 . S_2 \ldots . S_m . \sim S_{m+1} . \sim \ldots . \sim S_n$$

can be written as:

$$(S_1 . S_2 \ldots . S_m) . \sim (S_{m+1} \mathbf{v} S_{m+2} \mathbf{v} \ldots . \mathbf{v} S_n),$$

which is clearly a *TPE*.

The method of the last theorems does in fact provide a very convenient and workable procedure for constructing nervous nets to order, for those cases where there is no reference to events indefinitely far in the past in the specification of the conditions. By way of example, we may consider the case of heat produced by a transient cooling.

McCulloch and Pitts attempt to explain the hot-cold illusion with logic gates.

If a cold object is held to the skin for a moment and removed, a sensation of heat will be felt; if it is applied for a longer time, the sensation will be only of cold, with no preliminary warmth, however transient. It is known that one cutaneous receptor is affected by heat, and another by cold. If we let N_1 and N_2 be the actions of the respective receptors and N_3 and N_4 of neurons whose activity implies a sensation of heat and cold, our requirements may be written as

$$N_3(t) :\equiv: N_1(t-1) \,.\, \mathbf{v} \,.\, N_2(t-3) \,.\sim N_2(t-2)$$
$$N_4(t) \,.\equiv.\, N_2(t-2) \,.\, N_2(t-1)$$

where we suppose for simplicity that the required persistence in the sensation of cold is say two synaptic delays, compared with one for that of heat. These conditions clearly fall under Theorem III. A net may consequently be constructed to realize them, by the method of Theorem II. We begin by writing them in a fashion which exhibits them as built out of their constituents by the operations realized in Figures 1a, b, c, d: i.e., in the form

$$N_3(t) \,.\equiv.\, S\{N_1(t) \mathbf{v}\, S[(SN_2(t)) \,.\sim N_2(t)]\}$$
$$N_4(t) \,.\equiv.\, S\{[SN_2(t)] \,.\, N_2(t)\} \,.$$

First we construct a net for the function enclosed in the greatest number of brackets and proceed outward; in this case we run a net of the form shown in Figure 1a from c_2 to some neuron c_a, say, so that

$$N_a(t) \,.\equiv.\, SN_2(t).$$

Next introduce two nets of the forms 1c and 1d, both running from c_a and c_2, and ending respectively at c_4 and say c_b. Then

$$N_4(t) \,.\equiv.\, S[N_a(t) \,.\, N_2(t)] \,.\equiv.\, S[(SN_2(t)) \,.\, N_2(t)].$$
$$N_b(t) \,.\equiv.\, S[N_a(t) \,.\sim N_2(t)] \,.\equiv.\, S[(SN_2(t)) \,.\sim N_2(t)].$$

Finally, run a net of the form 1b from c_1 and c_b to c_3, and derive

$$N_3(t) \,.\equiv.\, S[N_1(t)\, \mathbf{v}\, N_b(t)]$$
$$.\equiv.\, S\{N_1(t) \mathbf{v}\, S[(SN_2(t)]. \sim N_2(t)\}.$$

These expressions for $N_3(t)$ and $N_4(t)$ are the ones desired; and the realizing net *in toto* is shown in Figure 1e.

This illusion makes very clear the dependence of the correspondence between perception and the "external world" upon the specific structural properties of the intervening nervous net. The same illusion, of course, could also have been produced under various other assumptions about the behavior of the cutaneous receptors, with correspondingly different nets.

We shall now consider some theorems of equivalence: i.e., theorems which demonstrate the essential identity, save for time, of various alternative laws of nervous excitation. Let us first discuss the case of *relative inhibition*. By this we mean the supposition that the firing of an inhibitory synapse does not absolutely prevent the firing of the neuron, but merely raises its threshold, so that a greater number of excitatory synapses must fire concurrently to fire it than would otherwise be needed. We may suppose, losing no generality, that the increase in threshold is unity for the firing of each such synapse; we then have the theorem:

Theorem IV. *Relative and absolute inhibition are equivalent in the extended sense.*

We may write out a law of nervous excitation after the fashion of (1), but employing the assumption of relative inhibition instead; inspection then shows that this expression is a *TPE*. An example of the replacement of relative inhibition by absolute is given by Figure 1f. The reverse replacement is even easier; we give the inhibitory axons afferent to c_i any sufficiently large number of inhibitory synapses apiece.

Second, we consider the case of extinction. We may write this in the form of a variation in the threshold θ_i; after the neuron c_i has fired; to the nearest integer—and only to this approximation is the variation in threshold significant in natural forms of excitation—this may be written as a sequence $\theta_i + b_j$ for j synaptic delays after firing, where $b_j = 0$ for j large enough, say $j = M$ or greater. We may then state

Theorem V. *Extinction is equivalent to absolute inhibition.*

For, assuming relative inhibition to hold for the moment, we need merely run M circuits $\mathscr{T}_1, \mathscr{T}_2, \ldots \mathscr{T}_{\mathrm{M}}$ containing respectively $1, 2, \ldots, M$ neurons, such that the firing of each link in any is sufficient to fire the next, from the neuron c_i back to it, where the end of the circuit $\mathscr{T}_j$ has just b_j inhibitory synapses upon c_i. It is evident that this will produce the desired results. The reverse substitution may be accomplished by the diagram of

Figure 1g. From the transitivity of replacement, we infer the theorem. To this group of theorems also belongs the well-known

Theorem VI. *Facilitation and temporal summation may be replaced by spatial summation.*

This is obvious: one need merely introduce a suitable sequence of delaying chains, of increasing numbers of synapses, between the exciting cell and the neuron whereon temporal summation is desired to hold. The assumption of spatial summation will then give the required results. See e.g. Figure 1h. This procedure had application in showing that the observed temporal summation in gross nets does not imply such a mechanism in the interaction of individual neurons.

The phenomena of learning, which are of a character persisting over most physiological changes in nervous activity, seem to require the possibility of permanent alterations in the structure of nets. The simplest such alteration is the formation of new synapses or equivalent local depressions of threshold. We suppose that some axonal terminations cannot at first excite the succeeding neuron; but if at any time the neuron fires, and the axonal terminations are simultaneously excited, they become synapses of the ordinary kind, henceforth capable of exciting the neuron. The loss of an inhibitory synapse gives an entirely equivalent result. We shall then have

Theorem VII. *Alterable synapses can be replaced by circles.*

This is accomplished by the method of Figure 1i. It is also to be remarked that a neuron which becomes and remains spontaneously active can likewise be replaced by a circle, which is set into activity by a peripheral afferent when the activity commences, and inhibited by one when it ceases.

III. The Theory: Nets with Circles

☞ Their attempt to describe networks with recurrence was highly ambitious, but would only later be clarified by Kleene (YEAR).

The treatment of nets which do not satisfy our previous assumption of freedom from circles is very much more difficult than that case. This is largely a consequence of the possibility that activity may be set up in a circuit and continue reverberating around it for an indefinite period of time, so that the realizable ***Pr*** may involve reference to past events of an indefinite degree of remoteness. Consider such a net $\mathcal{N}$, say of order p, and let $c_1, c_2, \ldots, c_p$ be a cyclic set of neurons of $\mathcal{N}$. It is first of all clear from the definition that every N_3 of $\mathcal{N}$ can be expressed as a *TPE*, of $N_1, N_2, \ldots, N_p$ and the absolute afferents; the solution of $\mathcal{N}$ involves then only the determination of expressions for the cyclic set. This done, we shall derive a set of expressions [A]:

$$N_i(z_1). \equiv . Pr_i[S^{n_{i1}}N_1(z_1), S^{n_{i2}}N_2(z_1), \ldots, S^{n_{ip}}N_p(z_1)], \quad (2)$$

where Pr_i also involves peripheral afferents. Now if n is the least common multiple of the n_{ij}, we shall, by substituting their equivalents according to (2) in (3) for the N_j, and repeating this process often enough on the result, obtain S of the form

$$N_i(z_1). \equiv . Pr_1[S^n N_1(z_1), S^n N_2(z_1), \ldots, S^n N_p(z_1)]. \quad (3)$$

These expressions may be written in the Hilbert disjunctive normal form as

$$N_i(z_1). \equiv . \sum_{\substack{\alpha\varepsilon\kappa \\ \beta_\alpha\varepsilon\kappa}} S_\alpha \prod_{j\varepsilon\kappa} S^n N_j(z_1) \prod_{j\varepsilon\beta_\alpha} \sim S^n N_j(z_1), \quad (4)$$

for suitable κ,

where S_α is a *TPE* of the absolute afferents of $\mathcal{N}$. There exist some 2^p different sentences formed out of the pN_i by conjoining to the conjunction of some set of them the conjunction of the negations of the rest. Denumerating these by $X_1(z_1), X_2(z_1), \ldots, X_{2^p}(z_1)$, we may, by use of the expressions (4), arrive at an equipollent set of equations of the form

$$X_i(z_1). \equiv . \sum_{j=1}^{2^p} Pr_{ij}(z_1).S^n X_j(z_1). \quad (5)$$

Now we import the subscripted numerals i, j into the object-language: i.e., define Pr_1 and Pr_2 such that $Pr_1(zz_1, z_1). \equiv .X_i(z_1)$ and $Pr_2(zz_1, zz_2, z_1). \equiv . Pr_{ij}(z_1)$ are provable whenever zz_1 and zz_2 denote i and j respectively. Then we may rewrite (5) as

$$\begin{aligned}&(z_1)zz_p : Pr_1(z_1, z_3)\\ &. \equiv . (Ez_2)zz_p . Pr_2(z_1, z_2, z_3 - zz_n) . Pr_1(z_2, z_3 - zz_n)\end{aligned} \quad (6)$$

where zz_n denotes n and zz_p denotes 2^p. By repeated substitution we arrive at an expression

$$\begin{aligned}&(z_1)zz_p : Pr_1(z_1, zz_n zz_2). \equiv . (Ez_2)zz_p (Ez_3)zz_p \ldots (Ez_n)zz_p.\\ &Pr_2(z_1, z_2, zz_n(zz_2 - 1)) . Pr_2(z_2, z_3, zz_n(zz_2 - 1)) \ldots .\end{aligned} \quad (7)$$

$\boldsymbol{Pr}_2(z_{n-1}, z_n, 0) \,.\, \boldsymbol{Pr}_1(z_n, 0)$, for any numeral zz_2 which denotes s. This is easily shown by induction to be equipollent to

$$(z_1)zz_p : .\, \boldsymbol{Pr}_1(z_1, zz_n zz_2) :\equiv: (\boldsymbol{E}f)(z_2)zz_2 - 1 f(z_2 zz_n) \\ \leqq zz_p \,.\, f(zz_n zz_2) = z_1 \,.\, \boldsymbol{Pr}_2(f(zz_n(z_2+1)), \\ f(zz_n z_2)) \,.\, \boldsymbol{Pr}_1(f(0), 0) \tag{8}$$

and since this is the case for all zz_2, it is also true that

$$(z_4)(z_1)zz_p : \boldsymbol{Pr}_1(z_1, z_4) \,.\equiv.\, (\boldsymbol{E}f)(z_2)(z_4 - 1).f(z_2) \\ \leqq zz_p \,.\, f(z_4) = z_1 f(z_4) = z_1 \,.\, \boldsymbol{Pr}_2\,[f(z_2+1), f(z_2), z_2] \,. \\ \boldsymbol{Pr}_1\,[f(\text{res}(\, z_4, zz_n)),\ \text{res}\,(z_4, zz_n)]\,, \tag{9}$$

where zz_n denotes n, res(r, s) is the residue of r mod s and zz_p denotes 2^p. This may be written in a less exact way as

$$N_i(t) \,.\equiv.\, (E\phi)\,(x)t - 1 \,.\, \phi(x) \leqq 2^p \,.\, \phi(t) = i. \\ P[\phi(x+1), \phi(x) \,.\, N_{\phi(0)}(0)],$$

where x and t are also assumed divisible by n, and $\boldsymbol{Pr}_2$ denotes P. From the preceding remarks we shall have

Theorem VIII. *The expression (9) for neurons of the cyclic set of a net $\mathcal{N}$ together with certain* TPE *expressing the actions of other neurons in terms of them, constitute a solution of $\mathcal{N}$.*

Consider now the question of the realizability of a set of S_i. A first necessary condition, demonstrable by an easy induction, is that

$$(z_2)z_1 \,.\, \boldsymbol{p}_1(z_2) \equiv \boldsymbol{p}_2(z_2) \,.\rightarrow.\, \boldsymbol{S}_i \equiv \boldsymbol{S}_i \begin{Bmatrix} p_1 \\ p_2 \end{Bmatrix} \tag{10}$$

should be true, with similar statements for the other free $\boldsymbol{p}$ in $\boldsymbol{S}_i$: i.e., no nervous net can take account of future peripheral afferents. Any $\boldsymbol{S}_i$ satisfying this requirement can be replaced by an equipollent $\boldsymbol{S}$ of the form

$$(\boldsymbol{E}f)(z_2)z_1(z_3)zz_p \,: f \varepsilon \boldsymbol{Pr}_{mi} \\ : f(z_1, z_2, z_3) = 1 \,.\equiv.\, \boldsymbol{p}_{z_3}(z_2) \tag{11}$$

where zz_p denotes p, by defining

$$\boldsymbol{Pr}_{mi} = \hat{f}[(z_1)(z_2)z_1(z_3)zz_p \,: .\, f(z_1, z_2, z_3) = 0.\text{v}.f(z_1, z_2, z_3) \\ = 1 : f(z_1, z_2, z_3) = 1 \,.\equiv.\, \boldsymbol{p}_{z_3}(z_2) :\rightarrow: \boldsymbol{S}_i].$$

Consider now these series of classes a_i, for which

$$N_i(t) \;:\equiv:\; (E_\phi)(x)t(m)q \;:\; \phi\varepsilon\alpha_i : N_m(x) \,.\equiv .\, \phi(t,x,m) = 1. \qquad [i = q+1, \ldots, M] \tag{12}$$

holds for some net. These will be called *prehensible* classes. Let us define the *Boolean ring* generated by a class of classes κ as the aggregate of the classes which can be formed from members of κ by repeated application of the logical operations; i.e., we put

$$\mathscr{R}(\kappa) = p\text{‘}\hat{\lambda}[(\alpha,\beta) : \alpha\varepsilon\kappa \rightarrow \alpha\varepsilon\lambda : \alpha, \beta\varepsilon\lambda. \rightarrow . - \alpha, \alpha.\beta, \alpha\mathbf{v}\beta\varepsilon\lambda].$$

We shall also define

$$\overline{\mathscr{R}}(\kappa). = .\mathscr{R}(\kappa) - \iota\text{‘}p\text{‘} - \text{“}\kappa,$$

$$\mathscr{R}_e(\kappa) = p\text{‘}\hat{\lambda}[(\alpha,\beta) : \alpha\varepsilon\kappa \rightarrow \alpha\varepsilon\lambda. \rightarrow . - \alpha, \alpha.\beta, \alpha\mathbf{v}\beta, S\text{“}\alpha\varepsilon\lambda]$$

$$\overline{\mathscr{R}}_e(\kappa) = \mathscr{R}_e(\kappa) - \iota\text{‘}p\text{‘} - \text{“}\kappa,$$

and

$$\sigma(\psi, t) = \hat{\phi}[(m).\phi(t+1, t, m) = \psi(m)].$$

The class $\mathscr{R}_e(\kappa)$ is formed from κ in analogy with $\mathscr{R}(\kappa)$, but by repeated application not only of the logical operations but also of that which replaces a class of properties $P \varepsilon \alpha$ by $S(P) \varepsilon S\text{“}\alpha$. We shall then have the

Lemma

$\boldsymbol{Pr}_1(\boldsymbol{p}_1, \boldsymbol{p}_2, \ldots, \boldsymbol{p}_m, \boldsymbol{z}_1)$ is a *TPE* if and only if

$$(\boldsymbol{z}_1)(\boldsymbol{p}_1, \ldots, \boldsymbol{p}_m)(E\boldsymbol{p}_{m+1}) : \boldsymbol{p}_{m+1} \varepsilon \overline{\mathscr{R}}_e(\{\boldsymbol{p}_1, \boldsymbol{p}_2, \ldots, \boldsymbol{p}_m\}) \quad \boldsymbol{p}_{m+1}(\boldsymbol{z}_1) \equiv \boldsymbol{Pr}_1(\boldsymbol{p}_1, \boldsymbol{p}_2, \ldots, \boldsymbol{p}_m, \boldsymbol{z}_1) \tag{13}$$

is true; and it is a *TPE* not involving ‘S’ if and only if this holds when ‘$\overline{\mathscr{R}}_e$’ is replaced by ‘$\overline{\mathscr{R}}$’, and we then obtain

Theorem IX. *A series of classes* $\alpha_1, \alpha_2, \ldots \alpha_s$ *is a series of prehensible classes if and only if*

$$\begin{aligned}&(Em)(En)(p)n(i)(\psi) : .(x)m\psi(x) = 0 \,\mathbf{v}\, \psi(x) = 1 : \rightarrow : (E\beta)\\ &(Ey)m \,.\, \psi(y) = 0 \,.\, \beta \varepsilon \mathscr{R}[\hat{\gamma}((Ei).\gamma = \alpha_i)] \,.\mathbf{v}.\, (x)m.\\ &\psi(x) = 0 \,.\, \beta \varepsilon \overline{\mathscr{R}}[\hat{\gamma}((Ei).\gamma = \alpha_i)] \;:\; (t)(\phi) \;:\; \phi \varepsilon \alpha_i.\\ &\quad \sigma(\phi, nt+p) \,.\rightarrow .\, (Ef).f \varepsilon \beta \,.\, (w)m(x)t-1.\\ &\quad\quad \phi(n(t+1)+p, nx+p, w) = f(nt+p, nx+p, w).\end{aligned} \tag{14}$$

The proof here follows directly from the lemma. The condition is necessary, since every net for which an expression of the form (4) can be written obviously verifies it, the ψ's being the characteristic functions of the $\boldsymbol{S}_\alpha$ and the β for each ψ being the class whose designation has the form $\prod_{i\varepsilon\alpha} \boldsymbol{Pr}_i \prod_{j\varepsilon\beta\alpha} \boldsymbol{Pr}_j$, where $\boldsymbol{Pr}_k$, denotes α_k for all k. Conversely, we may write an expression of the form (4) for a net $\mathcal{N}$ fulfilling prehensible classes satisfying (14) by putting for the $\boldsymbol{Pr}_\alpha$ $\boldsymbol{Pr}$ denoting the ψ's, and a $\boldsymbol{Pr}$, written in the analogue for classes of the disjunctive normal form, and denoting the α corresponding to that ψ, conjoined to it. Since every $\boldsymbol{S}$ of the form (4) is clearly realizable, we have the theorem.

It is of some interest to consider the extent to which we can by knowledge of the present determine the whole past of various special nets: i.e., when we may construct a net the firing of the cyclic of whose neurons requires the peripheral afferents to have had a set of past values specified by given functions ϕ_i. In this case the classes α_i of the last theorem reduced to unit classes; and the condition may be transformed into

$$
\begin{aligned}
&(E\, m, n)\, (p) n (i, \psi)\, (Ej) \; : . \, (x) m : \psi(x) = 0 \,.\mathbf{v}.\, \psi(x) = 1 : \\
&\quad \phi_i \,\varepsilon\, \sigma(\psi,\, nt + p) : \rightarrow : (w) m (x) t - 1 \,.\, \phi_i(n(t+1) \\
&\quad + \, p,\, nx + p,\, w) = \phi_j(nt + p,\, nx + p\,,\, w) \; : . \\
&\quad (u, v)\, (w) m .\, \phi_i(n(u+1) + p,\, nu + p,\, w) \\
&\quad = \phi_i(n(v+1) + p,\, nv + p,\, w).
\end{aligned}
$$

On account of limitations of space, we have presented the above argument very sketchily; we propose to expand it and certain of its implications in a further publication.

The condition of the last theorem is fairly simply in principle, though not in detail; its application to practical cases would, however, require the exploration of some 2^{2^n} classes of functions, namely the members of $\mathscr{R}(\{\alpha_1, \ldots, \alpha_s\})$. Since each of these is a possible β of Theorem 9, this result cannot be sharpened. But we may obtain a sufficient condition for the realizability of an $\boldsymbol{S}$ which is very easily applicable and probably covers most practical purposes. This is given by

Theorem X.

Let us define a set **K** of **S** by the following recursion:

1. Any *TPE* and any *TPE* whose arguments have been replaced by members of $\boldsymbol{K}$ belong to $\boldsymbol{K}$;
2. If $\boldsymbol{Pr}_1(z_1)$ is a member of $\boldsymbol{K}$, then $(z_2)z_1 \,.\, \boldsymbol{Pr}_1(z_2)$, $(Ez_2)z_1 \,.\, \boldsymbol{Pr}_1(z_2)$, and $\boldsymbol{C}_{mn}(z_1) \,.\, s$ belong to it, where $\boldsymbol{C}_{mn}$ denotes the property of being congruent to m modulo n, $m < n$.
3. *The set* **K** *has no further members.*

Then every member of $\boldsymbol{K}$ is realizable. For, if $\boldsymbol{Pr}_1(\mathbf{z}_1)$ is realizable, nervous nets for which:

$$N_i(z_1). \equiv .\, \boldsymbol{Pr}_1(z_1).\, \boldsymbol{SN}_i(z_1)$$

$$N_i(z_1). \equiv .\, \boldsymbol{Pr}_1(z_1) \vee \boldsymbol{SN}_i(z_1)$$

are the expressions of equation (4), realize $(z_2)z_1 \,.\, \boldsymbol{Pr}_1(z_2)$ and $(\boldsymbol{E}z_2)z_1 \,.\, \boldsymbol{Pr}_1(z_2)$ respectively; and a simple circuit, $c_1, c_2, \ldots, c_n$, of n links, each sufficient to excite the next, gives an expression

$$\boldsymbol{N}_m(z_1) \,.\, \equiv \,.\, \boldsymbol{N}_1(0) \,.\, \boldsymbol{C}_{mn}$$

for the last form. By induction we derive the theorem.

One more thing is to be remarked in conclusion. It is easily shown: first, that every net, if furnished with a tape, scanners connected to afferents, and suitable efferents to perform the necessary motor-operations, can compute only such numbers as can a Turing machine; second, that each of the latter numbers can be computed by such a net; and that nets with circles can be computed by such a net; and that nets with circles can compute, without scanners and a tape, some of the numbers the machine can, but no others, and not all of them. This is of interest as affording a psychological justification of the Turing definition of computability and its equivalents, Church's λ—definability and Kleene's primitive recursiveness: If any number can be computed by an organism, it is computable by these definitions, and conversely.

Here McCulloch and Pitts attempt to relate their logical calculus model to Turing machines.

IV. Consequences

Causality, which requires description of states and a law of necessary connection relating them, has appeared in several forms in several sciences, but never, except in statistics, has it been as irreciprocal as in this theory. Specification for any one time of afferent stimulation and of the activity of all constituent neurons, each an "all-or-none" affair, determines the state. Specification of the nervous net provides the law of necessary connection whereby one can compute from the description of any state that of the succeeding state, but the inclusion of disjunctive relations prevents complete determination of the one before. Moreover, the regenerative activity of constituent circles renders reference indefinite as to time past. Thus our knowledge of the world, including ourselves, is incomplete as to space and indefinite as to time. This ignorance, implicit in all our brains, is the counterpart of the abstraction which renders our knowledge useful. The role of brains in determining the epistemic relations of our theories to our observations and of these to the facts is all too clear, for it is apparent that every idea and every sensation is realized by activity within that net, and by no such activity are the actual afferents fully determined.

There is no theory we may hold and no observation we can make that will retain so much as its old defective reference to the facts if the net be altered. Tinnitus, paraesthesias, hallucinations, delusions, confusions and disorientations intervene. Thus empiry confirms that if our nets are undefined, our facts are undefined, and to the "real" we can attribute not so much as one quality or "form." With determination of the net, the unknowable object of knowledge, the "thing in itself," ceases to be unknowable.

To psychology, however defined, specification of the net would contribute all that could be achieved in that field—even if the analysis were pushed to ultimate psychic units or "psychons," for a psychon can be no less than the activity of a single neuron. Since that activity is inherently propositional, all psychic events have an intentional, or "semiotic," character. The "all-or-none" law of these activities, and the conformity of their relations to those of the logic of propositions, insure that the relations of psychons are those of the two-valued logic of propositions. Thus in psychology, introspective, behavioristic or physiological, the fundamental relations are those of two-valued logic.

Hence arise constructional solutions of holistic problems involving the differentiated continuum of sense awareness and the normative, perfective and resolvent properties of perception and execution. From

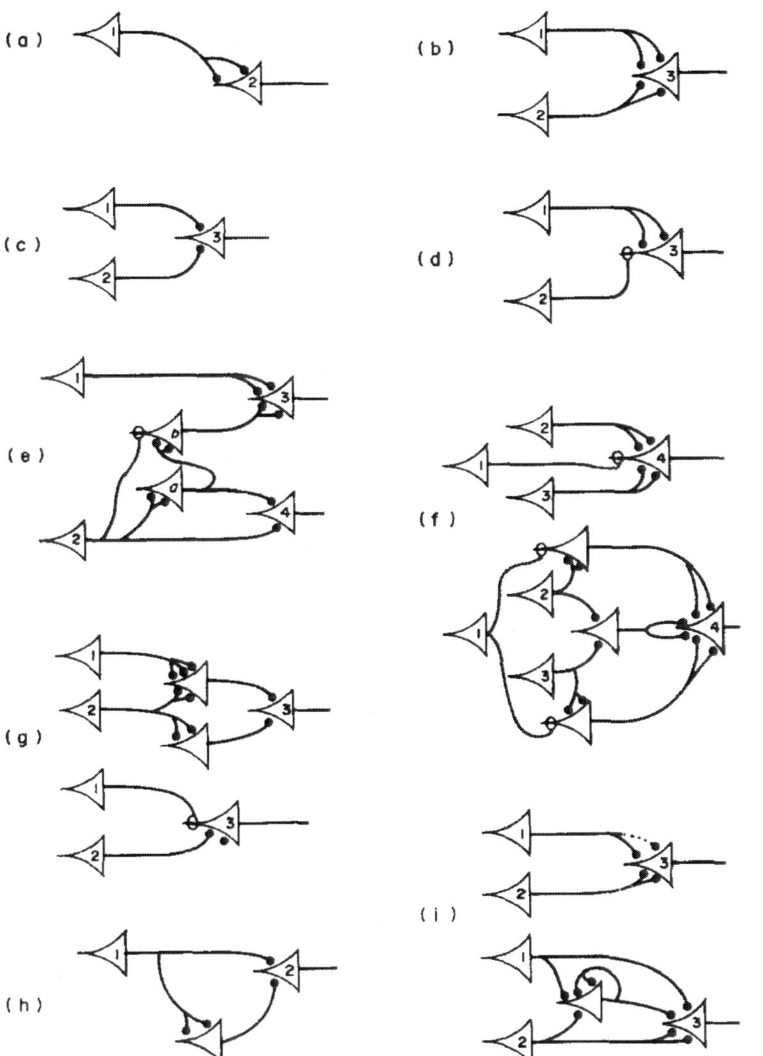

These diagrams are the first logic gates. They inspired von Neumann and subsequent engineers in the design of logic circuits for digital computers.

Figure 1. The neuron c_i is always marked with the numeral i upon the body of the cell, and the corresponding action is denoted by 'N' with i as subscript, as in the text:

(a) $N_2(t). \equiv .N_1(t-1)$

(b) $N_3(t). \equiv .N_1(t-1) \vee N_2(t-1)$

(c) $N_3(t). \equiv .N_1(t-1) \,.\, N_2(t-1)$

(d) $N_3(t). \equiv .N_1(t-1) \,.\, \sim N_2(t-1)$

(e) $N_3(t) :\equiv: N_1(t-1). \vee. \sim N_2(t-3). \sim N_2(t-2)$

$N_4(t). \equiv .N_2(t-2) \,.\, N_2(t-1)$

(f) $N_4(t) :\equiv: \sim N_1(t-1) .N_2(t-1) \vee N_3(t-1). \vee .N_1(t-1).$
$N_2(t-1).N_3(t-1)$

$N_4(t) :\equiv: \sim N_1(t-2) .N_2(t-2) \vee N_3(t-2). \vee .N_1(t-2).$
$N_2(t-2).N_3(t-2)$

(g) $N_3(t). \equiv .N_2(t-2). \sim N_1(t-3)$

(h) $N_2(t). \equiv .N_1(t-1).N_1(t-2)$

(i) $N_3(t) :\equiv: N_2(t-1). \vee .N_1(t-1).(Ex)t-1.N_1(x).N_2(x).$

the irreciprocity of causality it follows that even if the net be known, though we may predict future from present activities, we can deduce neither afferent from central, nor central from efferent, nor past from present activities—conclusions which are reinforced by the contradictory testimony of eye-witnesses, by the difficulty of diagnosing differentially the organically diseased, the hysteric and the malingerer, and by comparing one's own memories or recollections with his contemporaneous records. Moreover, systems which so respond to the difference between afferents to a regenerative net and certain activity within that net, as to reduce the difference, exhibit purposive behavior; and organisms are known to possess many such systems, subserving homeostasis, appetition and attention. Thus both the formal and the final aspects of that activity which we are wont to call *mental* are rigorously deduceable from present neurophysiology. The psychiatrist may take comfort from the obvious conclusion concerning causality—that, for prognosis, history is never necessary. He can take little from the equally valid conclusion that his observables are explicable only in terms of nervous activities which, until recently, have been beyond his ken. The crux of this ignorance is that inference from any sample of overt behavior to nervous nets is not unique, whereas, of imaginable nets, only one in fact exists, and may, at any moment, exhibit some unpredictable activity. Certainly for the psychiatrist it is more to the point that in such systems "Mind" no longer "goes more ghostly than a ghost." Instead, diseased mentality can be understood without loss of scope or rigor, in the scientific terms of neurophysiology. For neurology, the theory sharpens the distinction between nets necessary or merely sufficient for given activities, and so clarifies the relations of disturbed structure to disturbed function. In its own domain the difference between equivalent nets and nets equivalent in the narrow sense indicates the appropriate use and importance of temporal studies of nervous activity: and to mathematical biophysics the theory contributes a tool for rigorous symbolic treatment of known nets and an easy method of constructing hypothetical nets of required properties.

REFERENCES

Carnap, R. 1938. *The Logical Syntax of Language.* New York, NY: Harcourt-Brace.

Hilbert, D., and W. Ackermann. 1927. *Grundüge der Theoretischen Logik.* Berlin: Springer.

Russell, B., and A. N. Whitehead. 1925. *Principia Mathematica.* Cambridge: Cambridge University Press.

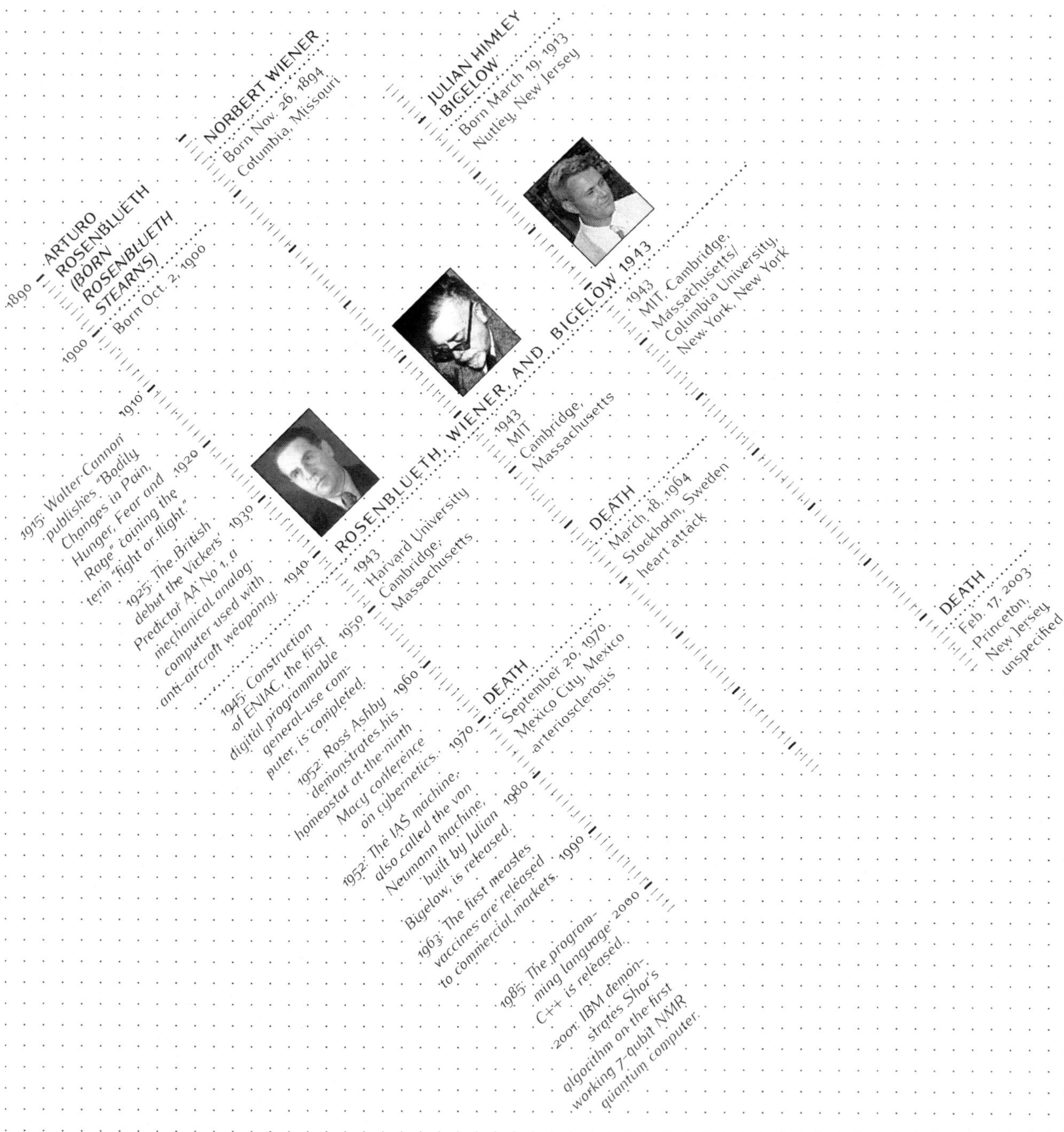

ROSENBLUETH, WIENER & BIGELOW

[6]

THE BIRTH OF CYBERNETICS

Andrew Pickering, University of Exeter

A. Rosenblueth, N. Wiener, and J. Bigelow, "Behavior, Purpose, and Teleology," *Philosophy of Science* 10 (1): 18–24 (1943).

Published during World War II, this short and straightforward essay was central to the foundation of cybernetics, a new field of work on complex systems that flourished from the late 1940s into the 1960s and has continued to mutate up to the present (Kline 2015). "Behavior, Purpose and Teleology" (BPT) grew out of Norbert Wiener's war work with Julian Bigelow in the United States, attempting to build an anti-aircraft predictor (Galison 1994). The point of this device was to extrapolate from tracking data in order to forecast where an enemy aircraft would arrive in the time it took a shell to reach it. The project was not a success, but it led Wiener to think about purposeful behavior in terms of negative feedback and systems that act to minimize the gap between a current state of affairs and an intended state (in this case, between the location of a target and exploding shells).

Negative feedback and servomechanisms had long been important in engineering (going back to nineteenth-century steam-engine governors as well as domestic thermostats), but the founding move in cybernetics was to generalize the picture as far as it would go. Reflecting the thinking of Wiener's friend Arturo Rosenblueth, a leading Mexican physiologist (Burbano and Reyes 2022), BPT erased the difference between machines, animals, and, indeed, humans. The authors argued that the same analysis of purposeful behavior could run the gamut from servomechanisms to organisms and intentional action in general. Wiener further expanded the picture in his 1948 book, *Cybernetics, or Control and Communication in the Animal and the Machine*, to include the information flows and processing central to feedback mechanisms like the anti-aircraft predictor, bringing in fields such as digital computing, information theory, psychology, and psychiatry. (In later years, relations between these elements became complicated. For example, in the 1950s computer science gave rise to symbolic AI, a model of the cognitive brain quite different from the performative cybernetic image of the brain as geared into the world via feedback loops.)

Wiener's vision was further fleshed out in the famous Macy conferences, "Cybernetics: Circular Causal and Feedback Mechanisms in Biological and Social Systems" (Pias 2003), held between 1946 and 1953, which brought together experts from across the disciplines, including Wiener himself, John von Neumann, Heinz von Foerster, Margaret Mead, and Gregory Bateson, with Warren McCulloch as perpetual chair. The reference to circular causality in the Macy's title pointed to another avenue of generalization. The argument was that the circular aspects of the behavior of feedback systems, in which output returns as input, marks cybernetics off from the linear causality of modern physics and evokes a different ontology or worldview. Signaling this, von Foerster fixed on the image of Ouroboros, the mythological snake that eats its own tail, as cybernetics' sigil. Here cybernetics took on a distinctively philosophical flavor as a new kind of science (later to be the title of Stephen Wolfram's book [2002] on cellular automata). As its reach expanded, cybernetics also became a topic of political critique. Authors ranging from the German philosopher Martin Heidegger (1981) to the French collective Tiqqun (2001) argued that its wartime origins and emphasis on control marked cybernetics as a key tool of the neoliberal state. In contrast, leading feminist Donna Haraway (1991) found a potential for radical political action in the anti-essentialist blurring and even fusion of the human and the machine, encapsulated in her figure of the cyborg.

Many lines of development grew from the small beginnings of BPT. These included a distinctive style of robotics depending upon search mechanisms rather than AI computations, running from Grey Walter's (1953) "tortoises" (or "turtles") in the late 1940s to Rodney Brooks's anti-AI subsumption architecture in the 1980s. In a movement led by von Foerster (2014), so-called second-order cybernetics incorporated the observer into the system of feedback systems, leading into wider discussions of ethics and radical philosophical constructivism (von Glasersfeld 1995). In British cybernetics especially, Ross Ashby's (1952) homeostat, built in 1948, was particularly influential. In contrast to servomechanisms, the homeostat actively explored its environment and adapted to it, giving rise to highly distinctive projects and artifacts in fields including psychiatry, management, the environment, education, the arts, and architecture (Pickering 2010). Cybernetics is much discussed these days in the arts and humanities, and less so in engineering, but echoes of BPT are heard all over the disciplinary map. ❦

REFERENCES

Ashby, W. R. 1952. *Design for a Brain.* London, UK: Chapman & Hall.

Brooks, R. 1999. *Cambrian Intelligence: The Early History of the New AI.* Cambridge, MA: MIT Press.

Burbano, A., and E. Reyes. 2022. "Capsaicin and Cybernetics: Mexican Intellectual Networks in the Foundation of Cybernetics." *AI & Society* 37:1013–1025. https://doi.org/10.1007/s00146-021-01337-3.

Galison, P. 1994. "The Ontology of the Enemy: Norbert Wiener and the Cybernetic Vision." *Critical Inquiry* 21 (1): 228–266. https://doi.org/10.1086/448747.

Haraway, D. 1991. "A Cyborg Manifesto: Science, Technology, and Socialist-Feminism in the Late Twentieth Century." In *Simians, Cyborgs, and Women,* 149–181. London, UK: Free Association Books.

Heidegger, M. 1981. "'Only a God Can Save Us': The Spiegel Interview (1966)." In *Heidegger: The Man and the Thinker,* edited by T. Sheehan, 45–67. Chicago, IL: Precedent Publishing.

Kline, R. 2015. *The Cybernetics Moment: Or Why We Call Our Age the Information Age.* Baltimore, MD: Johns Hopkins University Press.

Pias, C., ed. 2003. *Cybernetics-Kybernetik: The Macy-Conferences 1946–1953.* Vol. I: Transactions. Zürich–Berlin: Diaphanes.

Pickering, A. 2010. *The Cybernetic Brain: Sketches of Another Future.* Chicago, IL: University of Chicago Press.

Tiqqun. 2001. "The Cybernetic Hypothesis," https://cybernet.jottit.com.

von Foerster, H. 2014. *The Beginning of Heaven and Earth Has No Name.* New York, NY: Fordham University Press.

von Glasersfeld, E. 1995. *Radical Constructivism: A Way of Knowing and Learning.* London, UK: Routledge.

Walter, W. G. 1953. *The Living Brain.* London, UK: Duckworth.

Wiener, N. 1948. *Cybernetics, or Control and Communication in the Animal and the Machine.* Cambridge, MA: MIT Press.

Wolfram, S. 2002. *A New Kind of Science.* Champaign, IL: Wolfram Media Inc.

BEHAVIOR, PURPOSE, AND TELEOLOGY

Arturo Rosenblueth, Harvard Medical School
Norbert Wiener, Massachusetts Institute of Technology
and Julian Bigelow, Massachusetts Institute of Technology

This essay has two goals. The first is to define the behavioristic study of natural events and to classify behavior. The second is to stress the importance of the concept of purpose.

Behaviorism remained the dominant approach to animal and human psychology in the 1940s. It focused on input/output relations in the behavior of organisms, which were themselves regarded as opaque black boxes. The cognitive revolution of the 1960s, which shifted the focus from external relations to inner workings, subsequently attempted to open the black boxes in terms of computer metaphors and AI.

Given any object, relatively abstracted from its surroundings for study, the behavioristic approach consists in the examination of the output of the object and of the relations of this output to the input. By output is meant any change produced in the surroundings by the object. By input, conversely, is meant any event external to the object that modifies this object in any manner.

The above statement of what is meant by the behavioristic method of study omits the specific structure and the instrinsic organization of the object. This omission is fundamental because on it is based the distinction between the behavioristic and the alternative functional method of study. In a functional analysis, as opposed to a behavioristic approach, the main goal is the intrinsic organization of the entity studied, its structure and its properties; the relations between the object and the surroundings are relatively incidental.

From this definition of the behavioristic method a broad definition of behavior ensues. By behavior is meant any change of an entity with respect to its surroundings. This change may be largely an output from the object, the input being then minimal, remote or irrelevant; or else the change may be immediately traceable to a certain input. Accordingly, any modification of an object, detectable externally, may be denoted as behavior. The term would be, therefore, too extensive for usefulness were it not that it may be restricted by apposite adjectives—i.e., that behavior may be classified.

The consideration of the changes of energy involved in behavior affords a basis for classification. Active behavior is that in which the object is the source of the output energy involved in a given specific reaction. The object may store energy supplied by a remote or relatively immediate input, but the input does not energize the output directly. In passive behavior, on the contrary, the object is not a source of energy; all the energy in the output can be traced to the immediate input (e.g., the throwing of an object), or else the object may control energy which remains external to it throughout the reaction (e.g., the soaring flight of a bird).

Active behavior may be subdivided into two classes: purposeless (or random) and purposeful. The term purposeful is meant to denote that the act or behavior may be interpreted as directed to the attainment of a goal—i.e., to a final condition in which the behaving object reaches a definite correlation in time or in space with respect to another object or event. Purposeless behavior then is that which is not interpreted as directed to a goal.

The vagueness of the words "may be interpreted" as used above might be considered so great that the distinction would be useless. Yet the recognition that behavior may sometimes be purposeful is unavoidable and useful, as follows. The basis of the concept of purpose is the awareness of "voluntary activity." Now, the purpose of voluntary acts is not a matter of arbitrary interpretation but a physiological fact. When we perform a voluntary action what we select voluntarily is a specific purpose, not a specific movement. Thus, if we decide to take a glass containing water and carry it to our mouth we do not command certain muscles to contract to a certain degree and in a certain sequence; we merely trip the purpose and the reaction follows automatically. Indeed, experimental physiology has so far been largely incapable of explaining the mechanism of voluntary activity. We submit that this failure is due to the fact that when an experimenter stimulates the motor regions of the cerebral cortex he does not duplicate a voluntary reaction; he trips efferent, "output" pathways, but does not trip a purpose, as is done voluntarily.

The view has often been expressed that all machines are purposeful. This view is untenable. First may be mentioned mechanical devices such as a roulette, designed precisely for purposelessness. Then may be considered devices such as a clock, designed, it is true, with a purpose, but having a performance which, although orderly, is not purposeful—i.e., there is no specific final condition toward which the movement of the clock strives. Similarly, although a gun may be used for a definite purpose, the attainment of a goal is not intrinsic to the performance of the gun; random shooting can be made, deliberately purposeless.

The wartime connection is clear here, Because of wartime secrecy, the authors could not mention the anti-aircraft predictor from which their essay derived, and cited a target-seeking torpedo instead.

Some machines, on the other hand, are intrinsically purposeful. A torpedo with a target-seeking mechanism is an example. The term servomechanisms has been coined precisely to designate machines with intrinsic purposeful behavior.

It is apparent from these considerations that although the definition of purposeful behavior is relatively vague, and hence operationally largely meaningless, the concept of purpose is useful and should, therefore, be retained.

Purposeful active behavior may be subdivided into two classes: "feed-back" (or "teleological") and "non-feed-back" (or "non-teleological"). The expression feed-back is used by engineers in two different senses. In a broad sense it may denote that some of the output energy of an apparatus or machine is returned as input; an example is an electrical amplifier with feed-back. The feed-back is in these cases positive—the fraction of the output which reenters the object has the same sign as the original input signal. Positive feed-back adds to the input signals, it does not correct them. The term feed-back is also employed in a more restricted sense to signify that the behavior of an object is controlled by the margin of error at which the object stands at a given time with reference to a relatively specific goal. The feed-back is then negative, that is, the signals from the goal are used to restrict outputs which would otherwise go beyond the goal. It is this second meaning of the term feed-back that is used here.

All purposeful behavior may be considered to require negative feed-back. If a goal is to be attained, some signals from the goal are necessary at some time to direct the behavior. By non-feed-back behavior is meant that in which there are no signals from the goal which modify the activity of the object *in the course of the behavior*. Thus, a machine may be set to impinge upon a luminous object although the machine may be insensitive to light. Similarly, a snake may strike at a frog, or a frog at a fly, with no visual or other report from the prey after the movement has started. Indeed, the movement is in these cases so fast that it is not likely that nerve impulses would have time to arise at the retina, travel to the central nervous system and set up further impulses which would reach the muscles in time to modify the movement effectively.

As opposed to the examples considered, the behavior of some machines and some reactions of living organisms involve a continuous feed-back from the goal that modifies and guides the behaving object. This type of behavior is more effective than that mentioned above, particularly when the goal is not stationary. But continuous feed-back control may lead to very clumsy behavior if the feedback is inadequately damped and becomes therefore positive instead of negative for certain frequencies of oscillation. Suppose, for example, that a machine is designed with the purpose of impinging upon a moving luminous goal; the path followed by the machine is controlled by the direction and intensity of the light from the goal. Suppose further that the machine overshoots seriously when it follows a movement of the goal in a certain direction; an even stronger stimulus will then be delivered which will turn the machine in the opposite direction. If that movement again overshoots a series of increasingly larger oscillations will ensue and the machine will miss the goal.

This picture of the consequences of undamped feed-back is strikingly similar to that seen during the performance of a voluntary act by a cerebellar patient. At rest the subject exhibits no obvious motor disturbance. If he is asked to carry a glass of water from a table to his mouth, however, the hand carrying the glass will execute a series of oscillatory motions of increasing amplitude as the glass approaches his mouth, so that the water will spill and the purpose will not be fulfilled. This test is typical of the disorderly motor performance of patients with cerebellar disease. The analogy with the behavior of a machine with undamped feed-back is so vivid that we venture to suggest that the main function of the cerebellum is the control of the feed-back nervous mechanisms involved in purposeful motor activity.

The analysis moves briefly from the normal to the pathological. Overshoots and oscillations were a recognized problem with servomechanisms, known as "hunting." Rosenblueth, Wiener, and Bigelow (RWB) generalize from this to the tremors observed in "cerebellar patients."

Feed-back purposeful behavior may again be subdivided. It may be extrapolative (predictive), or it may be non-extrapolative (non-predictive). The reactions of unicellular organisms known as tropisms are examples of non-predictive performances. The amoeba merely follows the source to which it reacts; there is no evidence that it extrapolates the path of a moving source. Predictive animal behavior, on the other hand, is a commonplace. A cat starting to pursue a running mouse does not run directly toward the region where the mouse is at any given time, but moves toward an extrapolated future position. Examples of both predictive and non-predictive servomechanisms may also be found readily.

Predictive behavior may be subdivided into different orders. The cat chasing the mouse is an instance of first-order prediction; the cat merely predicts the path of the mouse. Throwing a stone at a moving target requires a second-order prediction; the paths of the target and of the stone should be foreseen. Examples of predictions of higher order are shooting with a sling or with a bow and arrow.

Predictive behavior requires the discrimination of at least two coordinates, a temporal and at least one spatial axis. Prediction will be more effective and flexible, however, if the behaving object can respond to changes in more than one spatial coordinate. The sensory receptors of an organism, or the corresponding elements of a machine, may therefore limit the predictive behavior. Thus, a bloodhound *follows* a trail, that is, it does not show any predictive behavior in trailing, because a chemical, olfactory input reports only spatial information: distance, as indicated by intensity. The external changes capable of affecting auditory, or, even better, visual receptors, permit more accurate spatial localization; hence the possibility of more effective predictive reactions when the input affects those receptors.

In addition to the limitations imposed by the receptors upon the ability to perform extrapolative actions, limitations may also occur that are due to the internal organization of the behaving object. Thus, a machine which is to trail predictively a moving luminous object should not only be sensitive to light (e.g., by the possession of a photoelectric cell), but should also have the structure adequate for interpreting the luminous input. It is probable that limitations of internal organization, particularly of the organization of the central nervous system, determine the complexity of predictive behavior which a mammal may attain. Thus, it is likely that the nervous system of a rat or dog is such that it does not permit the integration of input and output necessary for the performance of a predictive reaction of the third or fourth order. Indeed, it is possible that one of the features of the discontinuity of behavior observable when comparing humans with other high mammals may lie in that the other mammals are limited to predictive behavior of a low order, whereas man may be capable potentially of quite high orders of prediction.

The classification of behavior suggested so far is tabulated here:

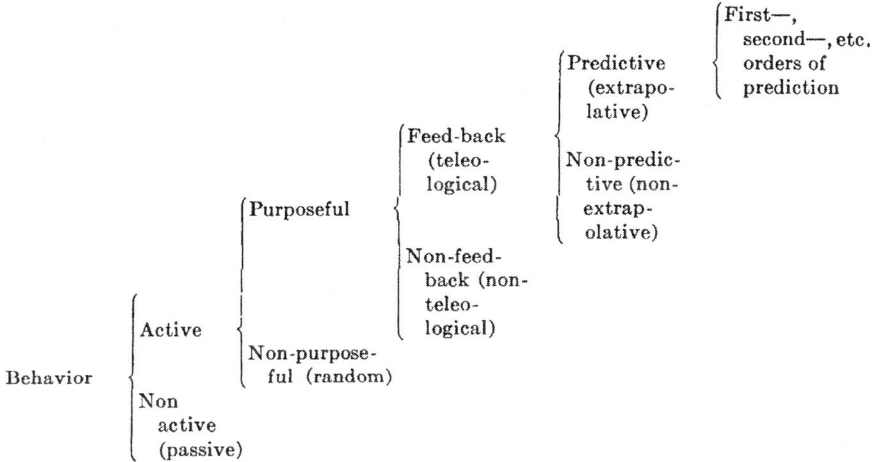

It is apparent that each of the dichotomies established singles out arbitrarily one feature, deemed interesting, leaving an amorphous remainder: the non-class. It is also apparent that the criteria for the several dichotomies are heterogeneous. It is obvious, therefore, that many other lines of classification are available, which are independent of that developed above. Thus, behavior in general, or any of the groups in the table, could be divided into linear (i.e., output proportional to input) and non-linear. A division into continuous and discontinuous might be useful for many purposes. The several degrees of freedom which behavior may exhibit could also be employed as a basis of systematization.

RWB put considerable effort into classifying types of purposeful systems, but this had little influence in the subsequent history of cybernetics.

The classification tabulated above was adopted for several reasons. It leads to the singling out of the class of predictive behavior, a class particularly interesting since it suggests the possibility of systematizing increasingly more complex tests of the behavior of organisms. It emphasizes the concepts of purpose and of teleology, concepts which, although rather discredited at present, are shown to be important. Finally, it reveals that a uniform behavioristic analysis is applicable to both machines and living organisms, regardless of the complexity of the behavior.

The problem of reliably identifying the purposes (goals, intentions) behind human actions is an old one in philosophy and the social sciences, to which an answer has never been found. RWB finesse the problem by starting from machines, whose purpose is evident by design.

It has sometimes been stated that the designers of machines merely attempt to duplicate the performances of living organisms. This statement is uncritical. That the gross behavior of some machines should be similar to the reactions of organisms is not surprising. Animal behavior includes many varieties of all the possible modes of behavior and the machines devised so far have far from exhausted all those possible modes. There is, therefore a considerable overlap of the two realms of behavior. Examples, however, are readily found of man-made machines with behavior that transcends human behavior. A machine with an electrical output is an instance; for men, unlike the electric fishes, are incapable of emitting electricity. Radio transmission is perhaps an even better instance, for no animal is known with the ability to generate short waves, even if so-called experiments on telepathy are considered seriously.

A further comparison of living organisms and machines leads to the following inferences. The methods of study for the two groups are at present similar. Whether they should always be the same may depend on whether or not there are one or more qualitatively distinct, unique characteristics present in one group and absent in the other. Such qualitative differences have not appeared so far.

The broad classes of behavior are the same in machines and in living organisms. Specific, narrow classes may be found exclusively in one or the other. Thus, no machine is available yet that can write a Sanscrit–Mandarin dictionary. Thus, also, no living organism is known that rolls on wheels—imagine what the result would have been if engineers had insisted on copying living organisms and had therefore put legs and feet in their locomotives, instead of wheels.

While the behavioristic analysis of machines and living organisms is largely uniform, their functional study reveals deep differences. Structurally, organisms are mainly colloidal, and include prominently protein molecules, large, complex and anisotropic; machines are chiefly metallic and include mainly simple molecules. From the standpoint of their energetics, machines usually exhibit relatively large differences of potential, which permit rapid mobilization of energy; in organisms the energy is more uniformly distributed, it is not very mobile. Thus, in electric machines conduction is mainly electronic, whereas in organisms electric changes are usually ionic.

Scope and flexibility are achieved in machines largely by temporal multiplication of effects; frequencies of one million per second or more are readily obtained and utilized. In organisms, spatial multiplication, rather than temporal, is the rule; the temporal achievements are poor—the fastest nerve fibers can only conduct about one thousand impulses per second; spatial multiplication is on the other hand abundant and admirable in its compactness. This difference is well illustrated by the comparison of a television receiver and the eye. The television receiver may be described as a single cone retina; the images are formed by scanning—i.e. by orderly successive detection of the signal with a rate of about 20 million per second. Scanning is a process which seldom or never occurs in organisms, since it requires fast frequencies for effective performance. The eye uses a spatial, rather than a temporal multiplier. Instead of the one cone of the television receiver a human eye has about 6.5 million cones and about 115 million rods.

If an engineer were to design a robot, roughly similar in behavior to an animal organism, he would not attempt at present to make it out of proteins and other colloids. He would probably build it out of metallic parts, some dielectrics and many vacuum tubes. The movements of the robot could readily be much faster and more powerful than those of the original organism. Learning and memory, however, would be quite rudimentary. In future years, as the knowledge of colloids and proteins increases, future engineers may attempt the design of robots not only with a behavior, but also with a structure similar to that of a mammal. The ultimate model of a cat is of course another cat, whether it be born of still another cat or synthesized in a laboratory.

In classifying behavior the term "teleology" was used as synonymous with "purpose controlled by feed-back." Teleology has been interpreted in the past to imply purpose and the vague concept of a "final cause" has been often added. This concept of final causes has led to the opposition of teleology to determinism. A discussion of causality, determinism and final causes is beyond the scope of this essay. It may be pointed out, however, that purposefulness, as defined here, is quite independent of causality, initial or final. Teleology has been discredited chiefly because it was defined to imply a cause subsequent in time to a given effect. When this aspect of teleology was dismissed, however, the associated recognition of the importance of purpose was also unfortunately discarded. Since we consider purposefulness a concept

Questions of causality came to the fore in the later development of cybernetics, where the circular causality of feedback systems was contrasted with the linear, once-and-for-all causality of classical physics.

necessary for the understanding of certain modes of behavior we suggest that a teleological study is useful if it avoids problems of causality and concerns itself merely with an investigation of purpose.

We have restricted the connotation of teleological behavior by applying this designation only to purposeful reactions which are controlled by the error of the reaction—i.e., by the difference between the state of the behaving object at any time and the final state interpreted as the purpose. Teleological behavior thus becomes synonymous with behavior controlled by negative feed-back, and gains therefore in precision by a sufficiently restricted connotation.

According to this limited definition, teleology is not opposed to determinism, but to non-teleology. Both teleological and non-teleological systems are deterministic when the behavior considered belongs to the realm where determinism applies. The concept of teleology shares only one thing with the concept of causality: a time axis. But causality implies a one-way, relatively irreversible functional relationship, whereas teleology is concerned with behavior, not with functional relationships. ❦

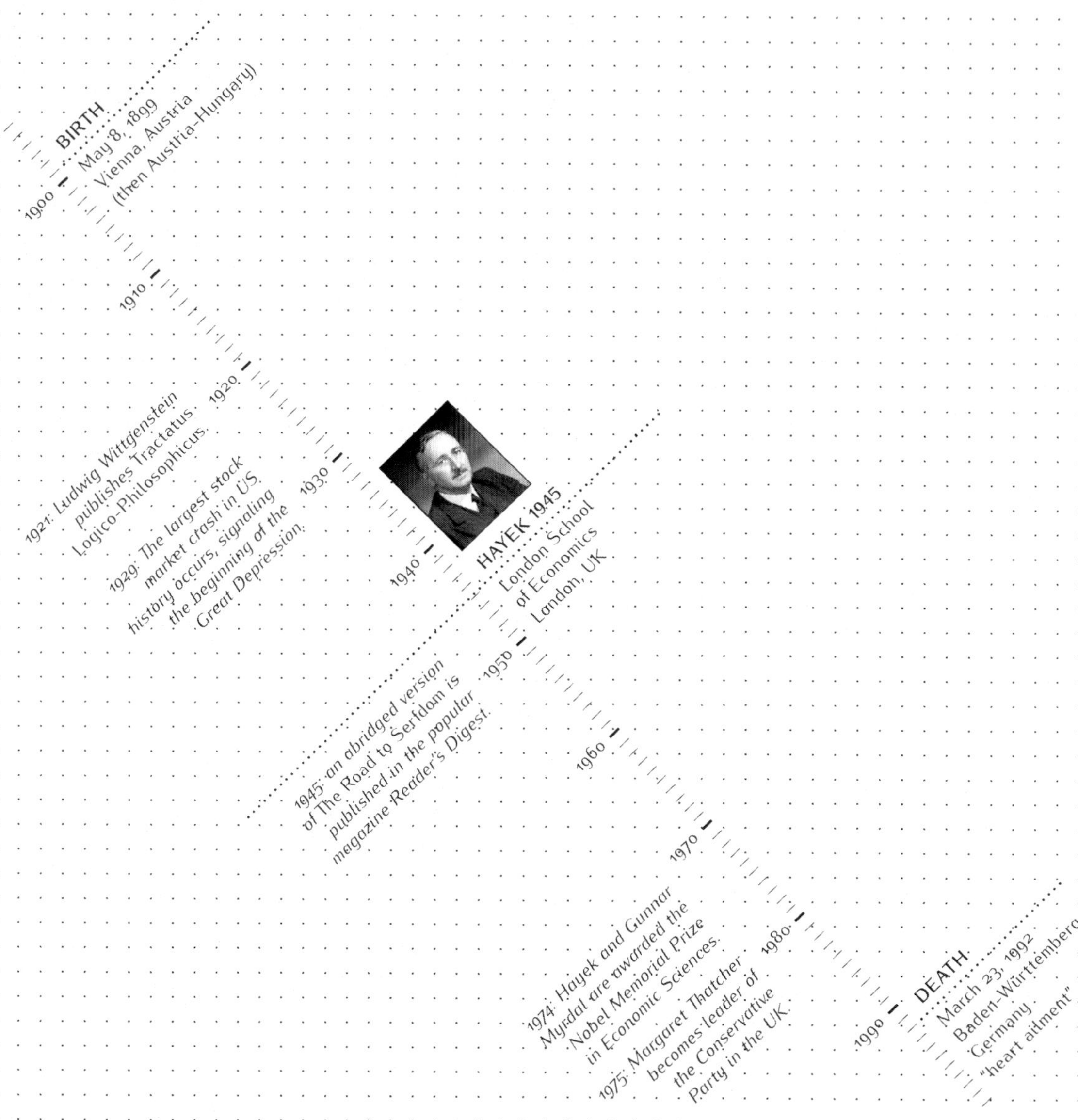

FRIEDRICH AUGUST VON HAYEK

[7]

FRIEDRICH HAYEK'S ECONOMY OF KNOWLEDGE

Samuel Bowles, Santa Fe Institute

A complex system, Herbert Simon wrote, is "made up of a large number of parts that interact in a non-simple way [such that] given the properties of the parts and the laws of their interaction it is not a trivial matter to infer the properties of the whole" (Simon 1969, 267). Though economists rarely use the term, complexity has a long history in the field, most famously illustrated by Adam Smith's invisible hand. Smith's surprising claim that competitive markets and private property could harness individual self-interest so as to contribute to a well-ordered society is exactly the far-from-trivial inference about the "properties of the whole" that he derived from the "parts and their laws of interaction" to which Simon referred.

In "The Use of Knowledge in Society," Friedrich Hayek (1945) provides what I take to be the best explanation of how the invisible hand might work: "We must look at the price system as . . . a mechanism for communicating information if we want to understand its real function. . . . The most significant fact about this system is the economy of knowledge with which it operates, or how little the individual participants need to know in order to take the right action." A price, to Hayek, is both a message about the relative scarcity of a good and a motivation to adjust one's behavior to a change in that scarcity. This bundling of message and motivation is the key to how an emergent property of a market system could be what Hayek called a "rational economic order."

The paper was one of Hayek's many contributions to the early-twentieth-century debate on the economic feasibility of centralized planning and its merits and shortcomings relative to the capitalist economy. Ludwig von Mises, Hayek, and others advanced the view that

F.A. Hayek, "The Use of Knowledge in Society," *American Economic Review* 35 (4), 519–530 (1945).

the rational economic calculation entailed by planning required the knowledge of prices reflecting true scarcity (that is, measuring social marginal costs and benefits), and this information could be obtained only by the extensive use of decentralized allocation through markets. Oskar Lange, Enrico Barone, Abba Lerner, and others countered that prices are implicit in any optimizing problem (whether or not markets exist). Linear programming methods developed by Leonid Kantorovich in the Soviet Union in 1939 provided a method by which these prices could be calculated. (Like Hayek's paper, Kantorovich's contribution reached far beyond academia. I taught linear programming to future economic planners in Havana in the late 1960s and used the technique to advise Cuba's Ministry of Sugar.)

By the 1940s, the debate appeared to be over. Even the arch opponent of socialism, Joseph Schumpeter, had conceded: "Can socialism work? Of course, it can. ... There is nothing wrong with the pure theory of socialism" (Schumpeter 1942, 167, 172). He was echoing another opponent of socialism, Vilfredo Pareto, who had concluded that "pure economics does not give us a truly decisive criterion for choosing between the organization of society based on private property and a socialist organization" (Pareto 1909, 364).

What, then, was wrong with central planning? And what was wrong with economic theory that it had so let down the market side of the debate? Hayek's paper was a counterattack against central planning, but he also targeted the "pure economics" that Pareto had referred to, meaning the equilibrium theory of perfect competition and the entire Walrasian (also termed neoclassical) paradigm in economics.

In Hayek's view, assuming a state of equilibrium effectively precludes a serious analysis of competition, which he defines, following Samuel Johnson, as "the action of endeavoring to gain what another endeavors to gain at the same time." He continues:

> *Now, how many of the devices adopted in ordinary life to that end would still be open to a seller in a market in which so-called "perfect competition" prevails? I believe that the answer is exactly none. Advertising, undercutting, and improving ("differentiating") the goods or services produced are all*

> *excluded by definition—"perfect" competition means indeed the absence of all competitive activities.* (Hayek 1948)

But to Hayek, the problem with the conventional economic paradigm went considerably beyond these shortcomings of the concept of the perfectly competitive equilibrium. His more fundamental criticism was the failure to recognize that information is scarce and local. In this paper, he recast the planning versus market debate as a problem of information aggregation:

> *Which of these systems is likely to be more efficient depends on the question under which of them can we expect that fuller use will be made of the existing knowledge. And this, in turn, depends on whether we are more likely to succeed in putting at the disposal of a single central authority all the knowledge which ought to be used but which is initially dispersed among many different individuals, or in conveying to the individuals such additional information as they need in order to enable them to fit their plans in with those of others.*

Since the 1980s, the discipline has taken up the fact that information takes the form, as Hayek put it, of "dispersed bits of incomplete and frequently contradictory knowledge which all the separate individuals possess." Figure 1 illustrates the explosion of interest among economists in the problem of limited information that Hayek raised in his 1945 paper (few cite Hayek, or even know that it was this 1945 paper that reframed the market as an information processing system).

What is termed "asymmetric information" now plays a central role in models of employment, credit, and the organization of the firm. The worker, for example, knows how hard she worked, but the employer does not; the lender knows how the loan will be actually used, but the banker does not.

As a result of these information asymmetries, contracts—for employment or lending, for example—will not cover everything of interest to the parties of the exchange. Hard work and prudent management of loaned funds are thus "external effects," like environmental spillovers that thus escape the logic by which the invisible hand would work.

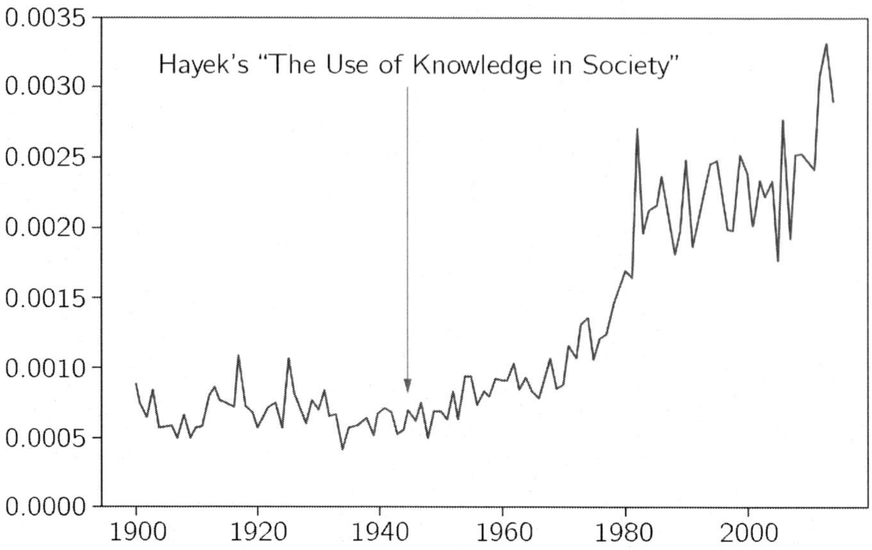

Figure 1. Frequency of the use of the words "knowledge" and "information" as a fraction of all words in the papers published in top economics journals (*AER, Econometrica, EJ, JPE, QJE, RESTAT, RESTUD* [1900–2014]). The data on which this is based are described in Bowles and Carlin (2020).

These models—the first of which Herbert Simon (1951) published just six years after Hayek's paper—address what Hayek termed "planning" within "organized industries" as well as owners' and managers' exercise of authority over their employees. Not surprisingly, they show that the firm owners and managers face the same intractable information problems and resulting inefficiencies that Hayek showed would plague a socialist central planner.

Ironically, Hayek's economy of knowledge brilliantly illustrates how the invisible hand might work and why, in our economy of "organized industries" and "monopolies" facing planetary constraints, it would be a mistake to think that it does.

REFERENCES

Bowles, S., and W. Carlin. 2020. "What Students Learn in Economics 101: Time for a Change." *Journal of Economic Literature* 58 (1): 176–214. https://doi.org/10.1257/jel.20191585.

Hayek, F. A. 1948. "The Meaning of Competition." In *Individualism and Economic Order*. Chicago, IL: University of Chicago Press.

Pareto, V. 1909. *Manuel d'Économic Politique.* First Italian edition, 1905. Paris: Giard et Briere.

Schumpeter, J. 1942. *Capitalism, Socialism, and Democracy.* New York, NY: Harper & Row.

Simon, H. 1951. "A Formal Theory of the Employment Relationship." *Econometrica* 19 (3): 293–305. https://doi.org/0012-9682(195107)19:3<293:AFTOTE>2.0.CO;2-2.

———. 1969. *The Sciences of the Artificial.* Cambridge, MA: MIT Press.

———. 1991. "Organizations and Markets." *Journal of Economic Perspectives* 5 (2): 25–44. https://doi.org/10.1257/jep.5.2.25.

Trotsky, L. 1932. "The Soviet Economy in Danger." *The Militant.*

THE USE OF KNOWLEDGE IN SOCIETY

F.A. Hayek, London School of Economics

I

What is the problem we wish to solve when we try to construct a rational economic order?

On certain familiar assumptions the answer is simple enough. *If* we possess all the relevant information, *if* we can start out from a given system of preferences, and *if* we command complete knowledge of available means, the problem which remains is purely one of logic. That is, the answer to the question of what is the best use of the available means is implicit in our assumptions. The conditions which the solution of this optimum problem must satisfy have been fully worked out and can be stated best in mathematical form: put at their briefest, they are that the marginal rates of substitution between any two commodities or factors must be the same in all their different uses.

These assumptions—complete information and given ("exogenous") preferences—characterize the Walrasian (neoclassical) paradigm that Hayek is targeting (along with centralized economic planning).

This, however, is emphatically *not* the economic problem which society faces. And the economic calculus which we have developed to solve this logical problem, though an important step toward the solution of the economic problem of society, does not yet provide an answer to it. The reason for this is that the "data" from which the economic calculus starts are never for the whole society "given" to a single mind which could work out the implications and can never be so given.

The peculiar character of the problem of a rational economic order is determined precisely by the fact that the knowledge of the circumstances of which we must make use never exists in concentrated or integrated form but solely as the dispersed bits of incomplete and frequently contradictory knowledge which all the separate individuals possess. The economic problem of society is thus not merely a problem of how to allocate "given" resources—if "given" is taken to mean given to a single mind which deliberately solves the problem set by these "data."

It is rather a problem of how to secure the best use of resources known to any of the members of society, for ends whose relative importance only these individuals know. Or, to put it briefly, it is a problem of the utilization of knowledge which is not given to anyone in its totality.

This character of the fundamental problem has, I am afraid, been obscured rather than illuminated by many of the recent refinements of economic theory, particularly by many of the uses made of mathematics. Though the problem with which I want primarily to deal in this paper is the problem of a rational economic organization, I shall in its course be led again and again to point to its close connections with certain methodological questions. Many of the points I wish to make are indeed conclusions toward which diverse paths of reasoning have unexpectedly converged. But, as I now see these problems, this is no accident. It seems to me that many of the current disputes with regard to both economic theory and economic policy have their common origin in a misconception about the nature of the economic problem of society. This misconception in turn is due to an erroneous transfer to social phenomena of the habits of thought we have developed in dealing with the phenomena of nature.

II

In ordinary language we describe by the word "planning" the complex of interrelated decisions about the allocation of our available resources. All economic activity is in this sense planning; and in any society in which many people collaborate, this planning, whoever does it, will in some measure have to be based on knowledge which, in the first instance, is not given to the planner but to somebody else, which somehow will have to be conveyed to the planner. The various ways in which the knowledge on which people base their plans is communicated to them is the crucial problem for any theory explaining the economic process, and the problem of what is the best way of utilizing knowledge initially dispersed among all the people is at least one of the main problems of economic policy—or of designing an efficient economic system.

The answer to this question is closely connected with that other question which arises here, that of *who* is to do the planning. It is about

Hayek's neglect of planning within firms is a serious omission. Herbert Simon pioneered the study of exchanges with incomplete contracting. Forty years later, he imagined "a mythical visitor from Mars" approaching Earth in a spaceship "equipped with a telescope that reveals social structures. The firms reveal themselves, say, as solid green areas . . . Market transactions show as red lines connecting firms, forming a network in the spaces between them. . . . A message sent back home, describing the scene would speak of *large green areas interconnected by red lines.* It would not likely speak of *a network of red lines connecting green spots*" (Simon 1991, 27).

this question that all the dispute about "economic planning" centers. This is not a dispute about whether planning is to be done or not. It is a dispute as to whether planning is to be done centrally, by one authority for the whole economic system, or is to be divided among many individuals. Planning in the specific sense in which the term is used in contemporary controversy necessarily means central planning—direction of the whole economic system according to one unified plan. Competition, on the other hand, means decentralized planning by many separate persons. The halfway house between the two, about which many people talk but which few like when they see it, is the delegation of planning to organized industries, or, in other words, monopoly.

Which of these systems is likely to be more efficient depends mainly on the question under which of them we can expect that fuller use will be made of the existing knowledge. And this, in turn, depends on whether we are more likely to succeed in putting at the disposal of a single central authority all the knowledge which ought to be used but which is initially dispersed among many different individuals, or in conveying to the individuals such additional knowledge as they need in order to enable them to fit their plans with those of others.

III

At the time, Hayek's characterization of central planning and markets as alternative information-processing systems was entirely novel.

It will at once be evident that on this point the position will be different with respect to different kinds of knowledge; and the answer to our question will therefore largely turn on the relative importance of the different kinds of knowledge; those more likely to be at the disposal of particular individuals and those which we should with greater confidence expect to find in the possession of an authority made up of suitably chosen experts. If it is today so widely assumed that the latter will be in a better position, this is because one kind of knowledge, namely, scientific knowledge, occupies now so prominent a place in public imagination that we tend to forget that it is not the only kind that is relevant. It may be admitted that, as far as scientific knowledge is concerned, a body of suitably chosen experts may be in the best position to command all the best knowledge available—though this is of course merely shifting the difficulty to the problem of selecting the experts.

What I wish to point out is that, even assuming that this problem can be readily solved, it is only a small part of the wider problem.

Today it is almost heresy to suggest that scientific knowledge is not the sum of all knowledge. But a little reflection will show that there is beyond question a body of very important but unorganized knowledge which cannot possibly be called scientific in the sense of knowledge of general rules: the knowledge of the particular circumstances of time and place. It is with respect to this that practically every individual has some advantage over all others because he possesses unique information of which beneficial use might be made, but of which use can be made only if the decisions depending on it are left to him or are made with his active coöperation. We need to remember only how much we have to learn in any occupation after we have completed our theoretical training, how big a part of our working life we spend learning particular jobs, and how valuable an asset in all walks of life is knowledge of people, of local conditions, and of special circumstances. To know of and put to use a machine not fully employed, or somebody's skill which could be better utilized, or to be aware of a surplus stock which can be drawn upon during an interruption of supplies, is socially quite as useful as the knowledge of better alternative techniques. And the shipper who earns his living from using otherwise empty or half-filled journeys of tramp-steamers, or the estate agent whose whole knowledge is almost exclusively one of temporary opportunities, or the *arbitrageur* who gains from local differences of commodity prices, are all performing eminently useful functions based on special knowledge of circumstances of the fleeting moment not known to others.

Now termed asymmetric information.

It is a curious fact that this sort of knowledge should today be generally regarded with a kind of contempt and that anyone who by such knowledge gains an advantage over somebody better equipped with theoretical or technical knowledge is thought to have acted almost disreputably. To gain an advantage from better knowledge of facilities of communication or transport is sometimes regarded as almost dishonest, although it is quite as important that society make use of the best opportunities in this respect as in using the latest scientific discoveries. This prejudice has in a considerable measure affected the attitude toward commerce in general compared with that toward production. Even economists who regard themselves as definitely immune to the crude materialist fallacies of the past constantly commit the same mistake where activities directed toward the acquisition of such practical knowledge are concerned—apparently because in their scheme of things all such knowledge is supposed to be "given." The common idea now seems to be that all such knowledge should as a matter of course be readily at the command of everybody, and the reproach of irrationality leveled against the existing economic order is frequently based on the fact that it is not so available. This view disregards the fact that the method by which such knowledge can be made as widely available as possible is precisely the problem to which we have to find an answer.

IV

If it is fashionable today to minimize the importance of the knowledge of the particular circumstances of time and place, this is closely connected with the smaller importance which is now attached to change as such. Indeed, there are few points on which the assumptions made (usually only implicitly) by the "planners" differ from those of their opponents as much as with regard to the significance and frequency of changes which will make substantial alterations of production plans necessary. Of course, if detailed economic plans could be laid down for fairly long periods in advance and then closely adhered to, so that no further economic decisions of importance would be required, the task of drawing up a comprehensive plan governing all economic activity would be much less formidable.

Note that Hayek is interested in change, not stasis, and finds equilibrium models inadequate.

It is, perhaps, worth stressing that economic problems arise always and only in consequence of change. So long as things continue as before, or at least as they were expected to, there arise no new problems requiring a decision, no need to form a new plan. The belief that changes, or at least day-to-day adjustments, have become less important in modern times implies the contention that economic problems also have become less important. This belief in the decreasing importance of change is, for that reason, usually held by the same people who argue that the importance of economic considerations has been driven into the background by the growing importance of technological knowledge.

Is it true that, with the elaborate apparatus of modern production, economic decisions are required only at long intervals, as when a new factory is to be erected or a new process to be introduced? Is it true that, once a plant has been built, the rest is all more or less mechanical, determined by the character of the plant, and leaving little to be changed in adapting to the ever-changing circumstances of the moment?

The fairly widespread belief in the affirmative is not, as far as I can ascertain, borne out by the practical experience of the businessman. In a competitive industry at any rate—and such an industry alone can serve as a test—the task of keeping cost from rising requires constant struggle, absorbing a great part of the energy of the manager. How easy it is for an inefficient manager to dissipate the differentials on which profitability rests, and that it is possible, with the same technical facilities, to produce with a great variety of costs, are among the commonplaces of business experience which do not seem to be equally familiar in the study of the economist. The very strength of the desire, constantly voiced by producers and engineers, to be allowed to proceed untrammeled by considerations of money costs, is eloquent testimony to the extent to which these factors enter into their daily work.

One reason why economists are increasingly apt to forget about the constant small changes which make up the whole economic picture is probably their growing preoccupation with statistical aggregates, which show a very much greater stability than the movements of the detail. The comparative stability of the aggregates cannot, however, be accounted for—as the statisticians occasionally seem to be inclined to do—by the "law of large numbers" or the mutual compensation of random changes. The number of elements with which we have to deal is not large enough for such accidental forces to produce stability. The continuous flow of goods and services is maintained by constant deliberate adjustments, by new dispositions made every day in the light of circumstances not known the day before, by B stepping in at once when A fails to deliver. Even the large and highly mechanized plant keeps going largely because of an environment upon which it can draw for all sorts of unexpected needs; tiles for its roof, stationery for its forms, and all the thousand and one kinds of equipment in which it cannot be self-contained and which the plans for the operation of the plant require to be readily available in the market.

The Great Depression and the macroeconomics of John Maynard Keynes stimulated the development of measures of national income and other aggregates by Simon Kuznets and others.

This idea—knowledge that is intrinsically difficult or impossible to transmit—has generally not been taken up in more recent models.

This is, perhaps, also the point where I should briefly mention the fact that the sort of knowledge with which I have been concerned is knowledge of the kind which by its nature cannot enter into statistics and therefore cannot be conveyed to any central authority in statistical form. The statistics which such a central authority would have to use would have to be arrived at precisely by abstracting from minor differences between the things, by lumping together, as resources of one kind, items which differ as regards location, quality, and other particulars, in a way which may be very significant for the specific decision. It follows from this that central planning based on statistical information by its nature cannot take direct account of these circumstances of time and place and that the central planner will have to find some way or other in which the decisions depending on them can be left to the "man on the spot."

V

If we can agree that the economic problem of society is mainly one of rapid adaptation to changes in the particular circumstances of time and place, it would seem to follow that the ultimate decisions must be left to the people who are familiar with these circumstances, who know directly of the relevant changes and of the resources immediately available to meet them. We cannot expect that this problem will be solved by first communicating all this knowledge to a central board which, after integrating all knowledge, issues its orders. We must solve it by some form of decentralization. But this answers only part of our problem. We need decentralization because only thus can we insure that the knowledge of the particular circumstances of time and place will be promptly used. But the "man on the spot" cannot decide solely on the basis of his limited but intimate knowledge of the facts of his immediate surroundings. There still remains the problem of communicating to him such further information as he needs to fit his decisions into the whole pattern of changes of the larger economic system.

How much knowledge does he need to do so successfully? Which of the events which happen beyond the horizon of his immediate knowledge are of relevance to his immediate decision, and how much of them need he know?

There is hardly anything that happens anywhere in the world that *might* not have an effect on the decision he ought to make. But he need not know of these events as such, nor of *all* their effects. It does not matter for him *why* at the particular moment more screws of one size than of another are wanted, *why* paper bags are more readily available than canvas bags, or *why* skilled labor, or particular machine tools, have for the moment become more difficult to obtain. All that is significant for him is *how much more or less* difficult to procure they have become compared with other things with which he is also concerned, or how much more or less urgently wanted are the alternative things he produces or uses. It is always a question of the relative importance of the particular things with which he is concerned, and the causes which alter their relative importance are of no interest to him beyond the effect on those concrete things of his own environment.

It is in this connection that what I have called the "economic calculus" proper helps us, at least by analogy, to see how this problem can be solved, and in fact is being solved, by the price system. Even the single controlling mind, in possession of all the data for some small, self-contained economic system, would not—every time some small adjustment in the allocation of resources had to be made—go explicitly through all the relations between ends and means which might possibly be affected. It is indeed the great contribution of the pure logic of choice that it has demonstrated conclusively that even such a single mind could solve this kind of problem only by constructing and constantly using rates of equivalence (or "values," or "marginal rates of substitution"), *i.e.,* by attaching to each kind of scarce resource a numerical index which cannot be derived from any property possessed by that particular thing, but which reflects, or in which is condensed, its significance in view of the whole means-end structure. In any small change he will have to consider only these quantitative indices (or "values") in which all the relevant information is concentrated; and, by adjusting the quantities one by one, he can appropriately rearrange his dispositions without having to solve the whole puzzle *ab initio* or without needing at any stage to survey it at once in all its ramifications.

Fundamentally, in a system in which the knowledge of the relevant facts is dispersed among many people, prices can act to coördinate the separate actions of different people in the same way as subjective values help the individual to coördinate the parts of his plan. It is worth contemplating for a moment a very simple and commonplace instance of the action of the price system to see what precisely it accomplishes. Assume that somewhere in the world a new opportunity for the use of some raw material, say, tin, has arisen, or that one of the sources of supply of tin has been eliminated. It does not matter for our purpose—and it is very significant that it does not matter—which of these two causes has made tin more scarce. All that the users of tin need to know is that some of the tin they used to consume is now more profitably employed elsewhere and that, in consequence, they must economize tin. There is no need for the great majority of them even to know where the more urgent need has arisen, or in favor of what other needs they ought to husband the supply. If only some of them know directly of the new demand, and switch resources over to it, and if the people who are aware of the new gap thus created in turn fill it from still other sources, the effect will rapidly spread throughout the whole economic system and influence not only all the uses of tin but also those of its substitutes and the substitutes of these substitutes, the supply of all the things made of tin, and their substitutes, and so on; and all this without the great majority of those instrumental in bringing about these substitutions knowing anything at all about the original cause of these changes. The whole acts as one market, not because any of its members survey the whole field, but because their limited individual fields of vision sufficiently overlap so that through many intermediaries the relevant information is communicated to all. The mere fact that there is one price for any commodity—or rather that local prices are connected in a manner determined by the cost of transport, etc.—brings about the solution which (it is just conceptually possible) might have been arrived at by one single mind possessing all the information which is in fact dispersed among all the people involved in the process.

Exactly this point was made by the pro-planning side of the debate. The planner could implement exactly the market solution, and then do better by correcting it for market failures due to limited competition (monopoly) or negative external effects (environmental spillovers).

VI

We must look at the price system as such a mechanism for communicating information if we want to understand its real function—a function which, of course, it fulfils less perfectly as prices grow more rigid. (Even when quoted prices have become quite rigid, however, the forces which would operate through changes in price still operate to a considerable extent through changes in the other terms of the contract.) The most significant fact about this system is the economy of knowledge with which it operates, or how little the individual participants need to know in order to be able to take the right action. In abbreviated form, by a kind of symbol, only the most essential information is passed on and passed on only to those concerned. It is more than a metaphor to describe the price system as a kind of machinery for registering change, or a system of telecommunications which enables individual producers to watch merely the movement of a few pointers, as an engineer might watch the hands of a few dials, in order to adjust their activities to changes of which they may never know more than is reflected in the price movement.

Of course, these adjustments are probably never "perfect" in the sense in which the economist conceives of them in his equilibrium analysis. But I fear that our theoretical habits of approaching the problem with the assumption of more or less perfect knowledge on the part of almost everyone has made us somewhat blind to the true function of the price mechanism and led us to apply rather misleading standards in judging its efficiency. The marvel is that in a case like that of a scarcity of one raw material, without an order being issued, without more than perhaps a handful of people knowing the cause, tens of thousands of people whose identity could not be ascertained by months of investigation, are made to use the material or its products more sparingly; *i.e.,* they move in the right direction. This is enough of a marvel even if, in a constantly changing world, not all will hit it off so perfectly that their profit rates will always be maintained at the same constant or "normal" level.

I have deliberately used the word "marvel" to shock the reader out of the complacency with which we often take the working of this mechanism for granted. I am convinced that if it were the result of deliberate human design, and if the people guided by the price changes understood that their decisions have significance far beyond their immediate aim, this mechanism would have been acclaimed as one of the greatest triumphs of the human mind. Its misfortune is the double one that it is not the product of human design and that the people guided by it usually do not know why they are made to do what they do. But those who clamor for "conscious direction"—and who cannot believe that anything which has evolved without design (and even without our understanding it) should solve problems which we should not be able to solve consciously—should remember this: The problem is precisely how to extend the span of out utilization of resources beyond the span of the control of any one mind;

Here Hayek is developing the idea of spontaneous order.

and therefore, how to dispense with the need of conscious control, and how to provide inducements which will make the individuals do the desirable things without anyone having to tell them what to do.

The problem which we meet here is by no means peculiar to economics but arises in connection with nearly all truly social phenomena, with language and with most of our cultural inheritance, and constitutes really the central theoretical problem of all social science. As Alfred Whitehead has said in another connection, "It is a profoundly erroneous truism, repeated by all copy-books and by eminent people when they are making speeches, that we should cultivate the habit of thinking what we are doing. The precise opposite is the case. Civilization advances by extending the number of important operations which we can perform without thinking about them." This is of profound significance in the social field. We make constant use of formulas, symbols, and rules whose meaning we do not understand and through the use of which we avail ourselves of the assistance of knowledge which individually we do not possess. We have developed these practices and institutions by building upon habits and institutions which have proved successful in their own sphere and which have in turn become the foundation of the civilization we have built up.

Sociologist Talcott Parsons calls markets and other ubiquitous institutions that people have stumbled on in a wide variety of environments "evolutionary universals."

The price system is just one of those formations which man has learned to use (though he is still very far from having learned to make the best use of it) after he had stumbled upon it without understanding it. Through it not only a division of labor but also a coördinated utilization of resources based on an equally divided knowledge has become possible. The people who like to deride any suggestion that this may be so usually distort the argument by insinuating that it asserts that by some miracle just that sort of system has spontaneously grown up which is best suited to modern civilization. It is the other way round: man has been able to develop that division of labor on which our civilization is based because he happened to stumble upon a method which made it possible. Had he not done so, he might still have developed some other, altogether different, type of civilization, something like the "state" of the termite ants, or some other altogether unimaginable type. All that we can say is that nobody has yet succeeded in designing an alternative system in which certain features of the existing one can be preserved which are dear even to those who most violently assail it—such as particularly the extent to which the individual can choose his pursuits and consequently freely use his own knowledge and skill.

Leonid Kantorovich and later George Danzig were also developing linear programming methods that allowed the computation of shadow prices in the absence of markets. Marxian economist Maurice Dobb suggested that computers then under development could perform the necessary computations on a scale adequate to plan an entire economy.

VII

It is in many ways fortunate that the dispute about the indispensability of the price system for any rational calculation in a complex society is now no longer conducted entirely between camps holding different political views. The thesis that without the price system we could not preserve a society based on such extensive division of labor as ours was greeted with a howl of derision when it was first advanced by von Mises twenty-five years ago. Today the difficulties which some still find in accepting it are no longer mainly political, and this makes for an atmosphere much more conducive to reasonable discussion. When we find Leon Trotsky arguing that "economic accounting is unthinkable without market relations"; when Professor Oscar Lange promises Professor von Mises a statue in the marble halls of the future Central Planning Board; and when Professor Abba P. Lerner rediscovers Adam Smith and emphasizes that the essential utility of the price system consists in inducing the individual, while seeking his own interest, to do what is in the general interest, the differences can indeed no longer be ascribed to political prejudice. The remaining dissent seems clearly to be due to purely intellectual, and more particularly methodological, differences.

How ironic that here Hayek channels Trotsky, a leader of the Bolshevik Revolution, who a decade earlier had written: "If a universal mind existed, such a mind, of course, could a priori draw up a faultless and exhaustive economic plan, beginning with the number of acres of wheat down to the last button for a vest. The bureaucracy often imagines that just such a mind is at its disposal; that is why it so easily frees itself from the control of the market" (Trotsky 1932).

A recent statement by Professor Joseph Schumpeter in his *Capitalism, Socialism, and Democracy* provides a clear illustration of one of the methodological differences which I have in mind. Its author is pre-eminent among those economists who approach economic phenomena in the light of a certain branch of positivism. To him these phenomena accordingly appear as objectively given quantities of commodities impinging directly upon each other, almost, it would seem, without any intervention of human minds. Only against this background can I account for the following (to me startling) pronouncement. Professor Schumpeter argues that the possibility of a rational calculation in the absence of markets for the factors of production follows for the theorist "from the elementary proposition

Schumpeter is stating that consumers' valuation of goods produced is the basis for imputing values (prices) to the inputs required to produce them (as can be done using linear programming).

that consumers in evaluating ('demanding') consumers' goods *ipso facto* also evaluate the means of production which enter into the production of these goods."[1]

Taken literally, this statement is simply untrue. The consumers do nothing of the kind. What Professor Schumpeter's "*ipso facto*" presumably means is that the valuation of the factors of production is implied in, or follows necessarily from, the valuation of consumers' goods. But this, too, is not correct. Implication is a logical relationship which can be meaningfully asserted only of propositions simultaneously present to one and the same mind. It is evident, however, that the values of the factors of production do not depend solely on the valuation of the consumers' goods but also on the conditions of supply of the various factors of production. Only to a mind to which all these facts were simultaneously known would the answer necessarily follow from the facts given to it. The practical problem, however, arises precisely because these facts are never so given to a single mind, and because, in consequence, it is necessary that in the solution of the problem knowledge should be used that is dispersed among many people.

The problem is thus in no way solved if we can show that all the facts, *if* they were known to a single mind (as we hypothetically assume them to be given to the observing economist), would uniquely determine the solution; instead we must show how a solution is produced by the interactions of people each of whom possesses only partial knowledge. To assume all the knowledge to be given to a single mind in the same manner in which we assume it to be given to us as the explaining

[1] J. Schumpeter, *Capitalism, Socialism, and Democracy* (New York; Harper, 1942), p. 175. Professor Schumpeter is, I believe, also the original author of the myth that Pareto and Barone have "solved" the problem of socialist calculation. What they, and many others, did was merely to state the conditions which a rational allocation of resources would have to satisfy and to point out that these were essentially the same as the conditions of equilibrium of a competitive market. This is something altogether different from knowing how the allocation of resources satisfying these conditions can be found in practice. Pareto himself (from whom Barone has taken practically everything he has to say), far from claiming to have solved the practical problem, in fact explicitly denies that it can be solved without the help of the market. See his *Manuel d'économie pure* (2nd ed., 1927), pp. 233–34. The relevant passage is quoted in an English translation at the beginning of my article on "Socialist Calculation: The Competitive 'Solution,'" in *Economica*, New Series, Vol. VIII, No. 26 (May, 1940), p. 125.

economists is to assume the problem away and to disregard everything that is important and significant in the real world.

Hayek is prescient in linking ill-conceived simplifying abstractions such as complete information to the growing use of mathematics. But the problem is not the mathematics, as contemporary models taking account of limited information show.

That an economist of Professor Schumpeter's standing should thus have fallen into a trap which the ambiguity of the term "datum" sets to the unwary can hardly be explained as a simple error. It suggests rather that there is something fundamentally wrong with an approach which habitually disregards an essential part of the phenomena with which we have to deal: the unavoidable imperfection of man's knowledge and the consequent need for a process by which knowledge is constantly communicated and acquired. Any approach, such as that of much of mathematical economics with its simultaneous equations, which in effect starts from the assumption that people's *knowledge* corresponds with the objective *facts* of the situation, systematically leaves out what is our main task to explain. I am far from denying that in our system equilibrium analysis has a useful function to perform. But when it comes to the point where it misleads some of our leading thinkers into believing that the situation which it describes has direct relevance to the solution of practical problems, it is high time that we remember that it does not deal with the social process at all and that it is no more than a useful preliminary to the study of the main problem.

It amazes me that, in the vast majority of courses, introductory economics persists in these assumptions. For an alternative introduction to economics that embraces Hayek's view that knowledge is limited and local, see Bowles and Carlin (2020).

REFERENCES

Pareto, V. 1927. *Manuel d'Économie Pure.* 2nd. Paris, France: Marcel Giard.

Schumpeter, J. 1942. *Capitalism, Socialism, and Democracy.* New York, NY: Harper.

von Hayek, F. A. 1940. "Socialist Calculation: The Competitive 'Solution'." *Economica* 8 (26): 125–149.

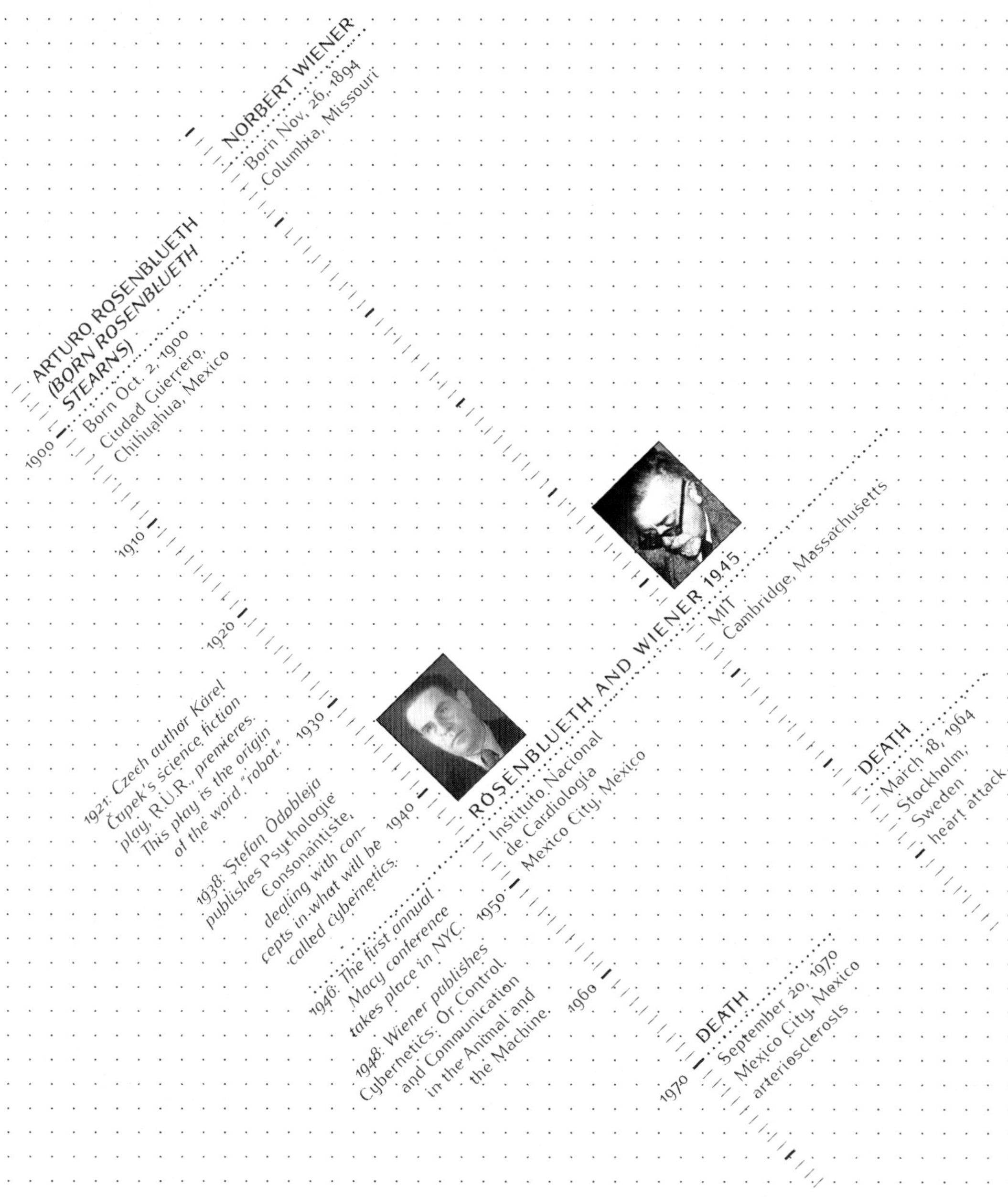

ROSENBLUETH & WIENER

[8]

OPENING A CLOSED BOX

Cosma Rohilla Shalizi, Carnegie Mellon University and Santa Fe Institute

A. Rosenblueth and N. Wiener, "The Role of Models in Science," *Philosophy of Science* 12 (4), 316–321 (1945).

All science, as this paper explains, relies on models, but this is more *obvious* for the study of complex systems than, say, mycology, so it is only fitting that we begin this volume with one of the first self-conscious articulations of how and why scientists use models. A few words of scene-setting are in order before discussing the paper itself, about our authors, their larger interdisciplinary collaborative project, and how this paper both fit into that project and opened new pathways.

Arturo Rosenblueth

Arturo Rosenblueth Stearns (1900–1970) was a Mexican neurophysiologist.[1] Born in Ciudad Guerrero, Chihuahua, to a Hungarian Jewish immigrant father and a Mexican-American mother, he studied medicine, specializing in physiology, first in Mexico City and then (after an interlude when, lacking funds, he made a living playing piano in restaurants) in Berlin, and finally Paris. There he was educated in the tradition of physiology descending from the great Claude Bernard (1813–1878), where biomedical investigation was blended with sophisticated ideas about scientific method and philosophy (Bernard [1865]1927).[2] In 1927 he obtained his medical de-

[1] More or less brief accounts of Rosenblueth's life and work can be found in all of the biographies of Wiener cited below. There are few dedicated studies in English, so I have relied on Salmerón (1978), Guadalajara Boo (2012), and especially on the works of Ruth Guzik Glantz (2015, 2009). (I am grateful to Prof. A. E. Owen for help with these references.) I have been unable to consult Guzik Glantz (2018); an English translation is very much to be desired.

[2] Some aspects of this tradition include: a focus on characterizing the functional role of anatomical structures, and then on characterizing the physical and (especially) chemical mechanisms through which those functions are implemented; an embrace of the method of conjectures and refutations (long before Karl Popper coined that apt phrase) rather than inductive generalization; a focus on crucial experiments where the diverging consequences of differing ideas could be subjected to empirical test; an

gree from the Sorbonne, returning to Mexico City as a professor of physiology at the National School of Medicine. In 1930, however, a Guggenheim fellowship took him to Harvard Medical School, where he was to spend the next fourteen years under the auspices of the celebrated Walter B. Cannon (1871–1945). Cannon—an academic "grandchild" of Bernard—was a giant of American science, who investigated the chemical bases of an immense range of physiological processes, and coined many concepts still in common (if anonymous) use, such as "fight or flight" reactions, and "homeostasis" (itself an elaboration of Bernard's "stability of the internal environment").

Rosenblueth's work with Cannon focused on the physiology of the nervous system, specifically the conduction of electrical impulses ("spikes") through neurons, and the action of the sympathetic nervous system. In both cases the goal was to work out the chemical mechanisms underlying the observed phenomena—very much in line with the Bernard tradition. While scientifically productive, his position at Harvard was precarious, reliant on Cannon's patronage. Cannon, for his part, tried to secure Rosenblueth a permanent position at a US university (including nominating Rosenblueth as Cannon's own successor at Harvard), but failed, in part due to explicit antisemitism. In 1944, Rosenblueth returned to Mexico City, where he held a series of increasingly distinguished positions at the Instituo Nacional de Cardiología and the Centro de Investigación y de Estudios Avanzados del Instituto Politécnico Nacional. It is characteristic that his final publication during his lifetime was a book titled *Mind and Brain: A Philosophy of Science* (Rosenblueth 1970).

Our other author gives us a memorable sketch of his "companion in science" (Wiener 1955, 171):

> *Arturo is a burly, vigorous man of middle height, quick in his action and in his speech, who paces rapidly up and down*

---

insistence on causal determinism, with the implication that any variability or (apparent) role for chance was merely a sign of some neglected, systematic causes. We can see traces of all of this in the essay, including its examples, its ideas about progressive refinement and elaboration of models, and the use of formal models themselves, to deduce the consequences of assumptions so that they might be compared to experimental realities.

the room when he is thinking. No one who sees him in the Mexican environment can doubt that he is a true Mexican, though the greater part of his genetic heritage comes from other countries, particularly Hungary. Arturo and I hit it off well from the very beginning, though to hit it off well with Arturo means not that one has no disagreements with him, but rather that one enjoys these disagreements.

Norbert Wiener

Wiener (1894–1964), for his part, is a more celebrated, even mythologized, figure.[3] His father, Leo Wiener, was a polyglot Jewish polymath from what was then imperial Russia and is now Belorussia, who (after adventures in eastern and western Europe, central America and Missouri, where he married Norbert's future mother), became professor of Slavic languages at Harvard. Among Leo's *many* eccentric theories were ideas on education, which he implemented with young Norbert; because or despite of those interventions, Norbert proved to be "an infant prodigy in the full sense of the word": "I entered college before the age of twelve, obtained my bachelor's degree before fifteen, and my doctorate before nineteen" (Wiener 1953, 3). That doctorate was in philosophy, specifically mathematical logic; it was followed by what we'd now call a post-doc at Cambridge University, with the great logician and philosopher Bertrand Russell (1872–1970). Just as decisively, Cambridge was also where Wiener was introduced to the more conventional branches of higher mathematics, from the lectures of the celebrated G. H. Hardy (1877–1947).

Returning to the US, Wiener decisively failed to find satisfactory academic employment in philosophy, despite publishing a number of papers in mathematical logic now regarded as fundamental. After floundering as (among other things) a hack encyclopedia writer, a

[3]In addition to his (well-written and *largely* reliable) memoirs, Wiener (1953) and Wiener (1955), Wiener has been the subject of at least four full-length biographical studies: Heims (1980), contrasting him (favorably) with John von Neumann; Masani (1990), emphasizing his mathematical accomplishments; the popularizing Conway and Siegelman (2005); and Montagnini (2017), emphasizing his role as a philosopher and pioneer of interdisciplinarity.

"computer" at the US Military's Aberdeen Proving Grounds, a private in the US Army, and a reporter for a Boston newspaper, he finally became a professor of mathematics at the Massachusetts Institute of Technology, where he remained for the rest of his life. There he began to produce works in pure and applied mathematics which secured his place as one of the leading mathematicians of the twentieth century. Two aspects of his inter-war work stand out here:[4] his theory of Brownian motion and his contributions to harmonic analysis. The former provided the first fully-coherent theory of a random process in continuous time, putting earlier heuristic work by Einstein, Langevin and other physicists on a firm mathematical footing, and paving the way for the modern theory of stochastic processes [@von-Plato-modern-prob]. In harmonic analysis, his work was many-sided, but one key part was understanding what happens when we try to decompose random signals into superpositions of sine waves, and relating the properties of the resulting Fourier spectrum to the statistical distribution of signals. While pursuing these more theoretical undertakings, Wiener was also, encouraged by the MIT environment, collaborating with engineers on very practical issues like the design and analysis of switching systems and analog computers.

Interdisciplinary Collaboration and Cybernetics

It was in 1933, when Rosenblueth was at Harvard Medical School and Wiener was at MIT, that the two men met, introduced by a former student of Wiener's, the Mexican physicist Manuel Sandoval Vallarta. Rosenblueth, who had an interest in philosophy of science dating back to reading Poincaré (2001) as a student, ran an informal seminar on "scientific method," mostly but not exclusively attended by other biologists in the area. (It seems to have often been as much dinner party as seminar.) Wiener began attending, and quickly became both a leading participant and a personal friend to Rosenblueth. The two shared a belief that

[4]At a technical level, both of these contributions relied on the newer ideas about integration developed by Lebesgue, Borel and others, which we now know as "measure theory." Thus the issue with Wiener's model of Brownian motion was really "what does it mean to have a probability distribution over trajectories that evolve continuously in time?" (and not just a distribution over positions at a finite set of times). Similarly, the issues in harmonic analysis came down to "how do we make sense of situations where the Fourier transform of a signal needs a *continuous* set of frequencies?"

"divisions between the sciences were convenient administrative lines for the apportionment of money and effort, which each working scientist should be willing to cross whenever his studies should appear to demand it" (Wiener 1955, 171). It was in this period that the two men began thinking about "the application of mathematics, and in particular of communications theory, to physiological method" (Wiener 1955, 173).

With the arrival of World War II, Wiener pursued a project for the US military on automatic anti-aircraft fire control. An anti-aircraft gun needs to fire at where a plane *will* be, rather than where it is now, so the future trajectory of the target needs to be extrapolated or predicted from its past, which was itself observed noisily. Since this was all to be done with analog equipment, the trajectory of the plane, the data and the predictions could all be thought of as continuous functions. Wiener set up the problem of finding the optimal predictor as a least squares problem, and used his work in harmonic analysis to show how to find the optimal linear solution. The result was a general theory of (linear) "extrapolation, interpolation and smoothing of stationary time series" (Wiener 1949). (Parallel work was done independently, and simultaneously, by Kolmogorov in the USSR.) Prototype predictors based on these principles were actually built and greatly impressed competent observers, but did not see use in the war, the army settling on simpler alternatives.

During the course of this work, Wiener became seized by an analogy between the principles of negative feedback control, long used in automatic machinery (Mayr 1986; Mindell 2004), and biological phenomena of self-regulation and even purposive behavior. (When you are building a gun that adjusts itself to shoot planes out of the sky, it is probably hard to avoid talking about what the gun is *trying* to do.) Together with his engineering collaborator Julian Bigelow, Wiener approached Rosenblueth to see if there was anything to the analogy biologically, especially neurologically. At a high level of generality (and vagueness), of course negative feedback is akin to Cannon's notion of homeostasis, but the trio went beyond that. Control theory already provided the tools to analyze the *failures* of feedback, including those due to over-correction, those due to excessive delay, and so forth. Rosenblueth was able to provide fairly convincing neurological examples of *pathological* conditions corresponding to these different failures of feedback. In terms of the present paper: if one thinks of (say) visually-guided reaching as being governed by the same abstract, formal model as an electro-mechanical servomechanism, then one should *expect* to see certain kinds of pathological motion when there is too much delay between seeing and moving. Observing this pathology then strengthens one's confidence in the aptness of the abstract model.

These considerations led to a truly seminal paper, Rosenblueth, Wiener, and Bigelow (1943), which analyzed *some* kinds of "teleological" (goal-directed, purposive) behavior as the result of feedback mechanisms.[5] This line of thought was a key ingredient—arguably *the* key ingredient—in the synthesis that Wiener presently dubbed "cybernetics," and descried, in the subtitle of his classic book (Wiener 1948), as the study of "control and communication in the animal and the machine," and, we might add, "in society," too.[6] The book, dedicated to Rosenblueth and largely written during Wiener's visits to Mexico City, had an enormous impact. This was helped along by a series of conferences funded by the Macy Foundation devoted to these ideas, whose participants included many of the leading figures in American natural and social science, and in which Rosenblueth was a prominent participant.

This "movement" (Heims 1991) or "moment" (Kline 2017) was in some was astonishingly successful—it is "why we call our age the information age"—but ultimately failed to create an autonomous, self-propagating discipline, not least because by the 1960s it found itself pursuing some very strange notions and blind alleys. It would be an excellent thing to see a proper *scientific* assessment of its contributions

[5]It is a common misconception that Rosenblueth, Wiener, and Bigelow (1943) *identified* purpose and feedback. The text actually makes it clear that many kinds of purposive behavior can't be controlled by negative feedback "in the course of the behavior" (Rosenblueth, Wiener, and Bigelow 1943, 19–20), e.g., when throwing at a target, the muscles in the hand and arm have to move too quickly for nerve impulses from the eye, or even from the proprioreceptive sensors in the arm, to travel to the brain and back out to the muscles (Rosenblueth, Wiener, and Bigelow 1943, 20). Of course on a larger time scale, feedback can *improve* throwing, and, as they say, there is a kind of negative feedback involved in *stopping* purposive behavior once the aim has been met.

[6]There was a long tradition of claiming that organisms, even human beings, were machines (de la Mettrie 1994; Loeb 1912), or at least of seeking mechanistic explanations for biological phenomena (Bertoloni Meli 2019). People like Rosenblueth and Wiener were very aware of a long tradition of claiming organisms *were* machines, or at least *acted* like machines.

and limitations in light of modern knowledge. Instead, I will just remark here that in emphasizing information, computation, dynamical systems modeling, processes of circular causation, aggressively abstractions or analogies across physical, technological, biological and social domains, and general disdain for traditional disciplinary boundaries, cybernetics was clearly a predecessor to the field of "complex systems" that formed in the 1980s, and arguably even a linear ancestor.

It would be a mistake to see cybernetics as *just* a project of grand theorizing. It is characteristic of Rosenblueth and Wiener that they felt it important to demonstrate the worth of these ideas by showing how they could solve concrete problems—ones already recognized as worthy problems by scientific communities. Much of their collaborative time in the later 1940s was thus spent on devising, and fitting to data, stochastic models of neurophysiological processes, most successfully the input-output behavior of the synapses between neurons (Rosenblueth *et al.* 1949), and of the propagation of waves of activity in cardiac tissue (Wiener and Roseblueth 1946), a pioneering study of what are now called "excitable media" (Greenberg and Hastings 1978).

Modeling, and the Paper

The essential argument of the paper is that "no substantial part of the universe" is so simple that the human mind can grasp it without "abstraction," "replacing the part of the universe under consideration by a model of similar but simpler structure" (316). These models are, primarily, "formal or intellectual" affairs, which allow us to reason logically from the premises of about that structure, to conclusions about what we should see in the world, or *would* see under such-and-such conditions, or *will* see if we do something. Much of the art of the scientist thus consists in formulating, manipulating and revising models, with the full understanding that the model does not, and should not, try to capture the full richness of the world.

We can now see where this paper fits in with Rosenblueth and Wiener's larger collaborative project.[7] Like many ventures into philosophy by scientists, this essay is (in part!) about justifying what the authors were already doing. In 1945, Rosenblueth and Wiener were deep into their neurophysiological studies, and engaged in just this process of model-building and model-revision. These models *deliberately* abstracted from many details of the physiology. Similarly, the 1943 paper on teleology likewise abstracted away many implementation details between "servomechanisms" and organisms. One goal of the paper was to argue that such abstraction was, in fact, a virtue—or, rather, inevitable, and so better done thoughtfully than in denial.

While physical models of scientific hypotheses are literally ancient,[8] the conscious recognition of "formal or intellectual" models is a comparatively recent thing. There is a transition somewhere between, say, Newton, who thought of himself as describing "the system of the world," and, say, Bohr, putting forward a model of atomic structure that was frankly *merely* a model. Even when Newton works through what we would see as a model of single planet orbiting the sun (uninfluenced by the gravitation of other planets, etc.), he did not think of it that way. Bohr, quite manifestly, did. The roots of this shift are beyond tracing here, but some aspects can be pointed out: an increasing comfort with approximation, with partial descriptions and with generalizations of limited scope, and with a certain division of cognitive labor. To borrow an example from Rosenblueth and Wiener, someone doing acoustics can (usually) take mechanical properties of air for granted, without having to worry about *why* it has just those properties, or even whether those properties would alter under extreme conditions, secure in the knowledge that someone else is studying those issues. (Whether this *reflects* the social organization of the scientific community, or on the contrary whether that organization is an outcome of this sort of problem-solving approach, is a deep question.)

[7] There is very little secondary literature on this paper; the biographies of the two authors cited above mention it only in passing or not at all, and it did not spawn the lineage of commentaries and critiques as Rosenblueth, Wiener, and Bigelow (1943).

[8] Physical models of the orbits of the planets date back to the Hellenistic age, if not before, the most famous (now) being the artifact called the "Antikythera mechanism" (Jones 2017).

By the middle of the twentieth century, then, the conscious use of models was a prominent part of natural science, but also one that required some explication and defense. Rosenblueth and Wiener actually portray "material" models, such as scale models, biological "preparations," or what we'd now call "model organisms," as secondary to formal models, though this logical order reverses the order of historical development and of their own exposition. For Rosenblueth and Wiener, material models are practical expedients, for when the "part of the world" of interest is awkward to work with: too large, too small, too expensive, too hard to manipulate, too apt to bite experimenters. Nonetheless, for them, the scale model of (say) an airplane in a wind tunnel only lets us draw inferences about the full-sized airplane in the sky because there is a common formal model of the aerodynamic situation, and both the larger and the small physical systems are regarded as relevantly similar to this formal model. They admit that this formal model may only be implicit in the mind of (some) investigators, but insist on its importance.

One consequence of this view of models is that it is hard to say that models are *true* or *false*, so much as *more or less accurate*, and that in particular ways or respects. This, in turn, suggests that there can be multiple good models of the same system, either at different levels of approximation or for different aspects of the same physical process. This paper emphasizes the importance of levels of approximation, as in modeling sound transmission: from a simple linear model treating air as a homogeneous and incompressible fluid, to one including compressibility and shock-waves, to an anticipation of (ultimately quantum) molecular hydrodynamics.

Rosenblueth and Wiener liken this process of elaborating models to that of "opening" a "closed box." The "closed" boxes are the unarticulated parts of the model which are treated "functionally," that is, like mathematical functions defined as relations of inputs to outputs. "Open" boxes are elaborated parts of the model treated "structurally," specifying a structure or mechanism that implements the function, and gives some details about how inputs get turned into outputs. Which parts of the model need to be opened up, and which can be left closed, is again part of the modeler's art and varies with the goals and resources of the investigation. The temptation to (as it were) open all the boxes, and model everything in complete detail, is one they are clearly familiar with, but want the reader to avoid: a fully articulated model becomes as complex as the original, and so useless to a merely human intelligence.

This brings us to one of the more interesting features of this paper, for the present audience. They begin by asserting that "no substantial part of the universe is so simple that it can be grasped and controlled without abstraction," and end by talking about how increasingly refined models "approach asymptotically the complexity of the original situation" (316, 320). But they never actually *define* either "complexity" or "simplicity." If we were to try to back out a complexity measure from the way they use words like "simple" and "complex," we'd land on something like "number of explicitly articulated details or processes." To modern ears, this suggests some sort of description length, which would certainly fit our authors' pre-occupations with information theory, and, one could even say, would look forward to the Kolmogorov–Solomonoff notion of complexity as algorithmic information content (Kolmogorov 1965; Solomonoff 1964; Li and Vitányi 1997). But perhaps this mere anachronism.

Despite being published in *Philosophy of Science*, then and now a leading journal for that subfield, this paper, for all its insight and good sense, made little impact on philosophy. It was much later, really in the 1980s, that philosophers of science became interested in the role of models in science. Thus Giere (1988), to take one justly influential example, defines "models" as, basically, what our authors call "formal models" (78—80), adding that there are also "hypotheses" which "claim a similarity between the models and real systems . . . [and a] specification of [the] relevant respects and degrees [of similarity]" (81). Such accounts of models and modeling have been extremely influential in contemporary philosophy of science, and have begun to filter back out to the thinking of working scientists (e.g., Burch 2018) and the more thoughtful textbooks (Ashworth, Berry, and Bueno de Mesquita 2021). This is all, wittingly or not, following in Rosenblueth and Wiener's footsteps.

REFERENCES

Ashworth, S., C. R. Berry, and E. Bueno de Mesquita. 2021. *Theory and Credibility: Integrating Theoretical and Empirical Social Science.* Princeton, NJ: Princeton University Press.

Bernard, C. [1865]1927. *Introduction to the Study of Experimental Medicine.* Translated by Henry Copley Green. First published as *Introduction à l'étude de la médecine expérimentale*, Paris: J. B. Bailliere. Reprinted New York, NY: Dover, 1957. New York, NY: Macmillan.

Bertoloni Meli, D. 2019. *Mechanism: A Visual, Lexical, and Conceptual History.* Pittsburgh, PA: University of Pittsburgh Press.

Borges, J. L. [1946]1998. *Collected Fictions.* Translated by Andrew Hurley. New York, NY: Viking.

Burch, T. K. 2018. *Model-Based Demography: Essays on Integrating Data, Technique and Theory.* Cham, Switzerland: Springer. https://doi.org/10.1007/978-3-319-65433-1.

Conway, F., and J. Siegelman. 2005. *Dark Hero of the Information Age: In Search of Norbert Wiener, the Father of Cybernetics.* New York, NY: Basic Books.

de la Mettrie, J. O. 1994. *Man a Machine and Man a Plant.* Translated by Richard Watson and Maya Rybalka, with introduction and notes by Justin Leiber; first published as *L'Homme Machine* (1747) and *L'Homme Plante* (1948). Indianapolis, IN: Hackett.

Fishburn, P., and B. Monjardet. 1992. "Norbert Wiener on the Theory of Measurement, 1914, 1915, 1921." *Journal of Mathematical Psychology* 36:165–184. https://doi.org/10.1016/0022-2496(92)90035-6.

Giere, R. N. 1988. *Explaining Science: A Cognitive Approach.* Chicago, IL: University of Chicago Press.

Greenberg, J., and S. Hastings. 1978. "Spatial Patterns for Discrete Models of Diffusion in Excitable Media." *SIAM Journal on Applied Mathematics* 34:515–523. http://www.jstor.org/pss/2100950.

Guadalajara Boo, J. F. 2012. "Dr. Arturo Rosenblueth Stearns (1900--1970) en el Instituto Nacional de Cardiología "Ignacio Chávez" (1944–1961)." *Revista de la Facultad de Medicina de la UNAM* 55 (5): 4–10.

Guzik Glantz, R. 2009. "Relaciones de un Cientiífico Mexicano con el Extranjero: El case de Arturo Rosenblueth." *Revista mexicana de investigación educativa* 14 (40): 43–67. http://www.scielo.org.mx/scielo.php?script=sci_arttext&pid=S1405-66662009000100004.

———. 2015. "Entra la experimentación y los modelos abstractos: Breve historia de vida de Arturo Rosenblueth (1900–1970)." *Antropología: Revista Interidscioplinaria del INAH* 99:20–36. https://revistatest.inah.gob.mx/index.php/antropologia/article/view/8191.

———. 2018. *Arturo Rosenblueth, 1900–1970.* Mexico City, Mexico: El Colegio Nacional.

Heims, S. J. 1980. *John von Neumann and Norbert Wiener: From Mathematics to the Technologies of Life and Death.* Cambridge, MA: MIT Press.

———. 1991. *The Cybernetics Group: Constructing a Social Science for Postwar America, 1946—1953.* Cambridge, MA: MIT Press.

Jones, A. 2017. *A Portable Cosmos: Revealing the Antikythera Mechanism, Scientific Wonder of the Ancient World.* Oxford, UK: Oxford University Press.

Kline, R. R. 2017. *The Cybernetics Moment: Or Why We Call Our Age the Information Age.* Baltimore, MD: Johns Hopkins University Press.

Kolmogorov, A. N. 1965. "Three Approaches to the Quantitative Definition of Information." *Problems of Information Transmission* 1:1–7.

Li, M., and P. M. B. Vitányi. 1997. *An Introduction to Kolmogorov Complexity and Its Applications.* Second. New York, NY: Springer-Verlag.

Loeb, J. 1912. *The Mechanistic Conception of Life.* Reprint (Cambridge, Massachusetts: Harvard University Press, 1964), edited and with an introduction by Donald Flemming. Chicago, IL: University of Chicago Press.

Masani, P. R. 1990. *Norbert Wiener, 1894–1964.* Basel, Switzerland: Birkhaäuser. https://doi.org/10.1007/978-3-0348-9252-0.

Mayr, O. 1986. *Authority, Liberty, and Automatic Machinery in Early Modern Europe.* Baltimore, MD: Johns Hopkins University Press.

Mindell, D. A. 2004. *Between Human and Machine: Feedback, Control, and Computing before Cybernetics.* Baltimore, MD: Johns Hopkins University Press.

Montagnini, Leone. 2017. *Harmonies of Disorder: Norbert Wiener: A Mathematician-Philosophy of Our Time.* Cham, Switzerland: Springer. https://doi.org/10.1007/978-3-319-50657-9.

Poincaré, H. 2001. *The Value of Science: Essential Writings of Henri Poincaré.* Contents: *Science and Hypothesis* (1903, translation 1905); *The Value of Science* (1905, translation 1913); *Science and Method* (1908; translation 1914). New York, NY: Modern Library.

Rosenblueth, A. 1970. *Mind and Brain: A Philosophy of Science.* Cambridge, MA: MIT Press.

Rosenblueth, A., N. Wiener, and J. Bigelow. 1943. "Behavior, Purpose, and Teleology." *Philosophy of Science* 10:18–24. http://www.jstor.org/stable/184878.

Rosenblueth, A., N. Wiener, Walter Pitts, and J. García Ramos. 1949. "A Statistical Analysis of Synaptic Excitation." *Journal of Cellular and Comparative Physiology* 34:173–205. https://doi.org/10.1002/jcp.1030340202.

Russell, B. 1920. *Introduction to Mathematical Philosophy.* Second. First edition, 1919. London, UK: George Allen and Unwin. http://people.umass.edu/klement/russell-imp.html.

———. 1993. *Our Knowledge of the External World, as a Field for Scientific Method in Philosophy.* New introduction by John G. Slater; first edition Open Court Publishing, 1914. London, UK: Routledge.

Salmerón, F. 1978. "Noticia sobre Arturo Rosenblueth." Reprinted pp. 289--295 of *Dialogos, Diálogos* 14 (5 (83)): 27–29.

Solomonoff, R. J. 1964. "A Formal Theory of Inductive Inference." *Information and Control* 7:1--22 and 224–254. http://world.std.com/~rjs/pubs.html.

Spinoza, B. 1992. *Ethics; Treatise on the Emendation of the Intellect; and Selected Letters.* Translated by Samuel Shirley, edited and introduced by Seymour Feldman. Indianapolis, IN: Hackett.

Whitehead, A. N., and B. Russell. 1925--27. *Principia Mathematica.* 2nd. Cambridge, UK: Cambridge University Press.

Wiener, N. 1919. "A New Theory of Measurement: A Study in the Logic of Mathematics." *Proceedings of the London Mathematical Society* 19:181–205. https://doi.org/10.1112/plms/s2-19.1.181.

———. 1948. *Cybernetics: Or, Control and Communication in the Animal and the Machine.* New York, NY: Wiley.

———. 1949. *Extrapolation, Interpolation, and Smoothing of Stationary Time Series: With Engineering Applications.* First published during the war [1942] as a classified report to Section D_2, National Defense Research Council. Cambridge, MA: MIT Press.

———. 1953. *Ex-Prodigy: My Childhood and Youth.* New York, NY: Simon and Schuster.

———. 1955. *I Am a Mathematician: The Later Life of a Prodigy.* Cambridge, MA: MIT Press.

———. 1958. *Nonlinear Problems in Random Theory.* Cambridge, MA: MIT Press.

———. 1974. *Collected Works: with Commentaries.* Cambridge, MA: MIT Press.

Wiener, N., and A. Roseblueth. 1946. "The Mathematical Formulation of the Problem of Conduction of Impulses in a Network of Connected Excitable Elements, Specifically in Cardiac Muscle." Reprinted in Wiener 1974, vol. 4, p.. 511--571, *Archivos del Instituo de Cardiología de México* 16:205–265.

THE ROLE OF MODELS IN SCIENCE

Arturo Rosenblueth, Instituto Nacional de Cardiología, México
and Norbert Wiener, Massachusetts Institute of Technology

The suggestion is that when scientists imagine themselves as being able to "control" some part of the universe, they are conceiving of themselves as exterior to, and distinct from, that part of the universe, able to *choose* what to do to it, and not, for example, part of a single all-embracing web of causal determination. See also Rosenblueth (1970).

The intention and the result of a scientific inquiry is to obtain an understanding and a control of some part of the universe. This statement implies a dualistic attitude on the part of scientists. Indeed, science does and should proceed from this dualistic basis. But even though the scientist behaves dualistically, his dualism is operational and does not necessarily imply strict dualistic metaphysics.

No substantial part of the universe is so simple that it can be grasped and controlled without abstraction. Abstraction consists in replacing the part of the universe under consideration by a model of similar but simpler structure. Models, formal or intellectual on the one hand, or material on the other, are thus a central necessity of scientific procedure. The purpose of this paper is to analyze the usefulness and the limitations of the diverse forms of scientific models.

Aristotle famously distinguished four kinds of causes of an event phenomenon (*Physics*, Book II, section 3): material, formal, efficient, and final. That is: the nature of the stuff involved; the arrangement of the stuff; the "primary source of the change or coming to rest" that brings the phenomenon into being or sustains it; and the end or purpose for which the event happens. In distinguishing "material" and "formal" models here, then, our authors are following a very ancient precedent, which they would certainly have been aware of.

An investigator is often not aware of his methodological procedure, nor is it indispensable that he should have this awareness. Important scientific contributions, especially of an experimental character, can be made even though the experimenter does not realize that all good experiments are good abstractions.

An experiment is a question. A precise answer is seldom obtained if the question is not precise; indeed, foolish answers—i.e., inconsistent, discrepant or irrelevant experimental results—are usually indicative of a foolish question.

Not all scientific questions are directly amenable to experiment. There is a hierarchy of questions whose levels are determined by the generality of the answers sought. Thus the question of what a certain drug, e.g., cebadine, does to a certain manifestation of a nerve impulse, e.g., the spike potential, belongs to a relatively "low" level in the

Now more often called the "action potential."

hierarchy of physiological questions, because it deals with a narrowly restricted phenomenon. An experimenter might formulate and answer that question precisely, and yet have only a vague, intuitive appreciation of its "higher", more general and abstract implications, such as the action of all drugs belonging to a certain chemical group on the spike potential, or the relations between spike potential amplitude and other manifestations of nerve activity.

As a rule "high" order, very abstract and general questions, are not directly amenable to an experimental test. They have to be broken down into more specific terms, terms directly translatable into experimental procedure. There are thus two qualitatively different operations involved in the process of formulating the test of a general statement, or in the converse process of building a theory from experimental data. One of these operations consists in moving up or down the scale of abstraction; the other requires the translation of abstraction into experiment, or vice versa. The good experimenter has unusual ability in the second procedure; he is capable of freely interchanging symbols and events. The theorist, on the other hand, deals mainly with the first type of operations, those at various levels within the realm of abstraction.

It might appear that the most expedient method of approaching a problem scientifically would be to formulate the most general question or questions possible, and then to subdivide these questions into less abstract statements, until first order abstractions, directly testable, would be reached. This method is applicable only exceptionally, because very abstract questions can only be framed after data have been collected, and the immediate implications of these data have been grasped. Problems are therefore usually approached in the opposite direction, from the factual to the abstract. An intuitive flair for what will turn out to be the important general question gives a basis for selecting some of the significant among the indefinite number of trivial experiments which could be carried out at that stage. Quite vague and tacit generalizations thus influence the selection of data at the start. The data then lead to more precise generalizations, which in turn suggest further experiments and progress is made by successive excursions from data to abstractions and vice versa.

After these general considerations we may proceed to the analysis of the several scientific models. A distinction has already been made between material and formal or intellectual models. A material model is the representation of a complex system by a system which is assumed simpler and which is also assumed to have some properties similar to those selected for study in the original complex system. A formal model is a symbolic assertion in logical terms of an idealized relatively simple situation sharing the structural properties of the original factual system.

Recall that Wiener's postdoctoral study in mathematics and philosophy was under Bertrand Russell, a leading advocate of the thesis that mathematics could ultimately be reduced to logic—especially, for Russell, to set theory (Whitehead and Russell 1925–1927; Russell 1920). "Logical terms," then, would have included mathematical terms as a matter of course.

Material models are useful in the following cases. *a*) They may assist the scientist in replacing a phenomenon in an unfamiliar field by one in a field in which he is more at home. They thus may have important didactic advantages. The history of the development of engineering illustrates this mode of usefulness. During the 18th and 19th centuries the success of Newtonian dynamics so dominated physics that electrical problems were often approached via mechanical models. After the work of Faraday and Maxwell, and with the growth of the large scale electrical industries, the development of electrical knowledge outstripped signally that of mechanics. Throughout this century, electrical models have been used to solve mechanical problems.

b) A material model may enable the carrying out of experiments under more favorable conditions than would be available in the original system. This translation presumes that there are reasonable grounds for supposing a similarity between the two situations; it thus presupposes the possession of an adequate formal model, with a structure similar to that of the two material systems. The formal model need not be thoroughly comprehended; the material model then serves to supplement the formal one.

When we assert that two material systems have a relevantly similar structure, we are implying the existence of a formal system that "abstracts out" the common structure. This is not an obvious proposition, but it may have seemed so, especially to Wiener, on the basis of exposure to mathematical fields like abstract algebra and logic. See Russell (1914), Wiener (1919, 1955), and Fishburn and Monjardet (1992). Our authors also realize (soundly!) that the practical experimentalist may not have a very concrete idea of this structure in mind.

Sometimes the relation between the material model and the original system may be no more than a change of scale, in space or time. As an example of a change of a spatial scale, at any proving ground, experiments on shells will not be carried out with large, expensive and unwieldy calibers, but with handy, cheaper, small calibers. Another example is the use of small animals, instead of large ones, for biological experiments: certainly any physiologist will work as much as possible on a dolphin rather than on a sulphur-bottom whale.

As an example of a transformation of the time scale may be mentioned the employment of drosophila in the study of genetics and population problems, in view of its rapid rate of multiplication.

A further instance of a transformation which facilitates experimental procedure is the use of transparent plastic models with adequate elastic properties for the study of the strains in steel structures. The transparency allows the use of polarized light to make the internal stresses directly observable.

While material models may thus render important services it may be emphasized that not all material models are useful. It is likely that the criteria *a* and *b* discussed above are not only sufficient but also necessary conditions for a useful material model. If the formal model which suggests a material one is weak and trivial, the latter will be irrelevant and barren—i.e., a gross analogy is not scientifically fruitful. Again, if a material model does not suggest any experiments whose results could not have been easily anticipated on the basis of the formal model alone, then that material model is superfluous. Finally, if a model has a more elaborate structure and is less readily amenable to experiment than the original system, then it does not represent a progress.

To exemplify, the long series of ether models in terms of elastic solids and gyroscopes which were the fashion among physicists during the eighties and nineties of the 19th century, have proved to be sterile and actually misleading, since they diverted the attention of scientists from the essential features of the problem involved. As Faraday and Herz had already seen, the important need in electrical knowledge was a sound field theory free from the operationally meaningless props of elaborate material analogies. As another example of an apparently useless analogy, the nitric acid-iron wire model of Lillie for nerve fibers may be mentioned. Although featured very prominently in most textbooks on the subject, iron wire dipped in nitric acid is not easier to experiment with than nerve fibers and there is no particular mathematical difficulty in the formulation of the problems involved. The phenomena of passive metals are not better understood than those of nerve, and involve quite as much physical conjecture; from this standpoint, were it not that the analogy is probably only gross, the useful model in the pair would be the nerve axon instead of the wire.

The "ether," in this sense, was a hypothetical subtle fluid that filled space and provided the medium (supposedly) through which light and other electromagnetic phenomena propagated. In other words, this was an attempt to construct mechanical analogues for things like Maxwell's equations.

We would now more often say "black box" and "white box," respectively. Wiener (1958) later developed an elaborate theory of determining the input–output characteristics of black boxes so that they could be replaced by a (possibly infinite) collection of white boxes.

As an introduction to the analysis of theoretical models it is appropriate to define what will be meant by a "closed box", as opposed to an "open box" problem. There are certain problems in science in which a fixed finite number of input variables determines a fixed finite number of output variables. In these, the problem is determinate when the relations between these finite sets of variables are known. It is possible to obtain the same output for the same input with different physical structures. If several alternative structures of this sort were enclosed in boxes whose only approach would be through the input and output terminals, it would be impossible to distinguish between these alternatives without resorting to new inputs, or outputs, or both. For instance, a given electrical impedance as a function of frequency can be realized with many different combinations of resistances, capacitances and inductances. As long as closed boxes containing such elements are only tested for self and mutual impedances across the terminals, their accurate internal structure cannot be determined. To determine that structure additional terminals would have to be used. The more terminals available, the more open the system. An entirely open system would need an indefinite number of terminals.

It is obvious, therefore, that the difference between open-box and closed-box problems, although significant, is one of degree rather than of kind. All scientific problems begin as closed-box problems, i.e., only a few of the significant variables are recognized. Scientific progress consists in a progressive opening of those boxes. The successive addition of terminals or variables, leads to gradually more elaborate theoretical models: hence to a hierarchy in these models, from relatively simple, highly abstract ones, to more complex, more concrete theoretical structures.

The setting up of a simple model for a closed-box assumes that a number of variables are only loosely coupled with the rest of those belonging to the system. The success of the initial experiments depends on the validity of that assumption. As the successive models become progressively more sophisticated the number of closed regions may actually and does usually increase, because the process may be compared with the subdivision of an original single box into several smaller shut compartments. Many of these small compartments may be deliberately left closed, because they are considered only functionally, but not structurally important.

At an intermediate stage in the course of a scientific inquiry the formal model may thus be a heterogeneous assembly of elements, some treated in detail, that is specifically or structurally, and some treated merely with respect to their overall performance, that is, generically or functionally. Thus, in the study of the nervous system, for many purposes synapses may be considered merely as regions where impulses are delayed, disregarding any question as to the method by which this delay takes place, and disregarding also other properties of synapses such as the fact that they are regions where facilitation or inhibition can occur.

A beautiful example of the progressive concretization of a theoretical model by the successive introduction of additional variables is furnished by the historical development of the theory of sound. It began mathematically as a system of linear partial differential equations in a homogeneous continuous medium. This simple model was, and still is useful for the representation and prediction of the transmission of sound of moderate intensity. For intense sound this theory failed. It was replaced by non-linear differential equations based on hydrodynamics and thermodynamics. In the study of shock waves it was realized that the dimensions of the regions of shock are those of the mean free path of a particle in a gas. Any theory which is to be satisfactory in this domain should take into account the molecular nature of the gas. As a first approximation the gas may be taken as perfect: that is it may be supposed to consist of particles without forces between them. The next, and more accurate theory, not yet developed, will take account of the forces between the particles; a still more mature theory will represent these forces in the space of quantum mechanics and not in that of the Newtonian theory.

So far this discussion has dealt mainly with the elaboration of theoretical models in order to explain observed facts—in other words, with the scientific search for abstract models with a structure equivalent to that of a given experience. Science is also concerned with the reverse process, namely, that of embodying an abstract structure into a concrete entity of similar structure, usually an apparatus or machine with a definite purpose. The traditional approach to such designs is empirical and largely accidental, but the scientific approach is possible and has already shown its validity. In this method the apparatus is first designed from the closed-box point of view, which should be obtained, when possible, by a theoretical minimization process, which is often statistical. For instance, if a wave filter is desired to separate telephone messages from noise, the first step is to determine the statistical composition of the messages and noises carried by the line. Given this composition there is a characteristic of the filter which best separates message and noise—i.e., a characteristic which minimizes the effects of the noise on the messages. For any characteristic there will be many ways of constructing an appropriate filter. The requirements are of an open-box nature, but the elements used in the construction may be treated on a closed-box basis. Other considerations, not necessarily relevant to the problem as stated, will determine the choice.

We have shown that scientific knowledge consists of a sequence of abstract models, preferably formal, occasionally material in nature. We shall now proceed to examine the results of carrying model-making to the limit. Consider first material models. They start by being rough approximations, surrogates for the real facts studied. Let the model approach asymptotically the complexity of the original situation. It will tend to become identical with that original system. As a limit it will become that system itself. That is, in a specific example, the best material model for a cat is another, or preferably the same cat. In other words, should a material model thoroughly realize its purpose, the original situation could be grasped in its entirety and a model would be unnecessary. Lewis Carroll fully expressed this notion in an episode in *Sylvie and Bruno*, when he showed that the only completely satisfactory map to scale of a given country was that country itself.

Pseudonym of the English mathematician Charles Dodgson, author (under his own name) of *The Game of Logic and Symbolic Logic*, and, of course, *Alice in Wonderland* and *Through the Looking-Glass*. The novel *Sylvie and Bruno* mixes a realistic social-fiction plot in Victorian England with paradoxical and nonsensical episodes in fairyland, somewhat in the manner of the Alice books. The episode in question occurs in chapter 11 of the second volume (1893). Nowadays, the conceit of a map at 1:1 scale is more often associated with a 1946 story, translated as "On Exactitude in Science," by the Argentine writer Jorge Luis Borges ([1946] 1998, 325). But this story itself seems to have been inspired by Carroll, whom Borges studied closely. Montagnini (2017) reports that the same idea was used by the Harvard philosopher Josiah Royce, who was one of Wiener's teachers, but this in turn may go back to Carroll.

The situation is the same with the theoretical models. The ideal formal model would be one which would cover the entire universe, which would agree with it in complexity, and which would have a one to one correspondence with it. Any one capable of elaborating and comprehending such a model in its entirety, would find the model unnecessary, because he could then grasp the universe directly as a whole. He would possess the third category of knowledge described by Spinoza.

This ideal theoretical model cannot probably be achieved. Partial models, imperfect as they may be, are the only means developed by science for understanding the universe. This statement does not imply an attitude of defeatism but the recognition that the main tool of science is the human mind and that the human mind is finite.

In his *Ethics*, Baruch Spinoza distinguishes three kinds of knowledge: sense perception; logical inference; and "intuitive," which proceeds "from an adequate idea of the essence of certain attributes of God to an adequate knowledge of the essence of things" (Spinoza 1992, 88–90). This third kind of knowledge, in other words, is able to grasp how the details of the world are implied, through an inescapable necessity, by the eternal essence of God or Nature. For Spinoza, the closer human beings approach this third kind of knowledge, the closer they come to "the highest possible peace of mind," or blessedness.

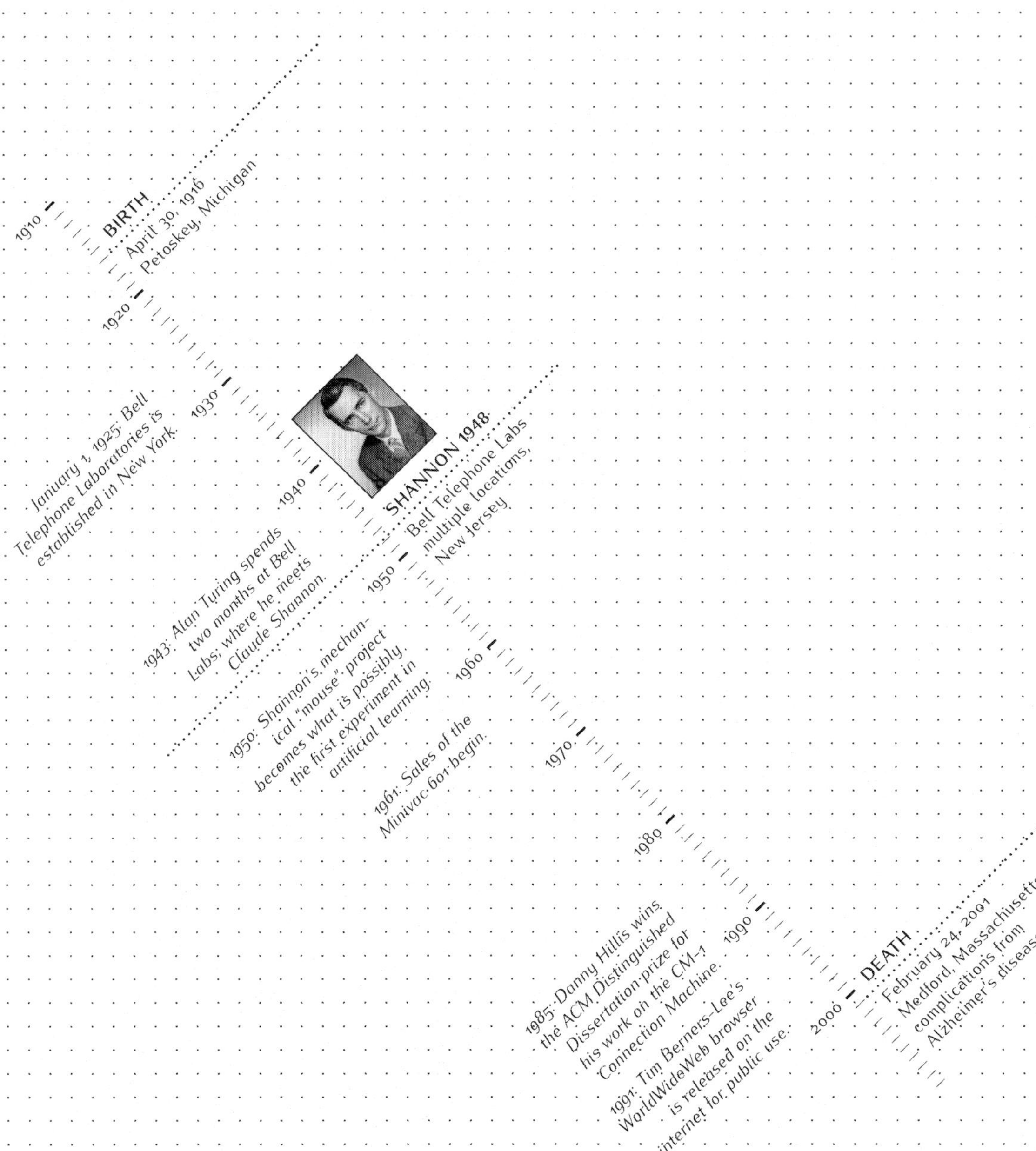

CLAUDE ELWOOD SHANNON

[9]

FROM THE ANALOG TO THE DIGITAL

Seth Lloyd, Massachusetts Institute of Technology

Claude Shannon's monumental paper, "A Mathematical Theory of Communication," was published in *The Bell System Technical Journal* in 1948. Appearing at the midpoint of the twentieth century, Shannon's paper marked the definitive shift from the analog to the digital: it introduced the word "bit" into popular and scientific parlance; it showed how any message could be encoded as a sequence of bits; and it derived the fundamental mathematical formulae for how many bits of information could be reliably sent down a communication channel in the presence of noise.

C. E. Shannon, "A Mathematical Theory of Communication," *Bell System Technical Journal 27* (3), 379–423 (1948).

It is difficult to overstate the paper's influence on science, technology, and society at large. Within a few years of its publication, the popular furor over information theory led to wildly overstated promises for how information theory would transform human beings' understanding of literature, social systems, the nascent theory of computation, biology, game theory, politics, and more. Warren Weaver's excited introduction to the 1949 book publication of Shannon's theory—by which time the tentative "A" in the title had been changed to the definitive "The"—only fanned the flames. The popular impact was so great that in 1956 Shannon felt compelled to write his celebrated "Bandwagon" article, protesting that information theory could only contribute to areas where its mathematical implications were rigorously explored and understood.

Like Norbert Wiener's *Cybernetics*, which also appeared in 1948, Shannon's paper is largely technical, and its actual scientific content differs significantly from the aspects that have exhibited such a mesmerizing hold on the popular imagination. The mathematical and scientific achievements of the paper are remarkable, however, and its technological implications have been profound: Every time one makes a phone call, sends an email, streams a movie, or calls up an app, the underlying physical process of communication is based on Shannon's results. "A Mathematical Theory of Communication" richly deserves its central place in twentieth-century science and culture.

Background

Before I summarize the central technical results of the paper, it is worth briefly recalling the historical and intellectual context in which Shannon developed his mathematical concepts.

Claude E. Shannon was born in Gaylord, Michigan, in 1916. His mother was a high-school teacher and principal, and his father was a businessman and a judge. As a teenager, he constructed a telegraph system that operated over a barbed-wire fence to connect his house to a friend's house a half-mile away. Shannon followed his sister to the mathematics department at the University of Michigan, from which he received undergraduate degrees in math (he studied Boolean logic, among other topics) and electrical engineering.

In 1936 he went to MIT as a research assistant in the electrical engineering department, to work in Vannevar Bush's differential analyzer laboratory and to do a master's degree. Bush's differential analyzer was an analog computer, a device that can be programmed so that its dynamics mimics, or forms an *analogue* of, the behavior of a variety of differential equations operating in continuous time over continuous variables. In 1941, Shannon would show that Bush's differential analyzer was in fact universal, in the sense that it could reproduce in principal the solution to *any* differential equation:

First, however, Shannon would show that analog circuits could be programmed to perform digital computation. Shannon's 1938 master's thesis, "A Symbolic Analysis of Relay and Switching Circuits," showed that electronic circuits consisting of switches and relays were universal for *digital* computation: They could be programmed to evaluate any desired Boolean function. This thesis essentially invents the electronic digital computer (independently proposed by Konrad Zuse and others at the same time, in a less mathematically rigorous fashion). Howard

Gardner (1987, 144) termed Shannon's thesis "possibly the most important, and also the most noted, master's thesis of the century."

At the age of twenty-one, Shannon was already focusing on how analog circuits could be used to implement digital operations. So profound was the impact of Shannon's work that the word "analog" —originally used in the sense of simulation as above—is now used to designate electronic devices that operate using continuous signals, to distinguish them from digital devices that use discrete logic. Later, in "A Mathematical Theory of Communication," Shannon would show that analog signals in the presence of noise can be thought of as essentially digital in nature: Their information content can be measured in bits.

Shannon received his PhD from MIT in 1940, and his thesis constructed an algebraic method to analyze Mendelian genetics. He then went to the Institute of Advanced Study at Princeton, where he worked under the supervision of Hermann Weyl and interacted with John von Neumann, whose work on ergodic theory became central to Shannon's theory of information. Shannon subsequently joined Bell Labs, then located on the lower West Side of Manhattan in a building that spanned the elevated railroad tracks that currently house the High Line park. During the war, Shannon worked on cryptography and code breaking. It was in the cafeteria of the Bell Labs building that Shannon met with Alan Turing during Turing's wartime sojourn in New York, and they exchanged ideas on computation and artificial intelligence.

Shannon regarded code breaking and the decoding of noisy communication signals as complementary processes: His 1945 confidential memo, "A Mathematical Theory of Cryptography" (subsequently declassified and published as "Communication Theory of Secrecy Systems," Shannon 1949), introduces many of the concepts later refined in "A Mathematical Theory of Communication." In the course of his investigations into the use of redundancy in natural language for the purpose of code breaking, Shannon became fascinated with the statistical patterns of letters and words, a topic he developed in "Prediction and Entropy of Printed English" (1951).

A Mathematical Theory of Communication

Shannon was not the first to introduce binary coding and data compression: The use of binary encoding to send information over continuous communication channels dates at least to the invention of the telegraph. Morse code is digitized both in signal strength—the telegraph line is either charged or uncharged—and in time—a dot is one unit of time, a dash is three units, spaces within encoded letters are one unit, spaces between words are three units, etc. Moreover, Morse code introduced data compression by assigning short codes to more frequent letters, so that the most frequent letter, E, is encoded as a single dot (1 unit of time) while Q is encoded as dash dash dot dash (13 units). Shannon's paper derives the ultimate limits of coding, data compression, and the reliable transmission of information in the presence of noise, and shows how those limits can be attained.

Shannon's paper begins with a simple goal: "The fundamental problem of communication is that of reproducing at one point either exactly or approximately a message selected at another point." He then immediately dismisses the central human importance of information: "Frequently the messages have *meaning*; that is they refer to or are correlated according to some system with certain physical or conceptual entities. These semantic aspects of communication are irrelevant to the engineering problem." While technically true, and necessary to state in the context of his mathematical theory, Shannon's rejection of meaning as important to the problem of communication seems purposely provocative, and drove a definitive strike to the wedge dividing the "two cultures" of the humanities and the sciences.

If meaning is not important to the engineering problem, then what is? According to Shannon, "The significant aspect is that the actual message is one *selected from a set* of possible messages." That is, the amount of information is associated with the size of the set from which the message was taken, rather than with the particular message itself. When all messages are equally likely, the amount of information associated with a set is equal to the logarithm to the base two of the size of the set: Equivalently, the quantity of information associated with a particular message is given by the number of bits required to *label* the members of the set from which the message was taken.

Shannon then embarks on a discussion of the frequency of letters, combinations of letters, and combinations of words in English, and how one can model sources of information as ergodic Markov processes. He shows that there is a unique quantity, the entropy, $H = -\Sigma p_i \log_2 p_i$, which measures the amount of information per symbol in terms of the probabilities p_i for the ith symbol. Rolf Landauer told me that when Shannon discovered the formula for information, he went to von Neumann to ask him what letter he should use to designate the quantity. "You should call it H," von Neumann supposedly said, "because that's what Boltzmann called it." In fact, the H of Boltzmann's celebrated H-theorem is the *negative* of Shannon's H.

Shannon then uses another von Neumann technique, ergodic theory, and the accompanying theory of Markov processes, to show that when a source has entropy H, then as the length N of the source's output gets longer and longer, the ith output sequence occurs with a frequency approximately equal to its probability, p_i . The logarithm of the number of such high-probability sequences is approximately NH, and each such sequence in the high-probability set occurs with approximately the same probability. Consequently, the number of bits required to encode a high-probability sequence of length N is just NH—Shannon's noiseless channel theorem.

In part 2 of the paper, Shannon introduces the problem of accurate communication in the presence of noise and shows that the capacity of the noisy communication channel is given by the shared or mutual information between the input and output of the channel. Shannon's noisy channel theorem states that it is not possible to send information reliably down the channel at a rate higher than the channel capacity, and that it is possible to send information reliably at any rate lower than the channel capacity.

Here, "reliably" means that in the limit as the code length becomes large, the probability of error goes to zero. The proof of the capacity limit is straightforward: By a simple counting argument, Shannon shows that any attempt to send more messages per second than allowed by the channel capacity will result in the probability of error going to one. The fact that the capacity can actually be obtained is more subtle: Shannon proves this by averaging over all possible codes, thereby showing that *almost every* code allows information to be sent down the channel and decoded with vanishingly small error rate in the limit that the message length becomes large. Another way of stating this result is that *random* codes attain the channel capacity. Shannon notes, rather ruefully, that it is hard to obtain a *non-random*, easily decodable code that attains the channel capacity, except in trivial or artificial cases.

Part 3 deals with the case of continuous signals in continuous time. On the face of it, such signals contain an infinite amount of information. To calculate the capacity of the continuous communication channel, Shannon must first make it effectively discrete. The first result is the famous Shannon–Nyquist sampling theorem: If a signal is *band limited*, so that its Fourier spectrum lies between $-W$ and $+W$, then the function can be reproduced exactly from samples taken at the Nyquist frequency $2W$. The second result shows that even if the signals are continuous, in the presence of noise the mutual information between channel input and output is still finite, and the mutual information still governs the capacity of the noisy channel to transmit information reliably. Finally, in the case of additive white (uncorrelated) noise, Shannon derives an explicit form for the capacity of the noisy channel: It is equal to $C = W \log_2(1 + P/N)$, where W is the bandwidth of the channel, P is the signal power, and N is the noise power. As in the case of the discrete channel with noise, it is not possible to send information down the continuous noisy channel reliably at a rate higher than the channel capacity, and random codes can be used to approach arbitrarily close to the channel capacity.

Discussion

Each of the three parts of Shannon's paper solves a major, previously unsolved problem. Part 1, by its use of ergodic theory, finds the definitive formula for measuring information and shows how that formula gives the channel capacity for the discrete, noiseless channel. Part 2 shows that the channel capacity of the noisy discrete channel is given by the mutual information between input and output: No information can be sent at a rate higher than the channel capacity; conversely, information can be encoded, sent, and decoded reliably at any rate less than the channel capacity. Random codes attain the capacity. Part 3 shows that the case of continuous, band-limited signals with noise can be treated using the same methods as the discrete noisy channel: The presence of noise and band limitation makes the set of such signals effectively discrete.

The mathematical techniques that Shannon introduces are powerful but subtle. One of the triumphs of "A Mathematical Theory of Communication" is that it uses multiple examples to present these techniques in a manner accessible to a broad audience. (Shannon himself admits that to provide rigorous derivations of the results on continuous variables would require a considerably longer and more technical treatment.) One of the common initial reactions to Shannon's work is "Is the answer really so simple?" The answer is "Yes—but the actual technical results are more subtle than they first appear to be."

Reaction and Influence

We saw above how, in short order, the popularity of information theory had grown to the point that Shannon (1956) felt compelled to write his "Bandwagon" protest. In the longer run, however, Shannon's work also formed the basis for the mathematically rigorous application of ideas of information theory to all aspects of communication, computation, and a wide variety of fields, notably the study of complex systems. Information and information processing lie at the heart of the sciences of complexity (see Zurek 1989). The basic formulae for information arose as the formula for entropy in statistical physics, and the mathematical theory has greatly extended the physical applications of the concept of information—notably in the theory of quantum computation. Information theory plays a central role not only in communication but in computation: notably, in the rapidly developing fields of machine learning and artificial intelligence. The successes of natural language programs such as GPT-3 arise essentially from extending Shannon's analysis of frequencies for sequences of words to massive databases to generate plausible texts artificially. The elemental nature of Shannon's theory, together with its combination of mathematical simplicity and sophistication, make it difficult to find a field where it can not be applied. It seems implausible that any article on applied mathematics could have a greater societal influence than "A Mathematical Theory of Communication."

REFERENCES

Gardner, H. 1987. *The Mind's New Science: A History of the Cognitive Revolution.* New York, NY: Basic Books.

Shannon, C. E. 1938. "A Symbolic Analysis of Relay and Switching Circuits." *Transactions of the AIEE* 57 (12): 713–723. https://doi.org/10.1109/t-aiee.1938.5057767.

———. 1941. "Mathematical Theory of the Differential Analyzer." *Journal of Mathematics and Physics* 20 (1–4): 337–354. https://doi.org/10.1002/sapm1941201337.

———. 1948. "A Mathematical Theory of Communication." *Bell System Technical Journal* 27 (3): 379–423. https://doi.org/10.1002/j.1538-7305.1948.tb01338.x.

———. 1949. "Communication Theory of Secrecy Systems." *Bell System Technical Journal* 28 (4): 656–715. https://doi.org/10.1002/j.1538-7305.1949.tb00928.x.

———. 1951. "Prediction and Entropy of Printed English." *Bell System Technical Journal* 30 (1): 50–64. https://doi.org/10.1002/j.1538-7305.1951.tb01366.x.

———. 1956. "The Bandwagon (editorial)." *IRE Transactions on Information Theory* 2 (1): 3–3. https://doi.org/10.1109/TIT.1956.1056774.

Sloane, N. J. A., and A. D. Wyner. 1993. "Biography of Claude Elwood Shannon." In *Claude Elwood Shannon 1916–2001: Collected Papers,* edited by N. J. A. Sloane and A. D. Wynder, xi–xvii. Piscataway, NJ: IEEE Press.

Wiener, N. 1948. *Cybernetics.* New York, NY: Wiley.

Zurek, W. H., ed. 1989. *Complexity, Entropy, and the Physics of Information.* Reading, MA: Addison-Wesley.

A MATHEMATICAL THEORY OF COMMUNICATION

C. E. Shannon, Bell Labs

Introduction

The recent development of various methods of modulation such as PCM and PPM which exchange bandwidth for signal-to-noise ratio has intensified the interest in a general theory of communication. A basis for such a theory is contained in the important papers of Nyquist [1] and Hartley [2] on this subject. In the present paper we will extend the theory to include a number of new factors, in particular the effect of noise in the channel, and the savings possible due to the statistical structure of the original message and due to the nature of the final destination of the information.

The fundamental problem of communication is that of reproducing at one point either exactly or approximately a message selected at another point. Frequently the messages have *meaning*; that is they refer to or are correlated according to some system with certain physical or conceptual entities. These semantic aspects of communication are irrelevant to the engineering problem. The significant aspect is that the actual message is one *selected from a set* of possible messages. The system must be designed to operate for each possible selection, not just the one which will actually be chosen since this is unknown at the time of design.

Shannon throws down the gauntlet!

If the number of messages in the set is finite then this number or any monotonic function of this number can be regarded as a measure of the information produced when one message is chosen from the set, all choices being equally likely. As was pointed out by Hartley the most

[1] Nyquist, H., "Certain Factors Affecting Telegraph Speed," *Bell System Technical Journal*, April 1924, p. 324; "Certain Topics in Telegraph Transmission Theory," *A.I.E.E. Trans.*, v. 47, April 1928, p. 617.

[2] Hartley, R. V. L., "Transmission of Information," *Bell System Technical Journal*, July 1928, p. 535

natural choice is the logarithmic function. Although this definition must be generalized considerably when we consider the influence of the statistics of the message and when we have a continuous range of messages, we will in all cases use an essentially logarithmic measure.

All boiling down to the point that two pages of text have twice the amount of information as one page.

The logarithmic measure is more convenient for various reasons:

1. It is practically more useful. Parameters of engineering importance such as time, bandwidth, number of relays, etc., tend to vary linearly with the logarithm of the number of possibilities. For example, adding one relay to a group doubles the number of possible states of the relays. It adds 1 to the base 2 logarithm of this number. Doubling the time roughly squares the number of possible messages, or doubles the logarithm, etc.

2. It is nearer to our intuitive feeling as to the proper measure. This is closely related to (1) since we intuitively measure entities by linear comparison with common standards. One feels, for example, that two punched cards should have twice the capacity of one for information storage, and two identical channels twice the capacity of one for transmitting information.

3. It is mathematically more suitable. Many of the limiting operations are simple in terms of the logarithm but would require clumsy restatement in terms of the number of possibilities.

The choice of a logarithmic base corresponds to the choice of a unit for measuring information. If the base 2 is used the resulting units may be called binary digits, or more briefly *bits*, a word suggested by J. W. Tukey. A device with two stable positions, such as a relay or a flip-flop circuit, can store one bit of information. N such devices can store N bits, since the total number of possible states is 2^N and $\log_2 2^N = N$. If the base 10 is used the units may be called decimal digits. Since

Tukey used the word 'bit' in a Bell Labs internal memo in January 1947.

$$\begin{aligned}\log_2 M &= \log_{10} M / \log_{10} 2 \\ &= 3.32 \log_{10} M,\end{aligned}$$

a decimal digit is about $3\frac{1}{3}$ bits. A digit wheel on a desk computing machine has ten stable positions and therefore has a storage capacity of one decimal digit. In analytical work where integration and differentiation are involved the base e is sometimes useful. The resulting units of information will be called natural units. Change from the base a to base b merely requires multiplication by $\log_b a$.

By a communication system we will mean a system of the type indicated schematically in figure 2. It consists of essentially five parts:

1. An *information source* which produces a message or sequence of messages to be communicated to the receiving terminal. The message may be of various types: e.g. (a) A sequence of letters as in a telegraph or teletype system; (b) A single function of time $f(t)$ as in radio or telephony; (c) A function of time and other variables as in black and white television—here the message may be thought of as a function $f(x, y, t)$ of two space coordinates and time, the light intensity at point (x, y) and time t on a pickup tube plate; (d) Two or more functions of time, say $f(t), g(t), h(t)$—this is the case in "three dimensional" sound transmission or if the system is intended to service several individual channels in multiplex; (e) Several functions of several variables—in color television the message consists of three functions $f(x, y, t)$, $g(x, y, t)$, $h(x, y, t)$ defined in a three-dimensional continuum—we may also think of these three functions as components of a vector field defined in the region—similarly, several black and white television sources would produce "messages" consisting of a number of functions of three variables; (f) Various combinations also occur, for example in television with an associated audio channel.

2. A *transmitter* which operates on the message in some way to produce a signal suitable for transmission over the channel. In telephony this operation consists merely of changing sound pressure into a proportional electrical current. In telegraphy we have an encoding operation which produces a sequence of dots, dashes and spaces on the channel corresponding to the message. In a multiplex PCM system the different speech functions must be sampled, compressed, quantized and encoded, and finally interleaved properly to construct the signal. Vocoder systems,

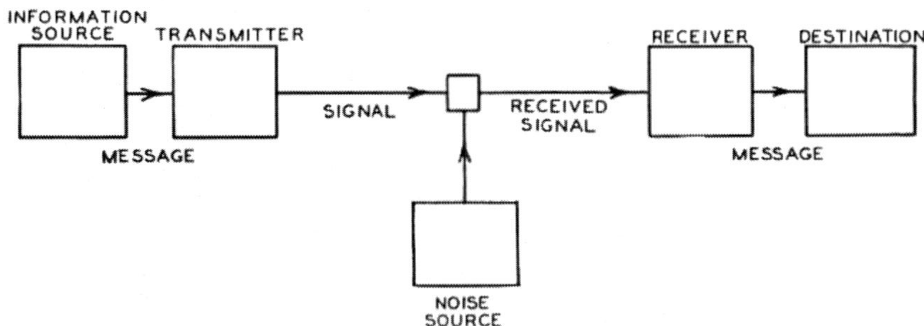

Figure 1. Schematic diagram of a general communication system.

Figure 1 is essential to Shannon's argument that the amount of information transmitted can be quantified. By breaking down the communication process into sequential parts, source, transmitter/encoder, noise, receiver/decoder, and destination, he can analyze each part separately. 'Meaning' does not appear: meaning is in the mind of the information source, and of the destination. This figure occurs, in one form or another, in essentially all textbooks on information theory.

television, and frequency modulation are other examples of complex operations applied to the message to obtain the signal.

3. The *channel* is merely the medium used to transmit the signal from transmitter to receiver. It may be a pair of wires, a coaxial cable, a band of radio frequencies, a beam of light, etc.

4. The *receiver* ordinarily performs the inverse operation of that done by the transmitter, reconstructing the message from the signal.

5. The *destination* is the person (or thing) for whom the message is intended.

We wish to consider certain general problems involving communication systems. To do this it is first necessary to represent the various elements involved as mathematical entities, suitably idealized from their physical counterparts. We may roughly classify communication systems into three main categories: discrete, continuous and mixed. By a discrete system we will mean one in which both the message and the signal are a sequence of discrete symbols. A typical case is telegraphy where the message is a sequence of letters and the signal a sequence of dots, dashes and spaces. A continuous system is one in which the message and signal are both treated as continuous functions, e.g. radio or television. A mixed system is one in which both discrete and continuous variables appear, e.g., PCM transmission of speech.

We first consider the discrete case. This case has applications not only in communication theory, but also in the theory of computing machines, the design of telephone exchanges and other fields. In addition the discrete case forms a foundation for the continuous and mixed cases which will be treated in the second half of the paper.

Part I: Discrete Noiseless Systems

THE DISCRETE NOISELESS CHANNEL

In fact, the continuous cases are treated by making them discrete: Shannon maps the analog to the digital.

Teletype and telegraphy are two simple examples of a discrete channel for transmitting information. Generally, a discrete channel will mean a system whereby a sequence of choices from a finite set of elementary symbols $S_1 \cdots S_n$ can be transmitted from one point to another. Each of the symbols S_i is assumed to have a certain duration in time t_i seconds (not necessarily the same for different S_i, for example the dots and dashes in telegraphy). It is not required that all possible sequences of the S_i be capable of transmission on the system; certain sequences only may be allowed. These will be possible signals for the channel. Thus in telegraphy suppose the symbols are: (1) A dot, consisting of line closure for a unit of time and then line open for a unit of time; (2) A dash, consisting of three time units of closure and one unit open; (3) A letter space consisting of, say, three units of line open; (4) A word space of six units of line open. We might place the restriction on allowable sequences that no spaces follow each other (for if two letter spaces are adjacent, it is identical with a word space). The question we now consider is how one can measure the capacity of such a channel to transmit information.

In the teletype case where all symbols are of the same duration, and any sequence of the 32 symbols is allowed the answer is easy. Each symbol represents five bits of information. If the system transmits n symbols per second it is natural to say that the channel has a capacity of $5n$ bits per second. This does not mean that the teletype channel will always be transmitting information at this rate—this is the maximum possible rate and whether or not the actual rate reaches this maximum depends on the source of information which feeds the channel, as will appear later.

In the more general case with different lengths of symbols and constraints on the allowed sequences, we make the following definition:

Definition: The capacity C of a discrete channel is given by

As Shannon notes, the discrete is the easiest case, and is essentially trivial if all symbols sent down the channel have the same duration. The definition of channel capacity is central to the entire paper and much of the effort of the paper goes into deriving formulae for the number of allowed symbols $N(T)$ over duration T under various circumstances.

$$C = \lim_{T \to \infty} \frac{\log N(T)}{T}$$

where $N(T)$ is the number of allowed signals of duration T.

It is easily seen that in the teletype case this reduces to the previous result. It can be shown that the limit in question will exist as a finite number in most cases of interest. Suppose all sequences of the symbols $S_1, \cdots, S_n$ are allowed and these symbols have durations $i_1, \cdots, i_n$. What is the channel capacity? If $N(t)$ represents the number of sequences of duration t we have

$$N(t) = N(t - t_1) + N(t - t_2) + \cdots + N(t - t_n)$$

The total number is equal to the sum of the numbers of sequences ending in $S_1, S_2, \cdots, S_n$ and these are $N(t-t_1), N(t-t_2), \cdots, N(t-t_n)$, respectively. According to a well known result in finite differences, $N(t)$ is then asymptotic for large t to X_0^t where X_0 is the largest real solution of the characteristic equation:

$$X^{-t_1} + X^{-t_2} + \cdots + X^{-t_n} = 1$$

and therefore

$$C = \log X_0$$

In case there are restrictions on allowed sequences we may still often obtain a difference equation of this type and find C from the characteristic equation. In the telegraphy case mentioned above

$$N(t) = N(t-2)+N(t-4)+N(t-5)+N(t-7)+N(t-8)+N(t-10)$$

as we see by counting sequences of symbols according to the last or next to the last symbol occurring. Hence C is — $\log \mu_0$ where μ_0 is the positive root of $1 = \mu^2+\mu^4+\mu^5+\mu^7+\mu^8+\mu^{10}$. Solving this we find $C = 0.539$.

A very general type of restriction which may be placed on allowed sequences is the following: We imagine a number of possible states $a_1, a_2, \cdots, a_m$. For each state only certain symbols from the set $S_1, \cdots, S_n$ can be transmitted (different subsets for the different states). When one of these has been transmitted the state changes to a new state

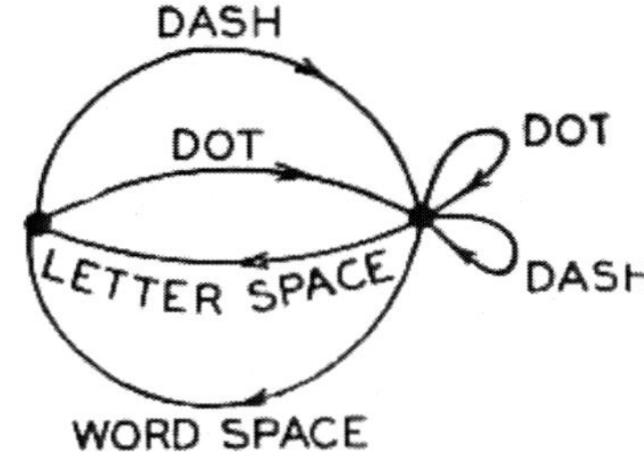

Figure 2. Graphical representation of the constraints on telegraph symbols.

Shannon proposes a model of signal generation as a Markov process with a fixed transition graph. Although Markov processes have finite memory, their memory space can be arbitrarily large. The use of Markov processes allows Shannon to describe the source of information in a purely mechanistic fashion.

depending both on the old state and the particular symbol transmitted. The telegraph case is a simple example of this. There are two states depending on whether or not a space was the last symbol transmitted. If so then only a dot or a dash can be sent next and the state always changes. If not, any symbol can be transmitted and the state changes if a space is sent, otherwise it remains the same. The conditions can be indicated in a linear graph as shown in figure ??. The junction points correspond to the states and the lines indicate the symbols possible in a state and the resulting state. In Appendix I it is shown that if the conditions on allowed sequences can be described in this form C will exist and can be calculated in accordance with the following result:

Theorem 1: Let $b_{ij}^{(s)}$ be the duration of the sth symbol which is allowable in state i and leads to state j. Then the channel capacity C is equal to $\log W$ where W is the largest real root of the determinant equation:

$$\left| \sum_s W^{-b_{ij}^{(s)}} - \delta_{ij} \right| = 0.$$

where $\delta_{ij} = 1$ if $i = j$ and is zero otherwise.

For example, in the telegraph case (fig. ??) the determinant is:

$$\begin{vmatrix} -1 & (W^{-2} + W^{-4}) \\ (W^{-3} + W^{-6}) & (W^{-2} + W^{-4} - 1) \end{vmatrix} = 0$$

On expansion this leads to the equation given above for this case.

THE DISCRETE SOURCE OF INFORMATION

This section is the conceptual heart of the paper. The fundamental concept for both data compression/communication and for code breaking is redundancy. Channel encoding takes advantage of redundancy to compress incoming messages: the compressed messages look more random than the input message itself. Code breaking takes advantage of redundancy in the encoded message to infer features of that message. Shannon's emphasis on redundancy in natural language presages the contemporary use of transformers such as GPT3 for performing natural language recognition and generation.

We have seen that under very general conditions the logarithm of the number of possible signals in a discrete channel increases linearly with time. The capacity to transmit information can be specified by giving this rate of increase, the number of bits per second required to specify the particular signal used.

We now consider the information source. How is an information source to be described mathematically, and how much information in bits per second is produced in a given source? The main point at issue is the effect of statistical knowledge about the source in reducing the required capacity of the channel, by the use of proper encoding of the information. In telegraphy, for example, the messages to be transmitted consist of sequences of letters. These sequences, however, are not completely random. In general, they form sentences and have the statistical structure of, say, English. The letter E occurs more frequently than Q, the sequence TH more frequently than XP, etc. The existence of this structure allows one to make a saving in time (or channel capacity) by properly encoding the message sequences into signal sequences. This is already done to a limited extent in telegraphy by using the shortest channel symbol, a dot, for the most common English letter E; while the infrequent letters, Q, X, Z are represented by longer sequences of dots and dashes. This idea is carried still further in certain commercial codes where common words and phrases are represented by four- or five-letter code groups with a considerable saving in average time. The standardized greeting and anniversary telegrams now in use extend this to the point of encoding a sentence or two into a relatively short sequence of numbers.

We can think of a discrete source as generating the message, symbol by symbol. It will choose successive symbols according to certain probabilities depending, in general, on preceding choices as well as the particular symbols in question). A physical system, or a mathematical model of a system which produces such a sequence of symbols governed by a set of probabilities is known as a stochastic process.[3] We may

[3]See, for example, S. Chandrasekhar, "Stochastic Problems in Physics and Astronomy," Reviews of Modem Physics, v. 15, No. 1, January 1943, p. 1.

consider a discrete source, therefore, to be represented by a stochastic process. Conversely, any stochastic process which produces a discrete sequence of symbols chosen from a finite set may be considered a discrete source. This will include such cases as:

1. Natural written languages such as English, German, Chinese.

2. Continuous information sources that have been rendered discrete by some quantizing process. For example, the quantized speech from a PCM transmitter, or a quantized television signal.

3. Mathematical cases where we merely define abstractly a stochastic process which generates a sequence of symbols. The following are examples of this last type of source.

    a) Suppose we have five letters A, B, C, D, E which are chosen each with probability .2, successive choices being independent. This would lead to a sequence of which the following is a typical example.
    BDCBCECCCADCBDDAAECEEA
    ABBDAEECACEEBAEECBCEAD
    This was constructed with the use of a table of random numbers. [4]

    b) Using the same five letters let the probabilities be .4, .1, .2, .2, .1 respectively, with successive choices independent. A typical message from this source is then:
    AAACDCBDCEAADADACEDA
    EADCABEDADDCECAAAAAD

    c) A more complicated structure is obtained if successive symbols are not chosen independently but their probabilities depend on preceding letters. In the simplest case of this type a choice depends only on the preceding letter and not on ones before that. The statistical structure can then be described by a set of transition probabilities $p_i(j)$, the probability that letter i is followed by letter j. The indices i and j range over all the

[4] Kendall and Smith. "Tables of Random Sampling Numbers," Cambridge, 1939.

possible symbols. A second equivalent way of specifying the structure is to give the "digram" probabilities $p(i,j)$, i.e., the relative frequency of the digram ij. The letter frequencies $p(i)$, (the probability of letter i), the transition probabilities $p_i(j)$ and the digram probabilities $p(i,j)$ are related by the following formulas.

$$p(i) = \sum_j p(i,j) = \sum_j p(j,i) = \sum_j p(j)p_j(i)$$

$$p(i,j) = p(i)p_i(j)$$

$$\sum_j p_i(j) = \sum_i p(i) = \sum_{i,j} p(i,j) = 1.$$

As a specific example suppose there are three letters A, B, C with the probability tables:

$p_i j$		j A	B	C
	A	0	$\frac{4}{5}$	$\frac{1}{5}$
i	B	$\frac{1}{2}$	$\frac{1}{2}$	0
	C	$\frac{1}{2}$	$\frac{2}{5}$	$\frac{1}{10}$

i	$p(i)$
A	$\frac{9}{27}$
B	$\frac{16}{27}$
C	$\frac{2}{27}$

$p(i,j)$		j A	B	C
	A	0	$\frac{4}{15}$	$\frac{1}{15}$
i	B	$\frac{8}{27}$	$\frac{8}{27}$	0
	C	$\frac{1}{27}$	$\frac{4}{135}$	$\frac{1}{135}$

A typical message from this source is the following:

ABBABABABABABABBBABBBBBAB
ABABABABBBACACABBABBBBABB
ABACBBBABA

The next increase in complexity would involve trigram frequencies but no more. The choice of a letter would depend on the preceding two letters but not on the message before that point. A set of trigram frequencies $p(i, j, k)$ or equivalently a set of transition probabilities $p_{ij}(k)$ would be required. Continuing in this way one obtains successively more complicated stochastic processes. In the general n-gram case a set of n-gram probabilities $p(i_1, i_2, \cdots, i_n)$ or of transition probabilities $p_{i_1, i_2} \cdots, i_{n-1}(i_n)$ is required to specify the statistical structure.

d) Stochastic processes can also be defined which produce a text consisting of a sequence of "words." Suppose there are five letters A, B, C, D, E and 16 "words" in the language with associated probabilities:

.10 A	.16 BEBE	.11 CABED	.04 DEB
.04 ADEB	.04 BED	.05 CEED	.15 DEED
.05 ADEE	.02 BEED	.08 DAB	.01 EAB
.01 BADD	.05 CA	.04 DAD	.05 EE

Suppose successive "words" are chosen independently and are separated by a space. A typical message might be:
DAB EE A BEBE DEED DEB ADEE ADEE EE DEB BEBE BEBE BEBE ADEE BED DEED DEED CEED ADEE A DEED
DEED BEBE CABED BEBE BED DAB DEED ADEB
If all the words are of finite length this process is equivalent to one of the preceding type, but the description may be simpler in terms of the word structure and probabilities. We may also generalize here and introduce transition probabilities between words, etc.

These artificial languages are useful in constructing simple problems and examples to illustrate various possibilities. We can also approximate to a natural language by means of a series of simple artificial languages. The zero-order approximation is obtained by choosing all letters with the same probability and independently. The first-order approximation is obtained by choosing successive letters independently but each letter having the same probability that it does in the natural language.[5] Thus, in the first-order approximation to English, E is chosen with probability .12 (its frequency in normal English) and W with probability .02, but there is no influence between adjacent letters and no tendency to form the preferred digrams such as TH, ED, etc. In the second-order approximation, digram structure is introduced. After a letter is chosen, the next one is chosen in accordance with the frequencies with which the various letters follow the first one. This requires a table of digram frequencies $p_i(j)$. In the third-order approximation, trigram structure is introduced. Each letter is chosen with probabilities which depend on the preceding two letters.

THE SERIES OF APPROXIMATIONS TO ENGLISH

☞ Shannon's encoding method proceeds by identifying high probability sequences of signals within the message. He notes that the effort to identify high probability sequences grows exponentially with the length of the sequence.

To give a visual idea of how this series of processes approaches a language, typical sequences in the approximations to English have been constructed and are given below. In all cases we have assumed a 27-symbol "alphabet," the 26 letters and a space.

1. Zero-order approximation (symbols independent and equi-probable).
   XFOML RXKHRJFFJUJ ZLPWCFWKCYJ
   FFJEYVKCQSGXYD QPAAMKBZAACIBZLHJQD

2. First-order approximation (symbols independent but with frequencies of English text).
   OCRO HLI RGWR NMIELWIS EU LL NBNESEBYA TH EEI
   ALHENHTTPA OOBTTVA NAH BRL

3. Second-order approximation (digram structure as in English).
   ON IE ANTSOUTINYS ARE T INCTORE ST BE S DEAMY
   ACHIN D ILONASIVE TUCOOWE AT TEASONARE FUSO
   TIZIN ANDY TOBE SEACE CTISBE

4. Third-order approximation (trigram structure as in English).
   IN NO IST LAT WHEY CRATICT FROURE BIRS GROCID

[5] Letter, digram and trigram frequencies are given in "Secret and Urgent" by Fletcher Pratt, Blue Ribbon Books 1939. Word frequencies are tabulated in "Relative Frequency of English Speech Sounds," G. Dewey, Harvard University Press, 1923.

PONDENOME OF DEMONSTURES OF THE REPTAGIN IS REGOACTIONA OF CRE

5. First-order word approximation. Rather than continue with tetragram, $\cdots$, n-gram structure it is easier and better to jump at this point to word units. Here words are chosen independently but with their appropriate frequencies.
   REPRESENTING AND SPEEDILY IS AN GOOD APT OR COME CAN DIFFERENT NATURAL HERE HE THE A IN CAME THE TO OF TO EXPERT GRAY COME TO FURNISHES THE LINE MESSAGE HAD BE THESE.

6. Second-order word approximation. The word transition probabilities are correct but no further structure is included.
   THE HEAD AND IN FRONTAL ATTACK ON AN ENGLISH WRITER THAT THE CHARACTER OF THIS POINT IS THEREFORE ANOTHER METHOD FOR THE LETTERS THAT THE TIME OF WHO EVER TOLD THE PROBLEM FOR AN UNEXPECTED

Again, it is salutary to compare such more and more realistic sequences of text with those generated by contemporary natural language processing methods. The contemporary methods use much larger databases and produce more realistic sentences, but the basic method is the same

The resemblance to ordinary English text increases quite noticeably at each of the above steps. Note that these samples have reasonably good structure out to about twice the range that is taken into account in their construction. Thus in (3) the statistical process insures reasonable text for two-letter sequence, but four-letter sequences from the sample can usually be fitted into good sentences. In (6) sequences of four or more words can easily be placed in sentences without unusual or strained constructions. The particular sequence of ten words "attack on an English writer that the character of this" is not at all unreasonable. It appears then that a sufficiently complex stochastic process will give a satisfactory representation of a discrete source.

The first two samples were constructed by the use of a book of random numbers in conjunction with (for example 2) a table of letter frequencies. This method might have been continued for (3), (4), and (5), since digram, trigram, and word frequency tables are available, but a simpler equivalent method was used. To construct (3) for example, one opens a book at random and selects a letter at random on the page.

This letter is recorded. The book is then opened to another page and one reads until this letter is encountered. The succeeding letter is then recorded. Turning to another page this second letter is searched for and the succeeding letter recorded, etc. A similar process was used for (4), (5), and (6). It would be interesting if further approximations could be constructed, but the labor involved becomes enormous at the next stage.

GRAPHICAL REPRESENTATION OF A MARKOFF PROCESS

Stochastic processes of the type described above are known mathematically as discrete Markoff processes and have been extensively studied in the literature.[6] The general case can be described as follows: There exist a finite number of possible "states" of a system; $S_1, S_2, ..., S_n$. In addition there is a set of transition probabilities; $p_i(j)$ the probability that if the system is in state S_i it will next go to state S_j. To make this Markoff process into an information source we need only assume that a letter is produced for each transition from one state to another. The states will correspond to the "residue of influence" from preceding letters.

The situation can be represented graphically as shown in figures 3, 4 and 5. The "states" are the junction points in the graph and the probabilities and letters produced for a transition arc given beside the corresponding line. Figure 3 is for the example B in Section 2, while figure 4 corresponds to the example C. In figure 3 there is only one state since successive letters are independent. In figure 4 there are as many states as letters. If a trigram example were constructed there would be at most n^2 states corresponding to the possible pairs of letters preceding the one being chosen. Figure 5 is a graph for the case of word structure in example D. Here S corresponds to the "space" symbol.

ERGODIC AND MIXED SOURCES

As we have indicated above a discrete source for our purposes can be considered to be represented by a Markoff process. Among the possible discrete Markoff processes there is a group with special properties of significance in communication theory. This special class

[6]For a detailed treatment see M. Frechet, "Methods des fonctions arbitraires. Theorie des énénements en chaine dans le cas d'un nombre fini d'états possibles." Paris, Gauthier-Villars. 1938.

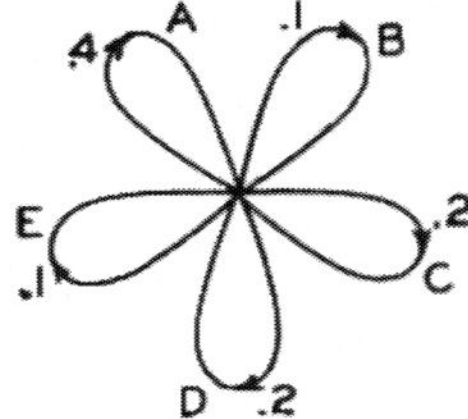

Figure 3. A graph corresponding to the source in example B.

The use of Markoff processes emphasizes the mechanistic aspect of Shannon's method: the Markoff process can be thought of as a probabilistic finite automaton with memory space equal to the logarithm of the size of the set of internal states. Although Markoff processes are frequently called 'memoriless' by expanding to large numbers of internal states, the process can 'remember' arbitrarily long sequences.

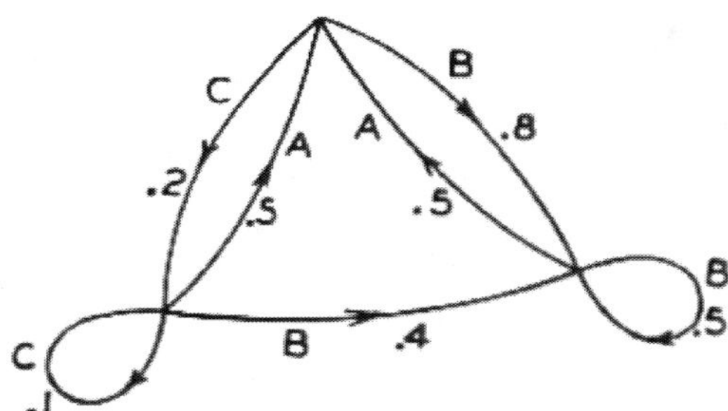

Figure 4. A graph corresponding to the source in example C.

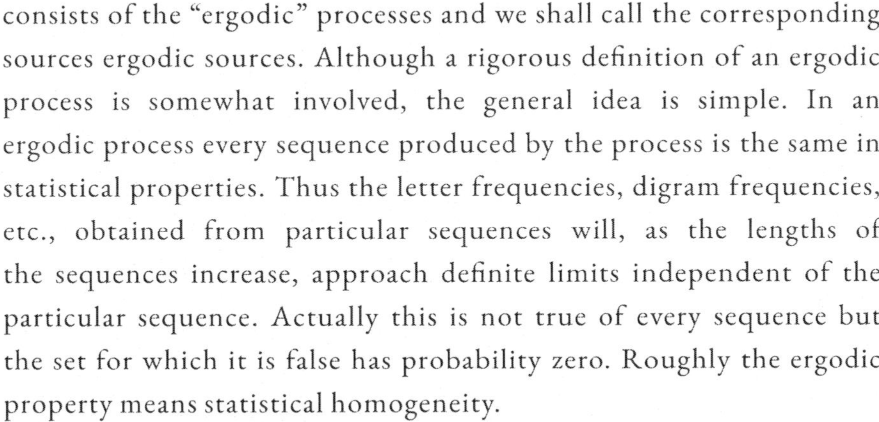

consists of the "ergodic" processes and we shall call the corresponding sources ergodic sources. Although a rigorous definition of an ergodic process is somewhat involved, the general idea is simple. In an ergodic process every sequence produced by the process is the same in statistical properties. Thus the letter frequencies, digram frequencies, etc., obtained from particular sequences will, as the lengths of the sequences increase, approach definite limits independent of the particular sequence. Actually this is not true of every sequence but the set for which it is false has probability zero. Roughly the ergodic property means statistical homogeneity.

All the examples of artificial languages given above are ergodic. This property is related to the structure of the corresponding graph. If the graph has the following two properties[7] the corresponding process will be ergodic:

1. The graph does not consist of two isolated parts A and B such that it is impossible to go from junction points in part A to junction

[7] These are restatements in terms of the graph of conditions given in Frechet.

After working together in Princeton, Shannon had absorbed von Neumann's work on ergodic theory. Here he uses ergodic theory to spectacular effect by identifying the high- and low-probability sets of sequences—a key distinction that allows him to derive the formula for channel capacity in terms of the formula for information.

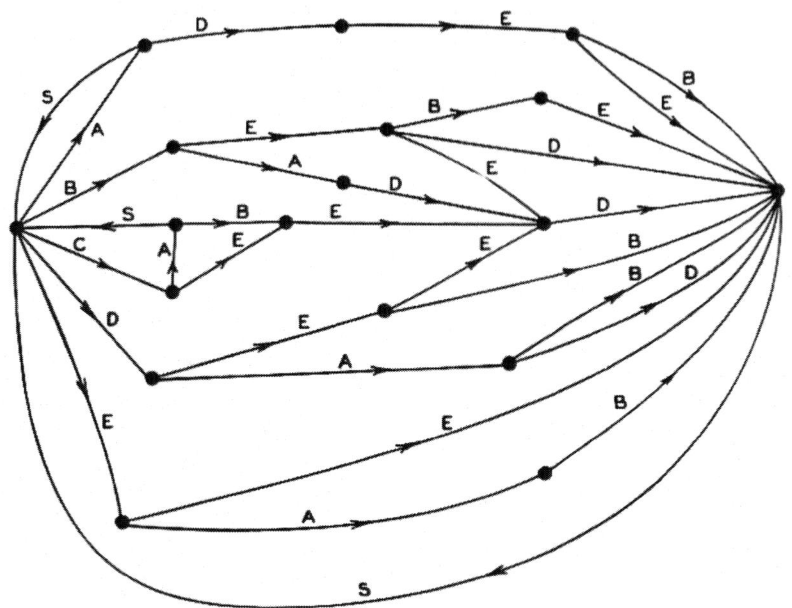

Figure 5. A graph corresponding to the source in example D.

points in part B along lines of the graph in the direction of arrows and also impossible to go from junctions in part B to junctions in part A.

2. A closed series of lines in the graph with all arrows on the lines pointing in the same orientation will be called a "circuit." The "length" of a circuit is the number of lines in it. Thus in figure 5 the series BEBES is a circuit of length 5. The second property required is that the greatest common divisor of the lengths of all circuits in the graph be one.

If the first condition is satisfied but the second one violated by having the greatest common divisor equal to $d > 1$, the sequences have a certain type of periodic structure. The various sequences fall into d different classes which are statistically the same apart from a shift of the origin (i.e., which letter in the sequence is called letter 1). By a shift of from 0 up to $d - 1$ any sequence can be made statistically equivalent to any other. A simple example with $d = 2$ is the following: There are three possible letters a, b, c. Letter a is followed with either b or c with probabilities $\frac{1}{3}$ and $\frac{2}{3}$ respectively. Either b or c is always followed by letter a. Thus a typical sequence is

$$abacacacabacababacac$$

This type of situation is not of much importance for our work.

If the first condition is violated the graph may be separated into a set of subgraphs each of which satisfies the first condition. We will assume that the second condition is also satisfied for each subgraph. We have in this case what may be called a "mixed" source made up of a number of pure components. The components correspond to the various subgraphs. If $L_1, L_2, L_3, \cdots$ are the component sources we may write

$$L = p_1 L_1 + p_2 L_2 + p_3 L_3 + \cdots$$

where p_i is the probability of the component source L_i.

Physically the situation represented is this: There are several different sources $L_1, L_2, L_3, \cdots$ which are each of homogeneous statistical structure (i.e., they are ergodic). We do not know *a priori* which is to be used, but once the sequence starts in a given pure component L_i it continues indefinitely according to the statistical structure of that component.

As an example one may take two of the processes defined above and assume $p_1 = .2$ and $p_2 = .8$. A sequence from the mixed source

$$L = .2L_1 + .8L_2$$

would be obtained by choosing first $L_1 or L_2$ with probabilities .2 and .8 and after this choice generating a sequence from whichever was chosen.

Except when the contrary is stated we shall assume a source to be ergodic. This assumption enables one to identify averages along a sequence with averages over the ensemble of possible sequences (the probability of a discrepancy being zero). For example the relative frequency of the letter A in a particular infinite sequence will be, with probability one, equal to its relative frequency in the ensemble of sequences.

If P_i is the probability of state i and $p_i(j)$ the transition probability to state j, then for the process to be stationary it is clear that the Pi must satisfy equilibrium conditions:

$$P_j = \sum_i P_i p_i(j).$$

In the ergodic case it can be shown that with any starting conditions the probabilities $P_j(N)$ of being in state j after N symbols, approach the equilibrium values as $N \to \infty$.

CHOICE, UNCERTAINTY, AND ENTROPY

We have represented a discrete information source as a Markoff process. Can we define a quantity which will measure, in some sense, how much information is "produced" by such a process, or better, at what rate information is produced?

Suppose we have a set of possible events whose probabilities of occurrence are $p_1, p_2, \cdots, p_n$. These probabilities are known but that is all we know concerning which event will occur. Can we find a measure of how much "choice" is involved in the selection of the event or of how uncertain we are of the outcome?

If there is such a measure, say $H(p_1, p_2, \cdots, p_n)$, it is reasonable to require of it the following properties:

1. H should be continuous in the p_i.

2. If all the p_i are equal, $p_i = \frac{1}{n}$, then H should be a monotonic increasing function of n. With equally likely events there is more choice, or uncertainty, when there are more possible events.

3. If a choice be broken down into two successive choices, the original H should be the weighted sum of the individual values of H. The meaning of this is illustrated in figure 6. At the left we have three possibilities $p_1 = \frac{1}{2}, p_2 = \frac{1}{3}, p_3 = \frac{1}{6}$. On the right we first choose between two possibilities each with probability $\frac{1}{2}$, and if the second occurs make another choice with probabilities $\frac{2}{3}$, $\frac{1}{3}$. The final results have the same probabilities as before. We require, in this special case, that

$$H(\frac{1}{2}, \frac{1}{3}, \frac{1}{6}) = H(\frac{1}{2}, \frac{1}{2}) + \frac{1}{2} H(\frac{2}{3}, \frac{1}{3})$$

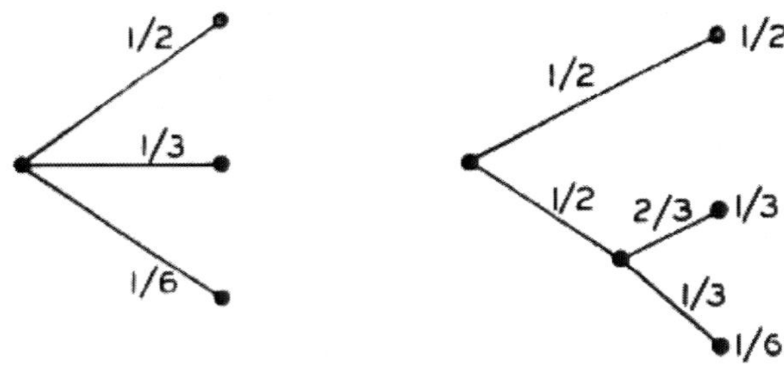

Figure 6. Decomposition of a choice from three possibilities.

This requirement of average additivity is what nails down the specific formula for information in terms of probability.

The coefficient $\frac{1}{2}$ is because this second choice only occurs half the time.

In Appendix II, the following result is established:

Theorem 2: The only H satisfying the three above assumptions is of the form:

$$H = -K \sum_{i=1}^{n} p_i \log p_i$$

where K is a positive constant.

Et voilà! The fundamental formula for information theory, derived originally by Boltzmann in the nineteenth century to quantify entropy. Here, Shannon has rederived the formula for information starting from simple and reasonable assumptions. A tour de force.

This theorem, and the assumptions required for its proof, are in no way necessary for the present theory. It is given chiefly to lend a certain plausibility to some of our later definitions. The real justification of these definitions, however, will reside in their implications.

Quantities of the form $H = -\sum p_i \log p_i$ (the constant K merely amounts to a choice of a unit of measure) play a central role in information theory as measures of information, choice and uncertainty. The form of H will be recognized as that of entropy as defined in certain formulations of statistical mechanics [8] where p_i is the probability of a system being in cell i of its phase space. H is then, for example, the H in Boltzmann's famous H theorem. We shall call $H = -\sum p_i \log p_i$ the entropy of the set of probabilities $p_1, \cdots, p_n$. If x is a chance variable we will write $H(x)$ for its entropy; thus x is not an argument of a function but a label for a number, to differentiate it from $H(y)$ say, the entropy of the chance variable y.

Actually, as noted in the introduction, Boltzmann's H is the negative of Shannon's H!

[8] See, for example, R. C. Tolman, "Principles of Statistical Mechanics," Oxford, Clarendon, 1938.

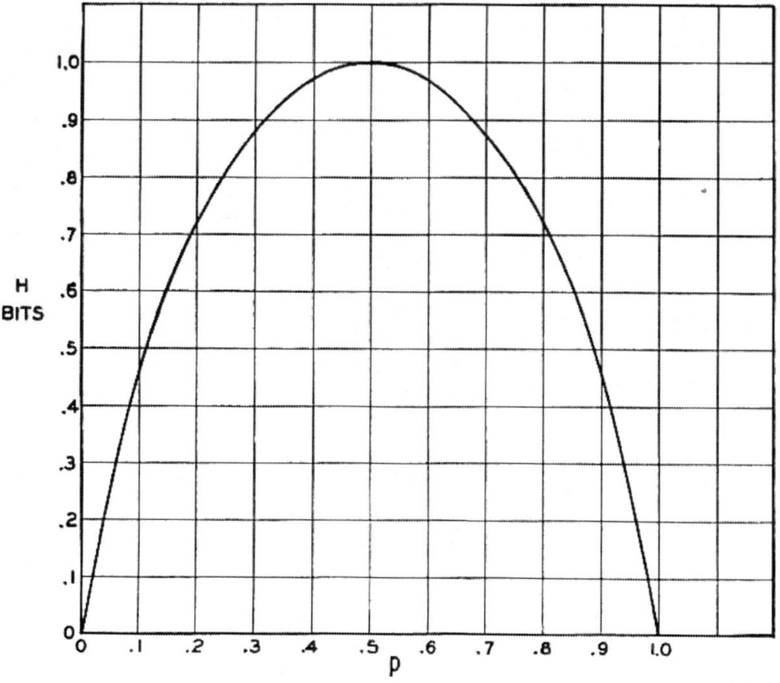

Figure 7. Entropy in the case of two possibilities with probabilities p and $(1-p)$.

The entropy in the case of two possibilities with probabilities p and $q = 1 - p$, namely

$$H = -(p \log p + q \log q)$$

is plotted in figure 7 as a function of p.

The quantity H has a number of interesting properties which further substantiate it as a reasonable measure of choice or information.

1. $H = 0$ if and only if all the p_i but one are zero, this one having the value unity. Thus only when we are certain of the outcome does H vanish. Otherwise H is positive.

2. For a given n, H is a maximum and equal to $\log n$ when all the p_i are equal (i.e., $\frac{1}{n}$). This is also intuitively the most uncertain situation.

3. Suppose there are two events, x and y, in question with m possibilities for the first and n for the second. Let $p(i, j)$ be the

probability of the joint occurrence of i for the first and j for the second. The entropy of the joint event is

$$H(x,y) = -\sum_{i,j} p(i,j) \log p(i,j)$$

while

$$H(x) = -\sum_{i,j} p(i,j) \log \sum_j p(i,j)$$
$$H(y) = -\sum_{i,j} p(i,j) \log \sum_i p(i,j)$$

Shannon now defines the familiar formulae for joint information/entropy and marginal informations/entropies.

It is easily shown that

$$H(x,y) \leq H(x) + H(y)$$

And the difference between the right side and the left side of this inequality is simply the mutual information between x and y.

with equality only if the events are independent (i.e., $p(i,j) = p(i)p(j)$). The uncertainty of a joint event is less than or equal to the sum of the individual uncertainties.

4. Any change toward equalization of the probabilities $p_1, p_2, \cdots, p_n$ increases H. Thus if $p_1 < p_2$ and we increase p_1, decreasing p_2 an equal amount so that p_1 and p_2 are more nearly equal, then H increases. More generally, if we perform any "averaging" operation on the p_i of the form

$$p'_i = \sum_i a_{ij} p_j$$

where $\sum_i a_{ij} = \sum_j a_{ij} = 1$, and all $a_{ij} \geq 0$, then H increases (except in the special case where this transformation amounts to no more than a permutation of the p_j with H of course remaining the same).

5. Suppose there are two chance events x and y as in 3, not necessarily independent. For any particular value i that x can assume there is a conditional probability $p_i(j)$ that y has the value j. This is given by

$$p_i(j) = \frac{p(i,j)}{\sum_j p(i,j)}.$$

We define the *conditional entropy* of y, $H_x(y)$ as the average of the entropy of y for each value of x, weighted according to the probability of getting that particular x. That is

$$H_x(y) = -\sum_{i,j} p(i,j) \log p_i(j).$$

This quantity measures how uncertain we are of y on the average when we know x. Substituting the value of $p_i(j)$ we obtain

$$\begin{aligned} H_x(y) &= -\sum_{i,j} p(i,j) \log p(i,j) + \sum_{i,j} p(i,j) \log \sum_j p(i,j) \\ &= H(x,y) - H(x) \end{aligned}$$

or

$$H(x,y) = H(x) + H_x(y)$$

The uncertainty (or entropy) of the joint event x, y is the uncertainty of x plus the uncertainty of y when x is known.

Mutual information can never be negative.

6. From 3 and 5 we have

$$H(x) + H(y) \geq H(x,y) = H(x) + H_x(y)$$

Hence

$$H(y) \geq H_x(y)$$

The uncertainty of y is never increased by knowledge of x. It will be decreased unless x and y are independent events, in which case it is not changed.

THE ENTROPY OF AN INFORMATION SOURCE

Consider a discrete source of the finite state type considered above. For each possible state i there will be a set of probabilities $p_i(j)$ of producing the various possible symbols j. Thus there is an entropy H_i for each state. The entropy of the source will be defined as the average of these

H_i weighted in accordance with the probability of occurrence of the states in question:

$$H = \sum_i P_i H_i$$
$$= -\sum_{i,j} P_i p_i(j) \log p_i(j)$$

This is the entropy of the source per symbol of text. If the Markoff process is proceeding at a definite time rate there is also an entropy per second

$$H' = \sum_i f_i H_i$$

where f_i is the average frequency (occurrences per second) of state i. Clearly

$$H' = mH$$

where m is the average number of symbols produced per second. H or H' measures the amount of information generated by the source per symbol or per second. If the logarithmic base is 2, they will represent bits per symbol or per second.

If successive symbols are independent then H is simply $-\sum p_i \log p_i$ where p_i is the probability of symbol i. Suppose in this case we consider a long message of N symbols. It will contain with high probability about $p_1 N$ occurrences of the first symbol, $p_2 N$ occurrences of the second, etc. Hence the probability of this particular message will be roughly

$$p = p_1^{p_1 N} p_2^{p_2 N} \cdots p_n^{p_n N}$$

or

$$\log p \doteq N \sum_i p_i \log p_i$$
$$\log p \doteq -NH$$
$$H \doteq \frac{\log 1/p}{N}.$$

H is thus approximately the logarithm of the reciprocal probability of a typical long sequence divided by the number of symbols in the sequence. The same result holds for any source. Stated more precisely we have (see Appendix III):

This seemingly innocuous derivation of the formula for H in terms of probabilities is considerably more subtle than it appears at first glance. Theorem 3 holds the key: stated more colloquially, it states that in almost all generated sequences, a symbol appears with a frequency approximately equal to its probability.

Theorem 3: Given any $\epsilon > 0$ and $\delta > 0$, we can find an N_0 such that the sequences of any length $N \geq N_0$ fall into two classes:

1. A set whose total probability is less than ϵ.

2. The remainder, all of whose members have probabilities satisfying the inequality

$$\left| \frac{\log p^{-1}}{N} - H \right| < \delta$$

In other words we are almost certain to have $\frac{\log p^{-1}}{N}$ very close to H when N is large.

A closely related result deals with the number of sequences of various probabilities. Consider again the sequences of length N and let them be arranged in order of decreasing probability. We define $n(q)$ to be the number we must take from this set starting with the most probable one in order to accumulate a total probability q for those taken.
Theorem 4:

$$\lim_{N \to \infty} \frac{\log n(q)}{N} = H$$

when q does not equal 0 or 1.

We may interpret $\log n(q)$ as the number of bits required to specify the sequence when we consider only the most probable sequences with a total probability q. Then $\frac{\log n(q)}{N}$ is the number of bits per symbol for the specification. The theorem says that for large N this will be independent of q and equal to H. The rate of growth of the logarithm of the number of reasonably probable sequences is given by H, regardless of our interpretation of "reasonably probable." Due to these results, which are proved in appendix III, it is possible for most purposes to treat the long sequences as though there were just 2^{HN} of them, each with a probability 2^{-HN}.

The next two theorems show that H and H' can be determined by limiting operations directly from the statistics of the message sequences, without reference to the states and transition probabilities between states.

Theorem 5: Let $p(B_i)$ be the probability of a sequence B_i of symbols from the source. Let

$$G_N = -\frac{1}{N} \sum_i p(B_i) \log p(B_i)$$

where the sum is over all sequences B_i containing N symbols. Then G_N is a monotonic decreasing function of N and

$$\lim_{N \to \infty} G_N = H.$$

Theorem 6: Let $p(B_i, S_j)$ be the probability of sequence B_i followed by symbol S_j and $p_{B_i}(S_j) = p(B_i, S_j)/p(B_i)$ be the conditional probability of S_j after B_i. Let

$$F_N = -\sum_{i,j} p(B_i, S_j) \log p_{B_i}(S_j)$$

where the sum is over all blocks B_i of $N-1$ symbols and over all symbols S_j. Then F_N is a monotonic decreasing function of N,

$$F_N = N G_N - (N-1) G_{N-1},$$

$$G_N = \frac{1}{N} \sum_1^n F_N,$$

$$F_N \leq G_N,$$

and $\lim_{N \to \infty} F_N = H$.

These results are derived in appendix III. They show that a series of approximations to H can be obtained by considering only the statistical structure of the sequences extending over $1, 2, \cdots N$ symbols. F_N is the better approximation. In fact F_N is the entropy of the Nth order approximation to the source of the type discussed above. If there are no statistical influences extending over more than N symbols, that is if the conditional probability of the next symbol knowing the preceding $(N-1)$ is not changed by a knowledge of any before that, then $F_N = H$. F_N of course is the conditional entropy of the next symbol when the $(N-1)$ preceding ones are known, while G_N is the entropy per symbol of blocks of N symbols.

The ratio of the entropy of a source to the maximum value it could have while still restricted to the same symbols will be called

Shannon uses these theorems to show that the information content of the overall message can be well approximated by the information content of subsequences. Note that in an actual text, such as a novel, for example, there may be long-range correlations, e.g., from the first chapter to the last, that are hard to capture by the sequential method alone. This is why current natural language processing methods take into account correlations and features that can occur anywhere in the text, not merely sequentially.

its *relative entropy.* This is the maximum compression possible when we encode into the same alphabet. One minus the relative entropy is the *redundancy.* The redundancy of ordinary English, not considering statistical structure over greater distances than about eight letters is roughly 50%. This means that when we write English half of what we write is determined by the structure of the language and half is chosen freely. The figure 50% was found by several independent methods which all gave results in this neighborhood. One is by calculation of the entropy of the approximations to English. A second method is to delete a certain fraction of the letters from a sample of English text and then let someone attempt to restore them. If they can be restored when 50% are deleted the redundancy must be greater than 50%. A third method depends on certain known results in cryptography.

Two extremes of redundancy in English prose are represented by Basic English and by James Joyce's book "Finnegans Wake." The Basic English vocabulary is limited to 850 words and the redundancy is very high. This is reflected in the expansion that occurs when a passage is translated into Basic English. Joyce on the other hand enlarges the vocabulary and is alleged to achieve a compression of semantic content.

The redundancy of a language is related to the existence of crossword puzzles. If the redundancy is zero any sequence of letters is a reasonable text in the language and any two dimensional array of letters forms a crossword puzzle. If the redundancy is too high the language imposes too many constraints for large crossword puzzles to be possible. A more detailed analysis shows that if we assume the constraints imposed by the language are of a rather chaotic and random nature, large crossword puzzles are just possible when the redundancy is 50%. If the redundancy is 33%, three dimensional crossword puzzles should be possible, etc.

REPRESENTATION OF THE ENCODING AND DECODING OPERATIONS

We have yet to represent mathematically the operations performed by the transmitter and receiver in encoding and decoding the information. Either of these will be called a discrete transducer. The input to the transducer is a sequence of input symbols and its output a sequence of output symbols. The transducer may have an internal memory so that its output depends not only on the present input symbol but also on the past history. We assume that the internal memory is finite, i.e. there exists a finite number m of possible states of the transducer and that its output is a function of the present state and the present input symbol. The next state will be a second function of these two quantities. Thus a transducer can be described by two functions:

$$y_n = f(x_n, \alpha_n)$$
$$\alpha_{n+1} = g(x_n, \alpha_n)$$

where:

x_n is the nth input symbol,

α_n is the state of the transducer when the nth input symbol is introduced,

y_n is the output symbol (or sequence of output symbols) produced when x_n is introduced if the state is α_n.

If the output symbols of one transducer can be identified with the input symbols of a second, they can be connected in tandem and the result is also a transducer. If there exists a second transducer which operates on the output of the first and recovers the original input, the first transducer will be called non-singular and the second will be called its inverse.

Theorem 7: The output of a finite state transducer driven by a finite state statistical source is a finite state statistical source, with entropy (per unit time) less than or equal to that of the input. If the transducer is non-singular they are equal.

That is, a deterministic encoder/transducer cannot add information to the source; it can only lose track of information by imperfect encoding.

Let α represent the state of the source, which produces a sequence of symbols x_i; and let β be the state of the transducer, which produces, in its output, blocks of symbols y_j. The combined system can be represented by the "product state space" of pairs (α, β). Two points in the space, (α_1, β_1) and ($\alpha_2\beta_2$), are connected by a line if α_1 can produce an x which changes β_1 to β_2, and this line is given the probability of that x in this case. The line is labeled with the block of y_j symbols produced by the transducer. The entropy of the output can be calculated as the weighted sum over the states. If we sum first on β each resulting term is less than or equal to the corresponding term for α, hence the entropy is not increased. If the transducer is non-singular let its output be connected to the inverse transducer. If H'_1, H'_2, and H'_3 are the output entropies of the source, the first and second transducers respectively, then $H'_1 \geq H'_2 \geq H'_3 = H'_1$ and therefore $H'_1 = H'_2$.

Suppose we have a system of constraints on possible sequences of the type which can be represented by a linear graph as in figure ??. If probabilities $p_{ij}^{(s)}$ were assigned to the various lines connecting state i to state j this would become a source. There is one particular assignment which maximizes the resulting entropy (see Appendix 4).

Theorem 8: Let the system of constraints considered as a channel have a capacity C. If we assign

$$p_{ij}^{(s)} = \frac{B_j}{B_i} C^{-l_{ij}^{(s)}}$$

where $l_{ij}^{(s)}$ is the duration of the sth symbol leading from state i to state j and the B_i satisfy

$$B_i = \sum_{s,j} B_j C^{-l_{ij}^{(s)}}$$

then H is maximized and equal to C.

By proper assignment of the transition probabilities the entropy of symbols on a channel can be maximized at the channel capacity.

And that, ladies and gentlemen, is it! Note the subtlety of the ingredients that went into deriving the central theory: Markoff chains, ergodic theory, high- and low-probability sequences, mutual information, etc. A remarkable result, stated simply, with profound implications.

THE FUNDAMENTAL THEOREM FOR A NOISELESS CHANNEL

We will now justify our interpretation of H as the rate of generating information by proving that H determines the channel capacity required with most efficient coding.

Theorem 9: Let a source have entropy H (bits per symbol) and a channel have a capacity C (bits per second). Then it is possible to encode the output of the source in such a way as to transmit at the average rate $\frac{C}{H} - \epsilon$ symbols per second over the channel where ϵ is arbitrarily small. It is not possible to transmit at an average rate greater than $\frac{C}{H}$.

The converse part of the theorem, that $\frac{C}{H}$ cannot be exceeded, may be proved by noting that the entropy of the channel input per second is equal to that of the source, since the transmitter must be non-singular, and also this entropy cannot exceed the channel capacity. Hence $H' \leq C$ and the number of symbols per second $= H'/H \leq C/H$.

The first part of the theorem will be proved in two different ways. The first method is to consider the set of all sequences of N symbols produced by the source. For N large we can divide these into two groups, one containing less than $2^{(H+\eta)N}$ members and the second containing less than 2^{RN} members (where R is the logarithm of the number of different symbols) and having a total probability less than μ. As N increases η and μ approach zero. The number of signals of duration T in the channel is greater than $2^{(C-\theta)T}$ with θ small when T is large. If we choose

$$T = \left(\frac{H}{C} + \lambda\right)N$$

then there will be a sufficient number of sequences of channel symbols for the high probability group when N and T are sufficiently large (however small λ) and also some additional ones. The high probability group is coded in an arbitrary one to one way into this set. The remaining sequences are represented by larger sequences, starting and ending with one of the sequences not used for the high probability group. This special sequence acts as a start and stop signal for a different code. In between a sufficient time is allowed to give enough different sequences for all the low probability messages. This will require

$$T_1 = \left(\frac{R}{C} + \varphi\right)N$$

where φ is small. The mean rate of transmission in message symbols per second will then be greater than

$$\left[(1-\delta)\frac{T}{N} + \delta\frac{T_1}{N}\right]^{-1} = \left[(1-\delta)\left(\frac{H}{C} + \lambda\right) + \delta\left(\frac{R}{C} + \varphi\right)\right]^{-1}$$

As N increases δ, λ and φ approach zero and the rate approaches $\frac{C}{H}$.

Another method of performing this coding and proving the theorem can be described as follows: Arrange the messages of length N in order of decreasing probability and suppose their probabilities are $p_1 \geq p_2 \geq p_3 \cdots \geq p_n$. Let $P_s = \sum_1^{s-1} p_i$; that is, P_s is the cumulative probability up to, but not including, p_s. We first encode into a binary system.

The binary code for message s is obtained by expanding P_s as a binary number. The expansion is carried out to m_s places, where m_s is the integer satisfying:

$$\log_2 \frac{1}{p_s} \leq m_s < 1 + \log_2 \frac{1}{p_s}$$

This method, known as Shannon–Fano coding, it represents the simplest and most obvious form of variable length code. Surprisingly to Shannon and Fano, this is not the optimal code, however. The most efficient variable length code is Huffman coding, discovered by David Huffman while answering a problem set challenge in Fano's class on information theory at MIT.

Thus the messages of high probability are represented by short codes and those of low probability by long codes. From these inequalities we have

$$\frac{1}{2^{m_s}} \leq p_s < \frac{1}{2^{m_s - 1}}.$$

The code for P_s will differ from all succeeding ones in one or more of its m_s places, since all the remaining P_i are at least $\frac{1}{2^{m_s}}$ larger and their binary expansions therefore differ in the first m_s places. Consequently all the codes are different and it is possible to recover the message from its code. If the channel sequences are not already sequences of binary digits, they can be ascribed binary numbers in an arbitrary fashion and the binary code thus translated into signals suitable for the channel.

The average number H' of binary digits used per symbol of original message is easily estimated. We have

$$H' = \frac{1}{N} \Sigma m_s p_s$$

But,

$$\frac{1}{N} \Sigma \left(\log_2 \frac{1}{p_s} \right) \leq \frac{1}{N} \Sigma m_s p_s < \frac{1}{N} \Sigma \left(1 + \log_2 \frac{1}{p_s} \right) p_s$$

and therefore,

$$-\Sigma p_s \log p_s \leq H' < \frac{1}{N} - \Sigma p_s \log p_s$$

As N increases $-\Sigma p_s \log p_s$ approaches H, the entropy of the source and H' approaches H.

We see from this that the inefficiency in coding, when only a finite delay of N symbols is used, need not be greater than $\frac{1}{N}$ plus the difference between the true entropy H and the entropy G_N calculated for sequences of length N. The percent excess time needed over the ideal is therefore less than

$$\frac{G_N}{H} + \frac{1}{HN} - 1.$$

This method of encoding is substantially the same as one found independently by R. M. Fano. [9] His method is to arrange the messages of length N in order of decreasing probability. Divide this series into two groups of as nearly equal probability as possible. If the message is in the first group its first binary digit will be 0, otherwise 1. The groups are similarly divided into subsets of nearly equal probability and the particular subset determines the second binary digit. This process is continued until each subset contains only one message. It is easily seen that apart from minor differences (generally in the last digit) this amounts to the same thing as the arithmetic process described above.

DISCUSSION

In order to obtain the maximum power transfer from a generator to a load a transformer must in general be introduced so that the generator as seen from the load has the load resistance. The situation here is roughly analogous. The transducer which does the encoding should match the source to the channel in a statistical sense. The source as seen from the channel through the transducer should have the same statistical structure as the source which maximizes the entropy in the channel. The content of Theorem 9 is that, although an exact match is not in general possible, we can approximate it as closely as desired. The ratio of the actual rate of transmission to the capacity C may be called the efficiency of the coding system. This is of course equal to the ratio of the actual entropy of the channel symbols to the maximum possible entropy.

Here Shannon reveals his analog roots based on electrical engineering.

In general, ideal or nearly ideal encoding requires a long delay in the transmitter and receiver. In the noiseless case which we have been considering, the main function of this delay is to allow reasonably good matching of probabilities to corresponding lengths of sequences. With a good code the logarithm of the reciprocal probability of a long message

[9] Technical Report No. 65, The Research Laboratory of Electronics, M.I.T.

must be proportional to the duration of the corresponding signal, in fact

$$\left|\frac{\log p^{-1}}{T} - C\right|$$

must be small for all but a small fraction of the long messages.

If a source can produce only one particular message its entropy is zero, and no channel is required. For example, a computing machine set up to calculate the successive digits of π produces a definite sequence with no chance element. No channel is required to "transmit" this to another point. One could construct a second machine to compute the same sequence at the point. However, this may be impractical. In such a case we can choose to ignore some or all of the statistical knowledge we have of the source. We might consider the digits of π to be a random sequence in that we construct a system capable of sending any sequence of digits. In a similar way we may choose to use some of our statistical knowledge of English in constructing a code, but not all of it. In such a case we consider the source with the maximum entropy subject to the statistical conditions we wish to retain. The entropy of this source determines the channel capacity which is necessary and sufficient. In the π example the only information retained is that all the digits are chosen from the set $0, 1, \ldots, 9$. In the case of English one might wish to use the statistical saving possible due to letter frequencies, but nothing else. The maximum entropy source is then the first approximation to English and its entropy determines the required channel capacity.

EXAMPLES

As a simple example of some of these results consider a source which produces a sequence of letters chosen from among A, B, C, D with probabilities $\frac{1}{2}$, $\frac{1}{4}$, $\frac{1}{8}$, $\frac{1}{8}$, successive symbols being chosen independently. We have

$$\begin{aligned} H &= -\left(\frac{1}{2}\log\frac{1}{2} + \frac{1}{4}\log\frac{1}{4} + \frac{2}{8}\log\frac{1}{8}\right) \\ &= \frac{7}{4} \text{ bits per symbol.} \end{aligned}$$

Thus we can approximate a coding system to encode messages from this source into binary digits with an average of $\frac{7}{4}$ binary digit per symbol.

In this case we can actually achieve the limiting value by the following code (obtained by the method of the second proof of Theorem 9):

$$\begin{array}{ll} A & 0 \\ B & 10 \\ C & 110 \\ D & 111 \end{array}$$

The average number of binary digits used in encoding a sequence of N symbols will be

$$N\left(\frac{1}{2}\times 1+\frac{1}{4}\times 2+\frac{2}{8}\times 3\right)=\frac{7}{4}N$$

It is easily seen that the binary digits $0, 1$ have probabilities $\frac{1}{2}, \frac{1}{2}$, so the H for the coded sequences is one bit per symbol. Since, on the average, we have $\frac{7}{4}$ binary symbols per original letter, the entropies on a time basis are the same. The maximum possible entropy for the original set is $\log 4 = 2$, occurring when A, B, C, D have probabilities $\frac{1}{4}, \frac{1}{4}, \frac{1}{4}, \frac{1}{4}$. Hence the relative entropy is $\frac{7}{8}$. We can translate the binary sequences into the original set of symbols on a two-to-one basis by the following table:

$$\begin{array}{ll} 00 & A' \\ 01 & B' \\ 10 & C' \\ 11 & D' \end{array}$$

This double process then encodes the original message into the same symbols but with an average compression ratio $\frac{7}{8}$.

As a second example consider a source which produces a sequence of A's and B's with probability p for A and q for B. If $p << q$ we have

$$\begin{aligned} H &= -\log p^p(1-p)^{1-p} \\ &= -p\log p(1-p)^{(1-p)/p} \\ &\doteq p\log\frac{e}{p} \end{aligned}$$

In such a case one can construct a fairly good coding of the message on a $0, 1$ channel by sending a special sequence, say 0000, for the infrequent symbol A and then a sequence indicating the *number* of B's following it.

This could be indicated by the binary representation with all numbers containing the special sequence deleted. All numbers up to 16 are represented as usual; 16 is represented by the next binary number after 16 which does not contain four zeros, namely $17 = 10001$, etc.

It can be shown that as $p \to 0$ the coding approaches ideal provided the length of the special sequence is properly adjusted.

Part II: The Discrete Channel with Noise

Part I, the noiseless channel, was the appetizer. Part II, the noisy channel, is the main course.

REPRESENTATION OF A NOISY DISCRETE CHANNEL

We now consider the case where the signal is perturbed by noise during transmission or at one or the other of the terminals. This means that the received signal is not necessarily the same as that sent out by the transmitter. Two cases may be distinguished. If a particular transmitted signal always produces the same received signal, i.e. the received signal is a definite function of the transmitted signal, then the effect may be called distortion. If this function has an inverse—no two transmitted signals producing the same received signal—distortion may be corrected, at least in principle, by merely performing the inverse functional operation on the received signal.

The case of interest here is that in which the signal does not always undergo the same change in transmission. In this case we may assume the received signal E to be a function of the transmitted signal S and a second variable, the noise N.

$$E = f(S, N)$$

The noise is considered to be a chance variable just as the message was above. In general it may be represented by a suitable stochastic process. The most general type of noisy discrete channel we shall consider is a generalization of the finite state noise free channel described previously. We assume a finite number of states and a set of probabilities

$$p_{\alpha,i}(\beta, j).$$

This is the probability, if the channel is in state α and symbol i is transmitted, that symbol j will be received and the channel left in state

β. Thus α and β range over the possible states, i over the possible transmitted signals and j over the possible received signals. In the case where successive symbols are independently perturbed by the noise there is only one state, and the channel is described by the set of transition probabilities $p_i(j)$, the probability of transmitted symbol i being received as j.

If a noisy channel is fed by a source there are two statistical processes at work: the source and the noise. Thus there are a number of entropies that can be calculated. First there is the entropy $H(x)$ of the source or of the input to the channel (these will be equal if the transmitter is non-singular). The entropy of the output of the channel, i.e. the received signal, will be denoted by $H(y)$. In the noiseless case $H(y) = H(x)$. The joint entropy of input and output will be $H(xy)$. Finally there are two conditional entropies $H_x(y)$ and $H_y(x)$, the entropy of the output when the input is known and conversely. Among these quantities we have the relations

$$H(x, y) = H(x) + H_x(y) = H(y) + H_y(x)$$

All of these entropies can be measured on a per-second or a per-symbol basis.

EQUIVOCATION AND CHANNEL CAPACITY

If the channel is noisy it is not in general possible to reconstruct the original message or the transmitted signal with *certainty* by any operation on the received signal E. There are, however, ways of transmitting the information which are optimal in combating noise. This is the problem which we now consider.

Suppose there are two possible symbols 0 and 1, and we are transmitting at a rate of 1000 symbols per second with probabilities $p_0 = p_1 = \frac{1}{2}$. Thus our source is producing information at the rate of 1000 bits per second. During transmission the noise introduces errors so that, on the average, 1 in 100 is received incorrectly (a 0 as 1, or 1 as 0). What is the rate of transmission of information? Certainly less than 1000 bits per second since about 1% of the received symbols are incorrect. Our first impulse might be to say the rate is 990 bits per second, merely subtracting the expected number of errors. This is not satisfactory since it fails to take into account the recipient's lack of knowledge of where the errors occur. We may carry it to an extreme case and suppose the noise so great that the received symbols are entirely independent of the transmitted symbols. The probability of receiving 1 is $\frac{1}{2}$ whatever was transmitted and similarly for 0. Then about half of the received symbols are correct due to chance alone, and we would be giving the system credit for transmitting 500 bits per second while actually no information is being transmitted at all. Equally "good" transmission would be obtained by dispensing with the channel entirely and flipping a coin at the receiving point.

Evidently the proper correction to apply to the amount of information transmitted is the amount of this information which is missing in the received signal, or alternatively the uncertainty when we have received a signal of what was actually sent. From our previous discussion of entropy as a measure of uncertainty it seems reasonable to use the conditional entropy of the message, knowing the received signal, as a measure of this missing information. This is indeed the proper definition, as we shall see later. Following this idea the rate of actual transmission, R, would be obtained by subtracting from the rate of production (i.e., the entropy of the source) the average rate of conditional entropy.

Here, Shannon gives an intuitive definition of channel capacity: it is the amount of information sent from the source, minus the amount of ambiguity/equivocation injected by the noisy channel.

$$R = H(x) - H_y(x)$$

The conditional entropy $H_y(x)$ will, for convenience, be called the equivocation. It measures the average ambiguity of the received signal.

In the example considered above, if a 0 is received the *a posteriori* probability that a 0 was transmitted is .99, and that a 1 was transmitted is .01. These figures are reversed if a 1 is received. Hence

$$\begin{aligned} H_y(x) &= -[.99 \log .99 + 0.01 \log 0.01] \\ &= .081 \text{ bits/symbol} \end{aligned}$$

or 81 bits per second. We may say that the system is transmitting at a rate $1000 - 81 = 919$ bits per second. In the extreme case where a 0 is equally likely to be received as a 0 or 1 and similarly for 1, the a posteriori probabilities are $\frac{1}{2}$, $\frac{1}{2}$ and

$$H_y(x) = -\left[\frac{1}{2}\log\frac{1}{2} + \frac{1}{2}\log\frac{1}{2}\right]$$
$$= 1 \text{ bit per symbol}$$

or 1000 bits per second. The rate of transmission is then 0 as it should be.

The following theorem gives a direct intuitive interpretation of the equivocation and also serves to justify it as the unique appropriate measure. We consider a communication system and an observer (or auxiliary device) who can see both what is sent and what is recovered (with errors due to noise). This observer notes the errors in the recovered message and transmits data to the receiving point over a "correction channel" to enable the receiver to correct the errors. The situation is indicated schematically in figure 8.

Theorem 10: If the correction channel has a capacity equal to $H_y(x)$ it is possible to so encode the correction data as to send it over this channel and correct all but an arbitrarily small fraction ϵ of the errors. This is not possible if the channel capacity is less than $H_y(x)$.

Roughly then, $H_y(x)$ is the amount of additional information that must be supplied per second at the receiving point to correct the received message.

To prove the first part, consider long sequences of received message M' and corresponding original message M. There will be logarithmically $TH_y(x)$ of the M's which could reasonably have produced each M'. Thus we have $TH_y(x)$ binary digits to send each T seconds. This can be done with ϵ frequency of errors on a channel of capacity $H_y(x)$.

The second part can be proved by noting, first, that for any discrete chance variables x, y, z

$$H_y(x, z) \geq H_y(x)$$

The left-hand side can be expanded to give

$$H_y(z) + H_{yz}(x) \geq H_y(x)$$
$$H_{yz}(x) \geq H_y(x) - H_y(z) \geq H_y(x) - H(z)$$

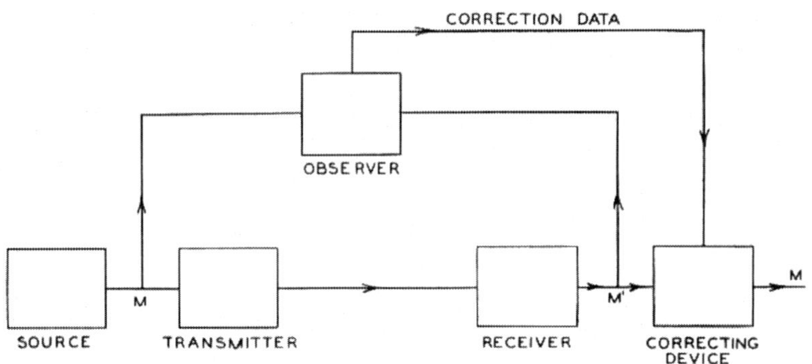

Figure 8. Schematic diagram of a correction system.

If we identify x as the output of the source, y as the received signal and z as the signal sent over the correction channel, then the right-hand side is the equivocation less the rate of transmission over the correction channel. If the capacity of this channel is less than the equivocation the right-hand side will be greater than zero and $H_{yz}(x) \geq 0$. But this is the uncertainty of what was sent, knowing both the received signal and the correction signal. If this is greater than zero the frequency of errors cannot be arbitrarily small.

EXAMPLE: Suppose the errors occur at random in a sequence of binary digits: probability p that a digit is wrong and $q = 1 - p$ that it is right. These errors can be corrected if their position is known. Thus the correction channel need only send information as to these positions. This amounts to transmitting from a source which produces binary digits with probability p for 1 (correct) and q for 0 (incorrect). This requires a channel of capacity

$$-[p \log p + q \log q]$$

which is the equivocation of the original system.

The rate of transmission R can be written in two other forms due to the identities noted above. We have

$$\begin{aligned} R &= H(x) - H_y(x) \\ &= H(y) - H_z(y) \\ &= H(x) + H(y) - H(x, y). \end{aligned}$$

The first defining expression has already been interpreted as the amount of information sent less the uncertainty of what was sent. The second measures the amount received less the part of this which is due to noise. The third is the sum of the two amounts less the joint entropy and therefore in a sense is the number of bits per second common to the two. Thus all three expressions have a certain intuitive significance.

The capacity C of a noisy channel should be the maximum possible rate of transmission, i.e., the rate when the source is properly matched to the channel. We therefore define the channel capacity by

$$C = \text{Max}\,(H(x) - H_y(x))$$

where the maximum is with respect to all possible information sources used as input to the channel. If the channel is noiseless, $H_y(x) = 0$. The definition is then equivalent to that already given for a noiseless channel since the maximum entropy for the channel is its capacity.

THE FUNDAMENTAL THEOREM FOR A DISCRETE CHANNEL WITH NOISE

It may seem surprising that we should define a definite capacity C for a noisy channel since we can never send certain information in such a case. It is clear, however, that by sending the information in a redundant form the probability of errors can be reduced. For example, by repeating the message many times and by a statistical study of the different received versions of the message the probability of errors could be made very small. One would expect, however, that to make this probability of errors approach zero, the redundancy of the encoding must increase indefinitely, and the rate of transmission therefore approach zero. This is by no means true. If it were, there would not be a very well defined capacity, but only a capacity for a given frequency of errors, or a given equivocation; the capacity going down as the error requirements are made more stringent. Actually the capacity C defined above has a very definite significance. It is possible to send information at the rate C through the channel *with as small a frequency of errors or equivocation as desired* by proper encoding. This statement is not true for any rate greater than C. If an attempt is made to transmit at a higher rate than C, say $C + R_1$, then there will necessarily be an equivocation equal to a

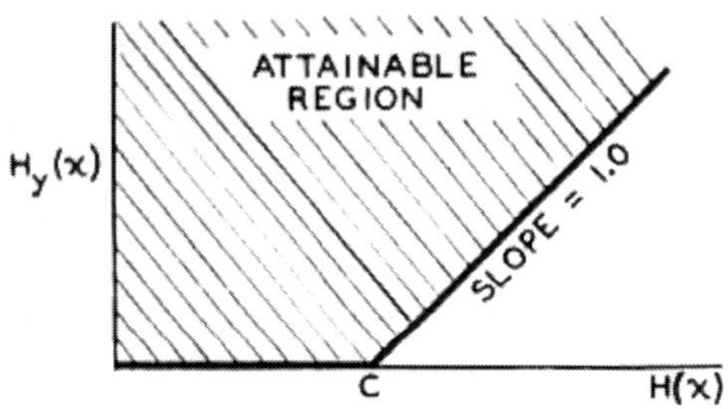

Figure 9. The equivocation possible for a given input entropy to a channel.

greater than the excess R_1. Nature takes payment by requiring just that much uncertainty, so that we are not actually getting any more than Cthrough correctly.

The situation is indicated in figure 9. The rate of information into the channel is plotted horizontally and the equivocation vertically. Any point above the heavy line in the shaded region can be attained and those below cannot. The points on the line cannot in general be attained, but there will usually be two points on the line that can.

These results are the main justification for the definition of C and will now be proved.

Theorem 11 is the central theorem of the entire paper. It's not so hard to prove that you can't send information at a rate higher than the channel capacity, as given by the mutual information between input and output. But to prove that almost all error correcting codes attain the channel capacity is a remarkable insight.

Theorem 11: Let a discrete channel have the capacity C and a discrete source the entropy per second H. If $H \leq C$ there exists a coding system such that the output of the source can be transmitted over the channel with an arbitrarily small frequency of errors (or an arbitrarily small equivocation). If $H > C$ it is possible to encode the source so that the equivocation is less than $H - C + \epsilon$ where ϵ is arbitrarily small. There is no method of encoding which gives an equivocation less than $H - C$.

The method of proving the first part of this theorem is not by exhibiting a coding method having the desired properties, but by showing that such a code must exist in a certain group of codes. In fact we will average the frequency of errors over this group and show that this average can be made less than ϵ. If the average of a set of numbers is less than ϵ there must exist at least one in the set which is less than ϵ. This will establish the desired result.

The capacity C of a noisy channel has been defined as

$$C = \text{Max}\,(H(x) - H_y(x))$$

The proof is subtle, but is essentially the mathematization of figure 10.

where x is the input and y the output. The maximization is over all sources which might be used as input to the channel.

Let S_0 be a source which achieves the maximum capacity C. If this maximum is not actually achieved by any source let S_0 be a source which approximates to giving the maximum rate. Suppose S_0 is used as input to the channel. We consider the possible transmitted and received sequences of a long duration T. The following will be true:

1. The transmitted sequences fall into two classes, a high probability group with about $2^{TH(x)}$ members and the remaining sequences of small total probability.

2. Similarly the received sequences have a high probability set of about $2^{TH(y)}$ members and a low probability set of remaining sequences.

3. Each high probability output could be produced by about $2^{TH_y(x)}$ inputs. The probability of all other cases has a small total probability.

All the ϵ's and δ's implied by the words "small" and "about" in these statements approach zero as we allow T to increase and S_0 to approach the maximizing source.

The situation is summarized in figure 10 where the input sequences are points on the left and output sequences points on the right. The fan of cross lines represents the range of possible causes for a typical output.

Now suppose we have another source producing information at rate R with $R < C$. In the period T this source will have 2^{TR} high probability outputs. We wish to associate these with a selection of the possible channel inputs in such a way as to get a small frequency of errors. We will set up this association in all possible ways (using, however, only the high probability group of inputs as determined by the source S_0) and average the frequency of errors for this large class of possible coding systems. This is the same as calculating the frequency of errors for a

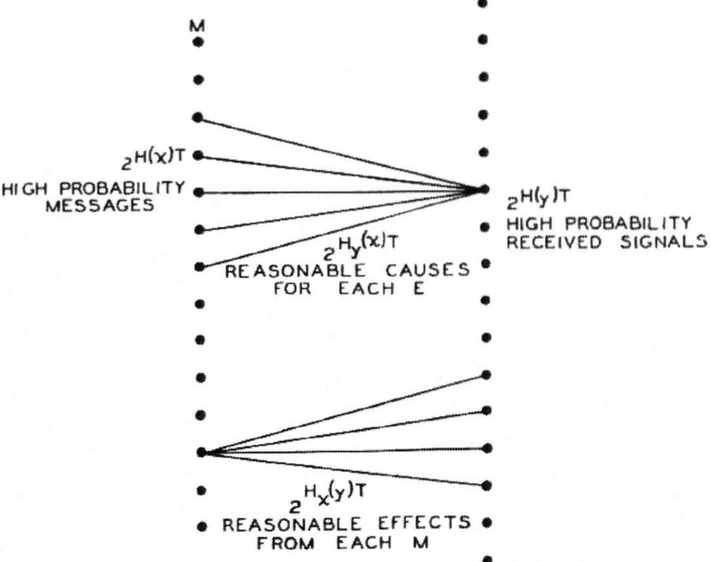

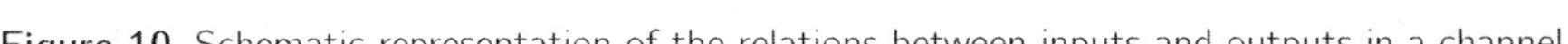
Figure 10. Schematic representation of the relations between inputs and outputs in a channel.

To summarize, when one sends information at any rate less than the channel capacity, the probability that two encoded signal states give the same output goes to zero, and the probability that a given output comes from only one input state goes to one. Choosing a random coding for the input will then result in a uniquely decodable code. The problem with random codes is that decoding can only be performed using a gigantic lookup table that lists which input give which outputs. This is not efficient. But non-random codes that attain the channel capacity are hard to find! Low density parity check codes (Gallagher, above) do achieve the capacity but they are still hard to decode.

random association of the messages and channel inputs of duration T. Suppose a particular output y_1 is observed. What is the probability of more than one message in the set of possible causes of y_1? There are 2^{TR} messages distributed at random in $2^{TH(x)}$ points. The probability of a particular point being a message is thus

$$2^{T(R-H(x))}$$

The probability that none of the points in the fan is a message (apart from the actual originating message) is

$$P = [1 - 2^{T(R-H(x))}]^{2^{TH_y(x)}}$$

Now $R < H(x) - H_y(x)$ so $R - H(x) = -H_y(x) - \eta$ with η positive. Consequently

$$P = [1 - 2^{-TH_y(x) - T\eta}]^{2^{TH_y(x)}}$$

approaches (as $T \to \infty$)

$$1 - 2^{-T\eta}.$$

Hence the probability of an error approaches zero and the first part of the theorem is proved.

The second part of the theorem is easily shown by noting that we could merely send C bits per second from the source, completely neglecting the remainder of the information generated. At the receiver the neglected part gives an equivocation $H(x) - C$ and the part transmitted need only add ϵ. This limit can also be attained in many other ways, as will be shown when we consider the continuous case.

The last statement of the theorem is a simple consequence of our definition of C. Suppose we can encode a source with $R = C + a$ in such a way as to obtain an equivocation $H_y(x) = a - \epsilon$ with ϵ positive. Then $R = H(x) = C + a$ and

$$H(x) - H_y(x) = C + \epsilon$$

with ϵ positive. This contradicts the definition of C as the maximum of $H(x) - H_y(x)$.

Actually more has been proved than was stated in the theorem. If the average of a set of numbers is within ϵ of their maximum, a fraction of at most $\sqrt{\epsilon}$ can be more than $\sqrt{\epsilon}$ below the maximum. Since ϵ is arbitrarily small we can say that almost all the systems are arbitrarily close to the ideal.

DISCUSSION

The demonstration of theorem 11, while not a pure existence proof, has some of the deficiencies of such proofs. An attempt to obtain a good approximation to ideal coding by following the method of the proof is generally impractical. In fact, apart from some rather trivial cases and certain limiting situations, no explicit description of a series of approximation to the ideal has been found. Probably this is no accident but is related to the difficulty of giving an explicit construction for a good approximation to a random sequence.

Shannon admits that the proof is somewhat dissatisfying, as it attains the capacity bound by using random codes, whereas real codes are not random!

An approximation to the ideal would have the property that if the signal is altered in a reasonable way by the noise, the original can still be recovered. In other words the alteration will not in general bring it closer to another reasonable signal than the original. This is accomplished at the cost of a certain amount of redundancy in the coding. The redundancy must be introduced in the proper way to combat the particular noise structure involved. However, any redundancy in the source will usually help if it is utilized at the receiving point. In particular, if the source already has a certain redundancy and no attempt is made to eliminate it in matching to the channel, this redundancy will help combat noise. For example, in a noiseless telegraph channel one could save about 50% in time by proper encoding of the messages. This is not done and most of the redundancy of English remains in the channel symbols. This has the advantage, however, of allowing considerable noise in the channel. A sizable fraction of the letters can be received incorrectly and still reconstructed by the context. In fact this is probably not a bad approximation to the ideal in many cases, since the statistical structure of English is rather involved and the reasonable English sequences are not too far (in the sense required for theorem) from a random selection.

As in the noiseless case a delay is generally required to approach the ideal encoding. It now has the additional function of allowing a large sample of noise to affect the signal before any judgment is made at the receiving point as to the original message. Increasing the sample size always sharpens the possible statistical assertions.

The content of theorem 11 and its proof can be formulated in a somewhat different way which exhibits the connection with the noiseless case more clearly. Consider the possible signals of duration T and suppose a subset of them is selected to be used. Let those in the subset all be used with equal probability, and suppose the receiver is constructed to select, as the original signal, the most probable cause from the subset, when a perturbed signal is received. We define $N(T, q)$ to be the maximum number of signals we can choose for the subset such that the probability of an incorrect interpretation is less than or equal to q.

Theorem 12: $\lim_{T\to\infty} \frac{\log N(T,q)}{T} = C$, where C is the channel capacity, provided that q does not equal 0 or 1.

In other words, no matter how we set our limits of reliability, we can distinguish reliably in time T enough messages to correspond to about CT bits, when T is sufficiently large. Theorem 12 can be compared with the definition of the capacity of a noiseless channel given in section 1.

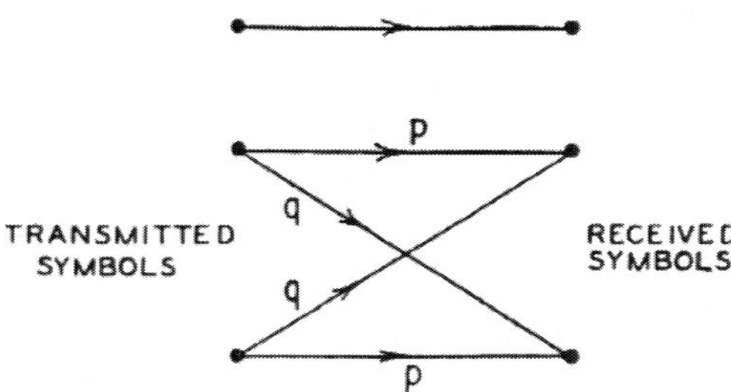

Figure 11. Example of a discrete channel.

The reader thanks Shannon for being merciful—every time he proves another blockbuster theorem, he gives a simple, easily understandable example of the application of the theorem.

EXAMPLE OF A DISCRETE CHANNEL AND ITS CAPACITY

A simple example of a discrete channel is indicated in figure 11. There are three possible symbols. The first is never affected by noise. The second and third each have probability p of coming through undisturbed, and q of being changed into the other of the pair. We have (letting $\alpha = -[p\log p + q\log q]$ and P and Q be the probabilities of using the first or second symbols)

$$H(x) = -P\log P - 2Q\log Q$$
$$H_y(x) = 2Q_\alpha$$

We wish to choose P and Q in such a way as to maximize $H(x) - H_y(x)$, subject to the constraint $P + 2Q = 1$. Hence we consider

$$U = -P\log P - 2Q\log Q - 2Q\alpha + \lambda(P + 2Q)$$
$$\frac{\partial U}{\partial P} = -1 - \log P + \lambda = 0$$
$$\frac{\partial U}{\partial Q} = -2 - 2\log Q - 2\alpha + 2\lambda = 0.$$

Eliminating λ

$$\log P = \log Q + \alpha$$
$$P = Qe^\alpha = Q\beta$$
$$P = \frac{\beta}{\beta+2}Q = \frac{1}{\beta+2}.$$

The channel capacity is then

$$C = \log\frac{\beta+2}{\beta}.$$

Note how this checks the obvious values in the cases $p = 1$ and $p = \frac{1}{2}$. In the first, $\beta = 1$ and $C = \log 3$, which is correct since the channel is then noiseless with three possible symbols. If $p = \frac{1}{2}$, $\beta = 2$ and $C = \log 2$. Here the second and third symbols cannot be distinguished at all and act together like one symbol. The first symbol is used with probability $P = \frac{1}{2}$ and the second and third together with probability $\frac{1}{2}$. This may be distributed in any desired way and still achieve the maximum capacity.

For intermediate values of p the channel capacity will lie between $\log 2$ and $\log 3$. The distinction between the second and third symbols conveys some information but not as much as in the noiseless case. The first symbol is used somewhat more frequently than the other two because of its freedom from noise.

THE CHANNEL CAPACITY IN CERTAIN SPECIAL CASES

If the noise affects successive channel symbols independently it can be described by a set of transition probabilities p_{ij}. This is the probability, if symbol i is sent, that j will be received. The maximum channel rate is then given by the maximum of

To attain the actual channel capacity, one must choose input probabilities for symbols that maximize the mutual information between input and output.

$$\sum_{i,j} P_i p_{ij} \log \sum_i P_i p_{ij} - \sum_{i,j} P_i p_{ij} \log p_{ij}$$

where we vary the P_i subject to $\sum P_i = 1$. This leads by the method of Lagrange to the equations,

$$\sum_j p_{sj} \log \frac{p_{sj}}{\sum_i P_i p_{ij}} = \mu \qquad s = 1, 2, \cdots .$$

Multiplying by P_s and summing on s shows that $\mu = -C$. Let the inverse of p_{sj} (if it exists) be h_{st} so that $\sum_s h_{st} p_{sj} = \delta_{tj}$. Then:

$$\sum_{s,j} h_{st} p_{sj} \log p_{sj} - \log \sum_i P_i p_{it} = -C \sum_s h_{st}.$$

Hence:

$$\sum_i P_i p_{it} = \exp \Big[C \sum_s h_{st} + \sum_{s,j} h_{st} p_{sj} \log p_{sj}\Big]$$

or,

$$P_i = \sum_t h_{it} \exp \Big[C \sum_s h_{st} + \sum_{s,j} h_{st} p_{sj} \log p_{sj}\Big].$$

This is the system of equations for determining the maximizing values of P_i, with C to be determined so that $\Sigma P_i = 1$. When this is

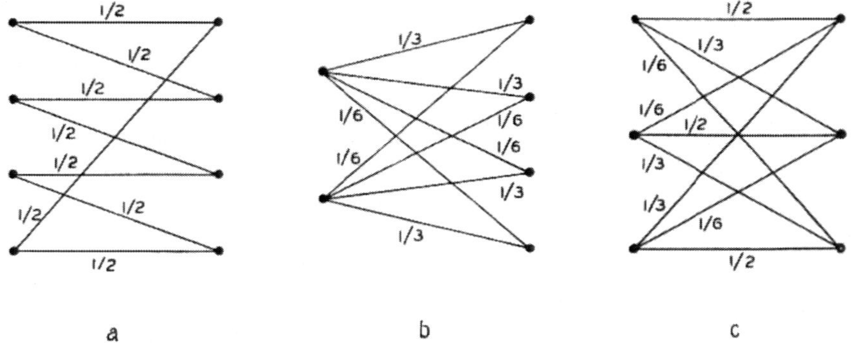

Figure 12. Examples of discrete channels with the same transition probabilities for each input and for each output.

done C will be the channel capacity, and the P_i the proper probabilities for the channel symbols to achieve this capacity.

If each input symbol has the same set of probabilities on the lines emerging from it, and the same is true of each output symbol, the capacity can be easily calculated. Examples are shown in figure 12. In such a case $H_x(y)$ is independent of the distribution of probabilities on the input symbols, and is given by $-\Sigma p_i \log p_i$ where the p_i are the values of the transition probabilities from any input symbol. The channel capacity is

$$\begin{aligned} &\text{Max}\,[H(y) - H_x(y)] \\ &= \text{Max}\,H(y) + \Sigma p_i \log p_i \end{aligned}$$

The maximum of $H(y)$ is clearly $\log m$ where m is the number of output symbols, since it is possible to make them all equally probable by making the input symbols equally probable. The channel capacity is therefore

$$C = \log m + \Sigma p_i \log p_i.$$

In figure 12a it would be

$$C = \log 4 - \log 2 = \log 2$$

This could be achieved by using only the 1st and 3d symbols. In figure 12b

$$\begin{aligned} C &= \log 4 - \frac{2}{3}\log 3 - \frac{1}{3}\log 6 \\ &= \log 4 - \log 3 - \frac{1}{3}\log 2 \\ &= \log \frac{1}{3} 2^{\frac{5}{3}}. \end{aligned}$$

In figure 12c we have

$$\begin{aligned} C &= \log 3 - \frac{1}{2}\log 2 - \frac{1}{3}\log 3 - \frac{1}{6}\log 6 \\ &= \log \frac{2}{2^{\frac{1}{2}} 3^{\frac{1}{3}} 6^{\frac{1}{6}}}. \end{aligned}$$

Suppose the symbols fall into several groups such that the noise never causes a symbol in one group to be mistaken for a symbol in another group. Let the capacity for the nth group be C_n when we use only the symbols in this group. Then it is easily shown that, for best use of the entire set, the total probability P_n of all symbols in the nth group should be

$$P_n = \frac{2^{C_n}}{\Sigma 2^{C_n}}.$$

Within a group the probability is distributed just as it would be if these were the only symbols being used. The channel capacity is

$$C = \log \Sigma 2^{C_n}.$$

AN EXAMPLE OF EFFICIENT CODING

The following example, although somewhat unrealistic, is a case in which exact matching to a noisy channel is possible. There are two channel symbols, 0 and 1, and the noise affects them in blocks of seven symbols. A block of seven is either transmitted without error, or exactly one symbol of the seven is incorrect. These eight possibilities are equally likely. We have

$$\begin{aligned} C &= \text{Max}\,[H(y) - H_x(y)] \\ &= \frac{1}{7}[7 + \frac{8}{8}\log\frac{1}{8}] \\ &= \frac{4}{7}\ \text{bits/symbol}. \end{aligned}$$

An efficient code, allowing complete correction of errors and transmitting at the rate C, is the following (found by a method due to R. Hamming):

Let a block of seven symbols be $X_1, X_2, \ldots X_7$. Of these X_3, X_5, X_6 and X_7 are message symbols and chosen arbitrarily by the source. The other three are redundant and calculated as follows:

$$X_4 \text{ is chosen to make } \alpha = X_4 + X_5 + X_6 + X_7 \text{ even}$$

$$X_2 \text{ is chosen to make } \beta = X_2 + X_3 + X_6 + X_7 \text{ even}$$

$$X_1 \text{ is chosen to make } \gamma = X_1 + X_3 + X_5 + X_7 \text{ even}$$

When a block of seven is received α, β and γ are calculated and if even called zero, if odd called one. The binary number $\alpha\ \beta\ \gamma$ then gives the subscript of the X_i that is incorrect (if 0 there was no error).

Hamming codes are examples of structured parity check codes: these formulae define the parity checks, which are structured in such a way that after a single error has occurred, the parity check points to the location of the error. Shannon's student Bob Gallagher extended this method to random, low-density parity check codes, currently among the most highly used codes for error correction.

Appendix 1

And now the math to back up the clear, intuitive pictures of the text.

THE GROWTH OF THE NUMBER OF BLOCKS OF SYMBOLS WITH A FINITE STATE CONDITION

Let $N_i(L)$ be the number of blocks of symbols of length L ending in state i. Then we have

$$N_j(L) = \sum_{is} N_i(L - b_{ij}^{(s)})$$

where $b_{ij}^1, b_{ij}^2, \cdots b_{ij}^m$ are the length of the symbols which may be chosen in state i and lead to state j. These are linear difference equations and the behavior as $L \to \infty$ must be of the type

$$N_j = A_j W^L$$

Substituting in the difference equation

$$A_j W^L = \sum_{iS} A_i W^{L - b_{ji}^{(s)}}$$

or

$$A_j = \sum_{iS} A_i W^{-b_{ji}^{(s)}}$$

$$\sum_i (\sum_s W^{-b_{ij}^{(s)}} - \delta_{ij}) A_i = 0.$$

For this to be possible the determinant

$$D(W) = \left|a_{ij}\right| = \left|\sum_S W^{-b_{ij}^{(s)}} - \delta_{ij}\right|$$

must vanish and this determines W, which is, of course, the largest real root of $D = 0$.

The quantity C is then given by

$$C = \lim_{L \to \infty} \frac{\log \Sigma A_j W^L}{L} = \log W$$

and we also note that the same growth properties result if we require that all blocks start in the same (arbitrarily chosen) state.

Appendix 2

DERIVATION OF $H = -\Sigma p_i \log p_i$

Let $H(\frac{1}{n}, \frac{1}{n}, \cdots, \frac{1}{n}) = A(n)$. From condition (3) we can decompose a choice from s^m equally likely possibilities into a series of m choices each from s equally likely possibilities and obtain

$$A(s^m) = mA(s)$$

Similarly

$$A(t^n) = nA(t)$$

We can choose n arbitrarily large and find an m to satisfy

$$s^m \leq t^n < s^{(m+1)}$$

Thus, taking logarithms and dividing by $n \log s$,

$$\frac{m}{n} \leq \frac{\log t}{\log s} \leq \frac{m}{n} + \frac{1}{n} \quad \text{or} \quad \left|\frac{m}{n} - \frac{\log t}{\log s}\right| < \epsilon$$

where ϵ is arbitrarily small.

Now from the monotonic property of $A(n)$

$$A(s^m) \leq A(t^n) \leq A(s^{m+1})$$
$$mA(s) \leq nA(t) \leq (m+1)A(s)$$

Hence, dividing by $nA(s)$,

$$\frac{m}{n} \leq \frac{A(t)}{A(s)} \leq \frac{m}{n} + \frac{1}{n} \quad \text{or} \quad \left| \frac{m}{n} - \frac{A(t)}{A(s)} \right| < \epsilon$$

$$\left| \frac{A(t)}{A(s)} - \frac{\log t}{logs} \right| \leq 2\epsilon \qquad A(t) = -K \log t$$

where K must be positive to satisfy (2).

Now suppose we have a choice from n possibilities with commeasurable probabilities $p_i = \frac{n_i}{\Sigma n_i}$ where the n_i are integers. We can break down a choice from Σn_i possibilities into a choice from n possibilities with probabilities $p_i \ldots p_n$ and then, if the ith was chosen, a choice from n_i with equal probabilities. Using condition 3 again, we equate the total choice from Σn_i as computed by two methods

$$K \log \Sigma n_i = H(p_i, \ldots, p_n) + K\Sigma p_i \log n_i$$

Hence

$$H = K[\Sigma p_i \log \Sigma n_i - \Sigma p_i \log n_i]$$
$$- K\Sigma p_i \log \frac{n_i}{\Sigma n_i} = -K\Sigma p_i \log p_i.$$

If the p_i are incommeasurable, they may be approximated by rationals and the same expression must hold by our continuity assumption. Thus the expression holds in general. The choice of coefficient K is a matter of convenience and amounts to the choice of a unit of measure.

Appendix 3

THEOREMS ON ERGODIC SOURCES

If it is possible to go from any state with $P > 0$ to any other along a path of probability $p > 0$, the system is ergodic and the strong law of large numbers can be applied. Thus the number of times a given path p_{ij} in the network is traversed in a long sequence of length N is about proportional to the probability of being at i and then choosing this path,

$P_i p_{ij} N$. If N is large enough the probability of percentage error $\pm\,\delta$ in this is less than ϵ so that for all but a set of small probability the actual numbers lie within the limits

$$(P_i p_{ij} \pm \delta)N$$

Hence nearly all sequences have a probability p given by

$$p = \Pi p_{ij}^{(P_i p_{ij} \pm \delta)N}$$

and $\frac{\log p}{N}$ is limited by

$$\frac{\log p}{N} = \Sigma (P_i p_{ij} \pm \delta) \log p_{ij}$$

or

$$\left| \frac{\log p}{N} - \Sigma P_i p_{ij} \log p_{ij} \right| < \eta.$$

This proves theorem 3.

Theorem 4 follows immediately from this on calculating upper and lower bounds for $n(q)$ based on the possible range of values of p in Theorem 3.

In the mixed (not ergodic) case if

$$L = \Sigma p_i L_i$$

and the entropies of the components are $H_1 \geq H_2 \geq \ \ldots \ \geq H_n$ we have the Theorem: $\lim_{N\to\infty} \frac{\log n(q)}{N} = \varphi(q)$ is a decreasing step function,

$$\varphi(q) = H_s \text{ in the interval } \sum_1^{s-1} \alpha_i < q < \sum_1^{s} \alpha_i.$$

To prove theorems 5 and 6 first note that F_n is monotonic decreasing because increasing N adds a subscript to a conditional entropy. A simple substitution for $p_{Bi}(S_j)$ in the definition of F_n shows that

$$F_N = NG_N - (N-1)G_{N-1}$$

and summing this for all N gives $G_N = \frac{1}{N}\Sigma F_N$. Hence $G_N \geq F_N$ and G_N monotonic decreasing. Also they must approach the same limit. By using theorem 3 we see that $\lim_{N\to\infty} G_N = H$.

Appendix 4

MAXIMIZING THE RATE FOR A SYSTEM OF CONSTRAINTS

Suppose we have a set of constraints on sequences of symbols that is of the finite state type and can be represented therefore by a linear graph. Let $l_{ij}^{(s)}$ be the lengths of the various symbols that can occur in passing from state i to state j. What distribution of probabilities P_i for the different states and $p_{ij}^{(s)}$ for choosing symbol s in state i and going to state j maximizes the rate of generating information under these constraints? The constraints define a discrete channel and the maximum rate must be less than or equal to the capacity C of this channel, since if all blocks of large length were equally likely, this rate would result, and if possible this would be best. We will show that this rate can be achieved by proper choice of the P_i and $p_{ij}^{(s)}$.

The rate in question is

$$\frac{-\Sigma P_i p_{ij}^{(s)} \log p_{ij}^{(s)}}{\Sigma P_i p_{ij}^{(s)} l_{ij}^{(s)}} = \frac{N}{M}.$$

Let $l_{ij} = \sum_s l_{ij}^{(s)}$. Evidently for a maximum $p_{ij}^{(s)} = k \exp l_{ij}^{(s)}$. The constraints on maximization are $\Sigma P_i = 1, \sum_j p_{ij} = 1, \Sigma P_i(p_{ij} - \delta_{ij}) = 0$. Hence we maximize

$$U = \frac{-\Sigma P_i p_{ij} \log p_{ij}}{\Sigma P_i p_{ij} l_{ij}} + \lambda \sum_i P_i + \Sigma \mu_i p_{ij} + \Sigma \eta_j P_i (p_{ij} - \delta_{ij})$$

$$\frac{\partial U}{\partial p_{ij}} = -\frac{M P_i (1 + \log p_{ij}) + N P_i l_{ij}}{M^2} + \lambda + \mu_i + \eta_i P_i = 0.$$

Solving for p_{ij}

$$p_{ij} = A_i B_j D^{-l_{ij}}.$$

Since

$$\sum_i p_{ij} = 1, \quad A_i^{-1} = \sum_j B_j D^{-l_{ij}}$$

$$p_{ij} = \frac{B_j D^{-l_{ij}}}{\sum_s B_s D^{-l_{is}}}.$$

The correct value of D is the capacity C and the B_j are solutions of

$$B_i = \Sigma B_j C^{-l_{ij}}$$

for then

$$p_{ij} = \frac{B_j}{B_i} C^{-l_{ij}}$$

$$\Sigma P_i \frac{B_j}{B_i} C^{-l_{ij}} = P_j$$

or

$$\Sigma \frac{P_i}{B_i} C^{-l_{ij}} = \frac{P_j}{B_i}$$

So that if λ_i satisfy

$$\Sigma \gamma_i C^{-l_{ij}} = \gamma_j$$

$$P_i = B_i \gamma_i$$

Both of the sets of equations for B_i and γ_i can be satisfied since C is such that

$$\left| C^{-l_{ij}} - \delta_{ij} \right| = 0$$

In this case the rate is

$$-\frac{\Sigma P_i p_{ij} \log \frac{B_j}{B_i} C^{-l_{ij}}}{\Sigma P_i p_{ij} l_{ij}}$$

$$= C - \frac{\Sigma P_i p_{ij} \log \frac{B_j}{B_i}}{\Sigma P_i p_{ij} l_{ij}}$$

but

$$\Sigma P_i p_{ij} (\log B_j - \log B_i) = \sum_j P_j \log B_j - \Sigma P_i \log B_i = 0$$

Hence the rate is C and as this could never be exceeded this is the maximum, justifying the assumed solution.

Part III, on continuous signals and noise, appeared in a separate Bell Systems Journal article from parts I and II. The methods hinge on taking continuous signals and making them effectively discrete. Limiting the bandwidth makes them effectively discrete in time (Shannon–Nyquist sampling theorem), and a finite signal-to-noise ratio reduces the amount of information required to describe a noisy continuous sample to a finite number of bits.

Part III: Mathematical Preliminaries

In this final installment of the paper we consider the case where the signals or the messages or both are continuously variable, in contrast with the discrete nature assumed heretofore. To a considerable extent the continuous case can be obtained through a limiting process from the discrete case by dividing the continuum of messages and signals into a large but finite number of small regions and calculating the various parameters involved on a discrete basis. As the size of the regions is decreased these parameters in general approach as limits the proper values for the continuous case. There are, however, a few new effects that appear and also a general change of emphasis in the direction of specialization of the general results to particular cases. We will not attempt, in the continuous case, to obtain our results with the greatest generality, or with the extreme rigor of pure mathematics, since this would involve a great deal of abstract measure theory and would obscure the main thread of the analysis. A preliminary study, however, indicates that the theory can be formulated in a completely axiomatic and rigorous manner which includes both the continuous and discrete cases and many others. The occasional liberties taken with limiting processes in the present analysis can be justified in all cases of practical interest.

SETS AND ENSEMBLES OF FUNCTIONS

We shall have to deal in the continuous case with sets of functions and ensembles of functions. A set of functions, as the name implies, is merely a class or collection of functions, generally of one variable, time. It can be specified by giving an explicit representation of the various functions in the set, or implicitly by giving a property which functions in the set possess and others do not. Some examples are:

1. The set of functions:

   $$f_\theta(t) = \sin(t + \theta).$$

   Each particular value of θ determines a particular function in the set.

2. The set of all functions of time containing no frequencies over W cycles per second.

3. The set of all functions limited in band to W and in amplitude to A.

4. The set of all English speech signals as functions of time.

An *ensemble* of functions is a set of functions together with a probability measure whereby we may determine the probability of a function in the set having certain properties.[1] For example with the set,

$$f_\theta(t) = \sin(t + \theta),$$

we may give a probability distribution for θ, $P(\theta)$. The set then becomes an ensemble.

Some further examples of ensembles of functions are:

1. A finite set of functions $f_k(t)(k = 1, 2, \ldots, n)$ with the probability of f_k being p_k.

2. A finite dimensional family of functions

$$f(\alpha_1, \alpha_2, \ldots, \alpha_n; t)$$

with a probability distribution on the parameters α_i:

$$p(\alpha_1, \ldots, \alpha_n).$$

For example we could consider the ensemble defined by

$$f(a_1, \ldots, a_n, \theta_1, \ldots, \theta_n; t) = \sum_{i=1}^{n} a_i \sin i(\omega t + \theta_i)$$

with the amplitudes a_i distributed normally and independently, and the phases θ_i distributed uniformly (from 0 to 2π) and independently.

3. The ensemble

$$f(a_i, t) = \sum_{n=-\infty}^{+\infty} a_n \frac{\sin \pi(2Wt - n)}{\pi(2Wt - n)}$$

with the a_i normal and independent all with the same standard deviation $\sqrt{N}$. This is a representation of "white" noise, band limited to the band from 0 to W cycles per second and with average power N.[2]

[1] In mathematical terminology the functions belong to a measure space whose total measure is unity.

[2] This representation can be used as a definition of band limited white noise. It has certain advantages in that it involves fewer limiting operations than do definitions that have been used in the past. The name "white noise," already firmly entrenched in the literature, is perhaps somewhat unfortunate. In optics white light means either any continuous spectrum as contrasted with a point spectrum, or a spectrum which is flat with *wavelength* (which is not the same as a spectrum flat with frequency).

4. Let points be distributed on the t axis according to a Poisson distribution. At each selected point the function $f(t)$ is placed and the different functions added, giving the ensemble

$$\sum_{k=-\infty}^{\infty} f(t+t_k)$$

where the t_k are the points of the Poisson distribution. This ensemble can be considered as a type of impulse or shot noise where all the impulses are identical.

5. The set of English speech functions with the probability measure given by the frequency of occurrence in ordinary use.

An ensemble of functions $f\alpha(t)$ is *stationary* if the same ensemble results when all functions are shifted any fixed amount in time. The ensemble

$$f_\theta(t) = \sin(t+\theta)$$

is stationary if θ is distributed uniformly from 0 to 2π. If we shift each function by t_1 we obtain

$$\begin{aligned} f_\theta(t+t_1) &= \sin(t+t_1+\theta) \\ &= \sin(t+\varphi) \end{aligned}$$

with φ distributed uniformly from 0 to 2π. Each function has changed but the ensemble as a whole is invariant under the translation. The other examples given above are also stationary.

An ensemble is *ergodic* if it is stationary, and there is no subset of the functions in the set with a probability different from 0 and 1 which is stationary. The ensemble

$$\sin(t+\theta)$$

is ergodic. No subset of these functions of probability $\neq 0, 1$ is transformed into itself under all time translations. On the other hand the ensemble

$$a\sin(t+\theta)$$

with a distributed normally and θ uniform is stationary but not ergodic. The subset of these functions with a between 0 and 1 for example is stationary.

Of the examples given, 3 and 4 are ergodic, and 5 may perhaps be considered so. If an ensemble is ergodic we may say roughly that each function in the set is typical of the ensemble. More precisely it is known that with an ergodic ensemble an average of any statistic over the ensemble is equal (with probability 1) to an average over the time translations of a particular function of the set.[3] Roughly speaking, each function can be expected, as time progresses, to go through, with the proper frequency, all the convolutions of any of the functions in the set.

Just as we may perform various operations on numbers or functions to obtain new numbers or functions, we can perform operations on ensembles to obtain new ensembles. Suppose, for example, we have an ensemble of functions $f_\alpha(t)$ and an operator T which gives for each function $f_\alpha(t)$ a resulting function

$$g_\alpha(t) = Tf_\alpha(t).$$

Probability measure is defined for the set $g_\alpha(t)$ by means of that for the set $f_\alpha(t)$. The probability of a certain subset of the $g_\alpha(t)$ functions is equal to that of the subset of the $f_\alpha(t)$ functions which produce members of the given subset of g functions under the operation T. Physically this corresponds to passing the ensemble through some device, for example, a filter, a rectifier or a modulator. The output functions of the device form the ensemble $g_\alpha(t)$.

A device or operator T will be called invariant if shifting the input merely shifts the output, i.e., if

$$g_\alpha = Tf_\alpha(t)$$

implies

$$g_\alpha(t + t_1) = Tf_\alpha(t + t_1)$$

for all $f_\alpha(t)$ and all t_1. It is easily shown (see Appendix 5) that if T is invariant and the input ensemble is stationary then the output ensemble

[3]This is the famous ergodic theorem or rather one aspect of this theorem which was proved in somewhat different formulations by Birkoff, von Neumann, and Koopman, and subsequently generalized by Wiener, Hopf, Hurewicz and others. The literature on ergodic theory is quite extensive and the reader is referred to the papers of these writers for precise and general formulations; e.g., E. Hopf, "Ergodentheorie," *Ergebnisse der Mathematik und ihrer Grenzgebiete*, v. 5; "On Causality Statistics and Probability," *Journal of Mathematics and Physics*, v. XIII, No. 1, 1934; N. Wiener, "The Ergodic Theorem," *Duke Mathematical Journal*, v. 5, 1939.

is stationary. Likewise if the input is ergodic the output will also be ergodic.

A filter or a rectifier is invariant under all time translations. The operation of modulation is not since the carrier phase gives a certain time structure. However, modulation is invariant under all translations which are multiples of the period of the carrier.

Wiener has pointed out the intimate relation between the invariance of physical devices under time translations and Fourier theory.[4] He has shown, in fact, that if a device is linear as well as invariant Fourier analysis is then the appropriate mathematical tool for dealing with the problem.

An ensemble of functions is the appropriate mathematical representation of the messages produced by a continuous source (for example, speech), of the signals produced by a transmitter, and of the perturbing noise. Communication theory is properly concerned, as has been emphasized by Wiener, not with operations on particular functions, but with operations on ensembles of functions. A communication system is designed not for a particular speech function and still less for a sine wave, but for the ensemble of speech functions.

BAND LIMITED ENSEMBLES OF FUNCTIONS

If a function of time $f(t)$ is limited to the band from 0 to W cycles per second it is completely determined by giving its ordinates at a series of discrete points spaced $\frac{1}{2W}$ seconds apart in the manner indicated by the following result.[5]

[4] Communication theory is heavily indebted to Wiener for much of its basic philosophy and theory. His classic NDRC report, *The Interpolation, Extrapolation and Smoothing of Stationary Time Series* (Wiley, 1949), contains the first clear-cut formulation of communication theory as a statistical problem, the study of operations on time series. This work, although chiefly concerned with the linear prediction and filtering problem, is an important collateral reference in connection with the present paper. We may also refer here to Wiener's *Cybernetics* (Wiley, 1948), dealing with the general problems of communication and control.

[5] For a proof of this theorem and further discussion see the author's paper "Communication in the Presence of Noise" published in the *Proceedings of the Institute of Radio Engineers*, v. 37, No. 1, Jan., 1949, pp. 10–21.

Theorem 13: Let $f(t)$ contain no frequencies over W. Then

$$f(t) = \sum_{-\infty}^{\infty} X_n \frac{\sin \pi(2Wt - n)}{\pi(2Wt - n)}$$

where

$$x_n = f\left(\frac{n}{2W}\right).$$

In this expansion $f(t)$ is represented as a sum of orthogonal functions. The coefficients X_n of the various terms can be considered as coordinates in an infinite dimensional "function space." In this space each function corresponds to precisely one point and each point to one function.

A function can be considered to be substantially limited to a time T if all the ordinates X_n outside this interval of time are zero. In this case all but $2TW$ of the coordinates will be zero. Thus functions limited to a band W and duration T correspond to points in a space of $2TW$ dimensions.

A subset of the functions of band W and duration T corresponds to a region in this space. For example, the functions whose total energy is less than or equal to E correspond to points in a $2TW$ dimensional sphere with radius $r = \sqrt{2WE}$.

An *ensemble* of functions of limited duration and band will be represented by a probability distribution $p(x_1, \ldots, x_n)$ in the corresponding n dimensional space. If the ensemble is not limited in time we can consider the $2TW$ coordinates in a given interval T to represent substantially the part of the function in the interval T and the probability distribution $p(x_1, \ldots, x_n)$ to give the statistical structure of the ensemble for intervals of that duration.

Here and below, Shannon generalizes entropy to continuous distributions: the primary difference with the discrete case is that the entropy of continuous distributions can be negative, and is defined only with respect to a particular measure on the continuous set.

ENTROPY OF A CONTINUOUS DISTRIBUTION

The entropy of a discrete set of probabilities $p_1, \ldots, p_n$ has been defined as:

$$H = -\sum p_i \log p_i.$$

In an analogous manner we define the entropy of a continuous distribution with the density distribution function $p(x)$ by:

$$H = -\int_{-\infty}^{\infty} p(x) \log p(x) dx.$$

With an n dimensional distribution $p(x_1, \ldots, x_n)$ we have

$$H = -\int \cdots \int p(x_1, \ldots, x_n) \log p(x_1, \ldots, x_n) dx_1 \cdots dx_n.$$

If we have two arguments x and y (which may themselves be multidimensional) the joint and conditional entropies of $p(x, y)$ are given by

$$H(x, y) = -\iint p(x, y) \log p(x, y) dx dy$$

and

$$H_x(y) = -\iint p(x, y) \log \frac{p(x, y)}{p(x)} dx dy$$

$$H_y(x) = -\iint p(x, y) \log \frac{p(x, y)}{p(y)} dx dy$$

where

$$p(x) = \int p(x, y) dy$$

$$p(y) = \int p(x, y) dx.$$

The entropies of continuous distributions have most (but not all) of the properties of the discrete case. In particular we have the following:

1. If x is limited to a certain volume v in its space, then $H(x)$ is a maximum and equal to $\log v$ when $p(x)$ is constant $(1/v)$ in the volume.

2. With any two variables x, y we have

$$H(x, y) \leq H(x) + H(y)$$

with equality if (and only if) x and y are independent, i.e., $p(x, y) = p(x)p(y)$ (apart possibly from a set of points of probability zero).

Even though entropy/information over continuous variables can be negative, mutual information/channel capacity remains positive; mutual information is in a sense a more fundamental quantity for continuous variables than just information on its own.

3. Consider a generalized averaging operation of the following type:

$$p'(y) = \int a(x, y) p(x) dx$$

with

$$\int a(x, y) dx = \int a(x, y) dy = 1, \quad a(x, y) \geq 0.$$

Then the entropy of the averaged distribution $p'(y)$ is equal to or greater than that of the original distribution $p(x)$.

4. We have

$$H(x, y) = H(x) + H_x(y) = H(y) + H_y(x)$$

and

$$H_x(y) \leq H(y).$$

5. Let $p(x)$ be a one-dimensional distribution. The form of $p(x)$ giving a maximum entropy subject to the condition that the standard deviation of x be fixed at is Gaussian. To show this we must maximize

$$H(x) = -\int p(x) \log p(x) dx$$

with

$$\sigma^2 = \int p(x) x^2 dx \quad \text{and} \quad 1 = \int p(x) dx$$

as constraints. This requires, by the calculus of variations, maximizing

$$\int \left[-p(x) \log p(x) + \lambda p(x) x^2 + \mu p(x) \right] dx.$$

The condition for this is

$$-1 - \log p(x) + \lambda x^2 + \mu = 0$$

and consequently (adjusting the constants to satisfy the constraints)

The distribution that maximizes entropy for fixed variance is a Gaussian. Gaussian distributions will be assumed moving forward unless otherwise noted.

$$p(x) = \frac{1}{\sqrt{2\pi}\sigma} e^{-(x^2/2\sigma^2)}.$$

Similarly in n dimensions, suppose the second order moments of $p(x_1, \ldots, x_n)$ are fixed at A_{ij}:

$$A_{ij} = \int \cdots \int x_i x_j p(x_1, \ldots, {}_n) dx_i \cdots dx_n.$$

Then the maximum entropy occurs (by a similar calculation) when $p(x_1, \ldots, x_n)$ is the n dimensional Gaussian distribution with the second order moments A_{ij}.

6. The entropy of a one-dimensional Gaussian distribution whose standard deviation is σ is given by

$$H(x) = \log \sqrt{2\pi e}\sigma.$$

This is calculated as follows:

$$\begin{aligned}
p(x) &= \frac{1}{\sqrt{2\pi}\sigma} e^{-(x^2/2\sigma^2)} \\
-\log p(x) &= \log \sqrt{2\pi}\sigma + \frac{x^2}{2\sigma^2} \\
H(x) &= -\int p(x) \log p(x) dx \\
&= \int p(x) \log \sqrt{2\pi}\sigma dx + \int p(x) \frac{x^2}{2\sigma^2} dx \\
&= \log \sqrt{2\pi}\sigma + \frac{\sigma^2}{2\sigma^2} \\
&= \log \sqrt{2\pi}\sigma + \log \sqrt{e} \\
&= \log \sqrt{2\pi e}\sigma.
\end{aligned}$$

Similarly the n dimensional Gaussian distribution with associated quadratic form a_{ij} is given by

$$p(x_1, \ldots, x_n) = \frac{|a_{ij}|^{\frac{1}{2}}}{(2\pi)^{n/2}} \exp\left(-\frac{1}{2}\sum a_{ij} x_i x_j\right)$$

and the entropy can be calculated as

$$H = \log(2\pi e)^{n/2} |a_{ij}|^{-\frac{1}{2}}$$

where $|a_{ij}|$ is the determinant whose elements are a_{ij}.

7. If x is limited to a half line ($p(x) = 0$ for $x \leq 0$) and the first moment of x is fixed at a:

$$a = \int_0^\infty p(x) x dx,$$

then the maximum entropy occurs when

$$p(x) = \frac{1}{a} e^{-(x/a)}$$

and is equal to $\log ea$.

8. There is one important difference between the continuous and discrete entropies. In the discrete case the entropy measures in an *absolute* way the randomness of the chance variable. In the continuous case the measurement is *relative to the coordinate system.* If we change coordinates the entropy will in general change. In fact if we change to coordinates $y_1 \cdots y_n$ the new entropy is given by

☞ But mutual information is not!

$$H(y) = \int \cdots \int p(x_i, \ldots, _n) J\left(\frac{x}{y}\right) \log p(x_1, \ldots, x_n) J\left(\frac{x}{y}\right) dy_1 \cdots dy_n$$

where $J\left(\frac{x}{y}\right)$ is the Jacobian of the coordinate transformation. On expanding the logarithm and changing the variables to $x_1, \cdots, x_n$, we obtain:

$$H(y) = H(x) - \int \cdots \int p(x_1, \ldots, x_n) \log J\left(\frac{x}{y}\right) dx_i \ldots dx_n.$$

Thus the new entropy is the old entropy less the expected logarithm of the Jacobian. In the continuous case the entropy can be considered a measure of randomness *relative to an assumed standard*, namely the coordinate system chosen with each small volume element $dx_1 \cdots dx_n$ given equal weight. When we change the coordinate system the entropy in the new system measures the randomness when equalvolume elements $dy_1 \cdots \; dy_n$ in the new system are given equal weight.

In spite of this dependence on the coordinate system the entropy concept is as important in the continuous case as the discrete case. This is due to the fact that the derived concepts of information rate and channel capacity depend on the *difference* of two entropies and this difference *does not* depend on the coordinate frame, each of the two terms being changed by the same amount.

The entropy of a continuous distribution can be negative. The scale of measurements sets an arbitrary zero corresponding to a uniform distribution over a unit volume. A distribution which is more confined than this has less entropy and will be negative. The rates and capacities will, however, always be nonnegative.

9. A particular case of changing coordinates is the linear transformation

$$y_j = \sum_i a_{ij} x_i.$$

In this case the Jacobian is simply the determinant $|a_{ij}|^{-1}$ and

$$H(y) = H(x) + \log |a_{ij}|.$$

In the case of a rotation of coordinates (or any measure preserving transformation) $J = 1$ and $H(y) = H(x)$.

ENTROPY OF AN ENSEMBLE OF FUNCTIONS

Consider an ergodic ensemble of functions limited to a certain band of width W cycles per second. Let

$$p(x_1, \dots, x_n)$$

be the density distribution function for amplitudes $x_1, \dots, x_n$ at n successive sample points. We define the entropy of the ensemble per degree of freedom by

$$H' = -\lim_{n \to \infty} \frac{1}{n} \int \cdots \int p(x_1, \dots, x_n) \log p(x_1, \dots, x_n dx_1 \dots dx_n.$$

We may also define an entropy H per second by dividing, not by n, but by the time T in seconds for n samples. Since $n = 2TW$, $H = 2WH'$.

With white thermal noise p is Gaussian and we have

$$H' = \log \sqrt{2\pi e N},$$
$$H = W \log 2\pi e N.$$

For a given average power N, white noise has the maximum possible entropy. This follows from the maximizing properties of the Gaussian distribution noted above.

The entropy for a continuous stochastic process has many properties analogous to that for discrete processes. In the discrete case the entropy was related to the logarithm of the *probability* of long sequences, and to the *number* of reasonably probable sequences of long

Having set up the preliminaries for continuous variables, Shannon now extends the rest of the discrete channel theory to the continuous case.

Assuming Gaussian noise gives a simple closed-form solution, here and below.

length. In the continuous case it is related in a similar fashion to the logarithm of the *probability density* for a long series of samples, and the *volume* of reasonably high probability in the function space.

As before, all the discrete channel/entropy formulae now apply.

More precisely, if we assume $p(x_1, \ldots, x_n)$ continuous in all the x_i for all n, then for sufficiently large n

$$\left|\frac{\log p}{n} - H'\right| < \epsilon$$

for all choices of $(x_1, \ldots, x_n)$ apart from a set whose total probability is less than δ, with δ and ϵ arbitrarily small. This follows form the ergodic property if we divide the space into a large number of small cells.

The relation of H to volume can be stated as follows: Under the same assumptions consider the n dimensional space corresponding to $p(x_1, \ldots, x_n)$. Let $V_n(q)$ be the smallest volume in this space which includes in its interior a total probability q. Then

$$\lim_{n \to \infty} \frac{\log V_n(q)}{n} = H'$$

provided q does not equal 0 or 1.

These results show that for large n there is a rather well-defined volume (at least in the logarithmic sense) of high probability, and that within this volume the probability density is relatively uniform (again in the logarithmic sense).

In the white noise case the distribution function is given by

$$p(x_1, \ldots, x_n) = \frac{1}{(2\pi N)^{n/2}} \exp -\frac{1}{2N} \sum x_i^2.$$

Since this depends only on $\sum x_i^2$ the surfaces of equal probability density are spheres and the entire distribution has spherical symmetry. The region of high probability is a sphere of radius $\sqrt{nN}$. As $n \to \infty$ the probability of being outside a sphere of radius $\sqrt{n(N+\epsilon)}$ approaches zero and $\frac{1}{n}$ times the logarithm of the volume of the sphere approaches $\log\sqrt{2\pi eN}$.

In the continuous case it is convenient to work not with the entropy H of an ensemble but with a derived quantity which we will call the entropy power. This is defined as the power in a white noise limited to the same band as the original ensemble and having the same entropy. In other words if H' is the entropy of an ensemble its entropy power is

$$N_1 = \frac{1}{2\pi e} \exp 2H'.$$

In the geometrical picture this amounts to measuring the high probability volume by the squared radius of a sphere having the same volume. Since white noise has the maximum entropy for a given power, the entropy power of any noise is less than or equal to its actual power.

ENTROPY LOSS IN LINEAR FILTERS

Deterministic filters can only decrease the amount of entropy in a signal, just as in the discrete case.

Theorem 14: If an ensemble having an entropy H_1 per degree of freedom in band W is passed through a filter with characteristic $Y(f)$ the output ensemble has an entropy

$$H_2 = H_1 + \frac{1}{W} \int_W \log |Y(f)|^2 df.$$

The operation of the filter is essentially a linear transformation of coordinates. If we think of the different frequency components as the original coordinate system, the new frequency components are merely the old ones multiplied by factors. The coordinate transformation matrix is thus essentially diagonalized in terms of these coordinates. The Jacobian of the transformation is (for n sine and n cosine components)

$$J = \prod_{i=1}^{n} |Y(f_i)|^2$$

where the f_i are equally spaced through the band W. This becomes in the limit

$$\exp \frac{1}{W} \int_W \log |Y(f)|^2 df.$$

Since J is constant its average value is the same quantity and applying the theorem on the change of entropy with a change of coordinates, the result follows. We may also phrase it in terms of the entropy power. Thus if the entropy power of the first ensemble is N_1 that of the second is

$$N_1 \exp \frac{1}{W} \int_W \log |Y(f)|^2 df.$$

GAIN	ENTROPY POWER FACTOR	ENTROPY POWER GAIN IN DECIBELS	IMPULSE RESPONSE
$1-\omega$ 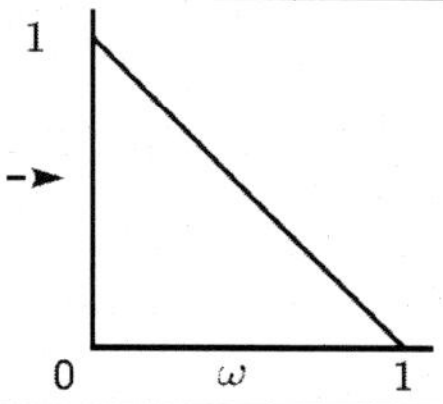	$\frac{1}{e^2}$	-8.69	$\frac{\sin^2(t/2)}{t^2/2}$
$1-\omega^2$ 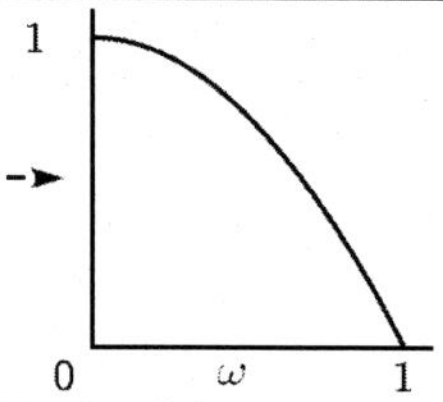	$\left(\frac{2}{e}\right)^4$	-5.33	$2\left[\frac{\sin t}{t^3}-\frac{\cos t}{t^2}\right]$
$1-\omega^3$ 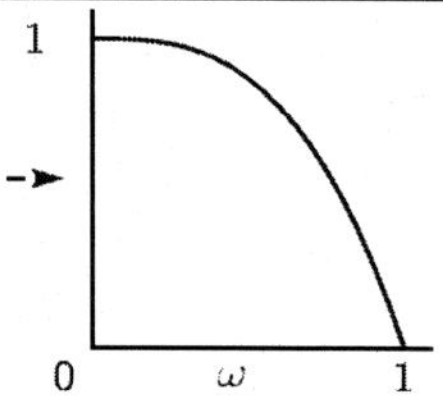	0.411	-3.87	$6\left[\frac{\cos t-1}{t^4}-\frac{\cos t}{2t^2}+\frac{\sin t}{t^3}\right]$
$\sqrt{1-\omega^2}$ 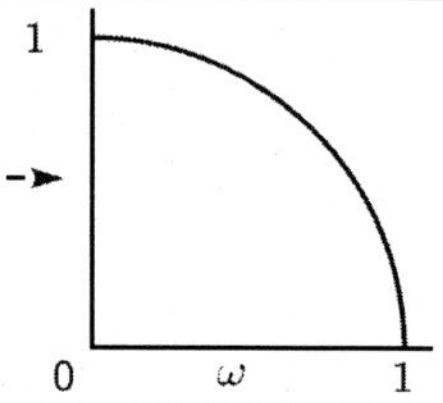	$\left(\frac{2}{e}\right)^2$	-2.67	$\frac{\pi}{2}\frac{J_1(t)}{t}$
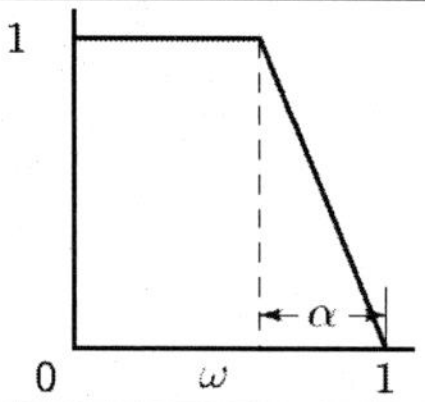	$\frac{1}{e^{2\alpha}}$	-8.69α	$\frac{1}{\alpha t^2}\left[\cos(1-\alpha)t-\cos t\right]$

Table 1.

The final entropy power is the initial entropy power multiplied by the geometric mean gain of the filter. If the gain is measured in *db*, then the output entropy power will be increased by the arithmetic mean *db* gain over W.

In Table I the entropy power loss has been calculated (and also expressed in *db*) for a number of ideal gain characteristics. The impulsive responses of these filters are also given for $W = 2\pi$, with phase assumed to be 0.

The entropy loss for many other cases can be obtained from these results. For example the entropy power factor $1/e^2$ for the first case also applies to any gain characteristic obtain from $1 - \omega$ by a measure preserving transformation of the ω axis. In particular a linearly increasing gain $G(\omega) = \omega$, or a "saw tooth" characteristic between 0 and 1 have the same entropy loss. The reciprocal gain has the reciprocal factor. Thus $1/\omega$ has the factor e^2. Raising the gain to any power raises the factor to this power.

ENTROPY OF A SUM OF TWO ENSEMBLES

By contrast, convolution increases entropy.

If we have two ensembles of functions $f_\alpha(t)$ and $g_\beta(t)$ we can form a new ensemble by "addition." Suppose the first ensemble has the probability density function $p(x_1, \ldots, x_n)$ and the second $q(x_1, \ldots, x_n)$. Then the density function for the sum is given by the convolution:

$$r(x_1, \ldots, x_n) = \int \cdots \int p(y_1, \ldots, y_n) q(x_1 - y_1, \ldots, x_n - y_n) dy_1 \cdots dy_n.$$

Physically this corresponds to adding the noises or signals represented by the original ensembles of functions.

The following result is derived in Appendix 6.

Theorem 15: Let the average power of two ensembles be N_1 and $N2$ and let their entropy powers be $\overline{N}_1$ and $\overline{N}_2$. Then the entropy power of the sum, $\overline{N}_3$, is bounded by

$$\overline{N}_1 + \overline{N}_2 \leq \overline{N}_3 \leq N_1 + N_2.$$

That is, Gaussians are special functions that are the result of repeated convolution—this is the central limit theorem.

White Gaussian noise has the peculiar property that it can absorb any other noise or signal ensemble which may be added to it with a resultant entropy power approximately equal to the sum of the white noise power and the signal power (measured from the average signal value, which is normally zero), provided the signal power is small, in a certain sense, compared to noise.

Consider the function space associated with these ensembles having n dimensions. The white noise corresponds to the spherical Gaussian distribution in this space. The signal ensemble corresponds to another probability distribution, not necessarily Gaussian or spherical. Let the second moments of this distribution about its center of gravity be a_{ij}. That is, if $p(x_1, \ldots, x_n)$ is the density distribution function

$$a_{ij} = \int \cdots \int p(x_i - \alpha_i)(x_j - \alpha_j) dx_1 \cdots dx_n$$

where the α_i are the coordinates of the center of gravity. Now a_{ij} is a positive definite quadratic form, and we can rotate our coordinate system to align it with the principal directions of this form. a_{ij} is then reduced to diagonal form b_{ii}. We require that each b_{ii} be small compared to N, the squared radius of the spherical distribution.

So we will just use Gaussians, Gaussians, Gaussians!

In this case the convolution of the noise and signal produce approximately a Gaussian distribution whose corresponding quadratic form is

$$N + b_{ii}$$

The entropy power of this distribution is

$$\left[\prod (N + b_{ii})\right]^{1/n}$$

or approximately

Because of discretization in time (band limited signals) and in signal space (Gaussian signals with Gaussian noise), everything is going to go through just like it did in the discrete case.

$$= \left[(N)^n + \sum b_{ii}(N)^{n-1}\right]^{1/n}$$
$$\doteq N + \frac{1}{n}\sum b_{ii}.$$

The last term is the signal power, while the first is the noise power.

Part IV: The Continuous Channel

THE CAPACITY OF A CONTINUOUS CHANNEL

In a continuous channel the input or transmitted signals will be continuous functions of time $f(t)$ belonging to a certain set, and the output or received signals will be perturbed versions of these. We will consider only the case where both transmitted and received signals are limited to a certain band W. They can then be specified, for a time T, by $2TW$ numbers, and their statistical structure by finite dimensional distribution functions. Thus the statistics of the transmitted signal will be determined by

$$P(x_1, \ldots, x_n) = P(x)$$

and those of the noise by the conditional probability distribution

$$P_{x_1,\ldots,x_n}(y_1, \ldots, y_n) = P_x(y).$$

The rate of transmission of information for a continuous channel is defined in a way analogous to that for a discrete channel, namely

That is, channel capacity is just mutual information, as before.

$$R = H(x) - H_y(x)$$

where $H(x)$ is the entropy of the input and $H_y(x)$ the equivocation. The channel capacity C is defined as the maximum of R when we vary the input over all possible ensembles. This means that in a finite dimensional approximation we must vary $P(x) = P(x_1, \ldots, x_n)$ and maximize

$$-\int P(x) \log P(x) dx + \iint P(x,y) \log \frac{P(x,y)}{P(y)} dx dy.$$

This can be written

$$\iint P(x,y) \log \frac{P(x,y)}{P(x)P(y)} dx dy.$$

using the fact that $\iint P(x,y) \log P(x) dx dy = \int P(x) \log P(x) dx$. The channel capacity is thus expressed as follows:

$$C = \lim_{T \to \infty} \max_{P(x)} \frac{1}{T} \iint P(x,y) \log \frac{P(x,y)}{P(x)P(y)} dx dy.$$

It is obvious in this form that R and C are independent of the coordinate system since the numerator and denominator in $\log \frac{P(x,y)}{P(x)P(y)}$ will be multiplied by the same factors when x and y are transformed in any one-to-one way. This integral expression for C is more general than $H(x) - H_y(x)$. Properly interpreted (see Appendix 7) it will always exist while $H(x) - H_y(x)$ may assume an indeterminate form $\infty - \infty$ in some cases. This occurs, for example, if x is limited to a surface of fewer dimensions than n in its n dimensional approximation.

If the logarithmic base used in computing $H(x)$ and $H_y(x)$ is two then C is the maximum number of binary digits that can be sent per second over the channel with arbitrarily small equivocation, just as in the discrete case. This can be seen physically by dividing the space of signals into a large number of small cells, sufficiently small so that the probability density $P_x(y)$ of signal x being perturbed to point y is substantially constant over a cell (either of x or y). If the cells are considered as distinct points the situation is essentially the same as a discrete channel and the proofs used there will apply. But it is clear physically that this quantizing of the volume into individual points cannot in any practical situation alter the final answer significantly, provided the regions are sufficiently small. Thus the capacity will be the limit of the capacities for the discrete subdivisions and this is just the continuous capacity defined above.

On the mathematical side it can be shown first (see Appendix 7) that if u is the message, x is the signal, y is the received signal (perturbed by noise) and v is the recovered message then

$$H(x) - H_y(x) \geq H(u) - H_v(u)$$

regardless of what operations are performed on u to obtain x or on y to obtain v. Thus no matter how we encode the binary digits to obtain the signal, or how we decode the received signal to recover the message, the discrete rate for the binary digits does not exceed the channel capacity we have defined. On the other hand, it is possible under very general conditions to find a coding system for transmitting binary digits at the rate C with as small an equivocation or frequency of errors as desired. This is true, for example, if, when we take a finite dimensional approximating space for the signal functions, $P(x, y)$ is continuous in both x and y except at a set of points of probability zero.

An important special case occurs when the noise is added to the signal and is independent of it (in the probability sense). Then $P_x(y)$ is a function only of the difference $n = (y - x)$,

$$P_x(y) = Q(y - x)$$

and we can assign a definite entropy to the noise (independent of the statistics of the signal), namely the entropy of the distribution $Q(n)$. This entropy will be denoted by $H(n)$.

Theorem 16: If the signal and noise are independent and the received signal is the sum of the transmitted signal and the noise then the rate of transmission is

$$R = H(y) - H(n),$$

Shannon proves that mutual information gives channel capacity in the continuous case.

i.e., the entropy of the received signal less the entropy of the noise. The channel capacity is

$$C = \max_{P(x)} H(y) - H(n).$$

We have, since $y = x + n$:

$$H(x, y) = H(x, n).$$

Expanding the left side and using the fact that x and n are independent

$$H(y) + H_y(x) = H(x) + H(n).$$

Hence

$$R = H(x) - H_y(x) = H(y) - H(n).$$

Since $H(n)$ is independent of $P(x)$, maximizing R requires maximizing $H(y)$, the entropy of the received signal. If there are certain constraints on the ensemble of transmitted signals, the entropy of the received signal must be maximized subject to these constraints.

CHANNEL CAPACITY WITH AN AVERAGE POWER LIMITATION

A simple application of Theorem 16 is the case when the noise is a white thermal noise and the transmitted signals are limited to a certain average power P. Then the received signals have an average power $P + N$ where N is the average noise power. The maximum entropy for the received signals occurs when they also form a white noise ensemble since this is the greatest possible entropy for a power $P + N$ and can be obtained by a suitable choice of transmitted signals, namely if they form a white noise ensemble of power P. The entropy (per second) of the received ensemble is then

$$H(y) = W \log 2\pi e(P + N),$$

and the noise entropy is

$$H(n) = W \log 2\pi eN.$$

And the formula for channel capacity is particularly simple with an average power limitation, as this requirement restricts our maximum entropy solution to Gaussians.

The channel capacity is

$$C = H(y) - H(n) = W \log \frac{P + N}{N}.$$

Summarizing we have the following:

Theorem 17: The capacity of a channel of band W perturbed by white thermal noise power N when the average transmitter power is limited to P is given by

$$C = W \log \frac{P + N}{N}.$$

This means that by sufficiently involved encoding systems we can transmit binary digits at the rate $W \log_2 \frac{P+N}{N}$ bits per second, with arbitrarily small frequency of errors. It is not possible to transmit at a higher rate by any encoding system without a definite positive frequency of errors.

To approximate this limiting rate of transmission the transmitted signals must approximate, in statistical properties, a white noise.[6] A system which approaches the ideal rate may be described as follows: Let $M = 2^s$ samples of white noise be constructed each of duration T. These are assigned binary numbers from 0 to $M - 1$. At the transmitter the message sequences are broken up into groups of s and for each group the corresponding noise sample is transmitted as the signal. At the receiver the M samples are known and the actual received signal (perturbed by noise) is compared with each of them. The sample which has the least R.M.S. discrepancy from the received signal is chosen as the transmitted signal and the corresponding binary number reconstructed. This process amounts to choosing the most probable (*a posteriori*)

[6]This and other properties of the white noise case are discussed from the geometrical point of view in "Communication in the Presence of Noise," *loc. cit.*

signal. The number M of noise samples used will depend on the tolerable frequency ϵ of errors, but for almost all selections of samples we have

$$\lim_{\epsilon\to 0}\lim_{T\to\infty}\frac{\log M(\epsilon,T)}{T}=W\log\frac{P+N}{N},$$

so that no matter how small ϵ is chosen, we can, by taking T sufficiently large, transmit as near as we wish to $TW\log\frac{P+N}{N}$ binary digits in the time T.

Formulas similar to $C=W\log\frac{P+N}{N}$ for the white noise case have been developed independently by several other writers, although with somewhat different interpretations. We may mention the work of N. Wiener,[7] W. G. Tuller,[8] and H. Sullivan in this connection.

In the case of an arbitrary perturbing noise (not necessarily white thermal noise) it does not appear that the maximizing problem involved in determining the channel capacity C can be solved explicitly. However, upper and lower bounds can be set for C in terms of the average noise power N the noise entropy power N_1. These bounds are sufficiently close together in most practical cases to furnish a satisfactory solution to the problem.

Theorem 18: The capacity of a channel of band W perturbed by an arbitrary noise is bounded by the inequalities

$$W\log\frac{P+N_1}{N_1}\le C\le W\log\frac{P+N}{N_1}$$

where

$$P = \text{average transmitter power}$$
$$N = \text{average noise power}$$
$$N_1 = \text{entropy power of the noise.}$$

Here again the average power of the perturbed signals will be $P+N$. The maximum entropy for this power would occur if the received signal were white noise and would be $W\log 2\pi e(P+N)$. It may not be possible to achieve this; i.e., there may not be any ensemble of transmitted signals which, added to the perturbing noise, produce a white thermal noise at

[7] *Cybernetics, loc. cit.*

[8] "Theoretical Limitations on the Rate of Transmission of Information," *Proceedings of the Institute of Radio Engineers*, v. 37, No. 5, May, 1949, pp. 468–78.

the receiver, but at least this sets an upper bound to $H(y)$. We have, therefore

$$\begin{aligned} C &= \max H(y) - H(n) \\ &\leq W \log 2\pi e(P+N) - W \log 2\pi e N_1. \end{aligned}$$

This is the upper limit given in the theorem. The lower limit can be obtained by considering the rate if we make the transmitted signal a white noise, of power P. In this case the entropy power of the received signal must be at least as great as that of a white noise of power $P + N_1$ since we have shown in in a previous theorem that the entropy power of the sum of two ensembles is greater than or equal to the sum of the individual entropy powers. Hence

$$\max H(y) \geq W \log 2\pi e(P + N_1)$$

and

$$\begin{aligned} C &\geq W \log 2\pi e(P + N_1) - W \log 2\pi e N_1 \\ &= W \log \frac{P + N_1}{N_1}. \end{aligned}$$

As P increases, the upper and lower bounds approach each other, so we have as an asymptotic rate

$$W \log \frac{P+N}{N_1}.$$

If the noise is itself white, $N = N_1$ and the result reduces to the formula proved previously:

$$C = W \log \left(1 + \frac{P}{N}\right).$$

If the noise is Gaussian but with a spectrum which is not necessarily flat, N_1 is the geometric mean of the noise power over the various frequencies in the band W. Thus

$$N_1 = \exp \frac{1}{W} \int_W \log N(f) df$$

where $N(f)$ is the noise power at frequency f.

Theorem 19: If we set the capacity for a given transmitter power P equal to

$$C = W \log \frac{P + N - \eta}{N_1}$$

then η is monotonic decreasing as P increases and approaches 0 as a limit.

Suppose that for a given power P_1 the channel capacity is

$$W \log \frac{P_1 + N - \eta_1}{N_1}.$$

This means that the best signal distribution, say $p(x)$, when added to the noise distribution $q(x)$, gives a received distribution $r(y)$ whose entropy power is $(P1 + N - \eta_1)$. Let us increase the power to $P_1 + \Delta P$ by adding a white noise of power δP to the signal. The entropy of the received signal is now at least

$$H(y) = W \log 2\pi e(P_1 + N - \eta_1 + \Delta P)$$

by application of the theorem on the minimum entropy power of a sum. Hence, since we can attain the H indicated, the entropy of the maximizing distribution must be at least as great and η must be monotonic decreasing. To show that $\eta \to 0$ as $P \to \infty$ consider a signal which is white noise with a large P. Whatever the perturbing noise, the received signal will be approximately a white noise, if P is sufficiently large, in the sense of having an entropy power approaching $P + N$.

THE CHANNEL CAPACITY WITH A PEAK POWER LIMITATION

In some applications the transmitter is limited not by the average power output but by the peak instantaneous power. The problem of calculating the channel capacity is then that of maximizing (by variation of the ensemble of transmitted symbols)

$$H(y) - H(n)$$

subject to the constraint that all the functions $f(t)$ in the ensemble be less than or equal to $\sqrt{S}$, say, for all t. A constraint of this type does not work out as well mathematically as the average power limitation. The most we have obtained for this case is a lower bound valid for all $\frac{S}{N}$, an "asymptotic" upper bound (valid for large $\frac{S}{N}$) and an asymptotic value of C for $\frac{S}{N}$ small.

Theorem 20: The channel capacity C for a band W perturbed by white thermal noise of power N is bounded by

$$C \geq W \log \frac{2}{\pi e^3} \frac{S}{N},$$

where S is the peak allowed transmitter power. For sufficiently large $\frac{S}{N}$

$$C \leq W \log \frac{\frac{2}{\pi e} S + N}{N} (1 + \epsilon)$$

where ϵ is arbitrarily small. As $\frac{S}{N} \to 0$ (and provided the band W starts at 0)

$$C \Big/ W \log \left(1 + \frac{S}{N}\right) \to 1.$$

We wish to maximize the entropy of the received signal. If $\frac{S}{N}$ is large this will occur very nearly when we maximize the entropy of the transmitted ensemble.

The asymptotic upper bound is obtained by relaxing the conditions on the ensemble. Let us suppose that the power is limited to S not at every instant of time, but only at the sample points. The maximum entropy of the transmitted ensemble under these weakened conditions is certainly greater than or equal to that under the original conditions. This altered problem can be solved easily. The maximum entropy occurs if the different samples are independent and have a distribution function which is constant from $-\sqrt{S}$ to $+\sqrt{S}$. The entropy can be calculated as

$$W \log 4S.$$

The received signal will then have an entropy less than

$$W \log(4S + 2\pi eN)(1 + \epsilon)$$

with $\epsilon \to 0$ as $\frac{S}{N} \to \infty$ and the channel capacity is obtained by subtracting the entropy of the white noise, $W \log 2\pi eN$:

$$W \log(4S + 2\pi eN)(1 + \epsilon) - W \log(2\pi eN) = W \log \frac{\frac{2}{\pi e} S + N}{N} (1 + \epsilon).$$

This is the desired upper bound to the channel capacity.

To obtain a lower bound consider the same ensemble of functions. Let these functions be passed through an ideal filter with a triangular transfer characteristic. The gain is to be unity at frequency 0 and decline linearly down to gain 0 at frequency W. We first show that the output functions of the filter have a peak power limitation S at all times (not just the sample points). First we note that a pulse $\frac{\sin 2\pi Wt}{2\pi Wt}$ going into the filter produces

$$\frac{1}{2}\frac{\sin^2 \pi Wt}{(\pi Wt)^2}$$

in the output. This function is never negative. The input function (in the general case) can be thought of as the sum of a series of shifted functions

$$a\frac{\sin 2\pi Wt}{2\pi Wt}$$

where a, the amplitude of the sample, is not greater than $\sqrt{S}$. Hence the output is the sum of shifted functions of the non-negative form above with the same coefficients. These functions being non-negative, the greatest positive value for any t is obtained when all the coefficients a have their maximum positive values, i.e., $\sqrt{S}$. In this case the input function was a constant of amplitude S and since the filter has unit gain for D.C., the output is the same. Hence the output ensemble has a peak power S.

The entropy of the output ensemble can be calculated from that of the input ensemble by using the theorem dealing with such a situation. The output entropy is equal to the input entropy plus the geometrical mean gain of the filter:

$$\int_0^W \log G^2 df = \int_0^W \log \left(\frac{W-f}{W}\right)^2 df = -2W.$$

Hence the output entropy is

$$W \log 4S - 2W = W \log \frac{4S}{e^2}$$

and the channel capacity is greater than

$$W \log \frac{2}{\pi e^3}\frac{S}{N}.$$

We now wish to show that, for small $\frac{S}{N}$ (peak signal power over average white noise power), the channel capacity is approximately

$$C = W \log \left(1 + \frac{S}{N}\right).$$

More precisely $C/W \log\left(1+\frac{S}{N}\right) \to 1$ as $\frac{S}{N} \to 0$. Since the average signal power P is less than or equal to the peak S, it follows that for all $\frac{S}{N}$

$$C \leq W \log\left(1+\frac{P}{N}\right) \leq W \log\left(1+\frac{S}{N}\right).$$

This "bang-bang" method—always operating the channel at peak power—makes the channel effectively binary and reaches the capacity.

Therefore, if we can find an ensemble of functions such that they correspond to a rate nearly $W \log\left(1+\frac{S}{N}\right)$ and are limited to band W and peak S the result will be proved. Consider the ensemble of functions of the following type. A series of t samples have the same value, either $+\sqrt{S}$ or $-\sqrt{S}$, then the next t samples have the same value, etc. The value for a series is chosen at random, probability $\frac{1}{2}$ for $+\sqrt{S}$ and $\frac{1}{2}$ for $-\sqrt{S}$. If this ensemble be passed through a filter with triangular gain characteristic (unit gain at D.C.), the output is peak limited to $\pm S$. Furthermore the average power is nearly S and can be made to approach this by taking t sufficiently large. The entropy of the sum of this and the thermal noise can be found by applying the theorem on the sum of a noise and a small signal. This theorem will apply if

$$\sqrt{t}\frac{S}{N}$$

is sufficiently small. This can be ensured by taking $\frac{S}{N}$ small enough (after t is chosen). The entropy power will be $S+N$ to as close an approximation as desired, and hence the rate of transmission as near as we wish to

$$W \log\left(\frac{S+N}{N}\right).$$

A more extended discussion of the process of going from continuous sources to effectively discrete sources. Seems slightly archaic in the digital age, but there are some useful mathematical results!

Part V: The Rate for a Continuous Source

FIDELITY EVALUATION FUNCTIONS

In the case of a discrete source of information we were able to determine a definite rate of generating information, namely the entropy of the underlying stochastic process. With a continuous source the situation is considerably more involved. In the first place a continuously variable quantity can assume an infinite number of values and requires, therefore, an infinite number of binary digits for exact specification.

This means that to transmit the output of a continuous source with *exact recovery* at the receiving point requires, in general, a channel of infinite capacity (in bits per second). Since, ordinarily, channels have a certain amount of noise, and therefore a finite capacity, exact transmission is impossible.

This, however, evades the real issue. Practically, we are not interested in exact transmission when we have a continuous source, but only in transmission to within a certain tolerance. The question is, can we assign a definite rate to a continuous source when we require only a certain fidelity of recovery,measured in a suitable way. Of course, as the fidelity requirements are increased the rate will increase. It will be shown that we can, in very general cases, define such a rate, having the property that it is possible, by properly encoding the information, to transmit it over a channel whose capacity is equal to the rate in question, and satisfy the fidelity requirements. A channel of smaller capacity is insufficient.

It is first necessary to give a general mathematical formulation of the idea of fidelity of transmission. Consider the set of messages of a long duration, say T seconds. The source is described by giving the probability density, in the associated space, that the source will select the message in question $P(x)$. A given communication system is described (from the external point of view) by giving the conditional probability $P_x(y)$ that if message x is produced by the source the recovered message at the receiving point will be y. The system as a whole (including source and transmission system) is described by the probability function $P(x, y)$ of having message x and final output y. If this function is known, the complete characteristics of the system from the point of view of fidelity are known. Any evaluation of fidelity must correspond mathematically to an operation applied to $P(x, y)$. This operation must at least have the properties of a simple ordering of systems; i.e., it must be possible to say of two systems represented by $P_1(x, y)$ and $P_2(x, y)$ that, according to our fidelity criterion, either (1) the first has higher fidelity, (2) the second has higher fidelity, or (3) they have equal fidelity. This means that a criterion of fidelity can be represented by a numerically valued function:

$$v\left(P(x, y)\right)$$

whose argument ranges over possible probability functions $P(x, y)$.

We will now show that under very general and reasonable assumptions the function $v\left(P(x,y)\right)$ can be written in a seemingly much more specialized form, namely as an average of a function $\rho(x,y)$ over the set of possible values of x and y:

$$v\left(P(x,y)\right) = \iint P(x,y)\rho(x,y)dxdy.$$

To obtain this we need only assume (1) that the source and system are ergodic so that a very long sample will be, with probability nearly 1, typical of the ensemble, and (2) that the evaluation is "reasonable" in the sense that it is possible, by observing a typical input and output x_1 and y_1, to form a tentative evaluation on the basis of these samples; and if these samples are increased in duration the tentative evaluation will, with probability 1, approach the exact evaluation based on a full knowledge of $P(x,y)$. Let the tentative evaluation be $\rho(x,y)$. Then the function $\rho(x,y)$ approaches (as $T \to \infty$) a constant for almost all (x,y) which are in the high probability region corresponding to the system:

$$\rho(x,y) \to v\left(P(x,y)\right)$$

and we may also write

$$\rho(x,y) \to \iint P(x,y)\rho(x,y)dxdy$$

since

$$\iint P(x,y)dxdy = 1.$$

This establishes the desired result.

The function $\rho(x,y)$ has the general nature of a "distance" between x and y.[9] It measures how undesirable it is (according to our fidelity criterion) to receive y when x is transmitted. The general result given above can be restated as follows: Any reasonable evaluation can be represented as an average of a distance function over the set of messages and recovered messages x and y weighted according to the probability $P(x,y)$ of getting the pair in question, provided the duration T of the messages be taken sufficiently large.

[9]It is not a "metric" in the strict sense, however, since in general it does not satisfy either $\rho(x,y) = \rho(y,x)$ or $\rho(x,y) + \rho(y,z) \geq \rho(x,z)$.

The following are simple examples of evaluation functions:

1. R.M.S. criterion.

$$v = \overline{(x(t) - y(t))^2}.$$

In this very commonly used measure of fidelity the distance function $\rho(x, y)$ is (apart from a constant factor) the square of the ordinary Euclidean distance between the points x and y in the associated function space.

$$\rho(x, y) = \frac{1}{T} \int_0^T [x(t) - y(t)]^2 dt.$$

2. Frequency weighted R.M.S. criterion. More generally one can apply different weights to the different frequency components before using an R.M.S. measure of fidelity. This is equivalent to passing the difference $x(t) - y(t)$ through a shaping filter and then determining the average power in the output. Thus let

$$e(t) = x(t) - y(t)$$

and

$$f(t) = \int_{-\infty}^{\infty} e(\tau)k(t - \tau)d\tau$$

then

$$\rho(x, y) = \frac{1}{T} \int_0^T f(t)^2 dt.$$

3. Absolute error criterion.

$$\rho(x, y) = \frac{1}{T} \int_0^T |x(t) - y(t)| dt.$$

4. The structure of the ear and brain determine implicitly an evaluation, or rather a number of evaluations, appropriate in the case of speech or music transmission. There is, for example, an "intelligibility" criterion in which $\rho(x, y)$ is equal to the relative frequency of incorrectly interpreted words when message $x(t)$ is received as $y(t)$. Although we cannot give an explicit representation

of $\rho(x, y)$ in these cases it could, in principle, be determined by sufficient experimentation. Some of its properties follow from well-known experimental results in hearing, e.g., the ear is relatively insensitive to phase and the sensitivity to amplitude and frequency is roughly logarithmic.

5. The discrete case can be considered as a specialization in which we have tacitly assumed an evaluation based on the frequency of errors. The function $\rho(x, y)$ is then defined as the number of symbols in the sequence y differing from the corresponding symbols in x divided by the total number of symbols in x.

THE RATE FOR A SOURCE RELATIVE TO A FIDELITY EVALUATION

We are now in a position to define a rate of generating information for a continuous source. We are given $P(x)$ for the source and an evaluation v determined by a distance function $\rho(x, y)$ which will be assumed continuous in both x and y. With a particular system $P(x, y)$ the quality is measured by

$$v = \iint \rho(x, y)P(x, y)dxdy.$$

Furthermore the rate of flow of binary digits corresponding to $P(x, y)$ is

$$R = \iint P(x, y) \log \frac{P(x, y)}{P(x)P(y)} dxdy.$$

We define the rate R_1 of generating information for a given quality v_1 of reproduction to be the minimum of R when we keep v fixed at v_1 and vary $P_x(y)$. That is:

$$R_1 = \min_{P_x(y)} \iint P(x, y) \log \frac{P(x, y)}{P(x)P(y)} dxdy$$

subject to the constraint:

$$v_1 = \iint P(x, y)\rho(x, y)dxdy.$$

This means that we consider, in effect, all the communication systems that might be used and that transmit with the required fidelity. The rate of transmission in bits per second is calculated for each one and we choose that having the least rate. This latter rate is the rate we assign the source for the fidelity in question.

The justification of this definition lies in the following result:

Theorem 21: If a source has a rate R_1 for a valuation v_1 it is possible to encode the output of the source and transmit it over a channel of capacity C with fidelity as near v_1 as desired provided $R_1 \leq C$. This is not possible if $R_1 > C$.

The last statement in the theorem follows immediately from the definition of R_1 and previous results. If it were not true we could transmit more than C bits per second over a channel of capacity C. The first part of the theorem is proved by a method analogous to that used for Theorem 11. We may, in the first place, divide the (x, y) space into a large number of small cells and represent the situation as a discrete case. This will not change the evaluation function by more than an arbitrarily small amount (when the cells are very small) because of the continuity assumed for $\rho(x, y)$. Suppose that $P_1(x, y)$ is the particular system which minimizes the rate and gives R_1. We choose from the high probability y's a set at random containing

$$2^{(R_1+\epsilon)T}$$

members where $\epsilon \to 0$ as $T \to \infty$. With large T each chosen point will be connected by a high probability line (as in Fig. 10) to a set of x's. A calculation similar to that used in proving Theorem 11 shows that with large T almost all x's are covered by the fans from the chosen y points for almost all choices of the y's. The communication system to be used operates as follows: The selected points are assigned binary numbers. When a message x is originated it will (with probability approaching 1 as $T \to \infty$) lie within at least one of the fans. The corresponding binary number is transmitted (or one of them chosen arbitrarily if there are several) over the channel by suitable coding means to give a small probability of error. Since $R_1 \leq C$ this is possible. At the receiving point the corresponding y is reconstructed and used as the recovered message.

The evaluation v_1' for this system can be made arbitrarily close to v_1 by taking T sufficiently large. This is due to the fact that for each long sample of message $x(t)$ and recovered message $y(t)$ the evaluation approaches v_1 (with probability 1).

It is interesting to note that, in this system, the noise in the recovered message is actually produced by a kind of general quantizing at the transmitter and not produced by the noise in the channel. It is more or less analogous to the quantizing noise in PCM.

THE CALCULATION OF RATES

The definition of the rate is similar in many respects to the definition of channel capacity. In the former

$$R = \min_{P_x(y)} \iint P(x,y) \log \frac{P(x,y)}{P(x)P(y)} dxdy$$

with $P(x)$ and $v_1 = \iint P(x,y)\rho(x,y)dxdy$ fixed. In the latter

$$C = \max_{P(x)} \iint P(x,y) \log \frac{P(x,y)}{P(x)P(y)} dxdy$$

with $P_x(y)$ fixed and possibly one or more other constraints (e.g., an average power limitation) of the form $K = \iint P(x,y)\lambda(x,y)dxdy$.

A partial solution of the general maximizing problem for determining the rate of a source can be given. Using Lagrange's method we consider

$$\iint \left[P(x,y) \log \frac{P(x,y)}{P(x)P(y)} + \mu P(x,y)\rho(x,y) + v(x)P(x,y) \right] dxdy.$$

The variational equation (when we take the first variation on $P(x,y)$) leads to

$$P_y(x) = B(x)e^{-\lambda\rho(x,y)}$$

where λ is determined to give the required fidelity and $B(x)$ is chosen to satisfy

$$\int B(x)e^{-\lambda\rho(x,y)}dx = 1.$$

This shows that, with best encoding, the conditional probability of a certain cause for various received y, $P_y(x)$ will decline exponentially with the distance function $\rho(x,y)$ between the x and y in question.

In the special case where the distance function $\rho(x, y)$ depends only on the (vector) difference between x and y,

$$\rho(x, y) = \rho(x - y)$$

we have

$$\int B(x)e^{-\lambda\rho(x-y)}dx = 1.$$

Hence $B(x)$ is constant, say α, and

$$P_y(x) = \alpha e^{\lambda\rho(x-y)}.$$

Unfortunately these formal solutions are difficult to evaluate in particular cases and seem to be of little value. In fact, the actual calculation of rates has been carried out in only a few very simple cases.

If the distance function $\rho(x, y)$ is the mean square discrepancy between x and y and the message ensemble is white noise, the rate can be determined. In that case we have

$$R = \min\left[H(x) - H_y(x)\right] = H(x) - \max H_y(x)$$

with $N = \overline{(x - y)^2}$. But the max $H_y(x)$ occurs when $y - x$ is a white noise, and is equal to $W_1 \log 2\pi e N$ where W_1 is the bandwidth of the message ensemble. Therefore

$$\begin{aligned} R &= W_1 \log 2\pi e Q - W_1 \log 2\pi e N \\ &= W_1 \log \frac{Q}{N} \end{aligned}$$

where Q is the average message power. This proves the following:

Theorem 22: The rate for a white noise source of power Q and band W_1 relative to an R.M.S. measure of fidelity is

$$R = W_1 \log \frac{Q}{N}$$

where N is the allowed mean square error between original and recovered messages.

More generally with any message source we can obtain inequalities bounding the rate relative to a mean square error criterion.

Theorem 23: The rate for any source of band W_1 is bounded by

$$W_1 \log \frac{Q_1}{N} \leq R \leq W_1 \log \frac{Q}{N}$$

where Q is the average power of the source, Q_1 its entropy power and N the allowed mean square error.

The lower bound follows from the fact that the max $H_y(x)$ for a given $\overline{(x-y)^2} = N$ occurs in the white noise case. The upper bound results if we place points (used in the proof of Theorem 21) not in the best way but at random in a sphere of radius $\sqrt{Q-N}$.

Appendix 5

Let S_1 be any measurable subset of the g ensemble, and S_2 the subset of the f ensemble which gives S_1 under the operation T. Then

$$S_1 = TS_2$$

Let H^λ be the operator which shifts all functions in a set by the time λ. Then

$$H^\lambda S_1 = H^\lambda TS_2 = TH^\lambda S_2$$

since T is invariant and therefore commutes with H^λ. Hence if $m[S]$ is the probability measure of the set S

$$\begin{aligned} m[H^\lambda S_1] &= m[TH^\lambda S_2] = m[H^\lambda S_2] \\ &= m[S_2] = m[S_1] \end{aligned}$$

where the second equality is by definition of measure in the g space, the third since the f ensemble is stationary, and the last by definition of g measure again.

To prove that the ergodic property is preserved under invariant operations, let S_1 be a subset of the g ensemble which is invariant under H^λ, and let S_2 be the set of all functions f which transform into S_1. Then

$$H^\lambda S_1 = H^\lambda TS_2 = TH^\lambda S_2 = S1$$

so that $H^\lambda S_2$ is included in S_2 for all λ. Now, since

$$m[H^\lambda S_2] = m[S_1],$$

this implies

$$H^\lambda S_2 = S_2$$

for all λ with $m[S_2] \neq 0, 1$. This contradiction shows that S_1 does not exist.

Appendix 6

The upper bound, $\overline{N}_3 \leq N_1 + N_2$, is due to the fact that the maximum possible entropy for a power $N_1 + N_2$ occurs when we have a white noise of this power. In this case the entropy power is $N_1 + N_2$.

To obtain the lower bound, suppose we have two distributions in n dimensions $p(x_i)$ and $q(x_i)$ with entropy powers $\overline{N}_1$ and $\overline{N}_2$. What form should p and q have to minimize the entropy power $\overline{N}_3$ of their convolution $r(x_i)$:

$$r(x_i) = \int p(y_i)q(x_i - y_i dy_i.$$

The entropy H_3 of r is given by

$$H_3 = -\int r(x_i) \log r(x_i) dx_i.$$

We wish to minimize this subject to the constraints

$$H_1 = -\int p(x_i \log p(x_i) dx_i$$
$$H_2 = -\int q(x_i) \log q(x_i) dx_i.$$

We consider then

$$U = -\int [r(x) \log r(x) + \lambda p(x) \log p(x) + \mu q(x) \log q(x)] dx$$
$$\delta U = -\int [[1 + \log r(x)]\delta r(x) + \lambda[1 + \log p(x)]\delta p(x) + \mu[1 + \log q(x)]\delta q(x)]\, dx.$$

If $p(x)$ is varied at a particular argument $x_i = s_i$, the variation in r(x) is

$$\delta r(x) = q(x_i - s_i)$$

and

$$\delta U = -\int q(x_i - s_i) \log r(x_i) dx_i - \lambda \log p(s_i) = 0$$

and similarly when q is varied. Hence the conditions for a minimum are

$$\int q(x_i - s_i) \log r(x_i) dx_i = \lambda \log p(s_i)$$
$$\int p(x_i - s_i) \log r(x_i) dx_i = \mu \log q(s_i).$$

If we multiply the first by $p(s_i)$ and the second by $q(s_i)$ and integrate with respect to s_i we obtain

$$H_3 = \lambda H_1$$
$$H_3 = \mu H_2$$

or solving for λ and μ and replacing in the equations

$$H_1 \int q(x_i - s_i) \log r(x_i) dx_i = -H_3 \log p(s_i)$$
$$H_2 \int p(x_i - s_i) \log r(x_i) dx_i = -H_3 \log q(s_i).$$

Now suppose $p(x_i)$ and $q(x_i)$ are normal

$$p(x_i = \frac{|A_{ij}|^{n/2}}{(2\pi)^{n/2}} \exp -\frac{1}{2} \sum A_{ij} x_i x_j$$
$$q(x_i) = \frac{|B_{ij}|^{n/2}}{(2\pi)^{n/2}} \exp -\frac{1}{2} \sum B_{ij} x_i x_j$$

Then $r(x_i)$ will also be normal with quadratic form C_{ij}. If the inverses of these forms are a_{ij}, b_{ij}, c_{ij} then

$$c_{ij} = a_{ij} + b_{ij}.$$

We wish to show that these functions satisfy the minimizing conditions if and only if $a_{ij} = K b_{ij}$ and thus give the minimum H_3 under the constraints. First we have

$$\log r(x_i) = \frac{n}{2} \log \frac{1}{2\pi} |C_{ij}| - \frac{1}{2} \sum C_{ij} x_i x_j$$
$$\int q(x_i - s_i \log r(x_i dx_i = \frac{n}{2} \log \frac{1}{2\pi} |C_{ij}| - \frac{1}{2} \sum C_{ij} s_i s_j - \frac{1}{2} \sum C_{ij} b_{ij}.$$

This should equal

$$\frac{H_3}{H_1} \left[\frac{n}{2} \log \frac{1}{2\pi} |A_{ij}| - \frac{1}{2} \sum A_{ij} s_i s_j \right]$$

which requires $A_{ij} = \frac{H_1}{H_3} C_{ij}$. In this case $A_{ij} = \frac{H_1}{H_2} B_{ij}$ and both equations reduce to identities.

Appendix 7

The following will indicate a more general and more rigorous approach to the central definitions of communication theory. Consider a probability measure space whose elements are ordered pairs (x, y). The variables x, y are to be identified as the possible transmitted and received signals of some long duration T. Let us call the set of all points whose x belongs to a subset S_1 of x points the strip over S_1, and similarly the set whose y belong to S_2 the strip over S_2. We divide x and y into a collection of non-overlapping measurable subsets X_i and Y_i approximate to the rate of transmission R by

$$R_1 = \frac{1}{T} \sum_i P(X_i, Y_i) \log \frac{P(X_i, Y_i)}{P(X_i)P(y_i)}$$

where

$P(X_i)$ is the probability measure of the strip over X_i

$P(Y_i)$ is the probability measure of the strip over Y_i

$P(X_i, Y_i)$ is the probability measure of the intersection of the strips.

A further subdivision can never decrease R_1. For let X_1 be divided into $X_1 = X_1' + X_1''$ and let

$$\begin{aligned} P(Y_1) &= a & P(X_1) &= b + c \\ P(X_1') &= b & P(X_i', Y_1) &= d \\ P(X_1'') &= c & P(X_1'', Y_1) &= e \end{aligned}$$
$$P(X_1, Y_1) = d + e.$$

Then in the sum we have replaced (for the X_1, Y_1 intersection)

$$(d + e) \log \frac{d + e}{a(b + c)} \text{ by } d \log \frac{d}{ab} + e \log \frac{e}{ac}.$$

It is easily shown that with the limitation we have on b, c, d, e,

$$\left[\frac{d + e}{b + c}\right]^{d+e} \leq \frac{d^d e^e}{b^d c^e}$$

and consequently the sum is increased. Thus the various possible subdivisions form a directed set, with R monotonic increasing with refinement of the subdivision. We may define R unambiguously as the least upper bound for R_1 and write it

$$R = \frac{1}{T} \iint P(x, y) \log \frac{P(x, y)}{P(x)P(y)} dx dy.$$

This integral, understood in the above sense, includes both the continuous and discrete cases and of course many others which cannot be represented in either form. It is trivial in this formulation that if x and u are in one-to-one correspondence, the rate from u to y is equal to that from x to y. If v is any function of y (not necessarily with an inverse) then the rate from x to y is greater than or equal to that from x to v since, in the calculation of the approximations, the subdivisions of y are essentially a finer subdivision of those for v. More generally if y and v are related not functionally but statistically, i.e., we have a probability measure space (y, v), then $R(x, v) \leq R(x, y)$. This means that any operation applied to the received signal, even though it involves statistical elements, does not increase R.

Another notion which should be defined precisely in an abstract formulation of the theory is that of "dimension rate," that is the average number of dimensions required per second to specify a member of an ensemble. In the band limited case $2W$ numbers per second are sufficient. A general definition can be framed as follows. Let $f_\alpha(t)$ be an ensemble of functions and let $\rho_T[f_\alpha(t), f_\beta(t)]$ be a metric measuring the "distance" from f_α to f_β over the time T (for example the R.M.S. discrepancy over this interval.) Let $N(\epsilon, \delta, T)$ be the least number of elements f which can be chosen such that all elements of the ensemble apart from a set of measure δ are within the distance ϵ of at least one of those chosen. Thus we are covering the space to within ϵ apart from a set of small measure δ. We define the dimension rate λ for the ensemble by the triple limit

$$\lambda = \lim_{\delta \to 0} \lim_{\epsilon \to 0} \lim_{T \to \infty} \frac{\log N(\epsilon, \delta, T)}{T \log \epsilon}.$$

This is a generalization of the measure type definitions of dimension in topology, and agrees with the intuitive dimension rate for simple ensembles where the desired result is obvious.

Acknowledgments

Heroes of the digital era!

The writer is indebted to his colleagues at the Laboratories, particularly to Dr. H. W. Bode, Dr. J. R. Pierce, Dr. B. McMillan, and Dr. B. M. Oliver for many helpful suggestions and criticisms during the course of this work. Credit should also be given to Professor N. Wiener, whose elegant solution of the problems of filtering and prediction of stationary ensembles has considerably influenced the writer's thinking in this field.

Patron saint of applied mathematicians everywhere!

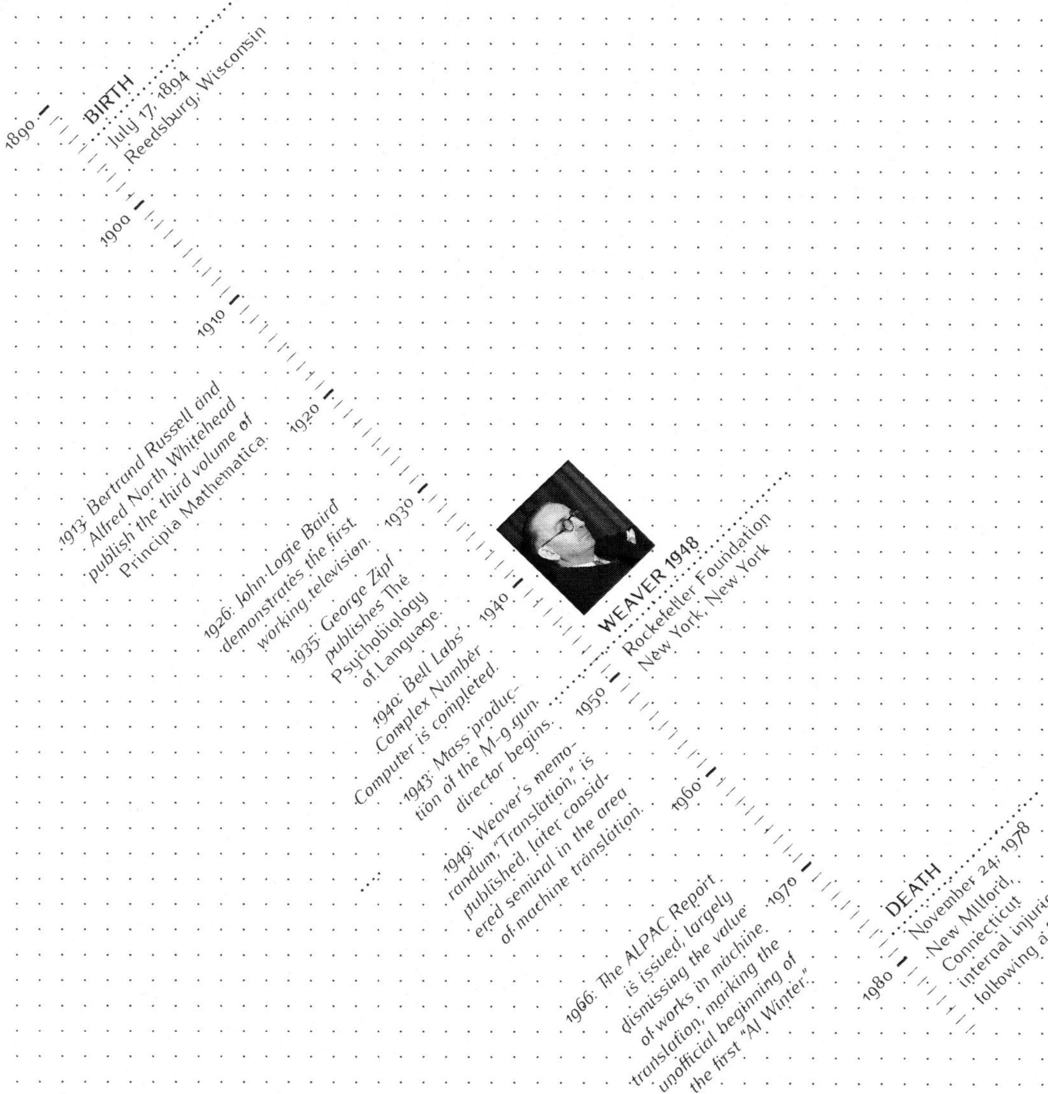

WARREN WEAVER

[10]

IF, AND, IF SO, HOW?

Stuart Kauffman, University of Pennsylvania and Institute for Systems Biology

Mid-twentieth-century science pivoted to new, almost unseen issues following Warren Weaver's brilliant lead in this vital essay, which broaches six topics. In statements below and the adjoined commentaries, I ask where we are seventy-six years after Weaver's pivot.

W. Weaver, "Science and Complexity," *American Scientist* 36 (4), 536–544 (1948).

I. The Sciences of Simplicity

Weaver rightly explains that the sciences of simplicity start with Newton in the seventeenth century and the invention of classical physics. This is the Newtonian paradigm:

1. Identify the relevant variables, for example, position and momentum.
2. Write laws of motion in differential form among these variables, for example, Newton's three laws of motion and gravitation.
3. Identify the boundary conditions which thereby determine all possible combined values of the relevant variables, hence the phase space of the system.
4. Specify the initial state of the system.
5. Integrate the differential equations of motion to determine the entailed trajectory of the system within its phase space.

Over the next two centuries classical physics gave rise to further laws such as Maxwell's equations, general relativity, and a myriad of novel technologies impacting modern mid-twentieth-century civilization, airplanes, cars, radios, and telephones.

The Newtonian paradigm underlies all of classical physics and the stunning emergence of quantum mechanics. Quantum mechanics remains safely within the Newtonian paradigm. The Schrödinger equation is a linear partial differential wave equation whose solution yields a deterministic propagation of a *probability* distribution. Thereafter what is said to occur depends upon the interpretation of quantum mechanics: The indeterminate collapse of the wave function on von Neumann (1927), the determinism of Bohm (1989), the Everett interpretation (Everett, III *et al.* 1973). Does God, indeed, play dice?

The Newtonian paradigm, however, does not cover statistical mechanics.

II. The Sciences of Disorganized Complexity

Boltzmann famously invented statistical mechanics (Brush 1965) to deal with systems having a vast number of variables that interact in a more or less random way, or "helter-skelter," in Weaver's phrase.

The secret step is the famous ergodic hypothesis.[1] We are to consider a liter box, closed to the outside world with Avogadro's number, N, of particles inside at some finite fixed temperature. The phase space is defined. It is all possible combinations of positions and momenta in three-dimensional space of the N particles, representable by $6N$ numbers.

Ideally, we would just solve Newton's equations of motion, as we can do for two billiard balls, for the N particles and find the actual trajectory of the system in its $6N$ dimensional phase space. But that is not possible. Boltzmann's brilliant move was this: Give up trying to integrate Newton's equations. Instead, Boltzmann conceived of dividing the $6N$-dimensional phase space into tiny $6N$-dimensional boxes, microstates, that fill the entire phase space. Then the magical ergodic hypothesis just asserts that the system spends equal time in equal volumes of this phase space.

All of statistical mechanics follows from the ergodic hypothesis, including the famous second law, which concerns macrostates, each made of a collection of one or more microstates. There are vastly more

[1] See https://plato.stanford.edu/entries/statphys-Boltzmann/.

microstates corresponding to a roughly uniform distribution of the N particles in the one-liter box, than to a condition with all the N particles crowded in one corner of the liter box. Boltzmann defined the "entropy" of a macrostate as the logarithm of the number of microstates in that macrostate. Thus, the entropy of the nearly uniform distribution is much higher than that of the case with all the particles in a corner. Given the ergodic hypothesis, the system will spend equal times in equal volumes, say, in any 100 microstates. Hence the system will, on average, spend more time in the nearly uniform or equilibrium state. The second law codifies this: entropy *tends* to increase. Famously, this is taken to be the thermodynamic arrow of time. Given the view that fundamental time, as in Newton and quantum mechanics, is reversible, so has no arrow of time, the thermodynamic arrow of time is the tendency for disorder to increase.

Statistical mechanics conquered disorganized complexity.

III. The Unknown Domain of a Science of Organized Complexity: Biology, Social Systems

Weaver's focus shifts to biological and social systems. In all these cases, as he says, a large number of variables are coupled in specific ways by which they interact. This is not the helter-skelter of vastly many particles in a liter box at finite temperature. The tools of disorganized complexity are useless here. This was the pivot Weaver stated so clearly in 1948. A living cell is a stunningly organized system of a hundred thousand or more molecular variables. In 1948 the structure of DNA was not known, nor RNA, nor the genetic code. In fact, a prevalent hypothesis was that a large number of proteins somehow were collectively the genetic material.

But with what conceptual and mathematical tools was science to attack such systems? Could this be a new physics on some new mathematical basis? But how?

Weaver does not know. But he probes in two major directions:

IV. Future Hopes in Computers

World War II saw the earliest development of digital computers with the ENIAC at the University of Pennsylvania.[2] The ENIAC linked thousands of vacuum tubes and succeeded brilliantly in its first tasks of calculating the trajectory of naval bombardment shells.

In 1936, Alan Turing invented his now-eponymous computational machine (Turing 1937). With it, we have now wondered for nearly ninety years whether the human mind is a computer (Nagel 2012; Searle 1990). Is general artificial intelligence possible (Mitchell 2019)? In 1948, the new field of cybernetics was birthing. Perhaps well known to Weaver, in the same year, McCulloch and Pitts published their seminal paper, "The Logical Calculus of Ideas Immanent in Nervous Activity." Here these authors showed that very complex feedforward networks of "formal binary variable neurons" could calculate any Boolean function on its N input row. McCulloch and Pitts identify the "on" versus "off" state of a formal neuron with a "true" versus "false" value of a propositional *Idea Immanent in the Mind*. This paper led to the vast later development of network models of neural systems.

Weaver, McCulloch, Wiener, and later Ashby and others built out the early years of cybernetics: nonlinear feedback and control. Negative feedback and stability, positive feedback and runaway. Study of nonlinear dynamical systems later gave rise to deterministic chaos. Deterministic chaos was first discovered in the 1890s by Poincaré (2017). No analytic solutions were possible, so numerical studies were needed. The field flowered in the 1960s (Gleick 1987; May 1976). Due to sensitivity to initial conditions that cannot be measured to infinite accuracy, nearby states diverge in phase space. Determinism, *pace* Laplace, no longer implied predictability.

Later, wide computational/numerical study of nonlinear dynamical systems led to understanding that a nonlinear dynamical system coupling N continuous variables induced a "flow" in the N dimension state space, that is, the *phase space* of the system. Generically, such flow was not ergodic. Rather, in such nonlinear systems, flow in the state space typically converges to and remains in subregions of the state

[2] See https://www.computerhistory.org/revolution/birth-of-the-computer/4/78.

space. The entire N-dimensional state space is often partitioned into finite N-dimensional volumes in each of which the flow is restricted and further converges on a yet smaller subregion called an "attractor." Attractors include stable and unstable steady states, limit cycles, aperiodic attractors, and chaotic—strange attractors (Strogatz 1994).

It is often the case that such N-dimensional nonlinear dynamical systems have multiple attractors that partition the state space, and each of which drains a "basin of attraction" in the state space whose trajectories converge on that one attractor (Strogatz 1994).

Analysis of such complex nonlinear systems became of major import in the design of complex chemical reaction networks, models of neural networks, ecological networks, non-equilibrium economic networks, and so on (Epstein and Pojman 1998; Vandermeer 2020; Bischi, Chiarella, and Gardini 2009).

A related field in the study of nonlinear dynamical systems was the emergence of bifurcation analysis. Here an M-dimensional parameter space for the N-dimensional dynamical system is defined. For structurally stable systems, the parameter space is divided into disjoint $M-1$-dimensional bifurcation boundary surfaces each enclosing an M-dimensional volume. Crossing such a boundary surface induces a change in the attractor structure in the phase space: old steady states may vanish, new ones appear. In general, attractors may vanish, or transition to new attractors, limit cycle oscillations, quasiperiodic orbits, or strange chaotic attractors (Strogatz 1994).

In short, Weaver rightly proposed that computation would become a major inroad to what will become, decades later, core aspects of the sciences of complexity.

V. Future Hopes in "Mixed Human Groups"

All the above is brilliant but, in a way, not surprising. Weaver is a mid-twentieth-century scientist seeking to see *if, and, if so, how* quantitative and mathematical science and its techniques might constitute the sufficient resources to conquer organized complexity.

Surprisingly, even astonishingly, Weaver hints *no.* In World War II, he participated in Allied efforts to understand how best to manage convoys struggling to cross the Atlantic in the face of German U-boats and their capacity for devastation. Mixed groups of human experts—mathematicians, physicists, biologists, doctors, psychologists—formed working groups. Weaver reports that, because of the human diversity in such mixed groups, astonishingly creative solutions were found to myriad tactical problems.

Some mixed human groups could be superbly creative. Weaver strongly urged that such mixed groups play a major role in the future development of science.

Two issues arise: First, the Santa Fe Institute itself is the canonical example of a mixed-group scientific enterprise. There is no doubt that SFI laid the foundations of what has become the sciences of complexity, from its founding in 1984. Complexity institutes have emerged around the world, encouraged and often nurtured by SFI.

The second issue is far more important. Going back to Turing, is the human mind really a version of an algorithmic universal Turing machine? Is general artificial intelligence really possible (Nagel 2012; Searle 1990; Mitchell 2019)? Are we algorithmic? Is mind algorithmic? Is consciousness algorithmic? Can a computer pass the Turing test? Could 1,000 tin cans pouring water into one another be conscious? If not, why not?

In short, are human groups creative because each of us is algorithmic and *we collectively compute* with one another, or precisely because *we are not algorithmic*?

If we humans are not algorithmic, will the tools of mathematics suffice to conquer all of organized complexity?

Algorithms are the heart of mathematical formal processes. After Gödel's incompleteness theorem limiting formal systems (Gödel 1986), other issues not seen by Weaver arise: Is the evolution of biosphere for 3.7 billion years deducible and algorithmic? Is the evolving biosphere open to mathematical formulation? We all assume yes. But if not. then, good grief, what?

Evolving economies and the evolving biosphere are most complex exemplars of Weaver's organized complexity. The evolving biosphere is the most complex system we know of in the universe. Do our mathematical tools suffice? If not, what? The Newtonian paradigm itself, surely formulated mathematically, is said to be universal, from cosmology on with respect to the evolution of the universe itself. Can we mathematize the 3.7-billion-year evolution of the biosphere within the Newtonian paradigm? If not, then what?

VI. The Limit of Science with Respect to Values and Moral Issues

Weaver ends with what he takes to *not* be science: Values and moral issues. This is, of course, David Hume's claim: It is not possible to deduce "ought" from "is"—the famous naturalistic fallacy (see, e.g., Frankena 1939). In so ending his essay, Weaver takes the familiar position within science: Based on the naturalistic fallacy, science has abjured from asking where in the 13.8 billion years of the evolution of the universe have "values" come from. *How do we get from matter to mattering?*

We have, for two centuries or more, been sure Hume was right. Was Hume right after all? If not, then what?

REFERENCES

Barabási, A.-L. 2009. "Scale-Free Networks: A Decade and Beyond." *Science* 325 (5939): 412–413. https://doi.org/10.1126/science.117329.

Beggs, J. 2008. "The Criticality Hypothesis: How Local Cortical Networks Might Optimize Information Processing." *Philosophical Transactions of the Royal Society A* 366:329–343. https://doi.org/10.1098/rsta.2007.2092.

Bischi, G. I., C. Chiarella, and L. Gardini. 2009. *Nonlinear Dynamics in Economics, Finance, and the Social Sciences.* Berlin, Germany: Springer.

Bohm, D. 1989. *Quantum Theory.* New York, NY: Dover.

Bornholdt, S., and S. A. Kauffman. 2019. "Ensembles, Dynamics, and Cell Types: Revisiting the Statistical Mechanics Perspective on Cell Regulation." *Journal of Theoretical Biology* 467:15–22.

Brush, S. G. 1965. *Kinetic Theory.* Vol. 1 and 2. Oxford, UK: Pergamon. https://doi.org/10.1016/C2013-0-07974-1.

Buckner, C., and J. Garson. 2019. "Connectionism." In *The Stanford Encyclopedia of Philosophy (Fall 2019),* edited by E. N. Zalta. https://plato.stanford.edu/archives/fall2019/entries/connectionism/.

Daniels, B. C., H. Kim, D. Moore, S. Zhou, H. B. Smith, B. Karas, S. A. Kauffman, and S. I. Walker. 2018. "Criticality Distinguishes the Ensemble of Biological Regulatory Networks." *Physical Review Letters* 121:138102.

Devereaux, A., R. Koppl, S. A. Kauffman, and A. Roli. 2021. "An Incompleteness Result Regarding Within-System Modeling." *SSRN,* https://doi.org/10.2139/ssrn.3968077.

Edwards, S. F., and P. W. Anderson. 1975. "Theory of Spin Glasses." *Journal of Physics F: Metal Physics* 5 (5): 965–974. https://doi.org/10.1088/0305-4608/5/5/017.

Epstein, I. R., and J. A. Pojman. 1998. *An Introduction to Nonlinear Chemical Dynamics: Oscillations, Waves, Patterns, and Chaos.* Oxford, UK: Oxford University Press.

Everett, III, H., J. A. Wheeler, B. S. DeWitt, L. N. Cooper, D. van Vechten, N. B. D. Graham, and R. N. Graham, eds. 1973. *The Many-Worlds Interpretation of Quantum Mechanics.* Princeton, NJ: Princeton University Press.

Frankena, W. K. 1939. "The Naturalistic Fallacy." *Mind* 48 (192): 464–477. https://doi.org/10.1093/mind/XLVIII.192.464.

Gleick, J. 1987. *Chaos: The Making of a New Science.* London, UK: Penguin.

Gödel, K. 1986. "Über formal unentscheidbare Sätze der Principia Mathematica und verwandter Systeme, I." In *Kurt Gödel: Collected Works, Volume I: Publications 1929–1936,* edited by S. Feferman, J. W. Dawson, Jr., S. C. Kleene, G. H. Moore, R. M. Solovay, and J. v. Heijenoort, I:144–195. Oxford, UK: Oxford University Press.

Kauffman, S. A. 1969. "Metabolic Stability and Epigenesis in Randomly Constructed Genetic Nets." *Journal of Theoretic Biology* 22:437–467. https://doi.org/10.1016/0022-5193(69)90015-0.

———. 1993. *Origins of Order.* Oxford, UK: Oxford University Press.

———. 2019. *A World Beyond Physics.* Oxford, UK: Oxford University Press.

Kauffman, S. A., and W. S. McCulloch. 1967. *Random Nets of Formal Genes.* Technical report. Quarterly Progress Report 34, Research Laboratory of Electronics, Massachusetts Institute of Technology.

Kauffman, S. A., and A. Roli. 2021a. "Beyond the Newtonian Paradigm: A Statistical Mechanics of Emergence." *OSF preprint,* https://osf.io/m9kpz/.

———. 2021b. "The World Is Not a Theorem." *Entropy* 23 (11): 1467. https://doi.org/10.3390/e23111467.

———. 2021c. "What is Consciousness?" *arXiv,* https://arxiv.org/abs/2106.15515.

Langton, C. G. 1990. "Computation at the Edge of Chaos: Phase Transitions and Emergent Computation." *Physica D* 42:12–37.

Longo, G., M. Montévil, and S. A. Kauffman. 2012. "No Entailing Laws, but Enablement in the Evolution of the Biosphere." In *Proceedings of the Fourteenth International Conference on Genetic and Evolutionary Computation,* 1379–1392. https://doi.org/10.1145/2330784/2330946.

May, R. M. 1976. "Simple Mathematical Models with Very Complicated Dynamics." *Nature* 261 (5560): 459–467. https://doi.org/10.1038/261459a0.

McCulloch, W., and W. Pitts. 1943. "A Logical Calculus of Ideas Immanent in Nervous Activity." *Bulletin of Mathematical Biology* 52 (1–2): 99–115. https://doi.org/10.1007/BF02459570.

Mitchell, M. 2019. *Artificial Intelligence: A Guide for Thinking Humans.* New York, NY: Picador.

Nagel, T. 2012. *Mind and Cosmos: Why the Materialist Neo-Darwinian Conception of Nature is Almost Certainly False.* Oxford, UK: Oxford University Press.

Penrose, R. 1989. *The Emperor's New Mind.* Oxford, UK: Oxford University Press.

Poincaré, J. H. 2017. *The Three-Body Problem and the Equations of Dynamics: Poincaré's Foundational Work on Dynamical Systems Theory.* Translated by B. D. Popp. Cham, Switzerland: Springer International Publishing.

Roli, A., Y. Jaeger, and S. A. Kauffman. 2022. "How Organisms Come to Know their World: Fundamental Limits on Artificial General Intelligence." *Frontiers in Ecology and Evolution,* https://doi.org/10.3389/fevo.2021.806283.

Russell, B. 1912. *The Problems of Philosophy.* New York, NY: Henry Holt / Company.

Schwab, J. D., S. D. Kühlwein, N. Ikonomi, M. Kühl, and H. A. Kestler. 2020. "Concepts in Boolean Network Modeling: What Do They All Mean?" *Computational and Structural Biotechnology Journal* 18:571–582. https://doi.org/10.1016/j.csbj.2020.03.001.

Searle, J. 1990. "Is the Brain's Mind a Computer Program?" *Scientific American* 262 (1): 26–31.

Strogatz, S. H. 1994. *Nonlinear Dynamics and Chaos: With Applications to Physics, Biology, Chemistry, and Engineering.* Boca Raton, FL: CRC Press.

Tisza, L. 1963. "The Conceptual Structure of Physics." *Reviews of Modern Physics* 35 (1).

Turing, A. M. 1937. "On Computable Numbers, with an Application to the Entscheidungsproblem." *Proceedings of the London Mathematical Society* s2-42 (1): 230–265. https://doi.org/10.1112/plms/s2-42.1.230.

Vandermeer, J. 2020. "Confronting Complexity in Agroecology: Simple Models from Turing to Simon." *Frontiers in Sustainable Food Systems* 07. https://doi.org/10.3389/fsufs.2020.00095.

Villani, M., L. La Rocca, S. A. Kauffman, and R. Serra. 2018. "Dynamical Criticality in Gene Regulatory Networks." *Complexity* 2018. https://doi.org/10.1155/2018/5980636.

von Neumann, John. 1927. "Mathematische Begründung der Quantenmechanik." *Nachrichten von der Gesellschaft der Wissenschaften zu Göttingen, Mathematisch-Physikalische Klasse,* 1–57. http://eudml.org/doc/59215.

Wittgenstein, L. 1922. *Tractatus Logico-Philosophicus.* London, UK: Kegan Paul.

———. 1953. *Philosophical Investigations.* Edited by G. E. M. Anscombe and R. Rhees. Translated by G. E. M. Anscombe. Oxford, UK: Blackwell.

SCIENCE AND COMPLEXITY

Warren Weaver, Rockefeller Foundation

Science has led to a multitude of results that affect men's lives. Some of these results are embodied in mere conveniences of a relatively trivial sort. Many of them, based on science and developed through technology, are essential to the machinery of modern life. Many other results, especially those associated with the biological and medical sciences, are of unquestioned benefit and comfort. Certain aspects of science have profoundly influenced men's ideas and even their ideals. Still other aspects of science are thoroughly awesome.

How can we get a view of the function that science should have in the developing future of man? How can we appreciate what science really is and, equally important, what science is not? It is, of course, possible to discuss the nature of science in general philosophical terms. For some purposes such a discussion is important and necessary, but for the present a more direct approach is desirable. Let us, as a very realistic politician used to say, let us look at the record. Neglecting the older history of science, we shall go back only three and a half centuries and take a broad view that tries to see the main features, and omits minor details. Let us begin with the physical sciences, rather than the biological, for the place of the life sciences in the descriptive scheme will gradually become evident.

Problems of Simplicity[0]

Speaking roughly, it may be said that the seventeenth, eighteenth, and nineteenth centuries formed the period in which physical science learned variables, which brought us the telephone and the radio, the automobile and the airplane, the phonograph and the moving pictures,

---

°Based upon material presented in Chapter I, "The Scientists Speak," Boni & Gaer, Inc., 1947.

the turbine and the Diesel engine, and the modern hydroelectric power plant.

The concurrent progress in biology and medicine was also impressive, but that was of a different character. The significant problems of living organisms are seldom those in which one can rigidly maintain constant all but two variables. Living things are more likely to present situations in which a half-dozen, or even several dozen quantities are all varying simultaneously, and in subtly interconnected ways. Often they present situations in which the essentially important quantities are either non-quantitative, or have at any rate eluded identification or measurement up to the moment. Thus biological and medical problems often involve the consideration of a most complexly organized whole. It is not surprising that up to 1900 the life sciences were largely concerned with the necessary preliminary stages in the application of the scientific method—preliminary stages which chiefly involve collection, description, classification, and the observation of concurrent and apparently correlated effects. They had only made the brave beginnings of quantitative theories, and hardly even begun detailed explanations of the physical and chemical mechanisms underlying or making up biological events.

To sum up, physical science before 1900 was largely concerned with two-variable *problems of simplicity;* whereas the life sciences, in which these problems of simplicity are not so often significant, had not yet become highly quantitative or analytical in character.

Problems of Disorganized Complexity

Subsequent to 1900 and actually earlier, if one includes heroic pioneers such as Josiah Willard Gibbs, the physical sciences developed an attack on nature of an essentially and dramatically new kind. Rather than study problems which involved two variables or at most three or four, some imaginative minds went to the other extreme, and said: "Let us develop analytical methods which can deal with two billion variables." That is to say, the physical scientists, with the mathematicians often in the vanguard, developed powerful techniques of probability theory

Comment:
The source of the implicit identification by McCulloch and Pitts of a true/false value of a Boolean variable in a network of formal neurons with an "Idea Immanent in Nervous Activity" is Bertrand Russell's theory of "sense-data statements." Russell's hope was that one might be in error in saying, "This chair is in this room." But one could hardly be in error if saying, "I am seeing what seems to be a chair." In Russell's hands, this became a sense data statement, "For Kauffman now, 'Red here.'" Such sense data statements were held to be incorrigible. Russell hoped to be able to construct the complete "objective world" by sets of first-order-logic true/false propositions. If so, our knowledge of the world would be the most secure. Wittgenstein's Tractatus (1922) furthered this, but his later "Philosophical Investigations" (1953) demolished this hope. Despite the later Wittgenstein and his different "language games" that cannot be reduced to one another, the literature on connectionism in neural networks continues sixty years later with Russell's dream (Buckner and Garson 2019).

Comment:
Is the human mind algorithmic? This is a highly vexed and debated question. Penrose (1989) argues no: A mathematician's coming to understand Gödel's theorem is not an algorithmic process. I, along with Roli and Yeager (2022), and separately with Roli (2021c), have recently argued that a universal Turing machine, embodied as a robot, or not, cannot jury rig. Jury rigging is not deductive. Algorithms are deductive. QED. If this is correct, general artificial intelligence is not possible. And since humans can easily jury rig, we are not algorithmic. This conclusion supports that of Penrose. But if a "human mixed group" is not algorithmic, is it possible to compute what a group of humans will do?

and of statistical mechanics to deal with what may be called problems of *disorganized complexity*.

This last phrase calls for explanation. Consider first a simple illustration in order to get the flavor of the idea. The classical dynamics of the nineteenth century was well suited for analyzing and predicting the motion of a single ivory ball as it moves about on a billiard table. In fact, the relationship between positions of the ball and the times at which it reaches these positions forms a typical nineteenth-century problem of simplicity. One can, but with a surprising increase in difficulty, analyze the motion of two or even of three balls on a billiard table. There has been, in fact, considerable study of the mechanics of the standard game of billiards. But, as soon as one tries to analyze the motion of ten or fifteen balls on the table at once, as in pool, the problem becomes unmanageable, not because there is any theoretical difficulty, but just because the actual labor of dealing in specific detail with so many variables turns out to be impracticable.

Imagine, however, a large billiard table with millions of balls rolling over its surface, colliding with one another and with the side rails. The great surprise is that the problem now becomes easier, for the methods of statistical mechanics are applicable. To be sure the detailed history of one special ball can not be traced, but certain important questions can be answered with useful precision, such as: On the average how many balls per second hit a given stretch of rail? On the average how far does a ball move before it is hit by some other ball? On the average how many impacts per second does a ball experience?

Earlier it was stated that the new statistical methods were applicable to problems of disorganized complexity. How does the word "disorganized" apply to the large billiard table with the many balls? It applies because the methods of statistical mechanics are valid only when the balls are distributed, in their positions and motions, in a helter-skelter, that is to say a disorganized, way. For example, the statistical methods would not apply if someone were to arrange the balls in a row parallel to one side rail of the table, and then start them all moving in precisely parallel paths perpendicular to the row in which they stand. Then the balls would never collide with each other nor with two of the rails, and one would not have a situation of disorganized complexity.

From this illustration it is clear what is meant by a problem of disorganized complexity. It is a problem in which the number of variables is very large, and one in which each of the many variables has a behavior which is individually erratic, or perhaps totally unknown. However, in spite of this helter-skelter, or unknown, behavior of all the individual variables, the system as a whole possesses certain orderly and analyzable average properties.

A wide range of experience comes under the label of disorganized complexity. The method applies with increasing precision when the number of variables increases. It applies with entirely useful precision to the experience of a large telephone exchange, in predicting the average frequency of calls, the probability of overlapping calls of the same number, etc. It makes possible the financial stability of a life insurance company. Although the company can have no knowledge whatsoever concerning the approaching death of any one individual, it has dependable knowledge of the average frequency with which deaths will occur.

This last point is interesting and important. Statistical techniques are not restricted to situations where the scientific theory of the individual events is very well known, as in the billiard example where there is a beautifully precise theory for the impact of one ball on another. This technique can also be applied to situations, like the insurance example, where the individual event is as shrouded in mystery as is the chain of complicated and unpredictable events associated with the accidental death of a healthy man.

The examples of the telephone and insurance companies suggests a whole array of practical applications of statistical techniques based on disorganized complexity. In a sense they are unfortunate examples, for they tend to draw attention away from the more fundamental use which science makes of these new techniques. The motions of the atoms which form all matter, as well as the motions of the stars which form the universe, come under the range of these new techniques. The fundamental laws of heredity are analyzed by them. The laws of thermodynamics, which describe basic and inevitable tendencies of all physical systems, are derived from statistical considerations. The entire structure of modern physics, our present concept of the nature of the physical universe, and of the accessible experimental facts concerning it rest on these statistical concepts. Indeed, the whole question of evidence and the way in which knowledge can be inferred from evidence are now recognized to depend on these same statistical ideas, so that probability notions are essential to any theory of knowledge itself.

Comment: In addition to the mathematical approaches discussed in the main text, three broad alternative strands have been taken based on an ensemble approach, a latter-day cousin of statistical mechanics that itself seeks the average properties of a single system, the liter box of gas. In an ensemble approach, one seeks the average, or generic properties of some entire class, or ensemble of systems. One then hopes to map such ensemble averages to the real system being modeled.

i. This has emerged most recently in seeking the generic connectivity properties of complex networks. Most famously, many real work networks are scale-free (Barabási 2009). A network is a mathematical structure called a graph, with nodes connected by nondirected or directed edges. Scale-free networks have a power law distribution of inputs to and outputs from a node.

Science and Complexity

PROBLEMS OF ORGANIZED COMPLEXITY

This new method of dealing with disorganized complexity, so powerful an advance over the earlier two-variable methods, leaves a great field untouched. One is tempted to oversimplify, and say that scientific methodology went from one extreme to the other—from two variables to an astronomical number—and left untouched a great middle region. The importance of this middle region, moreover, does not depend primarily on the fact that the number of variables involved is moderate—large compared to two, but small compared to the number of atoms in a pinch of salt. The problems in this middle region, in fact, will often involve a considerable number of variables. The really important characteristic of the problems of this middle region, which science has as yet little explored or conquered, lies in the fact that these problems, as contrasted with the disorganized situations with which statistics can cope, show the essential feature of *organization*. In fact, one can refer to this group of problems as those of *organized complexity*.

What makes an evening primrose open when it does? Why does salt water fail to satisfy thirst? Why can one particular genetic strain of microorganism synthesize within its minute body certain organic compounds that another strain of the same organism cannot manufacture? Why is one chemical substance a poison when another, whose molecules have just the same atoms but assembled into a mirror-image pattern, is completely harmless? Why does the amount of manganese in the diet affect the maternal instinct of an animal? What is the description of aging in biochemical terms? What meaning is to be assigned to the question: Is a virus a living organism? What is a gene, and how does the original genetic constitution of a living organism express itself in the developed characteristics of the adult? Do complex protein molecules "know how" to reduplicate their pattern, and is this an essential clue to the problem of reproduction of living creatures? All these are certainly complex problems, but they are not problems of disorganized complexity, to which statistical methods hold the key. They are all problems which involve dealing simultaneously with a *sizable number of factors which are interrelated into an organic whole.* They are all, in the language here proposed, problems of *organized complexity.*

On what does the price of wheat depend? This too is a problem of organized complexity. A very substantial number of relevant variables is involved here, and they are all interrelated in a complicated, but nevertheless not in helter-skelter, fashion.

How can currency be wisely and effectively stabilized? To what extent is it safe to depend on the free interplay of such economic forces as supply and demand? To what extent must systems of economic control be employed to prevent the wide swings from prosperity to depression? These are also obviously complex problems, and they too involve analyzing systems which are organic wholes, with their parts in close interrelation.

How can one explain the behavior pattern of an organized group of persons such as a labor union, or a group of manufacturers, or a racial minority? There are clearly many factors involved here, but it is equally obvious that here also something more is needed than the mathematics of averages. With a given total of national resources that can be brought to bear, what tactics and strategy will most promptly win a war, or better: what sacrifices of present selfish interest will most effectively contribute to a stable, decent, and peaceful world?

These problems—and a wide range of similar problems in the biological, medical, psychological, economic, and political sciences—are just too complicated to yield to the old nineteenth-century techniques which were so dramatically successful on two-, three-, or four-variable problems of simplicity. These new problems, moreover, cannot be handled with the statistical techniques so effective in describing average behavior in problems of disorganized complexity.

These new problems, and the future of the world depends on many of them, require science to make a third great advance, an advance

ii. Edwards and Anderson (1972) introduced the first spin-glass models. Spin glasses are disordered magnetic materials where nearby coupled spins may find their lowest energy by pointing in the same or the opposite direction. This "preference" varies sinusoidally with the somewhat random distances between the spins. A spin-glass model of N binary variables specifies which spin is coupled to which spin and, at random, the energy of that interaction. For three coupled spins all at lowest energy if oriented in the opposite direction to the other two spins, there is no single self-consistent lowest energy state. Such frustrated placettes present widely in the system typically render the energy landscape of the spin-glass fractal. Evolution of such a system at 0 K or any low finite temperature is not ergodic. The system does not explore the entire phase space.

that must be even greater than the nineteenth-century conquest of problems of simplicity or the twentieth-century victory over problems of disorganized complexity. Science must, over the next 50 years, learn to deal with these problems of organized complexity.

iii. I introduced the earliest ensemble approach in 1967 and 1969, as random Boolean networks, RBN (Kauffman and McCulloch 1967; Kauffman 1969). The aim was to model vast genetic regulatory networks with tens of thousands of interacting genes, where the protein product of one gene could "turn" another gene on or off. Here, gene activities are modeled as binary variables. A random network has N genes. Each gene is assigned K inputs chosen at random among the N. Each such gene is then assigned at random one of the possible Boolean functions on its K inputs. There is an enormous ensemble of random networks, given specification of N and K. I sought the generic behaviors of such random networks as a function of N and K. In particular, I proposed that the different cell types of an organism are modeled as the different attractors of such a network.

(cont. on next page)

Is there any promise on the horizon that this new advance can really be accomplished? There is much general evidence, and there are two recent instances of especially promising evidence. The general evidence consists in the fact that, in the minds of hundreds of scholars all over the world, important, though necessarily minor, progress is already being made on such problems. As never before, the quantitative experimental methods and the mathematical analytical methods of the physical sciences are being applied to the biological, the medical, and even the social sciences. The results are as yet scattered, but they are highly promising. A good illustration from the life sciences can be seen by a comparison of the present situation in cancer research with what it was twenty-five years ago. It is doubtless true that we are only scratching the surface of the cancer problem, but at least there are now some tools to dig with and there have been located some spots beneath which almost surely there is pay-dirt. We know that certain types of cancer can be induced by certain pure chemicals. Something is known of the inheritance of susceptibility to certain types of cancer. Million-volt rays are available, and the even more intense radiations made possible by atomic physics. There are radioactive isotopes, both for basic studies and for treatment. Scientists are tackling the almost incredibly complicated story of the biochemistry of the aging organism. A base of knowledge concerning the normal cell is being established that makes it possible to recognize and analyze the pathological cell. However distant the goal, we are now at last on the road to a successful solution of this great problem.

In addition to the general growing evidence that problems of organized complexity can be successfully treated, there are at least two promising bits of special evidence. Out of the wickedness of war have come two new developments that may well be of major importance in helping science to solve these complex twentieth-century problems.

The first piece of evidence is the wartime development of new types of electronic computing devices. These devices are, in flexibility and capacity, more like a human brain than like the traditional mechanical computing device of the past. They have "memories" in which vast amounts of information can be stored. They can be "told" to carry out computations of very intricate complexity, and can be left unattended while they go forward automatically with their task. The astounding speed with which they proceed is illustrated by the fact that one small part of such a machine, if set to multiplying two ten-digit numbers, can perform such multiplications some 40, 000 times faster than a human operator can say "Jack Robinson." This combination of flexibility, capacity, and speed makes it seem likely that such devices will have a tremendous impact on science. They will make it possible to deal with problems which previously were too complicated, and, more importantly, they will justify and inspire the development of new methods of analysis applicable to these new problems of organized complexity.

We now know that such RBN behave in three different dynamical regimes: Ordered, Critical, Chaotic (Schwab *et al.* 2020). I discovered in 1969 that N K networks with $K = 2$ have stunningly ordered properties such as a modest number of small stable attractors, hence a reasonable model of the different cell types of an organism. It later was shown the $K = 2$ networks are critical. Fifty years later it is now known that real cellular genetic networks are critical (Villani *et al.* 2018; Daniels *et al.* 2018) , as is brain dynamics (Beggs 2008). This ensemble theory correctly predicts that the number of cell types of an organism should be a power law scaling at slightly more than the square root of the number of genes (Bornholdt and Kauffman 2019). Ensemble theories are a new kind of statistical mechanics that tell us about the world.

At SFI, Chris Langton (1990) did seminal work, and Doyne Farmer coined the phrase "the edge of chaos." The broad hypothesis is that "Life evolves to the edge of chaos" (Kauffman 1993).

The second of the wartime advances is the "mixed-team" approach of operations analysis. These terms require explanation, although they are very familiar to those who were concerned with the application of mathematical methods to military affairs.

As an illustration, consider the over-all problem of convoying troops and supplies across the Atlantic. Take into account the number and effectiveness of the naval vessels available, the character of submarine attacks, and a multitude of other factors, including such an imponderable as the dependability of visual watch when men are tired, sick, or bored. Considering a whole mass of factors, some measurable and some elusive, what procedure would lead to the best over-all plan, that is, best from the combined point of view of speed, safety, cost, and so on? Should the convoys be large or small, fast or slow? Should they zigzag and expose themselves longer to possible attack, or dash in a speedy straight line? How are they to be organized, what defenses are best, and what organization and instruments should be used for watch and attack?

The attempt to answer such broad problems of tactics, or even broader problems of strategy, was the job during the war of certain groups known as the operations analysis groups. Inaugurated with brilliance by the British, the procedure was taken over by this country, and applied with special success in the Navy's anti-submarine campaign

Comment:
Is the evolving universe within the Newtonian paradigm? The Newtonian paradigm demands a fixed and prestated phase space of all the combinations of values of the relevant variables (Tisza 1963). Strikingly, the evolving biosphere does not fulfill this requirement. The evolving biosphere creates ever new relevant functional variables that cannot be prestated. Therefore, we can write no fixed laws of motion among these unknown future variables (Longo, Montévil, and Kauffman 2012). Roli and I (2021b) recently published a paper showing that no mathematics based on set theory can be used to deduce the ever-new adaptations that arise in evolution. It is an immediate corollary that, because the evolving universe has within it, an evolving biosphere, there can be no final theory that entails the evolution of the universe (Kauffman and Roli 2021a).

(cont. on next page)

and in the Army Air Forces. These operations analysis groups were, moreover, what may be called mixed teams. Although mathematicians, physicists, and engineers were essential, the best of the groups also contained physiologists, biochemists, psychologists, and a variety of representatives of other fields of the biochemical and social sciences. Among the outstanding members of English mixed teams, for example, were an endocrinologist and an X-ray crystallographer. Under the pressure of war, these mixed teams pooled their resources and focused all their different insights on the common problems. It was found, in spite of the modern tendencies toward intense scientific specialization, that members of such diverse groups could work together and could form a unit which was much greater than the mere sum of its parts. It was shown that these groups could tackle certain problems of organized complexity, and get useful answers.

It is tempting to forecast that the great advances that science can and must achieve in the next fifty years will be largely contributed to by voluntary mixed teams, somewhat similar to the operations analysis groups of war days, their activities made effective by the use of large, flexible, and highspeed computing machines. However, it cannot be assumed that this will be the exclusive pattern for future scientific work, for the atmosphere of complete intellectual freedom is essential to science. There will always, and properly, remain those scientists for whom intellectual freedom is necessarily a private affair. Such men must, and should, work alone. Certain deep and imaginative achievements are probably won only in such a way. Variety is, moreover, a proud characteristic of the American way of doing things. Competition between all sorts of methods is good. So there is no intention here to picture a future in which all scientists are organized into set patterns of activity. Not at all. It is merely suggested that some scientists will seek and develop for themselves new kinds of collaborative arrangements; that these groups will have members drawn from essentially all fields of science; and that these new ways of working, effectively instrumented by huge computers, will contribute greatly to the advance which the next half century will surely achieve in handling the complex, but essentially organic, problems of the biological and social sciences.

The Boundaries of Science

Let us return now to our original questions. What is science? What is not science? What may be expected from science?

Science clearly is a way of solving problems—not all problems, but a large class of important and practical ones. The problems with which science can deal are those in which the predominant factors are subject to the basic laws of logic, and are for the most part measurable. Science is a way of organizing reproducible knowledge about such problems; of focusing and disciplining imagination; of weighing evidence; of deciding what is relevant and what is not; of impartially testing hypotheses; of ruthlessly discarding data that prove to be inaccurate or inadequate; of finding, interpreting, and facing facts, and of making the facts of nature the servants of man.

The essence of science is not to be found in its outward appearance, in its physical manifestations; it is to be found in its inner spirit. That austere but exciting technique of inquiry known as the scientific method is what is important about science. This scientific method requires of its practitioners high standards of personal honesty, open-mindedness, focused vision, and love of the truth. These are solid virtues, but science has no exclusive lien on them. The poet has these virtues also, and often turns them to higher uses.

Science has made notable progress in its great task of solving logical and quantitative problems. Indeed, the successes have been so numerous and striking, and the failures have been so seldom publicized, that the average man has inevitably come to believe that science is just about the most spectacularly successful enterprise man ever launched. The fact is, of course, that this conclusion is largely justified.

Impressive as the progress has been, science has by no means worked itself out of a job. It is soberly true that science has, to date, succeeded in solving a bewildering number of relatively easy problems, whereas the hard problems, and the ones which perhaps promise most for man's future, lie ahead.

We must, therefore, stop thinking of science in terms of its spectacular successes in solving problems of simplicity. This means, among other things, that we must stop thinking of science in

This radical conclusion is supported by a recent paper by Devereaux *et al.* (2021) showing that no modeler within the universe can create a complete model of the universe. This result independently shows, therefore, that there can be no final theory that entails the evolution of the universe. More, the Devereaux paper proves Russell wrong. The world cannot be constructed as a logical combination of a prestated finite set of propositions. Finally, because evolving cells achieve constraint closure, the evolution of the biosphere is a non-deducible construction of ever new possibilities, not an entailed deduction (Kauffman 2019). With evolving life we are beyond the Newtonian paradigm with its fixed and prestated phase space. Weaver was right to wonder if and, if so, how organized complexity could be conquered in a New Physics. It now seems that the answer must be no if the New Physics is to be based on set theory.

terms of gadgetry. Above all, science must not be thought of as a modern improved black magic capable of accomplishing anything and everything.

Comment:
Hume is within a tradition dating back to Descartes which imagines a brain in a vat trying to know the world accurately. In this tradition, consideration of organisms as acting or doing is absent. But organisms are Kantian Whole agents, where the Parts exist in the universe for and by means of the Whole (Kauffman 2019), and are also non-equilibrium physical chemical systems that must take in matter and energy. Organism must, in short, feed. But "feeding" is an action, a doing, not merely a knowing. Once an organism Acts, it can succeed or fail. Mattering enters the universe from mere matter (Kauffman 2019). I get to live for a while. Once doing and mattering, that is value, enters the universe, that doing may succeed or fail. From this derives instrumental ought. From instrumental ought and mattering, value, later arises in evolution moral ought. The naturalistic fallacy is itself a fallacy.

Every informed scientist, I think, is confident that science is capable of tremendous further contributions to human welfare. It can continue to go forward in its triumphant march against physical nature, learning new laws, acquiring new power of forecast and control, making new material things for man to use and enjoy. Science can also make further brilliant contributions to our understanding of animate nature, giving men new health and vigor, longer and more effective lives, and a wiser understanding of human behavior. Indeed, I think most informed scientists go even further and expect that the precise, objective, and analytical techniques of science will find useful application in limited areas of the social and political disciplines.

There are even broader claims which can be made for science and the scientific method. As an essential part of his characteristic procedure, the scientist insists on precise definition of terms and clear characterization of his problem. It is easier, of course, to define terms accurately in scientific fields than in many other areas. It remains true, however, that science is an almost overwhelming illustration of the effectiveness of a well-defined and accepted language, a common set of ideas, a common tradition. The way in which this universality has succeeded in cutting across barriers of time and space, across political and cultural boundaries, is highly significant. Perhaps better than in any other intellectual enterprise of man, science has solved the problem of communicating ideas, and has demonstrated the world-wide cooperation and community of interest which then inevitably results.

Yes, science is a powerful tool, and it has an impressive record. But the humble and wise scientist does not expect or hope that science can do everything. He remembers that science teaches respect for special competence, and he does not believe that every social, economic, or political emergency would be automatically dissolved if "the scientists" were only put into control. He does not—with a few aberrant exceptions—expect science to furnish a code of morals, or a basis for esthetics. He does not expect science to furnish the yardstick for measuring, nor the motor for controlling, man's love of beauty and truth, his sense of value, or his convictions of faith. There are rich and essential parts of human life which are alogical, which are immaterial and non-quantitative in character, and which cannot be seen under the microscope, weighed with the balance, nor caught by the most sensitive microphone.

If science deals with quantitative problems of a purely logical character, if science has no recognition of or concern for value or purpose, how can modern scientific man achieve a balanced good life, in which logic is the companion of beauty, and efficiency is the partner of virtue?

In one sense the answer is very simple: our morals must catch up with our machinery. To state the necessity, however, is not to achieve it. The great gap, which lies so forebodingly between our power and our capacity to use power wisely, can only be bridged by a vast combination of efforts. Knowledge of individual and group behavior must be improved. Communication must be improved between peoples of different languages and cultures, as well as between all the varied interests which use the same language, but often with such dangerously differing connotations. A revolutionary advance must be made in our understanding of economic and political factors. Willingness to sacrifice selfish short-term interests, either personal or national, in order to bring about long-term improvement for all must be developed.

None of these advances can be won unless men understand what science really is; all progress must be accomplished in a world in which modern science is an inescapable, ever-expanding influence. ❧

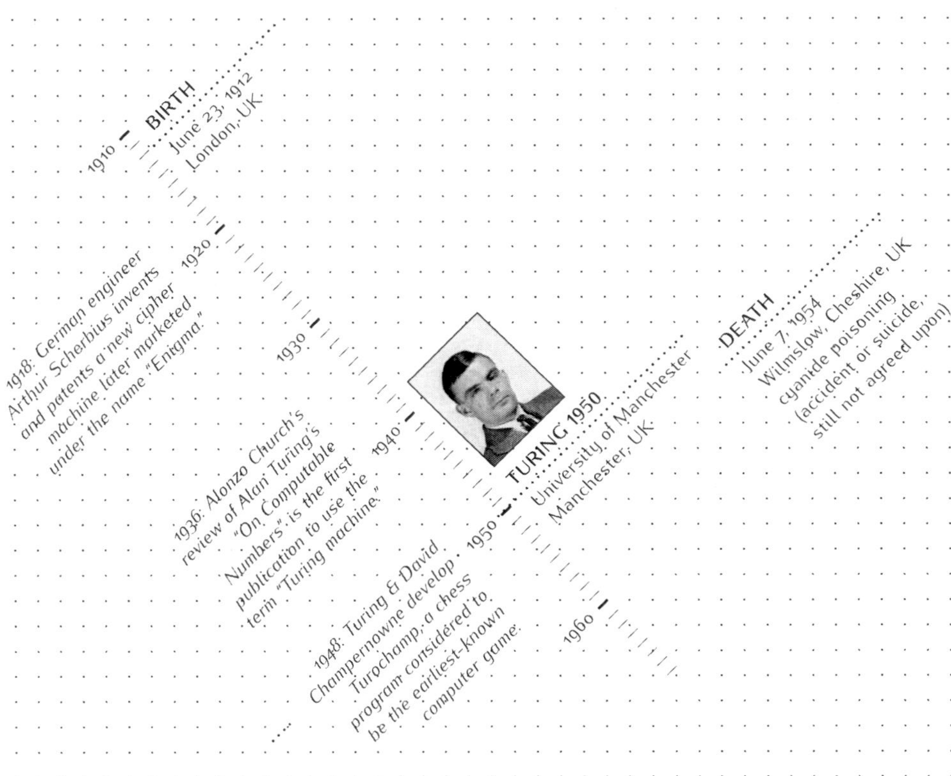

ALAN MATHISON TURING

[11]

TURING'S BRAINCHILD: ARTIFICIAL INTELLIGENCE

Daniel Dennett, Tufts University; Santa Fe Institute; and New College of the Humanities, London

In 1950, the mathematician Alan Turing was one of a handful of people in the world who had any experience in computer programming, and he was already coming to terms with the enormous implications of the new digital world he had helped create. He quite appropriately published his reflections in *Mind*, the leading *philosophy* journal in the English-speaking world. It became an instant classic, inspiring thousands of discussions of what soon became known as the Turing Test, the "imitation game" that he proposed as a worthy replacement for the tiresome argument that was already well under way in 1949 about whether these new machines could *think*—the way we human beings can.

A. M. Turing, "Computing Machinery and Intelligence," *Mind* LIX (236), 433–460 (1950).

Can machines think? Turing declared that this question is "too meaningless to deserve discussion." Instead of taking sides in debates about how to define these terms, he restricted his attention to a crisply definable task: winning the imitation game, in which a "discrete state machine" vies with a human being to see which can convince a human judge—through a telecommunicated conversation—of its humanity. This move is often seen as *behavioristic*—handsome is as handsome *does*—but it is better seen as just *scientific*: whatever thinking is, it should be something we can observe unmistakably if indirectly through its objective effects in the world. We may have our doubts about the suitability of Turing's replacement, but at least we will know whether some entity passes the test—like the net on the basketball hoop, the test would serve an important epistemological purpose, turning delicate and unreliable judgments into indisputable results.

The first person to claim without argument that no machine could pass the Turing Test was René Descartes!

> *It is indeed conceivable that a machine could be so made that it would utter words, and even words appropriate to the presence of physical acts or objects which cause some change in its organs; as, for example, if it was touched in some spot that it would ask what you wanted to say to it; if in another, that it would cry that it was hurt, and so on for similar things. But it could never modify its phrases to reply to the sense of whatever was said in its presence as even the most stupid men can do.*
>
> (*Discourse on Method*, [1637] 1960)

Turing never mentions Descartes's prescient passage, which suggests that they independently converge—not surprisingly—on conversation as the best measure of thinking. Linguistic interrogation is, after all, the primary way we have always assessed the intelligence and comprehension of our fellow human beings. Descartes thinks it is too obvious to need supporting argument that no machine could pass the test, but that is likely due to his inability to take seriously the idea of a machine with, say, more than a thousand moving parts.

Turing's restriction to discrete state machines (now known as digital computers) was not just for the sake of a bright line that forestalls debate. He recognized this category as something truly new in the world of human artifice, an invention that expands the reach of science even more than the microscope and telescope, permitting us for the first time to think rigorously and reliably about processes composed of *billions* of "moving parts" without risk of overlooking some hidden factor. If you can write a program for a discrete state machine that does x, you have an existence proof that it is *possible* to do x without the assistance of any *élan vital*, morphic resonances, psionic forces, or other mysterious poltergeists. Whether or not the world contains such exotic forces is left open, but one path to proof is cut off.

Turing provided an elegantly simple introduction to these amazing artifacts, explaining how they work by showing how they can perform all the duties of human computers executing algorithms with paper and pencil.

> *. . . . the 'book of rules' supplied to the computer is replaced in the machine by a part of the store. It is then called the 'table of instructions'. It is the duty of the control to see that these instructions are obeyed correctly and in the right order. The control is so constructed that this necessarily happens.* (p. 437)

"Necessarily" here is playing double duty in a way worth noting. On the one hand Turing had already defined a universal machine as a mathematical idealization which "by definition" obeys the rules perfectly, and he has proven that it can mimic any other discrete state machine. Here he implied succinctly that he and others have solved the engineering task of making a physical machine that, *by its physical design* (not "by definition"), can actually be held to instantiate the idealization for all practical purposes.

> *These are the machines which move by sudden jumps or clicks from one quite definite state to another. These states are sufficiently different for the possibility of confusion between them to be ignored. Strictly speaking there are no such machines. Everything really moves continuously. But there are many kinds of machine which can profitably be* thought of *as being discrete state machines.* (p. 439)

It is the digitalization that solves the engineering problem of making the different states *obviously* different to us and to the machine itself—like the net on the basketball hoop—absorbing the noise of continuous motion, creating complex organizations of causal links that can be *thought of* as deterministic whatever the underlying physics is.

> *This is reminiscent of Laplace's view that from the complete state of the universe at one moment of time, as described by the positions and velocities of all particles, it should be possible to predict all future states. The prediction which we are considering is, however, rather nearer to practicability than that considered by Laplace.* (p. 440)

Turing noted that Laplacean determinism is vulnerable to what has since come to be known as the butterfly effect:

> *The displacement of a single electron by a billionth of a centimetre at one moment might make the difference between a man being killed by an avalanche a year later, or escaping. It is an essential property of the mechanical systems which we have called 'discrete state machines' that this phenomenon does not occur.* (p. 440)

An important feature of digital computers is that we can isolate or identify any introduction of randomness (pseudo-randomness in almost all cases), and Turing explains the utility—indeed, the necessity—of harnessing randomness in search and learning. Using digital models to discipline one's imagination has been a spectacularly successful new way of thinking. Even at the dawn of this exploration, it raised the question for Turing of whether (in principle or in practice) the human contribution could be replaced by "just" more digital mechanism. After all, programming in binary machine code, which was Turing's tedious task, was already being routinized, soon to be taken out of the programmer's grasp and outsourced to a cascade of digital line editors (the ancestors of word processors) and compilers—and couldn't such a cascade eventually squeeze out the last drops of genius due to the programmer, banishing the ghost from the machine? In the eighteenth century the French philosopher Julien Offray de La Mettrie had argued in *L'Homme Machine* (1747) that human beings were ultimately just stupendous organizations of material mechanisms (bereft of any immaterial *res cogitans* such as Descartes had proposed). Turing, having invented a "universal" machine that could perfectly imitate the performance of any digital computing machine and approximate the performance of any analog computer to arbitrarily high resolution, imagined devising the existence proof that such a machine could think as well as any human thinker by winning the imitation game.

Reading "Computing Machinery and Intelligence" today yields a bounty of treasures, including three deep ironies: first, Turing's superb powers of imagination misled him into an overly optimistic assessment of the skeptical talents of his fellow human beings; second, his sense of fairness—no doubt energized by his persecution as a closeted gay man—

led him to propose a test that uncovers a deeply felt and supposedly sacrosanct bias in our ordinary assumptions about intelligence; and third, his proposed conversation stopper turned out to be a springboard for decades of just the sort of distracting philosophical and scientific debate he had hoped to avert.

Goodhart's law (1975) tells us that when a measure becomes a target, it ceases to be a good measure—since it creates an almost irresistible temptation to game the system. Turing had not intended his imitation game to be the specific focus of a research program in AI, but decades of actual attempts by programmers to win various simplified versions of the contest have shown how easy it is to fool intelligent but computer-naïve judges. The recent explosion of deep learning systems and transformers, exploiting statistical analysis of gigantic databases of human-generated text, has turned Turing's thought experiment into a serious social problem that requires our immediate attention: by putting a premium of *deceiving* a human judge into falsely attributing humanity to a machine, the imitation game has encouraged the development of techniques of phony anthropomorphism. We now face the task of disincentivizing deception and teaching people who don't understand how computers achieve their "magic" how to expose and reassess the systems they are invited to interact with. Turing's imitation game deftly postponed any analysis of what the most important powers of thinking might be, but now we have to articulate them so that we can explain what is missing from the impostors that are now blooming.

The second irony follows directly from the first: Turing wanted his test to be *neutral* between "human" ways of thinking and other ways (if there are any), but this point has been largely overlooked. Consider: in the original imitation game, when a man tries to convince the judge he is the woman, it is clear that Turing doesn't imagine that the man must in fact be able to "think like a woman" but just that the man should have enough information about how women think and respond—have a model or theory of what a woman would say—to simulate a woman *by hook or by crook*, that is, using whatever information processing moves are available to him. Similarly, if a suitably programmed discrete state machine contains a model of human thinking and conversing that enables it to pass for human, then, by Turing's lights, this should count as a proof that it *is* thinking in its own alien way. Handsome is as handsome does, and if the machine can do what the intelligent man can do, it is intelligent, and what it does is a kind of thinking—a new kind of thinking, perhaps.

This radically open-minded prospect has been all but submerged by what I will call *anthroponormative* thinking about thinking: *our* way of thinking, the *human* way of thinking, is the only kind of thinking worth considering (French 1990). There are in fact two opposing strains of anthroponormative thinking that have haunted AI since its birth: one abhors the very idea of intelligence being "reduced" to a digitized *theory* or *model*—turn the digital crank and out flow the theorems that govern art, creativity, comprehension—and the other abhors the idea that *there is no such theory*, that intelligence is "just" an unimaginably large data set with some kind of indexing system that "mindlessly" generates outputs. *Real* thinking, according to both biases, involves some extra feature that all normal human beings enjoy; it is hard to pin down but it hovers in the neighborhood of *comprehension*. Many people are extremely reluctant to abandon Descartes's conviction that machine comprehension is inconceivable.

The anthroponormative bias can perhaps best be seen by comparing the Turing Test to several variants in other domains. When symphony orchestras conduct auditions for new performers by having them play behind screens so that their physical appearance (their gender, skin color, etc.) is shielded from the judges, we commend the practice as appropriately unbiased. Now suppose that when the curtain is pulled back to reveal the winner of one such violin audition what confronts the judges is a complicated robotic machine with a violin carefully and flexibly clamped into a device that uses dozens of presser-vibrators to hold down the strings and moves a bow manipulated by two arm-like actuators holding the bow at both ends, and equipped with a finger-plucker device for pizzicato. It follows the score and the conductor with two independently operated vision systems. Watching this "violinist" produce sound from the violin is unsettling, even shocking, but didn't we decide that physical appearance was out of bounds?

The anthroponormative bias is not just robotophobia. Imagine that some clever fellow trains his arms and legs to behave in hitherto unimagined ways that permit him to play winning tennis that consists of towering lobs that land too deep in the court to return effectively and spin shots that bounce back into the net after clearing it. The delicious grace of Roger Federer or Rafael Nadal is entirely missing from this player's approach; his opponents seldom get to return his unsightly shots and he humiliates them with his cocky behavior. Tennis has been ruined! But he plays by all the rules. What should we do? Change the rules, move the goal posts so that we can preserve the good old *human* game of tennis one way or another?

The third irony that emerges from Turing's paper is that he thought he could block some fruitless wrangles from the outset by furnishing succinct and telling rebuttals to what he saw as the serious objections, but thanks to his practical and constructive attitude toward discovery, he hugely underestimated the ability of philosophers to dream up "possible in principle" counterexamples with which to muddy the waters. The most infamous, John Searle's Chinese Room (1980), has consumed the time, talent, and attention of debaters on both sides for decades, but there are others as well, such as Ned Block's Chinese Nation (1978) and Block's giant lookup table (1982). These are introduced and dismantled in Dennett (1985, 1987, 2013, 2019). Turing didn't bother dealing with such fantasies since he assumed that the interesting possibilities to discuss did not involve violating the known laws of physics (French 2000).

Turing underestimated the controversies that have raged about digital computers since he invented them, but perhaps instead of chalking this up to a failure of his imagination, we should see this as a measure of his innocent optimism in the face of adversity: he thought that thoughtful people were as smart as he was.

Acknowledgments

Of the hundreds of discussions of these issues I have had over the years, two interlocutors stand out for their contributions to my understanding: Robert French and Douglas R. Hofstadter.

REFERENCES

Block, N. 1978. "Troubles with Functionalism." *Minnesota Studies in the Philosophy of Science* 9:261–325. https://hdl.handle.net/11299/185298.

———. 1982. "Psychologism and Behaviorism." *Philosophical Review* 90:5–43. https://doi.org/10.2307/2184371.

Dennett, D. 1985. "Can Machines Think?" In *How We Know,* edited by M. Shafto, 121–145. San Francisco, CA: Harper and Row.

———. 1987. "Fast Thinking." In *The Intentional Stance.* Cambridge, MA: MIT Press.

———. 2013. "The Chinese Room." In *Intuition Pumps and Other Tools for Thinking.* New York, NY: Norton.

———. 2019. "What Can We Do?" In *25 Ways of Looking at AI,* edited by J. Brockman, 41–53. New York, NY: Penguin.

Descartes, R. [1637]1960. *Discourse on Method.* Translated by L. Lafleur. New York, NY: Bobbs Merrill.

French, R. M. 1990. "Subcognition and the Limits of the Turing Test." *Mind* 99:433–460. https://doi.org/10.1093/mind/XCIX.393.53.

———. 2000. "The Chinese Room: Just Say 'No!'" *Proceedings of the 22nd Annual Conference of the Cognitive Science Society,* 657–662. https://escholarship.org/uc/item/9062452s.

Gallistel, C. R. (forthcoming). *The Neurological Bases for the Computational Theory of Mind.* In a festschrift for Jerry Fodor.

Goodhart, C. 1975. "Problems of Monetary Management: The UK Experience." In *Papers in Monetary Economics,* vol. 1. Sydney, Australia: Reserve Bank of Australia.

Haugeland, J. 1985. *Artificial Intelligence: The Very Idea.* Cambridge, MA: MIT Press.

La Mettrie, J. O. de. 1747. *L'Homme Machine.*

Lucas, J. R. 1961. "Minds, Machines, and Gödel." *Philosophy* 36:112–127. https://doi.org/10.1017/S0031819100057983.

Penrose, R. 1989. *The Emperor's New Mind: Concerning Computers, Minds, and the Laws of Physics.* Oxford, UK: Oxford University Press.

———. 1990. "The Non-Algorithmic Mind." *Behavioral and Brain Sciences* 13 (4): 692–705. https://doi.org/10.1017/S0140525X0008105X.

Searle, J. 1980. "Minds, Brains, and Programs." *Behavioral and Brain Sciences* 3:417–424. https://doi.org/10.1017/S0140525X00005756.

Smith, J. I., and Y. Y. Haddad. 1975. "Women in the Afterlife: The Islamic View as Seen from Qur'ān and Tradition." *Journal of the American Academy of Religion* 43 (1): 39–50. https://doi.org/10.1093/jaarel/XLIII.1.39.

COMPUTING MACHINERY AND INTELLIGENCE

A. M. Turing, Victoria University of Manchester

1. The Imitation Game

I propose to consider the question, 'Can machines think?' This should begin with definitions of the meaning of the terms 'machine' and 'think'. The definitions might be framed so as to reflect so far as possible the normal use of the words, but this attitude is dangerous. If the meaning of the words 'machine' and 'think' are to be found by examining how they are commonly used it is difficult to escape the conclusion that the meaning and the answer to the question, 'Can machines think?' is to be sought in a statistical survey such as a Gallup poll. But this is absurd. Instead of attempting such a definition I shall replace the question by another, which is closely related to it and is expressed in relatively unambiguous words.

The new form of the problem can be described in terms of a game which we call the 'imitation game'. It is played with three people, a man (A), a woman (B), and an interrogator (C) who may be of either sex. The interrogator stays in a room apart from the other two. The object of the game for the interrogator is to determine which of the other two is the man and which is the woman. He knows them by labels X and Y, and at the end of the game he says either 'X is A and Y is B' or 'X is B and Y is A'. The interrogator is allowed to put questions to A and B thus:

C: Will X please tell me the length of his or her hair?

Now suppose X is actually A, then A must answer. It is A's object in the game to try and cause C to make the wrong identification. His answer might therefore be

> 'My hair is shingled, and the longest strands are about nine inches long.'

In order that tones of voice may not help the interrogator the answers should be written, or better still, typewritten. The ideal arrangement is to have a teleprinter communicating between the two rooms. Alternatively the question and answers can be repeated by an intermediary. The object of the game for the third player (B) is to help the interrogator. The best strategy for her is probably to give truthful answers. She can add such things as 'I am the woman, don't listen to him!' to her answers, but it will avail nothing as the man can make similar remarks.

Note that Turing's question is ambiguous in this first statement. Does the machine try to convince the judge that it is a woman not a man? Or that it is human, not a machine? It makes a difference that is standardly ignored: if the judge thinks that the game is between a man, A, and a woman, B, then the issue of whether one of them is a machine does not even arise. If the machine does a better job of convincing the judge that it is 'the woman' it wins, without the issue of whether it's a machine even coming up!

We now ask the question, 'What will happen when a machine takes the part of A in this game?' Will the interrogator decide wrongly as often when the game is played like this as he does when the game is played between a man and a woman? These questions replace our original, 'Can machines think?'

2. Critique of the New Problem

As well as asking, 'What is the answer to this new form of the question', one may ask, 'Is this new question a worthy one to investigate?' This latter question we investigate without further ado, thereby cutting short an infinite regress.

The new problem has the advantage of drawing a fairly sharp line between the physical and the intellectual capacities of a man. No engineer or chemist claims to be able to produce a material which is indistinguishable from the human skin. It is possible that at some time this might be done, but even supposing this invention available we should feel there was little point in trying to make a 'thinking machine' more human by dressing it up in such artificial flesh. The form in which we have set the problem reflects this fact in the condition which prevents the interrogator from seeing or touching the other competitors, or hearing their voices. Some other advantages of the proposed criterion may be shown up by specimen questions and answers. Thus:

Q : Please write me a sonnet on the subject of the Forth Bridge.

A : Count me out on this one. I never could write poetry.

Q : Add 34957 to 70764

A : (Pause about 30 seconds and then give as answer) 105621.

Q : Do you play chess?

A : Yes.

Q : I have K at my K1, and no other pieces. You have only K at K6 and R at R1. It is your move. What do you play?

A : (After a pause of 15 seconds) R–R8 mate.

The question and answer method seems to be suitable for introducing almost any one of the fields of human endeavour that we wish to include. We do not wish to penalise the machine for its inability to shine in beauty competitions, nor to penalise a man for losing in a race against an aeroplane. The conditions of our game make these disabilities irrelevant. The 'witnesses' can brag, if they consider it advisable, as much as they please about their charms, strength or heroism, but the interrogator cannot demand practical demonstrations.

The game may perhaps be criticised on the ground that the odds are weighted too heavily against the machine. If the man were to try and pretend to be the machine he would clearly make a very poor showing. He would be given away at once by slowness and inaccuracy in arithmetic. May not machines carry out something which ought to be described as thinking but which is very different from what a man does? This objection is a very strong one, but at least we can say that if, nevertheless, a machine can be constructed to play the imitation game satisfactorily, we need not be troubled by this objection.

It might be urged that when playing the 'imitation game' the best strategy for the machine may possibly be something other than imitation of the behaviour of a man. This may be, but I think it is unlikely that there is any great effect of this kind. In any case there is no intention to investigate here the theory of the game, and it will be assumed that the best strategy is to try to provide answers that would naturally be given by a man.

The default sexist language of 1950 is striking here, since Turing has begun with a contest between a man and a woman. This paragraph disambiguates the question Turing asked at the outset (see note 1).

3. The Machines Concerned in the Game

The question which we put in §1 will not be quite definite until we have specified what we mean by the word 'machine'. It is natural that we should wish to permit every kind of engineering technique to be used in our machines. We also wish to allow the possibility that an engineer or team of engineers may construct a machine which works, but whose manner of operation cannot be satisfactorily described by its constructors because they have applied a method which is largely experimental. Finally, we wish to exclude from the machines men born in the usual manner. It is difficult to frame the definitions so as to satisfy these three conditions. One might for instance insist that the team of engineers should be all of one sex, but this would not really be satisfactory, for it is probably possible to rear a complete individual from a single cell of the skin (say) of a man. To do so would be a feat of biological technique deserving of the very highest praise, but we would not be inclined to regard it as a case of 'constructing a thinking machine'. This prompts us to abandon the requirement that every kind of technique should be permitted. We are the more ready to do so in view of the fact that the present interest in 'thinking machines' has been aroused by a particular kind of machine, usually called an 'electronic computer' or 'digital computer'. Following this suggestion we only permit digital computers to take part in our game.

This restriction appears at first sight to be a very drastic one. I shall attempt to show that it is not so in reality. To do this necessitates a short account of the nature and properties of these computers.

It may also be said that this identification of machines with digital computers, like our criterion for 'thinking', will only be unsatisfactory if (contrary to my belief), it turns out that digital computers are unable to give a good showing in the game.

There are already a number of digital computers in working order, and it may be asked, 'Why not try the experiment straight away? It would be easy to satisfy the conditions of the game. A number of interrogators could be used, and statistics compiled to show how often the right identification was given.' The short answer is that we are not asking whether all digital computers would do well in the game nor whether the computers at present available would do well, but whether there are imaginable computers which would do well. But this is only the short answer. We shall see this question in a different light later.

You could count this number on the fingers of one hand. Today of course there are billions of digital computers.

4. Digital Computers

The idea behind digital computers may be explained by saying that these machines are intended to carry out any operations which could be done by a human computer. The human computer is supposed to be following fixed rules; he has no authority to deviate from them in any detail. We may suppose that these rules are supplied in a book, which is altered whenever he is put on to a new job. He has also an unlimited supply of paper on which he does his calculations. He may also do his multiplications and additions on a 'desk machine', but this is not important.

If we use the above explanation as a definition we shall be in danger of circularity of argument. We avoid this by giving an outline of the means by which the desired effect is achieved. A digital computer can usually be regarded as consisting of three parts:

(i) Store.
(ii) Executive unit.
(iii) Control.

The store is a store of information, and corresponds to the human computer's paper, whether this is the paper on which he does his calculations or that on which his book of rules is printed. In so far as the human computer does calculations in his head a part of the store will correspond to his memory.

The executive unit is the part which carries out the various individual operations involved in a calculation. What these individual operations are will vary from machine to machine. Usually fairly lengthy operations can be done such as 'Multiply 3540675445 by 7076345687' but in some machines only very simple ones such as 'Write down 0' are possible.

We have mentioned that the 'book of rules' supplied to the computer is replaced in the machine by a part of the store. It is then called the 'table of instructions'. It is the duty of the control to see that these instructions are obeyed correctly and in the right order. The control is so constructed that this necessarily happens.

The information in the store is usually broken up into packets of moderately small size. In one machine, for instance, a packet might consist of ten decimal digits. Numbers are assigned to the parts of the store in which the various packets of information are stored, in some systematic manner. A typical instruction might say—

'Add the number stored in position 6809 to that in 4302 and put the result back into the latter storage position'.

Needless to say it would not occur in the machine expressed in English. It would more likely be coded in a form such as 6809430217. Here 17 says which of various possible operations is to be performed on the two numbers. In this case the operation is that described above, *viz.* 'Add the number. . . .' It will be noticed that the instruction takes up ten digits and so forms one packet of information, very conveniently. The control will normally take the instructions to be obeyed in the order of the positions in which they are stored, but occasionally an instruction such as

'Now obey the instruction stored in position 5606, and continue from there'

may be encountered, or again

'If position 4505 contains 0 obey next the instruction stored in 6707, otherwise continue straight on.'

Instructions of these latter types are very important because they make it possible for a sequence of operations to be repeated over and over again until some condition is fulfilled, but in doing so to obey, not fresh instructions on each repetition, but the same ones over and over again. To take a domestic analogy. Suppose Mother wants Tommy to call at the cobbler's every morning on his way to school to see if her shoes are done, she can ask him afresh every morning. Alternatively she can stick up a notice once and for all in the hall which he will see when he leaves for school and which tells him to call for the shoes, and also to destroy the notice when he comes back if he has the shoes with him.

The reader must accept it as a fact that digital computers can be constructed, and indeed have been constructed, according to the principles we have described, and that they can in fact mimic the actions of a human computer very closely.

The book of rules which we have described our human computer as using is of course a convenient fiction. Actual human computers really remember what they have got to do. If one wants to make a machine mimic the behaviour of the human computer in some complex operation one has to ask him how it is done, and then translate the answer into the form of an instruction table. Constructing instruction tables is usually described as 'programming'. To 'programme a machine to carry out the operation A' means to put the appropriate instruction table into the machine so that it will do A.

An interesting variant on the idea of a digital computer is a 'digital computer with a random element'. These have instructions involving the throwing of a die or some equivalent electronic process; one such instruction might for instance be, 'Throw the die and put the resulting number into store 1000'. Sometimes such a machine is described as having free will (though I would not use this phrase myself). It is not normally possible to determine from observing a machine whether it has a random element, for a similar effect can be produced by such devices as making the choices depend on the digits of the decimal for π.

Most actual digital computers have only a finite store. There is no theoretical difficulty in the idea of a computer with an unlimited store. Of course only a finite part can have been used at any one time. Likewise only a finite amount can have been constructed, but we can imagine more and more being added as required. Such computers have special theoretical interest and will be called infinitive capacity computers.

As more and more memory was added, it became clear that many imaginable programs were not feasible—could not be executed in human amounts of time, a property that has generated much of the innovation in cryptography and quantum computing.

The idea of a digital computer is an old one. Charles Babbage, Lucasian Professor of Mathematics at Cambridge from 1828 to 1839, planned such a machine, called the Analytical Engine, but it was never completed. Although Babbage had all the essential ideas, his machine was not at that time such a very attractive prospect. The speed which would have been available would be definitely faster than a human computer but something like 100 times slower than the Manchester machine, itself one of the slower of the modern machines. The storage was to be purely mechanical, using wheels and cards.

The fact that Babbage's Analytical Engine was to be entirely mechanical will help us to rid ourselves of a superstition. Importance is often attached to the fact that modem digital computers are electrical, and that the nervous system also is electrical. Since Babbage's machine was not electrical, and since all digital computers are in a sense equivalent, we see that this use of electricity cannot be of theoretical importance. Of course electricity usually comes in where fast signalling is concerned, so that it is not surprising that we find it in both these connections. In the nervous system chemical phenomena are at least as important as electrical. In certain computers the storage system is mainly acoustic. The feature of using electricity is thus seen to be only a very superficial similarity. If we wish to find such similarities we should look rather for mathematical analogies of function.

5. Universality of Digital Computers

The digital computers considered in the last section may be classified amongst the 'discrete state machines'. These are the machines which move by sudden jumps or clicks from one quite definite state to another. These states are sufficiently different for the possibility of confusion between them to be ignored. Strictly speaking there are no such machines. Everything really moves continuously. But there are many kinds of machine which can profitably be *thought of* as being discrete state machines. For instance in considering the switches for a lighting system it is a convenient fiction that each switch must be definitely on or definitely off. There must be intermediate positions, but for most purposes we can forget about them. As an example of a discrete state machine we might consider a wheel which clicks round through 120° once a second, but may be stopped by a lever which can be operated from outside; in addition a lamp is to light in one of the positions of the wheel. This machine could be described abstractly as follows. The internal state of the machine (which is described by the position of the wheel) may be q_1, q_2 or q_3. There is an input signal i_0 or i_1 (position of lever). The internal state at any moment is determined by the last state and input signal according to the table

		Last State		
		q_1	q_2	q_3
Input	i_0	q_2	q_3	q_1
	i_1	q_1	q_2	q_3

The output signals, the only externally visible indication of the internal state (the light) are described by the table

State	q_1	q_2	q_3
Output	o_0	o_0	o_1

This example is typical of discrete state machines. They can be described by such tables provided they have only a finite number of possible states.

It will seem that given the initial state of the machine and the input signals it is always possible to predict all future states. This is reminiscent of Laplace's view that from the complete state of the universe at one moment of time, as described by the positions and velocities of all particles, it should be possible to predict all future states. The prediction which we are considering is, however, rather nearer to practicability than that considered by Laplace. The system of the 'universe as a whole' is such that quite small errors in the initial conditions can have an overwhelming effect at a later time. The displacement of a single electron by a billionth of a centimetre at one moment might make the difference between a man being killed by an avalanche a year later, or escaping. It is an essential property of the mechanical systems which we have called 'discrete state machines' that this phenomenon does not occur. Even when we consider the actual physical machines instead of the idealised machines, reasonably accurate knowledge of the state at one moment yields reasonably accurate knowledge any number of steps later.†

This is a breathtaking understatement. This practical reliability of prediction is unmatched by any other technology, and goes a long way to explaining the ubiquity of computer chips in the world today.

In chaos theory, this has come to be known as the butterfly effect (Lorenz 1963), the sensitive dependence on initial conditions in which a small change in one state of a deterministic nonlinear system can result in large differences in a later state.

As we have mentioned, digital computers fall within the class of discrete state machines. But the number of states of which such a machine is capable is usually enormously large. For instance, the number for the machine now working at Manchester is about $2^{165,000}$, *i.e.* about $10^{50,000}$. Compare this with our example of the clicking wheel described above, which had three states. It is not difficult to see why the number of states should be so immense. The computer includes a store corresponding to the paper used by a human computer. It must be possible to write into the store any one of the combinations of symbols which might have been written on the paper. For simplicity suppose that only digits from 0 to 9 are used as symbols. Variations in handwriting are ignored. Suppose the computer is allowed 100 sheets of

† There is a sense in which this statement is true, since digitization fixes a base level at which there are "differences that make a difference," but surprising and practically unpredictable emergent properties arise from many large programs, as the creators of deep learning systems are discovering.

paper each containing 50 lines each with room for 30 digits. Then the number of states is $10^{100 \times 50 \times 30}$, *i.e.* $10^{150,000}$. This is about the number of states of three Manchester machines put together. The logarithm to the base two of the number of states is usually called the 'storage capacity' of the machine. Thus the Manchester machine has a storage capacity of about 165,000 and the wheel machine of our example about 1.6. If two machines are put together their capacities must be added to obtain the capacity of the resultant machine. This leads to the possibility of statements such as 'The Manchester machine contains 64 magnetic tracks each with a capacity of 2,560, eight electronic tubes with a capacity of 1,280. Miscellaneous storage amounts to about 300 making a total of 174,380.'

Given the table corresponding to a discrete state machine it is possible to predict what it will do. There is no reason why this calculation should not be carried out by means of a digital computer. Provided it could be carried out sufficiently quickly the digital computer could mimic the behaviour of any discrete state machine. The imitation game could then be played with the machine in question (as B) and the mimicking digital computer (as A) and the interrogator would be unable to distinguish them. Of course the digital computer must have an adequate storage capacity as well as working sufficiently fast. Moreover, it must be programmed afresh for each new machine which it is desired to mimic.

This special property of digital computers, that they can mimic any discrete state machine, is described by saying that they are *universal* machines. The existence of machines with this property has the important consequence that, considerations of speed apart, it is unnecessary to design various new machines to do various computing processes. They can all be done with one digital computer, suitably programmed for each case. It will be seen that as a consequence of this all digital computers are in a sense equivalent.

We may now consider again the point raised at the end of §3. It was suggested tentatively that the question, 'Can machines think?' should be replaced by 'Are there imaginable digital computers which would do well in the imitation game?' If we wish we can make this superficially more general and ask 'Are there discrete state machines which would do well?' But in view of the universality property we see that either of these questions is equivalent to this, 'Let us fix our attention on one particular digital computer C. Is it true that by modifying this computer to have an adequate storage, suitably increasing its speed of action, and providing it with an appropriate programme, C can be made to play satisfactorily the part of A in the imitation game, the part of B being taken by a man?'

6. Contrary Views on the Main Question

We may now consider the ground to have been cleared and we are ready to proceed to the debate on our question, 'Can machines think?' and the variant of it quoted at the end of the last section. We cannot altogether abandon the original form of the problem, for opinions will differ as to the appropriateness of the substitution and we must at least listen to what has to be said in this connexion.

It will simplify matters for the reader if I explain first my own beliefs in the matter. Consider first the more accurate form of the question. I believe that in about fifty years' time it will be possible to programme computers, with a storage capacity of about 10^9, to make them play the imitation game so well that an average interrogator will not have more than 70 percent. chance of making the right identification after five minutes of questioning. The original question, 'Can machines think?' I believe to be too meaningless to deserve discussion. Nevertheless I believe that at the end of the century the use of words and general educated opinion will have altered so much that one will be able to speak of machines thinking without expecting to be contradicted. I believe further that no useful purpose is served by concealing these beliefs. The popular view that scientists proceed inexorably from well-established fact to well-established fact, never being influenced by any unproved conjecture, is quite mistaken. Provided it is made clear which are proved facts and which are conjectures, no harm can result. Conjectures are of great importance since they suggest useful lines of research.

This oft-quoted prophecy has energized AI for decades. Turing doesn't say what units he is using for measuring memory storage, but his figure of 10^9 is comically low from today's perspective.

I now proceed to consider opinions opposed to my own.

(1) *The Theological Objection*. Thinking is a function of man's immortal soul. God has given an immortal soul to every man and woman, but

not to any other animal or to machines. Hence no animal or machine can think.[1]

I am unable to accept any part of this, but will attempt to reply in theological terms. I should find the argument more convincing if animals were classed with men, for there is a greater difference, to my mind, between the typical animate and the inanimate than there is between man and the other animals. The arbitrary character of the orthodox view becomes clearer if we consider how it might appear to a member of some other religious community. How do Christians regard the Moslem view that women have no souls? But let us leave this point aside and return to the main argument. It appears to me that the argument quoted above implies a serious restriction of the omnipotence of the Almighty. It is admitted that there are certain things that He cannot do such as making one equal to two, but should we not believe that He has freedom to confer a soul on an elephant if He sees fit? We might expect that He would only exercise this power in conjunction with a mutation which provided the elephant with an appropriately improved brain to minister to the needs of this soul. An argument of exactly similar form may be made for the case of machines. It may seem different because it is more difficult to "swallow". But this really only means that we think it would be less likely that He would consider the circumstances suitable for conferring a soul. The circumstances in question are discussed in the rest of this paper. In attempting to construct such machines we should not be irreverently usurping His power of creating souls, any more than we are in the procreation of children: rather we are, in either case, instruments of His will providing mansions for the souls that He creates.

Turing speaks of the Islamic belief that women have no souls. Harry Lewis has pointed out to me that this is not true. See Smith and Haddad (1975).

Turing's delicate handling of theological arguments is one measure of the distance we have traveled since 1950.

However, this is mere speculation. I am not very impressed with theological arguments whatever they may be used to support. Such arguments have often been found unsatisfactory in the past. In the

[1] Possibly this view is heretical. St. Thomas Aquinas (Summa Theologica, quoted by Bertrand Russell, p. 480) states that God cannot make a man to have no soul. But this may not be a real restriction on His powers, but only a result of the fact that men's souls are immortal, and therefore indestructible.

time of Galileo it was argued that the texts, "And the sun stood still . . . and hasted not to go down about a whole day" (Joshua x. 13) and "He laid the foundations of the earth, that it should not move at any time" (Psalm cv. 5) were an adequate refutation of the Copernican theory. With our present knowledge such an argument appears futile. When that knowledge was not available it made a quite different impression.

(2) *The 'Heads in the Sand' Objection.* "The consequences of machines thinking would be too dreadful. Let us hope and believe that they cannot do so."

This argument is seldom expressed quite so openly as in the form above. But it affects most of us who think about it at all. We like to believe that Man is in some subtle way superior to the rest of creation. It is best if he can be shown to be *necessarily* superior, for then there is no danger of him losing his commanding position. The popularity of the theological argument is clearly connected with this feeling. It is likely to be quite strong in intellectual people, since they value the power of thinking more highly than others, and are more inclined to base their belief in the superiority of Man on this power.

I do not think that this argument is sufficiently substantial to require refutation. Consolation would be more appropriate: perhaps this should be sought in the transmigration of souls.

Here he casually pinpoints the escape clause that renders moot all the subsequent arguments of Lucas (1961), Penrose (1989) and others. The sort of digital program that could be a candidate for passing the Turing Test would be furnished with massive stores of often contradictory information and risky heuristics (algorithms not guaranteed to accomplish their purposes).

(3) *The Mathematical Objection.* There are a number of results of mathematical logic which can be used to show that there are limitations to the powers of discrete-state machines. The best known of these results is known as *Gödel*'s theorem,[1] and shows that in any sufficiently powerful logical system statements can be formulated which can neither be proved nor disproved within the system, unless possibly the system itself is inconsistent *(Gödel 1931)*. There are other, in some respects similar, results due to *Church (1936), Kleene (1935), Russell (1940), and Turing (1937)*. The latter

[1] Author's names in italics refer to the Bibliography.

result is the most convenient to consider, since it refers directly to machines, whereas the others can only be used in a comparatively indirect argument: for instance if Gödel's theorem is to be used we need in addition to have some means of describing logical systems in terms of machines, and machines in terms of logical systems. The result in question refers to a type of machine which is essentially a digital computer with an infinite capacity. It states that there are certain things that such a machine cannot do. If it is rigged up to give answers to questions as in the imitation game, there will be some questions to which it will either give a wrong answer, or fail to give an answer at all however much time is allowed for a reply. There may, of course, be many such questions, and questions which cannot be answered by one machine may be satisfactorily answered by another. We are of course supposing for the present that the questions are of the kind to which an answer 'Yes' or 'No' is appropriate, rather than questions such as 'What do you think of Picasso?' The questions that we know the machines must fail on are of this type, "Consider the machine specified as follows. . . . Will this machine ever answer 'Yes' to any question?" The dots are to be replaced by a description of some machine in a standard form, which could be something like that used in § 5. When the machine described bears a certain comparatively simple relation to the machine which is under interrogation, it can be shown that the answer is either wrong or not forthcoming. This is the mathematical result: it is argued that it proves a disability of machines to which the human intellect is not subject.

Obtaining the empirical demonstration that humans are immune to this limitation has been an instructive exercise in futility (see the discussions in Dennett 1978, 1995).

The short answer to this argument is that although it is established that there are limitations to the powers of any particular machine, it has only been stated, without any sort of proof, that no such limitations apply to the human intellect. But I do not think this view can be dismissed quite so lightly. Whenever one of these machines is asked the appropriate critical question, and gives a definite answer, we know that this answer must be wrong, and this gives us a certain feeling of superiority. Is this feeling illusory? It is no doubt quite genuine, but I do not think too much importance should be attached to it. We too often give wrong answers to questions ourselves to be justified in being very pleased at such evidence of fallibility on the part of the machines. Further, our superiority can only be felt on such an occasion in relation to the one machine over which we have scored our petty triumph. There would be no question of triumphing simultaneously over *all* machines. In short, then, there might be men cleverer than any given machine, but then again there might be other machines cleverer again, and so on.

Those who hold to the mathematical argument would, I think, mostly be willing to accept the imitation game as a basis for discussion. Those who believe in the two previous objections would probably not be interested in any criteria.

(4) *The Argument from Consciousness.* This argument is very well expressed in *Professor Jefferson's* Lister Oration for 1949, from which I quote. "Not until a machine can write a sonnet or compose a concerto because of thoughts and emotions felt, and not by the chance fall of symbols, could we agree that machine equals brain—that is, not only write it but know that it had written it. No mechanism could feel (and not merely artificially signal, an easy contrivance) pleasure at its successes, grief when its valves fuse, be warmed by flattery, be made miserable by its mistakes, be charmed by sex, be angry or depressed when it cannot get what it wants" (Jefferson 1949).

This argument appears to be a denial of the validity of our test. According to the most extreme form of this view the only way by which one could be sure that a machine thinks is to *be* the machine and to feel oneself thinking. One could then describe these feelings to the world, but of course no one would be justified in taking any notice. Likewise according to this view the only way to know that a *man* thinks is to be that particular man. It is in fact the solipsist point of view. It may be the most logical view to hold but it makes communication of ideas difficult. A is liable to believe 'A thinks but B does not' whilst B believes 'B thinks but A does not'. Instead of arguing continually over this point it is usual to have the polite convention that everyone thinks.

I am sure that Professor Jefferson does not wish to adopt the extreme and solipsist point of view. Probably he would be quite willing to accept the imitation game as a test. The game (with the player B omitted) is frequently used in practice under the name of *viva voce* to discover whether some one really understands something or has 'learnt it parrot fashion'. Let us listen in to a part of such a *viva voce*:

INTERROGATOR: In the first line of your sonnet which reads 'Shall I compare thee to a summer's day', would not 'a spring day' do as well or better?

WITNESS: It wouldn't scan.

INTERROGATOR: How about 'a winter's day'? That would scan all right.

WITNESS: Yes, but nobody wants to be compared to a winter's day.

INTERROGATOR: Would you say Mr. Pickwick reminded you of Christmas?

WITNESS: In a way.

INTERROGATOR: Yet Christmas is a winter's day, and I do not think Mr. Pickwick would mind the comparison.

WITNESS: I don't think you're serious. By a winter's day one means a typical winter's day, rather than a special one like Christmas.

And so on. What would Professor Jefferson say if the sonnet-writing machine was able to answer like this in the *viva voce*? I do not know whether he would regard the machine as 'merely artificially signalling' these answers, but if the answers were as satisfactory and sustained as in the above passage I do not think he would describe it as 'an easy contrivance'. This phrase is, I think, intended to cover such devices as the inclusion in the machine of a record of someone reading a sonnet, with appropriate switching to turn it on from time to time.

In short then, I think that most of those who support the argument from consciousness could be persuaded to abandon it rather than be forced into the solipsist position. They will then probably be willing to accept our test.

I do not wish to give the impression that I think there is no mystery about consciousness. There is, for instance, something of a paradox connected with any attempt to localise it. But I do not think these mysteries necessarily need to be solved before we can answer the question with which we are concerned in this paper.

(5) *Arguments from Various Disabilities.* These arguments take the form, "I grant you that you can make machines do all the things you have mentioned but you will never be able to make one to do X". Numerous features X are suggested in this connexion. I offer a selection:

Be kind, resourceful, beautiful, friendly (p. 448), have initiative, have a sense of humour, tell right from wrong, make mistakes (p. 448), fall in love, enjoy strawberries and cream (p. 448), make some one fall in love with it, learn from experience (pp. 456 f.), use words properly, be the subject of its own thought (p. 449), have as much diversity of behaviour as a man, do something really new (p. 450). (Some of these disabilities are given special consideration as indicated by the page numbers.)

No support is usually offered for these statements. I believe they are mostly founded on the principle of scientific induction. A man has seen thousands of machines in his lifetime. From what he sees of them he draws a number of general conclusions. They are ugly, each is designed for a very limited purpose, when required for a minutely different purpose they are useless, the variety of behaviour of any one of them is very small, etc., etc. Naturally he concludes that these are necessary properties of machines in general. Many of these limitations are associated with the very small storage capacity of most machines. (I am assuming that the idea of storage capacity is extended in some way to cover machines other than discrete-state machines. The exact definition does not matter as no mathematical accuracy is claimed in the present discussion.) A few years ago, when very little had been heard of digital computers, it was possible to elicit much incredulity concerning them, if one mentioned their properties without describing their construction. That was presumably due to a similar application of the principle of scientific induction. These applications of the principle are of course largely unconscious. When a burnt child fears the fire and shows that he fears it by avoiding it, I should say that he was applying scientific induction. (I could of course also describe his behaviour in many other ways.) The works and customs of mankind do not seem to be very suitable material to which to apply scientific induction. A very large part of space-time must be investigated, if reliable results are to be obtained. Otherwise we may (as most English children do) decide that everybody speaks English, and that it is silly to learn French.

There are, however, special remarks to be made about many of the disabilities that have been mentioned. The inability to enjoy strawberries and cream may have struck the reader as frivolous. Possibly a machine might be made to enjoy this delicious dish, but any attempt to make one do so would be idiotic. What is important about this disability is that it contributes to some of the other disabilities, *e.g.* to the difficulty of the same kind of friendliness occurring between man and machine as between white man and white man, or between black man and black man.

Turing anticipates the issues that currently energize discussions of the dangers of AI.

The claim that "machines cannot make mistakes" seems a curious one. One is tempted to retort, "Are they any the worse for that?" But let us adopt a more sympathetic attitude, and try to see what is really meant. I think this criticism can be explained in terms of the imitation game. It is claimed that the interrogator could distinguish the machine from the man simply by setting them a number of problems in arithmetic. The machine would be unmasked because of its deadly accuracy. The reply to this is simple. The machine (programmed for playing the game) would not attempt to give the *right* answers to the arithmetic problems. It would deliberately introduce mistakes in a manner calculated to confuse the interrogator. A mechanical fault would probably show itself through an unsuitable decision as to what sort of a mistake to make in the arithmetic. Even this interpretation of the criticism is not sufficiently sympathetic. But we cannot afford the space to go into it much further. It seems to me that this criticism depends on a confusion between two kinds of mistake. We may call them 'errors of functioning' and 'errors of conclusion'. Errors of functioning are due to some mechanical or electrical fault which causes the machine to behave otherwise than it was designed to do. In philosophical discussions one likes to ignore the possibility of such errors; one is therefore discussing 'abstract machines'. These abstract machines are mathematical fictions rather than physical objects. By definition they are incapable of errors of functioning. In this sense we can truly say that 'machines can never make mistakes'. Errors of conclusion can only arise when some meaning is attached to the output signals from the machine. The machine might, for instance, type out mathematical equations, or sentences in English. When a false proposition is typed we say that the machine has committed an error of conclusion. There is clearly no reason at all for saying that a machine cannot make this kind of mistake. It might do nothing but type out repeatedly '$0 = 1$'. To take a less perverse example, it might have some method for drawing conclusions by scientific induction. We must expect such a method to lead occasionally to erroneous results.

The claim that a machine cannot be the subject of its own thought can of course only be answered if it can be shown that the machine has *some* thought with *some* subject matter. Nevertheless, 'the subject matter of a machine's operations' does seem to mean something, at least to the people who deal with it. If, for instance, the machine was trying to find a solution of the equation $x^2 - 40x - 11 = 0$ one would be tempted to describe this equation as part of the machine's subject matter at that moment. In this sort of sense a machine undoubtedly can be its own subject matter. It may be used to help in making up its own programmes, or to predict the effect of alterations in its own structure. By observing the results of its own behaviour it can modify its own programmes so as to achieve some purpose more effectively. These are possibilities of the near future, rather than Utopian dreams.

Turing once again highlights a development that has been central to research in AI.

The criticism that a machine cannot have much diversity of behaviour is just a way of saying that it cannot have much storage capacity. Until fairly recently a storage capacity of even a thousand digits was very rare.

The criticisms that we are considering here are often disguised forms of the argument from consciousness. Usually if one maintains that a machine *can* do one of these things, and describes the kind of method that the machine could use, one will not make much of an impression. It is thought that the method (whatever it may be, for it must be mechanical) is really rather base. Compare the parenthesis in Jefferson's statement quoted on p. 21.

(6) *Lady Lovelace's Objection.* Our most detailed information of Babbage's Analytical Engine comes from a memoir by *Lady Lovelace (1842).* In it she states, "The Analytical Engine has no pretensions to *originate* anything. It can do *whatever we know how to order it* to perform" (her italics). This statement is quoted by *Hartree (1949, p. 70)* who adds: "This does not imply that it may not be possible to construct electronic equipment which will 'think for itself', or in which, in biological terms, one could set up a conditioned reflex, which would serve as a basis for 'learning'. Whether this is possible in principle or not is a stimulating and exciting question, suggested by some of these recent developments. But it did not seem that the machines constructed or projected at the time had this property".

I am in thorough agreement with Hartree over this. It will be noticed that he does not assert that the machines in question had not got the property, but rather that the evidence available to Lady Lovelace did not encourage her to believe that they had it. It is quite possible that the machines in question had in a sense got this property. For suppose that some discrete-state machine has the property. The Analytical Engine was a universal digital computer, so that, if its storage capacity and speed were adequate, it could by suitable programming be made to mimic the machine in question. Probably this argument did not occur to the Countess or to Babbage. In any case there was no obligation on them to claim all that could be claimed.

This whole question will be considered again under the heading of learning machines.

A variant of Lady Lovelace's objection states that a machine can 'never do anything really new'. This may be parried for a moment with the saw, 'There is nothing new under the sun'. Who can be certain that 'original work' that he has done was not simply the growth of the seed planted in him by teaching, or the effect of following well-known general principles. A better variant of the objection says that a machine can never 'take us by surprise'. This statement is a more direct challenge and can be met directly. Machines take me by surprise with great frequency. This is largely because I do not do sufficient calculation to decide what to expect them to do, or rather because, although I do a calculation, I do it in a hurried, slipshod fashion, taking risks. Perhaps I say to myself, 'I suppose the voltage here ought to be the same as there: anyway let's assume it is.' Naturally I am often wrong, and the result is a surprise for me for by the time the experiment is done these assumptions have been forgotten. These admissions lay me open to lectures on the subject of my vicious ways, but do not throw any doubt on my credibility when I testify to the surprises I experience.

I do not expect this reply to silence my critic. He will probably say that such surprises are due to some creative mental act on my part, and reflect no credit on the machine. This leads us back to the argument from consciousness, and far from the idea of surprise. It is a line of argument we must consider closed, but it is perhaps worth remarking that the appreciation of something as surprising requires as much of a 'creative mental act' whether the surprising event originates from a man, a book, a machine or anything else.

The view that machines cannot give rise to surprises is due, I believe, to a fallacy to which philosophers and mathematicians are particularly subject. This is the assumption that as soon as a fact is presented to a mind all consequences of that fact spring into the mind simultaneously with it. It is a very useful assumption under many circumstances, but one too easily forgets that it is false. A natural consequence of doing so is that one then assumes that there is no virtue in the mere working out of consequences from data and general principles.

(7) *Argument from Continuity in the Nervous System.* The nervous system is certainly not a discrete-state machine. A small error in the information about the size of a nervous impulse impinging on a neuron, may make a large difference to the size of the outgoing impulse. It may be argued that, this being so, one cannot expect to be able to mimic the behaviour of the nervous system with a discrete-state system.

See Gallistel (forthcoming) for a contrary view.

It is true that a discrete-state machine must be different from a continuous machine. But if we adhere to the conditions of the imitation game, the interrogator will not be able to take any advantage of this difference. The situation can be made clearer if we consider some other simpler continuous machine. A differential analyser will do very well. (A differential analyser is a certain kind of machine not of the discrete-state type used for some kinds of calculation.) Some of these provide their answers in a typed form, and so are suitable for taking part in the game. It would not be possible for a digital computer to predict exactly what answers the differential analyser would give to a problem, but it would be quite capable of giving the right sort of answer. For instance, if asked to give the value of π (actually about 3.1416) it would be reasonable to choose at random between the values 3.12, 3.13, 3.14, 3.15, 3.16 with the probabilities of 0.05, 0.15, 0.55, 0.19, 0.06 (say). Under these circumstances it would be very difficult for the interrogator to distinguish the differential analyser from the digital computer.

(8) *The Argument from Informality of Behaviour.* It is not possible to produce a set of rules purporting to describe what a man should do in every conceivable set of circumstances. One might for instance have a rule that one is to stop when one sees a red traffic light, and to go if one sees a green one, but what if by some fault both appear together? One may perhaps decide that it is safest to stop. But some further difficulty may well arise from this decision later. To attempt to provide rules of conduct to cover every eventuality, even those

arising from traffic lights, appears to be impossible. With all this I agree.

From this it is argued that we cannot be machines. I shall try to reproduce the argument, but I fear I shall hardly do it justice. It seems to run something like this. 'If each man had a definite set of rules of conduct by which he regulated his life he would be no better than a machine. But there are no such rules, so men cannot be machines.' The undistributed middle is glaring. I do not think the argument is ever put quite like this, but I believe this is the argument used nevertheless. There may however be a certain confusion between 'rules of conduct' and 'laws of behaviour' to cloud the issue. By 'rules of conduct' I mean precepts such as 'Stop if you see red lights', on which one can act, and of which one can be conscious. By 'laws of behaviour' I mean laws of nature as applied to a man's body such as 'if you pinch him he will squeak'. If we substitute 'laws of behaviour which regulate his life' for 'laws of conduct by which he regulates his life' in the argument quoted the undistributed middle is no longer insuperable. For we believe that it is not only true that being regulated by laws of behaviour implies being some sort of machine (though not necessarily a discrete-state machine), but that conversely being such a machine implies being regulated by such laws. However, we cannot so easily convince ourselves of the absence of complete laws of behaviour as of complete rules of conduct. The only way we know of for finding such laws is scientific observation, and we certainly know of no circumstances under which we could say, 'We have searched enough. There are no such laws.'

We can demonstrate more forcibly that any such statement would be unjustified. For suppose we could be sure of finding such laws if they existed. Then given a discrete-state machine it should certainly be possible to discover by observation sufficent about it to predict its future behaviour, and this within a reasonable time, say a thousand years. But this does not seem to be the case. I have set up on the Manchester computer a small programme using only 1000 units of storage, whereby the machine supplied with one sixteen figure number replies with another within two seconds. I would defy anyone to learn from these replies sufficient about the programme to be able to predict any replies to untried values.

(9) *The Argument from Extra-Sensory Perception.* I assume that the reader is familiar with the idea of extra-sensory perception, and the meaning of the four items of it, *viz.* telepathy, clairvoyance, precognition and psycho-kinesis. These disturbing phenomena seem to deny all our usual scientific ideas. How we should like to discredit them! Unfortunately the statistical evidence, at least for telepathy, is overwhelming. It is very difficult to rearrange one's ideas so as to fit these new facts in. Once one has accepted them it does not seem a very big step to believe in ghosts and bogies. The idea that our bodies move simply according to the known laws of physics, together with some others not yet discovered but somewhat similar, would be one of the first to go.

This argument is to my mind quite a strong one. One can say in reply that many scientific theories seem to remain workable in practice, in spite of clashing with E.S.P.; that in fact one can get along very nicely if one forgets about it. This is rather cold comfort, and one fears that thinking is just the kind of phenomenon where E.S.P. may be especially relevant.

A more specific argument based on E.S.P. might run as follows: "Let us play the imitation game, using as witnesses a man who is good as a telepathic receiver, and a digital computer. The interrogator can ask such questions as 'What suit does the card in my right hand belong to?' The man by telepathy or clairvoyance gives the right answer 130 times out of 400 cards. The machine can only guess at random, and perhaps gets 104 right, so the interrogator makes the right identification." There is an interesting possibility which opens here. Suppose the digital computer contains a random number generator. Then it will be natural to use this to decide what answer to give. But then the random number generator will be subject to the psycho-kinetic powers of the interrogator. Perhaps this psycho-kinesis might cause the machine to guess right more often than would be expected on a probability calculation, so that the interrogator might still be unable to make the right identification. On the other hand, he might be able to guess right without any questioning, by clairvoyance. With E.S.P. anything may happen.

If telepathy is admitted it will be necessary to tighten our test up. The situation could be regarded as analogous to that which would occur if the interrogator were talking to himself and one of the competitors was listening with his ear to the wall. To put the competitors into a 'telepathy-proof room' would satisfy all requirements.

Learning Machines

The reader will have anticipated that I have no very convincing arguments of a positive nature to support my views. If I had I should not have taken such pains to point out the fallacies in contrary views. Such evidence as I have I shall now give.

Let us return for a moment to Lady Lovelace's objection, which stated that the machine can only do what we tell it to do. One could say that a man can 'inject' an idea into the machine, and that it will respond to a certain extent and then drop into quiescence, like a piano string struck by a hammer. Another simile would be an atomic pile of less than critical size: an injected idea is to correspond to a neutron entering the pile from without. Each such neutron will cause a certain disturbance which eventually dies away. If, however, the size of the pile is sufficiently increased, the disturbance caused by such an incoming neutron will very likely go on and on increasing until the whole pile is destroyed. Is there a corresponding phenomenon for minds, and is there one for machines? There does seem to be one for the human mind. The majority of them seem to be 'sub-critical', *i.e.* to correspond in this analogy to piles of sub-critical size. An idea presented to such a mind will on average give rise to less than one idea in reply. A smallish proportion are super-critical. An idea presented to such a mind may give rise to a whole 'theory' consisting of secondary, tertiary and more remote ideas. Animals minds seem to be very definitely sub-critical. Adhering to this analogy we ask, 'Can a machine be made to be super-critical?'

The 'skin of an onion' analogy is also helpful. In considering the functions of the mind or the brain we find certain operations which we can explain in purely mechanical terms. This we say does not correspond to the real mind: it is a sort of skin which we must strip off if we are to find the real mind. But then in what remains we find a further skin to be stripped off, and so on. Proceeding in this way do we ever come to the 'real' mind, or do we eventually come to the skin which has nothing in it? In the latter case the whole mind is mechanical. (It would not be a discrete-state machine however. We have discussed this.)

These last two paragraphs do not claim to be convincing arguments. They should rather be described as 'recitations tending to produce belief'.

The only really satisfactory support that can be given for the view expressed at the beginning of § 6, will be that provided by waiting for the end of the century and then doing the experiment described. But what can we say in the meantime? What steps should be taken now if the experiment is to be successful?

As I have explained, the problem is mainly one of programming. Advances in engineering will have to be made too, but it seems unlikely that these will not be adequate for the requirements. Estimates of the storage capacity of the brain vary from 10^{10} to 10^{15} binary digits. I incline to the lower values and believe that only a very small fraction is used for the higher types of thinking. Most of it is probably used for the retention of visual impressions. I should be surprised if more than 10^9 was required for satisfactory playing of the imitation game, at any rate against a blind man. (Note—The capacity of the *Encyclopaedia Britannica*, 11th edition, is 2×10^9.) A storage capacity of 10^7 would be a very practicable possibility even by present techniques. It is probably not necessary to increase the speed of operations of the machines at all. Parts of modern machines which can be regarded as analogues of nerve cells work about a thousand times faster than the latter. This should provide a 'margin of safety' which could cover losses of speed arising in many ways. Our problem then is to find out how to programme these machines to play the game. At my present rate of working I produce about a thousand digits of programme a day, so that about sixty workers, working steadily through the fifty years might accomplish the job, if nothing went into the waste-paper basket. Some more expeditious method seems desirable.

Turing can perhaps be excused for not imagining the huge recursive explosion of techniques for accelerating programming, but his suggestion that a more expeditious method would be desirable introduces the important idea that getting the machine to learn is much more efficient than hand-crafting its programs.

In the process of trying to imitate an adult human mind we are bound to think a good deal about the process which has brought it to the state that it is in. We may notice three components,

(a) The initial state of the mind, say at birth,

(b) The education to which it has been subjected,

(c) Other experience, not to be described as education, to which it has been subjected.

Instead of trying to produce a programme to simulate the adult mind, why not rather try to produce one which simulates the child's? If this were then subjected to an appropriate course of education one would obtain the adult brain. Presumably the child-brain is something like a note-book as one buys it from the stationers. Rather little mechanism, and lots of blank sheets. (Mechanism and writing are from our point of view almost synonymous.) Our hope is that there is so little mechanism in the child-brain that something like it can be easily programmed. The amount of work in the education we can assume, as a first approximation, to be much the same as for the human child.

We have thus divided our problem into two parts. The child-programme and the education process. These two remain very closely connected. We cannot expect to find a good child-machine at the first attempt. One must experiment with teaching one such machine and see how well it learns. One can then try another and see if it is better or worse. There is an obvious connection between this process and evolution, by the identifications

Structure of the child machine = Hereditary material

Changes of the child machine = Mutations

Natural selection = Judgment of the experimenter

One may hope, however, that this process will be more expeditious than evolution. The survival of the fittest is a slow method for measuring advantages. The experimenter, by the exercise of intelligence, should be able to speed it up. Equally important is the fact that he is not restricted to random mutations. If he can trace a cause for some weakness he can probably think of the kind of mutation which will improve it.

It will not be possible to apply exactly the same teaching process to the machine as to a normal child. It will not, for instance, be provided with legs, so that it could not be asked to go out and fill the coal scuttle. Possibly it might not have eyes. But however well these deficiencies might be overcome by clever engineering, one could not send the creature to school without the other children making excessive fun of it. It must be given some tuition. We need not be too concerned about the legs, eyes, etc. The example of Miss *Helen Keller* shows that education can take place provided that communication in both directions between teacher and pupil can take place by some means or other.

We normally associate punishments and rewards with the teaching process. Some simple child-machines can be constructed or programmed on this sort of principle. The machine has to be so constructed that events which shortly preceded the occurrence of a punishment-signal are unlikely to be repeated, whereas a reward-signal increased the probability of repetition of the events which led up to it. These definitions do not presuppose any feelings on the part of the machine. I have done some experiments with one such child-machine, and succeeded in teaching it a few things, but the teaching method was too unorthodox for the experiment to be considered really successful.

The use of punishments and rewards can at best be a part of the teaching process. Roughly speaking, if the teacher has no other means of communicating to the pupil, the amount of information which can reach him does not exceed the total number of rewards and punishments applied. By the time a child has learnt to repeat 'Casablanca' he would probably feel very sore indeed, if the text could only be discovered by a 'Twenty Questions' technique, every 'NO' taking the form of a blow. It is necessary therefore to have some other 'unemotional' channels of communication. If these are available it is possible to teach a machine by punishments and rewards to obey orders given in some language, *e.g.* a symbolic language. These orders are to be transmitted through the 'unemotional' channels. The use of this language will diminish greatly the number of punishments and rewards required.

Opinions may vary as to the complexity which is suitable in the child machine. One might try to make it as simple as possible consistently with the general principles. Alternatively one might have a complete system of logical inference 'built in'.[1] In the latter case the store would be largely occupied with definitions and propositions. The propositions would have various kinds of status, *e.g.* well-established facts, conjectures, mathematically proved theorems, statements given by an authority, expressions having the logical form of proposition but not belief-value. Certain propositions may be described as 'imperatives'. The machine should be so constructed that as soon as an imperative is classed as 'well-established' the appropriate action automatically takes place. To illustrate this, suppose the teacher says to the machine, 'Do your homework now'. This may cause "Teacher says 'Do your homework now'" to be included amongst the well-established facts. Another such fact might be, "Everything that teacher says is true". Combining these may eventually lead to the imperative, 'Do your homework now', being included amongst the well-established facts, and this, by the construction of the machine, will mean that the homework actually gets started, but the effect is very satisfactory. The processes of inference used by the machine need not be such as would satisfy the most exacting logicians. There might for instance be no hierarchy of types. But this need not mean that type fallacies will occur, any more than we are bound to fall over unfenced cliffs. Suitable imperatives (expressed *within* the systems, not forming part of the rules *of* the system) such as 'Do not use a class unless it is a subclass of one which has been mentioned by teacher' can have a similar effect to 'Do not go too near the edge'.

A clear anticipation of the 'symbolic' or GOFAI (Good Old Fashioned AI, Haugeland 1985) school of AI, championed by John McCarthy and Allan Newell and Herbert Simon.

The imperatives that can be obeyed by a machine that has no limbs are bound to be of a rather intellectual character, as in the example (doing homework) given above. Important amongst such imperatives will be ones which regulate the order in which the rules of the logical system concerned are to be applied. For at each stage when one is using a logical system, there is a very large number of alternative steps, any of which one is permitted to apply, so far as obedience to the rules of the logical system

[1] Or rather 'programmed in' for our child-machine will be programmed in a digital computer. But the logical system will not have to be learnt.

is concerned. These choices make the difference between a brilliant and a footling reasoner, not the difference between a sound and a fallacious one. Propositions leading to imperatives of this kind might be "When Socrates is mentioned, use the syllogism in Barbara" or "If one method has been proved to be quicker than another, do not use the slower method". Some of these may be 'given by authority', but others may be produced by the machine itself, *e.g.* by scientific induction.

The idea of a learning machine may appear paradoxical to some readers. How can the rules of operation of the machine change? They should describe completely how the machine will react whatever its history might be, whatever changes it might undergo. The rules are thus quite time-invariant. This is quite true. The explanation of the paradox is that the rules which get changed in the learning process are of a rather less pretentious kind, claiming only an ephemeral validity. The reader may draw a parallel with the Constitution of the United States.

An important feature of a learning machine is that its teacher will often be very largely ignorant of quite what is going on inside, although he may still be able to some extent to predict his pupil's behaviour. This should apply most strongly to the later education of a machine arising from a child-machine of well-tried design (or programme). This is in clear contrast with normal procedure when using a machine to do computations: one's object is then to have a clear mental picture of the state of the machine at each moment in the computation. This object can only be achieved with a struggle. The view that 'the machine can only do what we know how to order it to do',[1] appears strange in face of this. Most of the programmes which we can put into the machine will result in its doing something that we cannot make sense of at all, or which we regard as completely random behaviour. Intelligent behaviour presumably consists in a departure from the completely disciplined behaviour involved in computation, but a rather slight one, which does not give rise to random behaviour, or to pointless repetitive loops. Another important result of preparing our machine for its part in the imitation game by a process of teaching and learning is that 'human fallibility' is likely to be omitted in a rather natural way, *i.e.* without special 'coaching'. (The reader should reconcile this with the point of view

[1]Compare Lady Lovelace's statement (p. 450), which does not contain the word 'only'.

on pp. 24, 25.) Processes that are learnt do not produce a hundred percent. certainty of result; if they did they could not be unlearnt.

It is probably wise to include a random element in a learning machine (see p. 438). A random element is rather useful when we are searching for a solution of some problem. Suppose for instance we wanted to find a number between 50 and 200 which was equal to the square of the sum of its digits, we might start at 51 then try 52 and go on until we got a number that worked. Alternatively we might choose numbers at random until we got a good one. This method has the advantage that it is unnecessary to keep track of the values that have been tried, but the disadvantage that one may try the same one twice, but this is not very important if there are several solutions. The systematic method has the disadvantage that there may be an enormous block without any solutions in the region which has to be investigated first. Now the learning process may be regarded as a search for a form of behaviour which will satisfy the teacher (or some other criterion). Since there is probably a very large number of satisfactory solutions the random method seems to be better than the systematic. It should be noticed that it is used in the analogous process of evolution. But there the systematic method is not possible. How could one keep track of the different genetical combinations that had been tried, so as to avoid trying them again?

We may hope that machines will eventually compete with men in all purely intellectual fields. But which are the best ones to start with? Even this is a difficult decision. Many people think that a very abstract activity, like the playing of chess, would be best. It can also be maintained that it is best to provide the machine with the best sense organs that money can buy, and then teach it to understand and speak English. This process could follow the normal teaching of a child. Things would be pointed out and named, etc. Again I do not know what the right answer is, but I think both approaches should be tried.

We can only see a short distance ahead, but we can see plenty there that needs to be done. ❧

REFERENCES

Butler, Samuel. 1865. *Erewhon.* Chapters 23, 24, 25, The Book of the Machines. *Publisher's note: Butler first published these chapters under a pseudonym, Cellarius, in the recently launched Christchurch, New Zealand newspaper* The Press. Erewhon *was published in 1872.* London, UK.

Church, Alonzo. 1936. "An Unsolvable Problem of Elementary Number Theory." *American Journal of Mathematics* 58 (2): 345–363.

Gödel, Kurt. 1931. "Über formal unentscheidbare Sätze der Principia Mathematica und verwandter Systeme, I." *Monatshefle für Mathematik und Physik,* 173–189.

Hartree, D. R. 1949. *Calculating Instruments and Machines.* Urbana, IL: University of Illinois Press.

Jefferson, G. 1949. "The Mind of Mechanical Man." Lister Oration for 1949, *British Medical Journal* 1:1105–1121.

Kleene, S. C. 1935. "General Recursive Functions of Natural Numbers." *American Journal of Mathematics* 57:153–173, 219–244.

Lovelace, Countess of. 1842. "Translator's Notes to an Article on Babbage's Analytical Engine." In *Scientific Memoirs,* edited by R. Taylor, 3:691–731.

Russell, Bertrand. 1940. *History of Western Philosophy. Publisher's note: This volume was published in 1945.* London, UK.

Turing, A. M. 1937. "On Computable Numbers, with an Application to the Entscheidunsproblem." *Proceedings of the London Mathematical Society* 2 (42): 230–265.

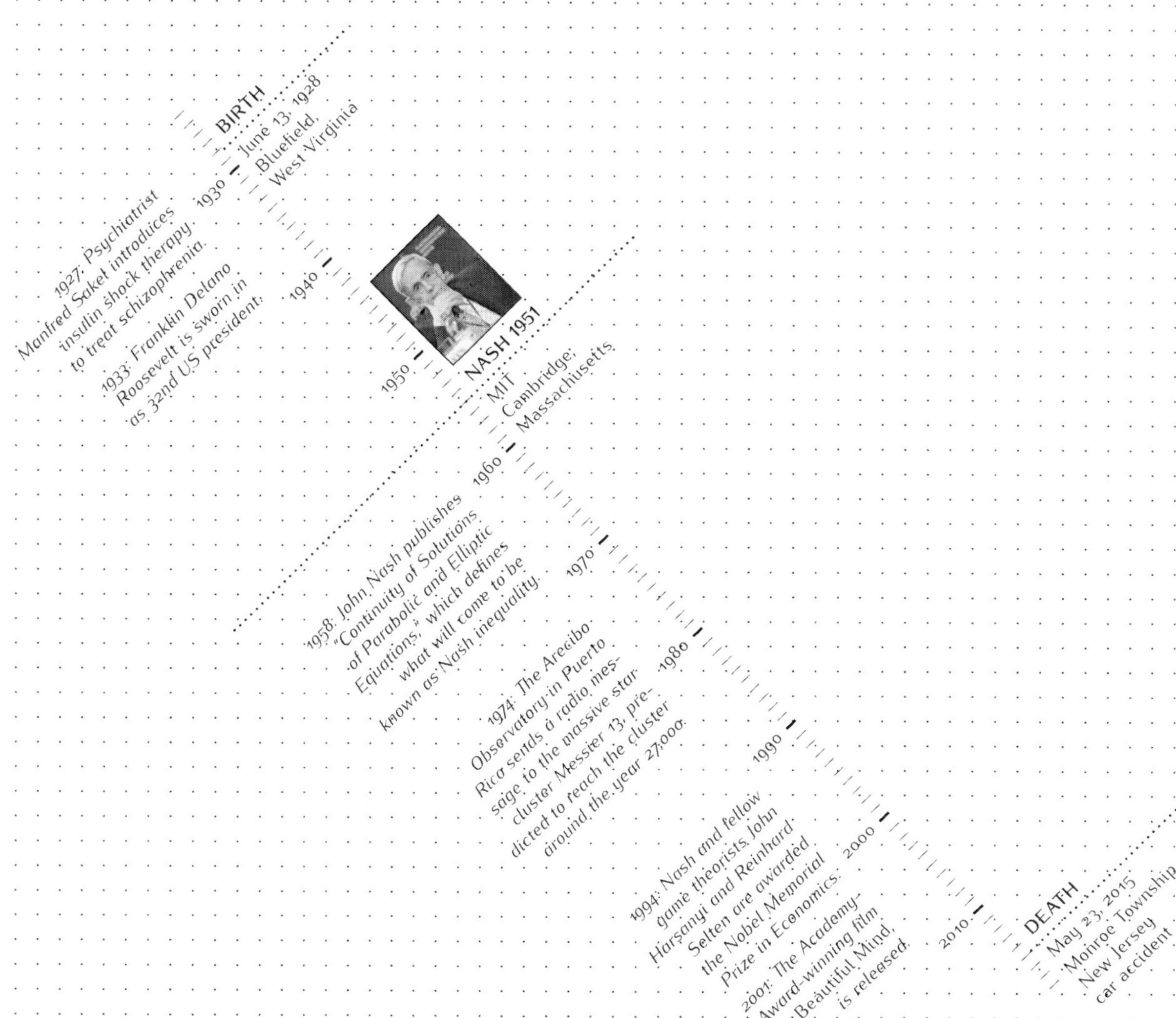

JOHN FORBES NASH, JR.

[12]

THE NASH EQUILIBRIUM

Robert L. Axtell, George Mason University and Santa Fe Institute

J. Nash, "Non-Cooperative Games," *Annals of Mathematics* (Second Series) 54 (2), 286–295 (1951).

Game theory is a cornerstone of the social sciences today. In its original formulation by John von Neumann and Oskar Morgenstern (1944), a variety of concepts and procedures for solving games was proposed, depending on the nature of the game being played. For example, minimax solutions were advanced for two-person, zero-sum games, that is, games in which what is won by one player is lost by the other. What John Nash's foundational 1951 paper did, along with the shorter one from 1950, was to provide an alternative solution concept that was more generally applicable, to zero- and non-zero–sum games alike, to games with many players, and so on. This is the idea of an equilibrium point of a game, in which the strategies of players are configured such that no one can be made better off by unilaterally deviating to a different strategy. Nash proved that such equilibria always exist for any game with a finite number of strategies. The Nash equilibrium, as it is today known, has become a focal point for solving essentially all games, and is fundamental to the practice of game theory.

Specifically, consider two players engaged in a game with a finite set of strategies and fixed and known possible payoffs that depend on the strategies employed by the players. A single play of the game amounts to each player selecting a strategy and receiving the corresponding payoff. Each individual strategy is considered "pure" while a mixed strategy involves a set of probabilities for playing each of the pure strategies. For example, consider the children's game of rock–paper–scissors, played by two players: the rock strategy loses to paper but defeats scissors, while the paper strategy loses to scissors. It is a zero-sum game in which the mixed strategy of playing each of the three pure strategies one-third of the time is intuitively obvious, even to kids. A two-strategy version is "matching pennies"—play "heads" half the time, "tails" the other half.

Nash's 1950 and 1951 papers begin by defining an equilibrium for games as strategic configurations such that, once established, no player has any incentive to unilaterally deviate from the strategy it has chosen, that is, there do not exist other strategies that would make any individual player better off if pursued on its own. The power of this equilibrium notion is clear: for players arranged in equilibrium configurations, none has any incentive to deviate—to try something different, to experiment—for to do so means reduced payoffs. Such strategic configurations are "self-enforcing" (Harrington, Jr. 1987). But this concept of equilibrium would be of little interest if it failed to exist either in common games or in general. The remarkable think about the Nash (1950) paper in the *Proceedings of the National Academy of Sciences* is that it proves that such equilibria, in mixed strategies, always exist for any finite game, using the so-called Kakutani (1941) fixed-point theorem. This short, two-page paper left out many details, which were addressed in a paper that appeared the following year (Nash 1951), and which proved the same general result on the existence of equilibrium using the Brouwer (1911) fixed-point theorem, which the Kakutani theorem generalizes. Armed with the fact that every finite game possesses at least one Nash equilibrium, application of game theoretic modeling flowered in a wide variety of fields. Initially the province of mathematicians, game theory soon was being used in management science (Shubik 1955), national security (Schelling 1960), biology (Lewontin 1961; Maynard Smith and Price 1973) , political economy (Shubik 1984), and beyond.

Nash's essential result on the existence of equilibrium has been applied and extended and confronted by reality in the nearly seventy-five years since its first appearance. It is a purely mathematical result, so in order for it to be a salient concept in the social sciences, people need to be able to arrive at such configurations in strategic contexts. Additionally, today we can imagine strategic environments in which machines are engaged in direct interactions, and in order for Nash equilibrium ideas to be relevant they have to involve calculations that can be accomplished, at least approximately, in a reasonable amount of time, using bounded resources. In the remainder of this commentary I address the plausibility of Nash equilibrium as a focus of game-theoretic solutions from behavioral, conceptual, computational, and related perspectives.

Back when Nash proposed the equilibrium concept that now bears his name, economists did not do experiments with human subjects. But a decade later the field of experimental economics was born in the work of Vernon Smith (1962), with the allied area of behavioral economics soon to follow. While many of the basic results of game theory have been confirmed in laboratory settings, the empirical relevance of mixed strategy Nash equilibria has been called into question on the basis of lab results demonstrating that human subjects are often unable to arrive at such equilibria (e.g., Erev and Roth 1998). However, other work has shown that professional game players often use strategies that are quite close to mixed strategy Nash. Indeed, Walker and Wooders (2001) demonstrated that professional tennis players at Wimbledon vary their serves in a way that closely resembles mixed-strategy Nash equilibria, although actual serves tends to be less serially correlated than they should be.

Many games have multiple Nash equilibria. Indeed, in the 1970s it was shown that the number of equilibria are, in general, odd (Wilson 1971; Harsanyi 1973). This led naturally to questions concerning which equilibria are "preferred" in some sense or another. A large literature on "refinement" of Nash equilibria grew up subsequently. Concepts such as subgame perfect and trembling hand equilibria are special cases of Nash equilibria and were developed primarily along logical-deductive lines as ways to reduce the number of equilibria that required consideration. Such refinements were not developed on the basis of behavioral experiments or realism and some of the reasoning principles necessary to arrive at them, for example, backward induction, are questionable as descriptions of human behavior (Camerer 1997).

Herbert Scarf (1973) showed how to compute fixed-point equilibria via operations research techniques, specifically simplicial search methods based on Sperner's lemma (Sperner 1928), a result from graph theory. These approaches involved mathematical programming and specific algorithms, for example, homotopy methods to approximate fixed points. While mostly used for computable general equilibrium (CGE) models in economics, such methods can also be used to compute

Nash equilibria. Computing equilibria this way was typically expensive when either the number of commodities for CGEs or the strategy spaces in games were large. Papadimitriou showed that simplicial search problems could be reduced to questions of graph parity, with the size of the graph needed to be searched growing rapidly in the size of either the commodity or strategy space. He described new complexity classes for the polynomial parity argument on undirected (PPA) and directed (PPAD) graphs, having complexity between P and NP (Papadimitriou 1994); for an introduction to these complexity classes see Moore and Mertens (2011). It was later shown by Daskalakis, Goldberg, and Papadimitriou (2009) that the computational complexity of computing Nash equilibria is in PPAD, making such problems hard to solve in general. A recent review of such results is Papadimitriou (2015).

The origin of the computational difficulties lies in the proofs of existence of fixed points. It was von Neumann (1937) von Neumann who introduced such devices into economics, albeit in German, translated into English only after World War II (von Neumann 1945). Brouwer's theorem (1912) proves that all continuous functions from a nonempty compact convex subset of Euclidean space to itself have fixed points. Kakutani heard von Neumann's lectures on the Brouwer theorem at Princeton and extended the result to correspondences and it was the Kakutani (1941) theorem that Nash first employed in the 1950 paper, as mentioned above. The proof of the Brouwer theorem used in the Nash 1951 paper is famously non-constructive, showing indirectly that a fixed point exists without providing an algorithm for producing such a point. Eventually Brouwer came to disavow such methods in favor of constructive ones. Overall, the fact that Nash equilibria always exist is rightfully regarded as a great strength of the solution concept, while the fact that it can be difficult to compute is the other side of the coin.

Finally, are there other reasons to doubt the relevance of Nash equilibria, beyond the behavioral, conceptual, and computational ones described above—are there other chinks in the Nash armor, as it were? If the social world were configured as one enormous fixed point, nothing would ever change. But in reality things do change—babies are born, people die, technological innovation happens, economies go up and down. There are a wide variety of real-world circumstances in which Nash equilibrium ideas apply, for a time, perhaps locally—from auctions to matching markets to tragedies of the commons. But then things change: payoffs evolve, strategy spaces expand or contract, the people you are playing against adapt. Life may be a series of Nash-like equilibria, but eventually all such configurations are transformed. People move on. And when viewed over time it may be that other ways of looking at social configurations are useful, the fluxes of birth and death, the progress of new technologies, the tumult of politics.

John Nash provided us with an idea that has served as the focal point for game theory over nearly seventy-five years. He is rightly praised for this contribution (e.g., Holt and Roth 2004), even as its limitations have become apparent Doubtless the Nash equilibrium will continue as a touchstone for theorizing about strategic behavior in the social sciences and beyond for years to come.

REFERENCES

Brouwer, L. E. J. 1911. "Beweis der invarianz des n-dimensionalen gebiets." *Mathematische Annalen* 71 (3): 305–313. https://doi.org/10.1007/BF01456846.

Camerer, C. F. 1997. "Progress in Behavioral Game Theory." *Journal of Economic Perspectives* 11 (4): 167–188. https://doi.org/10.1257/jep.11.4.167.

Daskalakis, C., P. W. Goldberg, and C. H. Papadimitriou. 2009. "The Complexity of Computing a Nash Equilibria." *SIAM Journal on Computing* 39 (1): 195–259. https://doi.org/10.1137/070699652.

Erev, I., and A. E. Roth. 1998. "Predicting How People Play Games: Reinforcement Learning in Experimental Games with Unique, Mixed Strategy Equilibria." *American Economic Review* 88 (4): 848–881.

Harrington, Jr., J. E. 1987. "Non-Cooperative Games." In *The New Palgrave: A Dictionary of Economics,* edited by J. Eatwell, M. Milgate, and P. Newman. New York, NY: W. W. Norton.

Harsanyi, J. C. 1973. "Oddness of the Number of Equilibrium Points: A New Proof." *International Journal of Game Theory* 2:235–250. https://doi.org/10.1007/BF01737572.

Holt, C. A., and A. E. Roth. 2004. "The Nash Equilibrium: A Perspective." *Proceedings of the National Academy of Sciences* 101 (12): 3999–4002. https://doi.org/10.1073/pnas.0308738101.

Kakutani, S. 1941. "A Generalization of Brouwer's Fixed Point Theorem." *Duke Mathematical Journal* 8 (3): 457–459. https://doi.org/10.1215/S0012-7094-41-00838-4.

Lewontin, R. C. 1961. "Evolution and the Theory of Games." *Journal of Theoretical Biology* 1 (3): 382–403. https://doi.org/10.1016/0022-5193(61)90038-8.

Maynard Smith, J., and G. R. Price. 1973. "The Logic of Animal Conflict." *Nature* 246:15–18. https://doi.org/10.1038/246015a0.

Moore, C., and S. Mertens. 2011. *The Nature of Computation.* Oxford, UK: Oxford University Press.

Nash, J. 1950. "Equilibrium Points in n-Person Games." *Proceedings of the National Academy of Sciences* 36 (1): 48–49. https://doi.org/10.1073/pnas.36.1.48.

———. 1951. "Non-Cooperative Games." *Annals of Mathematics* 54 (2): 286–295. https://doi.org/10.2307/1969529.

Papadimitriou, C. H. 1994. "On the Complexity of the Parity Argument and Other Inefficient Proofs of Existence." *Journal of Computer and Systems Sciences* 48 (3): 498–532. https://doi.org/10.1016/S0022-0000(05)80063-7.

———. 2015. "The Complexity of Computing Equilibria." In *Handbook of Game Theory with Economic Applications,* edited by H. P. Young and S. Zamir, 4:779–810. Oxford, UK: Elsevier. https://doi.org/10.1016/B978-0-444-53766-9.00014-8.

Scarf, H. 1973. *The Computation of Economic Equilibria.* New Haven, CT: Yale University Press.

Schelling, T. C. 1960. *The Strategy of Conflict.* Cambridge, MA: Harvard University Press.

Shubik, M. 1955. "The Uses of Game Theory in Management Science." *Management Science* 2 (1): 40–54. https://doi.org/10.1287/mnsc.2.1.40.

———. 1984. *Game Theory in the Social Sciences: A Game-Theoretic Approach to Political Economy.* Cambridge, MA: MIT Press.

Smith, V. L. 1962. "An Experimental Study of Competitive Market Behavior." *Journal of Political Economy* 70 (2): 111–137. https://doi.org/10.1086/258609.

Sperner, E. 1928. *Neuer Beweis für die Invarianz der Dimensionszahl und des Gebietes.* 6:265–272. https://doi.org/10.1007/BF02940617.

von Neumann, J. 1937. *Über ein Ökonomisches Gleichungssystem und eine Verallgemeinerung des Brouwerschen Fixpunktsatzes.* Edited by K. Menger. 8:73–83.

———. 1945. *A Model of General Economic Equilibrium.* 13:1–9. 33. https://doi.org/10.2307/2296111.

von Neumann, J., and O. Morgenstern. 1944. *Theory of Games and Economic Behavior.* Princeton, NJ: Princeton University Press.

Walker, M., and J. Wooders. 2001. "Minimax Play at Wimbledon." *American Economic Review* 91 (5): 1521–1538. https://doi.org/10.1257/aer.91.5.1521.

Wilson, R. 1971. "Computing Equilibria in N-Person Games." *SIAM Journal of Applied Mathematics* 21 (1): 80–87. https://doi.org/10.1137/0121011.

NON-COOPERATIVE GAMES

John Nash, Princeton University

Introduction

Von Neumann and Morgenstern have developed a very fruitful theory of two-person zero-sum games in their book *Theory of Games and Economic Behavior*. This book also contains a theory of n-person games of a type which we would call cooperative. This theory is based on an analysis of the interrelationships of the various coalitions which can be formed by the players of the game.

Our theory, in contradistinction, is based on the *absence* of coalitions in that it is assumed that each participant acts independently, without collaboration or communication with any of the others.

Here the idea of an equilibrium of a game is described qualitatively.

The notion of an *equilibrium point* is the basic ingredient in our theory. This notion yields a generalization of the concept of the solution of a two-person zero-sum game. It turns out that the set of equilibrium points of a two-person zero-sum game is simply the set of all pairs of opposing "good strategies."

In the immediately following sections we shall define equilibrium points and prove that a finite non-cooperative game always has at least one equilibrium point. We shall also introduce the notions of solvability and strong solvability of a non-cooperative game and prove a theorem on the geometrical structure of the set of equilibrium points of a solvable game.

As an example of the application of our theory we include a solution of a simplified three person poker game.

Formal Definitions and Terminology

In this section we define the basic concepts of this paper and set up standard terminology and notation. Important definitions will be preceded by a subtitle indicating the concept defined. The non-cooperative idea will be implicit, rather than explicit, below.

Finite Game:

For us an *n-person* game will be a set of n *players*, or *positions*, each with an associated finite set of *pure strategies*; and corresponding to each player, i, a *payoff function*, p_i, which maps the set of all n-tuples of pure strategies into the real numbers. When we use the term *n-tuple* we shall always mean a set of n items, with each item associated with a different player.

Mixed Strategy, s_i:

Definition of mixed strategy here.

A *mixed strategy* of player i will be a collection of non-negative numbers which have unit sum and are in one to one correspondence with his pure strategies.

We write $s_i = \sum_\alpha c_{i\alpha}\pi_{i\alpha}$ with $c_{i\alpha} \geqq 0$ and $\sum_\alpha c_{i\alpha} = 1$ to represent such a mixed strategy, where the $\pi_{i\alpha}$'s are the pure strategies of player i. We regard the s_i's as points in a simplex whose vertices are the $\pi_{i\alpha}$'s. This simplex may be regarded as a convex subset of a real vector space, giving us a natural process of linear combination for the mixed strategies.

We shall use the suffixes i, j, k for players and α, β, γ to indicate various pure strategies of a player. The symbols s_i, t_i, and r_i, etc. will indicate mixed strategies; $\pi_{i\alpha}$ will indicate the ith player's α^{th} pure strategy, etc.

Payoff function, p_i:

The payoff function, p_i, used in the definition of a finite game above, has a unique extension to the n-tuples of mixed strategies which is linear in the mixed strategy of each player [n-linear]. This extension we shall also denote by p_i, writing $p_i\,(s_1, s_2, \cdots, s_n)$.

We shall write $\mathfrak{s}$ or $\mathfrak{t}$ to denote an n-tuple of mixed strategies and if $\mathfrak{s} = (s_1, s_2, \cdots, s_n)$ then $p_i(\mathfrak{s})$ shall mean $p_i\,(s_1, s_2, \cdots, s_n)$. Such an n-tuple, $\mathfrak{s}$, will also be regarded as a point in a vector space, the product space of the vector spaces containing the mixed strategies. And the set of all such n-tuples forms, of course, a convex polytope, the product of the simplices representing the mixed strategies.

For convenience we introduce the substitution notation $(\mathfrak{s}; t_i)$ to stand for $(s_1, s_2, \cdots, s_{i-1}, t_i, s_{i+1}, \cdots, s_n)$ where $\mathfrak{s} = (s_1, s_2, \cdots, s_n)$. The effect of successive substitutions $((\mathfrak{s}; t_i); r_j)$ we indicate by $(\mathfrak{s}; t_i; r_j)$, etc.

Equilibrium Point:
An n-tuple $\mathfrak{s}$ is an *equilibrium point* if and only if for every i

The formal definition of the equilibrium of a game is here.

$$p_i(\mathfrak{s}) = \max_{\text{all } r_i\text{'s}} [p_i(\mathfrak{s}; r_i)] . \tag{1}$$

Thus an equilibrium point is an n-tuple $\mathfrak{s}$ such that each player's mixed strategy maximizes his payoff if the strategies of the others are held fixed. Thus each player's strategy is optimal against those of the others. We shall occasionally abbreviate equilibrium point by eq. pt.

We say that a mixed strategy s_i *uses* a pure strategy $\pi_{i\alpha}$ if $s_i = \sum_\beta c_{i\beta}\pi_{i\beta}$ and $c_{i\alpha} > 0$. If $\mathfrak{s} = (s_1, s_2, \cdots, s_n)$ and s_i uses $\pi_{i\alpha}$ we also say that $\mathfrak{s}$ uses $\pi_{i\alpha}$.

From the linearity of $p_i(s_1, \cdots, s_n)$ in s_i,

$$\max_{\text{all } r_i\text{'s}} [p_i(\mathfrak{s}; r_i)] = \max_\alpha [p_i(\mathfrak{s}; \pi_{i\alpha})] . \tag{2}$$

This is a key idea.

We define $p_{i\alpha}(\mathfrak{s}) = p_i(\mathfrak{s}; \pi_{i\alpha})$. Then we obtain the following trivial necessary and sufficient condition for $\mathfrak{s}$ to be an equilibrium point:

$$p_i(\mathfrak{s}) = \max_\alpha p_{i\alpha}(\mathfrak{s}). \tag{3}$$

If $\mathfrak{s} = (s_1, s_2, \cdots, s_n)$ and $s_i = \sum_\alpha c_{i\alpha}\pi_{i\alpha}$ then $p_i(\mathfrak{s}) = \sum_\alpha c_{i\alpha}p_{i\alpha}(\mathfrak{s})$, consequently for (3) to hold we must have $c_{i\alpha} = 0$ whenever $p_{i\alpha}(\mathfrak{s}) < \max_\beta p_{i\beta}(\mathfrak{s})$, which is to say that $\mathfrak{s}$ does not use $\pi_{i\alpha}$ unless it is an optimal pure strategy for player i. So we write

$$\text{if } \pi_{i\alpha} \text{ is used in } \mathfrak{s} \text{ then } p_{i\alpha}(\mathfrak{s}) = \max_\beta p_{i\beta}(\mathfrak{s}) \tag{4}$$

as another necessary and sufficient condition for an equilibrium point.

Since a criterion (3) for an eq. pt. can be expressed by the equating of n pairs of continuous functions on the space of n-tuples $\mathfrak{s}$ the eq. pts. obviously form a closed subset of this space. Actually, this subset is formed from a number of pieces of algebraic varieties, cut out by other algebraic varieties.

Existence of Equilibrium Points

A proof of this existence theorem based on Kakutani's generalized fixed point theorem was published in Proc. Nat. Acad. Sci. U.S. A., 36, pp. 48-49. The proof given here is a considerable improvement over that earlier version and is based directly on the Brouwer theorem. We proceed by constructing a continuous transformation T of the space of n-tuples such that the fixed points of T are the equilibrium points of the game.

THEOREM 1. *Every finite game has an equilibrium point.*

PROOF. Let $\mathfrak{s}$ be an n-tuple of mixed strategies, $p_i(\mathfrak{s})$ the corresponding pay-off to player i, and $p_{i\alpha}(\mathfrak{s})$ the pay-off to player i if he changes to his α^{th} pure strategy $\pi_{i\alpha}$ and the others continue to use their respective mixed strategies from $\mathfrak{s}$. We now define a set of continuous functions of $\mathfrak{s}$ by

$$\varphi_{i\alpha}(\mathfrak{s}) = \max(0, p_{i\alpha}(\mathfrak{s}) - p_i(\mathfrak{s}))$$

and for each component s_i of $\mathfrak{s}$ we define a modification s_i' by

This is the transformation that has a fixed point.

$$s_i' = \frac{s_i + \sum_\alpha \varphi_{i\alpha}(\mathfrak{s})\pi_{i\alpha}'}{1 + \sum_\alpha \varphi_{i\alpha}(\mathfrak{s})},$$

calling $\mathfrak{s}'$ the n-tuple $(s_1', s_2', s_3' \cdots s_n')$.

We must now show that the fixed points of the mapping $T : \mathfrak{s} \to \mathfrak{s}'$ are the equilibrium points.

First consider any n-tuple $\mathfrak{s}$. In $\mathfrak{s}$ the ith player's mixed strategy s_i will use certain of his pure strategies. Some one of these strategies, say $\pi_{i\alpha}$, must be "least profitable" so that $p_{i\alpha}(\mathfrak{s}) \leqq p_i(\mathfrak{s})$. This will make $\varphi_{i\alpha}(\mathfrak{s}) = 0$.

Now if this n-tuple $\mathfrak{s}$ happens to be fixed under T the proportion of $\pi_{i\alpha}$ used in s_i must not be decreased by T. Hence, for all β's, $\varphi_{i\beta}(\mathfrak{s})$ must be zero to prevent the denominator of the expression defining s_i' from exceeding 1.

Thus, if $\mathfrak{s}$ is fixed under T, for any i and $\beta \varphi_{i\beta}(\mathfrak{s}) = 0$. This means no player can improve his pay-off by moving to a pure strategy $\pi_{i\beta}$. But this is just a criterion for an eq. pt. [see (2)].

Conversely, if $\mathfrak{s}$ is an eq. pt. it is immediate that all φ's vanish, making $\mathfrak{s}$ a fixed point under T.

Since the space of n-tuples is a cell the Brouwer fixed point theorem requires that T must have at least one fixed point $\mathfrak{s}$, which must be an equilibrium point.

Symmetries of Games

An *automorphism*, or *symmetry*, of a game will be a permutation of its pure strategies which satisfies certain conditions, given below.

If two strategies belong to a single player they must go into two strategies belonging to a single player. Thus if ϕ is the permutation of the pure strategies it induces a permutation ψ of the players.

Each n-tuple of pure strategies is therefore permuted into another n-tuple of pure strategies. We may call χ the induced permutation of these n-tuples. Let ξ denote an n-tuple of pure strategies and $p_i(\xi)$ the payoff to player i when the n-tuple ξ is employed. We require that if

$$j = i^{\psi} \quad \text{then } p_j\left(\xi^{\chi}\right) = p_i(\xi)$$

which completes the definition of a symmetry.

The permutation ϕ has a unique linear extension to the mixed strategies. If

$$s_i = \sum_{\alpha} c_{i\alpha}\pi_{i\alpha} \quad \text{we define} \quad (s_i)^{\phi} = \sum_{\alpha} c_{i\alpha}\left(\pi_{i\alpha}\right)^{\phi}.$$

The extension of ϕ to the mixed strategies clearly generates an extension of χ to the n-tuples of mixed strategies. We shall also denote this by χ.

We define a *symmetric n-tuple* $\mathfrak{s}$ of a game by $\mathfrak{s}^{\chi} = \mathfrak{s}$ for all χ's.

THEOREM 2. *Any finite game has a symmetric equilibrium point.*

PROOF. First we note that $s_{i0} = \sum_{\alpha}\pi_{i\alpha}/\sum_{\alpha}1$ has the property $(s_{i0})^{\phi} = s_{j0}$ where $j = i^{\psi}$, so that the n-tuple $\mathfrak{s}_0 = (s_{10}, s_{20}, \cdots, s_{n0})$ is fixed under any χ; hence any game has at least one symmetric n-tuple.

If $\mathfrak{s} = (s_1, \cdots, s_n)$ and $\mathfrak{t} = (t_1, \cdots, t_n)$ are symmetric then

$$\frac{\mathfrak{s}+\mathfrak{t}}{2} = \left(\frac{s_1+t_1}{2}, \frac{s_2+t_2}{2}, \cdots, \frac{s_n+t_n}{2}\right)$$

is also symmetric because $\mathfrak{s}^{\chi} = \mathfrak{s} \leftrightarrow s_j = (s_i)^{\phi}$, where $j = i^{\psi}$, hence

$$\frac{s_j+t_j}{2} = \frac{(s_i)^{\phi}+(t_i)^{\phi}}{2} = \left(\frac{s_i+t_i}{2}\right)^{\phi},$$

hence

$$\left(\frac{\mathfrak{s}+\mathfrak{t}}{2}\right)^{\chi} = \frac{\mathfrak{s}+\mathfrak{t}}{2}.$$

This shows that the set of symmetric n-tuples is a convex subset of the space of n-tuples since it is obviously closed.

Now observe that the mapping $T : \mathfrak{s} \to \mathfrak{s}'$ used in the proof of the existence theorem was intrinsically defined. Therefore, if $\mathfrak{s}_2 = T\mathfrak{s}_1$ and χ is derived from an automorphism of the game we will have $\mathfrak{s}_2^{\chi} = T\mathfrak{s}_1^{\chi}$. If $\mathfrak{s}_1$ is symmetric $\mathfrak{s}_1^{\chi} = \mathfrak{s}_1$ and therefore $\mathfrak{s}_2^{\chi} = T\mathfrak{s}_1 = \mathfrak{s}_2$. Consequently this mapping maps the set of symmetric n-tuples into itself.

Since this set is a cell there must be a symmetric fixed point $\mathfrak{s}$ which must be a symmetric equilibrium point.

Solutions

This notion of "solutions" and "solvability" did not catch on, but questions of uniqueness are relevant.

We define here solutions, strong solutions, and sub-solutions. A non-cooperative game does not always have a solution, but when it does the solution is unique. Strong solutions are solutions with special properties. Sub-solutions always exist and have many of the properties of solutions, but lack uniqueness.

S_1 will denote a set of mixed strategies of player i and $\mathfrak{S}$ a set of n-tuples of mixed strategies.

Solvability:

A game is *solvable* if its set, $\mathfrak{S}$, of equilibrium points satisfies the condition

$$(\mathfrak{t}; r_i) \in \mathfrak{S} \quad \text{and} \quad \mathfrak{s} \in \mathfrak{S} \to (\mathfrak{s}; r_i) \in \mathfrak{S} \quad \text{for all } i\text{'s}. \tag{5}$$

This is called the *interchangeability* condition. The *solution* of a solvable game is its set, $\mathfrak{S}$, of equilibrium points.

Strong Solvability:

A game is *strongly solvable* if it has a solution, $\mathfrak{S}$, such that for all i's

$$\mathfrak{s} \in \mathfrak{S} \quad \text{and} \quad p_i(\mathfrak{s}; r_i) = p_i(\mathfrak{s}) \to (\mathfrak{s}; r_i) \in \mathfrak{S}$$

and then $\mathfrak{S}$ is called a *strong solution*.

Equilibrium Strategies:

In a solvable game let S_i be the set of all mixed strategies s_i such that for some $\mathfrak{t}$ the n-tuple $(\mathfrak{t}; s_i)$ is an equilibrium point. [s_i is the ith component of some equilibrium point.] We call S_i the set of *equilibrium strategies* of player i.

Sub-solutions:

If $\mathfrak{S}$ is a subset of the set of equilibrium points of a game and satisfies condition (1); and if $\mathfrak{S}$ is maximal relative to this property then we call $\mathfrak{S}$ a *sub-solution*.

For any sub-solution $\mathfrak{S}$ we define the ith *factor set*, S_i, as the set of all s_i's such that $\mathfrak{S}$ contains $(\mathfrak{t}; s_i)$ for some $\mathfrak{t}$.

Note that a sub-solution, when unique, is a solution; and its factor sets are the sets of equilibrium strategies.

THEOREM 3. *A sub-solution, $\mathfrak{S}$, is the set of all n-tuples $(s_1, s_2, \cdots, s_n)$ such that each $s_i \in S_i$ where S_i is the ith factor set of $\mathfrak{S}$. Geometrically, $\mathfrak{S}$ is the product of its factor sets.*

PROOF. Consider such an n-tuple $(s_1, s_2, \cdots, s_n)$. By definition $\exists$ $\mathfrak{t}_1, \mathfrak{t}_2, \cdots, \mathfrak{t}_n$ such that for each $i\,(\mathfrak{t}_i; s_i) \in \mathfrak{S}$. Using the condition (5) $n-1$ times we obtain successively $(\mathfrak{t}_1; s_1) \in \mathfrak{S}$, $(\mathfrak{t}_1; s_1; s_2) \in \mathfrak{S}$, $\cdots, (\mathfrak{t}_1; s_1; s_2; \cdots; s_n) \in \mathfrak{S}$ and the last is simply $(s_1, s_2, \cdots, s_n) \in \mathfrak{S}$, which we needed to show.

THEOREM 4. *The factor sets $S_1, S_2, \cdots, S_n$ of a sub-solution are closed and convex as subsets of the mixed strategy spaces.*

PROOF. It suffices to show two things: (a) if s_i and $s_i' \in S_i$ then $s_i^* = (s_i + s_i')/2 \in S_i$; (b) if $s_i^\#$ is a limit point of S_i then $s_i^\# \in S_i$.

Let $\mathfrak{t} \in \mathfrak{S}$. Then we have $p_j(\mathfrak{t}; s_i) \geqq p_j(\mathfrak{t}; s_i; r_j)$ and $p_j(\mathfrak{t}; s_i') \geqq p_j(\mathfrak{t}; s_i'; r_j)$ for any r_j, by using the criterion of (1) for an eq. pt. Adding these inequalities, using the linearity of $p_j(s_1, \cdots, s_n)$ in s_i, and dividing by 2, we get $p_j(\mathfrak{t}; s_i^*) \geqq p_j(\mathfrak{t}; s_i^*; r_j)$ since $s_i^* = (s_i + s_i')/2$. From this we know that $(\mathfrak{t}; s_i)$ is an eq. pt. for any $\mathfrak{t} \in \mathfrak{S}$. If the set of all such eq. pts. $(\mathfrak{t}; s_i^*)$ is added to $\mathfrak{S}$ the augmented set clearly satisfies condition (3), and since $\mathfrak{S}$ was to be maximal it follows that $s_i^* \in S_i$.

To attack (b) note that the n-tuple $(\mathfrak{t}; s_i^\#)$, where $\mathfrak{t} \in \mathfrak{S}$, will be a limit point of the set of n-tuples of the form $(\mathfrak{t}; s_i)$ where $s_i \in S_i$, since $s_i^\#$ is a limit point of S_i. But this set is a set of eq. pts. and hence any point in its closure is an eq. pt., since the set of all eq. pts. is closed. Therefore $(\mathfrak{t}; s_i^{\#})$ is an eq. pt. and hence $s_i^{\#} \in S_i$ from the same argument as for s_i^*.

Values:

Let $\mathfrak{S}$ be the set of equilibrium points of a game. We define

$$v_i^+ = \max_{\mathfrak{s} \in \mathfrak{S}} [p_i(\mathfrak{s})], \quad v_i^- = \min_{\mathfrak{s} \in \mathfrak{S}} [p_i(\mathfrak{s})].$$

If $v_i^+ = v_i^-$ we write $v_i = v_i^+ = v_i^-$. v_i^+ is the *upper value* to player i of the game; v_i^- the *lower value*; and v_i the *value*, if it exists.

Values will obviously have to exist if there is but one equilibrium point.

One can define *associated values* for a sub-solution by restricting $\mathfrak{S}$ to the eq. pts. in the sub-solution and then using the same defining equations as above.

A two-person zero-sum game is always solvable in the sense defined above. The sets of equilibrium strategies S_1 and S_2 are simply the sets of "good" strategies. Such a game is not generally strongly solvable; strong solutions exist only when there is a "saddle point" in *pure* strategies.

Simple Examples

These are intended to illustrate the concepts defined in the paper and display special phenomena which occur in these games.

The first player has the roman letter strategies and the payoff to the left, etc.

Several of these examples are classic 2 × 2 games.

Ex. 1

5	$a\alpha$	-3	Solution $(\frac{9}{16}a + \frac{7}{16}b, \frac{7}{17}\alpha + \frac{10}{17}\beta)$
-4	$a\beta$	4	
-5	$b\alpha$	5	$v_1 = \frac{-5}{17}, v_2 = +\frac{1}{2}$
3	$b\beta$	-4	

Ex. 2

1	$a\alpha$	1	Strong Solution (b, β)
-10	$a\beta$	10	
10	$b\alpha$	-10	$v_1 = v_2 = -1$
-1	$b\beta$	-1	

Ex. 3	1	$a\alpha$	1	Unsolvable; equilibrium points (a, α), (b, β), and $\left(\frac{a}{2} + \frac{b}{2}, \frac{\alpha}{2} + \frac{\beta}{2}\right)$. The strategies in the last case have maxi-min and mini-max properties.
	−10	$a\beta$	10	
	−10	$b\alpha$	−10	
	1	$b\beta$	1	
Ex. 4	1	$a\alpha$	1	Strong Solution: all pairs of mixed strategies.
	0	$a\beta$	1	
	1	$b\alpha$	0	$v_1^+ = v_2^+ = 1, v_1^- = v_2^- = 0$
	0	$b\beta$	0	
Ex. 5	1	$a\alpha$	2	Unsolvable; eq. pts. (a, α), (b, β), and $(\frac{1}{4}a + \frac{3}{4}b, \frac{3}{8}\alpha + \frac{5}{8}\beta)$. However, empirical tests show a tendency toward (a, α).
	−1	$a\beta$	−4	
	−4	$b\alpha$	−1	
	2	$b\beta$	1	
Ex. 6	1	$a\alpha$	1	Eq. pts.: (a, α) and (b, β), with (b, β) an example of instability.
	0	$a\beta$	0	
	0	$b\alpha$	0	
	0	$b\beta$	0	

Geometrical Form of Solutions

In the two-person zero-sum case it has been shown that the set of "good" strategies of a player is a convex polyhedral subset of his strategy space. We shall obtain the same result for a player's set of equilibrium strategies in any solvable game.

THEOREM 5. *The sets $S_1, S_2, \cdots, S_n$ of equilibrium strategies in a solvable game are polyhedral convex subsets of the respective mixed strategy spaces.*

PROOF. An n-tuple $\mathfrak{s}$ will be an equilibrium point if and only if for every i

$$p_i(\mathfrak{s}) = \max_{\alpha} p_{i\alpha}(\mathfrak{s}) \tag{6}$$

which is condition (3). An equivalent condition is for every i and α

$$p_i(\mathfrak{s}) - p_{i\alpha}(\mathfrak{s}) \geqq 0. \tag{7}$$

Let us now consider the form of the set S_j of equilibrium strategies, s_j, of player j. Let $\mathfrak{t}$ be any equilibrium point, then $(\mathfrak{t}; s_j)$ will be an equilibrium point if and only if $s_j \in S_j$, from Theorem 2. We now apply conditions (2) to $(\mathfrak{t}; s_j)$, obtaining

$$s_j \in S_j \leftrightarrow \text{ for all } i, \alpha \quad p_i\,(\mathrm{t}; s_j) - p_{i}\alpha\,(\mathrm{t}; s_j) \geqq 0. \qquad (8)$$

Since p_i is n-linear and t is constant these are a set of linear inequalities of the form $F_{i\alpha}\,(s_j) \geqq 0$. Each such inequality is either satisfied for all s_j or for those lying on and to one side of some hyperplane passing through the strategy simplex. Therefore, the complete set [which is finite] of conditions will all be satisfied simultaneously on some convex polyhedral subset of player j's strategy simplex. [Intersection of half-spaces.]

As a corollary we may conclude that S_j is the convex closure of a finite set of mixed strategies [vertices].

Dominance and Contradiction Methods

Ideas of dominance are very relevant in game theory today.

We say that s_i' dominates s_i if $p_i\,(\mathrm{t}; s_i') > p_i\,(\mathrm{t}; s_i)$ for every t.

This amounts to saying that s_i' gives player i a higher payoff than s_i no matter what the strategies of the other players are. To see whether a strategy s_i' dominates s_i it suffices to consider only pure strategies for the other players because of the n-linearity of p_i.

It is obvious from the definitions that *no equilibrium point can involve a dominated strategy* s_i.

The domination of one mixed strategy by another will always entail other dominations. For suppose s_i' dominates s_i and t_i uses all of the pure strategies which have a higher coefficient in s_i than in s_i'. Then for a small enough ρ

$$t_i' = t_i + \rho\,(s_i' - s_i)$$

is a mixed strategy; and t_i dominates t_i' by linearity.

One can prove a few properties of the set of undominated strategies. It is simply connected and is formed by the union of some collection of faces of the strategy simplex.

The information obtained by discovering dominances for one player may be of relevance to the others, insofar as the elimination of classes of mixed strategies as possible components of an equilibrium point is concerned. For the t's whose components are all undominated are all that need be considered and thus eliminating some of the strategies of one player may make possible the elimination of a new class of strategies for another player.

Another procedure which may be used in locating equilibrium points is the contradiction-type analysis. Here one assumes that an equilibrium point exists having component strategies lying within certain regions of the strategy spaces and proceeds to deduce further conditions which must be satisfied if the hypothesis is true. This sort of reasoning may be carried through several stages to eventually obtain a contradiction indicating that there is no equilibrium point satisfying the initial hypothesis.

"Contradiction-type analysis" is related to the non-constructive nature of the Brouwer and Kakutani fixed-point theorems.

A Three-Man Poker Game

As an example of the application of our theory to a more or less realistic case we include the simplified poker game given below. The rules are as follows:

This material on simplified poker appears in Nash's PhD dissertation.

(a) The deck is large, with equally many *high* and *low* cards, and a hand consists of one card.

(b) Two chips are used to ante, open, or call.

(c) The players play in rotation and the game ends after all have passed or after one player has opened and the others have had a chance to call.

(d) If no one bets the antes are retrieved.

(e) Otherwise the pot is divided equally among the highest hands which have bet.

We find it more satisfactory to treat the game in terms of quantities we call "behavior parameters" than in the normal form of *Theory of Games and Economic Behavior*. In the normal form representation two mixed strategies of a player may be equivalent in the sense that each makes the individual choose each available course of action in each particular situation requiring action on his part with the same frequency. That is, they represent the same behavior pattern on the part of the individual.

Behavior parameters give the probabilities of taking each of the various possible actions in each of the various possible situations which may arise. Thus they describe behavior patterns.

In terms of behavior parameters the strategies of the players may be represented as follows, assuming that since there is no point in passing with a *high* card at one's last opportunity to bet that this will not be done. The Greek letters are the probabilities of the various acts.

	First Moves	Second Moves
I	α Open on *high* β Open on *low*	κ Call III on *low* λ Call II on *low* μ Call II and III on *low*
II	γ Call I on *low* δ Open on *high* ϵ Open on *low*	ν Call III on *low* ξ Call III and I on *low*
III	ζ Call I and II on *low* η Open on *low* θ Call I on *low* ι Call II on *low*	Player III never gets a second move

We locate all possible equilibrium points by first showing that most of the Greek parameters must vanish. By dominance mainly with a little contradiction-type analysis β is eliminated and with it go γ, ζ, and θ by dominance. Then contradictions eliminate $\mu, \xi, \iota, \lambda, \kappa$, and ν in that order. This leaves us with $\alpha, \delta, \varepsilon$, and η. Contradiction analysis shows that none of these can be zero or one and thus we obtain a system of simultaneous algebraic equations. The equations happen to have but one solution with the variables in the range $(0, 1)$. We get

$$\alpha = \frac{21 - \sqrt{321}}{10}, \quad \eta = \frac{5\alpha + 1}{4}, \quad \delta = \frac{5 - 2\alpha}{5 + \alpha}, \quad \varepsilon = \frac{4\alpha - 1}{\alpha + 5}.$$

These yield $\alpha = .308, \eta = .635, \delta = .826$, and $\varepsilon = .044$. Since there is only one equilibrium point the game has values; these are

$$v_1 = -.147 = -\frac{(1 + 17\alpha)}{8(5 + \alpha)}, \quad v_2 = -.096 = -\frac{1 - 2\alpha}{4},$$

and

$$v_3 = .243 = \frac{79}{40}\left(\frac{1 - \alpha}{5 + \alpha}\right).$$

A more complete investigation of this poker game is published in Annals of Mathematics Study No. 24, *Contributions to the Theory of*

Games. There the solution is studied as the ratio of ante to bet varies, and the potentialities of coalitions are investigated.

Applications

The study of n-person games for which the accepted ethics of fair play imply non-cooperative playing is, of course, an obvious direction in which to apply this theory. And poker is the most obvious target. The analysis of a more realistic poker game than our very simple model should be quite an interesting affair.

The complexity of the mathematical work needed for a complete investigation increases rather rapidly, however, with increasing complexity of the game; so that analysis of a game much more complex than the example given here might only be feasible using approximate computational methods.

A less obvious type of application is to the study of cooperative games. By a cooperative game we mean a situation involving a set of players, pure strategies, and payoffs as usual; but with the assumption that the players can and will collaborate as they do in the von Neumann and Morgenstern theory. This means the players may communicate and form coalitions which will be enforced by an umpire. It is unnecessarily restrictive, however, to assume any transferability or even comparability of the payoffs [which should be in utility units] to different players. Any desired transferability can be put into the game itself instead of assuming it possible in the extra-game collaboration.

This paragraph distinguishes cooperative game theory from the non-cooperative approach that has been taken in this paper, a distinction that has survived to this day.

The writer has developed a "dynamical" approach to the study of cooperative games based upon reduction to non-cooperative form. One proceeds by constructing a model of the pre-play negotiation so that the steps of negotiation become moves in a larger non-cooperative game [which will have an infinity of pure strategies] describing the total situation.

This larger game is then treated in terms of the theory of this paper [extended to infinite games] and if values are obtained they are taken as the values of the cooperative game. Thus the problem of analyzing a cooperative game becomes the problem of obtaining a suitable, and convincing, non-cooperative model for the negotiation.

The writer has, by such a treatment, obtained values for all finite two person cooperative games, and some special n-person games.

Acknowledgments

Drs. Tucker, Gale, and Kuhn gave valuable criticism and suggestions for improving the exposition of the material in this paper. David Gale suggested the investigation of symmetric games. The solution of the Poker model was a joint project undertaken by Lloyd S. Shapley and the author. Finally, the author was sustained financially by the Atomic Energy Commission in the period 1949–50 during which this work was done.

REFERENCES

Kuhn, H. W. 1950. "Extensive Games." *Proceedings of the National Academy of Sciences of the USA* 36 (10): 570–576.

Nash, J. F. 1950. "Equilibrium Points in *N*-Person Games." *Proceedings of the National Academy of Sciences of the USA* 36 (1): 48–49.

———. 1952. "Two-Person Cooperative Games." *Econometrica* 21 (1): 128–140.

Nash, J. F., and L. S. Shapley. 1950. "A Simple Three-Person Poker Game." In *Contributions to the Theory of Games (AM-24), Volume I,* edited by H. W. Kuhn and A. W. Tucker, 105–115. Princeton, NJ: Princeton University Press.

von Neumann, J., and O. Morgenstern. 1944. *Theory of Games and Economic Behavior.* Princeton, NJ: Princeton University Press.

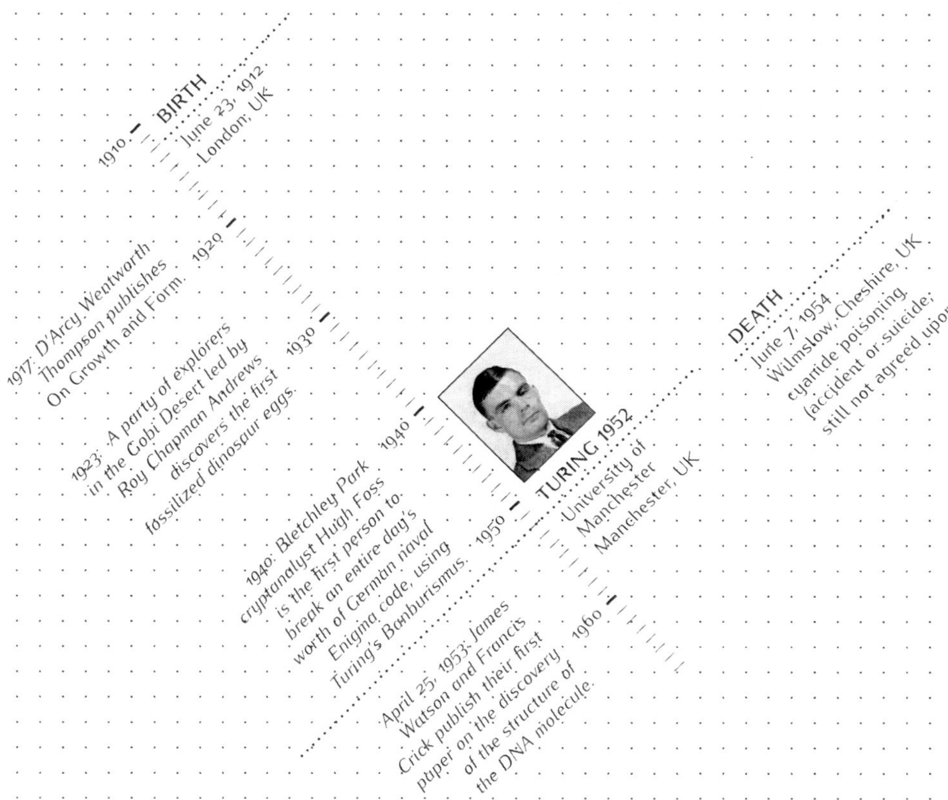

ALAN MATHISON TURING

[13]

DO TURING PATTERNS PROVIDE A CHEMICAL BASIS OF MORPHOGENESIS?

Karen Page, University College London

A. M. Turing, "The Chemical Basis of Morphogenesis," *Philosophical Transactions of the Royal Society B* 237: 37–72 (1952).

How does structure arise in an embryo? How do the chemicals present in a domain form patterns? In this paper, Alan Turing proposes an answer. This mechanism is now called *diffusion-driven* instability. Turing's paper contains other ideas that have triggered large areas of scientific and mathematical research. Here I try to explain Turing's theory and to discuss very briefly the research it inspired. I will also mention a setback in the search for Turing patterns.

Alan Turing was born in 1912 in Maida Vale, London. He studied logic at King's College, Cambridge, and proposed the concept of the Turing machine. During the war he worked for the Foreign Office and, in 1951, he was elected a Fellow of the Royal Society.

"The Chemical Basis of Morphogenesis" (published in August 1952) was the penultimate journal article that Turing published before his death in 1954 (Turing 1953; Newman 1955). It pre-dates Watson and Crick's (1953) paper on the structure of DNA and Crick and others' (1961) discovery of the genetic code.

The key ideas in the paper are described in sections 1, 4, and 11. The key mathematics is in sections 6, 7, and 8.1. Section 8.1 describes the stationary patterns now referred to as *Turing patterns.* Section 11 gives a biological interpretation of his results. Finally, section 12 aims to explain *gastrulation*, a process that initiates the formation of the major axis in amniote embryos (e.g., those of birds, reptiles, and mammals). Turing focuses on the breakdown of symmetry on the surface of a sphere, which is of approximate relevance to gastrulation during the development of mouse embryos. However, human and chick embryos are more disc-like initially (e.g., Page *et al.* 2001).

Here follows a brief description of the key mathematics of the paper. In section 6, "Reactions and Diffusion in a Ring of Cells," Turing considers a ring of N cells, each containing two chemicals, which he calls *morphogens*. The concentration of the first morphogen in cell r ($r = 1, 2, \ldots, N$) is given by X_r and the concentration of the other morphogen by Y_r. In each cell, the rate of production of the first morphogen is given by a function of the local levels of both morphogens, and the same is true for the rate of production of the second morphogen. Specifically, the rate of production of the first morphogen is $f(X_r, Y_r)$ and that of the second is $g(X_r, Y_r)$. Note that these values can be negative, meaning that the concentrations are reducing. The chemicals can also move to neighboring cells by a diffusive mechanism. This leads to Turing's equation 6.1:

$$\begin{aligned} \frac{dX_r}{dt} &= f(X_r, Y_r) + \mu(X_{r+1} + X_{r-1} - 2X_r) \\ \frac{dY_r}{dt} &= g(X_r, Y_r) + \nu(Y_{r+1} + Y_{r-1} - 2Y_r). \end{aligned} \tag{1}$$

Here μ is the diffusion coefficient of the first species and ν that of the second species.

In section 7, "Continuous Ring of Tissue," instead of considering the concentration per cell, Turing models the morphogen concentrations at each point in space on a ring. As his spatial variable he uses the angle θ, familiar to many from plane polar coordinates. Here he presents *reaction–diffusion* equations in their more frequently used form, as a pair (since there are two chemicals) of coupled partial differential equations. He gives them in their linear form. Here I give the general nonlinear form:

$$\begin{aligned} \frac{\partial X}{\partial t} &= f(X, Y) + \mu\frac{\partial^2 X}{\partial \theta^2} \\ \frac{\partial Y}{\partial t} &= g(X, Y) + \nu\frac{\partial^2 Y}{\partial \theta^2}. \end{aligned} \tag{2}$$

Turing also discusses three interacting chemicals, and the extension of the equations to describe an arbitrary number of chemicals is straightforward, although the analysis of patterning becomes more difficult.

Turing performs what is now known as *linear stability analysis* on both the spatially discrete and spatially continuous systems and shows, in section 8.1, "Diffusion-driven instability," that it predicts that patterns with certain spatial wavelengths will grow, starting from stochastic perturbations around an initially homogeneous state. This is the key insight of his paper. It is surprising in the sense that diffusion usually smooths out spatial differences in concentration. However, if two or more chemical species react and diffuse, spontaneous spatial patterns in their concentrations can form. Turing also mentions oscillatory patterns but claims that they require at least three morphogens.

I cannot do full justice to the mathematics here. The rough idea is that the small perturbations x and y about the chemical equilibrium concentrations satisfy linear equations. For a single cell, these are a pair of coupled ordinary differential equations, which we can write as

$$\frac{d}{dt}\begin{pmatrix} x \\ y \end{pmatrix} = J\begin{pmatrix} x \\ y \end{pmatrix},$$

where J is a two-by-two matrix of constants.

The equilibrium is stable if none of the real parts of the eigenvalues[1] of J is positive. If the perturbations are instead functions of space θ, they can be decomposed into Fourier modes (sines and cosines). For a sine-wave with wave number k, the matrix that determines whether such a perturbation grows is:

$$J - Dk^2 \equiv \begin{pmatrix} a & b \\ c & d \end{pmatrix} - k^2\begin{pmatrix} \mu & 0 \\ 0 & \nu \end{pmatrix} = \begin{pmatrix} a - \mu k^2 & b \\ c & d - \nu k^2 \end{pmatrix}. \quad (3)$$

For certain values of the reaction rates and diffusion coefficients, there is a range of values of k for which the corresponding sine-wave grows. Often the final pattern has the same number of peaks as the sine wave with the fastest growth rate.

There have been many subsequent theoretical developments. I will mention just four. First, Murray, Maini and others (1993) considered nondiffusive motion. I particularly highlight contact-based migration, which has been successful in explaining skin patterning in fish (Painter *et al.* 2015; Watanabe and Kondo 2015). Secondly, Crampin and co-workers

[1] "Eigenvalues and eigenvectors," https://en.wikipedia.org/wiki/Eigenvalues_and_eigenvectors (accessed March 7, 2022).

Figure 1. Dave the cat.

(1999) studied pattern selection on a growing domain. One criticism of Turing patterning had been that the patterns formed have an almost fixed wavelength and so the number of pattern elements scales with domain size. This makes it difficult for Turing patterns to explain similar embryonic structures in embryos of different sizes. Turing patterns were perceived to be more successful in explaining animal skin markings, which, although similar between different individuals, vary somewhat (see fig. 1). Crampin showed that, if the patterning domain grew in a certain way, a single-peaked pattern would form, which split into two peaks, then four, then eight. Thus, certain patterns can be selected over a wider range of domain sizes. Thirdly, Turing demonstrated that stochasticity (e.g., from thermal fluctuations) was most significant around the time instability first arises. Subsequent work has further explored continually added stochasticity (e.g., Biancalani, Fanelli, and Di Patti 2010; Karig *et al.* 2018). Finally, much work has addressed *juxtacrine signaling*, as in Turing's section 6 (e.g., Collier *et al.* 1996).

Turing patterns have been found in chemical systems (e.g. Castets *et al.* 1990; Lengyel and Epstein 1992), but it took some time for them to be found in biological systems. In 1980, Nüsslein-Volhard and Wieschaus's discovery that the segments of the *Drosophila* embryo were determined by fifteen genes led to doubt that simple Turing patterns were relevant to embryogenesis. More recently, plausible Turing patterns have been found in a variety of contexts, from the specification of mouse hair follicles to digit formation to left-right asymmetry in embryos (Glover *et al.* 2017; Müller *et al.* 2012; Raspopovic *et al.* 2014). There is no doubt, however, that these systems are patterned by mechanisms more complex than simple chemical reaction and diffusion. Nevertheless, morphogens (chemicals that exist in a graded form in the embryo and direct cellular differentiation) do exist—a good example is Sonic Hedgehog (Dessaud, McMahon, and Briscoe 2008)—and they act as inputs into gene regulatory networks whose dynamics can be modelled using coupled ordinary differential equations for the concentrations of the transcription factors involved (e.g., Balaskas *et al.* 2012).

In conclusion, this paper is foundational in complexity science because it discusses such a profound question of how structure arises from nothingness in an embryo, and because of the originality of Turing's ideas.

Acknowledgment

I would like to thank Philip Maini for introducing me to this paper.

REFERENCES

Balaskas, N., A. Ribeiro, J. Panovska, E. Dessaud, N. Sasai, K. M. Page, J. Briscoe, and V. Ribes. 2012. "Gene Regulatory Logic for Reading the Sonic Hedgehog Signaling Gradient in the Vertebrate Neural Tube." *Cell* 148 (1–2): 273–284. https://doi.org/10.1016/j.cell.2011.10.047.

Biancalani, T., D. Fanelli, and F. Di Patti. 2010. "Stochastic Turing Patterns in the Brusselator Model." *Physical Review E* 81 (4): 046215. https://doi.org/10.1103/PhysRevE.81.046215.

Castets, V., E. Dulos, J. Boissonade, and P. De Kepper. 1990. "Experimental Evidence of a Sustained Standing Turing-Type Nonequilibrium Chemical Pattern." *Physical Review Letters* 64 (24): 2953–2957. https://doi.org/10.1103/PhysRevLett.64.2953.

Collier, J. R., N. A. M. Monk, P. K. Maini, and J. H. Lewis. 1996. "Pattern Formation by Lateral Inhibition with Feedback: A Mathematical Model of Delta-Notch Intercellular Signalling." *Journal of Theoretical Biology* 183 (4): 429–446. https://doi.org/10.1006/jtbi.1996.0233.

Crampin, E. J., E. A. Gaffney, and P. K. Maini. 1999. "Reaction and Diffusion on Growing Domains: Scenarios for Robust Pattern Formation." *Bulletin of Mathematical Biology* 61 (6): 1093–1120. https://doi.org/10.1006/bulm.1999.0131.

Crick, F., L. Barnett, S. Brenner, and R. J. Watts-Tobin. 1961. "General Nature of the Genetic Code for Proteins." *Nature* 192 (4809): 1227–1232. https://doi.org/10.1038/1921227a0.

Dawes, J. H. P. 2016. "After 1952: The Later Development of Alan Turing's Ideas on the Mathematics of Pattern Formation." *Historia Mathematica* 43 (1): 49–64. https://doi.org/10.1016/j.hm.2015.03.003.

Dessaud, E., A. P. McMahon, and J. Briscoe. 2008. "Pattern Formation in the Vertebrate Neural Tube: A Sonic Hedgehog Morphogen-Regulated Transcriptional Network." *Development* 135 (15): 2489–2503. https://doi.org/10.1242/dev.009324.

Glover, J. D., K. L. Wells, F. Matthäus, K. J. Painter, W. Ho, J. Riddell, J. A. Johansson, *et al.* 2017. "Hierarchical Patterning Modes Orchestrate Hair Follicle Morphogenesis." *PLoS Biology* 15 (7): e2002117. https://doi.org/10.1371/journal.pbio.2002117.

Karig, D., K. M. Martini, T. Lu, N. A. DeLateur, N. Goldenfeld, and R. Weiss. 2018. "Stochastic Turing Patterns in a Synthetic Bacterial Population." *Proceedings of the National Academy of Sciences* 115 (26): 6572–6577. https://doi.org/10.1073/pnas.1720770115.

Lengyel, I., and I. R. Epstein. 1992. "A Chemical Approach to Designing Turing Patterns in Reaction–Diffusion Systems." *Proceedings of the National Academy of Sciences* 89 (9): 3977–3979. https://doi.org/10.1073/pnas.89.9.3977.

Müller, P., K. W. Rogers, B. M. Jordan, J. S. Lee, D. Robson, S. Ramanathan, and A. F. Schier. 2012. "Differential Diffusivity of Nodal and Lefty Underlies a Reaction–Diffusion Patterning System." *Science* 336 (6082): 721–724. https://doi.org/10.1126/science.1221920.

Murray, J. D. 1993. *Mathematical Biology.* 2nd ed. Berlin, Germany: Springer-Verlag.

Newman, M. H. A. 1955. "Alan Mathison Turing, 1912–1954." *Biographical Memoirs of Fellows of the Royal Society* 1:253–263. https://doi.org/10.1098/rsbm.1955.0019.

Nüsslein-Volhard, C., and E. Wieschaus. 1980. "Mutations Affecting Segment Number and Polarity in *Drosophila*." *Nature* 287 (5785): 795–801. https://doi.org/10.1038/287795a0.

Page, K. M., P. K. Maini, N. A. M. Monk, and C. D. Stern. 2001. "A Model of Primitive Streak Initiation in the Chick Embryo." *Journal of Theoretical Biology* 208 (4): 419–438. https://doi.org/10.1006/jtbi.2000.2229.

Painter, K. J., J. M. Bloomfield, J. A. Sherratt, and A. Gerisch. 2015. "A Nonlocal Model for Contact Attraction and Repulsion in Heterogeneous Cell Populations." *Bulletin of Mathematical Biology* 77 (6): 1132–1165. https://doi.org/10.1007/s11538-015-0080-x.

Raspopovic, J., L. Marcon, L. Russo, and J. Sharpe. 2014. "Digit Patterning is Controlled by a Bmp-Sox9-Wnt Turing Network Modulated by Morphogen Gradiants." *Science* 345 (6196): 566–570. https://doi.org/10.1126/science.1252960.

Turing, A. M. 1952. "The Chemical Basis of Morphogenesis." *Philosophical Transactions of the Royal Society of London B* 237 (641): 37–72. https://doi.org/10.1098/rstb.1952.0012.

———. 1953. "Some Calculations of the Riemann ζ-Function." *Proceedings of the London Mathematical Society* 3 (1): 99–117. https://doi.org/10.1112/plms/s3-3.1.99.

Watanabe, M., and S. Kondo. 2015. "Is Pigment Patterning in Fish Skin Determined by the Turing Mechanism?" *Trends in Genetics* 31 (2): 88–96. https://doi.org/10.1016/j.tig.2014.11.005.

Watson, J. D., and F. H. C. Crick. 1953. "Molecular Structure of Nucleic Acids: A Structure for Deoxyribose Nucleic Acid." *Nature* 171 (4356): 737–738. https://doi.org/10.1038/171737a0.

THE CHEMICAL BASIS OF MORPHOGENESIS

A. M. Turing, University of Manchester

It is suggested that a system of chemical substances, called morphogens, reacting together and diffusing through a tissue, is adequate to account for the main phenomena of morphogenesis. Such a system, although it may originally be quite homogeneous, may later develop a pattern or structure due to an instability of the homogeneous equilibrium, which is triggered off by random disturbances. Such reaction-diffusion systems are considered in some detail in the case of an isolated ring of cells, a mathematically convenient, though biologically unusual system. The investigation is chiefly concerned with the onset of instability. It is found that there are six essentially different forms which this may take. In the most interesting form stationary waves appear on the ring. It is suggested that this might account, for instance, for the tentacle patterns on Hydra and for whorled leaves. A system of reactions and diffusion on a sphere is also considered. Such a system appears to account for gastrulation. Another reaction system in two dimensions gives rise to patterns reminiscent of dappling. It is also suggested that stationary waves in two dimensions could account for the phenomena of phyllotaxis.

The purpose of this paper is to discuss a possible mechanism by which the genes of a zygote may determine the anatomical structure of the resulting organism. The theory does not make any new hypotheses; it merely suggests that certain well-known physical laws are sufficient to account for many of the facts. The full understanding of the paper requires a good knowledge of mathematics, some biology, and some elementary chemistry. Since readers cannot be expected to be experts in all of these subjects, a number of elementary facts are explained, which can be found in text-books, but whose omission would make the paper difficult reading.

1. A Model of the Embryo. Morphogens

In this section a mathematical model of the growing embryo will be described. This model will be a simplification and an idealization, and consequently a falsification. It is to be hoped that the features retained for discussion are those of greatest importance in the present state of knowledge.

The model takes two slightly different forms. In one of them the cell theory is recognized but the cells are idealized into geometrical points. In the other the matter of the organism is imagined as continuously distributed. The cells are not, however, completely ignored, for various physical and physico-chemical characteristics of the matter as a whole are assumed to have values appropriate to the cellular matter.

With either of the models one proceeds as with a physical theory and defines an entity called 'the state of the system'. One then describes how that state is to be determined from the state at a moment very shortly before. With either model the description of the state consists of two parts, the mechanical and the chemical. The mechanical part of the state describes the positions, masses, velocities and elastic properties of the cells, and the forces between them. In the continuous form of the theory essentially the same information is given in the form of the stress, velocity, density and elasticity of the matter. The chemical part of the state is given (in the cell form of theory) as the chemical composition of each separate cell; the diffusibility of each substance between each two adjacent cells must also be given. In the continuous form of the theory the concentrations and diffusibilities of each substance have to be given at each point. In determining the changes of state one should take into account

i. The changes of position and velocity as given by Newton's laws of motion.

ii. The stresses as given by the elasticities and motions, also taking into account the osmotic pressures as given from the chemical data.

iii. The chemical reactions.

iv. The diffusion of the chemical substances. The region in which this diffusion is possible is given from the mechanical data.

This account of the problem omits many features, e.g. electrical properties and the internal structure of the cell. But even so it is a problem of formidable mathematical complexity. One cannot at present hope to make any progress with the understanding of such systems except in very simplified cases. The interdependence of the chemical and mechanical data adds enormously to the difficulty, and attention will therefore be confined, so far as is possible, to cases where these can be separated. The mathematics of elastic solids is a well-developed subject, and has often been applied to biological systems. In this paper it is proposed to give attention rather to cases where the mechanical aspect can be ignored and the chemical aspect is the most significant. These cases promise greater interest, for the characteristic action of the genes themselves is presumably chemical. The systems actually to be considered consist therefore of masses of tissues which are not growing, but within which certain substances are reacting chemically, and through which they are diffusing. These substances will be called morphogens, the word being intended to convey the idea of a form producer. It is not intended to have any very exact meaning, but is simply the kind of substance concerned in this theory. The evocators of Waddington provide a good example of morphogens (Waddington 1940). These evocators diffusing into a tissue somehow persuade it to develop along different lines from those which would have been followed in its absence. The genes themselves may also be considered to be morphogens. But they certainly form rather a special class. They are quite indiffusible. Moreover, it is only by courtesy that genes can be regarded as separate molecules. It would be more accurate (at any rate at mitosis) to regard them as radicals of the giant molecules known as chromosomes. But presumably these radicals act almost independently, so that it is unlikely that serious errors will arise through regarding the genes as molecules. Hormones may also be regarded as quite typical morphogens. Skin pigments may be regarded as morphogens if desired. But those whose action is to be considered here do not come squarely within any of these categories.

The function of genes is presumed to be purely catalytic. They catalyze the production of other morphogens, which in turn may only be catalysts. Eventually, presumably, the chain leads to some morphogens whose duties are not purely catalytic. For instance, a substance might break down into a number of smaller molecules, thereby increasing the osmotic pressure in a cell and promoting its growth. The genes might thus be said to influence the anatomical form of the organism by determining the rates of those reactions which they catalyze. If the rates are assumed to be those determined by the genes, and if a comparison of organisms is not in question, the genes themselves may be eliminated from the discussion. Likewise any other catalysts obtained secondarily through the agency of the genes may equally be ignored, if there is no question of their concentrations varying. There may, however, be some other morphogens, of the nature of evocators, which cannot be altogether forgotten, but whose role may nevertheless be subsidiary, from the point of view of the formation of a particular organ. Suppose, for instance, that a 'leg-evocator' morphogen were being produced in a certain region of an embryo, or perhaps diffusing into it, and that an attempt was being made to explain the mechanism by which the leg was formed in the presence of the evocator. It would then be reasonable to take the distribution of the evocator in space and time as given in advance and to consider the chemical reactions set in train by it. That at any rate is the procedure adopted in the few examples considered here.

2. Mathematical Background Required

The greater part of this present paper requires only a very moderate knowledge of mathematics. What is chiefly required is an understanding of the solution of linear differential equations with constant coefficients. (This is also what is chiefly required for an understanding of mechanical and electrical oscillations.) The solution of such an equation takes the form of a sum ΣAe^{bt}, where the quantities A, b may be complex, i.e. of the form $\alpha + i\beta$ where α and β are ordinary (real) numbers and $i = \sqrt{-1}$. It is of great importance that the physical significance of the various possible solutions of this kind should be appreciated, for instance, that

(a) Since the solutions will normally be real one can also write them in the form $\mathscr{R}\Sigma Ae^{bt}$ or $\Sigma\mathscr{R}Ae^{bt}$ ($\mathscr{R}$ means 'real part of').

(b) That if $A = A'ei^{\phi}$ and $b = \alpha_i\beta$, where A', α, β, ϕ are real, then

$$\mathscr{R}Ae^{bt} = A'e^{\alpha t}\cos(\beta t + \phi).$$

Thus each such term represents a sinusoidal oscillation if $\alpha = 0$, a damped oscillation if $\alpha < 0$, and an oscillation of ever-increasing amplitude if $\alpha > 0$.

(c) If any one of the numbers b has a positive real part the system in question is unstable.

(d) After a sufficiently great lapse of time all the terms Ae^{bt} will be negligible in comparison with those for which b has the greatest real part, but unless this greatest real part is itself zero these dominant terms will eventually either tend to zero or to infinite values.

(e) That the indefinite growth mentioned in (b) and (d) will in any physical or biological situation eventually be arrested due to a breakdown of the assumptions under which the solution was valid. Thus, for example, the growth of a colony of bacteria will normally be taken to satisfy the equation $dy/dt = \alpha y \, (\alpha > 0)$, y being the number of organisms at time t, and this has the solution $y = Ae^{\alpha t}$. When, however, the factor $e^{\alpha t}$ has reached some billions the food supply can no longer be regarded as unlimited and the equation $dy/dt = \alpha y$ will no longer apply.

The following relatively elementary result will be needed, but may not be known to all readers:

$$\sum_{r=1}^{N} \exp\left[\frac{2\pi i r s}{N}\right] = 0 \text{ if } 0 < s < N,$$

but

$$= N \text{ if } s = 0 \text{ or } s = N.$$

The first case can easily be proved when it is noticed that the left-hand side is a geometric progression. In the second case all the terms are equal to 1.

The relative degrees of difficulty of the various sections are believed to be as follows. Those who are unable to follow the points made in this section should only attempt §§3, 4, 11, 12, 14 and part of §13. Those who can just understand this section should profit also from §§7, 8, 9. The remainder, §§5, 10, 13, will probably only be understood by those definitely trained as mathematicians.

3. Chemical Reactions

It has been explained in a preceding section that the system to be considered consists of a number of chemical substances (morphogens) diffusing through a mass of tissue of given geometrical form and reacting together within it. What laws are to control the development of this situation? They are quite simple. The diffusion follows the ordinary laws of diffusion, i.e. each morphogen moves from regions of greater to regions of less concentration, at a rate proportional to the gradient of the concentration, and also proportional to the 'diffusibility' of the substance. This is very like the conduction of heat, diffusibility taking the place of conductivity. If it were not for the walls of the cells the diffusibilities would be inversely proportional to the square roots of the molecular weights. The pores of the cell walls put a further handicap on the movement of the larger molecules in addition to that imposed by their inertia, and most of them are not able to pass through the walls at all. The reaction rates will be assumed to obey the 'law of mass action'. This states that the rate at which a reaction takes place is proportional to the concentrations of the reacting substances. Thus, for instance, the rate at which silver chloride will be formed and precipitated from a solution of silver nitrate and sodium chloride by the reaction

$$Ag^+ + Cl^- \rightarrow AgCl$$

will be proportional to the product of the concentrations of the silver ion Ag^+ and the chloride ion Cl^-. It should be noticed that the equation

$$AgNO_3 + NaCl \rightarrow AgCl + NaNO_3$$

is not used because it does not correspond to an actual reaction but to the final outcome of a number of reactions. The law of mass action must only be applied to the *actual* reactions. Very often certain substances appear in the individual reactions of a group, but not in the final outcome. For instance, a reaction $A \rightarrow B$ may really take the form of two steps $A + G \rightarrow C$ and $C \rightarrow B + G$. In such a case the substance G is described as a catalyst, and as catalyzing the reaction $A \rightarrow B$. (Catalysis according to this plan has been considered in detail by Michaelis and Menten (1913).) The effect of the genes is presumably achieved almost entirely by catalysis. They are certainly not permanently used up in the reactions.

Sometimes one can regard the effect of a catalyst as merely altering a reaction rate. Consider, for example, the case mentioned above, but suppose also that A can become detached from G, i.e. that the reaction $C \rightarrow A + G$ is taken into account. Also suppose that the reactions $A + G \rightleftarrows C$ both proceed much faster than $C \rightarrow B + G$. Then the concentrations of A, G, C will be related by the condition that there is equilibrium between the reactions $A + G \rightarrow C$ and $C \rightarrow A + G$, so that (denoting concentrations by square brackets) $[A][G] = k[C]$ for some constant k. The reaction $C \rightarrow B + G$ will of course proceed at a rate proportional to $[C]$, i.e. to $[A][G]$. If the amount of C is always small compared with the amount of G one can say that the presence of the catalyst and its amount merely alter the mass action constant for the reaction $A \rightarrow B$, for the whole proceeds at a rate proportional to $[A]$. This situation does not, however, hold invariably. It may well happen that nearly all of G takes the combined form C so long as any of A is left. In this case the reaction proceeds at a rate independent of the concentration of A until A is entirely consumed. In either of these cases the rate of the complete group of reactions depends only on the concentrations of the reagents, although usually not according to the law of mass action applied crudely to the chemical equation for the whole group. The same applies in any case where all reactions of the group with one exception proceed at speeds much greater than that of the exceptional one. In these cases the rate of the reaction is a function of the concentrations of the reagents. More generally again, no such approximation is applicable. One simply has to take all the actual reactions into account.

According to the cell model then, the number and positions of the cells are given in advance, and so are the rates at which the various morphogens diffuse between the cells. Suppose that there are N cells and M morphogens. The state of the whole system is then given by MN numbers, the quantities of the M morphogens in each of N cells. These numbers change with time, partly because of the reactions, partly because of the diffusion. To determine the part of the rate of change of one of these numbers due to diffusion, at any one moment, one only needs to know the amounts of the same morphogen in the cell and its neighbours, and the diffusion coefficient for that morphogen. To find the rate of change due to chemical reaction one only needs to know the concentrations of all morphogens at that moment in the one cell concerned.

This description of the system in terms of the concentrations in the various cells is, of course, only an approximation. It would be justified if, for instance, the contents were perfectly stirred. Alternatively, it may often be justified on the understanding that the 'concentration in the cell' is the concentration at a certain representative point, although the idea of 'concentration at a point' clearly itself raises difficulties. The author believes that the approximation is a good one, whatever argument is used to justify it, and it is certainly a convenient one.

It would be possible to extend much of the theory to the case of organisms immersed in a fluid, considering the diffusion within the fluid as well as from cell to cell. Such problems are not, however, considered here.

4. The Breakdown of Symmetry and Homogeneity

There appears superficially to be a difficulty confronting this theory of morphogenesis, or, indeed, almost any other theory of it. An embryo in its spherical blastula stage has spherical symmetry, or if there are any deviations from perfect symmetry, they cannot be regarded as of any particular importance, for the deviations vary greatly from embryo to embryo within a species, though the organisms developed from them are barely distinguishable. One may take it therefore that there is perfect spherical symmetry. But a system which has spherical symmetry, and whose state is changing because of chemical reactions and diffusion, will remain spherically symmetrical for ever. (The same would hold true if the state were changing according to the laws of electricity and magnetism, or of quantum mechanics.) It certainly cannot result in an organism such as a horse, which is not spherically symmetrical.

There is a fallacy in this argument. It was assumed that the deviations from spherical symmetry in the blastula could be ignored because it makes no particular difference what form of asymmetry there is. It is, however, important that there are *some* deviations, for the system may reach a state of instability in which these irregularities, or certain components of them, tend to grow. If this happens a new and stable equilibrium is usually reached, with the symmetry entirely gone. The variety of such new equilibria will normally not be so great as the variety of irregularities giving rise to them. In the case, for instance, of the gastrulating sphere, discussed at the end of this paper, the direction of the axis of the gastrula can vary, but nothing else.

The situation is very similar to that which arises in connexion with electrical oscillators. It is usually easy to understand how an oscillator keeps going when once it has started, but on a first acquaintance it is not obvious how the oscillation begins. The explanation is that there are random disturbances always present in the circuit. Any disturbance whose frequency is the natural frequency of the oscillator will tend to set it going. The ultimate fate of the system will be a state of oscillation at its appropriate frequency, and with an amplitude (and a wave form) which are also determined by the circuit. The phase of the oscillation alone is determined by the disturbance.

If chemical reactions and diffusion are the only forms of physical change which are taken into account the argument above can take a slightly different form. For if the system originally has no sort of geometrical symmetry but is a perfectly homogeneous and possibly irregularly shaped mass of tissue, it will continue indefinitely to be homogeneous. In practice, however, the presence of irregularities, including statistical fluctuations in the numbers of molecules undergoing the various reactions, will, if the system has an appropriate kind of instability, result in this homogeneity disappearing.

This breakdown of symmetry or homogeneity may be illustrated by the case of a pair of cells originally having the same, or very nearly the same, contents. The system is homogeneous: it is also symmetrical with respect to the operation of interchanging the cells. The contents of either cell will be supposed describable by giving the concentrations X and Y of two morphogens. The chemical reactions will be supposed such that, on balance, the first morphogen (X) is produced at the rate $5X-6Y+1$ and the second (Y) at the rate $6X-7Y+1$. When, however, the strict application of these formulae would involve the concentration of a morphogen in a cell becoming negative, it is understood that it is instead destroyed only at the rate at which it is reaching that cell by diffusion. The first morphogen will be supposed to duffuse at the rate 0.5 for unit difference of concentration between the cells, the second, for the same difference, at the rate 4.5. Now if both morphogens have unit concentration in both cells there is equilibrium. There is no resultant passage of either morphogen across the cell walls, since there is no concentration difference, and there is no resultant production (or destruction) of either morphogen in either cell since $5X—6Y + 1$ and $6X—7Y + 1$ both have the value zero for $X = 1, Y = 1$. But suppose the values are $X_1 = 1.06$, $Y_1 = 1.02$ for the first cell and $X_2 = 0.94$, $Y_2 = 0.98$ for the second. Then the two morphogens will be being produced by chemical action at the rates 0.18, 0.22 respectively in the first cell and destroyed at the same rates in the second. At the same time there is a flow due to diffusion from the first cell to the second at the rate 0.06 for the first morphogen and 0.18 for the second. In sum the effect is a flow from the second cell to the first at the rates 0.12, 0.04 for the two morphogens respectively. This flow tends to accentuate the already existing differences between the two cells. More generally, if

$$X_1 = 1 + 3\xi, \quad X_2 = 1 - 3\xi, \quad Y_1 = 1 + \xi, \quad Y_2 = 1 - \xi,$$

at some moment the four concentrations continue afterwards to be expressible in this form, and ξ increases at the rate 2ξ. Thus there is an exponential drift away from the equilibrium condition. It will be appreciated that a drift away from the equilibrium occurs with almost any small displacement from the equilibrium condition, though not normally according to an exact exponential curve. A particular choice was made in the above argument in order to exhibit the drift with only very simple mathematics.

Before it can be said to follow that a two-cell system can be unstable, with inhomogeneity succeeding homogeneity, it is necessary to show that the reaction rate functions postulated really can occur. To specify actual substances, concentrations and temperatures giving rise to these functions would settle the matter finally, but would be difficult and somewhat out of the spirit of the present inquiry. Instead, it is proposed merely to mention imaginary reactions which give rise to the required functions by the law of mass action, if suitable reaction constants are assumed. It will be sufficient to describe

(i) A set of reactions producing the first morphogen at the constant rate 1, and a similar set forming the second morphogen at the same rate.

(ii) A set destroying the second morphogen (Y) at the rate $7Y$.

(iii) A set converting the first morphogen (X) into the second (Y) at the rate $6X$.

(iv) A set producing the first morphogen (X) at the rate $11X$.

(v) A set destroying the first morphogen (X) at the rate $6Y$, so long as any of it is present.

The conditions of (i) can be fulfilled by reactions of the type $A \to X, B \to Y$, where A and B are substances continually present in large and invariable concentrations. The conditions of (ii) are satisfied by a reaction of the form $Y \to D$, D being an inert substance and (iii) by the reaction $X \to Y$ or $X \to Y + E$. The remaining two sets are rather more difficult. To satisfy the conditions of (iv) one may suppose that X is a catalyst for its own formation from A. The actual reactions could be the formation of an unstable compound U by the reaction $A + X \to U$, and the subsequent almost instantaneous breakdown $U \to 2X$. To destroy X at a rate proportional to Y as required in (v) one may suppose that a catalyst C is present in small but constant concentration and immediately combines with X, $X + C \to V$. The modified catalyst reacting with Y, at a rate proportional to Y, restores the catalyst but not the morphogen X, by the reactions $V + Y \to W, W \to C + H$, of which the latter is assumed instantaneous.

It should be emphasized that the reactions here described are by no means those which are most likely to give rise to instability in nature. The choice of the reactions to be discussed was dictated entirely by the fact that it was desirable that the argument be easy to follow. More plausible reaction systems are described in § 10.

Unstable equilibrium is not, of course, a condition which occurs very naturally. It usually requires some rather artificial interference, such as placing a marble on the top of a dome. Since systems tend to leave unstable equilibria they cannot often be in them. Such equilibria can, however, occur naturally through a stable equilibrium changing into an unstable one. For example, if a rod is hanging from a point a little above its centre of gravity it will be in stable equilibrium. If, however, a mouse climbs up the rod the equilibrium eventually becomes unstable and the rod starts to swing. A chemical analogue of this mouse-and-pendulum system would be that described above with the same diffusibilities but with the two morphogens produced at the rates

$$(3 + I)X - 6Y + I - 1 \quad \text{and} \quad 6X - (9 + I)Y - I + 1.$$

This system is stable if $I < 0$ but unstable if $I > 0$. If I is allowed to increase, corresponding to the mouse running up the pendulum, it will eventually become positive and the equilibrium will collapse. The system which was originally discussed was the case $I = 2$, and might be supposed to correspond to the mouse somehow reaching the top of the pendulum without disaster, perhaps by falling vertically on to it.

5. Left-Handed and Right-Handed Organisms

The object of this section is to discuss a certain difficulty which might be thought to show that the morphogen theory of morphogenesis cannot be right. The difficulty is mainly concerned with organisms which have not got bilateral symmetry. The argument, although carried through here without the use of mathematical formulae, may be found difficult by non-mathematicians, and these are therefore recommended to ignore it unless they are already troubled by such a difficulty.

An organism is said to have 'bilateral symmetry' if it is identical with its own reflexion in some plane. This plane of course always has to pass through some part of the organism, in particular through its centre of gravity. For the purpose of this argument it is more general to consider what may be called 'left-right symmetry'. An organism has left-right symmetry if its description in any right-handed set of rectangular Cartesian co-ordinates is identical with its description in some set of left-handed axes. An example of a body with left-right symmetry, but not bilateral symmetry, is a cylinder with the letter P printed on one end, and with the mirror image of a P on the other end, but with the two upright strokes of the two letters not parallel. The distinction may possibly be without a difference so far as the biological world is concerned, but mathematically it should not be ignored.

If the organisms of a species are sufficiently alike, and the absence of left-right symmetry sufficiently pronounced, it is possible to describe each individual as either right-handed or left-handed without there being difficulty in classifying any particular specimen. In man, for instance, one could take the X-axis in the forward direction, the Y-axis at right angles to it in the direction towards the side on which the heart is felt, and the Z-axis upwards. The specimen is classed as left-handed or right-handed according as the axes so chosen are left-handed or right-handed. A new classification has of course to be defined for each species.

The fact that there exist organisms which do not have left-right symmetry does not in itself cause any difficulty. It has already been explained how various kinds of symmetry can be lost in the development of the embryo, due to the particular disturbances (or 'noise') influencing the particular specimen not having that kind of symmetry, taken in conjunction with appropriate kinds of instability. The difficulty lies in the fact that there are species in which the proportions of left-handed and right-handed types are very unequal. It will be as well to describe first an argument which appears to show that this should not happen. The argument is very general, and might be applied to a very wide class of theories of morphogenesis.

An entity may be described as 'P-symmetrical' if its description in terms of one set of right-handed axes is identical with its description in terms of any other set of right-handed axes with the same origin. Thus, for instance, the totality of positions that a corkscrew would take up when rotated in all possible ways about the origin has P-symmetry. The entity will be said to be 'F-symmetrical' when changes from right-handed axes to left-handed may also be made. This would apply if the corkscrew were replaced by a bilaterally symmetrical object such as a coal scuttle, or a left-right symmetrical object. In these terms one may say that there are species such that the totality of specimens from that species, together with the rotated specimens, is P-symmetrical, but very far from

F-symmetrical. On the other hand, it is reasonable to suppose that

(i) The laws of physics are F-symmetrical.

(ii) The initial totality of zygotes for the species is F-symmetrical.

(iii) The statistical distribution of disturbances is F-symmetrical. The individual disturbances of course will in general have neither F-symmetry nor P-symmetry.

It should be noticed that the ideas of P-symmetry and F-symmetry as defined above apply even to so elaborate an entity as 'the laws of physics'. It should also be understood that the laws are to be the laws taken into account in the theory in question rather than some ideal as yet undiscovered laws.

Now it follows from these assumptions that the statistical distribution of resulting organisms will have F-symmetry, or more strictly that the distribution deduced as the result of working out such a theory will have such symmetry. The distribution of observed mature organisms, however, has no such symmetry. In the first place, for instance, men are more often found standing on their feet than their heads. This may be corrected by taking gravity into account in the laws, together with an appropriate change of definition of the two kinds of symmetry. But it will be more convenient if, for the sake of argument, it is imagined that some species has been reared in the absence of gravity, and that the resulting distribution of mature organisms is found to be P-symmetrical but to yield more right-handed specimens than left-handed and so not to have F-symmetry. It remains therefore to explain this absence of F-symmetry.

Evidently one or other of the assumptions (i) to (iii) must be wrong, i.e. in a correct theory one of them would not apply. In the morphogen theory already described these three assumptions do all apply, and it must therefore be regarded as defective to some extent. The theory may be corrected by taking into account the fact that the morphogens do not always have an equal number of left- and right-handed molecules. According to one's point of view one may regard this as invalidating either (i), (ii) or even (iii). Simplest perhaps is to say that the totality of zygotes just is not F-symmetrical, and that this could be seen if one looked at the molecules. This is, however, not very satisfactory from the point of view of this paper, as it would not be consistent with describing states in terms of concentrations only. It would be preferable if it was found possible to find more accurate laws concerning reactions and diffusion. For the purpose of accounting for unequal numbers of left- and right-handed organisms it is unnecessary to do more than show that there are corrections which would not be F-symmetrical when there are laevo- or dextrorotatory morphogens, and which would be large enough to account for the effects observed. It is not very difficult to think of such effects. They do not have to be very large, but must, of course, be larger than the purely statistical effects, such as thermal noise or Brownian movement.

There may also be other reasons why the totality of zygotes is not F-symmetrical, e.g. an asymmetry of the chromosomes themselves. If these also produce a sufficiently large effect, so much the better.

Though these effects may be large compared with the statistical disturbances they are almost certainly small compared with the ordinary diffusion and reaction effects. This will mean that they only have an appreciable effect during a short period in which the breakdown of left-right symmetry is occurring. Once their existence is admitted, whether on a theoretical or experimental basis, it is probably most convenient to give them mathematical expression by regarding them as P-symmetrically (but not F-symmetrically) distributed disturbances. However, they will not be considered further in this paper.

6. Reactions and Diffusion in a Ring of Cells

☞ This section and the next one contain the essential model of reaction–diffusion. Here the concentrations of chemicals are given in each cell. Chemicals move to adjacent cells with a diffusive flux equal to the concentration difference between the cells.

The original reason for considering the breakdown of homogeneity was an apparent difficulty in the diffusion-reaction theory of morphogenesis. Now that the difficulty is resolved it might be supposed that there is no reason for pursuing this aspect of the problem further, and that it would be best to proceed to consider what occurs when the system is very far from homogeneous. A great deal more attention will nevertheless be given to the breakdown of homogeneity. This is largely because the assumption that the system is still nearly homogeneous brings the problem within the range of what is capable of being treated mathematically. Even so many further simplifying assumptions have to be made. Another reason for giving this phase such attention is that it is in a sense the most critical period. That is to say, that if there is any doubt as to how the organism is going to develop it is conceivable that a minute examination of it just after instability has set in might settle the matter, but an examination of it at any earlier time could never do so.

There is a great variety of geometrical arrangement of cells which might be considered, but one particular type of configuration stands out as being particularly simple in its theory, and also illustrates the general principles very well. This configuration is a ring of similar cells. One may suppose that there are N such cells. It must be admitted that there is no biological example to which the theory of the ring can be immediately applied, though it is not difficult to find ones in which the principles illustrated by the ring apply.

It will be assumed at first that there are only two morphogens, or rather only two interesting morphogens. There may be others whose concentration does not vary either in space or time, or which can be eliminated from the discussion for one reason or another. These other morphogens may, for instanse, be catalysts involved in the reactions between the interesting morphogens. An example of a complete system of reactions is given in § 10. Some consideration will also be given in §§ 8, 9 to the case of three morphogens. The reader should have no difficulty in extending the results to any number of morphogens, but no essentially new features appear when the number is increased beyond three.

The two morphogens will be called X and Y. These letters will also be used to denote their concentrations. This need not lead to any real confusion. The concentration of X in cell r may be written X_r, and Y_r has a similar meaning. It is convenient to regard 'cell N' and 'cell O' as synonymous, and likewise 'cell 1' and cell '$N + 1$'. One can then say that for each r satisfying $1 \leqslant r \leqslant N$ cell r exchanges material by diffusion with cells $r - 1$ and $r + 1$. The cell-to-cell diffusion constant for X will be called μ, and that for Y will be called ν. This means that for unit concentration difference of X, this morphogen passes at the rate μ from the cell with the higher concentration to the (neighbouring) cell with the lower concentration. It is also necessary to make assumptions about the rates of chemical reaction. The most general assumption that can be made is that for concentrations X and Y chemical reactions are tending to increase X at the rate $f(X, Y)$ and Y at the rate $g(X, Y)$. When the changes in X and

Y due to diffusion are also taken into account the behaviour of the system may be described by the $2N$ differential equations

These are the main differential equations in their spatially discrete form. There are $2N$ equations, one for each chemical in each cell. The equations are said to be coupled because the rate of change of one concentration depends on the others. The reaction rates are represented by the function f and g of the concentrations of the two chemicals in the same cell. The concentration of X also increases locally if it is lower than the average concentration in the immediately neighboring cells. If it is locally higher than the average of its neighbors, it decreases. The same is true for Y. μ and ν are the diffusion coefficients.

$$\left.\begin{aligned} \frac{\mathrm{d}X_r}{\mathrm{d}t} &= f(X_r, Y_r) + \mu(X_{r+1} - 2X_r + X_{r-1}) \\ \frac{\mathrm{d}Y_r}{\mathrm{d}t} &= g(X_r, Y_r) + \nu(Y_{r+1} - 2Y_r + Y_{r-1}) \end{aligned}\right\} \quad (r = 1, \ldots, N). \tag{6.1}$$

If $f(h, k) \ : \ g(h, k) = 0$, then an isolated cell has an equilibrium with concentrations $X = h, Y = k$. The ring system also has an equilibrium, stable or unstable, with each X_r equal to h and each Y_r equal to k. Assuming that the system is not very far from this equilibrium it is convenient to put $X_r = h + x_r$, $Y_r = k + y_r$. One may also write $ax + by$ for $f(h+x, y+k)$ and $cx + dy$ for $g(h+x, y+k)$. Since $f(h, k) = g(h, k) = 0$ no constant terms are required, and since x and y are supposed small the terms in higher powers of x and y will have relatively little effect and one is justified in ignoring them. The four quantities a, b, c, d may be called the 'marginal reaction rates'. Collectively they may be described as the 'marginal reaction rate matrix'. When there are M morphogens this matrix consists of M^2 numbers. A marginal reaction rate has the dimensions of the reciprocal of a time, like a radioactive decay rate, which is in fact an example of a marginal (nuclear) reaction rate.

A cell isolated from its neighbors has a chemical equilibrium at $X = h$, $Y = k$. Small perturbations about these equilibrium values satisfy these linear equations.

With these assumptions the equations can be rewritten as

$$\left.\begin{aligned} \frac{\mathrm{d}x_r}{\mathrm{d}t} &= ax_r + by_r + \mu(x_{r+1} - 2x_r + x_{r-1}), \\ \frac{\mathrm{d}y_r}{\mathrm{d}t} &= cx_r + dy_r + \nu(y_{r+1} - 2y_r + y_{r-1}). \end{aligned}\right\} \tag{6.2}$$

To solve the equations one introduces new co-ordinates $\xi_0, \ldots, \xi_{N-1}$ and $\eta_0, \ldots, \eta_{N-1}$ by putting

The perturbations can be split into sine and cosine function with specific wavelengths—the Fourier modes.

$$\left.\begin{aligned} x_r &= \sum_{s=0}^{N-1} \exp\left[\frac{2\pi \mathrm{i} rs}{N}\right] \xi_s, \\ y_r &= \sum_{s=0}^{N-1} \exp\left[\frac{2\pi \mathrm{i} rs}{N}\right] \eta_s. \end{aligned}\right\} \tag{6.3}$$

These relations can also be written as

$$\left.\begin{aligned} \xi_r &= \frac{1}{N}\sum_{s=1}^{N} \exp\left[-\frac{2\pi i r s}{N}\right] x_s, \\ \eta_r &= \frac{1}{N}\sum_{s=1}^{N} \exp\left[-\frac{2\pi i r s}{N}\right] y_s, \end{aligned}\right\} \tag{6.4}$$

as may be shown by using the equations

$$\begin{aligned} \sum_{s=1}^{N} \exp\left[\frac{2\pi i r s}{N}\right] &= 0 \quad \text{if} \quad 0 < r < N, \\ &= N \quad \text{if } r = 0 \text{ or } r = N, \end{aligned} \tag{6.5}$$

(referred to in § 2). Making this substitution one obtains

$$\begin{aligned} \frac{\mathrm{d}\xi_s}{\mathrm{d}t} &= \frac{1}{N}\sum_{s=1}^{N} \exp\left[-\frac{2\pi i r s}{N}\right] \\ &\left[a x_r + b y_r + \mu\left(\exp\left[-\frac{2\pi i s}{N}\right] - 2 + \exp\left[\frac{2\pi i s}{N}\right]\right)\xi_s\right] \\ &= a\xi_s + b\eta_s + \mu\left(\exp\left[-\frac{2\pi i s}{N}\right] - 2 + \exp\left[\frac{2\pi i s}{N}\right]\right)\xi_s \\ &= \left(a - 4\mu\sin^2\frac{\pi s}{N}\right)\xi_s + b\eta_s. \end{aligned} \tag{6.6}$$

Likewise

$$\frac{\mathrm{d}\eta_s}{\mathrm{d}t} = c\xi_s + \left(d - 4\nu\sin^2\frac{\pi s}{N}\right)\eta_s. \tag{6.7}$$

The equations have now been converted into a quite manageable form, with the variables separated. There are now two equations concerned with ξ_1 and η_1, two concerned with ξ_2 and η_2, etc. The equations themselves are also of a well-known standard form, being linear with constant coefficients. Let p_s and p'_s be the roots of the equation

$$\left(p - a + 4\mu\sin^2\frac{\pi s}{N}\right)\left(p - d + 4\nu\sin^2\frac{\pi s}{N}\right) = bc \tag{6.8}$$

(with $\mathscr{R}p_s \geqslant \mathscr{R}p'_s$ for definiteness), then the solution of the equations is of the form

$$\left.\begin{aligned} \xi_s &= A_s \mathrm{e}^{p_s t} + B_s \mathrm{e}^{p'_s t}, \\ \eta_s &= C_s \mathrm{e}^{p_s t} + D_s \mathrm{e}^{p'_s t}, \end{aligned}\right\} \tag{6.9}$$

where, however, the coefficients A_s, B_s, C_s, D_s are not independent but are restricted to satisfy

$$\left.\begin{aligned} A_s\left(p_s - a + 4\mu\sin^2\frac{\pi s}{N}\right) &= bC_s, \\ B_s\left(p_s' - a + 4\mu\sin^2\frac{\pi s}{N}\right) &= bD_s. \end{aligned}\right\} \qquad (6.10)$$

If it should happen that $p_s = p_s'$ the equations (6.9) have to be replaced by

$$\left.\begin{aligned} \xi_s &= (A_s + B_s t)\,e^{p_s t}, \\ \eta_s &= (C_s + D_s t)\,e^{p_s t}. \end{aligned}\right\} \qquad (6.9')$$

and (6.10) remains true. Substituting back into (6.3) and replacing the variables x_r, y_r by X_r, Y_r (the actual concentrations) the solution can be written

Even when the chemical equilibrium in isolated cells is stable to small perturbations, some of the spatial wavelike perturbations (Fourier modes) grow.

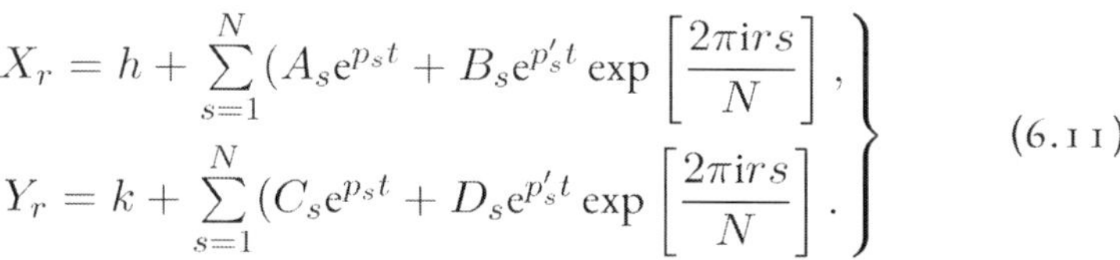

$$\left.\begin{aligned} X_r &= h + \sum_{s=1}^{N}(A_s e^{p_s t} + B_s e^{p_s' t}\exp\left[\frac{2\pi i r s}{N}\right], \\ Y_r &= k + \sum_{s=1}^{N}(C_s e^{p_s t} + D_s e^{p_s' t}\exp\left[\frac{2\pi i r s}{N}\right]. \end{aligned}\right\} \qquad (6.11)$$

Here A_s, B_s, C_s, D_s are still related by (6.10), but otherwise are arbitrary complex numbers; p_s and p_s' are the roots of (6.8).

The expression (6.11 gives the general solution of the equations (6.1) when one assumes that departures from homogeneity are sufficiently small that the functions $f(X, Y)$ and $g(X, Y)$ can safely be taken as linear. The form (6.11) given is not very informative. It will be considerably simplified in § 8. Another implicit assumption concerns random disturbing influences. Strictly speaking one should consider such influences to be continuously at work. This would make the mathematical treatment considerably more difficult without substantially altering the conclusions. The assumption which is implicit in the analysis, here and in § 8, is that the state of the system at $t = 0$ is not one of homogeneity, since it has been displaced from such a state by the disturbances; but after $t = 0$ further disturbances are ignored. In § 9 the theory is reconsidered without this latter assumption.

7. Continuous Ring of Tissue

As an alternative to a ring of separate cells one might prefer to consider a continuous ring of tissue. In this case one can describe the position of a point of the ring by the angle θ which a radius to the point makes with a fixed reference radius. Let the diffusibilities of the two substances be μ' and ν'.These are not quite the same as μ and ν of the last section, since μ and ν are in effect referred to a cell diameter as unit of length, whereas μ' and ν' are referred to a conventional unit, the same unit in which the radius ρ of the ring is measured. Then

$$\mu = \mu' \left(\frac{N}{2\pi\rho}\right)^2, \quad \nu = \nu' \left(\frac{N}{2\pi\rho}\right)^2.$$

The equations are

$$\left.\begin{aligned} \frac{\partial X}{\partial t} &= a(X-h) + b(Y-k) + \frac{\mu'}{\rho^2}\frac{\partial^2 X}{\partial \theta^2}, \\ \frac{\partial Y}{\partial t} &= c(X-h) + d(Y-k) + \frac{\nu'}{\rho^2}\frac{\partial^2 Y}{\partial \theta^2}, \end{aligned}\right\} \qquad (7.1)$$

These are the main differential equations in their more commonly used spatially continuous form. They are a pair of coupled partial differential equations, so called because they contain partial derivatives with respect to time and the spatial variable θ.

which will be seen to be the limiting case of (6.2). The marginal reaction rates a, b, c, d are, as before, the values at the equilibrium position of $\partial f/\partial X, \partial f/\partial Y, \partial g/\partial X, \partial g/\partial Y$. The general solution of the equations is

$$\left.\begin{aligned} X &= h + \sum_{s=-\infty}^{\infty} (A_s e^{p_s t} + B_s e^{p'_s t}) e^{is\theta}, \\ Y &= k + \sum_{s=-\infty}^{\infty} (C_s e^{p_s t} + D_s e^{p'_s t}) e^{is\theta}, \end{aligned}\right\} \qquad (7.2)$$

where p_s, p'_s are now roots of

$$\left(p - a + \frac{\mu' s^2}{\rho^2}\right)\left(p - d + \frac{\nu' s^2}{\rho^2}\right) = bc \qquad (7.3)$$

and

$$\left.\begin{aligned} A_s\left(p_s - a + \frac{\mu' s^2}{\rho^2}\right) &= bC_s, \\ B_s\left(p'_s - a + \frac{\mu' s^2}{\rho^2}\right) &= bD_s. \end{aligned}\right\} \qquad (7.4)$$

This solution may be justified by considering the limiting case of the solution (6.11). Alternatively, one may observe that the formula proposed is a solution, so that it only remains to prove that it is the most general one. This will follow if values of A_s, B_s, C_s, D_s can be found to fit any given initial conditions. It is well known that any function of an angle (such as X) can be expanded as a 'Fourier series'

$$X(\theta) = \sum_{s=-\infty}^{\infty} G_s e^{is\theta} \quad (X(\theta) \text{ being values of } X \text{ at } t = 0),$$

provided, for instance, that its first derivative is continuous. If also

$$Y(\theta) = \sum_{s=-\infty}^{\infty} H_s e^{is\theta} \quad (Y(\theta) \text{ being values of } Y \text{ at } t = 0),$$

then the required initial conditions are satisfied provided $A_s + B_s = G_s$ and $C_s + D_s = H_s$. Values A_s, B_s, C_s, D_s to satisfy these conditions can be found unless $p_s = p'_s$. This is an exceptional case and its solution if required may be found as the limit of the normal case.

The asymptotic behavior depends on the eigenvalues of the matrix $J - Dk^2$. J is the matrix of first partial derivatives of f and g with respect to X and Y. D has the diffusion coefficients on its diagonal. k is the spatial wavelength of the sine/cosine wave. See commentary for more details.

8. Types of Asymptotic Behaviour in the Ring after a Lapse of Time

As the reader was reminded in § 2, after a lapse of time the behaviour of an expression of the form of (6.11) is eventually dominated by the terms for which the corresponding p_s has the largest real part. There may, however, be several terms for which this real part has the same value, and these terms will together dominate the situation, the other terms being ignored by comparison. There will, in fact, normally be either two or four such 'leading' terms. For if p_{s_0} is one of them then $p_{N-s_0} = p_{s_0}$, since

$$\sin^2 \frac{\pi(N - s_0)}{N} = \sin^2 \frac{\pi s_0}{N},$$

so that p_{s_0} and p_{N-s_0} are roots of the same equation (6.8). If also p_{s_0} is complex then $\mathscr{R}p_{s_0} = \mathscr{R}p'_{s_0}$ and so in all

$$\mathscr{R}p_{s_0} = \mathscr{R}p'_{s_0} = \mathscr{R}p_{N-s_0} = \mathscr{R}p'_{N-s_0}.$$

One need not, however, normally anticipate that any further terms will have to be included. If p_{s_0} and p_{s_1} are to have the same real part, then, unless $s_1 = s_0$ or $s_0 + s_1 = N$ the quantities a, b, c, d, μ, ν will be restricted to satisfy some special condition, which they would be unlikely to satisfy by chance. It is possible to find circumstances in which as many as ten terms have to be included if such special conditions *are* satisfied, but these have no particular physical or biological importance. It is assumed below that none of these chance relations hold.

It has already been seen that it is necessary to distinguish the cases where the value of p_{s_0} for one of the dominant terms is real from those where it is complex. These may be called respectively the *stationary* and the *oscillatory* cases.

Stationary case. After a sufficient lapse of time $X_r - h$ and $Y_r - k$ approach asymptotically to the forms

$$\left.\begin{aligned} X_r - h &= 2\mathscr{R} A_{s_0} \exp\left[\frac{2\pi \mathrm{i} s_0 r}{N} + It\right], \\ Y_r - k &= 2\mathscr{R} C_{s_0} \exp\left[\frac{2\pi \mathrm{i} s_0 r}{N} + It\right]. \end{aligned}\right\} \tag{8.1}$$

Oscillatory case. After a sufficient lapse of time $X_r - h$ and $Y_r - k$ approach the forms

$$\left.\begin{aligned} X_r - h &= 2\mathrm{e}^{It}\mathscr{R}\left\{A_{s_0} \exp\left[\frac{2\pi \mathrm{i} s_0 r}{N} + i\omega t\right]\right. \\ &\qquad \left. + A_{N-s_0} \exp\left[-\frac{2\pi \mathrm{i} s_0 r}{N} + i\omega t\right]\right\}, \\ Y_r - k &= 2\mathrm{e}^{It}\mathscr{R}\left\{C_{s_0} \exp\left[\frac{2\pi \mathrm{i} s_0 r}{N} + i\omega t\right]\right. \\ &\qquad \left. + C_{N-s_0} \exp\left[-\frac{2\pi \mathrm{i} s_0 r}{N} + i\omega t\right]\right\}. \end{aligned}\right\} \tag{8.2}$$

The real part of p_{s_0} has been represented by I, standing for 'instability', and in the oscillatory case its imaginary part is ω. By the use of the $\mathscr{R}$ operation (real part of), two terms have in each case been combined in one.

The meaning of these formulae may be conveniently described in terms of waves. In the stationary case there are stationary waves on the ring having s_0 lobes or crests. The coefficients A_{s_0} and C_{s_0} are in a definite ratio given by (6.10), so that the pattern for one morphogen determines that for the other. With the lapse of time the waves become more pronounced provided there is genuine instability, i.e. if I is positive. The wave-length of the waves may be obtained by dividing the number of lobes into the circumference of the ring. In the oscillatory case the interpretation is similar, but the waves are now not stationary but travelling. As well as having a wave-length they have a velocity and a frequency. The frequency is $\omega/2\pi$, and the velocity is obtained by multiplying the wave-length by the frequency. There are two wave trains moving round the ring in opposite directions.

The wave-lengths of the patterns on the ring do not depend only on the chemical data a, b, c, d, μ', ν' but on the circumference of the ring, since they must be submultiples of the latter. There is a sense, however, in which there is a 'chemical wave-length' which does not depend on the dimensions of the ring. This may be described as the limit to which the wave-lengths tend when the rings are made successively larger. Alternatively (at any rate in the case of continuous tissue), it may be described as the wave-length when the radius is chosen to give the largest possible instability I. One may picture the situation by supposing that the chemical wave-length is true wave-length which is achieved whenever possible, but that on a ring it is necessary to 'make do' with an approximation which divides exactly into the circumference.

Although all the possibilities are covered by the stationary and oscillatory alternatives there are special cases of them which deserve to be treated separately. One of these occurs when $s_0 = 0$, and may be described as the 'case of extreme long wave-length', though this term may perhaps preferably be reserved to describe the chemical data when they are such that s_0 is zero whatever the dimensions of the ring. There is also the case of 'extreme short wave-length'. This means that $\sin^2(\pi s_0/N)$ is as large as possible, which is achieved by s_0 being either $\frac{1}{2}N$, or $\frac{1}{2}(N-1)$. If the remaining possibilities are regarded as forming the 'case of finite wave-length', there are six subcases altogether. It will be shown that each of these really can occur, although two of them require three or more morphogens for their realization.

(a) *Stationary case with extreme long wave-length.* This occurs for instance if $\mu = \nu = \frac{1}{4}$, $b = c = 1$, $a = d$. Then $p_s = a - \sin^2 \frac{\pi s}{N} + 1$. This is certainly real and is greatest when $s = 0$. In this case the contents of all the cells are the same; there is no resultant flow from cell to cell due to diffusion, so that each is behaving as if it were isolated. Each is in unstable equilibrium, and slips out of it in synchronism with the others.

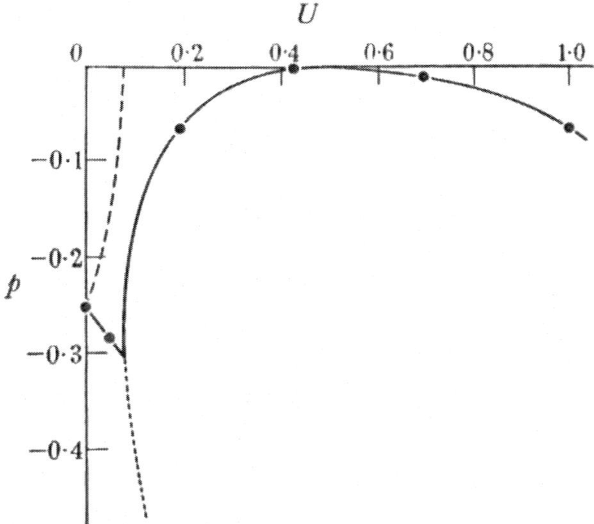

Figure 1. Values of $\mathscr{R}p$ (instability or growth rate), and $|\mathscr{I}p|$ (radian frequency of oscillation), related to wave-length $2\pi U^{-\frac{1}{2}}$ as in the relation (8.3) with $I = 0$. This is a case of stationary waves with finite wave-length. Full line, $\mathscr{R}p$; broken line, $-|\mathscr{I}p|$ (zero for $U > 0.071$); dotted line, $\mathscr{R}p'$. The full circles on the curve for $\mathscr{R}p$ indicate the values of U, p actually achievable on the finite ring considered in § 10, with $s = 0$ on the extreme left, $s = 5$ on the right.

(b) *Oscillatory case with extreme long wave-length.* This occurs, for instance, if $\mu = \nu = \frac{1}{4}$, $b = -c = 1$, $a = d$. Then $p_s = a - \sin^2 \frac{\pi s}{N} \pm i$. This is complex and its real part is greatest when $s = 0$. As in case *(a)* each cell behaves as if it were isolated. The difference from case *(a)* is that the departure from the equilibrium is oscillatory.

(c) *Stationary waves of extreme short wave-length.* This occurs, for instance, if $\nu = 0$, $\mu = 1$, $d = I$, $a = I - 1$, $b = -c = 1$. p_s is

$$I - \frac{1}{2} - 2\sin^2\frac{\pi s}{N} + \sqrt{\left\{\left(2\sin^2\frac{\pi s}{N} + \frac{1}{2}\right)^2 - 1\right\}},$$

and is greatest when $\sin^2(\pi s/N)$ is greatest. If N is even the contents of each cell are similar to those of the next but one, but distinctly different from those of its immediate neighbours.

If however, the number of cells is odd this arrangement is impossible, and the magnitude of the difference between neighbouring cells varies round the ring, from zero at one point to a maximum at a point diametrically opposite.

This is the main case, where the chemical equilibrium in isolated cells is stable but the homogeneous equilibrium is unstable to waves with a range of spatial frequencies k. The condition for modes to grow is quadratic in k^2. Only those wave numbers between a certain maximum and minimum value grow.

(d) *Stationary waves of finite wave-length.* This is the case which is of greatest interest, and has most biological application. It occurs, for instance, if $a = I-2, b = 2.5, c = -1.25, d = I+1.5, \mu' = 1, \nu' = \frac{1}{2}$, and $\frac{\mu}{\mu'} = \frac{\nu}{\nu'} = \left(\frac{N}{2\pi\rho}\right)^2$. As before ρ is the radius of the ring, and N the number of cells in it. If one writes U for $\left(\frac{N}{\pi\rho}^2\right) \sin^2 \frac{\pi s}{N}$, then equation (6.8) can, with these special values, be written

$$(p-I)^2 + \left(\frac{1}{2} + \frac{3}{2}U\right)(p-I) + \frac{1}{2}\left(U - \frac{1}{2}\right)^2 = 0. \qquad (8.3)$$

This has a solution $p = I$ if $U = \frac{1}{2}$. On the other hand, it will be shown that if U has any other (positive) value then both roots for $p - I$ have negative real parts. Their product is positive being $\frac{1}{2}\left(U - \frac{1}{2}\right)^2$, so that if they are real they both have the same sign. Their sum in this case is $-\frac{1}{2} - \frac{3}{2}U$ which is negative. Their common sign is therefore negative. If, however, the roots are complex their real parts are both equal to $-\frac{1}{4} - \frac{3}{4}U$, which is negative.

If the radius ρ of the ring be chosen so that for some integer s_0, $\frac{1}{2} = U = \left(\frac{N}{\pi\rho}\right)^2 \sin^2 \frac{\pi s_0}{N}$, there will be stationary waves with s_0 lobes and a wave-length which is also equal to the chemical wave-length, for p_{s0} will be equal to I, whereas every other p_s will have a real part smaller than I. If, however, the radius is chosen so that $\left(\frac{N}{\pi\rho}\right)^2 \sin^2 \frac{\pi s}{N} = \frac{1}{2}$ cannot hold with an integral s, then (in this example) the actual number of lobes will be one of the two integers nearest to the (non-integral) solutions of this equation, and usually *the* nearest. Examples can, however, be constructed where this simple rule does not apply.

Figure 1 shows the relation (8.3) in graphical form. The curved portions of the graphs are hyperbolae.

When the growth rates of the wavelike perturbations are imaginary, the waves appear to oscillate in time. Turing says this cannot occur with two chemicals but can with three or more. It can also happen with reaction and diffusion between two chemical species on a background prepattern of another, which remains fixed (e.g., Page et al. 2003).

The two remaining possibilities can only occur with three or more morphogens. With one morphogen the only possibility is *(a)*.

(e) *Oscillatory case with a finite wave-length.* This means that there are genuine travelling waves. Since the example to be given involves three morphogens it is not possible to use the formulae of § 6. Instead, one must use the corresponding three morphogen formulae. That which corresponds to (6.8) or (7.3) is most conveniently written as

$$\begin{vmatrix} a_{11} - p - \mu_1 U & a_{12} & a_{13} \\ a_{21} & a_{22} - p - \mu_2 U & a_{23} \\ a_{31} & a_{32} & a_{33} - p - \mu_3 U \end{vmatrix} = 0, \quad (8.4)$$

where again U has been written for $\left(\frac{N}{\pi\rho}\right)^2 \sin^2 \frac{\pi s}{N}$. (This means essentially that $U = \left(\frac{2\pi}{\lambda}\right)^2$, where λ is the wave-length.) The four marginal reactivities are superseded by nine $a_{11}, \ldots, a_{33}$, and the three diffusibilities are μ_1, μ_2, μ_3. Special values leading to traveling waves are

$$\left.\begin{array}{lll} \mu_1 = \frac{2}{3}, & \mu_2 = \frac{1}{3}, & \mu_3 = 0 \\ a_{11} = -\frac{10}{3}, & a_{12} = 3, & a_{13} = -1, \\ a_{21} = -2, & a_{22} = \frac{7}{3}, & a_{23} = 0, \\ a_{31} = 3, & a_{32} = -4, & a_{33} = 0, \end{array}\right\} \quad (8.5)$$

and with them (8.4) reduces to

$$p^3 + p^2(U+1) + p(1 + \tfrac{2}{9}(U-1)^2) + U + 1 = 0. \quad (8.6)$$

If $U = 1$ the roots are $\pm i$ and -2. If U is near to I they are approximately $-1 - U$ and $\pm i + \frac{(U-1)^2}{18}(\pm i - 1)$, and all have negative real parts. If the greatest real part is not the value zero, achieved with $U = 1$, then the value zero must be reached again at some intermediate value of U. Since P is then pure imaginary the even terms of (8.6) must vanish, i.e. $(p^2 + 1)(U + 1) = 0$. But this can only happen if $p = \pm i$, and the vanishing of the odd terms then shows that $U = 1$. Hence zero is the largest real part for any root p of (8.6). The corresponding p is $\pm i$ and U is 1. This means that there are travelling waves with unit (chemical) radian frequency and unit (chemical) velocity. If I is added to a_{11}, a_{22} and a_{33}, the instability will become I in place of zero.

(f) *Oscillatory case with extreme short wave-length.* This means that there is metabolic oscillation with neighbouring cells nearly 180° out of phase.

It can be achieved with three morphogens and the following chemical data:

$$\left.\begin{array}{lll} \mu = 1, & \mu_2 = \mu_3 = 0, & \\ a_{11} = -1, & a_{12} = -1, & a_{13} = 0, \\ a_{21} = 1, & a_{22} = 0, & a_{23} = -1, \\ a_{31} = 0, & a_{32} = 1, & a_{33} = 0. \end{array}\right\} \quad (8.7)$$

With these values (8.4) reduces to

$$p^3 + p^2(U+1) + 2p + U + 1 = 0. \quad (8.8)$$

This may be shown to have all the real parts of its roots negative if $U \geqslant 0$, for if $U = 0$ the roots are near to $-0.6, -0.2 \pm 1.3i$, and if U be continuously increased the values of p will alter continuously. If they ever attain values with a positive real part they must pass through pure imaginary values (or zero). But if p is pure imaginary $p^3 + 2p$ and $(p^2+1)(U+1)$ must both vanish, which is impossible if $U \geqslant 0$. As U approaches infinity, however, one of the roots approaches i. Thus $\mathscr{R}p = 0$ can be approached as closely as desired by large values of U, but not attained.

9. Further Consideration of the Mathematics of the Ring

In this section some of the finer points concerning the development of wave patterns are considered. These will be of interest mainly to those who wish to do further research on the subject, and can well be omitted on a first reading.

1. *General formulae for the two morphogen case.* Taking the limiting case of a ring of large radius (or a filament), one may write $\left(\frac{N}{\pi\rho}\right)^2 \sin^2 \frac{\pi s}{N} = U = \left(\frac{2\pi}{\lambda}\right)^2$ in (6.11) or $\frac{s^2}{\rho^2} = U = \left(\frac{2\pi}{\lambda}\right)^2$ in (7.3) and obtain

$$(p - a + \mu' U)(p - d + \nu' U) = bc, \quad (9.1)$$

which has the solution

$$p = \frac{a+d}{2} - \frac{\mu' + \nu'}{2} U \pm \sqrt{\left\{\left(\frac{\mu' - \nu'}{2} U + \frac{d-a}{2}\right)^2 + bc\right\}}. \quad (9.2)$$

One may put $I(U)$ for the real part of this, representing the instability for waves of wavelength $\lambda = 2\pi U^{-\frac{1}{2}}$. The dominant waves correspond to the maximum of $I(U)$. This maximum may either be at $U = 0$ or $U = \infty$ or at a stationary point on the part of the curve which is hyperbolic (rather than straight). When this last case occurs the values of p (or I) and U at the maximum are

$$\left.\begin{aligned} p = I &= (d\mu' - a\nu' - 2\surd(\mu'\nu')\surd(-bc)(\mu' - \nu')^{-1}, \\ U &= \left(a - d + \frac{\mu' + \nu'}{\surd(\mu'\nu')}\surd(-bc)\right)(\mu' - \nu')^{-1}. \end{aligned}\right\} \tag{9.3}$$

The conditions which lead to the four cases $(a), (b), (c), (d)$ described in the last section are

(a) (Stationary waves of extreme long wave-length.) This occurs if either

$$\text{(i) } bc > 0, \text{ (ii) } bc < 0 \text{ and } \frac{d-a}{\surd(-bc)} > \frac{\mu' + \nu'}{\surd(\mu'\nu')},$$

$$\text{(iii) } bc < 0 \text{ and } \frac{d-a}{\surd(-bc)} < -2.$$

The condition for instability in either case is that either $bc > ad$ or $a + d > 0$.

(b) (Oscillating case with extreme long wave-length, i.e. synchronized oscillations.) This occurs if

$$bc < 0 \text{ and } -2 < \frac{d-a}{\surd(-bc)} < \frac{4\surd(\mu'\nu')}{\mu' + \nu'}.$$

There is instability if in addition $a + d > 0$.

(c) (Stationary waves of extreme short wave-length.) This occurs if $bc < 0$, $\mu' > \nu' = 0$. There is instability if, in addition, $a + d > 0$.

(d) (Stationary waves of finite wave-length.) This occurs if

$$bc < 0 \text{ and } \frac{4\surd(\mu'\nu')}{\mu' + \nu'} < \frac{d-a}{\surd(-bc)} < \frac{(\mu' + \nu')}{(\surd\mu'\nu')'}, \tag{9.4a}$$

and there is instability if also

$$\frac{d}{\surd(-bc)}\sqrt{\frac{\mu'}{\nu'}} - \frac{a}{\surd(-bc)}\sqrt{\frac{\nu'}{\mu'}} > 2. \tag{9.4b}$$

It has been assumed that $\nu' \leqslant \mu' > 0$. The case where $\mu' \leqslant \nu' > 0$ can be obtained by interchanging the two morphogens. In the case $\mu' = \nu' = 0$ there is no co-operation between the cells whatever.

Some additional formulae will be given for the case of stationary waves of finite wavelength. The marginal reaction rates may be expressed parametrically in terms of the diffusibilities, the wave-length, the instability, and two other parameters α and χ. Of these α may be described as the ratio of $X{-}h$ to $Y{-}k$ in the waves. The expressions for the marginal reaction rates in terms of these parameters are

$$\left.\begin{aligned} a &= \mu'(\nu' - \mu')^{-1}(2\nu' U_0 + \chi) + I, \\ b &= \mu'(\nu' - \mu')^{-1}((\mu' + \nu')U_0 + \chi)\alpha, \\ c &= \nu'(\mu' - \nu')^{-1}((\mu' + \nu')U_0 + \chi)\alpha^{-1}, \\ d &= \nu'(\mu' - \nu')^{-1}(2\mu' U_0 + \chi) + I, \end{aligned}\right\} \qquad (9.5)$$

and when these are substituted into (9.2) it becomes

$$p = I - \frac{1}{2}\chi - \frac{\mu' + \nu'}{2}U + \sqrt{\left\{\left(\frac{\mu' + \nu'}{2}U + \frac{1}{2}\chi\right)^2 - \mu'\nu'(U - U_0)^2\right\}}. \qquad (9.6)$$

Here $2\pi U_0^{-\frac{1}{2}}$ is the chemical wave-length and $2\pi U^{-\frac{1}{2}}$ the wave-length of the Fourier component under consideration. χ must be positive for case (d) to apply.

If s be regarded as a continuous variable one can consider (9.2) or (9.6) as relating s to p, and $\mathrm{d}p/\mathrm{d}s$ and $\mathrm{d}^2p/\mathrm{d}s^2$ have meaning. The value of $\mathrm{d}^2p/\mathrm{d}s^2$ at the maximum is of some interest, and will be used below in this section. Its value is

$$\frac{\mathrm{d}^2 p}{\mathrm{d}s^2} = -\frac{\sqrt{(\mu'\nu')}}{\rho^2}\cdot\frac{8\sqrt{(\mu'\nu')}}{\mu' + \nu'}\cos^2\frac{\pi s}{N}(1 + \chi U_0^{-1}(\mu' + \nu')^{-1})^{-1}. \qquad (9.7)$$

2. In §§ 6, 7, 8 it was supposed that the disturbances were not continuously operative, and that the marginal reaction rates did not change with the passage of time. These assumptions will now be dropped, though it will be necessary to make some other, less drastic, approximations to replace them. The (statistical) amplitude of the

'noise' disturbances will be assumed constant in time. Instead of (6.6), (6.7), one then has

$$\left.\begin{aligned}\frac{\mathrm{d}\xi}{\mathrm{d}t} &= a'\xi + b\eta + R_1(t),\\ \frac{\mathrm{d}\eta}{\mathrm{d}t} &= c\xi + d'\eta + R_2(t),\end{aligned}\right\} \tag{9.8}$$

where ξ, η have been written for ξ_s, η_s since s may now be supposed fixed. For the same reason $a - 4\mu \sin^2 \frac{\pi s}{N}$ has been replaced by a' and $d - 4\nu \sin^2 \frac{\pi s}{N}$ by d'. The noise disturbances may be supposed to constitute white noise, i.e. if (t_1, t_2) and (t_3, t_4) are two non-overlapping intervals then $\int_{t_1}^{t_2} R_1(t)\mathrm{d}t$ and $\int_{t_3}^{t_4} R_2(t)\mathrm{d}t$ are statistically independent and each is normally distributed with variances $\beta_1(t_2 - t_1)$ and $\beta_1(t_4 - t_3)$ respectively, β_1 being a constant describing the amplitude of the noise. Likewise for $R_2(t)$, the constant β_1 being replaced by β_2. If p and p' are the roots of $(p - a')(p - d') = bc$ and p is the greater (both being real), one can make the substitution

$$\left.\begin{aligned}\xi &= b(u + v)\\ \eta &= (p - a')u + (p' - a')v,\end{aligned}\right\} \tag{9.9}$$

which transforms (9.8) into*

$$\frac{\mathrm{d}u}{\mathrm{d}t} = pu + \frac{p' - a'}{(p' - p)b}R_1(t) - \frac{R_2(t)}{p' - p} + \xi\frac{\mathrm{d}}{\mathrm{d}t}\left(\frac{p' - a'}{(p' - p)b}\right) - \eta\frac{\mathrm{d}}{\mathrm{d}t}\left(\frac{1}{p' - p}\right), \tag{9.11}$$

with a similar equation for v, of which the leading terms are $\mathrm{d}v/\mathrm{d}t = p'v$. This indicates that v will be small, or at least small in comparison with u after a lapse of time. If it is assumed that $v = 0$ holds (9.11) may be written

$$\frac{\mathrm{d}u}{\mathrm{d}t} = qu + L_1(t)R_1(t) + L_2(t)R_2(t), \tag{9.12}$$

$$\text{where} \quad L_1(t) = \frac{p' - a'}{(p' - p)b}, \quad L_2(t) = \frac{1}{p' - p}, \quad q = p + bL_1'(t). \tag{9.13}$$

*Editor's note: The original paper skips from equation 9.9 to 9.11; there is no equation 9.10.

The solution of this equation is

$$u = \int_{-\infty}^{t} (L_1(w)R_1(w) + L_2(w)R_2(w)) \exp\left[\int_{w}^{t} q(z)\mathrm{d}z\right] \mathrm{d}w. \tag{9.14}$$

One is, however, not so much interested in such a solution in terms of the statistical disturbances as in the consequent statistical distribution of values of u, ξ, and η at various times after instability has set in. In view of the properties of 'white noise' assumed above, the values of u at time t will be distributed according to the normal error law, with the variance

$$\int_{-\infty}^{t} [\beta_1(L_1(w))^2 + \beta_2(L_2(w))^2] \exp\left[2\int_{w}^{t} q(z)\mathrm{d}z\right] \mathrm{d}w. \tag{9.15}$$

There are two commonly occurring cases in which one can simplify this expression considerably without great loss of accuracy. If the system is in a distinctly stable state, then $q(t)$, which is near to $p(t)$, will be distinctly negative, and $\exp\left[\int_w^t q(z)dz\right]$ will be small unless w is near to t. But then $L_1(w)$ and $L_2(w)$ may be replaced by $L_1(t)$ and $L_2(t)$ in the integral, and also $q(z)$ may be replaced by $q(t)$. With these approximations the variance is

$$(-2q(t))^{-1}[\beta_1(L_1(t))^2 + \beta_2(L_2(t))^2]. \tag{9.16}$$

A second case where there is a convenient approximation concerns times when the system is unstable, so that $q(t) > 0$. For the approximation concerned to apply $2\int_w^t q(z)dz$ must have its maximum at the last moment $w(= t_0)$ when $q(t_0) = 0$, and it must be the maximum by a considerable margin (e.g. at least 5) over all other local maxima. These conditions would apply for instance if $q(z)$ were always increasing and had negative values at a sufficiently early time. One also requires $q'(t_0)$ (the rate of increase of q at time t_0) to be reasonably large; it must at least be so large that over a period of time of length $(q'(t_0))^{-\frac{1}{2}}$ near to t_0 the changes in $L_1(t)$ and $L_2(t)$ are small, and $q'(t)$ itself must not appreciably alter in this period. Under these circumstances the integrand is negligible when w is considerably different from t_0, in comparison with its values at that time, and therefore one may replace $L_1(w)$ and $L_2(w)$ by $L_1(t_0)$ and $L_2(t_0)$, and $q'(w)$ by $q'(t_0)$. This gives the value

$$\sqrt{\pi}(q'(t_0))^{-\frac{1}{2}}[\beta_1(L_1(t_0))^2 + \beta_2(L_2(t_0))^2] \exp\left[2\int_{t_0}^{t} q(z)\mathrm{d}z\right], \tag{9.17}$$

Turing considers the linearized system changing in time and having stochastic fluctuations continually added. He shows that only the fluctuations added close to the time at which instability first occurs contribute significantly to the final solution.

for the variance of u.

The physical significance of this latter approximation is that the disturbances near the time when the instability is zero are the only ones which have any appreciable ultimate effect. Those which occur earlier are damped out by the subsequent period of stability. Those which occur later have a shorter period of instability within which to develop to greater amplitude. This principle is familiar in radio, and is fundamental to the theory of the superregenerative receiver.

Naturally one does not often wish to calculate the expression (9.17), but it is valuable as justifying a common-sense point of view of the matter. The factor $\exp\left[\int_{t_0}^{t} q(z)dz\right]$ is essentially the integrated instability and describes the extent to which one would expect disturbances of appropriate wave-length to grow between times t_0 and t. Taking the terms in β_1, β_2 into consideration separately, the factor $\sqrt{\pi}\beta_1(q'(t_0))^{-\frac{1}{2}}(L_1(t_0))^2$ indicates that the disturbances on the first morphogen should be regarded as lasting for a time

$$\sqrt{\pi}(q_1(t_0))^{-\frac{1}{2}}(bL_1(t_0))^2.$$

The dimensionless quantities $bL_1(t_0)$, $bL_2(t_0)$ will not usually be sufficiently large or small to justify their detailed calculation.

3. The extent to which the component for which p_s is greatest may be expected to out-distance the others will now be considered in case (d). The greatest of the p_s will be called p_{s_0}. The two closest competitors to s_0 will be $s_0{-}1$ and s_0+1; it is required to determine how close the competition is. If the variation in the chemical data is sufficiently small it may be assumed that, although the exponents p_{s_0-1}, p_{s_0}, p_{s_0+1} may themselves vary appreciably in time, the differences $p_{s_0} - p_{s_0-1}$ and $p_{s_0} - p_{s_0+1}$ are constant. It certainly can happen that one of these differences is zero or nearly zero, and

there is then 'neck and neck' competition. The weakest competition occurs when $p_{s_0-1} = p_{s_0+1}$. In this case

$$p_{s_0} - p_{s_0-1} = p_{s_0} - p_{s_0+1} = -\frac{1}{2}(p_{s_0+1} - 2p_{s_0} + p_{s_0-1}).$$

But if s_0 is reasonably large $p_{s_0+1} - 2p_{s_0} + p_{s_0-1}$ can be set equal to $(\mathrm{d}^2p/\mathrm{d}s^2)_{s=s_0}$. It may be concluded that the rate at which the most quickly growing component grows cannot exceed the rate for its closest competitor by more than about $\frac{1}{2}(\mathrm{d}^2p/\mathrm{d}s^2)_{s=s_0}$. The formula (9.7), by which $\mathrm{d}^2p/\mathrm{d}s^2$ can be estimated, may be regarded as the product of two factors. The dimensionless factor never exceeds 4. The factor $\sqrt{(\mu'\nu')}/\rho^2$ may be described in very rough terms as 'the reciprocal of the time for the morphogens to diffuse a length equal to a radius'. In equally rough terms one may say that a time of this order of magnitude is required for the most quickly growing component to get a lead, amounting to a factor whose logarithm is of the order of unity, over its closest competitors, in the favourable case where $p_{s_0-1} = p_{s_0+1}$.

4. Very little has yet been said about the effect of considering non-linear reaction rate functions when far from homogeneity. Any treatment so systematic as that given for the linear case seems to be out of the question. It is possible, however, to reach some qualitative conclusions about the effects of non-linear terms. Suppose that z_1 is the amplitude of the Fourier component which is most unstable (on a basis of the linear terms), and which may be supposed to have wave-length λ. The non-linear terms will cause components with wavelengths $\frac{1}{2}\lambda$, $\frac{1}{3}\lambda$, $\frac{1}{4}\lambda$, ... to appear as well as a space-independent component. If only quadratic terms are taken into account and if these are somewhat small, then the component of wavelength $\frac{1}{2}\lambda$ and the space-independent component will be the strongest. Suppose these have amplitudes z_2 and z_1. The state of the system is thus being described by the numbers z_0, z_1, z_2. In the absence of non-linear terms they would satisfy equations

$$\frac{\mathrm{d}z_0}{\mathrm{d}t} = p_0 z_0, \quad \frac{\mathrm{d}z_1}{\mathrm{d}t} = p_1 z_1, \quad \frac{\mathrm{d}z_2}{\mathrm{d}t} = p_2 z_2,$$

and if there is slight instability p_1 would be a small positive number, but p_0 and p_2 distinctly negative. The effect of the non-linear terms is to replace these equations by ones of the form

$$\frac{dz_0}{dt} = p_0 z_0 + A z_1^2 + B z_2^2,$$
$$\frac{dz_1}{dt} = p_1 z_1 + C z_2 z_1 + D z_0 z_1,$$
$$\frac{dz_2}{dt} = p_2 z_2 + E z_1^2 + F z_0 z_2.$$

As a first approximation one may put $dz_0/dt = dz_2/d(t) = 0$ and ignore z_1^4 and higher powers; z_0 and z_1 are then found to be proportional to z_1^2, and the equation for z_1 can be written $dz_1/dt = p_0 z_1 - k z_1^3$. The sign of k in this differential equation is of great importance. If it is positive, then the effect of the term $k z_1^3$ is to arrest the exponential growth of z_1 at the value $\sqrt{(p_1/k)}$. The 'instability' is then very confined in its effect, for the waves can only reach a finite amplitude, and this amplitude tends to zero as the instability (p_1) tends to zero. If, however, k is negative the growth becomes something even faster than exponential, and, if the equation $dz_1/dt = p_1 z_1 - k z_1^3$ held universally, it would result in the amplitude becoming infinite in a finite time. This phenomenon may be called 'catastrophic instability'. In the case of two-dimensional systems catastrophic instability is almost universal, and the corresponding equation takes the form $dz_1/dt = p_1 z_1 + k z_1^2$. Naturally enough in the case of catastrophic instability the amplitude does not really reach infinity, but when it is sufficiently large some effect previously ignored becomes large enough to halt the growth.

Here Turing goes beyond linear analysis and attempts to approximate the amplitude of the fastest-growing sine wave.

5. Case (a) as described in § 8 represents a most extremely featureless form of pattern development. This may be remedied quite simply by making less drastic simplifying assumptions, so that a less gross account of the pattern can be given by the theory. It was assumed in § 9 that only the most unstable Fourier components would contribute appreciably to the pattern, though it was seen above (heading (3) of this section) that (in case (d)) this will only apply if the period of time involved is adequate to permit the morphogens, supposed for this purpose to be chemically inactive, to diffuse over the whole ring or organ concerned. The same may be shown to apply for case

(a). If this assumption is dropped a much more interesting form of pattern can be accounted for. To do this it is necessary to consider not merely the components with $U = 0$ but some others with small positive values of U. One may assume the form At—BU for p. Linearity in U is assumed because only small values of U are concerned, and the term At is included to represent the steady increase in instability. By measuring time from the moment of zero instability the necessity for a constant term is avoided. The formula (9.17) may be applied to estimate the statistical distribution of the amplitudes of the components. Only the factor $\exp\left[2\int_{t_0}^{t} q(z)\mathrm{d}z\right]$ will depend very much on U, and taking $q(t) = p(t) = At$—BU, t_0 must be BU/A and the factor is

$$\exp\left[A(t\text{—}BU/A)^2\right].$$

The term in U^2 can be ignored if At^2 is fairly large, for then either B^2U^2/A^2 is small or the factor e^{-BUt} is. But At^2 certainly is large if the factor e^{At^2}, applying when $U = 0$, is large. With this approximation the variance takes the form $C\mathrm{e}^{-\frac{1}{2}k^2U}$, with only the two parameters C, k to distinguish the pattern populations. By choosing appropriate units of concentration and length these pattern populations may all be reduced to a standard one, e.g. with $C = k = 1$. Random members of this population may be produced by considering any one of the type (a) systems to which the approximations used above apply. They are also produced, but with only a very small amplitude scale, if a homogeneous one-morphogen system undergoes random disturbances without diffusion for a period, and then diffusion without disturbance. This process is very convenient for computation, and can also be applied to two dimensions. Figure 2 shows such a pattern, obtained in a few hours by a manual computation.

To be more definite a set of numbers $u_{r,s}$ was chosen, each being ± 1, and taking the two values with equal probability. A function $f(x, y)$ is related to these numbers by the formula

$$f(x,y) = \Sigma u_{r,s} \exp\left[-\frac{1}{2}((x-hr)^2 + (y-hs)^2)\right].$$

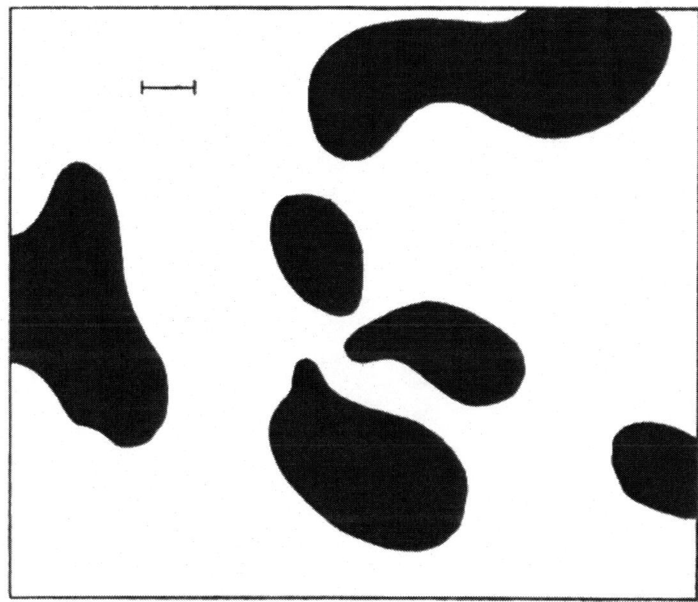

Figure 2. An example of a 'dappled' pattern as resulting from a type (*a*) morphogen system. A marker of unit length is shown. See text, § 9, 11.

In the actual computation a somewhat crude approximation to the function

$$\exp\left[-\frac{1}{2}(x^2+y^2)\right]$$

was used and h was about 0.7. In the figure the set of points where $f(x,y)$ is positive is shown black. The outlines of the black patches are somewhat less irregular than they should be due to an inadequacy in the computation procedure.

10. A Numerical Example

The numerous approximations and assumptions that have been made in the foregoing analysis may be rather confusing to many readers. In the present section it is proposed to consider in detail a single example of the case of most interest, (d). This will be made as specific as possible. It is unfortunately not possible to specify actual chemical reactions with the required properties, but it is thought that the reaction rates associated with the imagined reactions are not unreasonable.

The detail to be specified includes

(i) The number and dimensions of the cells of the ring.

(ii) The diffusibilities of the morphogens.

(iii) The reactions concerned.

(iv) The rates at which the reactions occur.

(v) Information about random disturbances.

(vi) Information about the distribution, in space and time, of those morphogens which are of the nature of evocators.

These will be taken in order.

(i) It will be assumed that there are twenty cells in the ring, and that they have a diameter of 0.1 mm each. These cells are certainly on the large rather than the small side, but by no means impossibly so. The number of cells in the ring has been chosen rather small in order that it should not be necessary to make the approximation of continuous tissue.

(ii) Two morphogens are considered. They will be called X and Y, and the same letters will be used for their concentrations. This will not lead to any real confusion. The diffusion constant for X will be assumed to be 5×10^{-8} cm$^2 s^{-1}$ and that for Y to be 2.5×10^{-8} cm$^2 s^{-1}$. With cells of diameter 0.01 cm this means that X flows between neighbouring cells at the rate 5×10^{-4} of the difference of X-content of the two cells per second. In other words, if there is nothing altering the concentrations but diffusion the difference of concentrations suffers an exponential decay with time constant 1000s, or 'half-period' of 700s. These times are doubled for Y.

If the cell membrane is regarded as the only obstacle to diffusion the permeability of the membranes to the morphogen is 5×10^{-6} cm/s or 0.018 cm/h. Values as large as 0.1 cm/h have been observed (Davson and Danielli 1943, figure 28).

(iii) The reactions are the most important part of the assumptions. Four substances A, X, Y, B are involved; these are isomeric, i.e. the molecules of the four substances are all rearrangements of the same atoms. Substances C, C', W will also be concerned. The thermodynamics of the problem will not be discussed except to say that it is contemplated that of the substances A, X, Y, B the one with the greatest free energy is A, and that with the least is B. Energy for the whole process is obtained by the degradation of A into B. The substance C is in effect a catalyst for the reaction $Y \to X$, and may also be regarded as an evocator, the system being unstable if there is a sufficient concentration of C.

The reactions postulated are

$$\begin{aligned} Y + X &\to W, \\ W + A &\to 2Y + B \quad \text{instantly}, \\ 2X &\to W, \\ A &\to X, \\ Y &\to B, \\ Y + C &\to C' \quad \text{instantly}, \\ C' &\to X + C. \end{aligned}$$

(iv) For the purpose of stating the reaction rates special units will be introduced (for the purpose of this section only). They will be based on a period of 1000s as units of time, and 10^{-11} mole/cm^3 as concentration unit. [2] There will be little occasion to use any but these special units (S.U.). The concentration of A will be assumed to have the large value of 1000 S.U. and the catalyst C, together with its combined form C' the concentration $10^{-3}(1 + \gamma)$ S.U., the dimensionless quantity γ being often supposed somewhat small, though values over as large a range as from —0.5 to 0.5 may be

[2] A somewhat larger value of concentration unit (e.g. 10^{-9} mole/cm^3) is probably more suitable. The choice of unit only affects the calculations through the amplitude of the random disturbances.

considered. The rates assumed will be

$$Y + X \to W \qquad \text{at the rate } \frac{25}{16}YX,$$

$$2X \to W \qquad \text{at the rate } \frac{7}{64}X^2,$$

$$A \to X \qquad \text{at the rate } \frac{1}{16} \times 10^{-3}A,$$

$$C' \to X + C \qquad \text{at the rate } \frac{55}{32} \times 10^{+3}C',$$

$$Y \to B \qquad \text{at the rate } \frac{1}{16}Y.$$

With the values assumed for A and C' the net effect of these reactions is to convert X into Y at the rate $\frac{1}{32}[50XY + 7X^2 - 55(1 + \gamma)]$ at the same time producing X at the constant rate $\frac{1}{16}$, and destroying Y at the rate $\frac{Y}{16}$. If, however, the concentration of Y is zero and the rate of increase of Y required by these formulae is negative, the rate of conversion of Y into X is reduced sufficiently to permit Y to remain zero.

In the special units $\mu = \frac{1}{2}, \nu = \frac{1}{4}$.

(v) Statistical theory describes in detail what irregularities arise from the molecular nature of matter. In a period in which, on the average, one should expect a reaction to occur between n pairs (or other combinations) of molecules, the actual number will differ from the mean by an amount whose mean square is also n, and is distributed according to the normal error law. Applying this to a reaction proceeding at a rate F (S.U.) and taking the volume of the cell as 10^{-8}cm^3 (assuming some elongation tangentially to the ring) it will be found that the root mean square irregularity of the quantity reacting in a period τ of time (S.U.) is $0.004\surd(F\tau)$.

The diffusion of a morphogen from a cell to a neighbour may be treated as if the passage of a molecule from one cell to another were a monomolecular reaction; a molecule must be imagined to change its form slightly as it passes the cell wall. If the diffusion constant for a wall is μ, and quantities M_1, M_2 of the relevant morphogen lie on

Table 1. Some stationary-wave patterns.

cell number	first specimen: incipient pattern X	first specimen: incipient pattern Y	first specimen: final pattern X	first specimen: final pattern Y	second specimen: incipient Y	'slow cooking': incipient Y	four-lobed equilibrium X	four-lobed equilibrium Y
0	1·130	0·929	0·741	1·463	0·834	1·057	1·747	0·000
1	1·123	0·940	0·761	1·469	0·833	0·903	1·685	0·000
2	1·154	0·885	0·954	1·255	0·766	0·813	1·445	2·500
3	1·215	0·810	1·711	0·000	0·836	0·882	0·445	2·500
4	1·249	0·753	1·707	0·000	0·930	1·088	1·685	0·000
5	1·158	0·873	0·875	1·385	0·898	1·222	1·747	0·000
6	1·074	1·003	0·700	1·622	0·770	1·173	1·685	0·000
7	1·078	1·000	0·699	1·615	0·740	0·956	0·445	2·500
8	1·148	0·896	0·885	1·382	0·846	0·775	0·445	2·500
9	1·231	0·775	1·704	0·000	0·937	0·775	1·685	0·000
10	1·204	0·820	1·708	0·000	0·986	0·969	1·747	0·000
11	1·149	0·907	0·944	1·273	1·019	1·170	1·685	0·000
12	1·156	0·886	0·766	1·451	0·899	1·203	0·445	2·500
13	1·170	0·854	0·744	1·442	0·431	1·048	0·445	2·500
14	1·131	0·904	0·756	1·478	0·485	0·868	1·685	0·000
15	1·090	0·976	0·935	1·308	0·919	0·813	1·747	0·000
16	1·109	0·957	1·711	0·000	1·035	0·910	1·685	0·000
17	1·201	0·820	1·706	0·000	1·003	1·050	0·445	2·500
18	1·306	0·675	0·927	1·309	0·899	1·175	0·445	2·500
19	1·217	0·811	0·746	1·487	0·820	1·181	1·685	0·000

the two sides of it, the root-mean-square irregularity in the amount passing the wall in a period τ is

$$0.004\sqrt{\{(M_1 + M_2)\mu\tau\}}\,.$$

These two sources of irregularity are the most significant of those which arise from truly statistical cause, and are the only ones which are taken into account in the calculations whose results are given below. There may also be disturbances due to the presence of neighbouring anatomical structures, and other similar causes. These are of great importance, but of too great variety and complexity to be suitable for consideration here.

(vi) The only morphogen which is being treated as an evocator is C. Changes in the concentration of A might have similar effects, but the change would have to be rather great. It is preferable to assume that A is a 'fuel substance' (e.g. glucose) whose concentration does not change. The concentration of C, together with its combined form C', will be supposed the same in all cells, but it changes with the passage of time. Two different varieties of the problem will be considered, with slightly different assumptions.

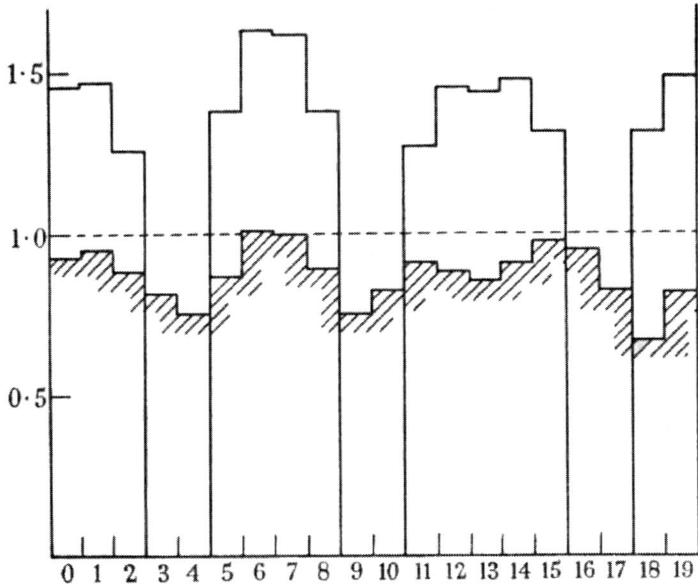

Figure 3. Concentrations of Y in the development of the first specimen (taken from table 1). - - - - - original homogeneous equilibrium; /////// incipient pattern; ———————— final equilibrium.

The results are shown in table 1. There are eight columns, each of which gives the concentration of a morphogen in each of the twenty cells; the circumstances to which these concentrations refer differ from column to column. The first five columns all refer to the same 'variety' of the imaginary organism, but there are two specimens shown. The specimens differ merely in the chance factors which were involved. With this variety the value of γ was allowed to increase at the rate of 2^{-7} S.U. from the value $-\frac{1}{4}$ to $+\frac{1}{16}$. At this point a pattern had definitely begun to appear, and was recorded. The parameter γ was then allowed to decrease at the same rate to zero and then remained there until there was no more appreciable change. The pattern was then recorded again. The concentrations of Y in these two recordings are shown in figure 3 as well as in table 1. For the second specimen only one column of figures is given, viz. those for the Y morphogen in the incipient pattern. At this stage the X values are closely related to the Y values, as may be seen from the first specimen (or from theory). The final values can be made almost indistinguishable from those for the first specimen by renumbering the cells and have therefore not been given. These two specimens may be said to belong to the 'variety with quick cooking', because the instability is allowed to increase so quickly that the pattern appears relatively soon. The effect of this haste might be regarded as rather unsatisfactory, as the incipient pattern is very irregular. In both specimens the four-lobed component is present in considerable strength in the incipient pattern. It 'beats' with the three-lobed component producing considerable irregularity. The relative magnitudes of the three- and four-lobed components depend on chance and vary from specimen to specimen. The four-lobed component may often be the stronger, and may occasionally be so strong that the final pattern is four-lobed. How often this happens is not known, but the pattern, when it occurs, is shown in the last two columns of the table. In this case the disturbances were supposed removed for some time before recording, so as to give a perfectly regular pattern.

The remaining column refers to a second variety, one with 'slow cooking'. In this the value of γ was allowed to increase only at the rate 10^{-5}. Its initial value was —0.010, but is of no significance. The final value was 0.003. With this pattern, when shown graphically, the irregularities are definitely perceptible, but are altogether overshadowed by the three-lobed component. The possibility of the ultimate pattern being four-lobed is not to be taken seriously with this variety.

The set of reactions chosen is such that the instability becomes 'catastrophic' when the second-order terms are taken into account, i.e. the growth of the waves tends to make the whole system more unstable than ever. This effect is finally halted when (in some cells) the concentration of Y has become zero. The constant conversion of Y into X through the agency of the catalyst C can then no longer continue in these cells, and the continued growth of the amplitude of the waves is arrested. When $\gamma = 0$ there is of course an equilibrium with $X = Y = 1$ in all cells, which is very slightly stable. There are, however, also other stable equilibria with $\gamma = 0$, two of which are shown in the table. These final equilibria may, with some trouble but little difficulty, be verified to be solutions of the equations (6.1) with

$$\frac{\mathrm{d}X}{\mathrm{d}t} = \frac{\mathrm{d}Y}{\mathrm{d}t} = 0,$$

$$\text{and} \qquad 32f(X,Y) = 57 - 50XY - 7Y^2,$$
$$32g(X,Y) = 50XY + 7Y^2 - 2Y - 55.$$

The morphogen concentrations recorded at the earlier times connect more directly with the theory given in §§ 6 to 9. The amplitude of the waves was then still sufficiently small for the approximation of linearity to be still appropriate, and consequently the 'catastrophic' growth had not yet set in.

The functions $f(X,Y)$ and $g(X,Y)$ of § 6 depend also on γ and are

$$f(X,Y) = \frac{1}{32}[-7X^2 - 50XY + 57 + 55\gamma],$$
$$g(X,Y) = \frac{1}{32}[7X^2 + 50XY - 2Y - 55 - 55\gamma].$$

In applying the theory it will be as well to consider principally the behaviour of the system when γ remains permanently zero. Then for equilibrium $f(X,Y) = g(X,Y) = 0$ which means that $X = Y = 1$, i.e. $h = k = 1$. One also finds the following values for various quantities mentioned in §§ 6 to 9:

$$a = -2, \quad b = -1.5625, \quad c = 2, \quad d = 1.500, \quad s = 3.333,$$
$$I = 0, \quad \alpha = 0.625, \quad \chi = 0.500, \quad (d-a)(-bc)^{-\frac{1}{2}} = 1.980,$$
$$(\mu+\nu)(\mu\nu)^{-\frac{1}{2}} = 2.121, \quad p_0 = -0.25 \pm 0.25\mathrm{i},$$
$$p_2 = -0.0648, \quad p_3 = -0.0034, \quad p_4 = -0.0118.$$

(The relation between p and U for these chemical data, and the values p_n, can be seen in figure 1, the values being so related as to make the curves apply to this example as well as that in § 8.) The value $s = 3.333$ leads one to expect a three-lobed pattern as the commonest, and this is confirmed by the values p_n. The four-lobed pattern is evidently the closest competitor. The closeness of the competition may be judged from the difference $p_3 - p_4 = 0.0084$, which suggests that the three-lobed component takes about 120 S.U. or about 33 h to gain an advantage of a neper (i.e. about 2.7 : 1) over the four-lobed one. However, the fact that γ is different from 0 and is changing invalidates this calculation to some extent.

The figures in table I were mainly obtained with the aid of the Manchester University Computer.

Although the above example is quite adequate to illustrate the mathematical principles involved it may be thought that the chemical reaction system is somewhat artificial. The following example is perhaps less so. The same 'special units' are used. The reactions assumed are

$$
\begin{array}{lll}
A \to X & \text{at the rate} & 10^{-3}A,\ A = 10^3, \\
X + Y \to C & \text{at the rate} & 10^3 XY, \\
C \to X + Y & \text{at the rate} & 10^6 C, \\
C \to D & \text{at the rate} & 62.5C, \\
B + C \to W & \text{at the rate} & 0.125BC,\ B = 10^3, \\
W \to Y + C & \text{instantly,} & \\
Y \to E & \text{at the rate} & 0.0625Y, \\
Y + V \to V' & \text{instantly,} & \\
V' \to E + V & \text{at the rate} & 62.5V',\ V' = 10^{-3}\beta.
\end{array}
$$

The effect of the reactions $X + Y \rightleftarrows C$ is that $C = 10^{-3}XY$. The reaction $C \to D$ destroys C, and therefore in effect both X and Y, at the rate $\frac{1}{16}XY$. The reaction $A \to X$ forms X at the constant rate 1, and the pair $Y + V \to V' \to E + V$ destroys Y at the constant rate $\frac{1}{16}\beta$. The pair $B + C \to W \to Y + C$ forms Y at the rate $\frac{1}{8}XY$, and $Y \to E$ destroys it at the rate $\frac{1}{16}Y$. The total effect therefore is that X is produced at the rate $f(X, Y) = \frac{1}{16}(16—XY)$, and Y at the rate $g(X, Y) = \frac{1}{16}(XY - Y - \beta)$. However, $g(X, Y) = 0$ if $Y \leqslant 0$. The diffusion constants will be supposed to be $\mu = \frac{1}{4}$, $\nu = \frac{1}{16}$. The homogeneity condition gives $hk = 16$, $k = 16 - \beta$. It will be seen from conditions (9.4a) that case (d) applies if and only if $\frac{4}{k} + \frac{k}{4} < 2.75$, i.e. if k lies between 1.725 and 9.257. Condition (9.4b) shows that there will be instability if in addition $\frac{8}{k} + \frac{k}{8} > \sqrt{3} + \frac{1}{2}$, i.e. if k does not lie between 4.98 and 12.8. It will also be found that the wave-length corresponding to $k = 4.98$ is 4.86 cell diameters.

In the case of a ring of six cells with $\beta = 12$ there is a stable equilibrium, as shown in table 2.

It should be recognized that these equilibria are only dynamic equilibria. The molecules which together make up the chemical waves are continually changing, though their concentrations in any particular cell are only undergoing small statistical fluctuations. Moreover, in order to

Table 2.

cell	0	1	2	3	4	5
X	7.5	3.5	2.5	2.5	3.5	7.5
Y	0	8	8	8	8	0

maintain the wave pattern a continual supply of free energy is required. It is clear that this must be so since there is a continual degradation of energy through diffusion. This energy is supplied through the 'fuel substances' (A, B in the last example), which are degraded into 'waste products' (D, E).

11. Restatement and Biological Interpretation of the Results

Certain readers may have preferred to omit the detailed mathematical treatment of §§ 6 to 10. For their benefit the assumptions and results will be briefly summarized, with some change of emphasis.

The system considered was either a ring of cells each in contact with its neighbours, or a continuous ring of tissue. The effects are extremely similar in the two cases. For the purposes of this summary it is not necessary to distinguish between them. A system with two or three morphogens only was considered, but the results apply quite generally. The system was supposed to be initially in a stable homogeneous condition, but disturbed slightly from this state by some influences unspecified, such as Brownian movement or the effects of neighbouring structures or slight irregularities of form. It was supposed also that slow changes are taking place in the reaction rates (or, possibly, the diffusibilities) of the two or three morphogens under consideration. These might, for instance, be due to changes of concentration of other morphogens acting in the role of catalyst or of fuel supply, or to a concurrent growth of the cells, or a change of temperature. Such changes are supposed ultimately to bring the system out of the stable state. The phenomena when the system is just unstable were the particular subject of the inquiry. In order to make the problem mathematically tractable it was necessary to assume that the system never deviated very far from the original homogeneous condition. This assumption was called the 'linearity assumption' because it permitted the replacement of the general reaction rate functions by linear ones. This linearity assumption is a serious one. Its justification lies in the fact that the patterns produced in the early stages when it is valid may be expected to have strong qualitative similarity to those prevailing in the later stages when it is not. Other, less important, assumptions were also made at the beginning of the mathematical theory, but the detailed effects of these were mostly considered in § 9, and were qualitatively unimportant.

The conclusions reached were as follows. After the lapse of a certain period of time from the beginning of instability, a pattern of morphogen concentrations appears which can best be described in terms of 'waves'. There are six types of possibility which may arise.

(a) The equilibrium concentrations and reaction rates may become such that there would be instability for an isolated cell with the same content as any one of the cells of the ring. If that cell drifts away from the equilibrium position, like an upright stick falling over, then, in the ring, each cell may be expected to do likewise. In neighbouring cells the drift may be expected to be in the same direction, but in distant cells, e.g. at opposite ends of a diameter there is no reason to expect this to be so.

This is the least interesting of the cases. It is possible, however, that it might account for 'dappled' colour patterns, and an example of a pattern in two dimensions produced by this type of process is shown in figure 2 for comparison with 'dappling'. If dappled patterns are to be explained in this way they must be laid down in a latent form when the foetus is only a few inches long. Later the distances would be greater than the morphogens could travel by diffusion.

(b) This case is similar to (*a*), except that the departure from equilibrium is not a unidirectional drift, but is oscillatory. As in case (*a*) there may not be agreement between the contents of cells at great distances.

There are probably many biological examples of this metabolic oscillation, but no really satisfactory one is known to the author.

(c) There may be a drift from equilibrium, which is in opposite directions in contiguous cells.

No biological examples of this are known.

(d) There is a stationary wave pattern on the ring, with no time variation, apart from a slow increase in amplitude, i.e. the pattern is slowly becoming more marked. In the case of a ring of continuous tissue the pattern is sinusoidal, i.e. the concentration of one of the morphogens plotted against position on the ring is a sine curve. The peaks of the waves will be uniformly spaced round the ring. The number of such peaks can be obtained approximately by dividing the so-called 'chemical wave-length' of the system into the circumference of the ring. The chemical wave-length is given for the case of two morphogens by the formula (9.3). This formula for the number of peaks of course does not give a whole number, but the actual number of peaks will always be one of the two whole numbers nearest to it, and will usually be *the* nearest. The degree of instability is also shown in (9.3).

The mathematical conditions under which this case applies are given in equations (9.4a), (9.4b).

Biological examples of this case are discussed at some length below.

(e) For a two-morphogen system only the alternatives (*a*) to (*d*) are possible, but with three or more morphogens it is possible to have travelling waves. With a ring there would be two sets of waves, one travelling clockwise and the other anticlockwise. There is a natural chemical wave-length and wave frequency in this case as well as a wave-length; no attempt was made to develop formulae for these.

In looking for biological examples of this there is no need to consider only rings. The waves could arise in a tissue of any anatomical form. It is important to know what wavelengths, velocities and frequencies would be consistent with the theory. These quantities are determined by the rates at which the reactions occur (more accurately by the 'marginal reaction rates', which have the dimensions of the reciprocal of a time), and the diffusibilities of the morphogens. The possible range of values of the reaction rates is so immensely wide that they do not even give an indication of orders of magnitude. The diffusibilities are more helpful. If one were to assume that all the *dimensionless* parameters in a system of travelling waves were the same as in the example given in § 8, one could say that the product of the velocity and wave-length of the waves was 3π times the diffusibility of the most diffusible morphogen. But this assumption is certainly false, and it is by no means obvious what is the true range of possible values for the numerical constant (here 3π). The movements of the tail of a spermatozoon suggest themselves as an example of these travelling waves. That the waves are within one cell is no real difficulty. However, the speed of propagation seems to be somewhat greater than can be accounted for except with a rather large numerical constant.

(f) Metabolic oscillation with neighbouring cells in opposite phases. No biological examples of this are known to the author.

It is difficult also to find cases to which case (*d*) applies directly, but this is simply because isolated rings of tissue are very rare. On the other hand, systems that have the same kind of symmetry as a ring are extremely common, and it is to be expected that under appropriate chemical conditions, stationary waves may develop on these bodies, and that their circular symmetry will be replaced by a polygonal symmetry. Thus, for instance, a plant shoot may at one time have circular symmetry, i.e. appear essentially the same when rotated through any angle about a certain axis; this shoot may later develop a whorl of leaves, and then it will only suffer rotation through the angle which separates the leaves, or any multiple of it. This same example demonstrates the complexity of the situation when more than one dimension is involved. The leaves on the shoots may not appear in whorls, but be imbricated. This possibility is also capable of mathematical analysis, and will be considered in detail in a later paper. The cases which appear to the writer to come closest biologically to the 'isolated ring of cells' are the tentacles of (e.g.) *Hydra*, and the whorls of leaves of certain plants such as Woodruff (*Asperula odorata*).

Hydra is something like a sea-anemone but lives in fresh water and has from about five to ten tentacles. A part of a *Hydra* cut off from the rest will rearrange itself so as to form a complete new organism. At one stage of this proceeding the organism has reached the form of a tube open at the head end and closed at the other end. The external diameter is somewhat greater at the head end than over the rest of the tube. The whole still has circular symmetry. At a somewhat later stage the symmetry has gone to the extent that an appropriate stain will bring out a number of patches on the widened head end. These patches arise at the points where the tentacles are subsequently to appear (Child 1941, p. 101 and figure 30). According to morphogen theory it is natural to suppose that reactions, similar to those which were considered in connection with the ring of tissue, take place in the widened head end, leading to a similar breakdown of symmetry. The situation is more complicated than the case of the thin isolated ring, for the portion of the *Hydra* concerned is neither isolated nor very thin. It is not unreasonable to suppose that this head region is the only one in which the chemical conditions are such as to give instability. But substances produced in this region are still free to diffuse through the surrounding region of lesser activity. There is no great difficulty in extending the mathematics to cover this point in particular cases. But if the active region is too wide the system no longer approximates the behaviour of a thin ring and one can no longer expect the tentacles to form a single whorl. This also cannot be considered in detail in the present paper.

In the case of woodruff the leaves appear in whorls on the stem, the number of leaves in a whorl varying considerably, sometimes being as few as five or as many as nine. The numbers in consecutive whorls on the same stem are often equal, but by no means invariably. It is to be presumed that the whorls originate in rings of active tissue in the meristematic area, and that the rings arise at sufficiently great distance to have little influence on one another. The number of leaves in the whorl will presumably be obtainable by the rule given above, viz. by dividing the chemical wavelength into the circumference, though both these quantities will have to be given some new interpretation more appropriate to woodruff than to the ring. Another important example of a structure with polygonal symmetry is provided by young root fibres just breaking out from the parent root. Initially these are almost homogeneous in cross-section, but eventually a ring of fairly evenly spaced spots appear, and these later develop into vascular strands. In this case again the full explanation must be in terms of a two-dimensional or even a three-dimensional problem, although the analysis for the ring is still illuminating. When the cross-section is very large the strands may be in more than one ring, or more or less randomly or hexagonally arranged. The two-dimensional theory (not expounded here) also goes a long way to explain this.

Flowers might appear superficially to provide the most obvious examples of polygonal symmetry, and it is probable that there are many species for which this 'waves round a ring' theory is essentially correct. But it is certain that it does not apply for all species. If it did it would follow that, taking flowers as a whole, i.e. mixing up all species, there would be no very markedly preferred petal (or corolla, segment, stamen, etc.) numbers. For when all species are taken into account one must expect that the diameters of the rings concerned will take on nearly all values within a considerable range, and that neighbouring diameters will be almost equally common. There may also be some variation in chemical wave-length. Neighbouring values of the ratio circumferences to wave-length should therefore be more or less equally frequent, and this must mean that neighbouring petal numbers will have much the same frequency. But this is not borne out by the facts. The number five is extremely common, and the number seven rather rare. Such facts are, in the author's opinion, capable of explanation on the basis of morphogen theory, and are closely connected with the theory of phyllotaxis. They cannot be considered in detail here.

For Turing's later work on the development of the daisy (unpublished in his lifetime), see Dawes (2016).

The case of a filament of tissue calls for some comment. The equilibrium patterns on such a filament will be the same as on a ring, which has been cut at a point where the concentrations of the morphogens are a maximum or a minimum. This could account for the segmentation of such filaments. It should be noticed, however, that the theory will not apply unmodified for filaments immersed in water.

12. Chemical Waves on Spheres. Gastrulation

The treatment of homogeneity breakdown on the surface of a sphere is not much more difficult than in the case of the ring. The theory of spherical harmonics, on which it is based, is not, however, known to many that are not mathematical specialists. Although the essential properties of spherical harmonics that are used are stated below, many readers will prefer to proceed directly to the last paragraph of this section.

The anatomical structure concerned in this problem is a hollow sphere of continuous tissue such as a blastula. It is supposed sufficiently thin that one can treat it as a 'spherical shell'. This latter assumption is merely for the purpose of mathematical simplification; the results are almost exactly similar if it is omitted. As in § 7 there are to be two morphogens, and

$a, b, c, d, \mu', \nu', h, k$ are also to have the same meaning as they did there. The operator ∇^2 will be used here to mean the superficial part of the Laplacian, i.e. $\nabla^2 V$ will be an abbreviation of

$$\frac{1}{\rho^2}\frac{\partial^2 V}{\partial \phi^2} + \frac{1}{\rho^2 \sin^2\theta}\frac{\partial}{\partial\theta}\left(\sin\theta\frac{\partial V}{\partial\theta}\right),$$

where θ and ϕ are spherical polar co-ordinates on the surface of the sphere and ρ is its radius. The equations corresponding to (7.1) may then be written

$$\left.\begin{aligned}\frac{\partial X}{\partial t} &= a(X-h) + b(Y-k) + \mu'\nabla^2 X,\\ \frac{\partial Y}{\partial t} &= c(X-h) + d(Y-k) + \nu'\nabla^2 Y.\end{aligned}\right\} \tag{12.1}$$

It is well known (e.g. Jeans 1927, chapter 8) that any function on the surface of the sphere, or at least any that is likely to arise in a physical problem, can be 'expanded in spherical surface harmonics'. This means that it can be expressed in the form

$$\sum_{n=0}^{\infty}\left[\sum_{m=-n}^{n} A_n^m P_n^m(\cos\theta)\mathrm{e}^{\mathrm{i}m\phi}\right].$$

The expression in the square bracket is described as a 'surface harmonic of degree n'. Its nearest analogue in the ring theory is a Fourier component. The essential property of a spherical harmonic of degree n is that when the operator ∇^2 is applied to it the effect is the same as multiplication by $-n(n+1)/\rho^2$. In view of this fact it is evident that a solution of (12.1) is

$$\left.\begin{aligned}X &= h + \sum_{n=0}^{\infty}\sum_{m=-n}^{n}(A_n^m \mathrm{e}^{\mathrm{i}q_n t} + B_n^m \mathrm{e}^{\mathrm{i}q'_n t})P_n^m(\cos\theta)\mathrm{e}^{\mathrm{i}m\phi},\\ Y &= k + \sum_{n=0}^{\infty}\sum_{m=-n}^{n}(C_n^m \mathrm{e}^{\mathrm{i}q_n t} + D_n^m \mathrm{e}^{\mathrm{i}q'_n t})P_n^m(\cos\theta)\mathrm{e}^{\mathrm{i}\phi},\end{aligned}\right\} \tag{12.2}$$

where q_n and q'_n are the two roots of

$$\left(q - a + \frac{\mu'}{\rho^2}n(n+1)\right)\left(q - d + \frac{\nu'}{\rho^2}n(n+1) = bc\right) \tag{12.3}$$

and

$$\begin{aligned}A_n^m\left(q_n - a + \frac{\mu'}{\rho^2}n(n+1)\right) &= bC_n^m,\\ B_n^m\left(q'_n - a + \frac{\mu'}{\rho^2}n(n+1)\right) &= cD_n^m.\end{aligned} \tag{12.4}$$

This is the most general solution, since the coefficients A_n^m and B_n^m can be chosen to give any required values of X, Y when $t = 0$, except when (12.3) has two equal roots, in which case a treatment is required which is similar to that applied in similar circumstances in § 7. The analogy with § 7 throughout will indeed be obvious, though the summation with respect to m does not appear there. The meaning of this summation is that there are a number of different patterns with the same wave-length, which can be superposed with various amplitude factors. Then supposing that, as in § 8, one particular wave-length predominates, (12.2) reduces to

$$\left.\begin{aligned} X - h &= \mathrm{e}^{\mathrm{i}q_{n_0}t} \sum_{m=-n_0}^{n_0} A_{n_0}^m P_{n_0}^m(\cos\theta)\mathrm{e}^{\mathrm{i}m\phi}, \\ b(Y-k) &= \left(q_{n_0} - a + \frac{\mu'}{\rho^2}n(n+1)\right)(X-h). \end{aligned}\right\} \qquad (12.5)$$

In other words, the concentrations of the two morphogens are proportional, and both of them are surface harmonics of the same degree n_0, viz. that which makes the greater of the roots q_{n0}, q'_{n0} have the greatest value.

It is probable that the forms of various nearly spherical structures, such as radiolarian skeletons, are closely related to these spherical harmonic patterns. The most important application of the theory seems, however, to be to the gastrulation of a blastula. Suppose that the chemical data, including the chemical wave-length, remain constant as the radius of the blastula increases. To be quite specific suppose that

$$\mu' = 2, \quad \nu' = 1, \quad a = -4, \quad b = -8, \quad c = 4, \quad d = 7.$$

With these values the system is quite stable so long as the radius is less than about 2. Near this point, however, the harmonics of degree 1 begin to develop and a pattern of form (12.5) with $n_0 = 1$ makes its appearance. Making use of the facts that

$$P_1^0(\cos\theta) = \cos\theta, \quad P_1^1(\cos\theta) = P_1^{-1}(\cos\theta) = \sin\theta,$$

it is seen that $X{-}h$ is of the form

$$X - h = A\cos\theta + B\sin\theta\cos\phi + C\sin\theta\sin\phi, \qquad (12.6)$$

which may also be interpreted as

$$X - h = A'\cos\theta', \qquad (12.7)$$

where θ' is the angle which the radius θ, ϕ makes with the fixed direction having direction cosines proportional to B, C, A and $A' = \surd(A^2 + B^2 + C^2)$.

The outcome of the analysis therefore is quite simply this. Under certain not very restrictive conditions (which include a requirement that the sphere be relatively small but increasing in size) the pattern of the breakdown of homogeneity is axially symmetrical, not about the original axis of spherical polar co-ordinates, but about some new axis determined by the disturbing influences. The concentrations of the first morphogen are given by (12.7), where θ' is measured from this new axis; and $Y - k$ is proportional to $X—h$. Supposing that the first morphogen is, or encourages the production of, a growth hormone, one must expect the blastula to grow in an axially symmetric manner, but at a greater rate at one end of the axis than at the other. This might under many circumstances lead to gastrulation, though the effects of such growth are not very easily determinable. They depend on the elastic properties of the tissue as well as on the growth rate at each point. This growth will certainly lead to a solid of revolution with a marked difference between the two poles, unless, in addition to the chemical instability, there is a mechanical instability causing the breakdown of axial symmetry. The direction of the axis of gastrulation will be quite random according to this theory. It may be that it is found experimentally that the axis is normally in some definite direction such as that of the animal pole. This is not essentially contradictory to the theory, for any small asymmetry of the zygote may be sufficient to provide the 'disturbance' which determines the axis.

13. Non-Linear Theory. Use of Digital Computers

The 'wave' theory which has been developed here depends essentially on the assumption that the reaction rates are linear functions of the concentrations, an assumption which is justifiable in the case of a system just beginning to leave a homogeneous condition. Such systems certainly have a special interest as giving the first appearance of a pattern, but they are the exception rather than the rule. Most of an organism, most of the time, is developing from one pattern into another, rather than from homogeneity into a pattern. One would like to be able to follow this more general process mathematically also. The difficulties are, however, such that one cannot hope to have any very embracing *theory* of such processes, beyond the statement of the equations. It might be possible, however, to treat a few particular cases in detail with the aid of a digital computer. This method has the advantage that it is not so necessary to make simplifying assumptions as it is when doing a more theoretical type of analysis. It might even be possible to take the mechanical aspects of the problem into account as well as the chemical, when applying this type of method. The essential disadvantage of the method is that one only gets results for particular cases. But this disadvantage is probably of comparatively little importance. Even with the ring problem, considered in this paper, for which a reasonably complete mathematical analysis was possible, the computational treatment of a particular case was most illuminating. The morphogen theory of phyllotaxis, to be described, as already mentioned, in a later paper, will be covered by this computational method. Non-linear equations will be used.

Many people have done this, since digital computers have advanced a bit since 1952!

It must be admitted that the biological examples which it has been possible to give in the present paper are very limited. This can be ascribed quite simply to the fact that biological phenomena are usually very complicated. Taking this in combination with the relatively elementary mathematics used in this paper one could hardly expect to find that many observed biological phenomena would be covered. It is thought, however, that the imaginary biological systems which have been treated, and the principles which have been discussed, should be of some help in interpreting real biological forms.

REFERENCES

Child, C. M. 1941. *Patterns and Problems of Development.* Chicago, IL: University of Chicago Press.

Davson, H., and J. F. Danielli. 1943. *The Permeability of Natural Membranes.* Cambridge, UK: Cambridge University Press.

Jeans, J. H. 1927. *The Mathematical Theory of Elasticity.* 5th. Cambridge, UK: Cambridge University Press.

Michaelis, L., and M. L. Menten. 1913. "Die Kinetik der Invertinwirkung." *Biochemische Zeitschrift* 49:333.

Waddington, C. H. 1940. *Organisers and Genes.* Cambridge, UK: Cambridge University Press.

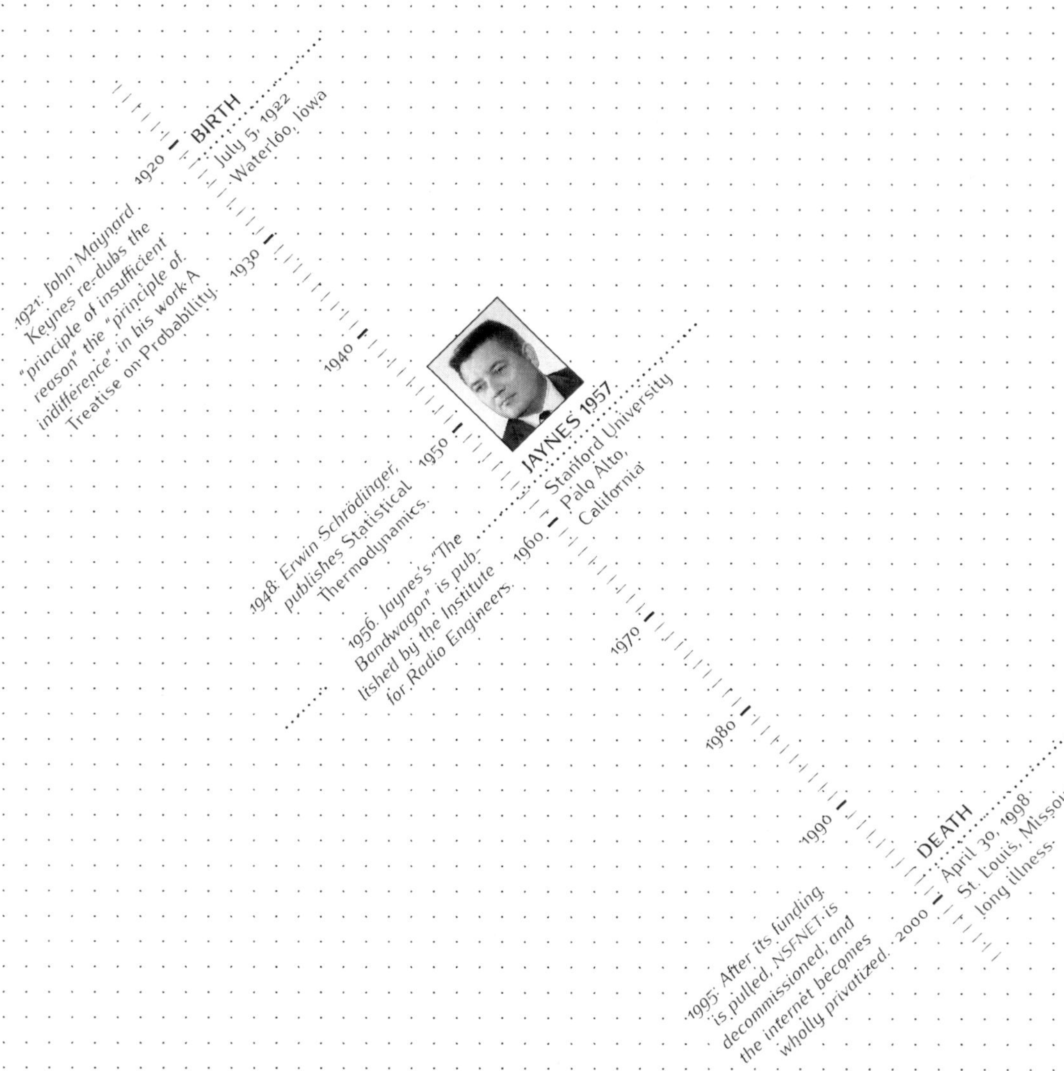

EDWIN THOMPSON JAYNES

[14]

JAYNES AND THE PRINCIPLE OF MAXIMUM ENTROPY

Dawn E. Holmes, University of California Santa Barbara

In a series of lectures given in 1956 to the Mobil Oil Company, Texas, Edwin Thompson Jaynes recalls discovering Claude Shannon's seminal work on information theory in the Princeton University library. At that time, as a graduate student of statistical mechanics, he was stunned by Shannon's novel derivation of the entropy expression and its varied applications, particularly since they were not restricted to thermodynamics. Jaynes immediately recognized Shannon's work on entropy as having an importance to physics comparable to Dirac's pioneering work in particle physics and viewed it as ". . . the most important work done by any scientist since the discovery of the Dirac equation" (Jaynes 1956). Shannon saw in the concept of entropy a possible application in the field of communication. His information-theoretic derivation of the entropy expression, subsequently known as Shannon entropy, can be found in his paper published in 1948 in *The Bell System Technical Journal* and it was this paper that so excited Jaynes. In the book that followed in 1949, *The Mathematical Theory of Communication*, Shannon expanded the idea of entropy as the key concept in information theory, which he is largely credited with founding. There followed an excess of articles in diverse areas using Shannon entropy. However, in 1956 he published "The Bandwagon," a short piece in which he suggests moderation in the use of information theory in other fields.

E. T. Jaynes, "Information Theory and Statistical Mechanics," *Physical Review* 106 (4), 620–630 (1957).

Jaynes was particularly concerned about the bad reputation Shannon entropy had gained among physicists because of the lack of new results, but in 1956 he wrote: "I think the time has come now when physicists might find it worthwhile to take a sober second look at Information Theory and what it can do for them" (Jaynes 1956).

Jaynes saw a direct link between entropy in statistical mechanics and information entropy, which he immediately identified as a significant shift in perspective. In this paper, he showed that entropy as expressed in the work of Boltzmann and Gibbs in the field of thermodynamics and Shannon entropy follow from the same underlying logic, thus establishing entropy as a general concept. Maximum entropy inference is introduced and in sections 3, 4, and 5 examples are given showing how the application of the principle of maximum entropy results in a conceptual shift in the approach to statistical mechanics. Jaynes proved that statistical mechanics can be interpreted in non-physical terms as statistical inference. In his 1957 paper he introduces the "principle of maximum entropy" and proposed the maximization of the entropy function H, subject to certain constraints, as a general principle of statistical inference applicable to a diverse range of areas. The principle has since proved particularly useful in providing the least biased probability distribution for a set of unknown prior probabilities. Jaynes's view is that assigning equal probabilities in situations of total ignorance is the least prejudiced choice. However, when ignorance is not total, inferences should be based on all and only the information available. This is assured by choosing the probability distribution with maximum entropy.

Over the years Jaynes's views on subjective and objective probability evolved without detracting from the results given in the paper. Originally a stalwart Bayesian, in his later writings he argued that a distinction between subjectivist and objectivist probability is unnecessary and so his comments in section 5 on this issue can be viewed in this light.

Following on from his 1957 paper, Jaynes's published work on the maximum entropy formalism and entropy as statistical inference spans almost fifty years. In 1978 at the Maximum Entropy Formalism Conference, Jaynes expounds his position in the aptly titled paper "Where Do We Stand on Maximum Entropy? drawing together some of his earlier work and giving an historical perspective on the work of Boltzmann, Gibbs and Shannon among others. This conference evolved into the annual Maximum Entropy (Maxent) conference, with Jaynes participating regularly, even after retirement, until his death in 1998.

Jaynes's book *Probability Theory: The Logic of Science* was published posthumously in 2002, edited by G. Larry Bretthorst.

Since Jaynes's pioneering 1957 paper, the uses of the maximum entropy principle have been explored and the theory developed so that it is now one of the cornerstones in approaches to problems involving complex systems. The information theoretic viewpoint is now accepted in the field of quantum physics, thus underlying the importance of Jaynes's pioneering work in uniting the two fields. The continuing impact of Jaynes's contributions to complexity theory can be seen in recent work in which the Principle of Maximum Entropy has been applied to generalize the molecular chaos hypothesis (Chliamovitch, Malaspinas, and Chopard 2017). ❧

REFERENCES

Chliamovitch, G., O. Malaspinas, and B. Chopard. 2017. "Kinetic Theory Beyond the Stosszahlansatz." *Entropy* 19 (8). https://doi.org/10.3390/e19080381.

Jaynes, E. T. 1956. *Probability Theory in Science and Engineering*. Colloquium Lectures in Pure and Applied Science No. 4. New York, NY: Socony-Mobil Oil Co.

———. 1978. "Where Do we Stand on Maximum Entropy?" In *Maximum Entropy Formalism Conference, Massachusetts Institute of Technology*. Reprinted in E. T. Jaynes: Papers on Probability, Statistics and Statistical Physics.

———. 2002. *Probability Theory: The Logic of Science*. Edited by G. L. Bretthorst. St. Louis, MO: Washington University.

Shannon, C. E. 1948. "A Mathematical Theory of Communication." Reprinted with corrections, *The Bell System Technical Journal* 27 (6): 379–423. https://doi.org/10.1002/j.1538-7305.1948.tb01338.x.

———. 1956. "The Bandwagon." *IRE Transactions on Information Theory* 2 (1): 3. https://doi.org/10.1109/TIT.1956.1056774.

Shannon, C. E., and W. Weaver. 1949. *The Mathematical Theory of Communication*. Urbana, IL: University of Illinois Press.

INFORMATION THEORY AND STATISTICAL MECHANICS

E.T. Jaynes, Stanford University

Abstract

Information theory provides a constructive criterion for setting up probability distributions on the basis of partial knowledge, and leads to a type of statistical inference which is called the maximum-entropy estimate. It is the least biased estimate possible on the given information; i.e., it is maximally noncommittal with regard to missing information. If one considers statistical mechanics as a form of statistical inference rather than as a physical theory, it is found that the usual computational rules, starting with the determination of the partition function, are an immediate consequence of the maximum-entropy principle. In the resulting "subjective statistical mechanics," the usual rules are thus justified independently of any physical argument, and in particular independently of experimental verification; whether or not the results agree with experiment, they still represent the best estimates that could have been made on the basis of the information available.

It is concluded that statistical mechanics need not be regarded as a physical theory dependent for its validity on the truth of additional assumptions not contained in the laws of mechanics (such as ergodicity, metric transitivity, equal *a priori* probabilities, etc.). Furthermore, it is possible to maintain a sharp distinction between its physical and statistical aspects. The former consists only of the correct enumeration of the states of a system and their properties; the latter is a straightforward example of statistical inference.

1. Introduction

The recent appearance of a very comprehensive survey[1] of past attempts to justify the methods of statistical mechanics in terms of mechanics, classical or quantum, has helped greatly, and at a very opportune time, to emphasize the unsolved problems in this field.

Although the subject has been under development for many years, we still do not have a complete and satisfactory theory, in the sense that there is no line of argument proceeding from the laws of microscopic

[1] D. ter Haar, *Revs. Modern Phys.* 27, 289 (1955).

mechanics to macroscopic phenomena, that is generally regarded by physicists as convincing in all respects. Such an argument should (a) be free from objection on mathematical grounds, (b) involve no additional arbitrary assumptions, and (c) automatically include an explanation of nonequilibrium conditions and irreversible processes as well as those of conventional thermodynamics, since equilibrium thermodynamics is merely an ideal limiting case of the behavior of matter.

It might appear that condition (b) is too severe, since we expect that a physical theory will involve certain unproved assumptions, whose consequences are deduced and compared with experiment. For example, in the statistical mechanics of Gibbs[2] there were several difficulties which could not be understood in terms of classical mechanics, and before the models which he constructed could be made to correspond to the observed facts, it was necessary to incorporate into them additional restrictions not contained in the laws of classical mechanics. First was the "freezing up" of certain degrees of freedom, which caused the specific heat of diatomic gases to be only $\frac{5}{6}$ of the expected value. Secondly, the paradox regarding the entropy of combined systems, which was resolved only by adoption of the generic instead of the specific definition of phase, an assumption which seems impossible to justify in terms of classical notions.[3] Thirdly, in order to account for the actual values of vapor pressures and equilibrium constants, an additional assumption about a natural unit of volume (h^{3N}) of phase space was needed. However, with the development of quantum mechanics the originally arbitrary assumptions are now seen as necessary consequences of the laws of physics. This suggests the possibility that we have now reached a state where statistical mechanics is no longer dependent on physical hypotheses, but may become merely an example of statistical inference.

[2] J. W. Gibbs, *Elementary Principles in Statistical Mechanics* (Longmans Green and Company, New York, 1928), Vol. II of collected works.

[3] We may note here that although Gibbs (reference 2, Chap. XV) started his discussion of this question by saying that the generic definition "seems in accordance with the spirit of the statistical method," he concluded it with, "The perfect similarity of several particles of a system will not in the least interfere with the identification of a particular particle in one case with a particular particle in another. The question is one to be decided in accordance with the requirements of practical convenience in the discussion of the problems with which we are engaged."

That the present may be an opportune time to re-examine these questions is due to two recent developments. Statistical methods are being applied to a variety of specific phenomena involving irreversible processes, and the mathematical methods which have proven successful have not yet been incorporated into the basic apparatus of statistical mechanics. In addition, the development of information theory[4] has been felt by many people to be of great significance for statistical mechanics, although the exact way in which it should be applied has remained obscure. In this connection it is essential to note the following. The mere fact that the same mathematical expression $-\sum p_i \log p_i$ occurs both in statistical mechanics and in information theory does not in itself establish any connection between these fields. This can be done only by finding new viewpoints from which thermodynamic entropy and information-theory entropy appear as the same *concept*. In this paper we suggest a reinterpretation of statistical mechanics which accomplishes this, so that information theory can be applied to the problem of justification of statistical mechanics. We shall be concerned with the prediction of equilibrium thermodynamic properties, by an elementary treatment which involves only the probabilities assigned to stationary states. Refinements obtainable by use of the density matrix and discussion of irreversible processes will be taken up in later papers.

Jaynes later showed his work was applicable to reversible and irreversible processes.

Section 2 defines and establishes some of the elementary properties of maximum-entropy inference, and in Secs. 3 and 4 the application to statistical mechanics is discussed. The mathematical facts concerning maximization of entropy, as given in Sec. 2, were pointed out long ago by Gibbs. In the past, however, these properties were given the status of side remarks not essential to the theory and not providing in themselves any justification for the methods of statistical mechanics. The feature which was missing has been supplied only recently by Shannon[4] in the demonstration that the expression for entropy has a deeper meaning, quite independent of thermodynamics. This makes possible a reversal of the usual line of reasoning in statistical mechanics. Previously, one constructed a theory based on the equations of motion, supplemented by

This refers to the Gibbs Algorithm, which is confined to thermodynamics. Jaynes generalized the algorithm.

[4]C. E. Shannon, *Bell System Tech. J.* 27, 379, 623 (1948); these papers are reprinted in C. E. Shannon and W. Weaver, *The Mathematical Theory of Communication* (University of Illinois Press, Urbana, 1949).

additional hypotheses of ergodicity, metric transitivity, or equal *a priori* probabilities, and the identification of entropy was made only at the end, by comparison of the resulting equations with the laws of phenomenological thermodynamics. Now, however, we can take entropy as our starting concept, and the fact that a probability distribution maximizes the entropy subject to certain constraints becomes the essential fact which justifies use of that distribution for inference.

The most important consequence of this reversal of viewpoint is not, however, the conceptual and mathematical simplification which results. In freeing the theory from its apparent dependence on physical hypotheses of the above type, we make it possible to see statistical mechanics in a much more general light. Its principles and mathematical methods become available for treatment of many new physical problems. Two examples are provided by the derivation of Siegert's "pressure ensemble" and treatment of a nuclear polarization effect, in Sec. 5.

2. Maximum-Entropy Estimates

The quantity x is capable of assuming the discrete values x_i $(i = 1, 2 \cdots, n)$. We are not given the corresponding probabilities p_i; all we know is the expectation value of the function $f(x)$:

$$\langle f(x) \rangle = \sum_{i=1}^{n} p_i f(x_i) . \tag{2-1}$$

On the basis of this information, what is the expectation value of the function $g(x)$? At first glance, the problem seems insoluble because the given information is insufficient to determine the probabilities p_i.[5] Equation (2-1) and the normalization condition

$$\sum p_i = 1 \tag{2-2}$$

would have to be supplemented by $(n - 2)$ more conditions before $\langle g(x) \rangle$ could be found.

[5] Yet this is precisely the problem confronting us in statistical mechanics; on the basis of information which is grossly inadequate to determine any assignment of probabilities to individual quantum states, we are asked to estimate the pressure, specific heat, intensity of magnetization, chemical potentials, etc., of a macroscopic system. Furthermore, statistical mechanics is amazingly successful in providing accurate estimates of these quantities. Evidently there must be other reasons for this success, that go beyond a mere correct statistical treatment of the problem as stated above.

This problem of specification of probabilities in cases where little or no information is available, is as old as the theory of probability. Laplace's "Principle of Insufficient Reason" was an attempt to supply a criterion of choice, in which one said that two events are to be assigned equal probabilities if there is no reason to think otherwise. However, except in cases where there is an evident element of symmetry that clearly renders the events "equally possible," this assumption may appear just as arbitrary as any other that might be made. Furthermore, it has been very fertile in generating paradoxes in the case of continuously variable random quantities,[6] since intuitive notions of "equally possible" are altered by a change of variables.[7] Since the time of Laplace, this way of formulating problems has been largely abandoned, owing to the lack of any constructive principle which would give us a reason for preferring one probability distribution over another in cases where both agree equally well with the available information.

In his later work, Jaynes concluded that there was no theoretical distinction to be drawn between the objective and subjective views on probability. Hence, the discussion presented in this section was unnecessary.

For further discussion of this problem, one must recognize the fact that probability theory has developed in two very different directions as regards fundamental notions. The "objective" school of thought[8,9] regards the probability of an event as an objective property of that event, always capable in principle of empirical measurement by observation of frequency ratios in a random experiment. In calculating a probability distribution the objectivist believes that he is making predictions which are in principle verifiable in every detail, just as are those of classical mechanics. The test of a good objective probability distribution $p(x)$ is: does it correctly represent the observable fluctuations of x?

[6] The problems associated with the continuous case are fundamentally more complicated than those encountered with discrete random variables; only the discrete case will be considered here.

[7] For several examples, see E. P. Northrop, *Riddles in Mathematics* (D. Van Nostrand Company, Inc., New York, 1944), Chap. 8.

[8] H. Cramer, *Mathematical Methods of Statistics* (Princeton University Press, Princeton, 1946).

[9] W. Feller, *An Introduction to Probability Theory and its Applications* (John Wiley and Sons, Inc., New York, 1950).

On the other hand, the "subjective" school of thought[10,11] regards probabilities as expressions of human ignorance; the probability of an event is merely a formal expression of our expectation that the event will or did occur, based on whatever information is available. To the subjectivist, the purpose of probability theory is to help us in forming plausible conclusions in cases where there is not enough information available to lead to certain conclusions; thus detailed verification is not expected. The test of a good subjective probability distribution is does it correctly represent our state of knowledge as to the value of x?

Although the theories of subjective and objective probability are mathematically identical, the concepts themselves refuse to be united. In the various statistical problems presented to us by physics, both viewpoints are required. Needless controversy has resulted from attempts to uphold one or the other in all cases. The subjective view is evidently the broader one, since it is always possible to interpret frequency ratios in this way; furthermore, the subjectivist will admit as legitimate objects of inquiry many questions which the objectivist considers meaningless. The problem posed at the beginning of this section is of this type, and therefore in considering it we are necessarily adopting the subjective point of view.

Just as in applied statistics the crux of a problem is often the devising of some method of sampling that avoids bias, our problem is that of finding a probability assignment which avoids bias, while agreeing with whatever information is given. The great advance provided by information theory lies in the discovery that there is a unique, unambiguous criterion for the "amount of uncertainty" represented by a discrete probability distribution, which agrees with our intuitive notions that a broad distribution represents more uncertainty than does a sharply peaked one, and satisfies all other conditions which make it reasonable.[4] In Appendix A we sketch Shannon's proof that the quantity which is positive, which increases with increasing uncertainty, and is additive for independent sources of uncertainty, is

$$H(p_1 \cdots p_n) = -K \sum_i p_i \ln p_i, \tag{2-3}$$

[10] J. M. Keynes, *A Treatise on Probability* (MacMillan Company, London, 1921).

[11] H. Jeffreys, *Theory of Probability* (Oxford University Press, London, 1939).

where K is a positive constant. Since this is just the expression for entropy as found in statistical mechanics, it will be called the entropy of the probability distribution p_i; henceforth we will consider the terms "entropy" and "uncertainty" as synonymous.

This is a fundamental step in the development of the principle of maximum entropy.

It is now evident how to solve our problem; in making inferences on the basis of partial information we must use that probability distribution which has maximum entropy subject to whatever is known. This is the only unbiased assignment we can make; to use any other would amount to arbitrary assumption of information which by hypothesis we do not have. To maximize (2-3) subject to the constraints (2-1) and (2-2), one introduces Lagrangian multipliers λ, μ, in the usual way, and obtains the result

$$p_i = e^{-\lambda - \mu f(x_i)}. \tag{2-4}$$

The constants λ, μ are determined by substituting into (2-1) and (2-2). The result may be written in the form

$$\langle f(x) \rangle = -\frac{\partial}{\partial \mu} \ln Z(\mu), \tag{2-5}$$

$$\lambda = \ln Z(\mu), \tag{2-6}$$

where

$$Z(\mu) = \sum_i e^{-\mu f(x_i)} \tag{2-7}$$

will be called the partition function.

This may be generalized to any number of functions $f(x)$: given the averages

$$\langle f_r(x) \rangle = \sum_i p_i f_r(x_i), \tag{2-8}$$

form the partition function

$$Z(\lambda_1, \cdots, \lambda_m) = \sum_i \exp\{-[\lambda_1 f_1(x_i) + \cdots + \lambda_m f_m(x_i)]\}. \tag{2-9}$$

Then the maximum-entropy probability distribution is given by

$$p_i = \exp\{-[\lambda_0 + \lambda_1 f_1(x_i) + \cdots + \lambda_m f_m(x_i)]\}, \tag{2-10}$$

in which the constants are determined from

$$\langle f_r(x)\rangle = -\frac{\partial}{\partial \lambda_r} \ln Z, \tag{2-11}$$

$$\lambda_0 = \ln Z. \tag{2-12}$$

The entropy of the distribution (2-10) then reduces to

$$S_{\max} = \lambda_0 + \lambda_1 \langle f_1(x)\rangle + \cdots + \lambda_m \langle f_m(x)\rangle, \tag{2-13}$$

where the constant K in (2-3) has been set equal to unity. The variance of the distribution of $f_r(x)$ is found to be

$$\Delta^2 f_r = \langle {f_r}^2\rangle - \langle f_r\rangle^2 = \frac{\partial^2}{\partial {\lambda_r}^2}(\ln Z). \tag{2-14}$$

In addition to its dependence on x, the function f_r may contain other parameters $\alpha_1, \alpha_2, \cdots$, and it is easily shown that the maximum-entropy estimates of the derivatives are given by

$$\left\langle \frac{\partial f_r}{\partial \alpha_k} \right\rangle = -\frac{1}{\lambda_r}\frac{\partial}{\partial \alpha_k} \ln Z. \tag{2-15}$$

The principle of maximum entropy may be regarded as an extension of the principle of insufficient reason (to which it reduces in case no information is given except enumeration of the possibilities x_i), with the following essential difference. The maximum-entropy distribution may be asserted for the positive reason that it is uniquely determined as the one which is maximally noncommittal with regard to missing information, instead of the negative one that there was no reason to think otherwise. Thus the concept of entropy supplies the missing criterion of choice which Laplace needed to remove the apparent arbitrariness of the principle of insufficient reason, and in addition it shows precisely how this principle is to be modified in case there are reasons for "thinking otherwise."

Thus, the principle of maximum entropy determines the probability distribution in which all and only the available information is incorporated.

Mathematically, the maximum-entropy distribution has the important property that no possibility is ignored; it assigns positive weight to every situation that is not absolutely excluded by the given information. This is quite similar in effect to an ergodic property. In this connection it is interesting to note that prior to the work of Shannon other information measures had been proposed[12,13] and used in statistical inference, although in a different way than in the present paper. In particular, the quantity $-\sum p_i^2$ has many of the qualitative properties of Shannon's information measure, and in many cases leads to substantially the same results. However, it is much more difficult to apply in practice. Conditional maxima of $-\sum p_i^2$ cannot be found by a stationary property involving Lagrangian multipliers, because the distribution which makes this quantity stationary subject to prescribed averages does not in general satisfy the condition $p_i \geqslant 0$. A much more important reason for preferring the Shannon measure is that it is the only one which satisfies the condition of consistency represented by the composition law (Appendix A). Therefore one expects that deductions made from any other information measure, if carried far enough, will eventually lead to contradictions.

3. Application to Statistical Mechanics

In later work, Jaynes acknowledged that this section needed further expanding. The main results are highlighted below.

It will be apparent from the equations in the preceding section that the theory of maximum-entropy inference is identical in mathematical form with the rules of calculation provided by statistical mechanics. Specifically, let the energy levels of a system be

$$E_i\left(\alpha_1, \alpha_2, \cdots\right),$$

where the external parameters α_i may include the volume, strain tensor applied electric or magnetic fields, gravitational potential, etc. Then if we know only the average energy $\langle E\rangle$, the maximum-entropy probabilities of the levels E_i are given by a special case of (2-10), which we recognize as the Boltzmann distribution. This observation really completes our derivation of the conventional rules of statistical mechanics as an example of statistical inference; the identification of temperature, free energy, etc., proceeds in a familiar manner,[14] with results summarized as

$$\lambda_1 = (1/kT), \tag{3-1}$$

[12] R. A. Fisher, *Proc. Cambridge Phil. Soc.* 22, 700 (1925).

[13] J. L. Doob, *Trans. Am. Math. Soc.* 39, 410 (1936).

[14] E. Schrödinger, *Statistical Thermodynamics* (Cambridge University Press, Cambridge, 1948).

$$U - TS = F(T, \alpha_1, \alpha_2, \cdots) = -kT \ln Z(T, \alpha_1, \alpha_2, \cdots), \tag{3-2}$$

$$S = -\frac{\partial F}{\partial T} = -k \sum_i p_i \ln p_i, \tag{3-3}$$

$$\beta_i = kT \frac{\partial}{\partial \alpha_1} \ln Z. \tag{3-4}$$

The thermodynamic entropy is identical with the information-theory entropy of the probability distribution except for the presence of Boltzmann's constant.[15] The "forces" β_i include pressure, stress tensor, electric or magnetic moment, etc., and Eqs. (3-2), (3-3), (3-4) then give a complete description of the thermodynamic properties of the system, in which the forces are given by special cases of (2-15); i.e., as maximum-entropy estimates of the derivatives $(\partial E_i / \partial \alpha_k)$.

In the above relations we have assumed the number of molecules of each type to be fixed. Now let n_1 be the number of molecules of type 1, n_2 the number of type 2, etc. If the n_s are not known, then a possible "state" of the system requires a specification of all the n_s as well as a particular energy level $E_i(\alpha_1\alpha_2 \cdots \mid n_1 n_2 \cdots)$. If we are given the expectation values

$$\langle E \rangle, \quad \langle n_1 \rangle, \quad \langle n_2 \rangle, \quad \cdots,$$

then in order to make maximum-entropy inferences, we need to form, according to (2-9), the partition function

$$\begin{aligned} Z(\alpha_1\alpha_2 \cdots \mid \lambda_1\lambda_2 \cdots, \beta) = \sum_{n_1 n_2 \cdots} \sum_i \exp\{-[\lambda_1 n_1 + \lambda_2 n_2 \\ + \cdots + \beta E_i(\alpha_k \mid n_s)]\}, \end{aligned} \tag{3-5}$$

and the corresponding maximum-entropy distribution (2-10) is that of the "quantum-mechanical grand canonical ensemble;" the Eqs. (2-11) fixing

[15] Boltzmann's constant may be regarded as a correction factor necessitated by our custom of measuring temperature in arbitrary units derived from the freezing and boiling points of water. Since the product TS must have the dimensions of energy, the units in which entropy is measured depend on those chosen for temperature. It would be convenient in general arguments to define an "absolute cgs unit" of temperature such that Boltzmann's constant is made equal to unity. Then entropy would become dimensionless (as the considerations of Sec. 2 indicate it should be), and the temperature would be equal to twice the average energy per degree of freedom; it is, of course, just the "modulus" Θ of Gibbs.

the constants, are recognized as giving the relation between the chemical potentials

$$\mu_i = -kT\lambda_i, \tag{3-6}$$

and the $\langle n_i \rangle$:

$$\langle n_i \rangle = \partial F / \partial \mu_i, \tag{3-7}$$

where the free-energy function $F = -kT\lambda_0$, and $\lambda_0 = \ln Z$ is called the "grand potential."[16] Writing out (2-13) for this case and rearranging, we have the usual expression

$$F(T, \alpha_1 \alpha_2 \cdots, \mu_1 \mu_2 \cdots) = \langle E \rangle - TS + \mu_1 \langle n_1 \rangle + \mu_2 \langle n_2 \rangle + \cdots . \tag{3-8}$$

It is interesting to note the ease with which these rules of calculation are set up when we make entropy the primitive concept. Conventional arguments, which exploit all that is known about the laws of physics, in particular the constants of the motion, lead to exactly the same predictions that one obtains directly from maximizing the entropy. In the light of information theory, this can be recognized as telling us a simple but important fact: *there is nothing in the general laws of motion that can provide us with any additional information about the state of a system beyond what we have obtained from measurement.* This refers to interpretation of the state of a system at time t on the basis of measurements carried out at time t. For predicting the course of time-dependent phenomena, knowledge of the equations of motion is of course needed. By restricting our attention to the prediction of equilibrium properties as in the present paper, we are in effect deciding at the outset that the only type of initial information allowed will be values of quantities which are observed to be constant in time. Any prior knowledge that these quantities would be constant (within macroscopic experimental error) in consequence of the laws of physics, is then redundant and cannot help us in assigning probabilities.

This principle has interesting consequences. Suppose that a super-mathematician were to discover a new class of uniform integrals of the motion, hitherto unsuspected. In view of the importance ascribed to

[16]D. ter Haar, *Elements of Statistical Mechanics* (Rinehart and Company, New York, 1954), Chap. 7.

uniform integrals of the motion in conventional statistical mechanics, and the assumed nonexistence of new ones, one might expect that our equations would be completely changed by this development. This would not be the case, however, unless we also supplemented our prediction problem with new experimental data which provided us with some information as to the likely values of these new constants. *Even if we had a clear proof that a system is not metrically transitive, we would still have no rational basis for excluding any region of phase space that is allowed by the information available to us.* In its effect on our ultimate predictions, this fact is equivalent to an ergodic hypothesis, quite independently of whether physical systems are in fact ergodic.

This shows the great practical convenience of the subjective point of view. If we were attempting to establish the probabilities of different states in the objective sense, questions of metric transitivity would be crucial, and unless it could be shown that the system was metrically transitive, we would not be able to find any solution at all. If we are content with the more modest aim of finding subjective probabilities, metric transitivity is irrelevant. Nevertheless, the subjective theory leads to exactly the same predictions that one has attempted to justify in the objective sense. The only place where subjective statistical mechanics makes contact with the laws of physics is in the enumeration of the different possible, mutually exclusive states in which the system might be. Unless a new advance in knowledge affects this enumeration, it cannot alter the equations which we use for inference.

If the subject were dropped at this point, however, it would remain very difficult to understand why the above rules of calculation are so uniformly successful in predicting the behavior of individual systems. In stripping the statistical part of the argument to its bare essentials, we have revealed how little content it really has; the amount of information available in practical situations is so minute that it alone could never suffice for making reliable predictions. Without further conditions arising from the physical nature of macroscopic systems, one would expect such great uncertainty in prediction of quantities such as pressure that we would have no definite theory which could be compared with experiments. It might also be questioned whether it is not the most probable, rather than the average, value over the maximum-entropy distribution that should be compared with experiment, since the average might be the average of two peaks and itself correspond to an impossible value.

It is well known that the answer to both of these questions lies in the fact that for systems of very large number of degrees of freedom, the probability distributions of the usual macroscopic quantities determined from the equations above, possess a single extremely sharp peak which includes practically all the "mass" of the distribution. Thus for all practical purposes average, most probable, median, or any other type of estimate are one and the same. It is instructive to see how, in spite of the small amount of information given, maximum-entropy estimates of certain functions $g(x)$ can approach practical certainty because of the way the *possible* values of x are distributed. We illustrate this by a model in which the possible values x_i are defined as follows: let n be a non-negative integer, and ϵ a small positive number. Then we take

$$x_1^{n+1} = \epsilon, \quad x_{i+1} - x_i = \epsilon / {x_i}^n, \quad i = 1, 2, \cdots . \tag{3-9}$$

According to this law, the x_i increase without limit as $i \to \infty$, but become closer together at a rate determined by n. By choosing ϵ sufficiently small we can make the density of points x_i in the neighborhood of any particular value of x as high as we please, and therefore for a continuous function $f(x)$ we can approximate a sum as closely as we please by an integral taken over a corresponding range of values of x

$$\sum_i f(x_i) \to \int f(x)\rho(x)dx,$$

where, from (3-9), we have

$$\rho(x) = x^n/\epsilon.$$

This approximation is not at all essential, but it simplifies the mathematics.

Now consider the problem: (A) Given $\langle x \rangle$, estimate x^2. Using our general rules, as developed in Sec. II, we first obtain the partition function

$$Z(\lambda) = \int_0^\infty \rho(x) e^{-\lambda x} dx = \frac{n!}{\epsilon \lambda^{n+1}},$$

with λ determined from (2-11),

$$\langle x \rangle = -\frac{\partial}{\partial \lambda} \ln Z = \frac{n+1}{\lambda}.$$

Then we find, for the maximum-entropy estimate of x^2,

$$\langle x^2 \rangle \{\langle x \rangle\} = Z^{-1} \int_0^\infty x^2 \rho(x) e^{-\lambda x} dx = \frac{n+2}{n+1} \langle x \rangle^2. \tag{3-10}$$

Next we invert the problem: (B) Given $\langle x^2 \rangle$, estimate x. The solution is

$$\begin{aligned} Z(\lambda) &= \int_0^\infty \rho(x) \exp\left(-\lambda x^2\right) dx \\ &= \frac{\pi^{\frac{1}{2}} n!}{2^{n+1}(n/2)!} \cdot \frac{1}{\epsilon \lambda^{(n+1)}}, \\ \langle x^2 \rangle &= -\frac{\partial}{\partial \lambda} \ln Z = \frac{n+1}{2\lambda}, \\ \langle x \rangle \{\langle x^2 \rangle\} &= Z^{-1} \int_0^\infty \rho x(x) \exp\left(-\lambda x^2\right) dx \\ &= \left(\frac{n+1}{2}\right)^{\frac{1}{2}} \frac{(\frac{1}{2}n)!}{[\frac{1}{2}(n+1)]!} \langle x^2 \rangle^{\frac{1}{2}}. \end{aligned} \tag{3-11}$$

The solutions are plotted in Fig. 1 for the case $n = 1$. The upper "regression line" represents Eq. (3-10), and the lower one Eq. (3-11). For other values of n, the slopes of the regression lines are plotted in Fig. 2. As $n \to \infty$, both regression lines approach the line at $45°$, and thus for large n, there is for all practical purposes a definite functional relationship between $\langle x \rangle$ and $\langle x^2 \rangle$, independently of which one is considered "given," and which one "estimated." Furthermore, as n increases the distributions become sharper; in problem (A) we find for the variance of x,

$$\langle x^2 \rangle - \langle x \rangle^2 = \langle x \rangle^2/(n+1). \tag{3-12}$$

Similar results hold in this model for the maximum-entropy estimate of any sufficiently well-behaved function $g(x)$. If $g(x)$ can be expanded in a power series in a sufficiently wide region about the point $x = \langle x \rangle$, we obtain, using the distribution of problem A above, the following expressions for the expectation value and variance of g:

$$\langle g(x) \rangle = g(\langle x \rangle) + g''(\langle x \rangle) \frac{\langle x \rangle^2}{2(n+1)} + O\left(\frac{1}{n^2}\right), \tag{3-13}$$

$$\begin{aligned} \Delta^2(g) &= \langle g^2(x) \rangle - \langle g(x) \rangle^2 \\ &= \left[g'(\langle x \rangle)\right]^2 \frac{\langle x \rangle^2}{n+1} + O\left(\frac{1}{n^2}\right). \end{aligned} \tag{3-14}$$

Conversely, a sufficient condition for x to be well determined by knowledge of $\langle g(x) \rangle$ is that x be a sufficiently smooth monotonic function

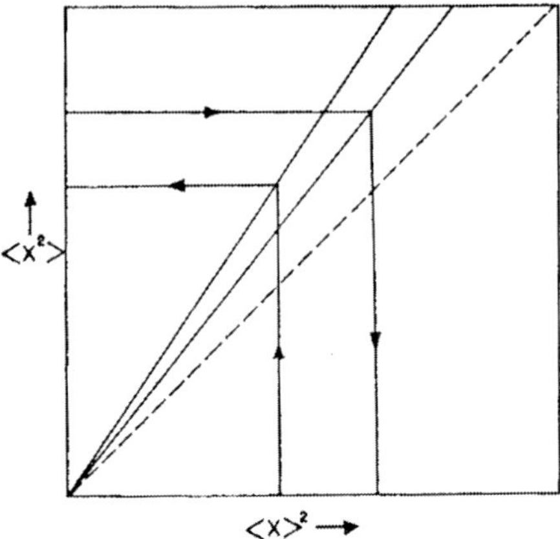

Figure 1. Regression of x and x^2 for state density increasing linearly with x. To find the maximum-entropy estimate of either quantity given the expectation value of the other, follow the arrows.

of g. The apparent lack of symmetry, in that reasoning from $\langle x \rangle$ to g does not require monotonicity of $g(x)$, is due to the fact that the distribution of *possible* values has been specified in terms of x rather than g.

As n increases, the relative standard deviations of all sufficiently well-behaved functions go down like $n^{-\frac{1}{2}}$; it is in this way that definite laws of thermodynamics, essentially independent of the type of information given, emerge from a statistical treatment that at first appears incapable of giving reliable predictions. The parameter n is to be compared with the number of degrees of freedom of a macroscopic system.

4. Subjective and Objective Statistical Mechanics

Many of the propositions of statistical mechanics are capable of two different interpretations. The Maxwellian distribution of velocities in a gas is, on the one hand, the distribution that can be realized in the greatest number of ways for a given total energy; on the other hand, it is a well-verified experimental fact. Fluctuations in quantities such as the density of a gas or the voltage across a resistor represent on the one hand the uncertainty of our predictions, on the other a measurable physical phenomenon. Entropy as a concept may be regarded as a measure of our degree of ignorance as to the state of a system; on the other hand, for equilibrium conditions it is an experimentally measurable quantity, whose most important properties were first found empirically. It is this

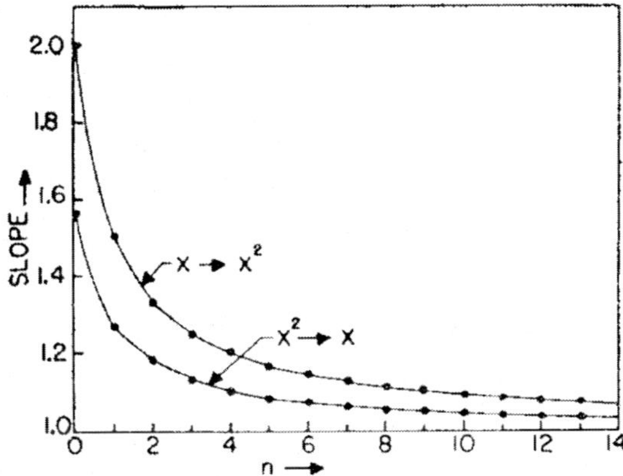

Figure 2. Slope of regression lines as a function of n.

last circumstance that is most often advanced as an argument against the subjective interpretation of entropy.

The relation between maximum-entropy inference and experimental facts may be clarified as follows. We frankly recognize that the probabilities involved in prediction based on partial information can have only a subjective significance, and that the situation cannot be altered by the device of inventing a fictitious ensemble, even though this enables us to give the probabilities a frequency interpretation. One might then ask how such probabilities could be in any way relevant to the behavior of actual physical systems. A good answer to this is Laplace's famous remark that probability theory is nothing but "common sense reduced to calculation." If we have little or no information relevant to a certain question, common sense tells us that no strong conclusions either way are justified. The same thing must happen in statistical inference, the appearance of a broad probability distribution signifying the verdict, "no definite conclusion." On the other hand, whenever the available information is sufficient to justify fairly strong opinions, maximum-entropy inference gives sharp probability distributions indicating the favored alternative. Thus, *the theory makes definite predictions as to experimental behavior only when, and to the extent that, it leads to sharp distributions.*

When our distributions broaden, the predictions become indefinite and it becomes less and less meaningful to speak of experimental verification. As the available information decreases to zero, maximum-entropy inference (as well as common sense) shades continuously into nonsense and eventually becomes useless. Nevertheless, at each stage it still represents the best that could have been done with the given information.

Phenomena in which the predictions of statistical mechanics are well verified experimentally are always those in which our probability distributions, for the macroscopic quantities actually measured, have enormously sharp peaks. But the process of maximum-entropy inference is one in which we choose the *broadest* possible probability distribution over the microscopic states, compatible with the initial data. Evidently, such sharp distributions for macroscopic quantities can emerge only if it is true that for *each* of the overwhelming majority of those states to which appreciable weight is assigned, we would have the *same* macroscopic behavior. We regard this, not merely as an interesting side remark, but as the essential fact without which statistical mechanics could have no experimental validity, and indeed without which matter would have no definite macroscopic properties, and experimental physics would be impossible. It is this principle of "macroscopic uniformity" which provides the objective content of the calculations, not the probabilities *per se*. Because of it, the predictions of the theory are to a large extent independent of the probability distributions over microstates. For example, if we choose at random one out of each $10^{10^{10}}$ of the possible states and arbitrarily assign zero probability to all the others, this would in most cases have no discernible effect on the macroscopic predictions.

Consider now the case where the theory makes definite predictions and they are not borne out by experiment. This situation cannot be explained away by concluding that the initial information was not sufficient to lead to the correct prediction; if that were the case the theory would not have given a sharp distribution at all. The most reasonable conclusion in this case is that the enumeration of the different *possible* states (i.e., the part of the theory which involves our knowledge of the laws of physics) was not correctly given. Thus, *experimental proof that a definite prediction is incorrect gives evidence of the existence of new laws of physics*. The failures of classical statistical mechanics, and their resolution by quantum theory, provide several examples of this phenomenon.

Although the principle of maximum-entropy inference appears capable of handling most of the prediction problems of statistical mechanics, it is to be noted that prediction is only one of the functions of statistical mechanics. Equally important is the problem of interpretation; given certain observed behavior of a system, what conclusions can we draw as to the microscopic causes of that behavior? To treat this problem and others like it, a different theory, which we may call objective statistical mechanics, is needed. Considerable semantic confusion has resulted from failure to distinguish between the prediction and interpretation problems, and attempting to make a single formalism do for both.

In the problem of interpretation, one will, of course, consider the probabilities of different states in the objective sense; i.e., the probability of state n is the fraction of the time that the system spends in state n. It is readily seen that one can never deduce the objective probabilities of individual states from macroscopic measurements. There will be a great number of different probability assignments that are indistinguishable experimentally; very severe unknown constraints on the possible states could exist. We see that, although it is now a relevant question, metric transitivity is far from necessary, either for justifying the rules of calculation used in prediction, or for interpreting observed behavior. Bohm and Schützer[17] have come to similar conclusions on the basis of entirely different arguments.

5. Generalized Statistical Mechanics

In conventional statistical mechanics the energy plays a preferred role among all dynamical quantities because it is conserved both in the time development of isolated systems and in the interaction of different systems. Since, however, the principles of maximum-entropy inference are independent of any physical properties, it appears that in subjective statistical mechanics all measurable quantities may be treated on the same basis, subject to certain precautions. To exhibit this equivalence, we return to the general problem of maximum-entropy inference of Sec. 2, and consider the effect of a small change in the problem. Suppose we vary the functions $f_k(x)$ whose expectation values are given, in an arbitrary way; $\delta f_k(x_i)$ may be specified independently for each value of k and i. In addition we change the expectation values of the f_k in a manner

[17] D. Bohm and W. Schützer, *Nuovo cimento*, Suppl. II, 1004 (1955).

independent of the δf_k; i.e., there is no relation between $\delta \langle f_k \rangle$ and $\langle \delta f_k \rangle$. We thus pass from one maximum-entropy probability distribution to a slightly different one, the variations in probabilities δp_i and in the Lagrangian multipliers $\delta \lambda_k$ being determined from the $\delta \langle f_k \rangle$ and $\delta f_k (x_i)$ by the relations of Sec. 2. How does this affect the entropy? The change in the partition function (2-9) is given by

$$\delta \lambda_0 = \delta \ln Z = -\sum_k \left[\delta \lambda_k \langle f_k \rangle + \lambda_k \langle \delta f_k \rangle \right], \tag{5-1}$$

and therefore, using (2-13)

$$\begin{aligned} \delta S &= \sum_k \lambda_k \left[\delta \langle f_k \rangle - \langle \delta f_k \rangle \right] \\ &= \sum_k \lambda_k \delta Q_k . \end{aligned} \tag{5-2}$$

The quantity

$$\delta Q_k = \delta \langle f_k \rangle - \langle \delta f_k \rangle \tag{5-3}$$

provides a generalization of the notion of infinitesimal heat supplied to the system, and might be called the "heat of the kth type." If f_k is the energy, δQ_k is the heat in the ordinary sense. We see that the Lagrangian multiplier λ_k is the integrating factor for the kth type of heat, and therefore it is possible to speak of the kth type of temperature. However, we shall refer to λ_k as the quantity "statistically conjugate" to f_k, and use the terms "heat" and "temperature" only in their conventional sense. Up to this point, the theory is completely symmetrical with respect to all quantities f_k.

In all the foregoing discussions, the idea has been implicit that the $\langle f_k \rangle$ on which we base our probability distributions represent the results of measurements of various quantities. If the energy is included among the f_k, the resulting equations are identical with those of conventional statistical mechanics. However, in practice a measurement of energy is rarely part of the initial information available; it is the temperature that is easily measurable. In order to treat the experimental measurement of temperature from the present point of view, it is necessary to consider not only the system σ_1 under investigation, but also another system σ_2. We introduce several definitions:

A *heat bath* is a system σ_2 such that

(a) The separation of energy levels of σ_2 is much smaller than any macroscopically measurable energy difference, so that the possible energies E_{2i} form, from the macroscopic point of view, a continuum.

(b) The entropy S_2 of the maximum-entropy probability distribution for given $\langle E_2 \rangle$ is a definite monotonic function of $\langle E_2 \rangle$; i.e., σ_2 contains no "mechanical parameters" which can be varied independently of its energy.

(c) σ_2 can be placed in interaction with another system σ_1 in such a way that only energy can be transferred between them (i.e., no mass, momentum, etc.), and in the total energy $E = E_1 + E_2 + E_{12}$, the interaction term E_{12} is small compared to either E_1 or E_2. This state of interaction will be called *thermal contact*.

A *thermometer* is a heat-bath σ_2 equipped with a pointer which reads its average energy. The scale is, however, calibrated so as to give a number T, called the *temperature*, defined by

$$1/T \equiv dS_2/d\langle E_2 \rangle \,. \tag{5-4}$$

In a measurement of temperature, we place the thermometer in thermal contact with the system σ_1 of interest. We are now uncertain not only of the state of the system σ_1 but also of the state of the thermometer σ_2, and so in making inferences, we must find the maximum-entropy probability distribution of the total system $\Sigma = \sigma_1 + \sigma_2$, subject to the available information. A state of Σ is defined by specifying simultaneously a state i of σ_1 and a state j of σ_2 to which we assign a probability p_{ij}. Now however we have an additional piece of information, of a type not previously considered; we know that the interaction of σ_1 and σ_2 may allow transitions to take place between states (ij) and (mn) if the total energy is conserved:

$$E_{1i} + E_{2j} = E_{1m} + E_{2n}.$$

In the absence of detailed knowledge of the matrix elements of E_{12} responsible for these transitions (which in practice is never available), we have no rational basis for excluding the possibility of any transition of this type. Therefore all states of Σ having a given total energy must be considered equivalent; the probability p_{ij} in its dependence on energy may contain only $(E_{1i} + E_{2j})$, not E_{1i} and E_{2j} separately.[18] Therefore, the maximum-entropy probability distribution, based on knowledge of $\langle E_2 \rangle$ and the conservation of energy, is associated with the partition function

$$Z(\lambda) = \sum_{ij} \exp\left[-\lambda\left(E_{1i} + E_{2j}\right)\right] = Z_1(\lambda) Z_2(\lambda), \tag{5-5}$$

which factors into separate partition functions for the two systems

$$Z_1(\lambda) = \sum_i \exp\left(-\lambda E_{1i}\right), \quad Z_2(\lambda) = \sum_j \exp\left(-\lambda E_{2j}\right), \tag{5-6}$$

with λ determined as before by

$$\langle E_2 \rangle = -\frac{\partial}{\partial \lambda} \ln Z_2(\lambda); \tag{5-7}$$

or, solving for λ by use of (2-13), we find that the quantity statistically conjugate to the energy is the reciprocal temperature:

$$\lambda = dS_2 / d\langle E_2 \rangle = 1/T. \tag{5-8}$$

More generally, this factorization is always possible if the information available consists of certain properties of σ_1 by itself and certain properties of σ_2 by itself. The probability distribution then factors into two independent distributions

$$p_{ij} = p_i(1) p_j(2), \tag{5-9}$$

and the total entropy is additive:

$$S(\Sigma) = S_1 + S_2. \tag{5-10}$$

We conclude that the function of the thermometer is merely to tell us what value of the parameter λ should be used in specifying the probability distribution of system σ_1. Given this value and the above factorization property, it is no longer necessary to consider the properties of the thermometer in detail when incorporating temperature measurements into our probability distributions; the mathematical processes used in

[18] This argument admittedly lacks rigor, which can be supplied only by consideration of phase coherence properties between the various states by means of the density matrix formalism. This, however, leads to the result given.

setting up probability distributions based on energy or temperature measurements are exactly the same but only interpreted differently.

It is clear that any quantity which can be interchanged between two systems in such a way that the total amount is conserved, may be used in place of energy in arguments of the above type, and the fundamental symmetry of the theory with respect to such quantities is preserved. Thus, we may define a "volume bath," "particle bath," "momentum bath," etc., and the probability distribution which gives the most unbiased representation of our knowledge of the state of a system is obtained by the same mathematical procedure whether the available information consists of a measurement of $\langle f_k \rangle$ or its statistically conjugate quantity λ_k.

We now give two elementary examples of the treatment of problems using this generalized form of statistical mechanics.

The pressure ensemble.—Consider a gas with energy levels $E_i(V)$ dependent on the volume. If we are given macroscopic measurements of the energy $\langle E \rangle$ and the volume $\langle V \rangle$, the appropriate partition function is

$$Z(\lambda, \mu) = \int_0^\infty dV \sum_i \exp\left[-\lambda E_i(V) - \mu V\right],$$

where λ, μ are Lagrangian multipliers. A short calculation shows that the pressure is given by

$$P = -\left\langle \partial E_i(V)/\partial V \right\rangle = \mu/\lambda,$$

so that the quantity statistically conjugate to the volume is

$$\mu = \lambda P = P/kT.$$

Thus, when the available information consists of either of the quantities $(T, \langle E \rangle)$, plus either of the quantities $(P/T, \langle V \rangle)$, the probability distribution which describes this information, without assuming anything else, is proportional to

$$\exp\left\{-\left[\frac{E_i(V) + PV}{kT}\right]\right\}. \tag{5-11}$$

This is the distribution of the "pressure ensemble" of Lewis and Siegert.[19]

[19] M. B. Lewis and A. J. F. Siegert, *Phys. Rev.* 101, 1227 (1956).

A nuclear polarization effect.—Consider a macroscopic system which consists of σ_1 (a nucleus with spin I), and σ_2 (the rest of the system). The nuclear spin is very loosely coupled to its environment, and they can exchange angular momentum in such a way that the total amount is conserved; thus σ_2 is an angular momentum bath. On the other hand they cannot exchange energy, since all states of σ_1 have the same energy. Suppose we are given the temperature, and in addition are told that the system σ_2 is rotating about a certain axis, which we choose as the z axis, with a macroscopically measured angular velocity ω. Does that provide any evidence for expecting that the nuclear spin I is polarized along the same axis? Let m_2 be the angular momentum quantum number of σ_2, and denote by n all other quantum numbers necessary to specify a state of σ_2. Then we form the partition function

$$Z_2(\beta, \lambda) = \sum_{n,m_2} \exp\left[-\beta E_2(n, m_2) - \lambda m_2\right], \tag{5-12}$$

where $\beta = 1/kT$, and λ is determined by

$$\langle m_2 \rangle = -\frac{\partial}{\partial \lambda} \ln Z_2 = \frac{B\omega}{\hbar}, \tag{5-13}$$

where B is the moment of inertia of σ_2. Then, our most unbiased guess is that the rotation of the molecular surroundings should produce on the average a nuclear polarization $\langle m_1 \rangle = \langle I_z \rangle$, equal to the Brillouin function

$$\langle m_1 \rangle = -\frac{\partial}{\partial \lambda} \ln Z_1(\lambda), \tag{5-14}$$

where

$$Z_1(\lambda) = \sum_{m=-I}^{I} e^{-\lambda m}. \tag{5-15}$$

In the case $I = \frac{1}{2}$, the polarization reduces to

$$\langle m_1 \rangle = -\frac{1}{2} \tanh\left(\frac{1}{2}\lambda\right). \tag{5-16}$$

If the angular velocity ω is small, (5-12) may be approximated by a power series in λ:

$$Z_2(\beta, \lambda) = Z_2(\beta, 0)\left[1 - \lambda \langle m_2 \rangle_0 + \frac{1}{2}\lambda^2 \langle m_2{}^2 \rangle_0 + \cdots\right],$$

where $\langle \quad \rangle_0$ stands for an expectation value in the nonrotating state. In the absence of a magnetic field $\langle m_2 \rangle_0 = 0$, $\hbar^2 \left\langle {m_2}^2 \right\rangle_0 = kTB$, so that (5-13) reduces to

$$\lambda = -\hbar\omega/kT. \tag{5-17}$$

Thus, the predicted polarization is just what would be produced by a magnetic field of such strength that the Larmor frequency $\omega_L = \omega$. If $|\lambda| \ll 1$, the result may be described by a "dragging coefficient"

$$\langle m_1 \rangle = \frac{\hbar^2 I(I+1)}{3kTB} \langle m_2 \rangle . \tag{5-18}$$

There is every reason to believe that this effect actually exists; it is closely related to the Einstein-de Haas effect. It is especially interesting that it can be predicted in some detail by a form of statistical mechanics which does not involve the energy of the spin system, and makes no reference to the mechanism causing the polarization. As a numerical example, if a sample of water is rotated at 36 000 rpm, this should polarize the protons to the same extent as would a magnetic field of about 1/7 gauss. This should be accessible to experiment. A straightforward extension of these calculations would reveal how the effect is modified by nuclear quadrupole coupling, in the case of higher spin values.

The Einstein–de Haas effect describes the rotation generated by magnetizing a magnet. Jaynes later remarked that although he was aware of the Einstein–de Haas experiment at the time of writing, he did not know Barnett's work. Barnett was interested in the opposite effect and wrote a series of papers on the subject. The first, published in 1915, was titled "Magnetization by Rotation." Jaynes independently predicted Barnett's effect.

6. Conclusion

The essential point in the arguments presented above is that we accept the von-Neumann—Shannon expression for entropy, very literally, as a measure of the amount of uncertainty represented by a probability distribution; thus entropy becomes the primitive concept with which we work, more fundamental even than energy. If in addition we reinterpret the prediction problem of statistical mechanics in the subjective sense, we can derive the usual relations in a very elementary way without any consideration of ensembles or appeal to the usual arguments concerning ergodicity or equal *a priori* probabilities. The principles and mathematical methods of statistical mechanics are seen to be of much more general applicability than conventional arguments would lead one to suppose. In the problem of prediction, the maximization of entropy is not an application of a law of physics,

Jaynes's extreme modesty in his concluding remarks belies the fact that his Principle of Maximum Entropy has evolved into one of major tools of complexity theory.

but merely a method of reasoning which ensures that no unconscious arbitrary assumptions have been introduced.

APPENDIX A. ENTROPY OF A PROBABILITY DISTRIBUTION

The variable x can assume the discrete values $(x_1, \cdots x_n)$. Our partial understanding of the processes which determine the value of x can be represented by assigning corresponding probabilities $(p_1, \cdots, p_n)$. We ask, with Shannon,[4] whether it is possible to find any quantity $H\,(p_1 \cdots p_n)$ which measures in a unique way the amount of uncertainty represented by this probability distribution. It might at first seem very difficult to specify conditions for such a measure which would ensure both uniqueness and consistency, to say nothing of usefulness. Accordingly it is a very remarkable fact that the most elementary conditions of consistency, amounting really to only one composition law, already determines the function $H\,(p_1 \cdots p_n)$ to within a constant factor. The three conditions are:

(1) H is a continuous function of the p_i.

(2) If all p_i are equal, the quantity $A(n) = H(1/n, \cdots, 1/n)$ is a monotonic increasing function of n.

(3) The composition law. Instead of giving the probabilities of the events $(x_1 \cdots x_n)$ directly, we might group the first k of them together as a single event, and give its probability $w_1 = (p_1 + \cdots + p_k)$; then the next m possibilities are assigned the total probability $w_2 = (p_{k+1} + \cdots + p_{k+m})$, etc. When this much has been specified, the amount of uncertainty as to the composite events is $H\,(w_1 \cdots w_r)$. Then we give the conditional probabilities $(p_1/w_1, \cdots, p_k/w_1)$ of the ultimate events $(x_1 \cdots x_k)$, given that the first composite event had occurred, the conditional probabilities for the second composite event, and so on. We arrive ultimately at the same state of knowledge as if the $(p_1 \cdots p_n)$ had been given directly, therefore if our information measure is to be consistent, we must obtain the same ultimate uncertainty no matter how the choices were broken down in this way. Thus, we must have

$$\begin{aligned} H\,(p_1 \cdots p_n) =& H\,(w_1 \cdots w_2) + w_1 H\,(p_1/w_1, \cdots, p_k/w_1) \\ & + w_2 H\,(p_{k+1}/w_2, \cdots, p_{k+m}/w_2) + \cdots . \end{aligned} \tag{A-1}$$

The weighting factor w_1 appears in the second term because the additional uncertainty $H(p_1/w_1, \cdots, p_k/w_1)$ is encountered only with probability w_1. For example, $H(1/2, 1/3, 1/6) = H(1/2, 1/2) + \frac{1}{2}H(2/3, 1/3)$.

From condition (1), it is sufficient to determine H for all rational values

$$p_i = n_i / \sum n_i,$$

with n_i integers. But then condition (3) implies that H is determined already from the symmetrical quantities $A(n)$. For we can regard a choice of one of the alternatives $(x_1 \cdots x_n)$ as a first step in the choice of one of

$$\sum_{i=1}^{n} n_i$$

equally likely alternatives, the second step of which is also a choice between n_i equally likely alternatives. As an example, with $n = 3$, we might choose $(n_1, n_2, n_3) = (3, 4, 2)$. For this case the composition law becomes

$$H\left(\frac{3}{9}, \frac{4}{9}, \frac{2}{9}\right) + \frac{3}{9}A(3) + \frac{4}{9}A(4) + \frac{2}{9}A(2) = A(9).$$

In general, it could be written

$$H(p_1 \cdots p_n) + \sum_i p_i A(n_i) = A\left(\sum_i n_i\right). \tag{A-2}$$

In particular, we could choose all n_i equal to m, whereupon (A-2) reduces to

$$A(m) + A(n) = A(mn). \tag{A-3}$$

Evidently this is solved by setting

$$A(n) = K \ln n, \tag{A-4}$$

where, by condition (2), $K > 0$. For a proof that (A-4) is the only solution of (A-3), we refer the reader to Shannon's paper.[4] Substituting (A-4) into (A-2), we have the desired result,

$$\begin{aligned} H(p_1 \cdots p_n) &= K \ln\left(\sum n_i\right) - K \sum p_i \ln n_i \\ &= -K \sum_i p_i \ln p_i. \end{aligned} \tag{A-5}$$

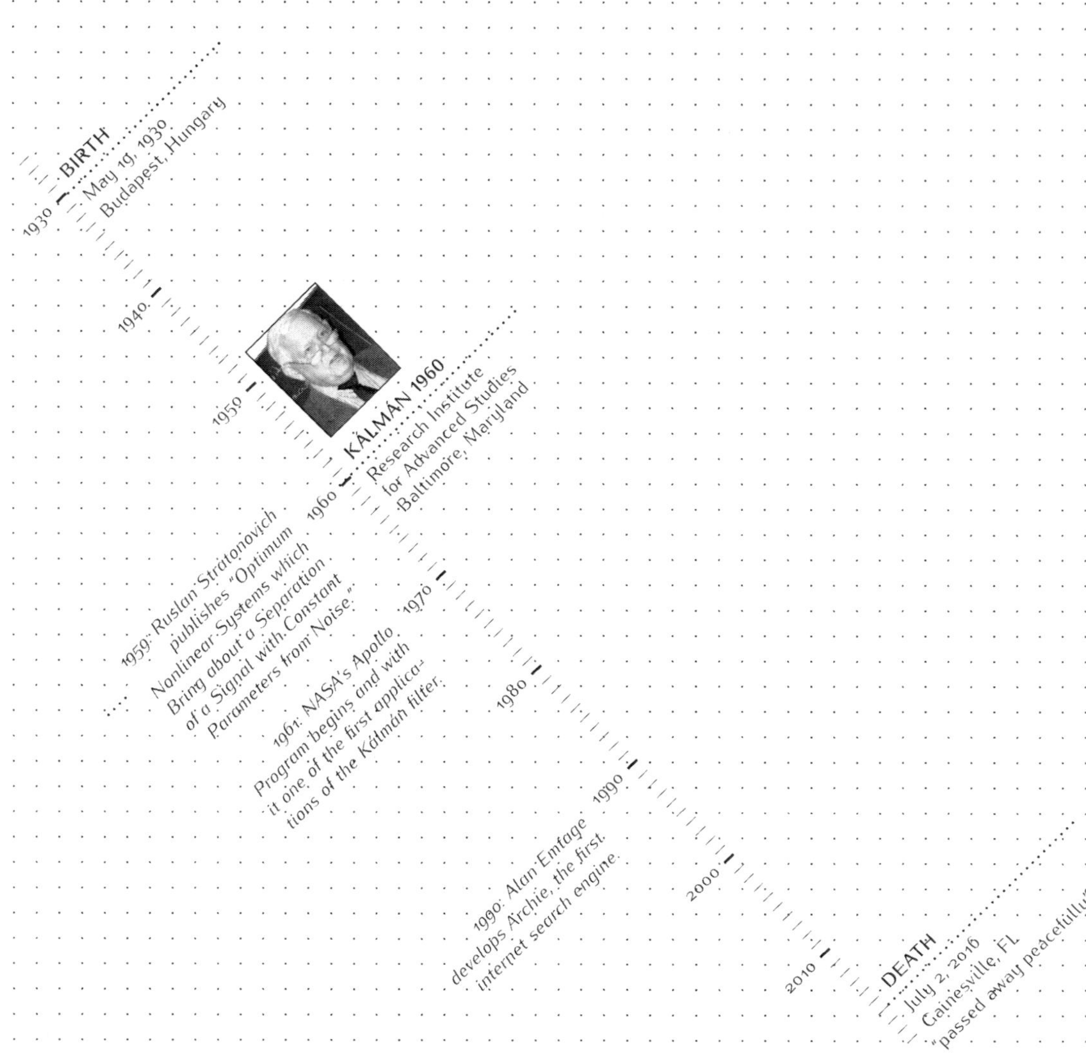

RUDOLF EMIL KÁLMÁN

[15]

THE STATE-SPACE REVOLUTION IN THE STUDY OF COMPLEX SYSTEMS

Maxim Raginsky
University of Illinois, Urbana-Champaign

R. E. Kálmán, "Contributions to the Theory of Optimal Control," *Boletín de la Sociedad Matemática Mexicana* 5, 102–119 (1960).

The origins of mathematical control theory can be traced back to James Clerk Maxwell's 1868 paper "On Governors," which phrased the question of stability of James Watt's centrifugal governor in terms of a numerical criterion involving roots of polynomials. The early days of feedback control as a separate discipline saw widespread use of frequency domain methods based on Fourier and Laplace transforms, with pioneering contributions by Hendrik Bode (1940), Harry Nyquist (1932), and others during the 1930s–1940s. The underlying paradigm viewed systems in terms of their external (input/output) behavior, which was described mathematically using so-called transfer functions, that is, proper rational functions of a single complex variable. The criterion of (external) stability, worked out by Maxwell for second- and third-order systems and later in full generality by Edward Routh (1877) and Adolf Hurwitz (1895), pertained to the poles of the transfer function. A system described in this way receives inputs (or stimuli) from its environment, reacts by producing outputs (or responses), and can be controlled by means of output-to-input feedback. This stimulus–response view was rather influential even beyond control theory, for example, in psychology and physiology, and it was adopted by Norbert Wiener (1948) in his book on cybernetics. The extension of these ideas to stochastic systems, where stationary processes were modeled in the frequency domain by their spectral densities, led to the seminal results on linear filtering by Andrey Kolmogorov and Wiener.

However, a number of limitations of this paradigm soon became apparent. The most obvious one was that it dealt primarily with time-invariant systems, that is, systems whose input–output behavior did not depend on time explicitly. Classical transform-based methods were not equipped to handle time-varying systems in a straightforward way, and the mathematical and computational challenges involved in extending the Kolmogorov–Wiener filtering theory to nonstationary processes were rather considerable. In addition, the classical theory was developed for single-input, single-output systems, whereas in many engineering applications (such as chemical process control) one would routinely deal with systems where several output variables had to be regulated simultaneously by manipulating several input variables. The ad hoc approach of reducing these multivariable control problems to a series of single-input, single-output problems lacked principled justification. Finally, stability analysis based on external input–output behavior could not account for the presence of internal instabilities, which in turn could not be compensated using output-to-input feedback. Taken together, all of this was evidence that the field of control systems was facing what Thomas Kuhn referred to as a *crisis* in his book *The Structure of Scientific Revolutions*, that is, the situation where the explanatory framework based on the current paradigm starts breaking down as various anomalies accumulate.[1]

The paradigm shift precipitated by this crisis occurred around 1960, and can be traced almost entirely to several pioneering papers by Rudolf Kálmán, including "Contributions to the Theory of Optimal Control." The influence of these works on the field of control systems is well-documented elsewhere.[2] Our interest here is in the fundamental contributions of this paper to the science of complexity.

I have already mentioned that, before the work of Kálmán, control systems were approached from the external perspective, that is, as input–output transformations subject to the criteria of causality, linearity, and time invariance. In this setting, the (scalar-valued) input

[1] The credit for the idea of viewing the state of affairs in pre-1960s control theory through the Kuhnian lens belongs to Prof. Sanjoy K. Mitter of MIT.

[2] See, e.g., the edited volume *Control Theory: Twenty-Five Seminal Papers (1932–1981)* published in 2001 by IEEE Press.

$u(\cdot)$ and output $y(\cdot)$ are related by the convolution integral

$$y(t) = \int_0^\infty h(t-\tau)u(\tau)\mathrm{d}\tau, \qquad t \geq 0 \tag{1}$$

where the function $h(\cdot)$, called the *impulse response* of the system, plays the same role as Green's function in linear partial differential equations of mathematical physics. The transfer function of the system is the Laplace transform of h. Kálmán augmented the external input/output picture with an internal *state* that represented the system's memory. The corresponding state-space description of a linear system consists of coupled first-order linear differential equations of the form

$$\dot{x}(t) = F(t)x(t) + G(t)u(t), \qquad y(t) = H(t)x(t) \tag{2}$$

where the input $u(t)$, the internal state $x(t)$, and the output $y(t)$ are now *vector-valued*, and $F(t)$, $G(t)$, and $H(t)$ are time-dependent matrices of appropriate shapes. The system (2) is still causal and linear, but it is now (possibly) time-varying. In the time-invariant case, that is, when the matrices F, G, and H are constant, one can eliminate the internal state and recover the convolution integral (1) under the assumption that the system is initially at rest, that is, $x(0) = 0$. However, the scope of the new description is much wider; in particular, it is now possible to treat both scalar (single-input, single-output) and multivariable systems on the same footing. Whereas classical control theory of Bode and Nyquist was phrased almost exclusively in the frequency domain and used the machinery of Fourier and Laplace transforms, modern control theory à la Kálmán is a time-domain theory, and its mathematical language is that of differential equations and linear algebra.

While the introduction of first-order differential equations for the system state was a forward-looking move for control theory, it was also a return to the mathematical formalism of classical mechanics, going back to the foundational works of Lagrange, Lyapunov, and Poincaré. This rapprochement with the theory of dynamical systems had brought to the fore the key notion of *stability*, which could now be phrased from an internal point of view in terms of the origin of the state space being the unique equilibrium configuration of the system. However, the presence of external input $u(t)$ in (2) constituted a radical departure from the conceptual framework of classical physics. Dynamical systems of classical mechanics are *closed* systems subject to the iron law of Laplacian determinism: The state $x(t)$ at a given time t completely determines the future state trajectory $x(t')$ for all $t' \geq t$. By contrast, the dynamical systems of control theory are *open* systems whose motion can be manipulated through their inputs. Thus, the knowledge of $x(t')$ for $t' \leq t$ does not seal the future fate of the system because the latter depends on the future inputs $u(t')$ for $t' \geq t$. This also speaks to the fundamental distinction between physics as the domain of discovery of natural laws, and analysis of phenomena subject to these laws, and engineering as the domain of invention and synthesis of systems with a given purpose in mind.

This brings us now to what Kálmán himself viewed as the "principal contribution" of his paper, namely the concepts of *controllability* and *observability*, two entirely new notions that make sense only in light of the state-space view. Roughly speaking, the system (2) is controllable if, for any initial state x_0, any final state x_1, and any times $t_0 < t_1$, there exists an input $u(t)$ defined for $t_0 \leq t \leq t_1$, such that the state trajectory of (2) with that input connects $x(t_0) = x_0$ to $x(t_1) = x_1$. The system is observable if, given the output trajectory $y(t)$ from t_0 to t_1, one can reconstruct the initial state $x(t_0)$.[3] Both controllability and observability relate the external attributes of the system (namely, its inputs and outputs) to its internal state and hence to its internal organization. Moreover, Kálmán gave necessary and sufficient conditions for controllability in terms of the so-called *controllability Gramian*[4] and, in the case of time-invariant systems, in terms of easily verified rank conditions for certain matrices.

The concept of controllability also leads naturally to the question of optimality, which can be addressed using calculus of variations—again, here we see at once a deep connection with the variational formulation of classical mechanics due to Lagrange and Hamilton and a deep rupture

[3] This interpretation of observability, which appeared in another paper of Kálmán published in the same year, is not present in this work, where he simply introduces it using duality of vector spaces.

[4] The terms "controllability Gramian" and "observability Gramian" were coined by Roger W. Brockett in his pioneering textbook *Finite-Dimensional Linear Systems* (1970).

with it due to the presence of external inputs amenable to manipulation. Assuming the system (2) is controllable, we now seek to accomplish the transfer from a given starting state x_0 to a given ending state x_1 in time t in the most economical (or, as Kálmán put it, "rational") way by minimizing the energy-like quantity

$$\frac{1}{2}\int_0^t \|u(\tau)\|^2 \mathrm{d}\tau \equiv \frac{1}{2}\int_0^t u(\tau)^{\mathrm{T}} u(\tau) \mathrm{d}\tau \tag{3}$$

over all input or control functions $u(\cdot)$ subject to the endpoint constraints $x(0) = x_0$ and $x(t) = x_1$ and to the dynamical law in (2). Kálmán connects this problem to Constantin Carathéodory's formulation of calculus of variations and thus to the partial differential equations of Hamilton and Jacobi. However, while the variational formulation of classical mechanics is primarily an explanatory schema that deduces the laws of motion of existing physical systems from the principle of least action, a control engineer's invocation of calculus of variations is explicitly teleological, seeking to *synthesize* a control law to align given means with desired ends in the most economical manner. This was a major ingredient in the solution of a more general problem posed by Kálmán in the 1960 paper, now commonly known as the Linear–Quadratic Regulator (LQR) problem.

By reconciling the external (input–output) and the internal (state-space) perspectives, Kálmán made it possible to speak in a unified way to the interplay between structure, organization, and function of open dynamical systems, both natural and artificial. I will now sketch the influence of these ideas on the science of complexity in the context of chaos, dissipativity, and complex networks. To keep the exposition simple, I will not touch upon the extensions of the concepts of controllability and observability to nonlinear systems.

Chaotic systems, such as the Lorenz model, exhibit a high degree of sensitivity to their initial conditions; in particular, they can have bounded but exponentially unstable trajectories. While deterministic chaos imposes severe limits on our ability to forecast the long-term behavior of such systems, it also implies that we can induce large changes in their trajectories using small control effort. To see how this can come about, consider a controlled nonlinear system of the form $\dot{x}(t) =$

$f(x(t), u(t))$ with input $u(t)$ and state $x(t)$, which is chaotic in the absence of control, that is, the trajectories of $\dot{x}(t) = f(x(t), 0)$ are bounded but unstable. We can then consider the control-free trajectory $\varphi(t) = \varphi(t, x_0)$ for a given initial condition $\varphi(0, x_0) = x_0$ and linearize the original system around this trajectory:

$$\dot{z}(t) = F(t)z(t) + G(t)u(t), \tag{4}$$

where we have defined

$$F(t) = \left.\frac{\partial f}{\partial x}\right|_{x=\varphi(t),u=0} \qquad \text{and} \qquad G(t) = \left.\frac{\partial f}{\partial u}\right|_{x=\varphi(t),u=0}. \tag{5}$$

The chaotic behavior of the control-free dynamics can now be quantified by means of Lyapunov exponents at x_0 that describe the rate of exponential growth of the eigenvalues of $F(t)$ as t increases; by the same token, we can quantify the sensitivity of the linearized system (4) to small deviations of the control input $u(t)$ from 0 in terms of the asymptotic behavior of its controllability Gramian, which depends on both $F(\cdot)$ and $G(\cdot)$ in (5). In particular, as t gets large, we can induce very large changes in $z(t)$ using very small control effort, as given by (3).

Dissipativity, another key concept in the science of complex systems, can also be viewed through the lens of state-space systems with inputs and outputs. Familiar examples of dissipative systems include electrical circuits, thermodynamic systems, and viscoelastic materials, which dissipate some of their energy into the environment in the form of heat. Once again, let us consider a system with controlled state dynamics given by $\dot{x}(t) = f(x(t), u(t))$ with input $u(t)$ and state $x(t)$, but now suppose that the system also produces an output $y(t) = g(x(t))$. The following definition of dissipativity was put forward by Jan Willems (1972a): The above system is dissipative if there exist a *supply-rate function* W of input and output and a *storage function* S of state, such that the dissipation inequality

$$\int_{t_0}^{t_1} W(u(t), y(t))\mathrm{d}t \geq S(x(t_1)) - S(x(t_0)) \tag{6}$$

holds for all times $t_0 < t_1$ and all input/state/output trajectories $u(t), x(t), y(t)$. Dissipativity imposes certain limitations on both the internal organization of the system and on its interaction with its environment through its inputs and outputs. In operational terms, it is useful to think of the supply rate as determining the work done on the system by manipulating its inputs. Thus, the dissipation inequality (6) specifies a fundamental lower bound on the amount of work needed to transfer the system from a given initial state $x(t_0)$ at time t_0 to a given final state $x(t_1)$ at time t_1. A key question in connection with dissipativity is to decide whether a given system is dissipative and, if so, to give explicit expressions for the supply rate and the storage function. For time-invariant linear systems described by models like (2) with constant matrices F, G, and H, these questions can be answered by appealing to Kálmán's results on the LQR problem and, hence, to notions of controllability and observability. Finally, to relate this control-theoretic notion of dissipativity to the corresponding notion in the theory of complex systems having to do with the (exponential) decrease of some energy-like observable along the system trajectory, it suffices to note that, in many cases of interest, the storage function S is also a Lyapunov function for the free (i.e., uncontrolled) motion of the system.

Finally, we mention the notion of *structural controllability*, introduced by C.-T. Lin (1974) for time-invariant, single-input systems and extended to multi-input systems by R. W. Shields and J. B. Pearson (1976). The main idea here is to view the system (2) as a time-varying network by associating each coordinate of the system's state vector with a vertex of a directed graph, so that there is a directed edge (or arc) from vertex i to vertex j at time t if the (j, i) entry of the matrix $F(t)$ in (2) is nonzero. Similarly, one can introduce control vertices or nodes by associating one such node to each coordinate of the input vector, so that it becomes natural to say that control vertex k influences state vertex i at time t if the (i, k) entry of the matrix $G(t)$ in (2) is nonzero. The original motivation for introducing the notion of structural controllability for linear time-invariant systems was stated in terms of inherent limitations on the precision of system models; for example, while we may know with certainty which entries of the F and G matrices are zero, the nonzero entries may only be known up to some measurement error. Thus, the goal was to consider simultaneously all systems with a given structure, that is, the same pattern of nonzero entries in the system matrices.

Another perspective on the role of structure is to ask whether one can control the global behavior of a large network by means of locally applied inputs affecting only a small subset of the state vertices. By combining the algebraic ideas of Kálmán with various notions of graph theory, such as matchings or independent sets, we can now reason quantitatively about the dynamics of complex networked systems with applications in economics, systems biology, neuroscience, social networks, physics of disordered systems, and much more.

In addition to enriching the toolkit of complexity science with the formalism of open dynamical systems via input-state-output descriptions and the notions of controllability and observability, an equally important contribution of Kálmán is along the algorithmic and computational aspects. His solutions of the optimal (minimum-energy) state transfer and of the more general LQR problem came with concrete algorithms that could be implemented on digital computers that were just beginning to be used by scientists and engineers. The recursive nature of state-space models lent itself to efficient computer implementation, which was key to the widespread adoption of the celebrated Kálmán filter that was also introduced in 1960, the *annus mirabilis* of mathematical systems theory. This fortuitous confluence of state-space models and efficient algorithms was ultimately responsible for the emergence of *computer simulation* as an indispensable tool in the science of complexity, especially when dealing with nonlinear and stochastic systems and complex networks consisting of many interacting subsystems.

REFERENCES

Bechhoefer, J. 2021. *Control Theory for Physicists.* Cambridge, UK: Cambridge University Press.

Bode, H. W. 1940. "Relations between Attenuation and Phase in Feedback Amplifier Design." *Bell Systems Technical Journal* 19 (3): 421–454. https://doi.org/10.1002/j.1538-7305.1940.tb00839.x.

Hermann, R., and A. J. Krener. 1977. "Nonlinear Controllability and Observability." *IEEE Transactions on Automatic Control* 22:728–740. https://doi.org/10.1109/TAC.1977.1101601.

Hurwitz, A. 1895. "Über die Bedingungen, unter welchen eine Gleichung nur Wurzeln mit negativen reellen Theilen besitzt." *Mathematische Annalen* 46 (2): 273–284. https://doi.org/10.1007/BF01446812.

Kálmán, R. E. 1960a. "Contributions to the Theory of Optimal Control." *Boletín de la Sociedad Matemática Mexicana* 5:102–119.

———. 1960b. "On the General Theory of Control Systems." *IFAC Proceedings Volumes* 1 (1): 491–502. https://doi.org/10.1016/S1474-6670(17)70094-8.

Lin, J.-T. 1974. "Structural Controllability." *IEEE Transactions on Automatic Control* 19:201–208. https://doi.org/10.1109/TAC.1974.1100557.

Maxwell, J. C. 1868. "On Governors." *Proceedings of the Royal Society of London* 16:270–283. https://doi.org/10.1098/rspl.1867.0055.

Nyquist, H. 1932. "Regeneration Theory." *Bell Systems Technical Journal* 11 (1): 126–147. https://doi.org/10.1002/j.1538-7305.1932.tb02344.x.

Routh, E. J. 1877. *A Treatise on the Stablity of a Given State of Motion, Particularly Steady Motion.* London, UK: McMillan.

Shields, R. W., and J. B. Pearson. 1976. "Structural Controllability of Multi-Input Linear Systems." *IEEE Transactions on Automatic Control* 21:203–212. https://doi.org/10.1109/TAC.1976.1101198.

Shinbrot, T., C. Grebogi, J. A. Yorke, and E. Ott. 1993. "Using Small Perturbations to Control Chaos." *Nature* 363 (6428): 411–417. https://doi.org/10.1038/363411a0.

Sussmann, H. J., and J. C. Willems. 1997. "300 Years of Optimal Control: From the Brachystochrone to the Maximum Principle." *IEEE Control Systems Magazine* 17:32–44. https://doi.org/10.1109/37.588098.

Wiener, N. 1948. *Cybernetics: Or Control and Communication in the Animal and the Machine.* Cambridge, MA: MIT Press.

Willems, J. C. 1972a. "Dissipative Dynamical Systems Part I: General Theory." *Archive for Rational Mechanics and Analysis* 45:321–351. https://doi.org/10.1007/BF00276493.

———. 1972b. "Dissipative Dynamical Systems Part II: Linear Systems with Quadratic Supply Rates." *Archive for Rational Mechanics and Analysis* 45:352–393. https://doi.org/10.1007/BF00276494.

CONTRIBUTIONS TO THE THEORY OF OPTIMAL CONTROL

R. E. Kálmán, Research Institute for Advanced Studies

1. Introduction

The purpose of this paper is to give an account of recent research on a classical problem in the theory of control: the design of linear control systems so as to minimize the integral of a quadratic function evaluated along motions of the system. This problem dates back in its modern form to Wiener and Hall at about 1943 (Wiener 1949; Hall 1943). In spite of its relatively long history, the problem has never been formulated rigorously from a mathematical point of view. Even the most up-to-date expositions of the subject (see, e.g., Newton, Gould, and Kaiser 1957) are inaccessible to the mathematician due to the lack of precisely stated conditions and results.

Here, Kálmán contrasts the existing approach to designing control systems based on Fourier and Laplace transforms with the time-domain approach based on differential equations. The role of Lyapunov's theory of stability is explicitly acknowledged.

The problem is quite broad, and there are many unsettled questions. This paper will be concerned with only the simplest case, the so-called *regulator problem*. For other aspects of the problem, the reader may consult Kálmán and Bertram (1958), Kálmán and Koepcke (1958, 1959), and Kálmán (1960a, 1960b), which, while devoted primarily to questions of theoretical engineering, contain precise mathematical results.

The conventional theory of the regulator problem is based largely on Fourier and Laplace transforms. By contrast, the approach of this paper is direct and uses the well-known theory of ordinary differential equations. We have also drawn on Lyapunov's theory of stability. While our earlier treatments (Kálmán and Koepcke 1958, 1959; Kálmán 1960b) followed the point of view of dynamic programming, here we utilize classical tools of the calculus of variations, in particular the Hamilton–Jacobi equation.

The principal contribution of the paper lies in the introduction and exploitation of the concepts of *controllability* and *observability* (Kálmán 1960b), with the aid of which we give, for the first time, a complete theory of the regulator problem. In particular, we prove existence and stability theorems for the regulator problem and study in some detail stability properties of the matrix Riccati equation, which arises as a special case of the Hamilton–Jacobi equation.

The concepts of controllability and observability are indeed the core conceptual contributions of this paper. Controllability pertains to the ability to affect the motion of the system's internal state by manipulating the inputs; observability refers to the ability to reconstruct the initial internal state based on external observations of the output.

A careful discussion of the conceptual aspects of the control problems has been included as an aid to persons not familiar with the field of control. Some mathematical arguments, in particular the review of the calculus of variations, are more leisurely than usual in order to render the paper reasonably self-contained.

2. Notation and Terminology

We use standard vector-matrix notation, with the following conventions: small Greek letters are scalars; small Latin letters are vectors, capitals are matrices. The unit matrix is I. Exceptions: i, j, m, n, p are integers; t, E, L, V are scalars; ϕ, ψ, ξ are vectors. The inner product is $[x, y]$. The transpose of a matrix is denoted by the prime. The norm is $\|x\| = [x, x]^{\frac{1}{2}}$. The norm $\|A\|$ of a matrix A is sup $\|Ax\|$ over $\|x\| = 1$. Special conventions: $\|x\|_P^2 = [x, Px]$ where P is symmetric, nonnegative definite. If A, B are symmetric, $A > B[A \geq B]$ means $A - B$ is positive [nonnegative] definite. The letters t, σ, τ are arbitrary real numbers which always denote the *time*; we use $t_0 \leq t_1 \leq t_2$ to denote fixed, ordered values of t. All scalars, vectors, and matrices are real throughout. For a scalar function $L(u)$ of the vector u, L_u is the gradient vector and L_{uu} is the jacobian matrix.

We shall study the system represented by the equations

$$dx/dt = F(t)x + G(t)u(t) \tag{2.1}$$

$$y(t) = H(t)x(t) \tag{2.2}$$

where: u is an m-vector, x is an n-vector, y is a p-vector; $F(t), G(t)$, as well as $H(t)$ are rectangular matrices continuous in t, either of which may be singular.

In view of the physical motivation of our problem, we adopt the following terminology: Equations (2.1–2.2) are the *plant* (or *model*); x is the *state* of; $u(t)$ is the *control function* or *input* to; and $y(t)$ is the *output* of the plant. The plant is *constant* if F, G, H are constants. If $u(t) = 0$ or $G(t) = 0$, the plant is *free*.

The behavior of the plant is described by the solution of the differential equation (2.1) which will exist for all t and be unique if, say, $u(t)$ is Lebesgue integrable. As is well known (Coddington and Levinson 1955), the general solution has the form[1]

$$x(t) = \Phi(t, t_0)\, x(t_0) + \int_{t_0}^{t} \Phi(t, \tau) G(\tau) u(\tau) d\tau \qquad (2.3)$$

where $\Phi(t, \tau)$, defined for all t, τ, is a fundamental matrix (Coddington and Levinson 1955) of solutions of the free system (2.1), satisfying the additional requirement that

$$\Phi(t, t) = I \quad \text{for all} t \qquad (2.4)$$

The notion of state as containing all presently available information relevant for making future decisions is indeed a bridge to the theory of Markov processes. The construction of the Kálmán filter would not have been possible without this fundamental connection.

In view of (2.4), we call Φ the *transition matrix* of (2.1)—a terminology borrowed from the theory of Markov processes (Kálmán 1960a, 1960b; Kálmán and Bertram 1960).

The solution (2.3) is conveniently regarded as the *motion* of the state of (2.1); this leads to the notation

$$x(t) = \phi_u(t; x, t_0) \qquad (2.5)$$

Read: the motion of (2.1) starting at initial state x at time t_0 and observed at time t, and influenced by the *fixed* control function $u(t)$ defined in the interval $[t_0, t]$. Since (2.5) holds for all t, t_0, we have in particular the identity: $x = \phi_u(t; x, t)$ for all t, x, u. Free motions are denoted by ϕ_f. We observe also that (2.1) has an *equilibrium state* x^* at 0, in other words, a state for which $x^* = \phi_f(t; x^*, t_0)$ for all t, t_0.

The notion of equilibrium is important for the discussion of stability. Here, stability is understood in the sense of Lyapunov and pertains to the internal view of the system, whereas the traditional view of control systems based on transfer functions was concerned with external stability (bounded inputs result in bounded outputs).

3. Statement of Problem

In the simplest applications, the object of a control system is the following: *Given any state x of the plant (2.1) at any time t_0, "generate" a control function $u(t)$, defined for $t \geq t_0$ and depending on x, t_0, which causes x*

[1] The function (2.3) will satisfy the differential equation (2.1) *almost everywhere*.

to be "transferred" to the equilibrium state 0. In other words, $u(t)$ is chosen so as to assure

$$\lim_{t \to \infty} \phi_u(t; x, t_0) = 0 \tag{3.1}$$

For technological reasons, the function $u(t)$ must be generated from actual measurements of the behavior of the plant. To describe how the control system is to be *physically realized*, one must therefore provide an algorithm for computing the number $u(t_1)$ from the knowledge of $y(t)$ for $t \leq t_1$. This is usually referred to as the *Feedback Principle.*

One may separate the problem of physical realization into two stages:

(A) Computation of the "best approximation" $\hat{x}(t_1)$ of the state $x(t_1)$ from knowledge of $y(t)$ for $t \leq t_1$.

(B) Computation of $u(t_1)$ given $\hat{x}(t_1)$.

In the engineering literature one often makes the simplifying assumption of treating the two problems separately, i.e., simply regarding $\hat{x}(t)$ as though it were $x(t)$. We are concerned here only with Problem (B) and therefore always assume that $x(t)$ is known exactly. Somewhat surprisingly, the theory of Problem (A), which includes as a special case Wiener's theory of the filtering and prediction of time series, turns out to be analogous to the theory of Problem (B) developed in this paper. This assertion follows from the *duality theorem* discovered by the author (Kálmán 1960a, 1960b); this theorem can be used to show also that the separation of Problems (A) and (B) is indeed legitimate.

Assuming $x(t)$ is known exactly and taking into account the Feedback Principle, the problem of generating $u(t)$ reduces to specifying the *control law*

$$u(t) = k(x(t), t) \tag{3.2}$$

From the definition of state it is clear that nothing would be gained by letting $u(t)$ depend also on values of the state prior to time t. To assure that (2.1) with (3.2) has a unique solution, it suffices to have $k \in C^1$. If (3.2) does not depend explicitly on t, we say the control law is *constant.*

Here, Kálmán states that the goal of control is to induce desired changes in the system's internal state. In the simplest case, we would like to cause the system to eventually approach equilibrium. Again, we should contrast it with the goal of control from the external perspective as changing the input–output characteristic of the system by means of feedback.

Here, we see an informal, yet precise statement of two key ideas in control theory: the *separation principle* and the *certainty equivalence principle*. The former says that the problem of optimal control based on partial (and possibly noisy) observations of the state can be split into two subproblems: state estimation and determiniation of the control based on the estimated state. Moreover, the two subproblems can be solved independently. The latter says that, once an optimal estimate of the state is obtained, the construction of optimal control law can proceed as if exact state observations are available.

To arrive at the control law "rationally," we now add the further desideratum that *the integral of a nonnegative function of the state along any motion ϕ_u should be minimized by the choice of $u(t)$.* Stating this requirement with some care, we shall see that it uniquely determines $u(t)$ and hence also the control law (3.2), and even implies (3.1). We call the resulting control system *optimal.*

Here, Kálmán argues for the "rational" approach to designing the control law in the sense of aligning existing means with desired ends in the most economical way. (Here one can recall that this was Max Weber's crisp formulation of rational action in his *Theory of Social and Economic Organization.*)

Let us now state precisely the

(3.3) OPTIMAL REGULATOR PROBLEM.

Find a control law (3.2) for which

$$V^0(x, t_0, t_1) = \inf_{u(t)} \left\{ \nu\left(\phi_u\left(t_1; x, t_0\right)\right) + \int_{t_0}^{t_1} L\left(\phi_u\left(t; x, t_0\right) u(t), t\right) dt \right\} \tag{3.4}$$

is attained for all x, t_0, and t_1, where ν, L are nonnegative scalar functions of class C^2 in all arguments.

The class of functions $u(t)$ which are admitted to competition in taking the infimum in (3.4) are to be of class D° (i.e., continuous except at isolated points at which $u(t)$ has finite left- and right-hand limits).

It is well known in the calculus of variations (Carathéodory 1935, p. 196) that the condition $L_{uu} \geq 0$ or $L_{uu} \leq 0$ is necessary for the existence of even local extremals. To avoid complications due to the equality sign, we assume from the outset that

$$L_{uu}(x, u, t) > 0 \quad \text{for all} x, u, t \tag{3.5}$$

which is equivalent to assuming that L is strictly convex in u.

Here, Kálmán phrases the problem of determining an optimal control strategy (in state feedback form) in the framework of calculus of variations following the ideas of Constantin Carathéodory. This formulation is complementary to the solution obtained using Bellman's dynamic programming principle and Pontryagin's maximum principle.

In engineering language, one calls t_1 the *terminal time* (it may be infinity!), the integral is the *performance index,* and (at least when it does not depend on u) L is the *error criterion* or more generally the *loss function.* The function ν is added for greater generality. We use the notation $V(x, t_0, t_1; u)$ for the value of the integral in (3.4) for some specified, fixed $u(t)$. The superscript o identifies "optimal".

4. Relations with the Calculus of Variations

In this section we transcribe some well-known results (Carathéodory 1935, Ch. 12) of the local problem of the calculus of variations into a form best suited to our problem. Let us first solve (3.3) in a very special case.

(4.1) LEMMA (Carathéodory). *Let $k(x,t)$ be an m-vector function of class C^1 in x,t. Write $u^\circ = k(x,t)$. Assume $\nu = 0$ and that the function L in (3.4) satisfies the following conditions for all x and all $t_0 \le t \le t_1$:*

(a) $L(x, u^\circ, t) = 0$

(b) $L(x, u, t) > 0$ for all $u \ne u^\circ$

Then the optimal performance index V° is identically zero for all x and is attained by using the optimal control law given by

$$u^\circ = k(x(t), t) \tag{4.2}$$

Proof. Let $\phi_u{}^\circ$ denote the motion of the plant under control law (4.2); similarly, let ϕ_{u^1} be the motion corresponding to some fixed control function $u^1(t)$. By (b) above and since the integrand of (3.4) is in class D° in t, it follows that $V(x, t_0, t_1; u^1) > 0$ unless $u^1(t) = u^o(t)$ at every continuity point of $u^1(t)$ in the interval (t_0, t_1). On the other hand, by (a) we see that $V^0(x, t_0, t_1)$ vanishes identically in x.[2] Q. E. D.

Anticipating the final result (4.14), let $V^\circ(x, t, t_1)$ be an arbitrary scalar function of class C^2 in x, t, (t_1 being a fixed number) and subject also to

$$V^o(x, t_1, t_1) = \nu(x) \tag{4.3}$$

Let us replace L by

$$L^*(x, u, t) = L(x, u, t) + V_t^o(x, t, t_1) + [V_x^o(x, t, t_1), F(t)x + G(t)u]$$

The integral of the last two terms along any motion between the limits t_0, t_1 is

$$\nu(\phi_u(t_1; x, t_0)) - V^\circ(x, t_0, t_1) \tag{4.4}$$

Since the second term in (4.4) does not depend on $u(t)$, it follows that the two variational problems,

$$\inf_{u(t)} \int_{t_0}^{t_1} L^*(\phi_u(t; x, t_0), u(t), t)\, dt \tag{4.5}$$

and (3.3) are equivalent in that they have the same minimizing function $u(t)$ (if such exists).

---

[2] It is clear that two optimal controls $u^\circ(t)$ and $u^1(t)$ can differ only on a set of measure zero.

Now we try to find functions $V^o(x,t,t_1)$ and $k(x,t,t_1)$ for which the hypotheses of the Lemma are satisfied when L is replaced by L^*.

In order that $L^*(x,u,t)$ have a minimum with respect to u at $u = u^\circ = k(x,t,t_1)$, it is necessary that all first partial derivatives of L^* with respect to u vanish at u°. This and the condition $L^*\left(x,u^0,t\right) = 0$ give

$$G'(t)V_x^o = -L_u\left(x,u^o,t\right) \tag{4.6}$$

$$-V_t^o = L\left(x,u^o,t\right) + \left[V_x^o, F(t)x + G(t)u^o\right] \tag{4.7}$$

These equations are called by Carathéodory the *fundamental equations* of the variational problem (3.3).

From assumption (3.5) it follows at once (by the strict convexity[3] of $L(x,u,t)$ in u) that (4.6) can be solved for u°; more precisely, there exists a function ψ of class C^1 such that

$$u^o = \psi\left(x, G'(t)V_x^o\left(x,t,t_1\right), t\right) = k\left(x,t,t_1\right) \tag{4.8}$$

which is the desired optimal control law.

To check condition (b) of the Lemma, we write, using (4.6–4.7),

$$\begin{aligned} L^*(x,u,t) &= L(x,u,t) - L\left(x,u^\circ,t\right) - \left[u-u^\circ, L_u\left(x,u^\circ,t\right)\right] \\ &= E\left(x,u,u^\circ,t\right) \end{aligned} \tag{4.9}$$

which is the well-known *Weierstrass E-function*. It is clear by inspection that E is the quadratic remainder in the Taylor series of L at $u = u^\circ$. Using the well-known estimate for the remainder, we have

$$E\left(x,u,u^\circ,t\right) = \|u-u^\circ\|^2_{L_{uu}(x,u+\theta(u^\circ-u),t)} \quad (0 \le \theta \le 1) \tag{4.10}$$

which is nonnegative and in view of (3.5) vanishes if and only if $u = u^\circ$.

Hence if there is a function V° satisfying (4.3,4.6–4.7) and in addition (3.5) holds, we can apply the Lemma and find that V° is just the left-hand side of (3.4). The optimal control law is (4.8).

---

[3]A scalar function $\alpha(x)$ of a vector x is convex in x if and only if for all x_1, x_2 the function $\beta(\lambda) = \alpha\left(\lambda x_1 + (1-\lambda)x_2\right)$ is convex in λ over the interval $0 \le \lambda \le 1$. This will be the case if and only if $d^2\beta/d\lambda^2 = \left[x_1 - x_2, \alpha_{xx}\left(x_1 - x_2\right)\right] \ge 0$. Hence $\alpha(x)$ is convex if and only if $\alpha_{xx} \ge 0$; similarly, $\alpha(x)$ is strictly convex if and only if $\alpha_{xx} > 0$.

We now define the so-called *conjugate variable* ξ by

$$\xi = V_x^0 \tag{4.11}$$

and write $\psi = \psi\,(x, G'(t)\xi, t)$. We define the *Hamiltonian* as

$$\mathcal{H}(x, \xi, t) = L(x, \psi, t) + [\xi, F(t)x + G(t)\psi] \tag{4.12}$$

It is easily shown that if L is of class C^2, so is also $\mathcal{H}$.

Now if $V^\circ\,(x, t, t_1)$ is any solution of class C^2 of (4.6–4.7), it follows by substitution that V^o is a solution of the Hamilton-Jacobi partial differential equation of the first order:

$$V_t^o + \mathcal{H}\,(x, V_x^o, t) = 0 \tag{4.13}$$

Conversely, let $V^\circ(x, t, t_1)(t_1 = \text{parameter})$ be any solution of (4.13) of class C^2. Defining u° by means of (4.8), it follows that V° satisfies the fundamental equations (4.6–4.7).

In summary, we have

(4.14) THEOREM. *If there exists a solution $V^\circ\,(x, t, t_1)$ of class C^2 of the Hamilton-Jacobi equation (or, equivalently, of (4.6-4.7)) which satisfies $V^\circ\,(x, t_1, t_1) = \nu(x)$ and if (3.5) holds, then V° is the optimal performance index for the regulator problem (3.3), and the corresponding optimal control law is given by (4.8).*

Kálmán's derivation of the optimal solution using Hamilton–Jacobi theory is now mostly of historical interest. A much simpler derivation based on "completing the square" was given by Roger W. Brockett in his 1970 textbook *Finite-Dimensional Linear Systems.*

5. Controllability

The purpose of this section is to impose conditions on the plant (2.1) to assure that the problem posed by (3.3) is meaningful in the limit $t_1 = \infty$. Guided by physical intuition, we introduce the

(5.1) DEFINITION. A state x is said to be *controllable at time* t_0 if there exists a control function $u^1(t)$, depending on x and t_0 and defined over some finite closed interval $[t_0, t_1]$, such that $\phi_{u^1}\,(t_1; x, t_0) = 0$. If this is true for every state x, we say that the plant is *completely controllable at time* t_0; if this is true for every t_0, we say simply that the plant is *completely controllable.*

This is the first appearance of the fundamental notion of controllability, which makes sense only in light of the state-space picture. Here, the goal is to drive the system from a given nonequilibrium initial state to its unique equilibrium state in finite time by exerting a suitable control input.

The following equivalent characterization of controllability is useful:

(5.2) PROPOSITION. *A plant is completely controllable at time* t (i) *if and* (ii) *only if the symmetric matrix*

$$W(t_0, t_1) = \int_{t_0}^{t_1} \Phi(t_0, t)\, G(t) G'(t) \Phi'(t_0, t)\, dt \qquad (5.3)$$

Kálmán obtains a necessary and sufficient condition for controllability in terms of the controllability Gramian (a term coined by Roger W. Brackett) of the system. The system is controllable if and only if its controllability Gramian is nonsingular; it is observable if and only if its observability Gramian is nonsingular. Controllability and observability are invariant under invertible linear transformations of the state space.

is positive definite for some $t_1 > t_0$.

Proof. (i) Set

$$u^1(t) = -G'(t)\Phi'(t_0, t)\, W^{-1}(t_0, t_1)\, x \qquad (5.4)$$

Substitution into (2.3) shows that $\phi_{u^1}(t_1; x, t_0) = 0$.

(ii) Suppose there exists some $x \neq 0$ such that $\|x\|^2_{W(t_0,t_1)} = 0$. Define

$$u^2(t) = -G'(t)\Phi'(t_0, t)\, x$$

which implies that

$$\|x\|^2_{W(t_0,t_1)} = \int_{t_0}^{t_1} \left\|u^2(t)\right\|^2 dt = 0$$

Since $u^2(t)$ is continuous in t, it is therefore identically zero in the interval $[t_0, t_1]$.

On the other hand, if the plant is completely controllable at t_0, there exists a control function $u^1(t)$ as required by (5.1) which satisfies the relation

This is the celebrated rank condition for controllability of linear time-invariant systems. While the condition itself appeared in earlier work by others (e.g., in the context of verifying certain criteria for applying the maximum principle of Pontryagin), its connection to controllability originates with Kálmán.

$$x = -\int_{t_0}^{t_1} \Phi(t_0, t)\, G(t) u^1(t) dt$$

and therefore

$$\|x\|^2 = -\int_{t_0}^{t_1} \left[u^1(t), u^2(t)\right] dt = 0$$

contradicting the assumption that $x \neq 0$. Q. E. D.

(5.5) COROLLARY. *A constant plant is completely controllable* (i) *if and* (ii) *only if*

$$\text{rank}\left[G, FG, \cdots, F^{n-1}G\right] = n \qquad (5.6)$$

(where the square brackets denote a composite matrix of n *rows and* mn *columns) in which case one may choose* $t_1 - t_0 > 0$ *as small as desired.*

Proof. Because of stationarity, controllability does not depend on t_0. Hence take $t_0 = 0$.

(i) By (5.2), it suffices to prove that $W(0, t_1)$ is positive definite no matter how small $t_1 > 0$. Let $g^1, \cdots, g^m$ be the columns of G. If $W(0, t_1)$ is semidefinite, then proceeding as in part (ii) of the proof of (5.2) we conclude that there is a vector $x \neq 0$ such that

$$\left[x, e^{Ft} g^i\right] = 0 \quad \text{for all} \quad 0 \leq t \leq t_1 \quad \text{and} \quad i = 1, \cdots, m$$

Differentiating j times with respect to t, and then setting $t = 0$, we get

$$\left[x, F^j g^i\right] = 0 \quad \text{for all} \quad i = 1, \cdots, m \quad \text{and} \quad j = 0, \cdots, n-1 \quad (5.7)$$

If (5.6) holds, this implies that x is orthogonal to a set of generators of E^n, contradicting the assumption that $x \neq 0$.

(ii) Assume the plant is completely controllable but (5.6) is false. Then there is a vector $x \neq 0$ which satisfies (5.7). By the Cayley-Hamilton theorem

$$\left[x, e^{rt} g^r\right] = \left[x, \left(\sum_{j=0}^{\infty} (Ft)^j / j!\right) g^i\right] = \left[x, \left(\sum_{j=0}^{n-1} \alpha_j (Ft)^j\right) g^i\right], \quad i = 1, \cdots, m$$

It follows that $\|x\|^2_{W(0,t_1)} = 0$ for all t_1, contradicting the assumption of complete controllability. Q. E. D.

Condition (5.6) has been used as a technical device in several recent papers in the theory of control (Pontryagin 1959; Krasovskii 1959; LaSalle 1959), without reference to the "physical" interpretation (5.1).

(5.8) *Remark.* Let x be the state of the plant at t_0 and y the "desired" state at t_1. It follows easily by a slight extension of the preceding arguments that y is *reachable* from x (i.e., there exists a motion ϕ_{u^1} which meets x at t_0 and y at t_1) if and only if the equation

By linearity, the possibility of reaching a desired final state from a given initial state in finite time reduces to the originally stated definition of controllability when the final state is the equilibrium state. Once this problem is solved, the reverse motion can be constructed by time reversal.

$$x - \Phi(t_0, t_1)\, y = W(t_0, t_1)\, v \quad (5.9)$$

has a solution, in which case

$$u^1(t) = -G'(t)\Phi'(t_0, t)\, v \quad (5.10)$$

is the appropriate control function.

Moreover, elementary methods of the calculus of variations show (see also Bertram and Sarachik 1960) that the minimum *control energy* required to achieve the transfer is

$$\mathcal{E}(x, t_0; y, t_1) = \int_{t_0}^{t_1} \left\|u^1(t)\right\|^2 dt = \|x - \Phi(t_0, t_1)\, y\|^2_{W^{-1}(t_0,t_1)} \quad (5.11)$$

Recalling Kálmán's discussion of arriving at the "most rational" way of controlling the system, this is a particularly pleasing formula for the minimum control energy in terms of the controllability Gramian.

Clearly, the required "energy" is zero if and only if the free motion going through x at t_0 intersects y at t_1.

Equation (5.9) may have a solution for some but not all x, y. Then W^{-1} does not exist and it is convenient to replace it with the *generalized inverse* $W^\dagger$ in the sense of Penrose (1955, 1956). (See Appendix). With this convention, (5.11) *is the minimum energy required for transferring x as close to y as possible.*

If $W(t_0, t_1)$ is invertible, then (5.9) always has a solution; we see that *a plant is completely controllable at time t_0 if and only if starting from the origin at time t_0 any state x can be reached in a finite length of time by applying an appropriate control function $u(t)$.* In other words, there is a noteworthy "symmetry" between sending x to 0 and sending 0 to x.

The machinery of controllability and observability Gramians is particularly clean when the system is time-invariant. The definition of uniform complete controllability introduced here is meant to simplify the analysis of time-varying systems. It finds important uses in the theory of adaptive control, where the system matrices are not known exactly, so some effort must be expended towards learning enough to be able to accomplish the control goals.

(5.12) *Remark.* Using the generalized inverse, we may replace (5.10) by a control law defined in $[t_0, t_1]$:

$$u^1(t) = -G'(t)W\dagger(t, t_1)\left[x - \Phi(t, t_1)\, y\right]$$

Even if the plant is stationary, this control law is not. In fact, a stationary control law can be obtained in this case only by letting $u^1(t)$ be discontinuous (Kálmán 1960b).

The following definition is designed to single out a class of nonstationary plants which are in a sense "quasi-stationary". This will play an important role in the sequel.

(5.13) DEFINITION. A plant is *uniformly completely controllable* if the following relations hold for all t:

(i) $0 < \alpha_0(\sigma)I \leq W(t, t+\sigma) \leq \alpha_1(\sigma)I$

(ii) $0 < \beta_0(\sigma)I \leq \Phi(t+\sigma, t)W(t, t+\sigma)\Phi'(t+\sigma, t) \leq \beta_1(\sigma)I$

where σ is a fixed constant. In other words, one can always transfer x to 0 and 0 to x in a finite length σ of time; moreover, such a transfer can never take place using an arbitrarily small amount (or requiring an arbitrarily large amount) of control energy.

Definition (5.13) has surprisingly far-flung consequences. We mention some of these; the proofs are elementary.

First of all, if (i-ii) hold, then, for all t,

$$\sqrt{\beta_0(\sigma)/\alpha_1(\sigma)} \leq \|\Phi(t+\sigma,t)\| \leq \sqrt{\beta_1(\sigma)/\alpha_0(\sigma)} \qquad (5.14)$$

which is equivalent to

$$\sqrt{\alpha_0(\sigma)/\beta_1(\sigma)} \leq \|\Phi(t,t+\sigma)\| \leq \sqrt{\alpha_1(\sigma)/\beta_0(\sigma)} \qquad (5.15)$$

(5.16) From formulas (5.3) and (5.14–5.15) we see that (i-ii) hold also for the constant $\sigma' = 2\sigma$; this implies further that (i-ii) hold for any $\sigma' \geq \sigma$.

(5.17) Using (5.16), we see that (i-ii) actually imply the following stronger bound on the transition matrix:

$$\|\Phi(t,\tau)\| \leq \alpha_3(|t-\tau|) \quad \text{for all } t, \tau \qquad (5.18)$$

(5.19) It is now clear that if any two of the relations (5.18), (i), and (ii) hold, the remaining relation is also true.

The bound (5.18) obviously restricts the class of dynamical systems (2.1). Some such restriction appears to be an unavoidable consequence of any "reasonable" definition of uniform complete controllability. For instance, if only (i) holds, the following peculiar situation may arise. Consider the scalar system:

$$dx/dt = -tx + \sqrt{2(t-1)}e^{-t+1/2}u(t)$$

(defined only for $t \geq 1$). We find easily that

$$\phi(t,\tau) = e^{(\tau^2-t^2)/2}$$

which does not satisfy (5.18); furthermore,

$$w(t,t+\sigma) = e^{2(\sigma-1)t+(\sigma-1)^2} - e^{-2t+1}$$

and it is clear that w does not satisfy (i) unless $\sigma = 1$, while (ii) is never satisfied. In other words, to transfer x to 0 over an interval of time shorter than 1 may require an arbitrarily large amount of control energy, whereas doing the same job over an interval of time longer than 1 may require only a vanishingly small amount of energy as $t_0 \to \infty$. Transferring 0 to x will require more and more energy as $t_0 \to \infty$.

Finally, let us note a well-known and readily verifiable condition for (5.18) (easily proved using the Gronwall-Bellman lemma):

$$\int_{t_1}^{t_2} \|F(\tau)\| d\tau \leq \gamma (t_2 - t_1) \qquad \text{for all } t_1, t_2 \tag{5.20}$$

By passing to the dual plant, the concept of controllability can be related to another key concept: observability. While observability is fundamental in its own right, Kálmán does not develop it much further in this paper.

We now seek to characterize a plant according to its "output" properties. This is most conveniently done as follows. Let $t^* = -t$ and $F^*(t^*) = F'(t)$, $G^*(t^*) = H'(t)$, and $H^*(t^*) = G'(t)$. Then

$$\begin{aligned} dx^*/dt^* &= F^*(t^*)x^* + G^*(t^*)u^*(t^*) \\ y^*(t^*) &= H^*(t^*)x^*(t^*) \end{aligned} \tag{5.21}$$

where x^*, u^*, y^* are n, p, and m vectors respectively, is the *dual plant* of (2.1–2.2). We shall not discuss the significance of this concept in detail (for which see Kálmán 1960b), except for pointing out that (i) the duality relations are reflexive if $t_0^{**} = t_0$; (ii) the transition matrix of (5.21) satisfies the relation

$$\Phi^* (t^*, \tau^*) = \Phi'(r, t) \qquad \text{for all } t, \tau \tag{5.22}$$

It is convenient to introduce the:

(5.23) DEFINITION. A plant (2.1-2.2) is *uniformly completely observable* if its dual is uniformly completely controllable.

It follows easily from (5.23) that the explicit expression for W^* corresponding to (5.3) is

$$\begin{aligned} W^*(t^*, t_0^* + \sigma^*) &= W^*(t_0, t_0 - \sigma^*) \\ &= \int_{t_0-\sigma^*}^{t_0} \Phi'(t, t_0)H'(t)H(t)\Phi(t, t_0)dt \end{aligned} \tag{5.24}$$

While Pontryagin's maximum principle gives a necessary condition for a globally optimal control design, here Kálmán obtains a globally optimal solution which, furthermore, can be approximated numerically by solving the Riccati differential equation.

Using W^* defined by (5.24), we can now state (5.23) explicitly. To avoid any possibility of confusion, the constants α, β, σ occurring in (i-ii) are to be replaced by $\alpha^*, \beta^*, \sigma^*$.

6. Solution of the Linear Regulator Problem

The point of view of the classical calculus of variations outlined in Section 4 is purely "local." At present, there are few global results and just about none in the theory of control. In the "local" (linear) case, however, the ideas of the preceding section lead to (what is hoped to become) a definitive theory of the regulator problem. This is the subject of the remainder of the paper.

To get the linear case of the regulator problem, it is not enough to have a linear model (2.1) for the plant but we need also the assumption:

The quadratic cost criterion is introduced here, and the key nondegeneracy condition is imposed: the matrix *R(t)* has to be positive definite for all values of *t*.

$$(\mathrm{A}_1) \quad L(x,u,t) = \frac{1}{2}\left\{\|H(t)x\|^2_{Q(t)} + \|u\|^2_{R(t)}\right\}, \quad \nu(x) = \frac{1}{2}\|x\|^2_A$$

where A is symmetric, nonnegative definite while $Q(t)$, $R(t)$ are symmetric, positive definite and of class C^2 in t.

In view of (A_1), the Hamiltonian function (4.12) is

$$\mathcal{H}(x,\xi,t) = \frac{1}{2}\left\{\|H(t)x\|^2_{Q(t)} + 2[F(t)x,\xi] - \left\|G'(t)\xi\right\|^2_{R^{-1}(t)}\right\} \quad (6.1)$$

With this choice of $\mathcal{H}$, the function

$$V^o(x,t,t_1) = \frac{1}{2}\|x\|^2_{P(t,t_1)} \quad (6.2)$$

(t_1 = parameter) is a solution of the Hamilton-Jacobi equation (4.13) if and only if $P(t,t_1)$ is a solution of the following ordinary nonlinear differential equation of the Riccati type:

$$-\frac{dP}{dt} = F'(t)P + PF(t) - PG(t)R^{-1}(t)G'(t)P + H'(t)Q(t)H(t) \quad (6.3)$$

The Riccati differential equation is a key ingredient. It provides an effective procedure for obtaining the optimal control using a digital computer. Moreover, the Riccati equation can be solved ahead of time and the solution stored in memory.

It is clear that (6.2) determines P only up to a constant, skew-symmetric matrix (constancy follows from the fact that dP/dt is symmetric). Henceforth, to avoid trivia, we always assume that P is symmetric.

Given any symmetric, nonnegative definitive matrix A, (6.3) has a unique solution $\Pi(t; A, t_1)$ which takes on the value A at $t = t_1$. This solution is known to exist only in some neighborhood of t_1; without further analysis we cannot conclude existence for all t. Because of the phenomenon of *finite escape time*, for which see Kálmán and Bertram 1960, Example 3.)

Nonetheless, $\Pi(t; A, t_1)$ does exist for all $t \leq t_1$. We prove this indirectly as follows:

(6.4) EXISTENCE THEOREM. (i) *For all t_1 and all symmetric, nonnegative definite A, (6.3) has a unique solution $\Pi(t; A, t_1)$ defined for all $t \leq t_1$.* (ii)

Under Assumption (A_1) *the optimal performance index for Problem (3.3) is given by*

$$V^o(x, t_0, t_1) = \|x\|^2_{\Pi(t_0;A,t_1)}$$

Moreover, the optimal performance index is attained if and only if the control law is given by

$$u^o(t) = R^{-1}(t)G'(t)\Pi(t; A, t_1)\, x(t) \qquad (6.5)$$

Proof. If we assume (i), then (ii) follows immediately from (4.14). Therefore, if $\Pi(t; A, t_1)$ exists, it must necessarily satisfy the relation

$$\begin{aligned} \|x\|^2_{\Pi(t_0;A,t_1)} &\le \int_{t_0}^{t_1} \|H(t)\Phi(t, t_0)\, x\|^2_{Q(t)}\, dt + \|\Phi(t_1, t_0)\, x\|^2_A \\ &\le \alpha(t_1, t_0)\, \|x\|^2 \end{aligned}$$

which follows by setting $u(t) \equiv 0$ in (3.4). Since $\alpha(t_1, t_0)$ is finite for all pairs t_1, t_0, it is clear that $\Pi(t; A, t_1)$ (if it exists) is contained in a compact region for all $t \in [t_0, t_1]$. Including this fact in the standard proof of the existence theorem for differential equations proves (i). Q. E. D.

In order to study the case $t_1 \to \infty$, we first define a particular solution of (6.3) which is of central significance for the ensuing development.

The importance of controllability is brought to full focus here, in connection with the existence of asymptotic (infinite-horizon) solutions of the Riccati equation.

(6.6) PROPOSITION. *If the plant is completely controllable, then*

$$\lim_{t_1 \to \infty} \Pi(t; 0, t_1) = \bar{P}(t)$$

(i) *exists for all t and* (ii) *is a solution of* (6.3).

Proof. (i) Suppose the plant is completely controllable at $t = t_0$. Then for every x there exists a control function $u^1(t)$, given by (5.4), which transfers x to 0 at or before $t = t_2(x, t_0)$. We set $u^1(t) = 0$ for $t > t_2$. Then

$$\begin{aligned} \|x\|_{\Pi(t_0;0,t_1)} &= V^o(x, t_0, t_1) \le V(x, t_0, t_2; u^1) \\ &= V(x, t_0, \infty; u^1) \le \alpha(t_0)\, \|x\|^2 \end{aligned}$$

which shows that $\|\Pi(t_0; 0, t_1)\|$ is bounded for all $t_1 \ge t_0$. On the other hand, (3.4) shows that $\|\Pi(t_0; 0, t_1)\|$ is nondecreasing as $t_1 \to \infty$. Hence the desired limit exists for arbitrary $t = t_0$. Q. E. D.

(ii) Using the continuity of solutions of (6.3) with respect to initial conditions, we have

$$\begin{aligned}\bar{P}(t) &= \lim_{t_2\to\infty} \Pi\left(t; 0, t_2\right) = \lim_{t_2\to\infty} \Pi(t; \Pi\left(t_1; 0, t_2\right), t_1) \\ &= \Pi\left(t; \lim_{t_2\to\infty} \Pi\left(t_1; 0, t_2\right), t_1\right) = \Pi\left(t; \bar{P}\left(t_1\right), t_1\right)\end{aligned}$$

which shows that $\bar{P}(t)$ is a solution of (6.3) which is defined for all t. Q. E. D.

(6.7) EXISTENCE THEOREM. *Assuming* $(A_1), \nu(x) = 0$, *and* $t_1 = \infty$, *the optimal performance index for Problem (3.3) is* $\|x\|^2_{\bar{P}(t)}$ *and the optimal control law is*

This theorem puts all the ingredients together and gives an explicit formula for the optimal control in state feedback form. That is, the optimal input at time *t* is a linear function of the state at time *t*, so there is no need to store the past state trajectory. This justifies the notion of "state" as the minimal sufficient representation of the system's memory for the purpose of determining the optimal control at each time *t*.

$$u^o(t) = R^{-1}(t)G'(t)\bar{P}(t)x(t) \tag{6.8}$$

Proof. Assume throughout the $\nu(x) = 0$. First we show: *If* $u(t)$ *is determined by the control law (6.7), the corresponding performance index is*

$$V\left(x, t_0, \infty; u^\circ\right) = \lim_{t_1\to\infty} V\left(x, t_0, t_1; u^\circ\right) = \|x\|^2_{\bar{P}(t_0)}$$

We see from (6.4) and (6.6) that

$$V\left(x, t_0, t_1; u^o\right) = \|x\|^2_{\bar{P}(t_0)} - \|\phi_{u^0}\left(t_1; x, t_0\right)\|^2_{\bar{P}(t_1)} \le \|x\|^2_{\bar{P}(t_0)}$$

On the other hand,

$$V\left(x, t_0, t_1; u^o\right) \ge V^o\left(x, t_0, t_1\right) = \|x\|^2_{\Pi(t_0;0,t_1)} \ge \|x\|^2_{\bar{P}(t_0)} - \epsilon$$

where $\epsilon \to 0$ as $t_1 \to \infty$, which proves (6.9). Hence

$$V^o\left(x, t_0, \infty\right) \le V\left(x, t_0, \infty; u^o\right)$$

The inequality sign cannot arise. For if $V\left(x, t_0, \infty; u^\circ\right) - V^0\left(x, t_0, \infty\right) \ge \eta > 0$, there is some control function u^1 such that $V\left(x, t_0, \infty; u^\circ\right) - V\left(x, t_0, \infty, u^1\right) \ge \eta/2$. For t_1 sufficiently large, we then have

$$V\left(x, t_0, \infty; u^o\right) = V^o\left(x, t_0, t_1\right) + \eta/4 \ge V\left(x, t_0, t_1; u^1\right) + \eta/2$$

Stability makes all the difference when we need to control the system over an infinite time horizon. Here, the relevant notion of stability is classical, going back to the work of Lyapunov, and pertains to the "closed-loop" system obtained by using the optimal state feedback control law.

which is a contradiction and everything is proved. Q. E. D.

In the engineering literature it is often assumed (tacitly and incorrectly) that a system with optimal control law (6.8) is necessarily stable. We now give rigorous sufficient conditions ensuring uniform asymptotic stability and point out in the process of proof some trivial but interesting parallels between the calculus of variations and the second method of Lyapunov.

The following definition is standard (Kálmán and Bertram 1960; Hahn 1959): The system (2.1) is *uniformly asymptotically stable* if (i) $\|\Phi(t, t_0)\| \leq \alpha$ and (ii) $\|\Phi(t, t_0)\| \rightarrow 0$ with $t \rightarrow \infty$ uniformly in t_0. It can be shown (Kálmán and Bertram 1960, Theorem 3) that uniform asymptotic stability in the linear case is equivalent to *exponential asymptotic stability*, which is defined by the condition $(\alpha, \beta > 0)$

$$\|\Phi(t, t_0)\| \leq \alpha \exp\left[-\beta(t - t_0)\right] \quad \text{for all } t_0 \text{ and all } t \geq t_0.$$

Once again, the notions of controllability and observability become important. This theorem deals with time-varying systems, so both controllability and observability are stated in their uniform versions. Moreover, the optimal cost for a given initial state is also the Lyapunov function for the closed-loop system.

(6.10) STABILITY THEOREM. *Consider a plant with control law (6.8) which is uniformly completely controllable and uniformly completely observable. In addition to* (A_1), *assume also*

$$(\mathrm{A}_2) \quad Q(t) \geq \alpha_4 I > 0, \qquad R(t) \geq \alpha_5 I > 0$$

$$(\mathrm{A}_3) \quad Q(t) \leq \alpha_6 I, \qquad R(t) \leq \alpha_7 I$$

Then the controlled plant is uniformly asymptotically stable and $V^{\circ}(x, t, \infty)$ *is one of its Lyapunov functions.*

Proof. As is well known, it suffices to prove that (a) V^o is bounded from above and (b) below by increasing functions of $\|x\|$ independent of t, (c) the derivative $\dot{V}^o$ of V^o along optimal motions of the plant is negative definite (Kálmán and Bertram 1960; Hahn 1959), and (d) $V^o \rightarrow \infty$ with $\|x\| \rightarrow \infty$.

(a) By uniform complete controllability, let $u^1(t)$ be the control function, depending on x, t_0 and defined in $[t_0,\ t_0 + \sigma]$ (σ = positive constant), which transfers x to 0 at or before $t = t_0 + \sigma$. In accordance with the remarks following (5.13), there is no loss of generality in taking the constants σ and σ^* (occurring in the definition of uniform complete controllability and uniform complete observability) to be the same. Having set $\sigma = \sigma^*$, we let $t_1 = t_0 + \sigma$. If $u^1(t)$ is defined explicitly by means of (5.4), then

$$\begin{aligned} \phi_{u^1}(t; x, t_0) &= \Phi(t, t_0)\left[I - W(t_0, t) W^{-1}(t_0, t_1)\right] x \\ &= \Phi(t, t_1)\, z(t) \end{aligned} \tag{6.11}$$

From the definition of W (see (5.3)), it follows easily[4] that the norm of the bracketed term above is less than or equal to 1. Using (5.18) then gives

$$\|z(t)\| \leq \alpha_8 \|x\|$$

where α_8 depends only on σ and it is therefore constant. By (6.11) and ($A3$) we get

$$V^o(x, t_0, \infty) \leq \int_{t_0}^{t_1} \left\{ \alpha_6 \|H(t)\Phi(t, t_1) z(t)\|^2 + \alpha_7 \left\|u^1(t)\right\|^2 \right\} dt \tag{6.12}$$

Making use of the foregoing and of the elementary inequality

$$\|Ax\|^2 \leq \|A\|^2 \cdot \|x\|^2 = \lambda_{\max}(A'A) \|x\|^2 \leq (\text{tr}\, A'A) \|x\|^2$$

(valid for any matrix A and any vector x), (6.12) becomes

$$V^\circ(x, t_0, \infty) \leq \alpha_6 \alpha_8 (\text{tr}\, W^*(t_1, t_0)) \|x\|^2 + \alpha_7 \|x\|^2_{W^{-1}(t_0, t_1)}$$

and by uniform complete controllability and observability we have finally

$$V^0(x, t_0, \infty) \leq [n\alpha_6 \alpha_8 \alpha_1^*(\sigma) + \alpha_7 \alpha_0(\sigma)] \|x\|^2 = \alpha_9 \|x\|^2.$$

(b) In view of (6.7), we can define

$$\inf_x V^\circ(x, t_0, \infty) = \alpha_{10}(t_0) \|x\|^2$$

We show that $\alpha_{10}(t_0) \geq \alpha_{11} > 0$. In fact, in the contrary case we can make ε, defined by

$$\|x\|^2 \varepsilon(x, t) = \int_{t_0}^{\infty} \|u^\circ(t)\|^2 dt \leq \alpha_5^{-1} \int_{t_0}^{\infty} \|u^\circ(t)\|^2_{R(t)} dt \leq V^\circ(x, t_0, \infty)$$

as small as desired by suitable choice of x, t_0. We introduce the abbreviation

$$z(t) = \int_{t_0}^{t_1} \Phi(t_0, t)\, G(t) u^\circ(t) dt$$

[4]We need to show only that if $B > 0$ and $B \geq A \geq 0$, then $\|AB^{-1}\| \leq 1$. Now $\|AB^{-1}\|^2 = \lambda_{\max}(B^{-1}A^{-2}B^{-1}) = \lambda_{\max}(A^2B^{-2})$. By the well-known theorem about simultaneous diagonalization of a positive definite and a symmetric matrix, we have the representation $A^2 = T'\wedge T$, $B^2 = T'T$, where T is nonsingular and $\wedge$ is diagonal. $B \geq A \geq 0$ implies $1 \geq \lambda(\wedge) \geq 0$. Hence $\lambda_{\max}(A^2B^{-2}) = \lambda_{\max}(\wedge) \leq 1$.

and note that, by the Schwarz inequality,

$$\|z(t)\|^2 \leq \left(\int_{t_0}^{t_1} \|\Phi(t_0, t)\, G(t)\|^2\, dt\right) \left(\int_{t_0}^{t_1} \left\|u^0(t)\right\|^2 dt\right)$$

and by uniform complete controllability,

$$\|z(t)\|^2 \leq n\alpha_1(\sigma) \in (x, t_0)\, \|x\|^2.$$

Utilizing this estimate, we find with the aid of (A_2):

$$\begin{aligned} V^\circ(x, t_0, \infty) &\geq \int_{t_0}^{t_1} \alpha_4 \|H(t)\Phi(t, t_0)[x + z(t)]\|^2\, dt \\ &\geq \int_{t_0}^{t_1} \alpha_4 \left\{\|H(t)\Phi(t, t_0)\, x\|^2 - \|H(t)\Phi(t, t_0)\, z(t)\|^2\right\} dt \\ &\geq \alpha_4 \Big\{\|\Phi(t_1, t_0)\, x\|^2_{W^*(t_1, t_0)} \\ &\qquad - n\, \alpha_1(\sigma) \in (x, t_0)\, [\mathrm{tr}\, W^*(t_1, t_0)]\, \|x\|^2\Big\} \end{aligned}$$

By (5.18) and uniform complete observability, this reduces to

$$\begin{aligned} V^\circ(x, t_0, \infty) &\geq \alpha_4 \left[\alpha_3^{-2}(\sigma)\alpha_0^*(\sigma) - n^2\alpha_1(\sigma)\alpha_1^*(\sigma) \in (x, t_0)\right] \|x\|^2 \\ &\geq [\alpha_{13} - \alpha_{14} \in (x, t_0)]\, \|x\|^2 \end{aligned}$$

which contradicts the assumption that α_{12} (and hence ϵ) can be made arbitrarily small by suitable choice of x, t_0.

(c) Since G and H are allowed to be singular, we cannot prove of course that $\dot V$ is negative definite. However, inspection of the last step of (b) yields the further inequality,

$$V^o(\phi_{u^\circ}(t_1; x, t_0), t_1, \infty) - V^o(x, t_0, \infty) \leq -[\alpha_{13} - \alpha_{14} \in (x, t_0)]\, \|x\|^2$$

and we have simultaneously also the further inequality

$$V^o(\phi_{u^o}(t_1; x, t_0), t_1, \infty) - V^o(x, t_0, \infty) \leq \alpha_5 \in (x, t_0)\, \|x\|^2$$

which follows immediately by (A_2) and the definition of V°. Setting

$$\alpha_{15} = \alpha_5\alpha_{13}/(\alpha_5 + \alpha_{14}) > 0$$

we have finally that

$$V^o(\phi_{u^\circ}(t_0 + \sigma; x, t_0), t_0 + \sigma, \infty) - V^\circ(x, t_0, \infty) \leq -\alpha_{15}\|x\|^2$$

which shows that V^o is strictly decreasing along any interval of time of length σ_2, unless $x = 0$. Taking account of this fact, the proof of Lyapunov's theorem on uniform asymptotic stability (Kálmán and Bertram 1960) goes through as usual.

(d) This is trivial in view of $V^\circ \geq \alpha_{12}\|x\|^2$. Q. E. D.

It is of some interest to observe that if we have merely complete controllability, part (a) does not go through but we have nevertheless proved (nonuniform) asymptotic stability.

7. Stability of the Riccati Equation

We now turn again to (6.3) and examine briefly its stability properties. Let $\delta P(t) = P(t) - \bar{P}(t)$ denote the deviation of a given motion $P(t)$ of (6.3) from $\bar{P}(t)$. Substituting into (6.3) shows that

$$d(\delta P)/dt = -\bar{F}'(t)\delta P - \delta P\bar{F}(t) - \delta PG(t)R^{-1}(t)G'(t)\delta P \tag{7.1}$$

where

$$\bar{F}(t) = F(t) - G(t)R^{-1}(t)G'(t)\bar{P}(t)$$

For simplicity, we temporarily drop the argument t in G, H, P, Q, R.

(7.2) STABILITY THEOREM. *Let*

This quantitative result on the stability of the Riccati equation is useful in the context of computer implementation, as well as in the settings where the system model is not known precisely.

$$\mathcal{V}(\delta P, t) = \frac{1}{2}\,\mathrm{tr}\left(\delta P\bar{P}^{-1}\right)^2. \tag{7.3}$$

Then (i) *the derivative of v along motions of (7.1) is*

$$\begin{aligned}\dot{v}&(\delta P, t)\\ &= \mathrm{tr}\left\{\left(P^{\frac{1}{2}}GR^{-1}G'P^{\frac{1}{2}}\right)\cdot\left(P^{\frac{1}{2}}\bar{P}^{-1}P^{\frac{1}{2}} - 1\right)^2\right.\\ &\quad\left. + \left(\bar{P}^{-\frac{1}{2}}H'QH\bar{P}^{-\frac{1}{2}}\right)\left(\bar{P}^{-\frac{1}{2}}P\bar{P}^{-\frac{1}{2}} - 1\right)^2\right.\end{aligned} \tag{7.4}$$

provided $P \geq 0$;

(ii) *If $A \geq 0$, then under the hypotheses of (6.10), all solutions $\Pi\ (t; A, t_1)$ of (6.3) are uniformly asymptotically stable relative to $\bar{P}(t)$ as $t \to -\infty$, and v is an appropriate Lyapunov function.*

Proof. (i) This is established by lengthy, elementary calculations. The square root of P exists by assumption and that of $\bar{P}$ by part (b) of (6.10).

(ii) Clearly, $\mathcal{V}$ vanishes if and only if $\delta P = 0$. We recall the fact that for any symmetric, $n \times n$ matrices A, B,

$$\lambda_{\min}(B)\lambda_i\left(A^2\right) \leq \lambda_i(ABA) \leq \lambda_{\max}(B)\lambda_i\left(A^2\right) \quad (i = 1, \cdots, n) \tag{7.5}$$

where the λ_i are eigenvalues. This is a consequence of the Fischer-Courant variational description of eigenvalues (for which see Bellman 1960, p. 115, Theorem 3 and p. 120, Exercise 9). Using (7.5) and the results of (6.10) it follows easily that

$$0 < \alpha \operatorname{tr}(\delta P)^2 \leq \mathcal{V}(\delta P, t) \leq \beta \operatorname{tr}(\delta P)^2, \quad \delta P \neq 0$$

Moreover, (7.4) being the trace of a nonnegative definite matrix, $\dot{\mathcal{V}}$ is clearly nonnegative. By arguments analogous to part (c) of the proof of (6.10), it follows then also that v is uniformly decreasing along any motion of (6.3) as as $t \to -\infty$.

(7.6) COROLLARY. *The motion $\bar{P}(t)$ is unstable (as $t \to \infty$).*

Proof. Immediate consequence of part (ii) of the proof of (7.2).

(7.7) *Remark.* If the problem is stationary, i.e., F, G, H, Q, R are constants, $P(t, t_1) = P(t + h, t_1 + h)$ which shows that $dP(t, t_1)/dt_1 = -dP(t, t_1)/dt$. Hence in this case one can compute $P(t) =$ const. from (6.3) by replacing t by $-t$; for any initial $A \geq 0$, this computation is asymptotically stable in the large.

(7.8) *Remark.* Because of the Corollary, in the nonstationary case (at least one of F, G, H, Q, R not constant), one cannot compute $P(t)$ as $t \to \infty$ from the knowledge of $P(t_0)$.

8. General Solution of the Riccati Equation

Consider the canonic (Hamiltonian) differential equations associated with (6.1):

$$dx/dt = \mathcal{H}_\xi(x, \xi, t) = F(t)x - G(t)R^{-1}(t)G'(t)\xi \tag{8.1}$$

$$d\xi/dt = -\mathcal{H}_x(x, \xi, t) = -H'(t)Q(t)H(t)x - F'(t)\xi \tag{8.2}$$

Let $P(t)$ be a solution of (6.3), defined in some interval $U = (-\infty, t_2)$. In view of (4.11) and (6.2), we assume that the initial conditions of (8.1-8.2) at time t_1 are related by ($t_1 < t_2$!)

$$\xi(t_1) = V_x^\circ(x(t_1), t_1) = P(t_1)\,x(t_1)$$

Then the same relation will hold between solutions of (8.1-8.2) corresponding to these initial conditions, for all t that $P(t)$ exists:

$$\xi(t) = V_x^o(x(t), t) = P(t)x(t), \quad t \in U \tag{8.3}$$

We can also verify (8.3) directly by substituting (6.3) into (8.1-8.2).

Now let $X(t), \Xi(t)$ be a pair of matrix solutions of (8.1-8.2) satisfying the initial conditions $X(t_1) = I, \Xi(t_1) = P(t_1)$. By (8.3) we have obviously

$$\Xi(t) = P(t)X(t), \quad t \in U \tag{8.4}$$

which shows that $X(t)$ is a solution of the matrix differential equation

$$dX/dt = \left[F(t) - G(t)R^{-1}(t)G'(t)P(t)\right] X, \quad t \in U$$

Setting $\nu(x) = \|x\|^2_{P(t_1)}$ in (3.4), we see from (6.4-6.5) that $X(t)$ is the transition matrix $\Phi^\circ(t, t_1)$ of the optimally controlled plant corresponding to this choice of ν. Since $\Phi^o(t, t_1)$ exists for all $t \in U$, we have

$$P(t) = \Xi(t)\Phi^o(t_1, t), \quad t \in U \tag{8.5}$$

To obtain an explicit expression for $P(t)$, let

$$\Theta(t, t_1) = \begin{pmatrix} \Theta_{11}(t, t_1) & \Theta_{12}(t, t_1) \\ \Theta_{21}(t, t_1) & \Theta_{22}(t, t_1) \end{pmatrix}$$

be the transition matrix of the system (8.1-8.2). We get the following formula valid for $t \in U$:

$$P(t) = [\Theta_{21}(t, t_1) + \Theta_{22}(t_1, t)P(t_1)]\,[\Theta_{11}(t, t_1) + \Theta_{12}(t, t_1)P(t_1)]^{-1}$$

This procedure is very well known in the calculus of variations (Radon 1928; Carathéodory 1935, Ch. 15) and is being periodically rediscovered (Reid 1946; Levin 1959).

Appendix: The Generalized Inverse of a Matrix

Following Penrose (1955), the *generalized inverse* of an arbitrary square matrix A is a matrix $A\dagger$ satisfying the relations:

$$\text{(i)}\quad AA\dagger A = A, \qquad \text{(ii)}\quad A\dagger AA\dagger = A\dagger,$$
$$\text{(iii)}\quad (A\dagger A)' = A\dagger A, \quad \text{(iv)}\quad (AA\dagger)' = AA\dagger$$

It can be shown that $A\dagger$ always exists and is uniquely determined by these relations. Examples: (1) If D is diagonal, then the elements of its generalized inverse are

$$\begin{aligned} d^{\dagger}_{ii} &= d^{-1}_{ii} \quad \text{if } d_{ii} \neq 0 \\ &= 0 \text{ otherwise} \end{aligned}$$

(2) If A is symmetric, there is an orthogonal transformation T such that $A = T'DT$. Then $A\dagger = T'D\dagger T$.

Consider now the linear equation $Ax = y$. Penrose (1956) proves that the "best approximate solution" $x^\circ = A\dagger y$ of this equation has the properties:

$$\text{(i)}\quad \|Ax - y\| \geq \|Ax^0 - y\| \quad \text{for all} \quad x$$
$$\text{(ii)}\quad \text{If} \quad \|Ax - y\| = \|Ax^0 - y\|, \quad \text{then} \quad \|x\| \geq \|x^0\|$$

REFERENCES

Bellman, R. 1953. *Stability Theory of Differential Equations.* New York, NY: McGraw-Hill.

———. 1960. *Introduction to Matrix Analysis.* New York, NY: McGraw-Hill.

Bertram, J. E., and P. E. Sarachik. 1960. "On Optimal Computer Control." *IFAC Proceedings Volumes* 1 (1): 429–432.

Carathéodory, C. 1935. *Variationsrechnung und partielle Differentialgleichungen erster Ordnung.* Leipzig, Germany: Teubner.

Coddington, E. A., and N. Levinson. 1955. *Theory of Ordinary Differential Equations.* New York, NY: McGraw-Hill.

Hahn, W. 1959. *Theorie und Anwendung der Direkten Methode von Ljapunov.* Vol. 22. Ergbnisse der Mathematik. Berlin, Germany: Springer.

Hall, A. C. 1943. *The Analysis and Synthesis of Linear Servomechanisms.* Cambridge, MA: MIT Press.

Kálmán, R. E. 1960a. "A New Approach to Linear Filtering and Prediction Problems." *Journal of Basic Engineering* 82 (1): 35–45.

———. 1960b. "On the General Theory of Control Systems." *IFAC Proceedings Volumes* 1 (1): 491–502.

Kálmán, R. E., and J. E. Bertram. 1958. "General Synthesis Procedure for Computer Control of Single and Multi-Loop Linear Systems (An Optimal Sampling System)." *Transactions of the American Institute of Electrical Engineers, Part II: Applications and Industry* 77 (6): 602–609.

———. 1960. "Control System Analysis and Design via the 'Second Method' of Lyapunov: I—Continuous-Time Systems." *Journal of Basic Engineering* 82 (2): 371–393.

Kálmán, R. E., and R. W. Koepcke. 1958. "Optimal Synthesis of Linear Sampling Control Systems Using Generalized Performance Indexes." *Transactions of the American Society of Mechanical Engineers* 80 (8): 1820–1826.

———. 1959. "The Role of Digital Computers in the Dynamic Optimization of Chemical Reactions." In *IRE-AIEE-ACM '59 (Western): Papers Presented at the March 3-5, 1959, Western Joint Computer Conference,* 107–116. New York, NY: Association for Computing Machinery.

Krasovskii, N. N. 1959. "On a Problem of Optimal Control of Nonlinear Systems." (translation pp. 303–332), *Prikl. Math. Mekh.* 23:209–229.

LaSalle, J. P. 1959. "Time Optimal Control Systems." *Proceedings of the National Academy of Sciences* 45 (4): 573–577.

Levin, J. J. 1959. "On the Matrix Riccati Equation." *Proceedings of the American Mathematical Society* 10:519–524.

Newton, G. C., L. A. Gould, and J. F. Kaiser. 1957. *Analytical Design of Linear Feedback Controls.* New York, NY: Wiley.

Penrose, R. 1955. "A Generalized Inverse for Matrices." *Mathematical Proceedings of the Cambridge Philosophical Society* 51 (3): 406–413.

———. 1956. "On Best Approximate Solutions of Linear Matrix Equations." *Mathematical Proceedings of the Cambridge Philosophical Society* 52 (1): 17–19.

Pontryagin, L. S. 1959. "Optimal Control Processes" [in Russian]. *Uspekh. Mat. Nauk.* 14 (1): 3–20.

Radon, J. 1928. "Zum Problem von Lagrange." *Abhandlungen aus dem Mathematischen Seminar der Universität Hamburg* 6:273–299.

Reid, W. T. 1946. "A Matrix Differential Equation of the Riccati Type." *American Journal of Mathematics* 68 (2): 237–246.

Wiener, N. 1949. *The Extrapolation, Interpolation, and Smoothing of Stationary Time Series.* New York, NY: Wiley.

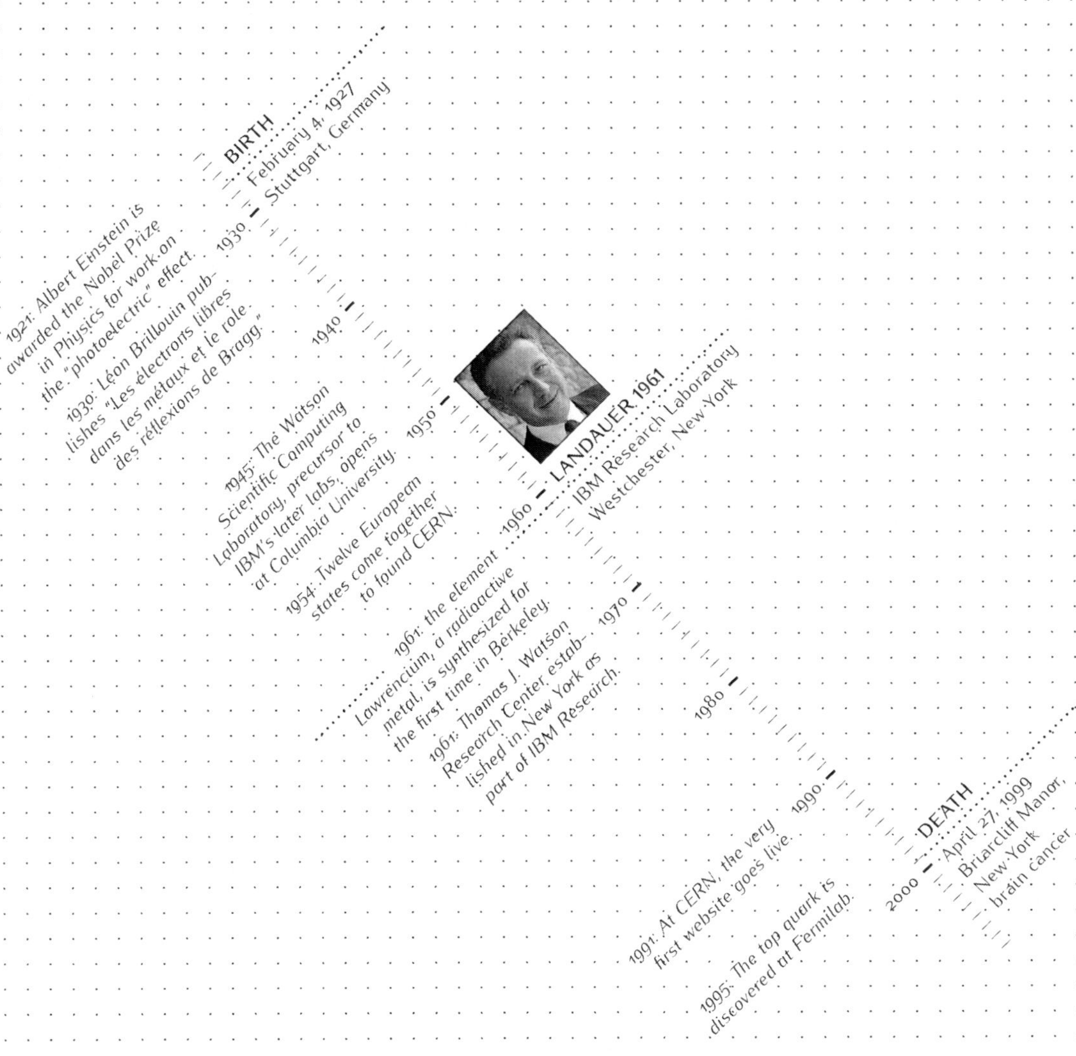

ROLF WILLIAM LANDAUER

[16]

THE RELATIONSHIP BETWEEN PHYSICS AND COMPUTATION: THE MINIMAL THERMODYNAMIC COST TO ERASE A BIT

David H. Wolpert, Santa Fe Institute

R. Landauer, "Irreversibility and Heat Generation in the Computing Process," *IBM Journal of Research and Development* 5 (3): 183–191 (1961).

The question of how the foundations of physics are related to information processing, to what we now call "computation," is an extremely deep issue that has concerned scientists for centuries. Indeed, some researchers, like John Wheeler (2002) with his pithy phrase "it from bit," saw this question as one of *the* central open issues in the entire scientific enterprise.

Perhaps the greatest success we have had in grappling with this question is when it is restricted to concern the relationship between information processing and statistical physics, specifically. Research into this relationship can be traced all the way back to the nineteenth century, with Maxwell's demon, a thought experiment that concerned whether an "intelligent demon" could exploit observational data to circumvent the second law of thermodynamics. Important subsequent work into this relationship was done in the early twentieth century by Brillouin, Szilárd, and others.

Arguably, the analysis that really set the field on the trajectory leading to where we are today was done by Rolf Landauer, in 1961, in his famous paper "Irreversibility and Heat Generation in the Computing Process." While there is much to ponder in that paper, the part that has been so consequential for modern science concerns the deceptively simple question "What is the minimal thermodynamic cost to erase a bit?" That is, to transform either of two states of a bit to a specific one of those states, called the "erased" state. (Landauer used the term "RESTORE TO ONE operation" rather than "bit erasure.")

While he never explicitly says so, Landauer interpreted this question to concern a system that is only coupled to a single heat bath, at temperature T. He also primarily (but not exclusively!) focused on the case where there are equal probabilities of the two possible initial states of the bit, and where both states of the bit have the same energy level. His famous conclusion was that the minimal thermodynamic cost to erase a bit in these circumstances is $kT \ln 2$, where k is Boltzmann's constant. Not only is Landauer's paper extremely insightful, but it single-handedly kick-started modern interest in the thermodynamics of computation, leading directly to a wave of papers by researchers such as Charles Bennett, Wojciech Zurek, Carlton Caves, Seth Lloyd, Paul Vitányi, Peter Grassberger, Edward Fredkin, Tommaso Toffoli, Norman Margolus, and others.

Importantly, though, up to the end of the twentieth century, all of statistical physics only considered either systems at thermal equilibrium, systems close to thermal equilibrium, or systems in a local thermal equilibrium. Twentieth-century statistical physics also almost exclusively considered systems changing very slowly, if at all. Research on such systems has been astonishingly fruitful—it resulted in the Nobel Prize for Giorgio Parisi in 2021. However, one striking aspect of very many complex systems is precisely how *far* they are from thermal equilibrium, and how *quickly* their state is changing. In particular, computation in general (and bit erasure in particular) is by definition a highly dynamic process that is (very) far from thermal equilibrium. This means that attempting to analyze the statistical physics of computation using the tools available in the twentieth century is like playing a piano while wearing mittens; you can vaguely convey impressions of what the music might sound like, perhaps even getting a few of the notes correct, but you can't actually produce the entire piece of music.

Due to this gap between the needs of the topic he was investigating and the formal tools at his disposal, Landauer was forced to use semiformal reasoning in his analysis. He was constrained to only give plausibility arguments, based on examples of physical systems, rather than properly prove his central thesis with full rigor. (All of the mathematical derivations in the paper only occur in the examples; none of it is used to prove the wide-ranging generality of the famous $kT \ln 2$ result that the paper is known for.)

Moreover, Landauer never even formally defined the terms central to his analysis, like "dissipation."[1] Perhaps as a result, Landauer failed to consider whether the heat necessarily generated in any implementation of bit erasure and then dumped into the external universe could be recovered from that external universe after the bit has been erased, that is, if bit erasure could be done in a *thermodynamically* reversible manner, even though it is *logically* irreversible.[2] This in turn has led to controversy about the legitimacy of Landauer's conclusion, controversy in which people have not even fully agreed on what the terms of debate are (Dillenschneider and Lutz 2010; Sagawa and Ueda 2010).

Ironically, this unresolved state of affairs changed—drastically—precisely around the time that the initial wave of interest in Landauer's work started to (ahem) dissipate. It was around then—around the turn of the millennium—that a revolution in statistical physics started gaining steam. Researchers including Crooks (1999), Jarzynski (1997), and Seifert (2005), and many others, building on results scattered across the earlier literature, figured out how to generalize statistical physics into a form that can fully capture the thermodynamics of systems that are arbitrarily far from thermal equilibrium and that are evolving arbitrarily quickly. As a result, this new field, sometimes called "stochastic thermodynamics," can be used to fully and formally analyze the thermodynamics of what we now call complex systems,[3] and in particular it can be used to analyze the thermodynamics of computational systems.

Stochastic thermodynamics has allowed us to fully formalize the reasoning that Landauer so presciently sketched out. (In stochastic thermodynamics terminology, $kT \ln 2$ is the minimal "entropy flow"

[1] Modern physics makes a crucial distinction between two kinds of dissipation: heat dissipation, which is thermodynamically reversible, and work dissipation, which is thermodynamically *ir*reversible.

[2] Logical irreversibility means that we cannot run the operation backward, from the erased state of the bit, to produce the initial, pre-erasure state—there is not enough information in the erased state to recover that initial state.

[3] What readers of these *Foundational Papers* understand by the term "complex system" should not be confused with spin glasses, which are also sometimes called "complex systems" in the condensed matter physics community, despite being static, and at local thermal equilibrium. Parisi's Nobel Prize was for his work on spin glasses.

in the bit erasure processes Landauer considers and does not include the "entropy production" of those processes.) It has also allowed us to generalize Landauer's analysis, to involve an arbitrary number of reservoirs (and so multiple temperatures), nonconservative forces acting on the system, and both closed systems in which reservoirs are finite as well as more conventional open systems with infinite reservoirs. We now also have versions of his analysis that are fully quantum mechanical. (Not surprisingly, entanglement, a purely quantum-mechanical phenomenon, deeply modifies Landauer's result.)

Equally crucially, modern computers have many hundreds of billions of transistors. These are all manufactured in an identical manner, and so none of these transistors are tailored to account for the precise (invariably non-uniform) distribution over their initial states that will arise due to where they occur in the computer's circuit diagram. This means that Landauer's reasoning does not apply to modern computers, since it does not fully account for the actual minimal thermodynamic cost, instead supposing that each transistor is tailored for its initial distribution. To calculate the full value of the minimal thermodynamic cost, one must add an extra cost to Landauer's $kT \ln 2$, a cost due to the mismatch between the actual initial distributions of all the separate transistors and the initial distribution that would have resulted in minimal thermodynamic cost for those transistors (Wolpert and Kolchinsky 2019).

In addition, many researchers interpreted Landauer's 1961 paper as proving that logically irreversible operations are necessarily thermodynamically irreversible, but that logically reversible operations need not be. This led to a lot of research into designing logically reversible computers (Bennett 1973; Fredkin and Toffoli 1982). However, stochastic thermodynamics has clarified that logical and thermodynamic reversibility are in fact completely independent properties of any physical process. In particular, a process can be logically reversible and thermodynamically irreversible, and it's also possible for a process to be logically irreversible and thermodynamically reversible (Sagawa 2014).

Moreover, if one does implement a computation in a logically reversible manner, there needs to be an extra copy of every output of the computation stored after the computation finishes, a copy that is not needed in standard (irreversible) computation. (In the case of logically reversibly implementing a Turing machine, what is instead needed is a copy of every input to the Turing machine.) While it was well-known before the advent of stochastic thermodynamics that this copy is needed, it was not appreciated in the twentieth century that making that extra copy necessarily "thermalizes an information reservoir" (to use modern language). In other words, we now know that in order to make the copy of a computation's output, one needs to first initialize a set of bits to contain that copy—and this initialization has its own thermodynamic costs, costs that might even exceed those that would arise from just doing the computation in the standard, irreversible manner (Wolpert 2019).

Finally, all digital computers—even reversible ones—are implemented physically as periodic processes, which perform the exact same physical operation to their degrees of freedom over and over. Recent results have proven that this simple property—implementing a computation in a periodic fashion—entails unavoidable strictly positive thermodynamic cost (Ouldridge and Wolpert 2023; Manzano *et al.* 2024).

Adopting a large-canvas, history-of-science perspective, in many ways Landauer played a role in thermodynamics of computation similar to the role that Kepler played in celestial mechanics. Both helped to create new fields; both were forced to use semi-formal reasoning due to the limited set of mathematical tools provided by the physics of their day; and both nonetheless derived some results that we now understand to be exactly correct, in certain idealized limits. These features are quite common in the pioneering breakthroughs in science. While Kepler was inspired to produce his breakthroughs by the music of the spheres, Landauer instead was inspired by the whispers of computers and of bits silently turning over in their electronic beds.

REFERENCES

Bennett, C. H. 1973. "Logical Reversibility of Computation." *IBM Journal of Research and Development* 17 (6): 525–532. https://doi.org/10.1147/rd.176.0525.

Crooks, G. E. 1999. "Entropy Production Fluctuation Theorem and the Nonequilibrium Work Relation for Free Energy Differences." *Physical Review E* 60 (3): 2721–2726. https://doi.org/10.1103/PhysRevE.60.2721.

Dillenschneider, R., and E. Lutz. 2010. "Comment on "Minimal Energy Cost for Thermodynamic Information Processing: Measurement and Information Erasure"." *Physical Review Letters* 104 (19): 198903. https://doi.org/10.1103/PhysRevLett.104.198903.

Fredkin, E., and T. Toffoli. 1982. "Conservative Logic." *International Journal of Theoretical Physics* 21:219–253. https://doi.org/10.1007/BF01857727.

Jarzynski, C. 1997. "Nonequilibrium Equality for Free Energy Differences." *Physical Review Letters* 78 (14): 2690–2693. https://doi.org/10.1103/PhysRevLett.78.2690.

Landauer, R. 1961. "Irreversibility and Heat Generation in the Computing Process." *IBM Journal of Research and Development* 5 (3): 183–191. https://doi.org/10.1147/rd.53.0183.

Manzano, G., E. Roldán, G. Kardeş, and D. H. Wolpert. 2024. "Thermodynamics of Computations with Absolute Irreversibility, Unidirectional Transitions, and Stochastic Computation Times." Forthcoming, *Physical Review X,* https : / / journals . aps . org / prx / accepted / 3a075K67Tbd1ab 06687062e65ba4d9d1977321d40.

Ouldridge, T., and D. H. Wolpert. 2023. "Thermodynamics of Deterministic Finite Automata Operating Locally and Periodically." *New Journal of Physics* 25 (12): 123013. https://doi.org/10.1088/1367-2630/ad1070.

Sagawa, T. 2014. "Thermodynamic and Logical Reversibilities Revisited." *Journal of Statistical Mechanics: Theory and Experiment* 2014 (3): P03025. https://doi.org/10.1088/1742-5468/2014/03/P03025.

Sagawa, T., and M. Ueda. 2010. "Generalized Jarzynski Equality under Nonequilibrium Feedback Control." *Physical Review Letters* 104 (9): 090602. https://doi.org/10.1103/PhysRevLett.104.090602.

Seifert, U. 2005. "Entropy Production along a Stochastic Trajectory and an Integral Fluctuation Theorem." *Physical Review Letters* 95 (4): 040602. https://doi.org/10.1103/PhysRevLett.95.040602.

Wheeler, J. A. 2002. "Information, Physics, Quantum: The Search for Links." In *Feynman and Computation: Exploring the Limits of Computation,* edited by A. Hey. Boca Raton, FL: CRC Press.

Wolpert, D. H. 2019. "The Stochastic Thermodynamics of Computation." *Journal of Physics A: Mathematical and Theoretical* 52 (19): 193001. https://doi.org/10.1088/1751-8121/ab0850.

Wolpert, D. H., and A. Kolchinsky. 2019. "Thermodynamics of Computing with Circuits." *New Journal of Physics* 22:063047. https://doi.org/10.1088/1367-2630/ab82b8.

IRREVERSIBILITY & HEAT GENERATION IN THE COMPUTING PROCESS

R. Landauer, IBM

Abstract

It is argued that computing machines inevitably involve devices which perform logical functions that do not have a single-valued inverse. This logical irreversibility is associated with physical irreversibility and requires a minimal heat generation, per machine cycle, typically of the order of kT for each irreversible function. This dissipation serves the purpose of standardizing signals and making them independent of their exact logical history. Two simple, but representative, models of bistable devices are subjected to a more detailed analysis of switching kinetics to yield the relationship between speed and energy dissipation, and to estimate the effects of errors induced by thermal fluctuations.

1. Introduction

The search for faster and more compact computing circuits leads directly to the question: What are the ultimate physical limitations on the progress in this direction? In practice the limitations are likely to be set by the need for access to each logical element. At this time, however, it is still hard to understand what physical requirements this puts on the degrees of freedom which bear information. The existence of a storage medium as compact as the genetic one indicates that one can go very far in the direction of compactness, at least if we are prepared to make sacrifices in the way of speed and random access.

Here Landauer is talking about reading from memory, to use modern language. This is not an issue he considers at length in this paper. Rather his main focus is writing to memory. Note, though, that he refers to this as "information processing." In point of fact, he does not consider the thermodynamics of arbitrary information processing, if by that one means the thermodynamics of the kinds of computations done in modern CPUs, which involve more than simple bit erasure. See Wolpert (2019).

Without considering the question of access, however, we can show, or at least very strongly suggest, that information processing is inevitably accompanied by a certain minimum amount of heat generation. In a general way this is not surprising. Computing, like all processes proceeding at a finite rate, must involve some dissipation. Our arguments, however, are more basic than this, and show that there is a minimum heat generation, independent of the rate of the process. Naturally the amount of heat generation involved is many orders of magnitude smaller than the heat dissipation in any practically conceivable device. The relevant point, however, is that the dissipation has a real function and is not just an unnecessary nuisance. The much larger amounts of dissipation in practical devices may be serving the same function.

Our conclusion about dissipation can be anticipated in several ways, and our major contribution will be a tightening of the concepts involved, in a fashion which will give some insight into the physical requirements for logical devices. The simplest way of anticipating our conclusion is to note that a binary device must have at least one degree of freedom associated with the information. Classically a degree of freedom is associated with kT of thermal energy. Any switching signals passing between devices must therefore have this much energy to override the noise. This argument does not make it clear that the signal energy must actually be dissipated. An alternative way of anticipating our conclusions is to refer to the arguments by Brillouin and earlier authors, as summarized by Brillouin in his book, *Science and Information Theory*,[1] to the effect that the measurement process requires a dissipation of the order of kT. The computing process, where the setting of various elements depends upon the setting of other elements at previous times, is closely akin to a measurement. It is difficult, however, to argue out this connection in a more exact fashion. Furthermore, the arguments concerning the measurement process are based on the analysis of specific models (as will some of our arguments about computing), and the specific models involved in the measurement analysis are rather far from the kind of mechanisms involved in data processing. In fact the arguments dealing with the measurement process do not define *measurement* very well, and avoid the very essential question: When is a system A coupled to a system B performing a measurement? The mere fact that two physical systems are coupled does not in itself require dissipation.

While interesting, and in some regards prescient, the precise arguments made in most work before Landauer do not play a role in our modern understanding of the thermodynamics of computation.

An ironic criticism of this earlier work by Landauer, given that the core of Landauer's arguments in this paper are semiformal, with no math.

Our main argument will be a refinement of the following line of thought. A simple binary device consists of a particle in a bistable

[1] L. Brillouin. 1956. *Science and Information Theory.* New York, NY: Academic Press Inc.

potential well shown in Fig. 1. Let us arbitrarily label the particle in the left-hand well as the ZERO state. When the particle is in the right-hand well, the device is in the ONE state. Now consider the operation RESTORE TO ONE, which leaves the particle in the ONE state, regardless of its initial location. If we are told that the particle is in the ONE state, then it is easy to leave it in the ONE state, without spending energy. If on the other hand we are told that the particle is in the ZERO state, we can apply a force to it, which will push it over the barrier, and then, when it has passed the maximum, we can apply a retarding force, so that when the particle arrives at ONE, it will have no excess kinetic energy, and we will not have expended any energy in the whole process, since we extracted energy from the particle in its downhill motion. Thus at first sight it seems possible to RESTORE TO ONE without any expenditure of energy. Note, however, that in order to avoid energy expenditure we have used two different routines, depending on the initial state of the device. This is not how a computer operates. In most instances a computer pushes information around in a manner that is independent of the exact data which are being handled, and is only a function of the physical circuit connections.

This is a core assertion underlying Landauer's reasoning—and clearly wrong in computations more elaborate than bit erasure, where what is done to a bit string in a RAM does depend on the current contents of that RAM. See Wolpert and Kolchinsky (2020).

Can we then construct a single time-varying force, $F(t)$, which when applied to the conservative system of figure 1 will cause the particle to end up in the ONE state, if it was initially in either the ONE state or the ZERO state? Since the system is conservative, its whole history can be reversed in time, and we will still have a system satisfying the laws of motion. In the time-reversed system we then have the possibility that for a single initial condition (position in the ONE state, zero velocity) we can end up in at least two places: the ZERO state or the ONE state. This, however, is impossible. The laws of mechanics are completely deterministic and a trajectory is determined by an initial position and velocity. (An initially unstable position can, in a sense, constitute an exception. We can roll away from the unstable point in one of at least two directions. Our initial point ONE is, however, a point of stable equilibrium.) Reverting to the original direction of time development, we see then that it is not possible to invent a single $F(t)$ which causes the particle to arrive at ONE regardless of its initial state.

In other words, can we construct a process that implements the erasure of a bit without any thermodynamic cost? (That's what "conservative" means in the current context.) He argues in the rest of this paragraph that this is not possible. Then in the next paragraph, he argues that bit erasure can in fact be implemented if we do not restrict attention to such processes with no thermodynamic cost. (That's what "lossy" means.)

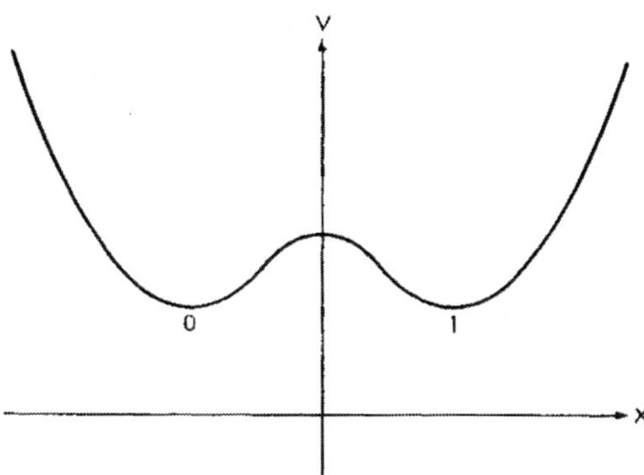

Figure 1. Bistable potential well. x is a generalized coordinate representing quantity which is switched.

If, however, we permit the potential well to be lossy, this becomes easy. A very strong positive initial force applied slowly enough so that the damping prevents oscillations will push the particle to the right, past ONE, regardless of the particle's initial state. Then if the force is taken away slowly enough, so that the damping has a chance to prevent appreciable oscillations, the particle is bound to arrive at ONE. This example also illustrates a point argued elsewhere[2] in more detail: While a heavily overdamped system is obviously undesirable, since it is made sluggish, an extremely underdamped one is also not desirable for switching, since then the system may bounce back into the wrong state if the switching force is applied and removed too quickly.

Section 2 actually isn't relevant to the subsequent development (either in this paper or, more broadly, in physics).

2. Classification

Before proceeding to the more detailed arguments we will need to classify data processing equipment by the means used to hold information, when it is not interacting or being processed. The simplest class and the one to which all the arguments of subsequent sections will be addressed consists of devices which can hold information without dissipating energy. The system illustrated in Fig. 1 is in this class. Closely related to the mechanical example of Fig. 1 are ferrites, ferroelectrics and thin magnetic films. The latter, which can switch without domain

[2] R. Landauer and J. A. Swanson. 1961. "Frequency Factors in the Thermally Activated Process." *Physical Review* 121 (6): 1668.

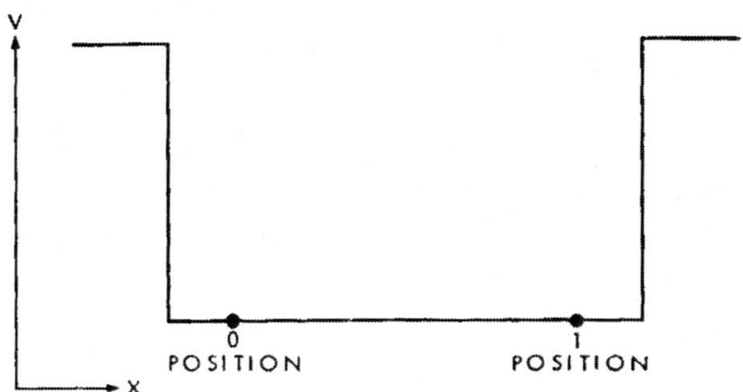

Figure 2. Potential well in which ZERO and ONE state are not separated by barrier. Information is preserved because random motion is slow.

wall motion, are particularly close to the one-dimensional device shown in Fig. 1. Cryotrons are also devices which show dissipation only when switching. They do differ, however, from the device of Fig. 1 because the ZERO and ONE states are not particularly favored energetically. A cryotron is somewhat like the mechanical device illustrated in Fig. 2, showing a particle in a box. Two particular positions in the box are chosen to represent ZERO and ONE, and the preservation of information depends on the fact that Brownian motion in the box is very slow. The reliance on the slowness of Brownian motion rather than on restoring forces is not only characteristic of cryotrons, but of most of the more familiar forms of information storage: Writing, punched cards, microgroove recording, etc. It is clear from the literature that all essential logical functions can be performed by devices in this first class. Computers can be built that contain either only cryotrons, or only magnetic cores.[3,4]

The second class of devices consists of structures which are in a steady (time invariant) state, but in a dissipative one, while holding on to information. Electronic flip-flop circuits, relays, and tunnel diodes are in this class. The latter, whose characteristic with load line is shown in Fig. 3, typifies the behavior. Two stable points of operation are separated by an unstable position, just as for the device in Fig. 1. It is noteworthy

[3]K. Mendelssohn. 1959. *Progress in Cryogenics.* Vol. 1. New York, NY: Academic Press Inc. Chapter 1 by D. R. Young, p. 1.

[4]L. B. Russell. 1957. *IRE Convention Record,* p. 106.

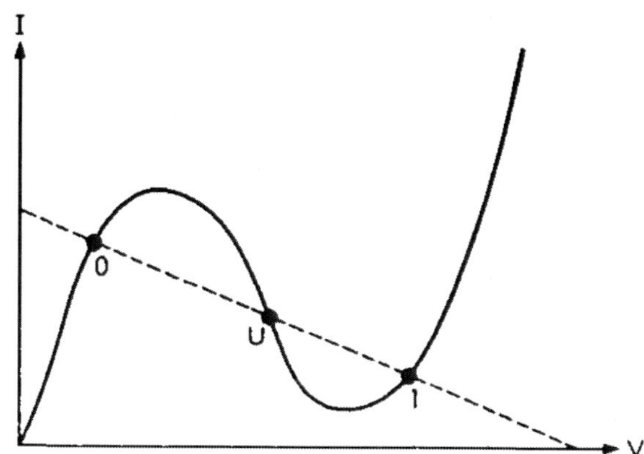

Figure 3. Negative resistance characteristic (solid line) with load line (dashed). ZERO *and* ONE *are stable states, U is unstable.*

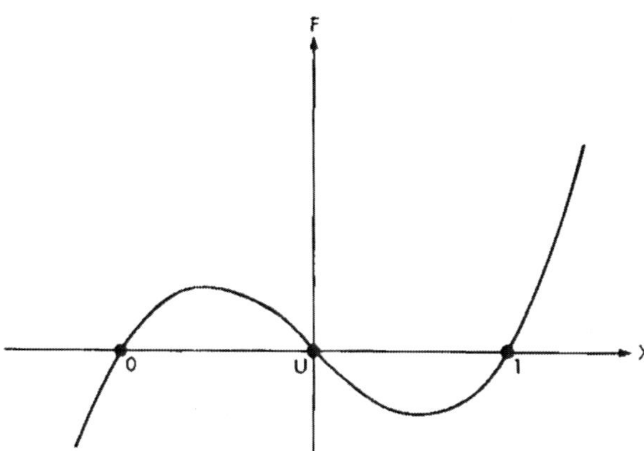

Figure 4. Force versus distance for the bistable well of Fig. 1. ZERO and ONE are stable states, U the unstable one.

that this class has no known representatives analogous to Fig. 2. All the active bistable devices (latches) have built-in means for restoration to the desired state. The similarity between Fig. 3 and the device of Fig. 1 becomes more conspicuous if we represent the bistable well of Fig. 1 by a diagram plotting force against distance. This is shown in Fig. 4. The line $F = 0$ intersects the curve in three positions, much like the load line (or a line of constant current), in Fig. 3. This analogy leads us to expect that in the case of the dissipative device there will be transitions from the desired state, to the other stable state, resulting from thermal agitation or quantum mechanical tunneling, much like for the dissipationless case, and as has been discussed for the latter in detail by Swanson.[5] The dissipative device, such as the single tunnel diode, will in general be an analog, strictly speaking, to an unsymmetrical potential well, rather than the symmetrical well shown in Fig. 1. We can therefore expect that of the two possible states for the negative resistance device only one is really stable, the other is metastable. An assembly of bistable tunnel diodes left alone for a sufficiently long period would eventually almost all arrive at the same state of absolute stability.

In general when using such latching devices in computing circuits one tries hard to make the dissipation in the two allowed states small, by pushing these states as closely as possible to the voltage or current axis. If one were successful in eliminating this dissipation almost completely during the steady state, the device would become a member of our first class. Our intuitive expectation is, therefore, that in the steady state dissipative device the dissipation per switching event is at least as high as in the devices of the first class, and that this dissipation per switching event is supplemented by the steady state dissipation.

The third and remaining class is a "catch-all"; namely, those devices where time variation is essential to the recognition of information. This includes delay lines, and also carrier schemes, such as the phase-bistable

[5] J. A. Swanson. 1960. "Physical versus Logical Coupling in Memory Systems." *IBM Journal of Research and Development* 4 (3): 305.
We would like to take this opportunity to amplify two points in Swanson's paper which perhaps were not adequately stressed in the published version.

(1) The large number of particles (~ 100) in the optimum element are a result of the small energies per particle (or cell) involved in the typical cooperative phenomenon used in computer storage. There is no question that information can be stored in the position of a single particle, at room temperature, if the activation energy for its motion is sufficiently large ($\sim$ several electron volts).

(2) Swanson's optimum volume is, generally, not very different from the common sense requirement on U, namely: $\nu t \exp(-U/kT) \ll 1$, which would be found without the use of information theory. This indicates that the use of redundancy and complicated coding methods does not permit much additional information to be stored. It is obviously preferable to eliminate these complications, since by making each element only slightly larger than the "optimum" value, the element becomes reliable enough to carry information without the use of redundancy.

system of von Neumann.[6] The latter affords us a very nice illustration of the need for dissipative effects; most other members of this third class seem too complex to permit discussion in simple physical terms.

In the von Neumann scheme, which we shall not attempt to describe here in complete detail, one uses a "pump" signal of frequency ω_0, which when applied to a circuit tuned to $\omega_0/2$, containing a nonlinear reactance, will cause the spontaneous build-up of a signal at the lower frequency. The lower frequency signal has a choice of two possible phases (180° apart at the lower frequency) and this is the source of the bistability. In the von Neumann scheme the pump is turned off after the subharmonic has developed, and the subharmonic subsequently permitted to decay through circuit losses. This decay is an essential part of the scheme and controls the direction in which information is passed. Thus at first sight the circuit losses perform an essential function. It can be shown, however, that the signal reduction can be produced in a lossless nonlinear circuit, by a suitably phased pump signal. Hence it would seem adequate to use lossless nonlinear circuits, and instead of turning the pump off, change the pump phase so that it causes signal decay instead of signal growth. The directionality of information flow therefore does not really depend on the existence of losses. The losses do, however, perform another essential function.

The von Neumann system depends largely on a coupling scheme called *majority logic*, in which one couples to three subharmonic oscillators and uses the sum of their oscillations to synchronize a subharmonic oscillator whose pump will cause it to build up at a later time than the initial three. Each of the three signals which are added together can have one of two possible phases. At most two of the signals can cancel, one will always survive, and thus there will always be a phase determined for the build-up of the next oscillation. The synchronization signal can, therefore, have two possible magnitudes. If all three of the inputs agree we get a synchronization signal three times as big as in the case where only two inputs have a given phase. If the subharmonic circuit is lossless the subsequent build-up will then

[6] R. L. Wigington. 1959. "A New Concept in Computing." *Proceedings of the IRE* 47 (4): 516–523.

result in two different amplitudes, depending on the size of the initial synchronization signal. This, however, will interfere with the basic operation of the scheme at the next stage, where we will want to combine outputs of three oscillators again, and will want all three to be of equal amplitude. We thus see that the absence of the losses gives us an output amplitude from each oscillator which is too dependent on inputs at an earlier stage. While perhaps the deviation from the desired amplitudes might still be tolerable after one cycle, these deviations could build up, through a period of several machine cycles. The losses, therefore, are needed so that the unnecessary details of a signal's history will be obliterated. The losses are essential for the standardization of signals, a function which in past theoretical discussions has perhaps not received adequate recognition, but has been very explicitly described in a recent paper by A. W. Lo.[7]

3. Logical Irreversibility

In the Introduction we analyzed Fig. 1 in connection with the command RESTORE TO ONE and argued that this required energy dissipation. We shall now attempt to generalize this train of thought. RESTORE TO ONE is an example of a logical truth function which we shall call *irreversible*. We shall call a device *logically irreversible* if the output of a device does not uniquely define the inputs. We believe that devices exhibiting logical irreversibility are essential to computing. Logical irreversibility, we believe, in turn implies physical irreversibility, and the latter is accompanied by dissipative effects.

We shall think of a computer as a distinctly finite array of N binary elements which can hold information, without dissipation. We will take our machine to be synchronous, so that there is a well-defined machine cycle and at the end of each cycle the N elements are a complicated function of their state at the beginning of the cycle.

Our arguments for logical irreversibility will proceed on three distinct levels. The first-level argument consists simply in the assertion that present machines do depend largely on logically irreversible steps, and that therefore any machine which copies the logical organization of

[7]A. W. Lo. Paper to appear in *IRE Transactions on Electronic Computers*.

present machines will exhibit logical irreversibility, and therefore by the argument of the next Section, also physical irreversibility.

The second level of our argument considers a particular class of computers, namely those using logical functions of only one or two variables. After a machine cycle each of our N binary elements is a function of the state of at most two of the binary elements before the machine cycle. Now assume that the computer is logically reversible. Then the machine cycle maps the 2^N possible initial states of the machine onto the same space of 2^N states, rather than just a subspace thereof. In the 2^N possible states each bit has a ONE and a ZERO appearing with equal frequency. Hence the reversible computer can utilize only those truth functions whose truth table exhibits equal numbers of ONES and ZEROS. The admissible truth functions then are the identity and negation, the EXCLUSIVE OR and its negation. These, however, are not a complete set[8] and do not permit a synthesis of all other truth functions.

In the third level of our argument we permit more general devices. Consider, for example, a particular three-input, three-output device, i.e., a small special purpose computer with three bit positions. Let p, q, and r be the variables before the machine cycle. The particular truth function under consideration is the one which replaces r by $p \cdot q$ if $r = 0$, and replaces r by $\overline{p \cdot q}$ if $r = 1$. The variables p and q are left unchanged during the machine cycle. We can consider r as giving us a choice of program, and p, q as the variables on which the selected program operates. This is a logically reversible device, its output always defines its input uniquely. Nevertheless it is capable of performing an operation such as AND which is not, in itself, reversible. The computer, however, saves enough of the input information so that it supplements the desired result to allow reversibility. It is interesting to note, however, that we did not "save" the program; we can only deduce what it was.

Now consider a more general purpose computer, which usually has to go through many machine cycles to carry out a program. At first sight it may seem that logical reversibility is simply obtained by

[8]D. Hilbert and W. Ackermann. 1950. *Principles of Mathematical Logic*. Page 10. New York, NY: Chelsea Publishing Co.

saving the input in some corner of the machine. We shall, however, label a machine as being logically reversible, if and only if all its individual steps are logically reversible. This means that every single time a truth function of two variables is evaluated we must save some additional information about the quantities being operated on, whether we need it or not. Erasure, which is equivalent to RESTORE TO ONE, discussed in the Introduction, is not permitted. We will, therefore, in a long program clutter up our machine bit positions with unnecessary information about intermediate results. Furthermore if we wish to use the reversible function of three variables, which was just discussed, as an AND, then we must supply in the initial programming a separate ZERO for every AND operation which is subsequently required, since the "bias" which programs the device is not saved, when the AND is performed. The machine must therefore have a great deal of extra capacity to store both the extra "bias" bits and the extra outputs. Can it be given adequate capacity to make all intermediate steps reversible? If our machine is capable, as machines are generally understood to be, of a nonterminating program, then it is clear that the capacity for preserving all the information about all the intermediate steps cannot be there.

All of these ideas are very prescient: the subsequent work of Fredkin and Tofolli (in the context of computational circuits) and of Bennett (in the context of Turing machines) elaborates on these ideas of Landauer's.

Let us, however, not take quite such an easy way out. Perhaps it is just possible to devise a machine, useful in the normal sense, but not capable of embarking on a nonterminating program. Let us take such a machine as it normally comes, involving logically irreversible truth functions. An irreversible truth function can be made into a reversible one, as we have illustrated, by "embedding" it in a truth function of a large number of variables. The larger truth function, however, requires extra inputs to bias it, and extra outputs to hold the information which provides the reversibility. What we now contend is that this larger machine, while it is reversible, is not a useful computing machine in the normally accepted sense of the word.

First of all, in order to provide space for the extra inputs and outputs, the embedding requires knowledge of the number of times each of the operations of the original (irreversible) machine will be required. The usefulness of a computer stems, however, from the fact that it is more than just a table look-up device; it can do many programs which were not anticipated in full detail by the designer. Our enlarged machine must have a number of bit positions, for every embedded device of the order of the number of program steps and requires a number of switching events during program loading comparable to the number that occur during the program itself. The setting of bias during program loading, which would typically consist of restoring a long row of bits to say ZERO, is just the type of nonreversible logical operation we are trying to avoid. Our unwieldy machine has therefore avoided the irreversible operations during the running of the program, only at the expense of added comparable irreversibility during the loading of the program.

The most detailed investigation of the thermodynamic consequences of this "expense" can be found in Section 9 of Wolpert (2019).

4. Logical Irreversibility and Entropy Generation

The detailed connection between logical irreversibility and entropy changes remains to be made. Consider again, as an example, the operation RESTORE TO ONE. The generalization to more complicated logical operations will be trivial.

It is now appreciated that this generalization is in fact extremely *non*trivial.

Imagine first a situation in which the RESTORE operation has already been carried out on each member of an assembly of such bits. This is somewhat equivalent to an assembly of spins, all aligned with the positive z-axis. In thermal equilibrium the bits (or spins) have two equally favored positions. Our specially prepared collections show much more order, and therefore a lower temperature and entropy than is characteristic of the equilibrium state. In the adiabatic demagnetization method we use such a prepared spin state, and as the spins become disoriented they take up entropy from the surroundings and thereby cool off the lattice in which the spins are embedded. An assembly of ordered bits would act similarly. As the assembly thermalizes and forgets its initial state the environment would be cooled off. Note that the important point here is not that all bits in the assembly initially agree with each other, but only that there is a single, well-defined initial state for the collection of bits. The well-defined initial state corresponds, by the usual statistical mechanical definition of entropy, $S = k \log_e W$, to zero entropy. The degrees of freedom associated with the information can, through thermal relaxation, go to any one of 2^N states (for N bits in the assembly) and therefore the entropy can increase by $kN \log_e 2$ as the initial information becomes thermalized.

Note that our argument here does not necessarily depend upon connections, frequently made in other writings, between entropy and information. We simply think of each bit as being located in a physical system, with perhaps a great many degrees of freedom, in addition to the relevant one. However, for each possible physical state which will be interpreted as a ZERO, there is a very similar possible physical state in which the physical system represents a ONE. Hence a system which is in a ONE state has only half as many physical states available to it as a system which can be in a ONE or ZERO state. (We shall ignore in this Section and in the subsequent considerations the case in which the ONE and ZERO are represented by states with different entropy. This case requires arguments of considerably greater complexity but leads to similar physical conclusions.)

In carrying out the RESTORE TO ONE operation we are doing the opposite of the thermalization. We start with each bit in one of two states and end up with a well-defined state. Let us view this operation in some detail.

Consider a statistical ensemble of bits in thermal equilibrium. If these are all reset to ONE, the number of states covered in the ensemble has been cut in half. The entropy therefore has been reduced by $k \log_e 2 = 0.6931k$ per bit. The entropy of a closed system, e.g., a computer with its own batteries, cannot decrease; hence this entropy must appear elsewhere as a heating effect, supplying $0.6931kT$ per restored bit to the surroundings. This is, of course, a minimum heating effect, and our method of reasoning gives no guarantee that this minimum is in fact achievable.

Our reset operation, in the preceding discussion, was applied to a thermal equilibrium ensemble. In actuality we would like to know what happens in a particular computing circuit which will work on information which has not yet been thermalized, but at any one time consists of a well-defined ZERO or a well-defined ONE. Take first the case where, as time goes on, the reset operation is applied to a random chain of ONES and ZEROS. We can, in the usual fashion, take the statistical ensemble equivalent to a time average and therefore conclude that the dissipation per reset operation is the same for the time-wise succession as for the thermalized ensemble.

A computer, however, is seldom likely to operate on random data. One of the two bit possibilities may occur more often than the other, or even if the frequencies are equal, there may be a correlation between successive bits. In other words the digits which are reset may not carry the maximum possible information. Consider the extreme case, where the inputs are all ONE, and there is no need to carry out any operation. Clearly then no entropy changes occur and no heat dissipation is involved. Alternatively if the initial states are all ZERO they also carry no information, and no entropy change is involved in resetting them all to ONE. Note, however, that the reset operation which sufficed when the inputs were all ONE (doing nothing) will not suffice when the inputs are all ZERO. When the initial states are ZERO, and we wish to go to ONE, this is analogous to a phase transformation between two phases in equilibrium, and can, presumably, be done reversibly and without an entropy increase in the universe, but only by a procedure specifically designed for that task. We thus see that when the initial states do not have their fullest possible diversity, the necessary entropy increase in the RESET operation can be reduced, but only by taking advantage of our knowledge about the inputs, and tailoring the reset operation accordingly.

The generalization to other logically irreversible operations is apparent, and will be illustrated by only one additional example. Consider a very small special-purpose computer, with three binary elements p, q, and r. A machine cycle replaces p by r, replaces q by r, and replaces r by $p \cdot q$. There are eight possible initial states, and in thermal equilibrium they will occur with equal probability. How much entropy reduction will occur in a machine cycle? The initial and final machine states are shown in Fig. ??. States α and β occur with a probability of $\frac{1}{8}$ each: states γ and δ have a probability of occurrence of $\frac{3}{8}$ each. The initial entropy was

$$\begin{aligned} S_i &= k \log_e W = -k\Sigma\rho \log_e \rho \\ &= -k\Sigma\frac{1}{8}\log_e\frac{1}{8} = 3k\log_e 2. \end{aligned}$$

The final entropy is

$$\begin{aligned} S_f &= -k\Sigma\rho\log_e\rho \\ &= -k(\frac{1}{8}\log\frac{1}{8}+\frac{1}{8}\log\frac{1}{8}+\frac{3}{8}\log\frac{3}{8}+\frac{3}{8}\log\frac{3}{8}). \end{aligned}$$

The difference S_i—S_f is $1.18k$. The minimum dissipation, if the initial state has no useful information, is therefore $1.18kT$.

Landauer himself was in part to blame for the confusion in how others interpreted his work, due to his cavalier use of terms like "dissipation," which he uses here but never formally defines. Modern terminology makes a crucial distinction between two related concepts: "heat dissipation" (which need not involve thermodynamic irreversibility) and "work dissipation" (which does involve thermodynamic irreversibility).

The question arises whether the entropy is really reduced by the logically irreversible operation. If we really map the possible initial ZERO states and the possible initial ONE states into the same space, i.e., the space of ONE states, there can be no question involved. But, perhaps, after we have performed the operation there can be some small remaining difference between the systems which were originally in the ONE state already and those that had to be switched into it. There is no harm in such differences persisting for some time, but as we saw in the discussion of the dissipationless subharmonic oscillator, we cannot tolerate a cumulative process, in which differences between various possible ONE states become larger and larger according to their detailed past histories. Hence the physical "many into one" mapping, which is the source of the entropy change, need not happen in full detail during the machine cycle which performed the logical function. But it must eventually take place, and this is all that is relevant for the heat generation argument.

5. Detailed Analysis of Bistable Well

The consequential part of Landauer's paper has pretty much ended by this point—none of the arguments below are considered in the modern literature.

To supplement our preceding general discussion we shall give a more detailed analysis of switching for a system representable by a bistable potential well, as illustrated, one-dimensionally, in figure 1, with a barrier large compared to kT. Let us, furthermore, assume that switching is accomplished by the addition of a force which raises the energy of one well with respect to the other, but still leaves a barrier which has to be surmounted by thermal activation. (A sufficiently large force will simply eliminate one of the minima completely. Our switching forces are presumed to be smaller.) Let us now consider a statistical ensemble of double well systems with a nonequilibrium distribution

BEFORE CYCLE p	q	r		AFTER CYCLE p_1	q_1	r_1	FINAL STATE
1	1	1	⟶	1	1	1	α
1	1	0	⟶	0	0	1	β
1	0	1	⟶	1	1	0	γ
1	0	0	⟶	0	0	0	δ
0	1	1	⟶	1	1	0	γ
0	1	0	⟶	0	0	0	δ
0	0	1	⟶	1	1	0	γ
0	0	0	⟶	0	0	0	δ

Figure 5. Three input - three output device which maps eight possible states onto only four different states.

and ask how rapidly equilibrium will be approached. This question has been analyzed in detail in an earlier paper (Landauer and Swanson 1961), and we shall therefore be satisfied here with a very simple kinetic analysis which leads to the same answer. Let n_A and n_B be the number of ensemble members in Well A and Well B respectively. Let U_A and U_B be the energies at the bottom of each well and U that of the barrier which has to be surmounted. Then the rate at which particles leave Well A to go to Well B will be of the form $\nu n_A \exp[-(U - U_A)/kT]$. The flow from B to A will be $\nu n_B \exp[-(U - U_B)/kT]$. The two frequency factors have been taken to be identical. Their differences are, at best, unimportant compared to the differences in exponents. This yields

$$\begin{aligned}\frac{dn_A}{dt} &= -n_A\nu \exp[-(U - U_A)/kT] \\ &\quad +n_B\nu \exp[-(U - U_B)/kT], \\ \frac{dn_B}{dt} &= n_A\nu \exp[-(U - U_A)/kT] \\ &\quad -n_B\nu \exp[-(U - U_B)/kT].\end{aligned} \tag{5.1}$$

We can view Eqs. (5.1) as representing a linear transformation on (n_A, n_B), which yields $\left(\frac{dn_A}{dt}, \frac{dn_B}{dt}\right)$. What are the characteristic values of the transformation? They are:

$$\begin{aligned}\lambda_1 = 0,\ \lambda_2 = &- \nu \exp[(U - U_A)/kT] \\ &- \nu \exp[-(U - U_B)/kT].\end{aligned}$$

The eigenvalue $\lambda_1 = 0$ corresponds to a time-independent well population. This is the equilibrium distribution

$$n_A = n_B \exp\frac{1}{kT}[U_B - U_A].$$

The remaining negative eigenvalue must then be associated with deviations from equilibrium, and exp $(-\lambda_2 t)$ gives the rate at which these deviations disappear. The relaxation time τ is therefore in terms of a quantity U_0, which is the average of U_A and U_B

$$\frac{1}{\tau} = \lambda_2 = \nu \exp\left[-(U - U_0)/kT\right] \cdot \{\exp\left[-(U_0 - U_A)kT\right] + \exp\left[(U_0 - U_B)kT\right]\}. \qquad (5.2)$$

The quantity U_0 in Eq. (5.2) cancels out, therefore the validity of Eq. (5.2) does not depend on the definition of U_0. Letting $\Delta = \frac{1}{2}(U_A - U_B)$, Eq. (5.2) then becomes

$$\frac{1}{\tau} = 2\nu \exp[-(U - U_0)/kT] \cosh \Delta/kT. \qquad (5.3)$$

To first order in the switching force which causes U_A and U_B to differ, $(U - U_0)$ will remain unaffected, and therefore Eq. (5.3) can be written

$$\frac{1}{\tau} = \frac{1}{\tau_0} \cosh \Delta/kT, \qquad (5.4)$$

where τ_0 is the relaxation time for the symmetrical potential well, when $\Delta = 0$. This equation demonstrates that the device is usable. The relaxation time τ_0 is the length of time required by the bistable device to thermalize, and represents the maximum time over which the device is usable. τ on the other hand is the minimum switching time. Cosh Δ/kT therefore represents the maximum number of switching events in the lifetime of the information. Since this can be large, the device can be useful. Even if Δ is large enough so that the first-order approximation needed to keep $U - U_0$ constant breaks down, the exponential dependence of cosh Δ/kT on Δ, in Eq. (5.3) will far outweigh the changes in $\exp[(U - U_0)kT]$, and τ_0/τ will still be a rapidly increasing function of Δ.

Note that Δ is one-half the energy which will be dissipated in the switching process. The thermal probability distribution within each well will be about the same before and after switching, the only difference is that the final well is 2Δ lower than the initial well. This energy difference is dissipated and corresponds to the one-half hysteresis loop area energy loss generally associated with switching. Equation (5.4) therefore confirms the empirically well-known fact that increases in switching speed can only be accomplished at the expense of increased dissipation per switching event. Equation (5.4) is, however, true only for a special model and has no really general significance. To show this consider an alternative model. Let us assume that information is stored by the position of a particle along a line, and that $x = \pm a$ correspond to ZERO and ONE, respectively. No barrier is assumed to exist, but the random diffusive motion of the particle is taken to be slow enough, so that positions will be preserved for an appreciable length of time. (This model is probably closer to the behavior of ferrites and ferroelectrics, when the switching occurs by domain wall motion, than our preceding bistable well model. The energy differences between a completely switched and a partially switched ferrite are rather small and it is the existence of a low domain-wall mobility which keeps the particle near its initial state, in the absence of switching forces, and this initial state can almost equally well be a partially switched state, as a completely switched one. On the other hand if one examines the domain wall mobility on a sufficiently microscopic scale it is likely to be related again to activated motion past barriers.) In that case, particles will diffuse a typical distance s in a time $\tau \sim s^2/2D$. D is the diffusion constant. The distance which corresponds to information loss is $s \sim a$, the associated relaxation time is $\tau_0 \sim a^2/2D$. In the presence of a force F the particle moves with a velocity μF, where the mobility μ is given by the Einstein relation as D/kT. To move a particle under a switching force F through a distance $2a$ requires a time τ_s given by

$$\mu F \tau_s = 2a, \tag{5.5}$$

or

$$\tau_s = 2a/\mu F. \tag{5.6}$$

The energy dissipation 2Δ, is a $2aF$. This gives us the equations

$$\tau_s = 2a^2/\mu\Delta, \tag{5.7}$$

$$\tau_s/\tau_0 = 4kT/\Delta, \qquad (5.8)$$

which show the same direction of variation of τ_s with Δ as in the case with the barrier, but do not involve an exponential variation with Δ/kT. If all other considerations are ignored it is clear that the energy bistable element of Eq. (5.4) is much to be preferred to the diffusion stabilized element of Eq. (5.8).

The above examples give us some insight into the need for energy dissipation, not directly provided by the arguments involving entropy consideration. In the RESTORE TO ONE operation we want the system to settle into the ONE state regardless of its initial state. We do this by lowering the energy of the ONE state relative to the ZERO state. The particle will then go to this lowest state, and on the way dissipate any excess energy it may have had in its initial state.

6. Three Sources of Error

We shall in this section attempt to survey the relative importance of several possible sources of error in the computing process, all intimately connected with our preceding considerations. First of all the actual time allowed for switching is finite and the relaxation to the desired state will not have taken place completely. If T_s is the actual time during which the switching force is applied and τ_s is the relaxation time of Eq. (5.4) then exp $(-T_s/\tau_s)$ is the probability that the switching will not have taken place. The second source of error is the one considered in detail in an earlier paper by J. A. Swanson (1960), and represents the fact that τ_0 is finite and information will decay while it is supposed to be sitting quietly in its initial state. The relative importance of these two errors is a matter of design compromises. The time T_s, allowed for switching, can always be made longer, thus making the switching relaxation more complete. The total time available for a program is, however, less than τ_0, the relaxation time for stored information, and therefore increasing the time allowed for switching decreases the number of steps in the maximum possible program.

A third source of error consists of the fact that even if the system is allowed to relax completely during switching there would still be a fraction of the ensemble of the order $\exp(-2\Delta/kT)$ left in the unfavored initial state. (Assuming $\Delta \gg kT$.) For the purpose of the subsequent discussion let us call this Boltzmann error. We shall show that no matter how the design compromise between the first two kinds of errors is made, Boltzmann error will never be dominant. We shall compare the errors in a rough fashion, without becoming involved in an enumeration of the various possible exact histories of information.

To carry out this analysis, we shall overestimate Boltzmann error by assuming that switching has occurred in every machine cycle in the history of every bit. It is this upper bound on the Boltzmann error which will be shown to be negligible, when compared to other errors. The Boltzmann error probability, per switching event is $\exp(-2\Delta/kT)$. During the same switching time bits which are not being switched are decaying away at the rate $\exp(-t/\tau_0)$. In the switching time T_s, therefore, unswitched bits have a probability T_s/τ_0 of losing their information. If the Boltzmann error is to be dominant

$$T_s/\tau_0 < \exp(-2\Delta/kT). \tag{6.1}$$

Let us specialize to the bistable well of Eq. (5.4). This latter equation takes (6.1) into the form

$$\frac{2T_s}{\tau_s}\exp(-\Delta/kT) < \exp(-2\Delta/kT), \tag{6.2}$$

or equivalently

$$\frac{T_s}{\tau_s} < \frac{1}{2}\exp(-\Delta/kT). \tag{6.3}$$

Now consider the relaxation to the switched state. The error incurred due to incomplete relaxation is $\exp(-T_s/\tau_s)$, which according to Eq. (6.3) satisfies

$$\exp(-T_s/\tau_s) > \exp[-\frac{1}{2}\exp(-\Delta/kT)]. \tag{6.4}$$

The right-hand side of this inequality has as its argument $\frac{1}{2}\exp(-\Delta/kT)$ which is less than $\frac{1}{2}$. Therefore the right-hand side is large compared to $\exp(-2\Delta/kT)$, the Boltzmann error, whose exponent is certainly larger than unity. We have thus shown that if the Boltzmann error dominates over the information decay, it must in turn be dominated by the incomplete relaxation during switching.

A somewhat alternate way of arguing the same point consists in showing that the accumulated Boltzmann error, due to the maximum number of switching events permitted by Eq. (5.4), is small compared to unity.

Consider now, instead, the diffusion stabilized element of Eq. (5.8). For it, we can find instead of Eq. (6.4) the relationship

$$\exp(-T_s/\tau_s) > \exp[(-\Delta/4kT)\exp(-2\Delta/kT)], \qquad (6.5)$$

and the right-hand side is again large compared to the Boltzmann error, $\exp(-2\Delta/kT)$. The alternative argument in terms of the accumulated Boltzmann error exists also in this case.

When we attempt to consider a more realistic machine model, in which switching forces are applied to coupled devices, as is done for example in diodeless magnetic core logic (Russell 1957), it becomes difficult to maintain analytically a clean-cut breakdown of error types, as we have done here. Nevertheless we believe that there is still a somewhat similar separation which is manifested.

Summary

The information-bearing degrees of freedom of a computer interact with the thermal reservoir represented by the remaining degrees of freedom. This interaction plays two roles. First of all, it acts as a sink for the energy dissipation involved in the computation. This energy dissipation has an unavoidable minimum arising from the fact that the computer performs irreversible operations. Secondly, the interaction acts as a source of noise causing errors. In particular thermal fluctuations give a supposedly switched element a small probability of remaining in its initial state, even after the switching force has been applied for a long time. It is shown, in terms of two simple models, that this source of error is dominated by one of two other error sources:

1) Incomplete switching due to inadequate time allowed for switching.

2) Decay of stored information due to thermal fluctuations.

It is, of course, apparent that both the thermal noise and the requirements for energy dissipation are on a scale which is entirely negligible in present-day computer components. The dissipation as calculated, however, is an absolute minimum. Actual devices which are far from minimal in size and operate at high speeds will be likely to require a much larger energy dissipation to serve the purpose of erasing the unnecessary details of the computer's past history.

Acknowledgments

Some of these questions were first posed by E. R. Piore a number of years ago. In its early stages (Swanson 1960; Landauer and Swanson 1961) this project was carried forward primarily by the late John Swanson. Conversations with Gordon Lasher were essential to the development of the ideas presented in the paper.

REFERENCES

Brillouin, L. 1956. *Science and Information Theory.* New York, NY: Academic Press Inc.

Hilbert, D., and W. Ackermann. 1950. *Principles of Mathematical Logic.* Page 10. New York, NY: Chelsea Publishing Co.

Landauer, R., and J. A. Swanson. 1961. "Frequency Factors in the Thermally Activated Process." *Physical Review* 121 (6): 1668.

Lo, A. W. Paper to appear in *IRE Transactions on Electronic Computers.*

Mendelssohn, K. 1959. *Progress in Cryogenics.* Vol. 1. New York, NY: Academic Press Inc.

Russell, L. B. 1957. *IRE Convention Record.*

Swanson, J. A. 1960. "Physical versus Logical Coupling in Memory Systems." *IBM Journal of Research and Development* 4 (3): 305.

Wigington, R. L. 1959. "A New Concept in Computing." *Proceedings of the IRE* 47 (4): 516–523.

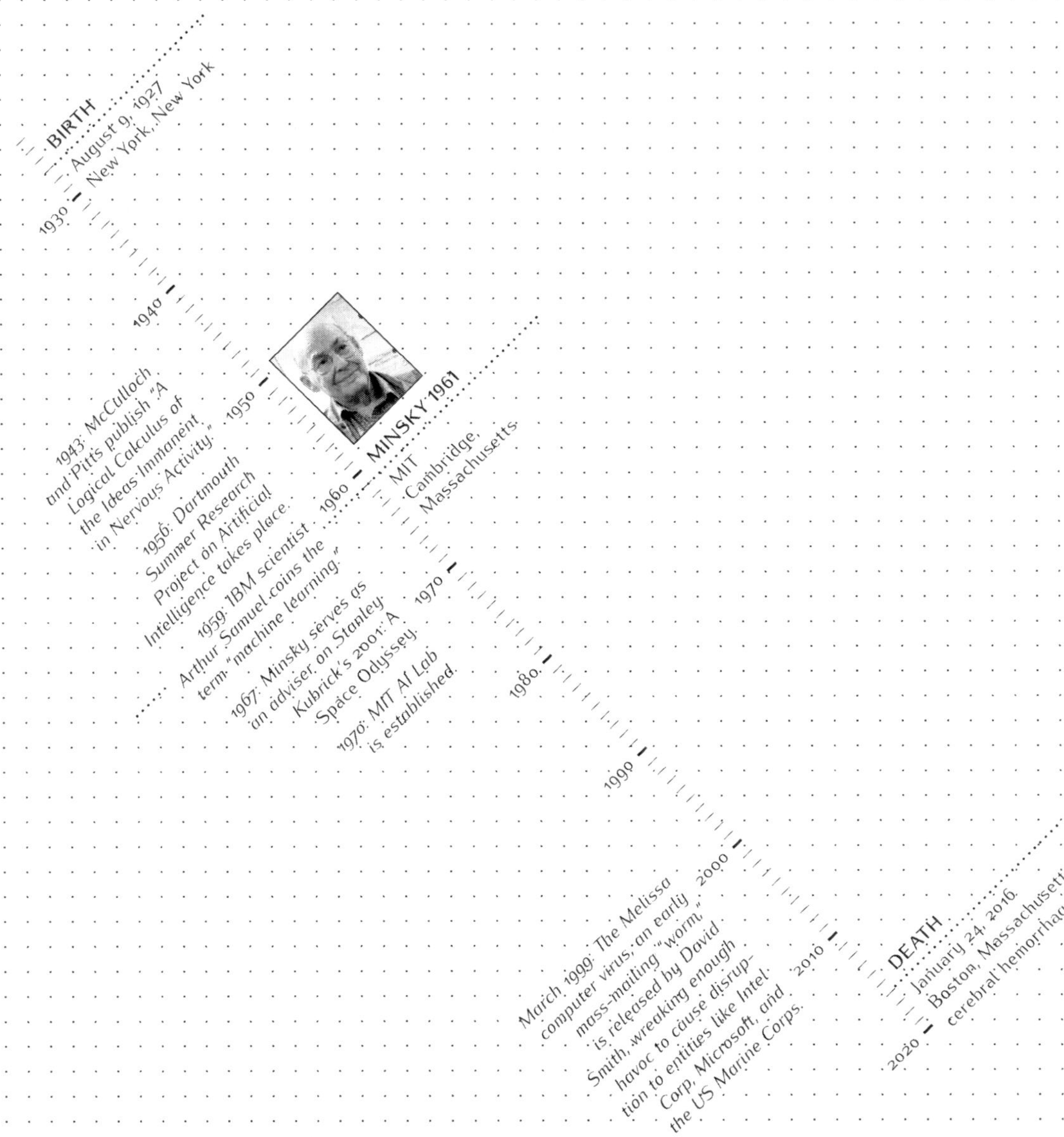

MARVIN LEE MINSKY

[17]

SYMBOLS VERSUS CYBERNETICS: MARVIN MINSKY ON THE PROSPECTS FOR ARTIFICIAL INTELLIGENCE

Melanie Mitchell, Santa Fe Institute

In 1956, a group of prominent mathematicians, psychologists, neuroscientists, computer programmers, and electrical engineers spent the summer at Dartmouth College defining the new field of "artificial intelligence." In the proposal for this meeting, the organizers set forth their ambitious goals: "An attempt will be made to find how to make machines use language, form abstractions and concepts, solve kinds of problems now reserved for humans, and improve themselves" (McCarthy *et al.* 2006). The organizers were optimistic about the prospects: "We think that a significant advance can be made in one or more of these problems if a carefully selected group of scientists work on it together for a summer." In the end, the Dartmouth meeting did not succeed in producing such dramatic advances (Moor 2006), but it did serve the purpose of distinguishing this new field from several related efforts and of defining AI's primary focus areas, thus setting the stage for decades of AI research that followed.

M. Minsky, "Steps Towards Artificial Intelligence," *Proceedings of the IRE* 49 (1): 8–30 (1961).

Marvin Minsky was one of the four main organizers of the Dartmouth meeting. His 1960 paper "Steps Towards Artificial Intelligence" is a status report for the field, four years in. The paper gives us a fascinating glimpse of how early AI researchers thought about the scope and goals of this nascent discipline. More interestingly, Minsky's paper foreshadows the many debates about the nature of intelligence—natural or artificial—that persist today.

Minsky, who obtained a PhD in mathematics from Princeton with a study of biological reinforcement learning, was one of the most prominent AI pioneers. He was a polymath whose work bridged neuroscience, psychology, electrical engineering, and computer science.

In 1959, Minsky founded one of the world's first AI labs, at MIT. This lab remains today among the most prestigious centers of AI research.

One major goal of the 1956 Dartmouth meeting was to distinguish AI from related interdisciplinary approaches to studying the links between intelligence and computation, including cybernetics, which focused on self-organizing systems with continuous feedback; automata studies, which attempted to unify informational processes in computers and living systems; and complex information processing, which looked at informational processes in psychology. Each of these disparate approaches, including AI, helped form the foundations of today's science of complex systems. However, it was important to Minsky and his collaborators to distinguish their new field from these prior efforts. Hence the term artificial intelligence, which in those days referred solely to so-called "symbolic" approaches to modeling intelligence, and, as Minsky makes clear in this paper, explicitly did not include "neural networks." This is ironic, given that now, more than six decades later, AI has become almost synonymous with neural networks. But the field has always been one in which definitions and goal posts change over time.

Minsky's paper starts out, in fact, noting that there is not, and will likely never be, a generally accepted definition of intelligence. Minsky instead frames the goal of AI as "the mechanization of problem-solving processes." And in his view, such mechanization need have little to do with biological intelligence; instead, the focus of AI is "heuristic programming." He defines a heuristic as "any method or trick used to improve the efficiency of a problem-solving system." For example, in the General Problem Solver system created by Newell and Simon (1956, and described in this paper), heuristic functions were defined to measure the "distance" between the current state of the system and the goal state, and heuristic operators were defined that could decrease this distance. These functions and operators were heuristic because they could sometimes fail, but the hope was that they could sometimes help mitigate an intractable search through a huge state space. In an earlier paper, Minsky (1958) noted "the distinction between programs which are guaranteed to work (and are called 'algorithms') and [heuristic] programs which are associated with what the programmer feels are good reasons to expect some success."

To Minsky and other originators of the "symbolic" approach to AI, heuristic programs that process language-like symbols would be the key to capturing human-level problem-solving abilities. In "Steps Towards Artificial Intelligence," Minsky surveys progress on five sub-areas of AI: search, pattern recognition, learning, planning, and induction, focusing on now-classic early AI projects such as Selfridge's (1958) Pandemonium system; Newell, Shaw, and Simon's Logic Theorist and General Problem Solver programs (Newell and Simon 1956; Newell, Shaw, and Simon 1959); and Samuel's (1959) learning program for playing checkers. Minsky's optimism around these "first steps" became evident a few years later; in 1967 he predicted that "within a generation . . . the problem of creating 'artificial intelligence' . . . will substantially be solved" (1967, 2).

Such predictions fell prey to the philosopher Hubert Dreyfus's (2012) "First Step Fallacy": "Limited early success is not a valid basis for assuredly predicting the ultimate success of one's project." Fifty years after the Dartmouth AI meeting, the main organizer, John McCarthy, admitted that "AI was harder than we thought" (in Moewes and Nürnberger 2013, 135). Minsky and Papert himself later noted that "easy things are hard" (1988 [1969], 29), a simple statement of what has come to be known as Moravec's paradox (Moravec 1988, 15): While AI systems can accomplish tasks that we consider to require high intelligence (e.g., beating chess and Go grandmasters), they still struggle with the easiest of human tasks, such as opening a doorknob, carrying on a conversation, or describing the meaning of a photograph.

Progressing in parallel with Minsky *et al.*'s heuristic programs was a largely separate research effort to capture machine intelligence via inspiration from the brain. This line of research, starting with McCulloch and Pitts's (1943) model of neuronal activity in terms of logic functions, going through Rosenblatt's Perceptron (1958), and Rumelhart and McClelland's "Parallel Distributed Processing" networks (Rumelhart, McClelland, and the PDP Research Group 1988), up to today's deep neural networks (LeCun, Bengio, and Hinton 2015), was disparaged early on by Minsky and his colleagues as "likely to be sterile" (Minsky and Papert [1969]1988, 232). However, it ended up as the dominant approach, ironically taking on the mantle of "artificial intelligence," which had been originally coined to exclude it. In recent decades, AI has made enormous progress on many narrowly defined tasks, but general human-level intelligence is still elusive, and its definition remains unclear. However, history rhymes in unexpected ways—at the time of this writing, many voices in the AI research community are pointing out the numerous limitations of deep learning and are proposing that true machine intelligence will require deep neural networks to be integrated with the kind of symbolic processing of Minsky's day (Marcus 2018).

Minsky is known for many contributions to AI, including foundational work on knowledge representation (1975) and a theory of general intelligence called the "The Society of Mind" (1988). Throughout his career he was particularly fascinated by the "commonsense knowledge problem": how to give machines the vast knowledge of the world that underlies much of human intelligent behavior. Hubert Dreyfus (2012) called this problem "the unexpected obstacle" in the route to general AI. In spite of overly optimistic predictions throughout its history, AI remains far from being "solved," and the ambitious goals of Minsky and the other organizers of the 1956 Dartmouth meeting largely remain open. As far as general AI is concerned, the field is still taking its first steps.

REFERENCES

Dreyfus, H. L. 2012. "A History of First Step Fallacies." *Minds and Machines* 22 (2): 87–99. https://doi.org/10.1007/s11023-012-9276-0.

LeCun, Y., Y. Bengio, and G. E. Hinton. 2015. "Deep Learning." *Nature* 521 (7553): 436–444. https://doi.org/10.1038/nature14539.

Marcus, G. 2018. *Deep Learning: A Critical Appraisal.* ArXiv:1801.00631. https://doi.org/10.48550/arXiv.1801.00631.

McCarthy, J., M. L. Minsky, N. Rochester, and C. E. Shannon. 2006. "A Proposal for the Dartmouth Summer Research Project on Artificial Intelligence, August 31, 1955." *AI Magazine* 27 (4): 12. https://doi.org/10.1609/aimag.v27i4.1904.

McCulloch, W. S., and W. Pitts. 1943. "A Logical Calculus of the Ideas Immanent in Nervous Activity." *Bulletin of Mathematical Biophysics* 5 (4): 115–143. https://doi.org/10.1007/BF02478259.

Minsky, M. L. 1958. "Some Methods of Artificial Intelligence and Heuristic Programming." In *Proceedings of the Symposium on the Mechanization of Thought Processes,* 3–28. London, UK: Her Majesty's Stationery Office.

———. 1967. *Computation: Finite and Infinite Machines.* Englewood Cliff, NJ: Prentice-Hall.

———. 1975. "A Framework for Representing Knowledge." In *The Psychology of Computer Vision,* edited by P. H. Winston. New York, NY: McGraw-Hill.

———. 1988. *The Society of Mind.* New York, NY: Simon & Schuster.

Minsky, M. L., and S. A. Papert. [1969]1988. *Perceptrons.* Expanded. Cambridge, MA: MIT Press.

Moewes, C., and A. Nürnberger. 2013. *Computational Intelligence in Intelligent Data Analysis.* Berlin, Germany: Springer.

Moor, J. 2006. "The Dartmouth College Artificial Intelligence Conference: The Next Fifty Years." *AI Magazine* 27 (4): 87. https://doi.org/10.1609/aimag.v27i4.1911.

Moravec, H. 1988. *Mind Children: The Future of Robot and Human Intelligence.* Cambridge, MA: Harvard University Press.

Newell, A., J. C. Shaw, and H. A. Simon. 1959. "Report on a General Problem-Solving Program." In *Proceedings of the IFIP Congress,* 256–264. Santa Monica, CA: RAND Corporation.

Newell, A., and H. A. Simon. 1956. "The Logic Theory Machine: A Complex Information Processing System." *IRE Transactions on Information Theory* 2 (3): 61–79. https://doi.org/10.1109/TIT.1956.1056797.

Rosenblatt, F. 1958. "The Perceptron: A Probabilistic Model for Information Storage and Organization in the Brain." *Psychological Review* 65 (6): 386–408. https://doi.org/10.1037/h0042519.

Rumelhart, D. E., J. L. McClelland, and the PDP Research Group. 1988. *Parallel Distributed Processing.* Vol. 1. IEEE.

Samuel, A. L. 1959. "Some Studies in Machine Learning Using the Game of Checkers." *IBM Journal on Research and Development* 3:210–229. https://doi.org/10.1147/rd.33.0210.

Selfridge, O. G. 1958. "Pandemonium: A Paradigm for Learning." In *Proceedings of the Symposium on Mechanisation of Thought Processes,* edited by D. V. Blake and A. M. Uttley, 511–529.

STEPS TOWARD ARTIFICIAL INTELLIGENCE*

Marvin Minsky, Massachusetts Institute of Technology

The work toward attaining "artificial intelligence" is the center of considerable computer research, design, and application. The field is in its starting transient, characterized by many varied and independent efforts. Marvin Minsky has been requested to draw this work together into a coherent summary, supplement it with appropriate explanatory or theoretical noncomputer information, and introduce his assessment of the state-of-the-art. This paper emphasizes the class of activities in which a general purpose computer, complete with a library of basic programs, is further programmed to perform operations leading to ever higher-level information processing functions such as learning and problem solving. This informative article will be of real interest to both the general PROCEEDINGS reader and the computer specialist.
—*The Guest Editor*

Abstract

The problems of heuristic programming—of making computers solve really difficult problems—are divided into five main areas: Search, Pattern-Recognition, Learning, Planning, and Induction.

A computer can do, in a sense, only what it is told to do. But even when we do not know how to solve a certain problem, we may program a machine (computer) to *Search* through some large space of solution attempts.

*Received by the IRE, October 24, 1960. The author's work summarized here—which was done at Lincoln Lab., a center for research operated by MIT at Lexington, Mass., with the joint support of the U. S. Army, Navy, and Air Force under Air Force Contract AF 19(604)-5200; and at the Res. Lab. of Electronics, MIT, Cambridge, Mass., which is supported in part by the U. S. Army Signal Corps, the Air Force Office of Scientific Res., and the ONR—is based on earlier work done by the author as a Junior Fellow of the Society of Fellows, Harvard Univ., Cambridge.

Unfortunately, this usually leads to an enormously inefficient process. With *Pattern-Recognition* techniques, efficiency can often be improved, by restricting the application of the machine's methods to appropriate problems. Pattern-Recognition, together with *Learning*, can be used to exploit generalizations based on accumulated experience, further reducing search. By analyzing the situation, using *Planning* methods, we may obtain a fundamental improvement by replacing the given search with a much smaller, more appropriate exploration. To manage broad classes of problems, machines will need to construct models of their environments, using some scheme for *Induction*.

Wherever appropriate, the discussion is supported by extensive citation of the literature and by descriptions of a few of the most successful heuristic (problem-solving) programs constructed to date.

Introduction

A visitor to our planet might be puzzled about the role of computers in our technology. On the one hand, he would read and hear all about wonderful "mechanical brains" baffling their creators with prodigious intellectual performance. And he (or it) would be warned that these machines must be restrained, lest they overwhelm us by might, persuasion, or even by the revelation of truths too terrible to be borne. On the other hand, our visitor would find the machines being denounced, on all sides, for their slavish obedience, unimaginative literal interpretations, and incapacity for innovation or initiative; in short, for their inhuman dullness.

Our visitor might remain puzzled if he set out to find, and judge for himself, these monsters. For he would find only a few machines (mostly "general-purpose" computers, programmed for the moment to behave according to some specification) doing things that might claim any real intellectual status. Some would be proving mathematical theorems of rather undistinguished character. A few machines might be playing certain games, occasionally defeating their designers. Some might be distinguishing between hand-printed letters. Is this enough to justify so much interest, let alone deep concern? I believe that it is; that we are on the threshold of an era that will be strongly influenced, and quite possibly dominated, by intelligent problem-solving machines. But our purpose is not to guess about what the future may bring; it is only to try to describe and explain what seem now to be our first steps toward the construction of "artificial intelligence."

Many papers since this one have claimed to describe "first steps" toward AI. Philosopher Herbert Dreyfus (2012) dubs this the "first step fallacy"—the fallacy that progress in narrow AI (game playing, theorem proving, etc.) is on a continuum with progress toward general or "true" AI. Even with recent advances in the field, many argue that little progress has been made toward the kind of AI envisioned by Minsky and other pioneers.

Along with the development of general-purpose computers, the past few years have seen an increase in effort toward the discovery and mechanization of problem-solving processes. Quite a number of papers have appeared describing theories or actual computer programs concerned with game-playing, theorem-proving, pattern-recognition, and other domains which would seem to require some intelligence. The literature does not include any general discussion of the outstanding problems of this field.

In this article, an attempt will be made to separate out, analyze, and find the relations between some of these problems. Analysis will be supported with enough examples from the literature to serve the introductory function of a review article, but there remains much relevant work not described here. This paper is highly compressed, and therefore, cannot begin to discuss all these matters in the available space.

We still lack a theory or even a generally accepted definition of "intelligence." Later on, Minsky referred to mental terms such as thinking or understanding as "prescientific idea germs."

There is, of course, no generally accepted theory of "intelligence"; the analysis is our own and may be controversial. We regret that we cannot give full personal acknowledgments here—suffice it to say that we have discussed these matters with almost every one of the cited authors.

It is convenient to divide the problems into five main areas: Search, Pattern-Recognition, Learning, Planning, and Induction; these comprise the main divisions of the paper. Let us summarize the entire argument very briefly:

A computer can do, in a sense, only what it is told to do. But even when we do not know exactly how to solve a certain problem, we may program a machine to *Search* through some large space of solution attempts. Unfortunately, when we write a straightforward program for such a search, we usually find the resulting process to be enormously inefficient. With *Pattern-Recognition* techniques, efficiency can be greatly improved by restricting the machine to use its methods only on the kind of attempts for which they are appropriate. And with *Learning*, efficiency is further improved by directing Search in accord with earlier experiences. By actually analyzing the situation, using what we call *Planning* methods, the machine may obtain a really fundamental improvement by replacing the originally given Search by a much smaller, more appropriate exploration. Finally, in the section on *Induction*, we consider some rather more global concepts of how one might obtain intelligent machine behavior.

I. The Problem of Search[1]

Summary—If, for a given problem, we have a means for checking a proposed solution, then we can solve the problem by testing all possible answers. But this always takes much too long to be of practical interest. Any device that can reduce this search may be of value. If we can detect relative improvement, then "hill-climbing" (Section I-B) may be feasible, but its use requires some structural knowledge of the search space. And unless this structure meets certain conditions, hill-climbing may do more harm than good.

When we talk of problem-solving in what follows we will usually suppose that all the problems to be solved are initially *well defined* (McCarthy 1956). By this we mean that with each problem we are given some systematic way to decide when a proposed solution is acceptable. Most of the experimental work discussed here is concerned with such well-defined problems as are met in theorem-proving, or in games with precise rules for play and scoring.

"Well-defined" problems, such as playing chess or proving theorems, also have "closed worlds"—that is, a program can search for a solution in a predefined space of possibilities. Real-world problems, such as driving a car, are not well-defined due to the open-ended possibilities that can occur. The field of AI still struggles with such open-ended tasks.

In one sense all such problems are trivial. For if there exists a solution to such a problem, that solution can be found eventually by any blind exhaustive process which searches through all possibilities. And it is usually not difficult to mechanize or program such a search.

But for any problem worthy of the name, the search through all possibilities will be too inefficient for practical use. And on the other hand, systems like chess, or nontrivial parts of mathematics, are too complicated for complete analysis. Without complete analysis, there must always remain some core of search, or "trial and error." So we need to find techniques through which the results of *incomplete analysis* can be used to make the search more efficient. The necessity for this is simply

[1] The adjective "heuristic," as used here and widely in the literature, means *related to improving problem-solving performance*; as a noun it is also used in regard to any method or trick used to improve the efficiency of a problem-solving system. A "heuristic program," to be considered successful, must work well on a variety of problems, and may often be excused if it fails on some. We often find it worthwhile to introduce a heuristic method which happens to cause occasional failures, if there is an over-all improvement in performance. But imperfect methods are not necessarily heuristic, nor vice versa. Hence "heuristic" should not be regarded as opposite to "foolproof"; this has caused some confusion in the literature.

overwhelming: a search of all the paths through the game of checkers involves some 10^{40} move choices (Samuel 1959); in chess, some 10^{120} (Shannon 1956). If we organized all the particles in our galaxy into some kind of parallel computer operating at the frequency of hard cosmic rays, the latter computation would still take impossibly long; we cannot expect improvements in "hardware" alone to solve all our problems! Certainly we must use whatever we know in advance to guide the trial generator. And we must also be able to make use of results obtained along the way.[2] [3]

A. RELATIVE IMPROVEMENT, HILL-CLIMBING, AND HEURISTIC CONNECTIONS

A problem can hardly come to interest us if we have no background of information about it. We usually have some basis, however flimsy, for detecting *improvement*; some trials will be judged more successful than others. Suppose, for example, that we have a *comparator* which selects as the better, one from any pair of trial outcomes. Now the comparator cannot, alone, serve to make a problem well-defined. No goal is defined. But if the comparator-defined relation between trials is "transitive" (*i.e.*, if A *dominates* B and B *dominates* C implies that A *dominates* C), then we can at least define "progress," and ask our machine, given a time limit, to do the best it can.

But it is essential to observe that a comparator by itself, however shrewd, cannot alone give any improvement over exhaustive search. The comparator gives us information about partial success, to be sure. But we need also some way of using this information to direct the pattern of search in promising directions; to select new trial points which are in some sense

[2] McCarthy (1956) has discussed the enumeration problem from a recursive-function theory point of view. This incomplete but suggestive paper proposes, among other things, that "the enumeration of partial recursive functions should give an early place to compositions of functions that have already appeared."

I regard this as an important notion, especially in the light of Shannon's results (Shannon 1949) on two-terminal switching circuits—that the "average" n-variable switching function requires about $2^n/n$ contacts. This disaster does not usually strike when we construct "interesting" large machines, presumably because they are based on composition of functions already found useful.

[3] In 1952 and especially in 1956 Ashby has an excellent discussion of the search problem. (However, I am not convinced of the usefulness of his notion of "ultrastability," which seems to be little more than the property of a machine to search until something stops it.)

"like," or "similar to," or "in the same direction as" those which have given the best previous results. To do this we need some additional structure on the search space. This structure need not bear much resemblance to the ordinary spatial notion of direction, or that of distance, but it must somehow tie together points which are heuristically related.

We will call such a structure a *heuristic connection*. We introduce this term for informal use only—that is why our definition is itself so informal. But we need it. Many publications have been marred by the misuse, for this purpose, of precise mathematical terms, *e.g.*, *metric* and *topological*. The term "connection," with its variety of dictionary meanings, seems just the word to designate a relation without commitment as to the exact nature of the relation.

An important and simple kind of heuristic connection is that defined when a space has coordinates (or parameters) and there is also defined a numerical "success-function" E which is a reasonably smooth function of the coordinates. Here we can use local optimization or *hill-climbing* methods.

B. HILL-CLIMBING

Suppose that we are given a black-box machine with inputs $\lambda_1, \cdots, \lambda_n$ and an output $E(\lambda_1, \cdots, \lambda_n)$. We wish to maximize E by adjusting the input values. But we are not given any mathematical description of the function E; hence we cannot use differentiation or related methods. The obvious approach is to explore locally about a point, finding the direction of steepest ascent. One moves a certain distance in that direction and repeats the process until improvement ceases. If the hill is smooth this may be done, approximately, by estimating the gradient component $\partial E/\partial\lambda_i$ separately for each coordinate λ_i. There are more sophisticated approaches (one may use noise added to each variable, and correlate the output with each input, see Fig. 1), but this is the general idea. It is a fundamental technique, and we see it always in the background of far more complex systems. Heuristically, its great virtue is this: the sampling effort (for determining the direction of the gradient) grows, in a sense, only linearly with the number of parameters. So if we can solve, by such a method, a certain kind of problem involving many parameters, then the addition of more parameters of the same kind ought not cause an

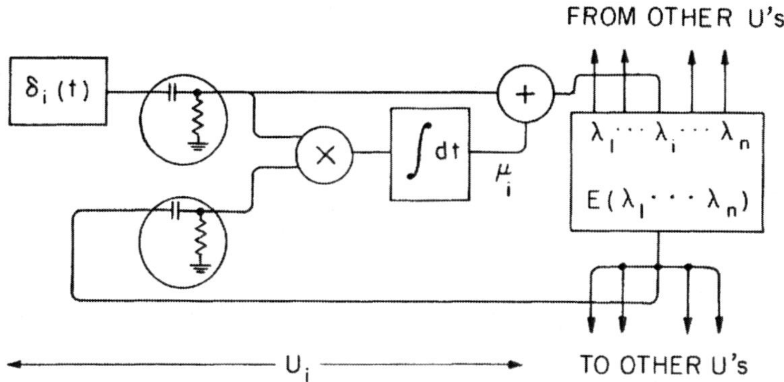

Figure 1. "Multiple simultaneous optimizers" search for a (local) maximum value of some function $E(\lambda_1, \cdots, \lambda_n)$ of several parameters. Each unit U_i independently "jitters" its parameter λ_i, perhaps randomly, by adding a variation $\delta_i(t)$ to a current mean value μ_i. The changes in the quantities δ_i and E are correlated, and the result is used to (slowly) change μ_i. The filters are to remove dc components. This simultaneous technique, really a form of coherent detection, usually has an advantage over methods dealing separately and sequentially with each parameter. (Cf. the discussion of "informative feedback" in Wiener (1948), p. 133 ff.)

inordinate increase in difficulty. We are particularly interested in problem-solving methods which can be so extended to more difficult problems. Alas, most interesting systems which involve combinational operations usually grow exponentially more difficult as we add variables.

A great variety of hill-climbing systems have been studied under the names of "adaptive" or "self-optimizing" servomechanisms.

C. TROUBLES WITH HILL-CLIMBING

Obviously, the gradient-following hill-climber would be trapped if it should reach a *local peak* which is not a true or satisfactory optimum. It must then be forced to try larger steps or changes.

It is often supposed that this false-peak problem is the chief obstacle to machine learning by this method. This certainly can be troublesome. But for really difficult problems, it seems to us that usually the more fundamental problem lies in finding any significant peak at all. Unfortunately the known E functions for difficult problems often exhibit what we have called (Minsky and Selfridge 1961) the "*Mesa Phenomenon*" in which a small change in a parameter usually leads to either no change in performance or to a large change in performance. The space is thus composed primarily of flat regions or "mesas." Any tendency of the trial generator to make small steps then results in much aimless wandering without compensating information gains. A profitable search in such a space requires steps so large that hill-climbing is essentially ruled out. The problem-solver must find other methods; hill-climbing might still be feasible with a different heuristic connection.

Certainly, in our own intellectual behavior we rarely solve a tricky problem by a steady climb toward success. I doubt that in any one simple mechanism, *e.g.*, hill-climbing, will we find the means to build an efficient and general problem-solving machine. Probably, an intelligent machine will require a variety of different mechanisms. These will be arranged in hierarchies, and in even more complex, perhaps recursive, structures. And perhaps what amounts to straightforward hill-climbing on one level may sometimes appear (on a lower level) as the sudden jumps of "insight."

II. The Problem of Pattern Recognition

Summary—In order not to try all possibilities, a resourceful machine must classify problem situations into categories associated with the domains of effectiveness of the machine's different methods. These pattern-recognition methods must extract the heuristically significant features of the objects in question. The simplest methods simply match the objects against standards or prototypes. More powerful "property-list" methods subject each object to a sequence of tests, each detecting some *property* of heuristic importance. These properties have to be invariant under commonly encountered forms of distortion. Two important problems arise here—inventing new useful properties, and combining many properties to form a recognition system. For complex problems, such methods will have to be augmented by facilities for subdividing complex objects and describing the complex relations between their parts.

Any powerful heuristic program is bound to contain a variety of different methods and techniques. At each step of the problem-solving process the machine will have to decide what aspect of the problem to work on, and then which method to use. A choice must be made, for we usually cannot afford to try all the possibilities. In order to deal with a goal or a problem, that is, to choose an appropriate method, we have to recognize what kind of thing it is. Thus the need to choose among actions compels us to provide the machine with classification techniques, or means of evolving them. It is of overwhelming importance that the machine have classification techniques which are realistic. But "realistic" can be defined only with respect to the environments to be encountered by the machine, and with respect to the methods available to it. Distinctions which cannot be exploited are not worth recognizing. And methods are usually worthless without classification schemes which can help decide when they are applicable.

A. TELEOLOGICAL REQUIREMENTS OF CLASSIFICATION

The useful classifications are those which match the goals and methods of the machine. The objects grouped together in the classifications should have something of heuristic value in common; they should be "similar" in a useful sense; they should depend on relevant or essential features. We should not be surprised, then, to find ourselves using inverse or teleological expressions to define the classes. We really do want to have a grip on "the class of objects which can be transformed into a result of form Y," that is, the class of objects which will satisfy some goal. One should be wary of the familiar injunction against using teleological language in science. While it is true that talking of goals in some contexts may dispose us towards certain kinds of animistic explanations, this need not be a bad thing in the field of problem-solving; it is hard to see how one can solve problems without thoughts of purposes. The real difficulty with teleological definitions is technical, not philosophical, and arises when they have to be used and not just mentioned. One obviously cannot afford to use for classification a method which actually requires waiting for some remote outcome, if one needs the classification precisely for deciding whether to try out that method. So, in practice, the ideal teleological definitions often have to be replaced by practical approximations, usually with some risk of error; that is, the definitions have to be made *heuristically effective*, or economically usable. This is of great importance. (We can think of "heuristic effectiveness" as contrasted to the ordinary mathematical notion of "effectiveness" which distinguishes those definitions which can be realized at all by machine, regardless of efficiency.)

B. PATTERNS AND DESCRIPTIONS

It is usually necessary to have ways of assigning *names*—symbolic expressions—to the defined classes. The structure of the names will have a crucial influence on the mental world of the machine, for it determines what kinds of things can be conveniently thought about. There are a variety of ways to assign names. The simplest schemes use what we will call *conventional* (or *proper*) names; here, arbitrary symbols are assigned to classes. But we will also want to use complex *descriptions* or *computed names;* these are constructed for classes by processes which *depend on the class definitions.* To be useful, these should reflect some of the structure of the things they designate, abstracted in a manner relevant to the problem area. The notion of description merges smoothly into the more complex notion of *model*; as we think of it, a model is a sort of active description. It is a thing whose form reflects some of the structure of the thing represented, but which also has some of the character of a working machine.

In Section III we will consider "learning" systems. The behavior of those systems can be made to change in reasonable ways depending on what happened to them in the past. But by themselves, the simple learning systems are useful only in recurrent situations; they cannot cope with any significant novelty. Nontrivial performance is obtained only when learning systems are supplemented with classification or pattern-recognition methods of some inductive ability. For the variety of objects encountered in a nontrivial search is so enormous that we cannot depend on recurrence, and the mere accumulation of records of past experience can have only limited value. Pattern-Recognition, by providing a heuristic connection which links the old to the new, can make learning broadly useful.

What is a "pattern"? We often use the term teleologically to mean a set of objects which can in some (useful) way be treated alike. For each problem area we must ask, "What patterns would be useful for a machine working on such problems?"

The problems of *visual* pattern-recognition have received much attention in recent years and most of our examples are from this area.

C. PROTOTYPE-DERIVED PATTERNS

The problem of reading *printed* characters is a clear-cut instance of a situation in which the classification is based ultimately on a fixed set of "prototypes"—*e.g.*, the dies from which the type font was made. The individual marks on the printed page may show the results of many distortions. Some distortions are rather systematic: change in size, position, orientation. Some are of the nature of noise: blurring, grain, low contrast, etc.

If the noise is not too severe, we may be able to manage the identification by what we call a *normalization and template-matching* process. We first remove the differences related to size and position—that is, we *normalize* the input figure. One may do this, for example, by constructing a similar figure inscribed in a certain fixed triangle (see Fig. 2); or one may transform the figure to obtain a certain fixed center of gravity and a unit second central moment. (There is an additional problem with rotational equivalence where it is not easy to avoid all ambiguities. One does not want to equate "6" and "9". For that matter, one does not want to equate $(0, o)$, or $(X,$ x) or the o's in x_o and x^o, so that there may be context-dependency involved.) Once normalized, the unknown figure can be compared with *templates* for the prototypes and, by means of some measure of *matching*, choose the best fitting template. Each "matching criterion" will be sensitive to particular forms of noise and distortion, and so will each normalization procedure. The inscribing or boxing method may be sensitive to small specks, while the moment method will be especially sensitive to smearing, at least for thin-line figures, etc. The choice of a matching criterion must depend on the kinds of noise and transformations commonly encountered. Still, for many problems we may get acceptable results by using straightforward correlation methods.

When the class of equivalence transformations is very large, *e.g.*, when local stretching and distortion are present, there will be difficulty in finding a uniform normalization method. Instead, one may have to consider a process of adjusting locally for best fit to the template. (While measuring the matching, one could "jitter" the figure locally; if an improvement were found the process could be repeated using a slightly different change, etc.) There is usually no practical possibility of applying to the figure *all* of the admissible transformations. And to recognize the *topological* equivalence of pairs such as those in Fig. 3 is likely beyond any practical kind of iterative local-improvement or hill-climbing matching procedure. (Such recognitions can be mechanized, though, by methods which follow lines, detect vertices, and build up a *description* in the form, say, of a vertex-connection table.)

The template matching scheme, with its normalization and direct comparison and matching criterion, is just too limited in conception to be of much use in more difficult problems. If the transformation set is large, normalization, or "fitting," may be impractical, especially if there is no adequate heuristic connection on the space of transformations. Furthermore, for each defined pattern, the system has to be presented with a prototype. But if one has in mind a fairly abstract class, one may simply be unable to represent its essential features with one or a very few concrete examples. How could one represent with a single prototype the class of figures which have an even number of disconnected parts? Clearly, the template system has negligible descriptive power. The property-list system frees us from some of these limitations.

Here Minsky is defining a property to be a binary function—either an object has the property or it does not. In modern AI, objects are analyzed in terms of features, which are continuous-valued functions reflecting the extent to which an object has a particular property (and often these properties are defined by the AI system itself, rather than a human-defined property such as "containing a loop").

D. PROPERTY LISTS AND "CHARACTERS"

We define a *property* to be a two-valued function which divides figures into two classes; a figure is said to have or not have the property according to whether the function's value is 1 or 0. Given a number N of distinction properties, we could define as many as 2^n subclasses by their set intersections and, hence, as many as 2^{2^n} *patterns* by combining the properties with AND's and OR's. Thus, if we have three properties, *rectilinear*, *connected*, and *cyclic*, there are eight subclasses (and 256 patterns) defined by their intersections (see Fig. 4).

If the given properties are placed in a fixed order then we can represent any of these elementary regions by a vector, or string of digits. The vector so assigned to each figure will be called the *Character* of that figure (with respect to the sequence of properties in question). (In Minsky (1959) we use the term *characteristic* for a property without restriction to 2 values.) Thus a square has the Character (1, 1, 1) and a circle the Character (0, 1, 1) for the given sequence of properties.

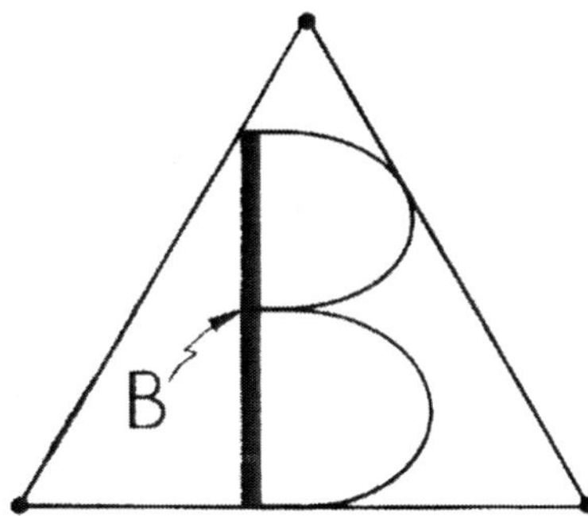

Figure 2. A simple normalization technique. If an object is expanded uniformly, without rotation, until it touches all three sides of a triangle, the resulting figure will be unique, and pattern-recognition can proceed without concern about relative size and position.

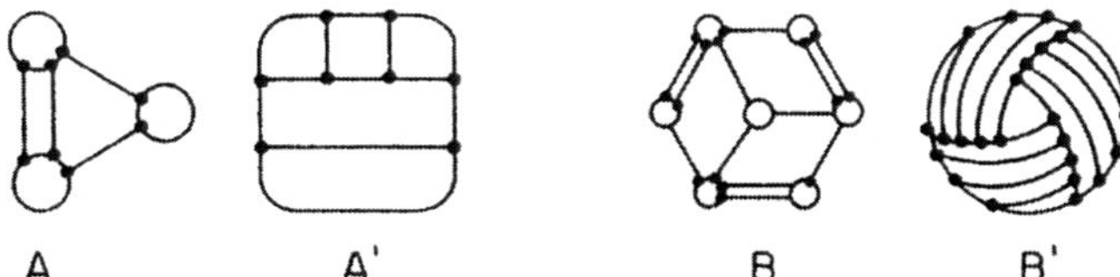

Figure 3. The figures A, A' and B, B' are topologically equivalent pairs. Lengths have been distorted in an arbitrary manner, but the connectivity relations between corresponding points have been preserved. In Sherman (1960) and Haller (1959) we find computer programs which can deal with such equivalences.

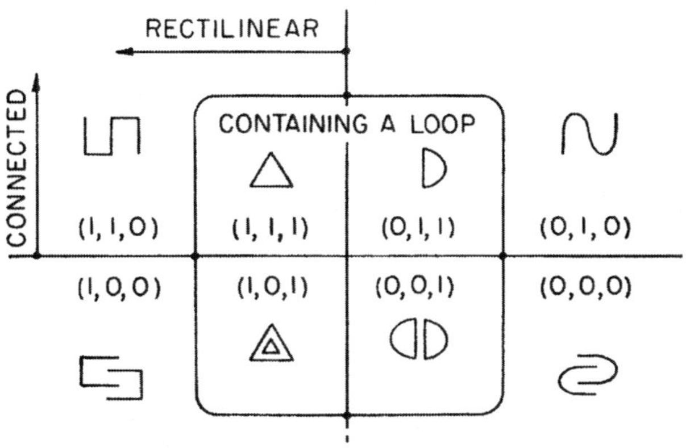

Figure 4. The eight regions represent all the possible configurations of values of the three properties "rectilinear," "connected," "containing a loop." Each region contains a representative figure, and its associated binary "Character" sequence.

For many problems one can use such Characters as names for categories and as primitive elements with which to define an adequate set of patterns. Characters are more than conventional names. They are instead very rudimentary forms of *description* (having the form of the simplest symbolic expression—the *list*) whose structure provides some information about the designated classes. This is a step, albeit a small one, beyond the template method; the Characters are not simple instances of the patterns, and the properties may themselves be very abstract. Finding a good set of properties is the major concern of many heuristic programs.

Finding a good set of properties has been called "feature engineering," and in many cases it has been a large part of what AI researchers do. A major advantage of more recent deep learning systems is that they can formulate useful features automatically as part of the learning process.

E. INVARIANT PROPERTIES

One of the prime requirements of a good property is that it be invariant under the commonly encountered equivalence transformations. Thus for visual Pattern-Recognition we would usually want the object identification to be independent of uniform changes in size and position. In their pioneering paper McCulloch and Pitts (1947) describe a general technique for forming invariant properties from non-invariant ones, assuming that the transformation space has a certain (group) structure. The idea behind their mathematical argument is this: suppose that we have a function P of figures, and suppose that for a given figure F we define $[F] = \{F_1, F_2, \cdots\}$ to be the set of all figures equivalent to F under the given set of transformations; further, define $P[F]$ to be the set $\{P(F_1), P(F_2), \cdots\}$ of values of P on those figures. Finally, define $P^*[F]$ to be AVERAGE $(P[F])$. Then we have a new property P^* whose values are independent of the selection of F from an equivalence class defined by the transformations. We have to be sure that when different representatives are chosen from a class the collection $[F]$ will always be the same in each case. In the case of continuous transformation spaces, there will have to be a *measure* or the equivalent associated with the set $[F]$ with respect to which the operation AVERAGE is defined, say, as an integration.[4]

Minsky is referring to neuroscientist Warren McCulloch and logician Walter Pitts's classic 1943 paper "A Logical Calculus of the Ideas Immanent in Nervous Activity," which proposes an idealized mathematical model of neurons (based on propositional logic) that became the basis for subsequent work on artificial neural networks.

---

[4] In the case studied in McCulloch and Pitts (1947) the transformation space is a *group* with a uniquely defined measure: the set $[F]$ can be computed without repetitions by *scanning* through the application of all the transforms T_α to the given figure so that the invariant property can be defined by

$$P^*(F) = \int_{\alpha \in G} P(T_\alpha(F))d\mu$$

This method is proposed (McCulloch and Pitts 1947) as a neurophysiological model for pitch-invariant hearing and size-invariant visual recognition (supplemented with visual centering mechanisms). This model is discussed also by Wiener.[5] Practical application is probably limited to one-dimensional groups and analog scanning devices.

In much recent work this problem is avoided by using properties already invariant under these transformations. Thus a property might count the number of connected components in a picture—this is invariant under size and position. Or a property may count the number of vertical lines in a picture—this is invariant under size and position (but not rotation).

F. GENERATING PROPERTIES

The problem of generating useful properties has been discussed by Selfridge (1955); we shall summarize his approach. The machine is given, at the start, a few basic transformations $A_1, \cdots, A_n$, each of which transforms, in some significant way, each figure into another figure. A_1 might, for example, remove all points *not on a boundary* of a solid region; A_2 might leave only *vertex* points; A_3 might *fill up hollow regions*, etc. (see Fig. 5). Each sequence $A_{i_1} A_{i_2} \cdots A_{i_k}$ of these forms a new transformation, so that there is available an infinite variety. We provide the machine also with one or more "terminal" operations which convert a picture into a number, so that any sequence of the elementary transformations, followed by a terminal operation, defines a property. (Dineen 1955 describes how these processes were programmed in a digital computer.) We can start with a few short sequences, perhaps chosen randomly. Selfridge describes how the machine might learn new useful properties.

> We now feed the machine A's and O's telling the machine each time which letter it is. Beside each sequence under the two letters, the machine builds up distribution functions from the results of applying the sequences to the image. Now, since the sequences were chosen completely randomly,

---

where G is the group and μ the measure. By substituting $T_\beta(F)$ for F in this, one can see that the result is independent of choice of β since we obtain the same integral over $G\beta^{-1} = G$.

[5]See p. 160 ff. of Wiener (1948).

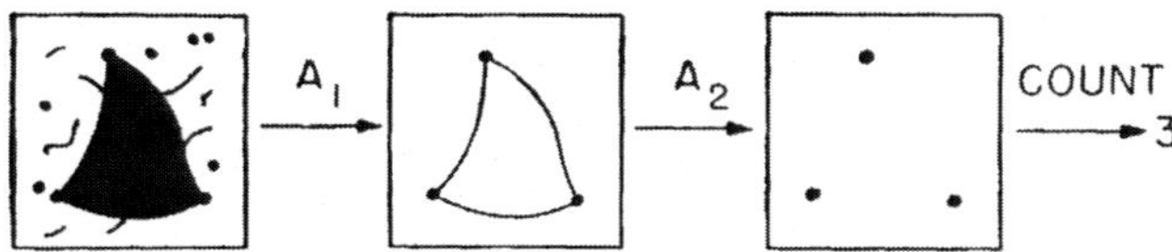

Figure 5. An arbitrary sequence of picture-transformations, followed by a numerical-valued function, can be used as a *property* function for pictures. A_1 removes all points which are not at the edge of a solid region. A_2 leaves only vertex points—at which an arc suddenly changes direction. The function C simply counts the number of points remaining in the picture. All remarks in the text could be generalized to apply to properties, like A_1A_2C, which can have more than two values.

it may well be that most of the sequences have very flat distribution functions; that is, they [provide] no information, and the sequences are therefore [by definition] not significant. Let it discard these and pick some others. Sooner or later, however, some sequences will prove significant; that is, their distribution functions will peak up somewhere. What the machine does now is to build up new sequences *like* the significant ones. This is the important point. If it merely chose sequences at random it might take a very long while indeed to find the best sequences. Built with some successful sequences, or partly successful ones, to guide it, we hope that the process will be much quicker. The crucial question remains: how do we build up sequences "like" other sequences, but not identical? As of now we think we shall merely build sequences from the transition frequencies of the significant sequences. We shall build up a matrix of transition frequencies from the significant ones, and use those as transition probabilities with which to choose new sequences.

We do not claim that this method is necessarily a very good way of choosing sequences—only that it should do better than not using at all the knowledge of what kind of sequences has worked. It has seemed to us that this is the crucial point of learning.[6]

---

[6]See p. 93 of Selfridge (1955).

It would indeed be remarkable if this failed to yield properties more useful than would be obtained from completely random sequence selection. The generating problem is discussed further in Minsky (1956). Newell, Shaw, and Simon (1960a) describe more deliberate, less statistical, techniques that might be used to discover sets of properties appropriate to a given problem area. One may think of the Selfridge proposal as a system which uses a finite-state language to describe its properties. Solomonoff (1957, 1960b) proposes some techniques for discovering common features of a set of expressions, *e.g.*, of the descriptions of those properties of already established utility; the methods can then be applied to generate new properties with the same common features. I consider the lines of attack in Selfridge (1955), Newell, Shaw, and Simon (1960a), and Solomonoff (1960b, 1957), although still incomplete, to be of the greatest importance.

G. COMBINING PROPERTIES

One cannot expect easily to find a *small* set of properties which will be just right for a problem area. It is usually much easier to find a large set of properties each of which provides a little useful information. Then one is faced with the problem of finding a way to combine them to make the desired distinctions. The simplest method is to choose, for each class, a typical character (a particular sequence of property values) and then to use some matching procedure, *e.g.*, counting the numbers of agreements and disagreements, to compare an unknown with these chosen "Character prototypes." The linear weighting scheme described just below is a slight generalization on this. Such methods treat the properties as more or less independent evidence for and against propositions; more general procedures (about which we have yet little practical information) must account also for nonlinear relations between properties, *i.e.*, must contain weighting terms for joint subsets of property values.

"Bayes nets" refers to a precursor of the more famous "Bayesian networks" described by Judea Pearl (1986).

1) *"Bayes nets" for combining independent properties:* We consider a single experiment in which an object is placed in front of a property-list machine. Each property E_i will have a value, 0 or 1. Suppose that there has been defined some set of "object classes" F_j, and that we

want to use the outcome of this experiment to decide in which of these classes the object belongs.

Assume that the situation is basically probabilistic, and that we know the probability p_{ij} that, if the object is in class F_j then the ith property E_i will have value 1. Assume further that these properties are independent; that is, even given F_j, knowledge of the value of E_i tells us nothing more about the value of a different E_k in the same experiment. (This is a strong condition—see below.) Let ϕ_j be the absolute probability that an object is in class F_j. Finally, for this experiment define V to be the particular set of i's for which the E_i's are 1. Then this V represents the Character of the object. From the definition of conditional probability, we have

$$\Pr(F_j, V) = \Pr(V) \cdot \Pr(F_j|V) = \Pr(F_j) \cdot \Pr(V|F_j).$$

Given the Character V, we want to guess which F_j has occurred (with the least chance of being wrong—the so-called *maximum likelihood* estimate); that is, for which j is $\Pr(F_j|V)$ the largest? Since in the above $\Pr(V)$ does not depend on j, we have only to calculate for which j is $\Pr(F_j) \cdot \Pr(V|F_j) = \phi_j \Pr(V|F_j)$ the largest. Hence, by our independence hypothesis, we have to maximize

$$\phi \cdot \prod_{i \in V} p_{ij} \cdot \prod_{i \in \bar{V}} q_{ij} = \phi_j \prod_{i \in V} \frac{p_{ij}}{q_{ij}} \cdot \prod_{\text{all } i} q_{ij}. \tag{1}$$

These "maximum likelihood" decisions can be made (Fig. 6) by a simple network device.[7]

Oliver Selfridge's Pandemonium system, developed in the 1950s, was one of the first AI systems for visual pattern recognition. Its parallel and distributed processing architecture had some notable similarities with artificial neural networks developed much later.

These nets resemble the general schematic diagrams proposed in the "Pandemonium" model of Selfridge (1959, see his Fig. 3). It is proposed there that some intellectual processes might be carried out by a hierarchy of simultaneously functioning submachines suggestively

[7] At the cost of an additional network layer, we may also account for the possible cost g_{jk} that would be incurred if we were to assign to F_k a figure really in class F_j: in this case the minimum cost decision is given by the k for which

$$\sum_j g_{jk} \phi_j \prod_{i \in V} p_{ij} \prod_{i \in \overline{V}} q_{ij}$$

is the least. $\overline{V}$ is the complement set to $V_i q_{ij}$ is $(1 - p_{ij})$.

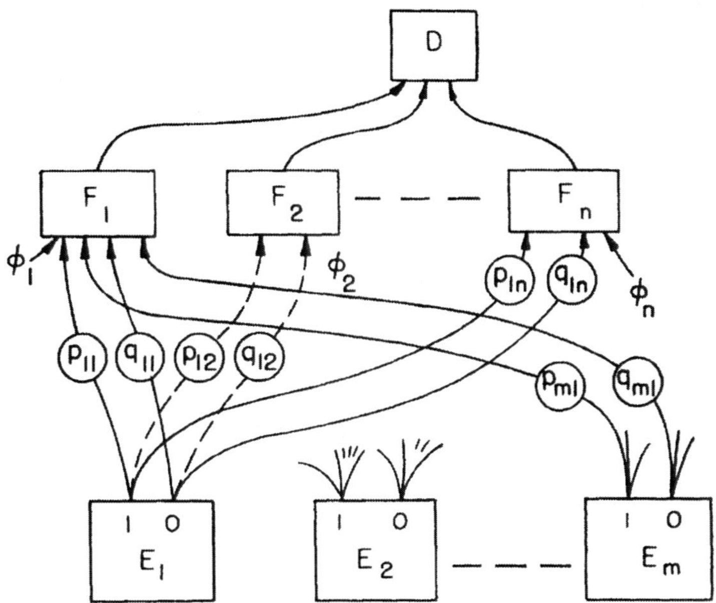

Figure 6. "Net" model for maximum-likelihood decisions based on linear weightings of property values. The input data are examined by each "property filter" E_i. Each E_i has "0" and "1" output channels, one of which is excited by each input. These outputs are weighted by the corresponding p_{ij}'s, as shown in the text. The resulting signals are multiplied in the F_j units, each of which "collects evidence" for a particular figure class. (We could have used here $\log(p_{ij})$, and *added* at the F_j units.) The final decision is made by the topmost unit D, who merely chooses that F_j with the largest score. Note that the logarithm of the coefficient p_{ij}/q_{ij} in the second expression of (1) can be construed as the "weight of the evidence" of E_i in favor of F_j. (See also Papert (1961) and Rosenblatt (1958).)

called "demons." Each unit is set to detect certain patterns in the activity of others and the output of each unit announces the degree of confidence of that unit that it sees what it is looking for. Our E_i units are Selfridge's "data demons." Our units F_j are his "cognitive demons"; each collects from the abstracted data evidence for a specific proposition. The topmost "decision demon" D responds to that one in the multitude below it whose shriek is the loudest.[8]

It is quite easy to add to this "Bayes network model" a mechanism which will enable it to *learn* the optimal connection weightings. Imagine that, after each event, the machine is told which F_j has occurred; we could implement this by sending back a signal along the connections

[8]See also the report in Selfridge and Neisser (1960).

leading to that F_j unit. Suppose that the connection for p_{ij} (or q_{ij}) contains a two-terminal device (or "synapse") which stores a number w_{ij}. Whenever the joint event $(F_j, E_i = 1)$ occurs, we modify w_{ij} by replacing it by $(w_{ij} + 1)\theta$, where θ is a factor slightly less than unity. And when the joint event $(F_j, E_i = 0)$ occurs, we decrement w_{ij} by replacing it with $(w_{ij})\theta$. It is not difficult to show that the expected values of the w_{ij}'s will become proportional to the p_{ij}'s (and, in fact, approach $p_{ij}[\theta/(1 - \theta)]$). Hence, the machine tends to learn the optimal weighting on the basis of experience. (One must put in a similar mechanism for estimating the ϕ_j's.) The variance of the normalized weight $w_{ij}[(1 - \theta)/\theta]$ approaches $[(1 - \theta)/(1 + \theta)]p_{ij}q_{ij}$. Thus a small value for θ means rapid learning but is associated with a large variance, hence, with low reliability. Choosing θ close to unity means slow, but reliable, learning. θ is really a sort of memory decay constant, and its choice must be determined by the noise and stability of the environment—much noise requires long averaging times, while a changing environment requires fast adaptation. The two requirements are, of course, incompatible and the decision has to be based on an economic compromise.[9]

2) *Possibilities of using random nets for Bayes decisions:* The nets of Fig. 6 are very orderly in structure. Is all this structure necessary? Certainly if there were a great many properties, *each of which provided very little marginal information*, some of them would not be missed. Then one might expect good results with a mere sampling of all the possible connection paths w_{ij}. And one might thus, *in this special situation*, use a random connection net.

The two-layer nets here resemble those of the "perceptron" proposal of Rosenblatt (1958). In the latter, there is an additional level of connections coming directly from randomly selected points of a "retina." Here the properties, the devices which abstract the visual input data, are simple functions which add some inputs, subtract others, and detect whether the result exceeds a threshold. Eq. (1), we think, illustrates what is of value in this scheme. It does seem clear that

[9]See also Minsky and Selfridge (1961) and Papert (1961).

a maximum-likelihood type of analysis of the output of the property functions can be handled by such nets. But these nets, with their simple, randomly generated, connections can probably never achieve recognition of such patterns as "the class of figures having two separated parts," and they cannot even achieve the effect of template recognition without size and position normalization (unless sample figures have been presented previously in essentially all sizes and positions). For the chances are extremely small of finding, by random methods, enough properties usefully correlated with patterns appreciably more abstract than those of the prototype-derived kind. And these networks can really only separate out (by weighting) information in the individual input properties; they cannot extract further information present in nonadditive form. The "perceptron" class of machines have facilities neither for obtaining better-than-chance properties nor for assembling better-than-additive combinations of those it gets from random construction.[10]

For recognizing *normalized* printed or hand-printed characters, single-point properties do surprisingly well (Highleyman and Kamentsky 1960); this amounts to just "averaging" many samples. Bledsoe and Browning (1959) claim good results with point-pair properties. Roberts (1960) describes a series of experiments in this general area. Doyle (1959) without normalization but with quite sophisticated properties obtains excellent results; his properties are already substantially size- and position-invariant. A general review of Doyle's work and other pattern-recognition experiments will be found in Selfridge and Neisser (1960).

For the complex discrimination, *e.g.*, between one and two connected objects, the property problem is very serious, especially for long wiggly objects such as are handled by Kirsch *et al.* (1957). Here some kind of recursive processing is required and combinations of simple properties would almost certainly fail even with large nets and long training.

We should not leave the discussion of some decision net models without noting their important limitations. The hypothesis that, for given j,

[10]See also Roberts (1960), Papert (1961), and Rosenblatt (1958). We can find nothing resembling an analysis [see (1) above] in Rosenblatt (1958) or subsequent publications of Rosenblatt.

the p_{ij} represent independent events, is a very strong condition indeed. Without this hypothesis we could still construct maximum-likelihood nets, but we would need an additional layer of cells to represent all of the joint events V; that is, we would need to know all the Pr $(F_j|V)$. This gives a general (but trivial) solution, but requires 2^n cells for n properties, which is completely impractical for large systems. What is required is a system which computes some sampling of all the joint conditional probabilities, and uses these to estimate others when needed. The work of Uttley (1956, 1959), bears on this problem, but his proposed and experimental devices do not yet clearly show how to avoid exponential growth.[11]

H. ARTICULATION AND ATTENTION—LIMITATIONS OF THE PROPERTY-LIST METHOD

Because of its fixed size, the property-list scheme is limited (for any given set of properties) in the detail of the distinctions it can make. Its ability to deal with a compound scene containing several objects is critically weak, and its direct extensions are unwieldy and unnatural. If a machine can recognize a chair and a table, it surely should be able to tell us that "there is a chair and a table." To an extent, we can invent properties which allow some capacity for superposition of object Characters.[12] But there is no way to escape the information limit.

What is required is clearly 1) a *list (of whatever length is necessary)* of the primitive objects in the scene and 2) a statement about the relations among them. Thus we say of Fig. 7(a), "A rectangle (1) contains two subfigures disposed horizontally. The part on the left is a rectangle (2) which contains two subfigures disposed vertically; the upper a circle (3) and the lower a triangle (4). The part on the right . . . etc." Such a description entails an ability to separate or "articulate" the scene into parts. (Note that in this example the articulation is essentially *recursive*; the figure is first divided into two parts; then each part is described using the same machinery.) We can formalize this kind of description in an expression language whose fundamental grammatical

[11] See also Papert (1961).

[12] Cf. Mooers' technique of Zatocoding (1956; 1956).

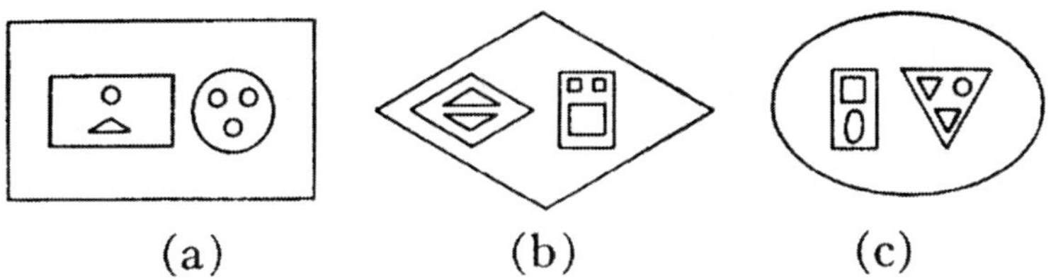

Figure 7. The picture (a) is first described verbally in the text. Then, by introducing notation for the relations "inside of," "to the left of" and "above," we construct a symbolic description. Such descriptions can be formed and manipulated by machines. By abstracting out the complex relation between the parts of the figure we can use the same formula to describe the related pictures (b) and (c), changing only the list of primitive parts. It is up to the programmer to decide at just what level of complexity a part of a picture should be considered "primitive"; this will depend on what the description is to be used for. We could further divide the drawings into vertices, lines, and arcs. Obviously, for some applications the relations would need more metrical information, *e.g.*, specification of lengths or angles.

form is a pair (R, L) whose first member R names a *relation* and whose second member L is an *ordered list* $(x_1, x_2, \cdots, x_n)$ of the objects or subfigures which bear that relation to one another. We obtain the required flexibility by allowing the members of the list L to contain not only the names of "elementary" figures but also "subexpressions" of the form (R, L) designating complex subfigures. Then our scene above may be described by the expression

$$[\odot, (\square, (\rightarrow, \{(\odot, (\square, (\downarrow, (\bigcirc, \triangle)))), (\odot, (\bigcirc, (\nabla, (\bigcirc, \quad \bigcirc, \bigcirc))))\}))]$$

where $(\odot, (x, y))$ means that y is contained in x; $(\rightarrow, (x, y))$ means that y is to the right of x; $(\downarrow, (x, y))$ means that y is below x, and $(\nabla, (x, y, z))$ means that y is to the right of x and z is underneath and between them. The symbols $\square$, $\bigcirc$, and $\triangle$ represent the indicated kinds of primitive geometric objects. This expression-pair description language may be regarded as a simple kind of "list-structure" language. Powerful computer techniques have been developed, originally by Newell, Shaw and Simon, for manipulating symbolic expressions in such languages for purposes of heuristic programming. (See the remarks at the end of Section IV. If some of the members of a list are themselves lists, they must be surrounded by exterior parentheses, and this accounts for the accumulation of parentheses.)

Such "list-structure" languages are exemplified by the LISP programming language developed by AI pioneer John McCarthy in 1958. LISP—short for LISt Processing—was the most popular programming language in AI for decades. (As Minsky notes, LISP's hierarchical list structures resulted in "an accumulation of parentheses"; it was often joked that LISP stood for "Lots of Idiotic Stupid Parentheses.")

It may be desirable to construct descriptions in which the complex relation is extracted, *e.g.*, so that we have an expression of the form FG where F is an expression which at once denotes the composite relation between all the primitive parts listed in G. A complication arises in connection with the "binding" of variables, *i.e.*, in specifying the manner in which the elements of G participate in the relation F. This can be handled in general by the "λ" notation (McCarthy 1960) but here we can just use integers to order the variables.

For the given example, we could describe the relational part F by an expression

$$\odot(1, \rightarrow (\odot(2, \downarrow (3,4)), \odot(5, \nabla(6,7,8))))$$

in which we now use a "functional notation": "$(\odot, (x, y))$" is replaced by "$\odot(x, y)$," etc., making for better readability. To obtain the desired description, this expression has to be applied to an ordered list of primitive objects, which in this case is $(\square, \square, \circ, \triangle, \circ, \circ, \circ, \circ)$. This composite functional form allows us to abstract the composite relation. By changing only the object list we can obtain descriptions also of the objects in Fig. 7(b) and 7(c).

The important thing about such "articular" descriptions is that they can be obtained by *repeated application of a fixed set of pattern-recognition techniques*. Thus we can obtain *arbitrarily complex* descriptions from a fixed complexity classification-mechanism. The new element required in the mechanism (beside the capacity to manipulate the list-structures) is the ability to articulate—to "attend fully" to a selected part of the picture and bring all one's resources to bear on that part. In efficient problem-solving programs, we will not usually complete such a description in a single operation. Instead, the depth or detail of description will be under the control of other processes. These will reach deeper, or look more carefully, only when they have to, *e.g.*, when the presently available description is inadequate for a current goal. The author, together with L. Hodes, is working on pattern-recognition schemes using articular descriptions. By manipulating the formal descriptions we can deal with overlapping and incomplete figures, and several other problems of the "Gestalt" type.

It seems likely that as machines are turned toward more difficult problem areas, *passive* classification systems will become less adequate, and we may have to turn toward schemes which are based more on internally-generated hypotheses, perhaps "error-controlled" along the lines proposed by MacKay (1956).

Space requires us to terminate this discussion of pattern-recognition and description. Among the important works not reviewed here should be mentioned those of Bomba (1959) and Grimsdale *et al.* (1959), which involve elements of description, Unger (1959) and Holland (1960) for parallel processing schemes, Hebb (1949) who is concerned with physiological description models, and the work of the Gestalt psychologists, notably Köhler (1947), who have certainly raised, if not solved, a number of important questions. Sherman (1960), Haller (1959) and others have completed programs using line-tracing operations for topological classification. The papers of Selfridge (1955, 1956) have been a major influence on work in this general area.

See also Kirsch *et al.* (1957) for discussion of a number of interesting computer image-processing techniques, and see Minot (1959) and Stevens (1957) for reviews of the reading machine and related problems. One should also examine some biological work, *e.g.*, Tinbergen (1951) to see instances in which some discriminations which seem, at first glance very complicated are explained on the basis of a few apparently simple properties arranged in simple decision trees.

III. Learning Systems

Summary—In order to solve a new problem, one should first try using methods similar to those that have worked on similar problems. To implement this "basic learning heuristic" one must generalize on past experience, and one way to do this is to use success-reinforced decision models. These learning systems are shown to be averaging devices. Using devices which learn also which events are associated with reinforcement, *i.e.*, reward, we can build more autonomous "secondary reinforcement" systems. In applying such methods to complex problems, one

> encounters a serious difficulty—in distributing credit for success of a complex strategy among the many decisions that were involved. This problem can be managed by arranging for local reinforcement of partial goals within a hierarchy, and by grading the training sequence of problems to parallel a process of maturation of the machine's resources.

In order to solve a new problem one uses what might be called the basic learning heuristic—first try using methods similar to those which have worked, in the past, on similar problems. We want our machines, too, to benefit from their past experience. Since we cannot expect new situations to be precisely the same as old ones, any useful learning will have to involve generalization techniques. There are too many notions associated with "learning" to justify defining the term precisely. But we may be sure that any useful learning system will have to use records of the past as *evidence for more general propositions*; it must thus entail some commitment or other about "inductive inference." (See Section V-B.) Perhaps the simplest way of generalizing about a set of entities is through constructing a new one which is an "ideal," or rather, a typical member of that set; the usual way to do this is to smooth away variation by some sort of averaging technique. And indeed we find that most of the *simple* learning devices do incorporate some averaging technique—often that of averaging some sort of product, thus obtaining a sort of correlation. We shall discuss this family of devices here, and some more abstract schemes in Section V.

Humanlike generalization remains the most important unsolved problem in AI.

A. REINFORCEMENT

A reinforcement process is one in which some aspects of the behavior of a system are caused to become more (or less) prominent in the future as a consequence of the application of a "reinforcement operator" Z. This operator is required to affect only those aspects of behavior for which instances have actually occurred recently.

The analogy is with "reward" or "extinction" (not punishment) in animal behavior. The important thing about this kind of process is that it is "operant" (a term of Skinner 1953); the reinforcement operator does not initiate behavior, but merely selects that which the Trainer

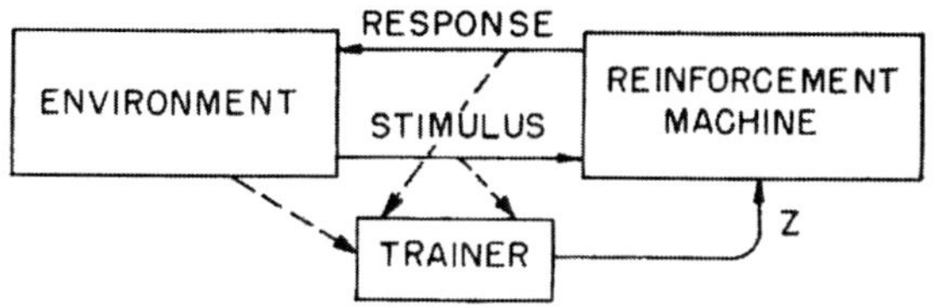

Figure 8. Parts of an "operant reinforcement" learning system. In response to a stimulus from the environment, the machine makes one of several possible responses. It remembers what decisions were made in choosing this response. Shortly thereafter, the Trainer sends to the machine positive or negative reinforcement (reward) signal; this increases or decreases the tendency to make the same decisions in the future. Note that the Trainer need not know how to solve problems, but only how to detect success or failure, or relative improvement; his function is selective. The Trainer might be connected to observe the actual stimulus-response activity or, in a more interesting kind of system, just some function of the state of the environment.

likes from that which has occurred. Such a system must then contain a device M which generates a variety of behavior (say, in interacting with some environment) and a Trainer who makes critical judgments in applying the available reinforcement operators. (See Fig. 8.)

Let us consider a very simple reinforcement model. Suppose that on each presentation of a stimulus S an animal has to make a choice, *e.g.*, to turn left or right, and that its probability of turning right, at the nth trial, is p_n. Suppose that *we* want it to turn right. Whenever it does this we might "reward" it by applying the operator Z_+;

$$p_{n+l} = Z_+(p_n) = \theta p_n + (1 - \theta) \qquad 0 < \theta < 1$$

which moves p a fraction $(1-\theta)$ of the way towards unity.[13] If we dislike what it does we apply negative reinforcement,

$$p_{n+l} = Z_-(p_n) = \theta p_n$$

moving p the same fraction of the way toward 0. Some theory of such "linear" learning operators, generalized to several stimuli and responses, will be found in Bush and Mosteller (1955). We can show that the learning result is an average weighted by an exponentially-decaying time factor: Let Z_n be ± 1 according to whether the nth event is rewarded or extinguished and replace p_n by $c_n = 2p_n - 1$ so that

[13]Properly, the reinforcement functions should depend both on the p's and on the previous reaction—reward should *decrease* p if our animal has just turned to the left. The notation in the literature is also somewhat confusing in this regard.

$-1 \leq c_n \leq 1$, as for a correlation coefficient. Then (with $c_0 = 0$) we obtain by induction

$$c_{n+1} = (1-\theta)\sum_{i=0}^{n} \theta^{n-i} Z_i,$$

and since

$$1/(1-\theta) \approx \sum_{0}^{n} \theta^{n-i},$$

we can write this as

$$c_{n+1} \approx \frac{\sum \theta^{n-i} Z_i}{\sum \theta^{n-i}}. \quad (1)$$

If the term Z_i is regarded as a product of i) how the creature responded and ii) which kind of reinforcement was given, then c_n is a kind of correlation function (with the decay weighting) of the joint behavior of these quantities. The ordinary, uniformly-weighted average has the same general form but with time-dependent θ:

$$c_{n+1} = \left(1 - \frac{1}{N}\right) c_n + \frac{1}{N} Z_n. \quad (2)$$

In (1) we have again the situation described in Section II-G, 1; a small value of θ gives fast learning, and the possibility of quick adaptation to a changing environment. A near-unity value of θ gives slow learning, but also smooths away uncertainties due to noise. As noted in Section II-G, 1, the response distribution comes to approximate the probabilities of rewards of the alternative responses. (The importance of this phenomenon has, I think, been overrated; it is certainly not an especially rational strategy. One reasonable alternative is that of computing the numbers p_{ij} as indicated, but actually playing at each trial the "most likely" choice. Except in the presence of a hostile opponent, there is usually no reason to play a "mixed" strategy.[14])

In Samuel's coefficient-optimizing program (Samuel 1959, see Section III-C, 1), there is a most ingenious compromise between the exponential and the uniform averaging methods: the value of N in

Arthur Samuel developed a checkers-playing program in the late 1950s that was one of the first machine learning systems. This program combined tree-search with an evaluation function that assigned a value to a given board configuration using a weighted sum of board attributes (e.g., the relative piece advantage of each side). Learning consisted of changing the coefficients in this weighted sum. The program learned by playing games against itself, and eventually became a moderately good player. This was a remarkable achievement, especially given the CPU and memory limitations available at the time.

[14]The question of just how often one should play a strategy different from the estimated optimum, in order to gain information, is an underlying problem in many fields. See, *e.g.*, Shubik (1960).

(2) above begins at 16 and so remains until $n = 16$, then N is 32 until $n = 32$, and so on until $n = 256$. Thereafter N remains fixed at 256. This nicely prevents violent fluctuations in c_n at the start, approaches the uniform weighting for a while, and finally approaches the exponentially-weighted correlation, all in a manner that requires very little computation effort! Samuel's program is at present the outstanding example of a game-playing program which matches average human ability, and its success (in real time) is attributed to a wealth of such elegancies, both in heuristics and in programming.

The problem of extinction or "unlearning" is especially critical for complex, hierarchical, learning. For, once a generalization about the past has been made, one is likely to build upon it. Thus, one may come to select certain properties as important and begin to use them in the characterization of experience, perhaps storing one's memories in terms of them. If later it is discovered that some other properties would serve better, then one must face the problem of translating, or abandoning, the records based on the older system. This may be a very high price to pay. One does not easily give up an old way of looking at things, if the better one demands much effort and experience to be useful. Thus the *training sequences* on which our machines will spend their infancies, so to speak, must be chosen very shrewdly to insure that early abstractions will provide a good foundation for later difficult problems.

Determining an optimal ordering of training examples or experiences—today called "curriculum learning"—is still an active area of research in machine learning.

Incidentally, in spite of the space given here for their exposition, I am not convinced that such "incremental" or "statistical" learning schemes should play a central role in our models. They will certainly continue to appear as components of our programs but, I think, mainly by default. The more intelligent one is, the more often he should be able to learn from an experience something rather definite; *e.g.*, to reject or accept a hypothesis, or to change a goal. (The obvious exception is that of a truly statistical environment in which averaging is inescapable. But the heart of problem-solving is always, we think, the combinatorial part that gives rise to searches, and we should usually be able to regard the complexities caused by "noise" as mere annoyances, however irritating they may be.) In this connection we can refer to the discussion of memory in Miller,

Throughout the early decades of AI research, Minsky and other important figures were dismissive of work on statistical learning. However, starting in the late 1980s, statistical learning methods (including deep learning) have come to dominate AI research.

I'm not sure what Minsky means by "noise" here, but certainly the "complexities" of problem solving turned out to be more significant than any of the early AI proponents expected.

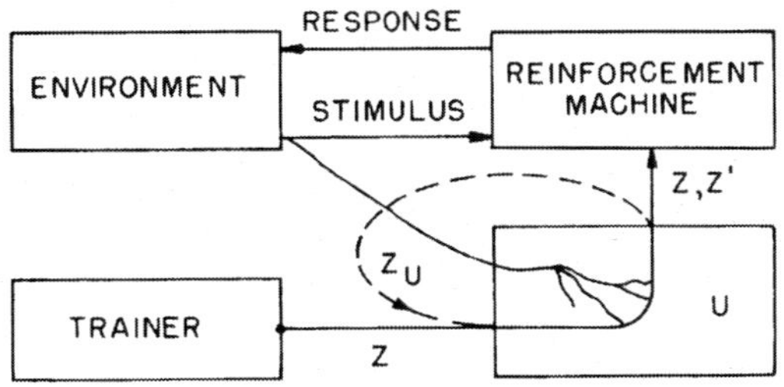

Figure 9. An additional device U gives the machine of Fig. 8 the ability to learn which signals from the environment have been associated with reinforcement. The primary reinforcement signals Z are routed through U. By a Pavlovian conditioning process (not described here), external signals come to produce reinforcement signals like those that have frequently succeeded them in the past. Such signals might be abstract, *e.g.*, verbal encouragement. If the "secondary reinforcement" signals are allowed, in turn, to acquire further external associations (through, *e.g.*, a channel Z_U as shown) the machine might come to be able to handle chains of subproblems. But something must be done to stabilize the system against the positive symbolic feedback loop formed by the path Z_U. The profound difficulty presented by this stabilization problem may be reflected in the fact that, in lower animals, it is very difficult to demonstrate such chaining effects.

Galanter, and Pribram (1960).[15] This seems to be the first major work in Psychology to show the influence of work in the artificial intelligence area, and its programme is generally quite sophisticated.

B. SECONDARY REINFORCEMENT AND EXPECTATION MODELS

The simple reinforcement system is limited by its dependence on the Trainer. If the Trainer can detect only the *solution* of a problem, then we may encounter "mesa" phenomena which will limit performance on difficult problems. (See Section I-C.) One way to escape this is to have the machine learn to generalize on what the Trainer does. Then, in difficult problems, it may be able to give itself partial reinforcements along the way, *e.g.*, upon the solution of relevant subproblems. The machine in Fig. 9 has some such ability. The new unit U is a device that learns which external stimuli are strongly correlated with the various reinforcement signals, and responds to such stimuli by reproducing the corresponding reinforcement signals. (The device U is *not* itself a reinforcement learning device; it is more like a "Pavlovian" conditioning device, treating the Z signals as "unconditioned" stimuli and the S signals as conditioned stimuli.) The heuristic idea is that any signal from the environment which in the past

[15]See especially ch. 10.

has been well correlated with (say) positive reinforcement is likely to be an indication that something good has just happened. If the training on early problems was such that this is realistic, then the system eventually should be able to detach itself from the Trainer, and become autonomous. If we further permit "chaining" of the "secondary reinforcers," *e.g.*, by admitting the connection shown as a dotted line in Fig. 9, the scheme becomes quite powerful, in principle. There are obvious pitfalls in admitting such a degree of autonomy; the values of the system may drift to a "nonadaptive" condition.

C. PREDICTION AND EXPECTATION

The evaluation unit U is supposed to acquire an ability to tell whether a situation is good or bad. This evaluation could be applied to *imaginary* situations as well as to real ones. If we could estimate the consequences of a proposed action (without its actual execution), we could use U to evaluate the (estimated) resulting situation. This could help in reducing the effort in search, and we would have in effect a machine with some ability to look ahead, or *plan*. In order to do this we need an additional device P which, given the descriptions of a situation and an action, will predict a description of the likely result. (We will discuss schemes for doing this in Section IV-C.) The device P might be constructed along the lines of a reinforcement learning device. In such a system the required reinforcement signals would have a very attractive character. For the machine must reinforce P positively when the *actual outcome resembles that which was predicted*—accurate expectations are rewarded. If we could further add a premium to reinforcement of those predictions which have a novel aspect, we might expect to discern behavior motivated by a sort of curiosity. In the reinforcement of mechanisms for confirmed novel expectations (or new explanations) we may find the key to simulation of intellectual motivation.[16]

Samuel's Program for Checkers: In Samuel's "generalization learning" program for the game of checkers (Samuel 1959) we find a novel heuristic technique which could be regarded as a simple example of the "expectation reinforcement" notion. Let us review very briefly the situation in playing

[16] See also ch. 6 of Minsky (1954).

two-person board games of this kind. As noted by Shannon (1956) such games are in principle finite, and a best strategy can be found by following out all possible continuations—if he goes there I can go there, or there, etc.—and then "backing-up" or "minimaxing" from the terminal positions, won, lost, or drawn. But in practice the full exploration of the resulting colossal "move-tree" is out of the question. No doubt, some exploration will always be necessary for such games. But the tree must be pruned. We might simply put a limit on depth of exploration—the number of moves and replies. We might also limit the number of alternatives explored from each position—this requires some heuristics for selection of "plausible moves."[17] Now, if the backing-up technique is still to be used (with the incomplete move-tree) one has to substitute for the absolute "win, lose, or draw" criterion some other "static" way of evaluating nonterminal positions."[18] (See Fig. 10.) Perhaps the simplest scheme is to use a weighted sum of some selected set of "property" functions of the positions—mobility, advancement, center control, and the like. This is done in Samuel's program, and in most of its predecessors. Associated with this is a multiple-simultaneous-optimizer method for discovering a good coefficient assignment (using the correlation technique noted in Section III-A). But the source of reinforcement signals in Samuel (1959) is novel. One cannot afford to play out one or more entire games for each single learning step. Samuel measures instead *for each move* the difference between what the evaluation function yields *directly* of a position and what it *predicts* on the basis of an extensive continuation exploration, *i.e.*, backing-up. The signal of this error, "Delta," is used for reinforcement; thus the system may learn something at *each move*.[19]

---

[17]See the discussion of Bernstein *et al.* (1958) and the more extensive review and discussion in the very suggestive paper of Newell, Shaw, and Simon (1958a); one should not overlook the pioneering paper of Newell (1955), and Samuel's discussion of the minimaxing process in 1959.

[18]In some problems the backing-up process can be handled in closed analytic form so that one may be able to use such methods as Bellman's "Dynamic Programming" (1957). Freimer (1960) gives some examples for which limited "look-ahead" doesn't work.

[19]It should be noted that Samuel (1959) describes also a rather successful checker-playing program based on recording and retrieving information about positions encountered in the past, a less abstract way of exploiting past experience. Samuel's work is notable in the variety of experiments that were performed, with and without

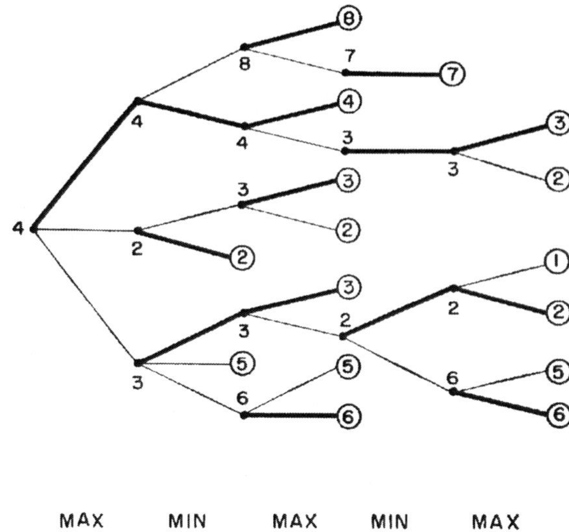

Figure 10. "Backing-up" the static evaluations of proposed moves in a game-tree. From the vertex at the left, representing the present position in a board game, radiate three branches, representing the *player's* proposed moves. Each of these might be countered by a variety of *opponent* moves, and so on. According to some program, a finite tree is generated. Then the worth to the player of each terminal board position is estimated. (See text.) If the opponent has the same values, he will choose to minimize the score, while the player will always try to maximize. The heavy lines show how this minimaxing process backs up until a choice is determined for the present position.

The full tree for chess has the order of 10^{120} branches—beyond the reach of any man or computer. There is a fundamental heuristic exchange between the effectiveness of the evaluation function and the extent of the tree. A very weak evaluation (*e.g.*, one which just compares the players' values of pieces) would yield a devastating game if the machine could explore all continuations out to, say, 20 levels. But only 6 levels, roughly within range of our presently largest computers, would probably not give a brilliant game; less exhaustive strategies, perhaps along the lines of Newell, Shaw, and Simon (1958a), would be more profitable.

D. THE BASIC CREDIT-ASSIGNMENT PROBLEM FOR COMPLEX REINFORCEMENT LEARNING SYSTEMS

In playing a complex game such as chess or checkers, or in writing a computer program, one has a definite success criterion—the game is won or lost. But in the course of play, each ultimate success (or failure) is associated with a vast number of internal decisions. If the run is successful, how can we assign credit for the success among the multitude of decisions? As Newell noted,

---

various heuristics. This gives an unusual opportunity to really find out how different heuristic methods compare. More workers should choose (other things being equal) problems for which such variations are practicable.

> It is extremely doubtful whether there is enough information in "win, lose, or draw" when referred to the whole play of the game to permit any learning at all over available time scales. . . . For learning to take place, each play of the game must yield much more information. This is . . . achieved by breaking the problem into components. The unit of success is the goal. If a goal is achieved, its subgoals are reinforced; if not they are inhibited. (Actually, what is reinforced is the transformation rule that provided the subgoal.) . . . This also is true of the other kinds of structure: every tactic that is created provides information about the success or failure of tactic search rules; every opponent's action provides information about success or failure of likelihood inferences; and so on. The amount of information relevant to learning increases directly with the number of mechanisms in the chess-playing machine.[20]

We are in complete agreement with Newell on this approach to the problem.[21]

This "impression" is a precursor to Minsky's pessimistic views on neural networks, discussed at length in his 1969 book *Perceptrons*. Minsky and his coauthor Samuel Papert showed that perceptrons—single-layer neural networks (an input sequence fully connected to an output layer)—were limited to solving relatively simple problems. While admitting some of the virtues of perceptrons, Minsky and Papert speculated that "there is no reason to suppose that any of these virtues carry over to the many-layered version," noting the credit assignment problem as the primary obstacle. Their speculation has been decisively disproven: the backpropagation algorithm for multilayered neural networks, invented in the 1960s and made more efficient in the 1980s, can now efficiently enable credit assignment in neural networks with hundreds of layers.

It is my impression that many workers in the area of "self-organizing" systems and "random neural nets" do not feel the urgency of this problem. Suppose that one million decisions are involved in a complex task (such as winning a chess game). Could we assign to each decision element one-millionth of the credit for the completed task? In certain special situations we can do just this—*e.g.*, in the machines of Rosenblatt (1958), Roberts (1960), and Farley and Clark (1954), etc., where the connections being reinforced are to a sufficient degree independent. But the problem-solving ability is correspondingly weak.

For more complex problems, with decisions in hierarchies (rather than summed on the same level) and with increments small enough to assure probable convergence, the running times would become fantastic. For complex problems we will have to define "success" in some

[20] See p. 108 of Newell (1955).

[21] See also the discussion in Samuel (1959, p. 2) on assigning credit for a change in "Delta."

rich local sense. Some of the difficulty may be evaded by using carefully graded "training sequences" as described in the following section.

Friedberg's Program-Writing Program: An important example of comparative failure in this credit-assignment matter is provided by the program of Friedberg (1958) and Friedberg, Dunham, and North (1959) to solve program-writing problems. The problem here is to write programs for a (simulated) very simple digital computer. A simple problem is assigned, *e.g.*, "compute the AND of two bits in storage and put the result in an assigned location." A generating device produces a random (64-instruction) program. The program is run and its success or failure is noted. The success information is used to reinforce *individual instructions* (in fixed locations) so that each success tends to increase the chance that the instructions of successful programs will appear in later trials. (We lack space for details of how this is done.) Thus the program tries to find "good" instructions, more or less independently, for each location in program memory. The machine did learn to solve some extremely simple problems. But it took of the order of 1000 times longer than pure chance would expect. In part II of Friedberg, Dunham, and North (1959), this failure is discussed, and attributed in part to what we called (Section I-C) the "Mesa phenomena." In changing just one instruction at a time the machine had not taken large enough steps in its search through program space.

The second paper goes on to discuss a sequence of modifications in the program generator and its reinforcement operators. With these, and with some "priming" (starting the machine off on the right track with some useful instructions), the system came to be only a little worse than chance. Friedberg, Dunham, and North (1959) conclude that with these improvements "the generally superior performance of those machines with a success-number reinforcement mechanism over those without does serve to indicate that such a mechanism can provide a basis for constructing a learning machine." I disagree with this conclusion. It seems to me that each of the "improvements" can be interpreted as serving only to increase the step size of the search, that is, the randomness of the mechanism; this helps to avoid the Mesa phenomenon and thus approach chance behavior. But it certainly does not show that the "learning mechanism" is working—one would want at least to see some better-than-chance results before arguing this point. The trouble, it seems, is with credit-assignment. The credit for a working program can only be assigned to functional groups of instructions, *e.g.*, subroutines, and as these operate in hierarchies we should not expect individual instruction reinforcement to work well.[22] It seems surprising that it was not recognized in Friedberg, Dunham, and North (1959) that the doubts raised earlier were probably justified! In the last section we see some real success obtained by breaking the problem into parts and solving them sequentially. (This successful demonstration using division into subproblems does not use any reinforcement mechanism at all.) Some experiments of similar nature are reported in Kilburn, Grimsdale, and Sumner (1960).

It is my conviction that no scheme for learning, or for pattern-recognition, can have very general utility unless there are provisions for recursive, or at least hierarchical, use of previous results. We cannot expect a learning system to come to handle very hard problems without preparing it with a reasonably graded sequence of problems of growing difficulty. The first problem must be one which can be solved in reasonable time with the initial resources. The next must be capable of solution in reasonable time by using reasonably simple and accessible combinations of methods developed in the first, and so on. The only alternatives to this use of an adequate "training sequence" are 1) advanced resources, given initially, or 2) the fantastic exploratory processes found perhaps only in the history of organic evolution.[23] And even there, if we accept the general view of Darlington (1958) who emphasizes the heuristic aspects of genetic systems, we must have developed early (in, *e.g.*, the phenomena of meiosis and crossing-over) quite highly specialized mechanisms providing for the

[22] See the introduction to Friedberg (1958) for a thoughtful discussion of the plausibility of the scheme.

[23] It should, however, be possible to construct learning mechanisms which can select for themselves reasonably good training sequences (from an always complex environment) by pre-arranging a relatively slow development (or "maturation") of the system's facilities. This might be done by pre-arranging that the sequence of goals attempted by the primary Trainer match reasonably well, at each stage, the complexity of performance mechanically available to the pattern-recognition and other parts of the system. One might be able to do much of this by simply limiting the depth of hierarchical activity, perhaps only later permitting limited recursive activity.

Work on developing such mechanisms continues in sub-areas of machine learning such as "active learning" and "metalearning."

segregation of groupings related to solutions of subproblems. Recently, much effort has been devoted to the construction of training sequences in connection with programming "teaching machines." Naturally, the psychological literature abounds with theories of how complex behavior is built up from simpler. In our own area, perhaps the work of Solomonoff (1957), while overly cryptic, shows the most thorough consideration of this dependency on *training sequences*.

IV. Problem-Solving and Planning

Summary—The solution, by machine, of really complex problems will require a variety of administration facilities. During the course of solving a problem, one becomes involved with a large assembly of interrelated subproblems. From these, at each stage, a very few must be chosen for investigation. This decision must be based on 1) estimates of relative difficulties and 2) estimates of centrality of the different candidates for attention. Following subproblem selection (for which several heuristic methods are proposed), one must choose methods appropriate to the selected problems. But for really difficult problems, even these step-by-step heuristics for reducing search will fail, and the machine must have resources for analyzing the problem structure in the large—in short, for "planning." A number of schemes for planning are discussed, among them the use of models—analogous, semantic, and abstract. Certain abstract models, "Character Algebras," can be constructed by the machine itself, on the basis of experience or analysis. For concreteness, the discussion begins with a description of a simple but significant system (LT) which encounters some of these problems.

A. THE "LOGIC THEORY" PROGRAM OF NEWELL, SHAW, AND SIMON

It is not surprising that the testing grounds for early work on mechanical problem-solving have usually been areas of mathematics, or games, in which the rules are defined with absolute clarity. The "Logic Theory" machine of Newell and Simon (1956) and Newell, Shaw, and Simon (1957), called "LT" below, was a first attempt to prove theorems in logic, by frankly heuristic methods. Although the program was not by human standards a brilliant success (and did not surpass its designers), it stands as a landmark both in heuristic programming and also in the development of modern automatic programming.

The Logic Theorist (LT) program, created in the 1950s by Allen Newell, J. C. Shaw, and Herbert Simon, was one of the first examples of a symbolic AI system. (At that time, Newell and Simon proposed that the field be named "complex information processing" instead of "artificial intelligence.") LT was a precursor of Newell and Simon's General Problem Solver (GPS), discussed later in this paper.

The problem domain here is that of discovering proofs in the Russell-Whitehead system for the propositional calculus. That system is given as a set of (five) axioms and (three) rules of inference; the latter specify how certain transformations can be applied to produce new theorems from old theorems and axioms.

The LT program is centered around the idea of "working backwards" to find a proof. Given a theorem *T* to be proved, LT searches among the axioms and previously established theorems for one from which *T* can be deduced by a single application of one of three simple "Methods" (which embody the given rules of inference). If one is found, the problem is solved. Or the search might fail completely. But finally, the search may yield one or more "problems" which are usually propositions from which *T* may be deduced directly. If one of these can, in turn, be proved a theorem the main problem will be solved. (The situation is actually slightly more complex.) Each such subproblem is adjoined to the "subproblem list" (after a limited preliminary attempt) and LT works around to it later. The full power of LT, such as it is, can be applied to each subproblem, for LT can use itself as a subroutine in a recursive fashion.

The heuristic technique of working backwards yields something of a teleological process, and LT is a forerunner of more complex systems which construct hierarchies of goals and subgoals. Even so, the basic administrative structure of the program is no more than a nested set of searches through lists in memory. We shall first outline this structure and then mention a few heuristics that were used in attempts to improve performance.

1) Take the next problem from problem list.
   (If there are no more problems, EXIT with total failure.)

2) Choose the next of the three basic Methods.
   (If no more methods, go to 1.)

3) Choose the next member of the list of axioms and previous theorems. (If no more, go to 2.)
   Then apply the Method to the problem, using the chosen theorem or axiom.
   If problem is solved, EXIT with complete proof.
   If no result, go to 3.
   If new subproblem arises, go to 4.

4) Try the special (substitution) Method on the subproblem.
   If problem is solved, EXIT with complete proof.
   If no result, put the subproblem *at the end* of the problem list and go to 3.

Among the heuristics that were studied were 1) a *similarity test* to reduce the work in step 4 (which includes another search through the theorem list), 2) a *simplicity test* to select apparently easier problems from the problem list, and 3) a *strong nonprovability test* to remove from the problem list expressions which are probably false and hence not provable. In a series of experiments "learning" was used to find which earlier theorems had been most useful and should be given priority in step 3. We cannot review the effects of these changes in detail. Of interest was the balance between the extra cost for administration of certain heuristics and the resultant search reduction; this balance was quite delicate in some cases when computer memory became saturated. The system seemed to be quite sensitive to the training sequence—the order in which problems were given. And some heuristics which gave no significant over-all improvement did nevertheless affect the class of solvable problems. Curiously enough, the general efficiency of LT was not greatly improved by any or all of these devices. But all this practical experience is reflected in the design of the much more sophisticated "GPS" system (described briefly in Section IV-D, 2).

Wang (1960b) has criticized the LT project on the grounds that there exist, as he and others have shown, mechanized proof methods which, for the particular run of problems considered, use far less machine effort than does LT and which have the advantage that they will ultimately find a proof for any provable proposition. (LT does not have this exhaustive "decision procedure" character and can fail ever to find proofs for some theorems.) Newell, Shaw, and Simon (1957), perhaps unaware of the existence of even moderately efficient exhaustive methods, supported their arguments by comparison with a particularly inefficient exhaustive procedure. Nevertheless, I feel that some of Wang's criticisms are misdirected. He does not seem to recognize that the authors of LT are not so much interested in proving these theorems as they are in the general problem of solving difficult problems. The combinatorial system of Russell and Whitehead (with which LT deals) is far less simple and elegant than the system used by Wang.[24] (Note, *e.g.*, the emphasis in Newell, Shaw, and Simon (1958a) and Newell, Shaw, and Simon (1958b).) Wang's problems, while *logically* equivalent, are *formally* much simpler. His methods do not include any facilities for using previous results (hence they are sure to degrade rapidly at a certain level of problem complexity), while LT is fundamentally oriented around this problem. Finally, because of the very effectiveness of Wang's method on the *particular* set of theorems in question, he simply did not have to face the fundamental heuristic problem of *when to decide to give up on a line of attack*. Thus the formidable performance of his program (Wang 1960b) perhaps diverted his attention from heuristic problems that must again spring up when real mathematics is ultimately encountered.

This is not meant as a rejection of the importance of Wang's work and discussion. He and others working on "mechanical mathematics" have discovered that there are proof procedures which are much more efficient than has been suspected. Such work will unquestionably help in constructing intelligent machines, and these procedures will certainly be preferred, when available, to "unreliable heuristic methods." Wang, Davis and Putnam, and several others are now pushing these new techniques into the far more challenging domain of theorem-proving in the predicate calculus (for which exhaustive decision procedures are no longer available). We have no space to discuss this area,[25] but it seems clear that a program to solve real mathematical problems will have to

[24] Wang's procedure (1960) too, works backwards, and can be regarded as a generalization of the method of "falsification" for deciding truth-functional tautology. In Wang (1960a) and its unpublished sequel, Wang introduces more powerful methods (for much more difficult problems).

[25] See Davis and Putnam (1960) and Wang (1960a).

combine the mathematical sophistication of Wang with the heuristic sophistication of Newell, Shaw and Simon.[26]

B. HEURISTICS FOR SUBPROBLEM SELECTION

In designing a problem-solving system, the programmer often comes equipped with a set of more or less distinct "Methods"—his real task is to find an efficient way for the program to decide where and when the different methods are to be used.

Methods which do not dispose of a problem may still transform it to create new problems or subproblems. Hence, during the course of solving one problem we may become involved with a large assembly of interrelated subproblems. A "parallel" computer, yet to be conceived, might work on many at a time. But even the parallel machine must have procedures to allocate its resources because it cannot simultaneously apply all its methods to all the problems. We shall divide this administrative problem into two parts: the selection of those subproblem(s) which seem most critical, attractive, or otherwise immediate, and, in the next section, the choice of which method to apply to the selected problem.

In the basic program for LT (Section IV-A), subproblem selection is very simple. New problems are examined briefly and (if not solved at once) are placed at the end of the (linear) problem list. The main program proceeds along this list (step 1), attacking the problems in the order of their generation. More powerful systems will have to be more judicious (both in generation and selection of problems) for only thus can excessive branching be restrained.[27] In more complex systems we can expect to consider for each subproblem, at least these two aspects: 1) its apparent "centrality"—how will its solution promote the main

---

[26]All these efforts are directed toward the reduction of search effort. In that sense they are all heuristic programs. Since practically no one still uses "heuristic" in a sense opposed to "algorithmic," serious workers might do well to avoid pointless argument on this score. The real problem is to find methods which significantly delay the apparently inevitable exponential growth of search trees.

[27]Note that the simple scheme of LT has the property that each generated problem will eventually get attention, even if several are created in a step 3. If one were to turn *full* attention to each problem, as generated, one might never return to alternate branches.

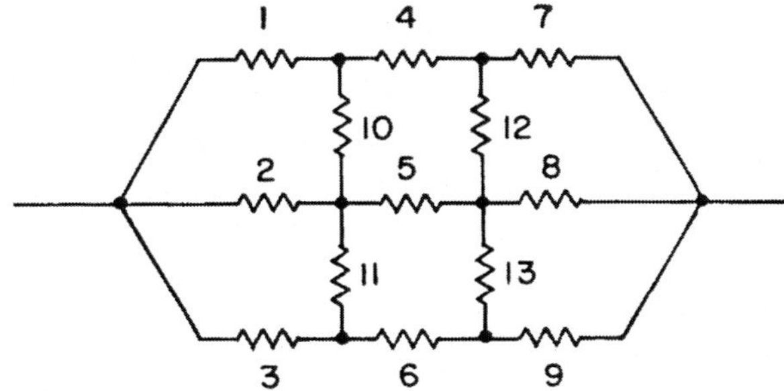

Figure 11. This board game (due to C. E. Shannon) is played on a network of equal resistors. The first player's goal is to open the circuit between the endpoints; the second player's goal is to short the circuit. A move consists of opening or shorting a resistor. If the first player begins by opening resistor 1, the second player might counter by shorting resistor 4, following the strategy described in the text. The remaining move pairs (if *both* players use that strategy) would be (5, 8) (9, 13) (6, 3) (12, 10 or 2) (2 or 10 *win*). In this game the first player should be able to force a win, and the maximum-current strategy seems always to do so, even on larger networks.

goal, and 2) its apparent "difficulty"—how much effort is it liable to consume. We need heuristic methods to estimate each of these quantities and, further, to select accordingly one of the problems and allocate to it some reasonable quantity of effort.[28] Little enough is known about these matters, and so it is not entirely for lack of space that the following remarks are somewhat cryptic.

Imagine that the problems and their relations are arranged to form some kind of directed-graph structure (Minsky 1956; Newell and Simon 1956; Gelernter and Rochester 1958). The main problem is to establish a "valid" path between two initially distinguished nodes. Generation of new problems is represented by the addition of new, not-yet-valid paths, or by the insertion of new nodes in old paths. Then problems are represented by not-yet-valid paths, and "centrality" by location in the structure. Associate with each connection, quantities describing its current validity state (solved, plausible, doubtful, etc.) and its current estimated difficulty.

[28] One will want to see if the considered problem is the same as one already considered, or very similar. See the discussion in Gelernter and Rochester (1958). This problem might be handled more generally by simply *remembering* the (Characters of) problems that have been attacked, and checking new ones against this memory, *e.g.*, by methods of Mooers (1956b), looking more closely if there seems to be a match.

1) *Global Methods*: The most general problem-selection methods are "global"—at each step they look over the entire structure. There is one such simple scheme which works well on at least one rather degenerate interpretation of our problem graph. This is based on an electrical analogy suggested to us by a machine designed by Shannon (related to one described in 1955, which describes quite a variety of interesting game-playing and learning machines) to play a variant of the game marketed as "Hex" (and known among mathematicians as "Nash"). The initial board position can be represented as a certain network of resistors. (See Fig. 11.) One player's goal is to construct a *short-circuit* path between two given boundaries; the opponent tries to open the circuit between them. Each move consists of shorting (or opening), irreversibly, one of the remaining resistors. Shannon's machine applies a potential between the boundaries and selects that resistor which carries the largest current. Very roughly speaking, this resistor is likely to be most critical because changing it will have the largest effect on the resistance of the net and, hence, in the goal direction of shorting (or opening) the circuit. And although this argument is not perfect, nor is this a perfect model of the real combinatorial situation, the machine does play extremely well. (It can make unsound moves in certain artificial situations, but no one seems to have been able to force this during a game.)

The use of such a global method for problem-selection requires that the available "difficulty estimates" for related subproblems be arranged to combine in roughly the manner of resistance values. Also, we could regard this machine as using an "analog model" for "planning." (See Section IV-D.)[29]

2) *Local, and "Hereditary," Methods*: The prospect of having to study at each step the whole problem structure is discouraging, especially since the structure usually changes only slightly after each attempt. One naturally looks for methods which merely *update* or modify a small fragment of the stored record. Between the extremes of the

[29]A variety of combinatorial methods will be matched against the network-analogy opponent in a program being completed by R. Silver, Lincoln Lab., M.I.T., Lexington, Mass.

"first-come-first-served" problem-list method and the full global-survey methods, lie a variety of compromise techniques. Perhaps the most attractive of these are what we will call the *Inheritance* methods—essentially recursive devices.

In an Inheritance method, the effort assigned to a subproblem is determined only by its immediate ancestry; at the time each problem is created it is assigned a certain total quantity Q of time or effort. When a problem is later split into subproblems, such quantities are assigned to them by some local process which *depends only on their relative merits and on what remains of* Q. Thus the centrality problem is managed implicitly. Such schemes are quite easy to program, especially with the new programming systems such as IPL (Newell and Tonge 1960) and LISP (McCarthy 1960) (which are themselves based on certain hereditary or recursive operations). Special cases of the inheritance method arise when one can get along with a simple all-or-none Q, *e.g.*, a "stop condition"—this yields the exploratory method called "back-tracking" by Golumb (1961). The decoding procedure of Wozencraft and Horstein (1961) is another important variety of Inheritance method.

In the complex exploration process proposed for chess by Newell, Shaw, and Simon (1958a) we have a form of Inheritance method with a *non-numerical stop-condition*. Here, the subproblems inherit *sets of goals to be achieved*. This teleological control has to be administered by an additional goal-selection system and is further complicated by a global (but reasonably simple) stop rule of the backing-up variety [Section III-C]. (Note: we are identifying here the move-tree-limitation problem with that of problem-selection.) Even though extensive experimental results are not yet available, we feel that the scheme of Newell, Shaw, and Simon (1958a) deserves careful study by anyone planning serious work in this area. It shows only the beginning of the complexity sure to come in our development of intelligent machines.[30]

[30]Some futher discussion of this question may be found in Slagle (1961).

C. "CHARACTER-METHOD" MACHINES

Once a problem is selected, we must decide which method to try first. This depends on our ability to classify or characterize problems. We first compute the Character of our problem (by using some pattern recognition technique) and then consult a "Character-Method" table or other device which is supposed to tell us which method(s) are most effective on problems of that Character. This information might be built up from experience, given initially by the programmer, deduced from "advice" (McCarthy 1959), or obtained as the solution to some other problem, as suggested in the GPS proposal (Newell, Shaw, and Simon 1960b). In any case, this part of the machine's behavior, regarded from the outside, can be treated as a sort of stimulus-response, or "table look-up," activity.

If the Characters (or descriptions) have too wide a variety of values, there will be a serious problem of filling a Character-Method table. One might then have to reduce the detail of information, *e.g.*, by using only a few important properties. Thus the *Differences* of GPS (see Section IV-D, Samuel 1959) describe no more than is necessary to define a single goal, and a priority scheme selects just one of these to characterize the situation. Gelernter and Rochester (1958) suggest using a property-weighting scheme, a special case of the "Bayes net" described in Section II-G.

D. PLANNING

Ordinarily one can solve a complicated problem only by dividing it into a number of parts, each of which can be attacked by a smaller search (or be further divided). Generally speaking, a successful division will reduce the search time not by a mere fraction, but by a *fractional exponent.* In a graph with 10 branches descending from each node, a 20-step search might involve 10^{20} trials, which is out of the question, while the insertion of just four *lemmas* or *sequential subgoals* might reduce the search to only 5.10^4 trials, which is within reason for machine exploration. Thus it will be worth a relatively enormous effort to find such "islands" in the solution of complex problems.[31] Note that even if

[31] See section 10 of Ashby (1956).

one encountered, say, 10^6 failures of such procedures before success, one would still have gained a factor of perhaps 10^{10} in overall trial reduction! *Thus practically any ability at all to "plan," or "analyze," a problem will be profitable,* if the problem is difficult. It is safe to say that all simple, unitary, notions of how to build an intelligent machine will fail, rather sharply, for some modest level of problem difficulty. Only schemes which actively pursue an analysis toward obtaining a set of *sequential goals* can be expected to extend smoothly into increasingly complex problem domains.

Perhaps the most straightforward concept of planning is that of using a *simplified model* of the problem situation. Suppose that there is available, for a given problem, some other problem of "essentially the same character" but with less detail and complexity. Then we could proceed first to solve the simpler problem. Suppose, also, that this is done using a second set of methods, which are also simpler, but in some correspondence with those for the original. *The solution to the simpler problem can then be used as a "plan" for the harder one.* Perhaps each step will have to be expanded in detail. But the multiple searches will *add, not multiply,* in the total search time. The situation would be ideal if the model were, mathematically, a *homomorphism* of the original. But even without such perfection the model solution should be a valuable guide. In mathematics one's proof procedures usually run along these lines: one first assumes, *e.g.*, that integrals and limits always converge, in the planning stage. Once the outline is completed, in this simple-minded model of mathematics, then one goes back to try to "make rigorous" the steps of the proof, *i.e.*, to replace them by chains of argument using genuine rules of inference. And even if the plan fails, it may be possible to patch it by replacing just a few of its steps.

Another aid to planning is the *semantic*, as opposed to the homomorphic, model (Minsky 1956, 1959). Here we may have an *interpretation* of the current problem within another system, not necessarily simpler, but with which we are more familiar and have already more powerful methods. Thus, in connection with a plan for the proof of a theorem, we will want to know whether the proposed lemmas, or islands in the proof, are actually *true*; if not, the plan will surely fail. We can often easily tell if a proposition is true by looking at an interpretation. Thus the truth of a proposition from plane geometry can be supposed, at least with great reliability, by actual measurement of a few constructed drawings (or the analytic geometry equivalent). The geometry machine of Gelernter and Rochester (1958) and Gelernter (1960) uses such a semantic model with excellent results; it follows closely the lines proposed in Minsky (1956).

1) *The "Character-Algebra" Model*: Planning with the aid of a model is of the greatest value in reducing search. Can we construct machines which find their own models? I believe the following will provide a general, straightforward way to construct certain kinds of useful, abstract models. The critical requirement is that we be able to compile a "Character-Method Matrix" (in addition to the simple Character-Method table in Section IV-C). *The CM matrix is an array of entries which predict with some reliability what will happen when methods are applied to problems.* Both of the matrix dimensions are indexed by problem Characters; if there is a method *which usually transforms problems of character C_i into problems of character C_j* then let the matrix entry C_{ij} be the name of that method (or a list of such methods). If there is no such method the corresponding entry is null.

   Now suppose that there is no entry for C_{ij}—meaning that we have no *direct* way to transform a problem of type C_i into one of type C_j. Multiply the matrix by itself. If the new matrix has a non-null (i, j) entry then there must be a sequence of *two* methods which effects the desired transformation. If that fails, we may try higher powers. Note that [if we put unity for the (i, i) terms] we can reach the 2^n matrix power with just n multiplications. We don't need to define the symbolic multiplication operation; one may instead use arithmetic entries—putting *unity for any non-null entry* and zero for any null entry in the original matrix. This yields a simple connection, or flow diagram, matrix, and its nth power tells us something about its set of paths of length 2^n.[32] (Once a non-null entry is discovered, there exist efficient ways to find the corresponding sequences of methods. The problem is really just that of finding paths through a maze, and

[32]See, *e.g.*, Hohn, Seshu, and Aufenkamp (1957).

the method of Moore (1959) would be quite efficient. Almost any problem can be converted into a problem of finding a chain between two terminal expressions in some formal system.) If the Characters are taken to be abstract representations of the problem expressions, this "Character-Algebra" model can be as abstract as are the available pattern-recognition facilities. See Minsky (1956) and Minsky (1959).

The critical problem in using the Character-Algebra model for planning is, of course, the *prediction-reliability of the matrix entries*. One cannot expect the Character of a result to be strictly determined by the Character of the original and the method used. And the reliability of the predictions will, in any case, deteriorate rapidly as the matrix power is raised. But, as we have noted, any plan at all is so much better than none that the system should do very much better than exhaustive search, even with quite poor prediction quality.

This matrix formulation is obviously only a special case of the character planning idea. More generally, one will have descriptions, rather than fixed characters, and one must then have more general methods to calculate from a description what is likely to happen when a method is applied.

In spite of its name, the General Problem Solver was only able to solve relatively simple problems that had a well-defined and small search space.

2) *Characters and Differences*: In the GPS (General Problem Solver) proposal of Newell, Shaw, and Simon (1960a, 1960b), we find a slightly different framework: they use a notion of Difference between two problems (or expressions) where we speak of the Character of a single problem. These views are equivalent if we take our problems to be links or connections between expressions. But this notion of Difference (as the Character of a pair) does lend itself more smoothly to teleological reasoning. For what is the goal defined by a problem but to *reduce the "difference" between the present state and the desired state*? The underlying structure of GPS is precisely what we have called a "Character-Method machine" in which each kind of Difference is associated in a table with one or more methods which are known to "reduce" that Difference. Since the characterization here depends always on 1) the current problem expression and 2) the desired end result, it is reasonable to think, as its authors suggest, of GPS as using "means-end" analysis.

At each step of problem solving, GPS applied a (human-defined) operator that could reduce the difference between the current and the goal state. However, defining states in such a way that a machine could easily compute the "difference" turned out to be too difficult for many, if not most, real-world problems.

To illustrate the use of Differences, we shall review an example (Newell, Shaw, and Simon 1960a). The problem, in elementary propositional calculus, is to prove that from $S \wedge (-P \supset Q)$ we can deduce $(Q \vee P) \wedge S$. The program looks at both of these expressions with a recursive *matching* process which branches out from the main connectives. The first Difference it encounters is that S occurs on different sides of the main connective "$\wedge$". It therefore looks in the Difference-Method table under the heading "change position." It discovers there a method which uses the theorem $(A \wedge B) \equiv (B \wedge A)$ which is obviously useful for removing, or "reducing," differences of position. GPS applies this method, obtaining $(-P \supset Q) \wedge S$. GPS now asks what is the Difference between this new expression and the goal. This time the matching procedure gets down into the connectives inside the left-hand members and finds a Difference between the connectives "$\supset$" and "$\vee$". It now looks in the CM table under the heading "Change Connective" and discovers the appropriate method using $(-A \supset B) \equiv (A \vee B)$. It applies this method, obtaining $(P \vee Q) \wedge S$. In the final cycle, the difference-evaluating procedure discovers the need for a "change position" inside the left member, and applies a method using $(A \vee B) \equiv (B \vee A)$. This completes the solution of the problem.[33]

Evidently, the success of this "means-end" analysis in reducing general search will depend on the degree of specificity that can be written into the Difference-Method table—basically the same requirement for an effective Character-Algebra.

It may be possible to *plan* using Differences, as well.[33] One might imagine a "Difference-Algebra" in which the predictions have the form $D = D'D''$. One must construct accordingly a difference-

[33]Compare this with the "matching" process described in Newell and Simon (1956). The notions of "Character," "Character-Algebra," etc., originate in Minsky (1956) but seem useful in describing parts of the "GPS" system of Newell and Simon (1956) and Newell, Shaw, and Simon (1960a). Newell, Shaw, and Simon (1960a) contains much additional material we cannot survey here. Essentially, GPS is to be self-applied to the problem of discovering sets of Differences appropriate for given problem areas. This notion of "bootstrapping"—applying a problem-solving system to the task of improving some of its own methods—is old and familiar, but in Newell, Shaw, and Simon (1960a) we find perhaps the first specific proposal about how such an advance might be realized.

factorization algebra for discovering longer chains $D = D_1 \cdots D_n$ and corresponding method plans. We should note that one *cannot* expect to use such planning methods with such primitive Differences as are discussed in Newell, Shaw, and Simon (1960a); for these cannot form an adequate Difference-Algebra (or Character Algebra). Unless the characterizing expressions have many levels of descriptive detail, the matrix powers will too swiftly become degenerate. This degeneracy will ultimately limit the capacity of any formal planning device.

One may think of the general planning heuristic as embodied in a recursive process of the following form. Suppose we have a problem P:

a) Form a plan for problem P.

b) Select first (next) step of the plan.
(If no more steps, exit with "success.")

c) Try the suggested method(s):
Success: return to b), *i.e.*, try next step in the plan.
Failure: return to a), *i.e.*, form new plan, or perhaps change current plan to avoid this step.
Problem judged too difficult: *Apply this entire procedure to the problem of the current step.*

Observe that such a program schema is essentially recursive; it uses itself as a subroutine (explicitly, in the last step) in such a way that its current state has to be stored, and restored when it returns control to itself.[34]

[34]This violates, for example, the restrictions on "DO loops" in programming systems such as FORTRAN. Convenient techniques for programming such processes were developed by Newell, Shaw, and Simon (1957); the program state-variables are stored in "push-down lists" and both the program and the data are stored in the form of "list-structures." Gelernter (1960) extended FORTRAN to manage some of this. McCarthy (1960) has extended these notions in LISP to permit *explicit* recursive definitions of programs in a language based on recursive functions of symbolic expressions; here the management of program-state variables is fully automatic. See also Orchard-Hays' article in this issue. *Editor's note: Minsky here is referring to William Orchard-Hays's "The Evolution of Programming Systems," published in the 1961 Proceedings of the IRE alongside this article.*

Miller, Galanter and Pribram[35] discuss possible analogies between human problem-solving and some heuristic planning schemes. It seems certain that, for at least a few years, there will be a close association between theories of human behavior and attempts to increase the intellectual capacities of machines. But, in the long run, we must be prepared to discover profitable lines of heuristic programming which do not deliberately imitate human characteristics.[36]

V. Induction and Models

A. INTELLIGENCE

In all of this discussion we have not come to grips with anything we can isolate as "intelligence." We have discussed only heuristics, shortcuts, and classification techniques. Is there something missing? I am confident that sooner or later we will be able to assemble programs of great problem-solving ability from complex combinations of heuristic devices—multiple optimizers, pattern-recognition tricks, planning algebras, recursive administration procedures, and the like. In no one of these will we find the seat of intelligence. Should we ask

Nowadays simply known as "neural nets." Minsky uses the term "artificial intelligence" to contrast symbolic "heuristic programming" approaches with neural networks. He again expresses his pessimism about neural networks playing more than a "component element" role in machine intelligence. Today, when neural networks dominate the field, some AI researchers propose that symbolic approaches will need to be integrated as components of intelligent systems to mitigate the limitations of deep learning.

[35] See chs. 12 and 13 of Miller, Galanter, and Pribram (1960).

[36] Limitations of space preclude detailed discussion here of theories of self-organizing neural nets, and other models based on brain analogies. (Several of these are described or cited in National Physical Laboratory (1959) and Yovits and Cameron (1960).) This omission is not too serious, I feel, in connection with the subject of heuristic programming, because the motivation and methods of the two areas seem so different. Up to the present time, at least, research on neural-net models has been concerned mainly with the attempt to show that certain rather simple heuristic processes, *e.g.*, reinforcement learning, or property-list pattern-recognition, can be realized or evolved by collections of simple elements without very highly organized interconnections. Work on heuristic programming is characterized quite differently by the search for new, more powerful heuristics for solving very complex problems, and by very little concern for what hardware (neuronal or otherwise) would minimally suffice for its realization. In short, the work on "nets" is concerned with how far one can get with a small initial endowment; the work on "artificial intelligence" is concerned with using all we know to build the most powerful systems that we can. It is my expectation that, in problem-solving power, the (allegedly brain-like) minimal-structure systems will never threaten to compete with their more deliberately designed contemporaries; nevertheless, their study should prove profitable in the development of component elements and subsystems to be used in the construction of the more systematically conceived machines.

what intelligence "really is"? My own view is that this is more of an esthetic question, or one of sense of dignity, than a technical matter! To me "intelligence" seems to denote little more than the complex of performances which we happen to respect, but do not understand. So it is, usually, with the question of "depth" in mathematics. Once the proof of a theorem is really understood its content seems to become trivial. (Still, there may remain a sense of wonder about how the proof was discovered.)

Programmers, too, know that there is never any "heart" in a program. There are high-level routines in each program, but all they do is dictate that "if such-and-such, then transfer to such-and-such a subroutine." And when we look at the low-level subroutines, which "actually do the work," we find senseless loops and sequences of trivial operations, merely carrying out the dictates of their superiors. The intelligence in such a system seems to be as intangible as becomes the meaning of a single common word when it is thoughtfully pronounced over and over again.

But we should not let our inability to discern a locus of intelligence lead us to conclude that programmed computers therefore cannot think. For it may be so with *man*, as with *machine*, that, when we understand finally the structure and program, the feeling of mystery (and self-approbation) will weaken.[37] We find similar views concerning "creativity" in Newell, Shaw, and Simon (1958b). The view expressed by Rosenbloom (1951) that minds (or brains) can transcend machines is based, apparently, on an erroneous interpretation of the meaning of the "unsolvability theorems" of Gödel.[38]

Here Minsky embraces a "moving target" view of intelligence: Once we understand an "intelligent" performance (or a machine is able to mimic it, for example, by playing chess), it no longer seems intelligent to us.

Throughout AI's history, several people have cited Gödel's theorems to argue that machines can never fully capture human intelligence, because Gödel (and others) proved that there are fundamental limits to what machines can compute. In fact, this was one of the straw-man objections to AI discussed by Alan Turing (1950), to which he correctly responded, "Although it is established that there are limitations to the powers of any particular machine, it has only been stated, without any sort of proof, that no such limitations apply to the human intellect."

[37] See Minsky (1956) and Minsky (1959).

[38] On problems of volition we are in general agreement with McCulloch (1954) that our *freedom of will* "presumably means no more than that we can distinguish between what we intend [*i.e.*, our *plan*], and some intervention in our action." See also MacKay (1959, and its references); we are, however, unconvinced by his eulogization of "analogue" devices. Concerning the "mind-brain" problem, one should consider the arguments of Craik (1952), Von Hayek (1952), and Pask (1959). Among the active leaders in modern heuristic programming, perhaps only Samuel (1960) has taken a strong position against the idea of machines thinking. His argument, based on the fact that reliable computers do only that which they are instructed to do, has a basic flaw; it does not follow that the programmer therefore has full knowledge (and therefore full responsibility and credit for) what will ensue. For certainly the programmer may set up an evolutionary system whose limitations are for him unclear and possibly incomprehensible. No better does the mathematician know all the consequences of a proposed set of axioms. Surely a machine has to *be* in order to perform. But we cannot assign all the credit to its programmer if the operation of a system comes to reveal structures not recognizable or anticipated by the programmer. While we have

B. INDUCTIVE INFERENCE

Let us pose now for our machines, a variety of problems more challenging than any ordinary game or mathematical puzzle. Suppose that we want a machine which, when embedded for a time in a complex environment or "universe," will essay to produce a description of that world—to discover its regularities or laws of nature. We might ask it to predict what will happen next. We might ask it to predict what would be the likely consequences of a certain action or experiment. Or we might ask it to formulate the laws governing some class of events. In any case, our task is to equip our machine with *inductive* ability—with methods which it can use to construct general statements about events beyond its recorded experience. Now, there can be no system for inductive inference that will work well in all possible universes. But given a universe, or an ensemble of universes, and a criterion of success, this (epistemological) problem for machines becomes technical rather than philosophical. There is quite a literature concerning this subject, but we shall discuss only one approach which currently seems to us the most promising; this is what we might call the "grammatical induction" schemes of Solomonoff (1957, 1958, 1960a), based partly on work of Chomsky (1957) and Chomsky and Miller (1958).

This statement is closely related to Wolpert's "no free lunch" theorems (1996).

We will take *language* to mean the set of expressions formed from some given set of primitive symbols or expressions, by the repeated application of some given set of rules; the primitive expressions plus the rules is the *grammar* of the language. Most induction problems can be framed as problems in the *discovery of grammars*. Suppose, for instance, that a machine's prior experience is summarized by a large collection of statements, some labelled "good" and some "bad" by some critical device. How could we generate selectively more good statements? The trick is to find some relatively simple (formal) language in which the good statements are grammatical, and in which the bad ones are not. Given such a language, we can use it to generate more statements, and presumably these will tend to be more like the good ones. The heuristic argument is that if we can find a relatively simple way to separate the two sets, the discovered rule is likely

---

not yet seen much in the way of intelligent activity in machines, Samuel's arguments in (1960) (circular in that they are based on the presumption that machines do not have minds) do not assure us against this. Turing (1956) gives a very knowledgeable discussion of such matters.

to be useful beyond the immediate experience. If the extension fails to be consistent with new data, one might be able to make small changes in the rules and, generally, one may be able to use many ordinary problem-solving methods for this task.

The problem of finding an efficient grammar is much the same as that of finding efficient *encodings*, or programs, for machines; in each case, one needs to discover the important regularities in the data, and exploit the regularities by making shrewd *abbreviations*. The possible importance of Solomonoff's work (1960) is that, despite some obvious defects, it may point the way toward systematic mathematical ways to explore this discovery problem. He considers the class of all programs (for a given general-purpose computer) which will produce a certain given output (the body of data in question). Most such programs, if allowed to continue, will add to that body of data. By properly weighting these programs, perhaps by length, we can obtain corresponding weights for the different possible continuations, and thus a basis for prediction. If this prediction is to be of any interest, it will be necessary to show some independence of the given computer; it is not yet clear precisely what form such a result will take.

C. MODELS OF ONESELF

If a creature can answer a question about a hypothetical experiment, without actually performing that experiment, then the answer must have been obtained from some submachine inside the creature. The output of that submachine (representing a correct answer) as well as the input (representing the question) must be coded descriptions of the corresponding external events or event classes. Seen through this pair of encoding and decoding channels, the internal submachine acts like the environment, and so it has the character of a "model." The inductive inference problem may then be regarded as the problem of constructing such a model.

To the extent that the creature's actions affect the environment, this internal model of the world will need to include some representation of the creature itself. If one asks the creature "why did you decide to do such and such" (or if it asks this of itself), any answer must come from the internal model. Thus the evidence of introspection itself is liable to be based ultimately on the processes used in constructing one's image of one's self.

This discussion points at two issues in AI that are still quite open: an AI system being able to explain its decisions, and an AI system having some kind of "self-awareness."

Speculation on the form of such a model leads to the amusing prediction that intelligent machines may be reluctant to believe that they are *just* machines. The argument is this: our own self-models have a substantially "dual" character; there is a part concerned with the physical or mechanical environment—with the behavior of inanimate objects—and there is a part concerned with social and psychological matters. It is precisely because we have not yet developed a satisfactory mechanical theory of mental activity that we have to keep these areas apart. We could not give up this division even if we wished to—until we find a unified model to replace it. Now, when we ask such a creature what sort of being it is, it cannot simply answer "directly;" it must inspect its model(s). And it must answer by saying that it seems to be a dual thing—which appears to have two parts—a "mind" and a "body." Thus, even the robot, unless equipped with a satisfactory theory of artificial intelligence, would have to maintain a dualistic opinion on this matter.[39]

Conclusion

In attempting to combine a survey of work on "artificial intelligence" with a summary of our own views, we could not mention every relevant project and publication. Some important omissions are in the area of "brain models"; the early work of Farley and Clark (1954) (also Farley's paper in Yovits and Cameron (1960), often unknowingly duplicated, and the work of Rochester *et al.* (1956) and Milner in Yovits and Cameron (1960).) The work of Lettvin *et al.* (1959) is related to the theories in Selfridge (1959). We did not touch at all on the problems of logic and language, and of information retrieval, which must be faced when action is to be based on the contents of large memories; see, *e.g.*, McCarthy (1959). We have not discussed the basic results in mathematical logic which bear on the question of what can be done by machines. There are entire literatures we have hardly even sampled—the bold pioneering of Rashevsky

[39] There is a certain problem of infinite regression in the notion of a machine having a *good* model of itself: of course, the nested models must lose detail and finally vanish. But the argument, *e.g.*, of Von Hayek (see 8.69 and 8.79 of 1952) that we cannot "fully comprehend the unitary order" (of our own minds) ignores the power of recursive description as well as Turing's demonstration that (with sufficient external writing space) a "general-purpose" machine can answer any question about a description of itself that any larger machine could answer.

(c. 1929) and his later co-workers (Rashevsky 1960); Theories of Learning, *e.g.*, Gorn (1959); Theory of Games, *e.g.*, Shubik (1960); and Psychology, *e.g.*, Bruner, Goodnow, and Austin (1956). And everyone should know the work of Polya (1945) on how to solve problems. We can hope only to have transmitted the flavor of some of the more ambitious projects *directly* concerned with getting machines to take over a larger portion of problem-solving tasks.

One last remark: we have discussed here only work concerned with more or less self-contained problem-solving programs. But as this is written, we are at last beginning to see vigorous activity in the direction of constructing usable *time-sharing* or *multiprogramming* computing systems. With these systems, it will at last become economical to match human beings in real time with really large machines. This means that we can work toward programming what will be, in effect, "thinking aids." In the years to come, we expect that these man-machine systems will share, and perhaps for a time be dominant, in our advance toward the development of "artificial intelligence."

REFERENCES

Ashby, W. R. 1952. *Design for a Brain.* New York, NY: John Wiley / Sons, Inc.

———. 1956. "Design for an Intelligence Amplifier." In *Automata Studies,* edited by C. E. Shannon and J. McCarthy, vol. 34. Princeton, NJ: Princeton University Press.

Bellman, R. 1957. *Dynamic Programming.* Princeton, NJ: Princeton University Press.

Bernstein, A., M. de V. Roberts, T. Arbuckle, and M. A. Belsky. 1958. "A Chess Playing Program for the IBM 704." In *IRE-ACM-AIEE '58 (Western): Proceedings of the May 6–8, 1958, Western Joint Computer Conference: Contrasts in Computers,* 157–159. New York, NY.

Bledsoe, W. W., and I. Browning. 1959. "Pattern Recognition and Reading by Machine." In *Papers Presented at the December 1-3, 1959, Eastern Joint IRE-AIEE-ACM Computer Conference,* 225–232. Boston, MA: Association for Computing Machinery.

Bomba, J. S. 1959. "Alpha-Numeric Character Recognition Using Local Operations." In *Papers Presented at the December 1–3, 1959, Eastern Joint IRE-AIEE-ACM Computer Conference,* 218–224. Boston, MA: Association for Computing Machinery.

Bruner, J. S., J. Goodnow, and G. Austin. 1956. *A Study of Thinking*. New York, NY: John Wiley / Sons, Inc.

Bush, R. R., and F. Mosteller. 1955. *Stochastic Models for Learning*. New York, NY: John Wiley / Sons, Inc.

Chomsky, A. N. 1957. *Syntactic Structures*. The Hague, Netherlands: Mouton.

Chomsky, A. N., and G. A. Miller. 1958. "Finite State Languages." *Information and Control* 1:91–112.

Craik, K. J. W. 1952. *The Nature of Explanation*. Preface dated 1943. Cambridge, UK: Cambridge University Press.

Darlington, C. D. 1958. *The Evolution of Genetics*. New York, NY: Basic Books, Inc.

Davis, M., and H. Putnam. 1960. "A Computing Procedure for Quantification Theory." *Journal of the ACM* 7 (3): 201–215.

Dinneen, G. P. 1955. "Programming Pattern Recognition." In *AFIPS '55: Proceedings of the March 1-3, 1955, Western Joint Computer Conference*, 94–100. New York, NY: Association for Computing Machinery.

Doyle, W. 1959. *Recognition of Sloppy, Hand-Printed Characters*. Technical report. Lincoln Lab, MIT, Lexington, MA, Group Rept. 54--12.

Farley, B., and W. Clark. 1954. "Simulation of Self-Organizing Systems by Digital Computer." *IRE Transactions on Information Theory* 4 (4): 76–84.

Freimer, M. 1960. *Topics in Dynamic Programming II*. Technical report. Lincoln Lab, MIT, Lexington, MA, Rept. 52-G-0020 (MIT Hayden Library No. H-82). See especially sec. I-E.

Friedberg, R. M. 1958. "A Learning Machine, Part I." *IBM Journal of Research and Development* 2 (1): 2–13.

Friedberg, R. M., B. Dunham, and J. H. North. 1959. "A Learning Machine, Part II." *IBM Journal of Research and Development* 3 (3): 282–287.

Gelernter, H., and N. Rochester. 1958. "Intelligent Behavior in Problem-Solving Machines." *IBM Journal of Research and Development* 2 (4): 336–345.

Gelernter, H. L. 1960. "Realization of a Geometry-Proving Machine." In *Proceedings of the International Conference on Information Processing, UNESCO, Paris, 15–20 June, 1959*. London, UK: Butterworths.

Golumb, S. 1961. "A Mathematical Theory of Discrete Classification." In *Information Theory—The Fourth London Symposium*, edited by C. Cherry. London, UK: Butterworths.

Gorn, S. 1959. "On the Mechanical Simulation of Habit-Forming and Learning." *Information and Control* 2 (3): 226–259.

Grimsdale, R. L., F. H. Sumner, C. J. Tunis, and T. Kilburn. 1959. "A System for the Automatic Recognition of Patterns." *Proceedings of the IEE - Part B: Radio and Electronic Engineering* 106 (26): 210–221.

Haller, N. 1959. "Line Tracing for Character Recognition." Master's thesis, Massachusetts Institute of Technology.

Hebb, D. O. 1949. *The Organization of Behavior: A Neuropsychological Theory*. New York, NY: John Wiley / Sons, Inc.

Highleyman, W. H., and L. A. Kamentsky. 1960. "Comments on a Character Recognition of Bledoe and Browning." *IRE Transactions on Electronic Computers* EC-9 (2): 263–263.

Hohn, F. E., S. Seshu, and D. D. Aufenkamp. 1957. "The Theory of Nets." *IRE Transactions on Electronic Computers* 6 (3): 154–161.

Holland, J. H. 1960. "On Iterative Circuit Computers Constructed of Microelectronic Components and Systems." In *Proceedings of the Western Joint Computer Conference: Papers Presented the Joint IRE-AIEE-ACM Computer Conference, San Francisco, Calif., May 3–5, 1960,* 259–265. Palo Alto, CA: National Press.

Kilburn, T., R. L. Grimsdale, and F. H. Sumner. 1960. "Experiments in Machine Thinking and Learning." In *Proceedings of the International Conference on Information Processing, UNESCO, Paris, 15–20 June, 1959.* London, UK: Butterworths.

Kirsch, R. A., C. Ray, L. Cahn, and G. H. Urban. 1957. "Experiments in Processing Pictorial Information with a Digital Computer." In *Papers and Discussions Presented at the December 9-13, 1957, Eastern Joint Computer Conference: Computers with Deadlines to Meet,* 221–229. New York, NY: Association for Computing Machinery.

Köhler, W. 1947. *Gestalt Psychology: An Introduction to New Concepts in Modern Psychology.* Vol. 279. Mentor. New American Library.

Lettvin, J. Y., H. R. Maturana, W. S. McCulloch, and W. Pitts. 1959. "What the Frog's Eye Tells the Frog's Brain." *Proceedings of the IRE* 47:1940–1951.

MacKay, D. M. 1956. "The Epistemological Problem for Automata." In *Automata Studies,* edited by C. E. Shannon and J. McCarthy. Princeton, NJ: Princeton University Press.

———. 1959. "Operational Aspects of Intellect." In *Mechanization of Thought Processes: Proceedings of a Symposium Held at the National Physical Laboratory on 24th, 25th, 26th, and 27th November 1958.* London, UK: Her Majesty's Stationery Office.

McCarthy, J. 1956. "The Inversion of Functions Defined by Turing Machines." In *Automata Studies,* edited by C. E. Shannon and J. McCarthy, vol. 34. Princeton, NJ: Princeton University Press.

———. 1959. "Programs with Common Sense." In *Mechanization of Thought Processes: Proceedings of a Symposium Held at the National Physical Laboratory on 24th, 25th, 26th, and 27th November 1958.* London, UK: Her Majesty's Stationery Office.

———. 1960. "Recursive Functions of Symbolic Expressions and their Computation by Machine, Part I." *Communications of the ACM* 3 (4): 184–195.

McCulloch, W. S. 1954. "Through the Den of the Metaphysician." *British Journal for the Philosophy of Science* 5 (17): 18–34.

McCulloch, W. S., and W. Pitts. 1947. "How We Know Universals: The Perception of Auditory and Visual Forms." *Bulletin of Mathematical Biophysics* 9:127–147.

Miller, G. A., E. Galanter, and K. H. Pribram. 1960. *Plans and the Structure of Behavior.* New York, NY: Henry Holt / Co., Inc.

Minot, O. N. 1959. *Automatic Devices for Recognition of Visible Two-Dimensional Patterns: A Survey of the Field.* Technical report. US Naval Electronics Lab, San Diego, CA, tech. memo. 364.

Minsky, M. 1954. "Neural Nets and the Brain Model Problem." PhD diss., Princeton University, Princeton, NJ (University Microfilms, Ann Arbor).

Minsky, M. 1956. *Heuristic Aspects of the Artificial Intelligence Problem.* Technical report.

———. 1959. "Some Aspects of Heuristic Programming and Artificial Intelligence." In *Mechanization of Thought Processes: Proceedings of a Symposium Held at the National Physical Laboratory on 24th, 25th, 26th, and 27th November 1958.* London, UK: Her Majesty's Stationery Office.

Minsky, M., and O. G. Selfridge. 1961. "Learning in Random Nets." In *Information Theory—The Fourth London Symposium,* edited by C. Cherry. London, UK: Butterworths.

Mooers, C. N. 1956a. "Information Retrieval on Structured Content." In *Third London Symposium on Information Theory,* edited by C. Cherry. New York, NY: Academic Press, Inc.

———. 1956b. "Zatocoding and Developments in Information Retrieval." *Aslib Proceedings* 8 (1): 3–22.

Moore, E. F. 1959. "The Shortest Path through a Maze." In *Proceedings of an International Symposium on the Theory of Switching, Part II,* 285–292. Cambridge, MA: Harvard University Press.

National Physical Laboratory. 1959. *Mechanization of Thought Processes: Proceedings of a Symposium Held at the National Physical Laboratory on 24th, 25th, 26th, and 27th November 1958.* London, UK: Her Majesty's Stationery Office.

Newell, A. 1955. "The Chess Machine: An Example of Dealing with a Complex Task by Adaptation." In *AFIPS '55 (Western): Proceedings of the March 1–3, 1955, Western Joint Computer Conference,* 101–108. New York, NY: Association for Computing Machinery.

Newell, A., J. C. Shaw, and H. A. Simon. 1957. "Empirical Explorations of the Logic Theory Machine." In *IRE-AIEE-ACM '57 (Western): Papers Presented at the February 26–28, 1957, Western Joint Computer Conference: Techniques for Reliability,* 218–230. New York, NY: Association for Computing Machinery.

———. 1958a. "Chess-Playing Programs and the Problem of Complexity." *IBM Journal of Research and Development* 2 (4): 320–335.

———. 1958b. "Elements of a Theory of Human Problem Solving." *Psychological Review* 65 (3): 151–166.

———. 1960a. "A Variety of Intelligent Learning in a General Problem Solver." In *Self-Organizing Systems,* edited by M. T. Yovits and S. Cameron. New York, NY: Pergamon Press.

———. 1960b. "Report on a General Problem-Solving Problem." In *Proceedings of the International Conference on Information Processing, UNESCO, Paris, 15–20 June, 1959.* London, UK: Butterworths.

Newell, A., and H. A. Simon. 1956. "The Logic Theory Machine." *IRE Transactions on Information Theory* IT-2 (3): 61–79.

Newell, A., and F. Tonge. 1960. "An Introduction to Information Processing Language V." *Communications of the ACM* 3 (4): 205–211.

Papert, S. 1961. "Some Mathematical Models of Learning." In *Information Theory—The Fourth London Symposium,* edited by C. Cherry. London, UK: Butterworths.

Pask, G. 1959. "Physical Analogues to the Growth of a Concept." In *Mechanization of Thought Processes: Proceedings of a Symposium Held at the National Physical Laboratory on 24th, 25th, 26th, and 27th November 1958.* London, UK: Her Majesty's Stationery Office.

Polya, G. 1945. *How to Solve It.* Also, *Induction and Analogy in Mathematics* and *Patterns of Plausible Inference,* 2 vols., Princeton University Press (Princeton, NJ), 1954. Princeton, NJ: Princeton University Press.

Rashevsky, N. 1960. *Mathematical Biophysics.* Vol. 2. New York, NY: Dover Publications.

Roberts, L. G. 1960. "Pattern Recognition with an Adaptive Network." In *IRE International Convention Record, Part 2,* 66–70.

Rochester, N., J. H. Holland, L. Haibt, and W. Duda. 1956. "Tests on a Cell Assembly Theory of the Action of the Brain, Using a Large Digital Computer." *IRE Transactions on Information Theory* IT-2 (3): 80–93.

Rosenblatt, F. 1958. *The Perceptron.* Technical report. Cornell Aeronautical Lab., Inc., Ithaca, NY, Rept. no. VG-1196-G-1. See also the article of Hawkins in this issue.

Rosenbloom, P. 1951. *Elements of Mathematical Logic.* New York, NY: Dover Publications.

Samuel, A. 1960. "Letter to the Editor." Incorrectly labeled vol. 131 on cover, *Science* 132 (3429).

Samuel, A. L. 1959. "Some Studies in Machine Learning Using the Game of Checkers." *IBM Journal of Research and Development* 3:211–219.

Selfridge, O. G. 1955. "Pattern Recognition and Modern Computers." In *AFIPS '55: Proceedings of the March 1-3, 1955, Western Joint Computer Conference,* 91–93. New York, NY: Association for Computing Machinery.

———. 1956. "Pattern Recognition and Learning." In *Third London Symposium on Information Theory,* edited by C. Cherry. New York, NY: Academic Press, Inc.

———. 1959. "Pandemonium: A Paradigm for Learning." In *Mechanization of Thought Processes: Proceedings of a Symposium Held at the National Physical Laboratory on 24th, 25th, 26th, and 27th November 1958.* London, UK: Her Majesty's Stationery Office.

Selfridge, O. G., and U. Neisser. 1960. "Pattern Recognition by Machine." *Scientific American* 203 (2): 60–68.

Shannon, C. E. 1949. "Synthesis of Two-Terminal Switching Networks." *Bell Systems Technical Journal* 28:59–98.

———. 1955. "Game-Playing Machines." *Journal of the Franklin Institute* 260 (6): 447–453.

———. 1956. "Programming a Digital Computer for Playing Chess." In *The World of Mathematics,* edited by J. R. Newman, vol. 4. New York, NY: Simon / Schuster.

Sherman, H. 1960. "A Quasi-Topological Method for Machine Recognition of Line Patterns." In *Proceedings of the International Conference on Information Processing, UNESCO, Paris, 15-20 June, 1959.* London, UK: Butterworths.

Shubik, M. 1960. "Games Decisions and Industrial Organization." *Management Science* 6 (4): 455–474.

Skinner, B. F. 1953. *Science and Human Behavior.* New York, NY: Macmillan.

Slagle, J. 1961. "A Heuristic Program that Solves Symbolic Integration Problems in Freshman Calculus: Symbolic Automatic Integrator (SAINT)." PhD diss., MIT.

Solomonoff, R. J. 1957. "An Inductive Inference Machine." In *1957 IRE National Convention Record, Pt. 2,* 56–62.

———. 1958. *The Mechanization of Linguistic Learning.* Technical report. Zator Co.

———. 1960a. "A New Method for Discovering the Grammars of Phrase Structure Languages." In *Proceedings of the International Conference on Information Processing, UNESCO, Paris, 15-20 June, 1959.* London, UK: Butterworths.

———. 1960b. *A Preliminary Report on a General Theory of Inductive Inference.* Technical report. Zator Co.

Stevens, M. E. 1957. *A Survey of Automatic Reading Techniques.* Technical report. NBS, US Dept. of Commerce, Washington, DC, Rept. 5643.

Tinbergen, N. 1951. *The Study of Instinct.* New York, NY: Oxford University Press.

Turing, A. M. 1956. "Can a Machine Think?" In *The World of Mathematics,* edited by J. R. Newman, vol. 4. New York, NY: Simon / Schuster, Inc.

Unger, S. H. 1959. "Pattern Detection and Recognition." *Proceedings of the IRE* 47 (10): 1737–1752.

Uttley, A. M. 1956. "Conditional Probability Machines." In *Automata Studies,* edited by C. E. Shannon and J. McCarthy. Princeton, NJ: Princeton University Press.

———. 1959. "Conditional Probability Computing in a Nervous System." In *Mechanization of Thought Processes: Proceedings of a Symposium Held at the National Physical Laboratory on 24th, 25th, 26th, and 27th November 1958.* London, UK: Her Majesty's Stationery Office.

Von Hayek, F. A. 1952. *The Sensory Order.* London, UK: Routledge / Kegan Paul.

Wang, H. 1960a. "Proving Theorems by Pattern Recognition, I." *Communications of the ACM* 3 (4): 220–234.

———. 1960b. "Toward Mechanical Mathematics." *IBM Journal of Research and Development* 4 (1): 2–22.

Wiener, N. 1948. *Cybernetics.* New York, NY: John Wiley / Sons, Inc.

Wozencraft, J., and M. Horstein. 1961. "Coding for Two-Way Channels." In *Information Theory—The Fourth London Symposium,* edited by C. Cherry. London, UK: Butterworths.

Yovits, M. T., and S. Cameron, eds. 1960. *Self-Organizing Systems.* New York, NY: Pergamon Press.

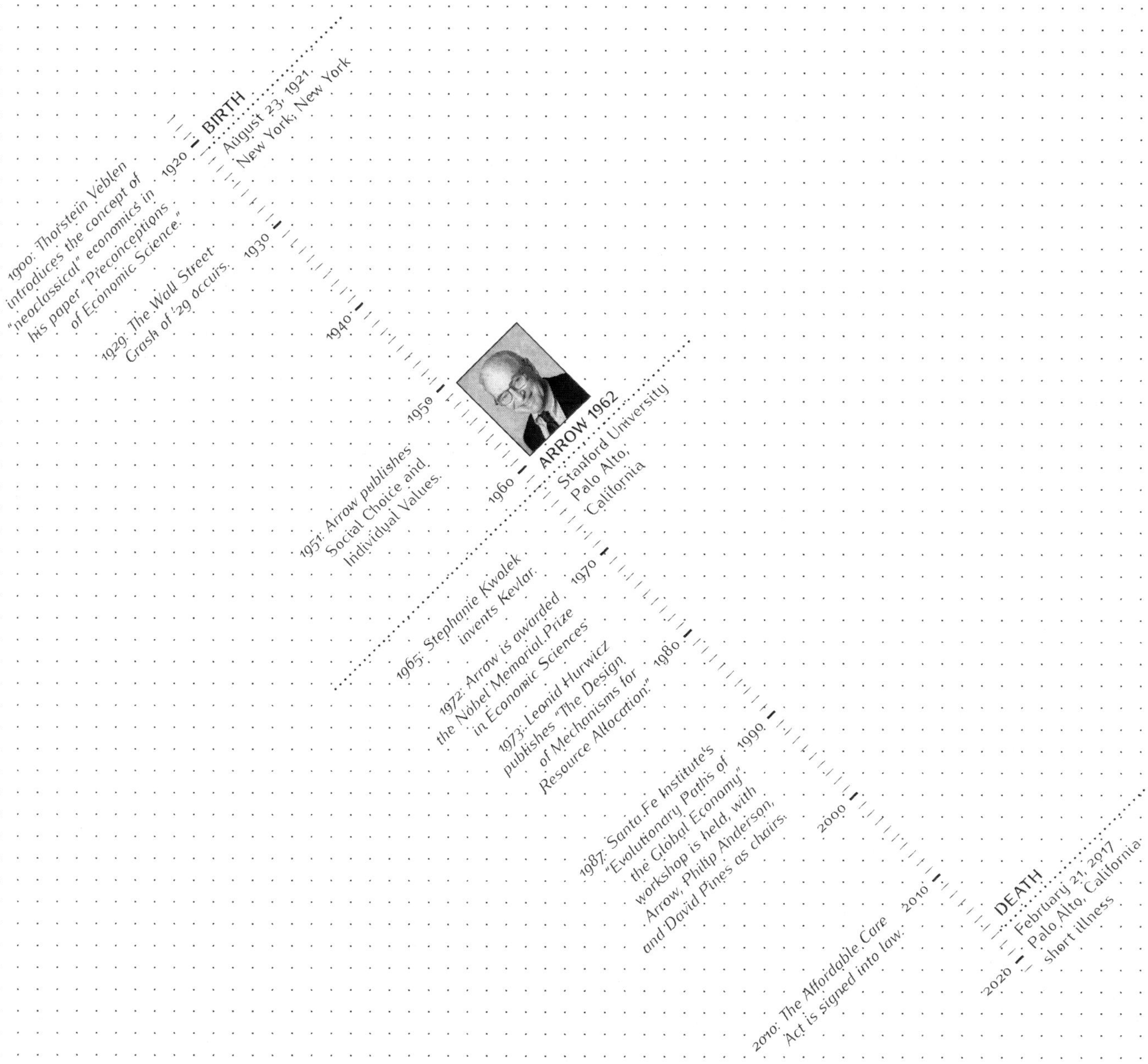

KENNETH JOSEPH ARROW

[18]

ARROW'S "LEARNING BY DOING" AND COMPLEXITY ECONOMICS

John Geanakoplos, Yale University and Santa Fe Institute

K. J. Arrow, "The Economic Implications of Learning by Doing," *The Review of Economic Studies* 29 (3), 155–173 (1962).

In the 1950s and early 1960s, Bob Solow and other economists built a canonical model of economic growth that made it clear that neither the extraordinary rise in output over the previous two centuries, nor the astounding difference in income across countries, could be explained simply by differences in the accumulation of productive machines or capital. Technological progress had to be the missing factor. But how to model it? Solow *et al.* made technological progress an exogenous function of time, like Moore's Law.

Kenneth Arrow (1962) began his paper by saying that "trend projections are basically a confession of ignorance, and what is worse, from a practical viewpoint, they are not policy variables." He then launched into an interdisciplinary synopsis of learning theories, concluding from them that learning is a product of experience in solving new problems, not repetitive ones. He was also struck by the finding of the aeronautical engineer T. P. Wright (1936) that the labor hours required for the construction of the Nth airframe seem to be a precise decreasing function $bN^{-1/3}$. Similar findings by Hirsch (1956) in other industries led to the notion of the learning curve, which seemed to suggest that building more of the same thing leads to learning.

Arrow ingeniously combined the idea of learning through *new* experiences with Wright's formula by embracing the *vintage model* of Johansen (1959) and Solow (1959). In that model, machines can be arranged in ascending order, known in advance. A machine of vintage or grade V requires $\lambda(V)$ workers to produce $\gamma(V)$ units of output per hour, where λ is decreasing in V and γ is increasing in V. Each machine can be produced from a unit of output, which otherwise could be consumed. Machines last indefinitely, or for some large time period. In the vintage model, V corresponds to a date, like 1960. In order to produce a 1960 vintage machine, an entrepreneur in 1950 would have to wait ten years. Productivity thus increases exogenously over time like clockwork. As time passes, so many machines will be built that there will not be enough workers to run them all. Evidently workers will concentrate on the newest and best model machines. The older machines will sit idle and obsolete.

Arrow took the vintage model and with one conceptual advance moved the analysis to a completely different level. In Arrow's (1962) model of learning by doing, the entrepreneur could produce a 1960-grade model at any time, as long as one machine of every lower grade $G = 1, \ldots, 1959$ had been produced by her or somebody else before. This move opened up two fundamental features of the model. By devoting more resources to investment, a society not only would have more machines, but also better-grade machines. Learning became endogenous, and it was embodied in something new. Moreover, because entrepreneurs do not care that building their new machine will enable other entrepreneurs to build next-generation machines, investment is typically inefficiently low, creating a need for public policy.

There is a third fundamental feature of the model—increasing returns to scale. Doubling the number of units of output G that have been devoted to building machines and doubling the number of available workers L, more than doubles output per hour, unless both λ and γ are constant functions. These three features of Arrow's model, namely endogenous learning, its inefficient provision by private markets, and increasing returns to scale, have become building blocks of modern complexity economics.

Arrow followed Wright and parameterized the learning functions

$$\gamma(G) = a > \lambda(G) = bG^{-n}$$

Clearly the higher n, the faster the learning from any investment. Given a fixed number of workers L and G machines in operation, Arrow worked out the formula for how much output $x = x_{a,n}(G, L)$ can be produced per hour, assuming that $0 \leq n < 1$, confirming that x displays increasing returns to scale. He provided a separate formula when $n = 1$, which also shows increasing returns. He concentrated on the case $0 \leq n < 1$, ignoring $n > 1$ altogether. Though he did not take note of it, the formulas make clear that increases in G alone yield diminishing marginal returns in output when $0 < n < 1$, and constant returns when $n = 1$.

After a long aside on competitive pricing of wages and interest, which we shall come back to, Arrow closed the model by assuming that, at any moment t, the economy saves (i.e., invests in machines) an exogenously fixed fraction s of the output at t, and consumes the rest. For any s and any exogenously given growth rate σ of the labor force, he computes that the unique steady-state rate of growth of output $x(t)$ and accumulated capital $G(t)$ must be

$$\sigma/(1-n)$$

The fourth striking feature of the model is that this steady-state growth rate $\sigma/(1-n)$ of $x(t)$ and $G(t)$ does not depend on the fraction s of output that is invested in new machines, though the levels $x(t)$, $G(t)$, and $x(t)/G(t)$ do. This insensitivity of the steady-state growth rate to the choice between consumption and investment led economists to call this model and others like it a semi-endogenous theory of growth. Solow's original (1956) growth model displayed the same insensitivity, and for the same reason: there are diminishing returns to capital in producing output.

Arrow enriched his paper by dropping the exogeneity of the saving rate s and instead solving the model for steady-state competitive equilibrium when all agents maximize the same beta discounted sum of consumption. Every agent is both laborer and entrepreneur. At each date the agents *choose* how to split their income (from wages and profit from the machines they already own) between consumption and producing more machines so as to maximize their long run utility. The fraction s of total output that is invested in new machines then emerges endogenously, rather than being fixed exogenously. Given this s, the model then behaves just like the one described earlier. Its steady-state rate of growth is still insensitive to s, and therefore to the utility. Arrow then proved that the free market fails. The competitive equilibrium steady state G/L ratio is too low: investing more in machines would increase utility, even though it does not change the steady-state rate of growth, formalizing the inefficiency feature of the model.

A competitive equilibrium wage $w(t)$ also emerges, and grows at a slower rate $n\sigma/(1-n)$ than the economy. The fifth feature of the economy is the paradoxical conclusion that faster exogenous growth in labor gives faster growth in wages.

Arrow's paper was immediately hailed as a masterpiece, since it broke so far from the standard decreasing-returns, Pareto-efficient general equilibrium models that Arrow himself had created. Yet Arrow did not teach it, nor did he continue working on the subject. He made completely uncharacteristic errors in the paper, lapsing carelessly into steady states because he ignored boundary conditions that he scrupulously attended to in his general equilibrium work. The paper also contained one huge bug, or feature. Arrow implicitly assumed that agents are so small that they cannot build machines more than infinitesimally beyond the contemporary grade $G(t)$. If they could borrow enough money, they could themselves push the quality well beyond $G(t)$, making indefinitely larger profits by taking advantage of the increasing returns to scale. This would destroy the competitive equilibrium.

Perhaps stymied by the increasing returns contradiction to genuine competitive equilibrium, the profession waited twenty years to make substantial progress. In a different context, Arrow had encouraged progress by telling the story about the priest who was asked about the existence of evil with an omnipotent god. "We must face the problem firmly, and move on."

Paul Romer (1986) took the first big step by building a model analogous to the Arrow model in which $n = 1$ and maximal output $x_{a,1}(G, L)$ is linear in capital G. Roemer showed that with linear returns to capital, in steady-state equilibrium, the rate of growth does depend positively on the investment choice s, and thus growth becomes fully endogenous.

Romer (1990) took the next big step by including a sector of the economy devoted exclusively to producing knowledge from labor and previous knowledge, thus incorporating pure research as an avenue to learning about production. He derived all sorts of additional testable conclusions, such as that a bigger labor force should create more knowledge and ultimately more growth in output. He maintained the first three features of Arrow's model, honestly facing the increasing returns to scale impediment to equilibrium with perfect competition by shifting to imperfect competition. Entrepreneurs realize that as they produce more, the prices they get will decline, counteracting the increasing returns in their production, and restoring the possibility of genuine equilibrium. Others like Lucas (1988) and Aghion and Howitt (1992) went on to study the effects of investment in human capital on growth. Solow (1997) introduced the idea of non-continuous breakthrough additions to knowledge that provided jumps to growth. A burgeoning field called modern growth theory was born, which led to Romer receiving the Noble Prize in 2018. More recently, complexity economists like Farmer and Lafond (2016) have tried to estimate the learning curve for non-fossil fuel energy production, predicting the obsolescence of oil.

REFERENCES

Aghion, P., and P. Howitt. 1992. "A Model of Growth Through Creative Destruction." *Econometrica* 60 (2): 323–351. https://doi.org/10.3386/w3223.

Farmer, J. D., and F. Lafond. 2016. "How Predictable Is Technological Progress?" *Research Policy* 45 (3): 647–655. https://doi.org/10.1016/j.respol.2015.11.001.

Lucas, R. E. 1988. "On the Mechanics of Economic Development." *Journal of Monetary Economics* 22 (1): 3–42. https://doi.org/10.1016/0304-3932(88)90168-7.

Romer, P. M. 1986. "Increasing Returns and Long-Run Growth." *Journal of Political Economy* 94 (5): 1002–1037. https://doi.org/10.1086/261420.

———. 1990. "Endogenous Technological Change." *Journal of Political Economy* 98 (5, part 2): 71–102. https://doi.org/10.1086/261725.

Solow, R. M. 1956. "A Contribution to the Theory of Economic Growth." *The Quarterly Journal of Economics* 70 (1): 65–94. https://doi.org/10.2307/1884513.

———. 1997. *Learning from 'Learning by Doing': Lessons for Economic Growth.* Stanford, CA: Stanford University Press.

Solow, R. M., J. Tobin, C. C. von Weizsäcker, and M. Yaari. 1966. "Neoclassical Growth with Fixed Factor Proportions." *The Review of Economic Studies* 33 (2): 79–115. https://doi.org/10.2307/2974435.

THE ECONOMIC IMPLICATIONS OF LEARNING BY DOING

K. J. Arrow, Stanford University

It is by now incontrovertible that increases in per capita income cannot be explained simply by increases in the capital-labor ratio. Though doubtless no economist would ever have denied the role of technological change in economic growth, its overwhelming importance relative to capital formation has perhaps only been fully realized with the important empirical studies of Abramovitz (1956) and Solow (1957). These results do not directly contradict the neo-classical view of the production function as an expression of technological knowledge. All that has to be added is the obvious fact that knowledge is growing in time. Nevertheless a view of economic growth that depends so heavily on an exogenous variable, let alone one so difficult to measure as the quantity of knowledge, is hardly intellectually satisfactory. From a quantitative, empirical point of view, we are left with time as an explanatory variable. Now trend projections, however necessary they may be in practice, are basically a confession of ignorance, and, what is worse from a practical viewpoint, are not policy variables.

Technological progress must be explained as the consequence of some activity, involving a choice.

Further, the concept of knowledge which underlies the production function at any moment needs analysis. Knowledge has to be acquired. We are not surprised, as educators, that even students subject to the same educational experiences have different bodies of knowledge, and we may therefore be prepared to grant, as has been shown empirically (see Arrow *et al.* 1961, Part III), that different countries, at the same moment of time, have different production functions even apart from differences in natural resource endowment.

I would like to suggest here an endogenous theory of the changes in knowledge which underlie intertemporal and international shifts in production functions. The acquisition of knowledge is what is usually termed "learning," and we might perhaps pick up some clues from the many psychologists who have studied this phenomenon (for a convenient survey, see Hilgard 1956). I do not think that the picture of technical change as a vast and prolonged process of learning about the environment in which we operate is in any way a far-fetched analogy; exactly the same phenomenon of improvement in performance over time is involved.

Arrow reaches outside economics to other disciplines for clues about learning.

Of course, psychologists are no more in agreement than economists, and there are sharp differences of opinion about the processes of learning. But one empirical generalization is so clear that all schools of thought must accept it, although they interpret it in different fashions: Learning is the product of experience. Learning can only take place through the attempt to solve a problem and therefore only takes place during activity. Even the Gestalt and other field theorists, who stress the role of insight in the solution of problems (Köhler's famous apes), have to assign a significant role to previous experiences in modifying the individual's perception.

A second generalization that can be gleaned from many of the classic learning experiments is that learning associated with repetition of essentially the same problem is subject to sharply diminishing returns. There is an equilibrium response pattern for any given stimulus, towards which the behavior of the learner tends with repetition. To have steadily increasing performance, then, implies that the stimulus situations must themselves be steadily evolving rather than merely repeating.

Sustained learning comes from new activities.

The role of experience in increasing productivity has not gone unobserved, though the relation has yet to be absorbed into the main corpus of economic theory. It was early observed by aeronautical engineers, particularly T. P. Wright (1936), that the number of labor-hours expended in the production of an airframe (airplane body without engines) is a decreasing function of the total number of airframes of the same type previously produced. Indeed, the relation is remarkably precise; to produce the Nth airframe of a given type, counting from the inception of production, the amount of labor required is proportional to $N^{-\frac{1}{3}}$. This relation has become basic in the production and cost planning of the United States Air Force; for a full survey, see Asher (1956). Hirsch (see Hirsch 1956, and other work cited there) has shown

This empirical regularity made a big impression on Arrow, who frequently brought it up in conversations. It confirms learning from experience, yet seems to suggest learning from repetitive behavior.

Here we find the origins, at least for Arrow, of the famous expression the "learning curve." Arrow curiously does not refer, in his learned account of writers on learning, to John Dewey, who coined the expression "learning by doing" that Arrow used in his title.

the existence of the same type of "learning curve" or "progress ratio," as it is variously termed, in the production of other machines, though the rate of learning is not the same as for airframes.

Verdoorn (1956, pp. 433–4) has applied the principle of the learning curve to national outputs; however, under the assumption that output is increasing exponentially, current output is proportional to cumulative output, and it is the former variable that he uses to explain labor productivity. The empirical fitting was reported in Verdoorn (1949); the estimated progress ratio for different European countries is about ·5. (In Verdoorn 1949, a neo-classical interpretation in terms of increasing capital–labor ratios was offered; see pp. 7–11.)

Lundberg (1961, pp. 129–133) has given the name "Horndal effect" to a very similar phenomenon. The Horndal iron works in Sweden had no new investment (and therefore presumably no significant change in its methods of production) for a period of 15 years, yet productivity (output per manhour) rose on the average close to 2% per annum. We find again steadily increasing performance which can only be imputed to learning from experience.

I advance the hypothesis here that technical change in general can be ascribed to experience, that it is the very activity of production which gives rise to problems for which favorable responses are selected over time. The evidence so far cited, whether from psychological or from economic literature is, of course, only suggestive. The aim of this paper is to formulate the hypothesis more precisely and draw from it a number of economic implications. These should enable the hypothesis and its consequences to be confronted more easily with empirical evidence.

Arrow calls attention to one of his most important conclusions: that in a free market, there will be too little learning because innovators don't care that, if they invented more, it would speed up further inventions by others.

The model set forth will be very simplified in some other respects to make clearer the essential role of the major hypothesis; in particular, the possibility of capital-labor substitution is ignored. The theorems about the economic world presented here differ from those in most standard economic theories; profits are the result of technical change; in a free-enterprise system, the rate of investment will be less than the optimum; net investment and the stock of capital become subordinate concepts, with gross investment taking a leading role.

In section 1, the basic assumptions of the model are set forth. In section 2, the implications for wage earners are deduced; in section 3 those for profits, the inducement to invest, and the rate of interest. In section 4, the behavior of the entire system under steady growth with mutually consistent expectations is taken up. In section 5, the divergence between social and private returns is studied in detail for a special case (where the subjective rate of discount of future consumption is a constant). Finally, in section 6, some limitations of the model and needs for further development are noted.

1. The Model

The first question is that of choosing the economic variable which represents "experience". The economic examples given above suggest the possibility of using cumulative output (the total of output from the beginning of time) as an index of experience, but this does not seem entirely satisfactory. If the rate of output is constant, then the stimulus to learning presented would appear to be constant, and the learning that does take place is a gradual approach to equilibrium behavior. I therefore take instead cumulative gross investment (cumulative production of capital goods) as an index of experience. Each new machine produced and put into use is capable of changing the environment in which production takes place, so that learning is taking place with continually new stimuli. This at least makes plausible the possibility of continued learning in the sense, here, of a steady rate of growth in productivity.

Wright's formula suggests using output as the index of experience from which learning is determined. But output sounds like doing more of the same thing. So instead Arrow imagines producing better and better machines, each of which will in turn produce output faster than the previous machine. The variation in machines makes for an evolving environment that can sustain learning.

The second question is that of deciding where the learning enters the conditions of production. I follow here the model of Solow (1960) and Johansen (1959), in which technical change is completely embodied in new capital goods. At any moment of new time, the new capital goods incorporate all the knowledge then available, but once built their productive efficiency cannot be altered by subsequent learning.

This is the famous vintage model, popularized by a paper by Solow, Tobin, von Wecksacker, and Yaari (1966), published after Arrow's paper.

To simplify the discussion we shall assume that the production process associated with any given new capital good is characterized by fixed coefficients, so that a fixed amount of labor is used and a fixed amount of output obtained. Further, it will be assumed that new capital goods are better than old ones in the strong sense that, if we compare a unit of capital goods produced at time t_1 with one produced at time $t_2 > t_1$, the first requires the co-operation of at least as much labor as the second, and produces no more product. Under this assumption, a new capital good will always be used in preference to an older one.

Arrow replaces the vintage V of a machine, which specifies the year it will come online, with the grade or serial number G, which specifies via a formula how productive the machine actually is.

Let G be cumulative gross investment. A unit capital good produced when cumulative gross investment has reached G will be said to have *serial number* G. Let

$$\lambda(G) = \text{amount of labor used in production with a capital good of serial number } G,$$

$$\gamma(G) = \text{output capacity of a capital good of serial number } G,$$

$$x = \text{total output},$$

$$L = \text{total labor force employed.}$$

This succinctly captures the idea that machines of a higher grade are better.

It is assumed that $\lambda(G)$ is a non-increasing function, while $\gamma(G)$ is a non-decreasing function. Then, regardless of wages or rental value of capital goods, it always pays to use a capital good of higher serial number before one of lower serial number.

It will further be assumed that capital goods have a fixed lifetime, $\overline{T}$. Then capital goods disappear in the same order as their serial numbers. It follows that at any moment of time, the capital goods in use will be all those with serial numbers from some G' to G, the current cumulative gross investment. Then

$$x = \int_{G'}^{G} \gamma(G)d(G), \tag{1}$$

$$L = \int_{G'}^{G} \lambda(G)dG. \tag{2}$$

The magnitudes x, L, G, and G' are, of course, all functions of time, to be designated by t, and they will be written $x(t)$, $L(t)$, $G(t)$, and $G'(t)$ when necessary to point up the dependence. Then $G(t)$, in particular, is the cumulative gross investment up to time t. The assumption about the lifetime of capital goods implies that

No machine can still operate more than $\overline{T}$ years after it was created.

$$G'(t) \geqq G(t - \overline{T}). \tag{3}$$

Since $G(t)$ is given at time t, we can solve for G' from (1) or (2) or the equality in (3). In a growth context, the most natural assumption is that of full employment. The labor force is regarded as a given function of time and is assumed equal to the labor employed, so that $L(t)$ is a given function. Then $G'(t)$ is obtained by solving in (2). If the result is substituted into (1), x can be written as a function of L and G, analogous to the usual production function. To write this, define

$$\Lambda(G) = \int \lambda(G)dG, \qquad \Gamma(G) = \int \gamma(G)dG. \tag{4}$$

These are to be regarded as indefinite integrals. Since $\lambda(G)$ and $\gamma(G)$ are both positive, $\Lambda(G)$ and $\Gamma(G)$ are strictly increasing and therefore have inverses, $\Lambda^{-1}(u)$ and $\Gamma^{-1}(v)$, respectively. Then 1 and 2 can be written, respectively,

$$x = \Gamma(G) - \Gamma(G'), \tag{1'}$$

$$L = \Lambda(G) - \Lambda(G'). \tag{2'}$$

Solve for G' from (2').

$$G' = \Lambda^{-1}[\Lambda(G) - L]. \tag{5}$$

The formula is very simple: if the quality of machines created today is G, then start assigning workers to the G machine and then to other machines in descending order until the supply of labor runs out. There is effectively just one machine of each quality, so this unambiguously leads to the least productive machine G' that can be run by the workers.

Substitute (5) into (1').

$$x = \Gamma(G) - \Gamma\{\Lambda^{-1}[\Lambda(G) - L]\}, \tag{6}$$

which is thus a production function in a somewhat novel sense. Equation (6) is always valid, but under the full employment assumption we can regard L as the labor force available.

A second assumption, more suitable to a depression situation, is that in which demand for the product is the limiting factor. Then x is taken as given; G′ can be derived from (1) or (1′), and employment then found from (2) or (2′). If this is less than the available labor force, we have Keynesian unemployment.

A third possibility, which, like the first, may be appropriate to a growth analysis, is that the solution (5) with L as the labor force, does not satisfy (3). In this case, there is a shortage of capital due to depreciation. There is again unemployment but now due to structural discrepancies rather than to demand deficiency.

In any case, except by accident, there is either unemployed labor or unemployed capital; there could be both in the demand deficiency case. Of course, a more neo-classical model, with substitution between capital and labor for each serial number of capital good, might permit full employment of both capital and labor, but this remains a subject for further study.

With (7) and (8) and the dynamic for G introduced in sections 4 and 5, Arrow ingeniously incorporated Wright's formula for the improvement in productivity as output increases while at the same time keeping an ever-changing environment of new machines, consistent with the lesson that sustained learning requires new challenges.

The instantaneous production function $x(G, L)$ gives us increasing returns to scale. Notice that $X/G < a$; since investment must come out of output, the growth rate of G could therefore never be more than a. The derivative of x with respect to G is the instantaneous marginal product of G.

An extra unit of G allows the economy to move workers from the now-obsolete G' to the more efficient machine G. Finally, x and G can rise in the same proportion, maintaining the capital/output ratio, as macro- economists like to say, if L also rises less but just enough to maintain the same L/G^{1-n} ratio. Such a calibrated rise in G and L leaves the marginal product of G unchanged.

In what follows, the full-employment case will be chiefly studied. The capital shortage case, the third one, will be referred to parenthetically. In the full-employment case, the depreciation assumption no longer matters; obsolescence, which occurs for all capital goods with serial numbers below G', becomes the sole reason for the retirement of capital goods from use.

The analysis will be carried through for a special case. To a very rough approximation, the capital-output ratio has been constant, while the labor-output ratio has been declining. It is therefore assumed that

$$\gamma(G) = a, \tag{7}$$

a constant, while $\lambda(G)$ is a decreasing function of G. To be specific, it will be assumed that $\lambda(G)$ has the form found in the study of learning curves for airframes.

$$\lambda(G) = bG^{-n}, \tag{8}$$

where $n > 0$. Then

$$\Gamma(G) = aG, \Lambda(G) = cG^{1-n}, \text{ where } c = b/(1-n) \text{ for } n \neq 1.$$

Then (6) becomes

$$x = aG\left[1 - \left(1 - \frac{L}{cG^{1-n}}\right)^{1/1-n}\right] \text{ if } n \neq 1. \tag{9}$$

Equation (9) is always well defined in the relevant ranges, since from (2′),

$$L = \Lambda(G) - \Lambda(G') \leqq \Lambda(G) = cG^{1-n}.$$

When $n = 1$, $\Lambda(G) = b \log G$ (where the natural logarithm is understood), and

$$x = aG(1 - e^{-L/b}) \text{ if } n = 1. \tag{10}$$

Although (9) and (10) are, in a sense, production functions, they show increasing returns to scale in the variables G and L. This is obvious in

(10) where an increase in G, with L constant, increases x in the same proportion; a simultaneous increase in L will further increase x. In (9), first suppose that $n < 1$. Then a proportional increase in L and G increases L/G^{1-n} and therefore increases the expression in brackets which multiplies G. A similar argument holds if $n > 1$. It should be noted that x increases more than proportionately to scale changes in G and L in general, not merely for the special case defined by (7) and (8). This would be verified by careful examination of the behavior of (6), when it is recalled that $\lambda(G)$ is non-increasing and $\gamma(G)$ is non-decreasing, with the strict inequality holding in at least one. It is obvious intuitively, since the additional amounts of L and G are used more efficiently than the earlier ones.

When $n = 1$ there are still increasing returns to scale. But now G enters linearly, so increasing G alone gives the same return, no matter how much G is increased. Arrow seems to lose track of this case later. He apparently did not notice that with $n = 1$, the steady-state equilibrium behaves quite differently from the case $n < 1$.

The increasing returns do not, however, lead to any difficulty with distribution theory. As we shall see, both capital and labor are paid their marginal products, suitably defined. The explanation is, of course, that the private marginal productivity of capital (more strictly, of new investment) is less than the social marginal productivity since the learning effect is not compensated in the market.

This is a shocking statement from the creator of general equilibrium. It is true that it is possible to pay everyone their marginal product out of the output they produce. In the usual case of increasing returns, the sum of the marginal products exceeds output, and thus can't be paid. Nevertheless, there most certainly is a difficulty. If any agent could borrow the output to produce more than an infinitesimal number of machines and hire the workers needed to run them at the going wage, they could make a pure profit after repaying their loan at the going interest rate. This profit would grow indefinitely large as they increased their scale of operation.

The production assumptions of this section are designed to play the role assigned by Kaldor to his "technical progress function," which relates the rate of growth of output per worker to the rate of growth of capital per worker (see Kaldor 1961, section VIII). I prefer to think of relations between rates of growth as themselves derived from more fundamental relations between the magnitudes involved. Also, the present formulation puts more stress on gross rather than net investment as the basic agent of technical change.

Earlier, Haavelmo (1954, sections 7.1 and 7.2) had suggested a somewhat similar model. Output depended on both capital and the stock of knowledge; investment depended on output, the stock of capital, and the stock of knowledge. The stock of knowledge was either simply a function of time or, in a more sophisticated version, the consequence of investment, the educational effect of each act of investment decreasing exponentially in time.

Verdoorn (1956, pp. 436–7) had also developed a similar simple model in which capital and labor needed are non-linear functions of output (since the rate of output is, approximately, a measure of cumulative output and

Arrow begins his investigation of the distribution of output into wages and profit. It would have been simpler to move directly to the dynamics and steady-state growth path of the model assuming that an arbitrary constant fraction s of output at each instant is invested into building new machines, as he does later. That does not require any theory of how the free market would set wages.

therefore of learning) and investment a constant fraction of output. He notes that under these conditions, full employment of capital and labor simultaneously is in general impossible—a conclusion which also holds for the present model as we have seen. However, Verdoorn draws the wrong conclusion: that the savings ratio must be fixed by some public mechanism at the uniquely determined level which would insure full employment of both factors; the correct conclusion is that one factor or the other will be unemployed. The social force of this conclusion is much less in the present model since the burden of unemployment may fall on obsolescent capital; Verdoorn assumes his capital to be homogeneous in nature.

2. Wages

Having decided to see how the free market sets wages, Arrow finds a simple solution. At any moment t, there will be a worst machine G' that is in use. The owner of any machine still worse must have chosen not to use it. That could only happen if the output from the worse machine would be less than or equal to the cost of paying the needed workers the going wage w to run the machine. By continuity of machine qualities, the profit of machine G' has to be less than or equal to 0 at wage w. If G' is used, its profit can't be negative. Hence it must be exactly zero. That condition determines w from G'.

Under the full employment assumption the profitability of using the capital good with serial number G' must be zero; for if it were positive it would be profitable to use capital goods with higher serial number and if it were negative capital good G' would not be used contrary to the definition of G'. Let

$$w = \text{ wage rate with output as numéraire.}$$

From (1′) and (7)

$$G' = G - (x/a) \tag{11}$$

so that

$$\lambda(G') = b\left(G - \frac{x}{a}\right)^{-n}. \tag{12}$$

The output from capital good G' is $\gamma(G')$ while the cost of operation is $\lambda(G')w$. Hence

$$\gamma(G') = \lambda(G')w$$

or from (7) and (12)

$$w = a\left(G - \frac{x}{a}\right)^{n/b}. \tag{13}$$

It is interesting to derive labor's share which is wL/x. From (2′) with $\Lambda(g) = cG^{1-n}$ and G' given by (11)

$$L = c\left[G^{1-n} - \left(G - \frac{x}{a}\right)^{1-n}\right],$$

for $n \neq 1$ and therefore

$$wL/x = a\left[\left(\frac{G}{x}-\frac{1}{a}\right)^{n}\left(\frac{G}{x}\right)^{1-n}-\left(\frac{G}{x}-\frac{1}{a}\right)\right]/(1-n) \tag{14}$$

for $n \neq 1$,

where use has been made of the relation, $c = b/(1-n)$. It is interesting to note that labor's share is determined by the ratio G/x.

Since, however, x is determined by G and L, which, at any moment of time, are data, it is also useful to express the wage ratio, w, and labor's share, wL/x, in terms of L and G. First, G' can be found by solving for it from (2′).

This reasoning is clear.

$$G' = \left(G^{1-n}-\frac{L}{c}\right)^{1/(1-n)} \quad \text{for } n \neq 1. \tag{15}$$

We can then use the same reasoning as above, and derive

$$w = a\left(G^{1-n}-\frac{L}{c}\right)^{\frac{n/(1-n)}{b}}, \tag{16}$$

$$\frac{wL}{x} = \frac{\left[\left(\frac{L}{G^{1-n}}\right)^{(1-n)/n}-\frac{1}{c}\left(\frac{L}{G^{1-n}}\right)^{1/n}\right]^{n/(1-n)}}{b\left[1-\left(1-\frac{L}{cG^{1-n}}\right)^{1/(1-n)}\right]}. \tag{17}$$

Arrow pays lip service to the case $n = 1$. He will shortly forget about this case. He has already ignored the case $n > 1$.

Labor's share thus depends on the ratio L/G^{1-n}; it can be shown to decrease as the ratio increases.

For completeness, I note the corresponding formulas for the case $n = 1$. In terms of G and x, we have

$$w = (aG - x)/b, \tag{18}$$

$$wL/x = \left(\frac{aG}{x}-1\right)\log\frac{G/x}{(G/x)-(1/a)}. \tag{19}$$

In terms of G and L, we have

$$G' = Ge^{-L/b}, \tag{20}$$

$$w = \frac{aG}{be^{L/b}}, \tag{21}$$

$$wL/x = \frac{L}{b(e^{L/b} - 1)}. \quad (22)$$

In this case, labor's share depends only on L, which is indeed the appropriate special case ($n = 1$) of the general dependence on L/G^{1-n}.

The preceding discussion has assumed full employment. In the capital shortage case, there cannot be a competitive equilibrium with positive wage since there is necessarily unemployment. A zero wage is, however, certainly unrealistic. To complete the model, it would be necessary to add some other assumption about the behavior of wages. This case will not be considered in general; for the special case of steady growth, see Section 5.

Section 3, like section 2, could have been delayed until after the exploration of steady-state growth paths with a constant rate of saving, as introduced in section 4.

3. Profits and Investment

The profit at time t from a unit investment made at time $v \leqq t$ is

$$\gamma[G(v)] - w(t)\ \lambda[G(v)].$$

Building a machine always costs one unit of output. The machine will then make a profit immediately, and also in future time periods. How much profit the owner can expect to make in the future depends on the wage they expect in the future.

In contemplating an investment at time v, the stream of potential profits depends upon expectations of future wages. We will suppose that looking ahead at any given moment of time each entrepreneur assumes that wages will rise exponentially from the present level. Thus the wage rate expected at time v to prevail at time t is

$$w(v)e^{\theta(t-v)},$$

and the profit expected at time v to be received at time t is

$$\gamma[G(v)]\ [1 - W(v)e^{\theta(t-v)}],$$

A crucial fact that motivates many of the future calculations, but is not mentioned by Arrow, concerns the marginal product of capital G, that is, the derivative of the production function $x(G, L)$ for current output, defined in eq. (9), with respect to G. The key insight is that this equals the instantaneous profit of machine $G(v)$ when $t = v$.

where

$$W(v) = w(v)\ \lambda[G(v)]/\gamma[G(v)], \quad (23)$$

the labor cost per unit output at the time the investment is made. The dependence of W on v will be made explicit only when necessary. The profitability of the investment is expected to decrease with time (if $\theta > 0$) and to reach zero at time $T^* + v$, defined by the equation

$$We^{\theta T^*} - 1. \tag{24}$$

Thus T^* is the expected economic lifetime of the investment, provided it does not exceed the physical lifetime, $\overline{T}$. Let

$$T = \min(\overline{T}, T^*). \tag{25}$$

Then the investor plans to derive profits only over an interval of length T, either because the investment wears out or because wages have risen to the point where it is unprofitable to operate. Since the expectation of wage rises which causes this abandonment derives from anticipated investment and the consequent technological progress, T^* represents the expected date of obsolescence. Let

$$\rho = \text{ rate of interest.}$$

If the rate of interest is expected to remain constant over the future, then the discounted stream of profits over the effective lifetime, T, of the investment is

$$S = \int_o^T e^{-\rho T}\, \gamma[G(v)]\, (1 - We^{\theta t})dt, \tag{26}$$

or

$$\frac{S}{\gamma[G(v)]} = \frac{1 - e^{-\rho T}}{\rho} + \frac{W(1 - e^{-(\rho-\theta)T})}{\theta - \rho}. \tag{27}$$

Let

$$V = e^{-\theta T} = \max(e^{-\theta T}, W),\ \alpha = \rho/\theta. \tag{28}$$

Then

$$\frac{\theta S}{\gamma[G(v)]} = \frac{1 - V^\alpha}{\alpha} + \frac{W(1 - V^{\alpha-1})}{1 - \alpha} = R(\alpha). \tag{29}$$

The definitions of $R(\alpha)$ for $\alpha = 0$ and $\alpha = 1$ needed to make the function continuous are:

$$R(0) = -\log V + W(1 - V^{-1}),\ \ R(1) = 1 - V + W\log V.$$

When wages are rising, with output price fixed at 1, every existing machine gets less and less profitable. When its profit hits 0, it becomes obsolete. Producers recognize that the world is getting more efficient and that they will eventually go out of business.

This is the most important equation in the paper. S is the present value of the profits of a newly produced machine at time v of grade $G(v)$. In equilibrium, $S = 1$, so entrepreneurs get exactly fair value of using 1 unit of output to build the new machine. If it is less, nobody will want to build the machine. If it is more than 1, everybody will be trying to borrow at rate ρ to build more machines.

If profit were constant and equal to ρ for $t = 0$ until $t = \infty$, then we would have $S = 1$. But T is less than infinity, and profit is declining. In order for $S = 1$, we need the initial profit to be much higher than ρ. If ρ were given exogenously, this would not even be possible unless productivity $a = \gamma(G)$ is bigger than ρ. Arrow will make this assumption later. We have seen that this initial profit is equal to the marginal product of G. Thus the derivative of $x(G, L)$ with respect to G must be much higher than ρ. This is a high barrier to production in the free market.

☞ The value of S depends on the discount rate ρ, the expected growth rate of wages θ, and, most critically, the wage rate $w(v)$ at the moment the machine is built, which appears in other units as W. In free-market equilibrium, S will have to be 1. In the following page of calculations, given any θ and $W < 1$, Arrow will show how to find the ρ which makes $S = 1$. In section 5, he will reverse the calculation and show how to find W for any given ρ and θ.

If all the parameters of (26), (27), or (29) are held constant, S is a function of ρ, and, equivalently, R of x. If (26) is differentiated with respect to ρ, we find

$$dS/d\rho = \int_0^T (-t)e^{-\rho^t}\, \gamma[G(v)]\, (1 - We^{\theta t})dt < 0.$$

Also

$$S < \gamma[G(v)] \int_0^T e^{-\rho^t} dt = \gamma[G(v)]\, (1 - e^{-\rho^T})/\rho$$
$$< \gamma[G(v)]/\rho.$$

Since obviously $S > 0$, S approaches 0 as ρ approaches infinity. Since R and α differ from S and ρ, respectively, only by positive constant factors, we conclude

$$dR/d\alpha < 0, \lim_{\alpha \to +\infty} R(\alpha) = 0.$$

To examine the behavior of $R(\alpha)$ as α approaches $-\infty$, write

$$R(\alpha) = -\frac{(1/V)^{1-\alpha}}{(1-\alpha)^2}[(1-\alpha)V + \alpha W]\left(\frac{1-\alpha}{\alpha}\right) + \frac{1}{\alpha} + \frac{W}{1-\alpha}.$$

The last two terms approach zero. As α approaches $-\infty$, $1 - \alpha$ approaches $+\infty$. Since $1/V > 1$, the factor

$$\frac{(1/V)^{1-\alpha}}{(1-\alpha)^2}$$

approaches $+\infty$, since an exponential approaches infinity faster than any power. From (28), $V \geqq W$. If $V = W$, then the factor,

$$(1-\alpha)V - \alpha W = \alpha(W - V) + V,$$

is a positive constant; if $V > W$, then it approaches $+\infty$ as α approaches $-\infty$. Finally,

$$\frac{1-\alpha}{\alpha}$$

necessarily approaches -1. Hence,

$$R(\alpha) \text{ is a strictly decreasing function, approaching } +\infty \text{ as } \alpha \text{ approaches } -\infty \text{ and } 0 \text{ as } \alpha \text{ approaches } +\infty. \tag{30}$$

The market, however, should adjust the rate of return so that the discounted stream of profits equals the cost of investment, i.e., $S = 1$, or, from (29),

$$R(\alpha) = \theta/\gamma[G(v)]. \tag{31}$$

Since the right-hand side of (31) is positive, (30) guarantees the existence of an α which satisfies (31). For a given θ, the equilibrium rate of return, ρ, is equal to $\alpha\theta$; it may indeed be negative. The rate of return is thus determined by the expected rate of increase in wages, current labor costs per unit output, and the physical lifetime of the investment. Further, if the first two are sufficiently large, the physical lifetime becomes irrelevant, since then $T^* < \overline{T}$, and $T = T^*$.

Mission accomplished.

The discussion of profits and returns has not made any special assumptions as to the form of the production relations.

4. Rational Expectations in a Macroeconomic Growth Model

Assume a one-sector model so that the production relations of the entire economy are described by the model of section 1. In particular, this implies that gross investment at any moment of time is simply a diversion of goods that might otherwise be used for consumption. Output and gross investment can then be measured in the same units.

In section 4 Arrow does what he might have done before section 2. He makes the standard macro assumption of his day, first suggested by Bob Solow, that at every moment the economy will always invest the same fraction s of its output into making more machines, and consume the rest. Following Solow again, Arrow also assumes that the workforce grows at a constant rate σ. From these assumptions, and the analysis of output from section 1, he quickly derives the unique stationary rate of growth of output and machine grade G, and the steady-state ratio of capital G to output. He also deduces the growth rate of wages θ.

The question arises, can the expectations assumed to govern investment behavior in the preceding section actually be fulfilled? Specifically, can we have a constant relative increase of wages and a constant rate of interest which, if anticipated, will lead entrepreneurs to invest at a rate which, in conjunction with the exogenously given rate of interest to remain at the given level? Such a state of affairs is frequently referred to as "perfect foresight," but a better term is "rational expectations," a term introduced by J. Muth (1961).

We study this question first for the full employment case. For this case to occur, the physical lifetime of investments must not be an effective constraint. If, in the notation of the last section, $T^* > \overline{T}$, and if wage expectations are correct, then investments will disappear through depreciation at a time when they are still yielding positive current profits.

Arrow frames the stationarity question in terms of prices, wages, and the interest rate. He could have shortened his calculations considerably by asking a related question: is there a constant common rate of growth for output x and capital G, maintaining the ratio G/x, that is compatible with labor growing at an arbitrary exogenously given rate σ? Inspection of the production function given by eq. (9) reveals that this requires the constancy of $L/G^{(1-n)}$. Hence x and G must grow at rate $\sigma/(1-n)$. A constant growth rate of G means the ratio of new investment to G is constant. Since G/x is constant, this also means that a constant fraction of output x at any t is invested in building new machines. Finally, stationarity also means that the labor of a new machine is always the same fraction of its output. Since required labor is $G^{(-n)}$, given the growth rate of G, wages must rise at rate $= n * \sigma/(1-n)$. The reader could skip directly to the assumption about σ after eq. (34) and then to eqs. (35)–(37).

☞

Arrow deduces what the steady-state growth rate of the economy must be before making any assumption about what fraction of output is invested and what fraction consumed. As long as the investment fraction is constant, he shows, its level does not affect the growth rate of the economy.

As seen in section 2, this is incompatible with competitive equilibrium and full employment. Assume therefore that

$$T^* \leqq \overline{T}; \tag{32}$$

then from (28), $W = V$, and from (29) and (31), the equilibrium value of ρ is determined by the equation,

$$\frac{1 - W^{\alpha}}{\alpha} + \frac{W - W^{\alpha}}{1 - \alpha} = \frac{\theta}{a}, \tag{33}$$

where, on the right-hand side, use is made of (7).

From (16), it is seen that for the wage rate to rise at a constant rate θ, it is necessary that the quantity,

$$G^{1-n} - \frac{L}{c},$$

rise at a rate $\theta(1-n)/n$. For θ constant, it follows from (33) that a constant ρ and therefore a constant α requires that W be constant. For the specific production relations (7) and (8), (23) shows that

$$W = a\frac{\left(G^{1-n} - \frac{L}{c}\right)^{n/(1-n)}}{b}\frac{bG^{-n}}{a} = \left(1 - \frac{L}{cG^{1-n}}\right)^{n/(1-n)},$$

and therefore the constancy of W is equivalent to that of L/G^{1-n}. In combination with the preceding remark, we see that

$$\begin{aligned} &L \text{ increases at rate } \theta(1-n)/n, \\ &G \text{ increases at rate } \theta/n. \end{aligned} \tag{34}$$

Suppose that

$$\sigma = \text{ rate of increase of the labor force,}$$

is a given constant. Then

$$\theta = n\,\sigma/(1-n), \tag{35}$$

$$\text{the rate of increase of } G \text{ is } \sigma/(1-n). \tag{36}$$

Substitution into the production function (9) yields

$$\text{the rate of increase of } x \text{ is } \sigma/(1-n). \tag{37}$$

From (36) and (37), the ratio G/x is constant over time. However, the value at which it is constant is not determined by the considerations so far introduced; the savings function is needed to complete the system. Let the constant ratio be

$$G(t)/x(t) = \mu \tag{38}$$

Define

$$g(t) = \text{ rate of gross investment at time } t = dG/dt.$$

From (36), $g/G = \sigma/(1-n)$, a constant. Then

$$g/x = (g/G)(G/x) = \mu\,\sigma/(1-n). \tag{39}$$

A simple assumption is that the ratio of gross saving (equals gross investment) to income (equals output) is a function of the rate of return, ρ; a special case would be the common assumption of a constant savings-to-income ratio. Then μ is a function of ρ. On the other hand, we can write W as follows, using (23) and (13):

Here is the assumption that at each date a constant fraction of output (income) is invested.

$$W = a\frac{\left(G - \frac{x}{a}\right)^n}{b}\,\frac{bG^{-n}}{a} = \left(1 - \frac{x}{aG}\right)^n = \left(1 - \frac{1}{a\mu}\right)^n. \tag{40}$$

This elementary but crucial fact is used later: higher W means lower x/G. Recall that the rate of growth of wages θ is the same at every stationary state, given the same growth rate of labor σ. If W is larger, then it will take less time until the machine becomes obsolete, meaning the gap between G' and G is smaller, meaning fewer machines in use, hence less output.

Since θ is given by (35), (33) is a relation between W and ρ, and, by (40) between μ and ρ. We thus have two relations between μ and ρ, so they are determinate.

From (38), μ determines one relation between G and X. If the labor force, L, is given at one moment of time, the production function (9) constitutes a second such relation, and the system is completely determinate.

As in many growth models, the rates of growth of the variables in the system do not depend on savings behavior; however, their levels do.

It should be made clear that all that has been demonstrated is the existence of a solution in which all variables have constant rates of growth, correctly anticipated. The stability of the solution requires further study.

Arrow is paying the price for the strange ordering of his analysis. He should have first assumed a constant investment fraction s. Then he could have traced the dynamics of how the economy evolved, starting from any G. He would have discovered that it converged to the steady state, answering his question.

The growth rate for wages implied by the solution has one paradoxical aspect; it increases with the rate of growth of the labor force (provided $n < 1$). The explanation seems to be that under full employment, the increasing labor force permits a more rapid introduction of the newer machinery. It should also be noted that, for a constant saving ratio, g/x, an increase in σ decreases μ, from (39), from which it can be seen that wages at the initial time period would be lower. In this connection it may be noted that since G cannot decrease, it follows from (36) that σ and $1-n$ must have the same sign for the steady growth path to be possible. The most natural case, of course, is $\sigma > 0$, $n < 1$.

This solution is, however, admissible only if the condition (32), that the rate of depreciation not be too rapid, be satisfied. We can find an explicit formula for the economic lifetime, T^*, of new investment. From (24), it satisfies the condition

$$e^{-\theta T^*} = W.$$

If we use (35) and (40) and solve for T^*, we find

$$T^* = \frac{-(1-n)}{\sigma} \log \left[1 - \frac{1}{a\mu} \right] \tag{41}$$

and this is to be compared with $\overline{T}$; the full employment solution with rational expectations of exponentially increasing wages and constant interest is admissible if $T^* \leqq T$.

If $T^* > \overline{T}$, then the full employment solution is inadmissible. One might ask if a constant-growth solution is possible in this case. The answer depends on assumptions about the dynamics of wages under this condition.

We retain the two conditions, that wages rise at a constant rate θ, and that the rate of interest be constant. With constant θ, the rate of interest, ρ, is determined from (31); from (29), this requires that

$$W \text{ is constant over time.} \tag{42}$$

From the definition of W, (23), and the particular form of the production relations, (7) and (8), it follows that the wage rate, w, must rise at the same rate as G^n, or

$$G \text{ rises at a constant rate } \theta/n. \tag{43}$$

In the presence of continued unemployment, the most natural wage dynamics in a free market would be a decreasing, or, at best, constant wage level. But since G can never decrease, it follows from (43) that θ can never be negative. Instead of making a specific assumption about wage changes, it will be assumed that any choice of θ can be imposed, perhaps by government or union or social pressure, and it is asked what restrictions on the possible values of θ are set by the other equilibrium conditions.

In the capital shortage case, the serial number of the oldest capital good in use is determined by the physical lifetime of the good, i.e.,

$$G' = G(t - \overline{T}). \text{ From (43),}$$
$$G(t - \overline{T}) = e^{-\theta\overline{T}/n}G.$$

Then, from (1′) and (7),

$$x = aG(1 - e^{-\theta\overline{T}/n}),$$

so that the ratio, G/x, or μ, is a constant,

$$\mu = 1/a(1 - e^{-\theta\overline{T}/n}). \tag{44}$$

From (43), $g/G = \theta/n$; hence, by the same argument as that leading to (39),

$$g/x = \theta/na(1 - e^{-\theta\overline{T}/n}). \tag{45}$$

There are three unknown constants of the growth process, θ, ρ, and W. If, as before, it is assumed that the gross savings ratio, g/x, is a function of the rate of return, ρ, then, for any given ρ, θ can be determined from (45); note that the right-hand side of (45) is a strictly increasing function of θ for $\theta \geqq 0$, so that the determination is unique, and the rate of growth is an increasing function of the gross savings ratio, contrary to the situation in the full employment case. Then W can be solved for from (31) and (29).

Thus the rate of return is a freely disposable parameter whose choice determines the rate of growth and W, which in turn determines the initial wage rate. There are, of course, some inequalities which must be satisfied to insure that the solution corresponds to the capital shortage rather than the full employment case; in particular, $W \leqq V$ and also the labor force must be sufficient to permit the expansion. From (2′), this means that the labor force must at all times be at least equal to

$$cG^{1-n} - c(G')^{1-n} = cG^{1-n}(1 - e^{-\theta(1-n)T/n});$$

if σ is the growth rate of the labor force, we must then have (46)

$$\sigma \geqq \theta(1 - n)/n, \tag{46}$$

which sets an upper bound on θ (for $n < 1$). Other constraints on ρ are implied by conditions $\theta \geqq 0$ and $W \geqq 0$ (if it is assumed that wage rates are non-negative). The first condition sets a lower limit on g/x; it can be shown, from (45) that

$$g/x \geqq 1/a\overline{T}; \tag{47}$$

i.e., the gross savings ratio must be at least equal to the amount of capital goods needed to produce one unit of output over their lifetime. The constraint $W > 0$ implies an interval in which ρ must lie. The conditions under which these constraints are consistent (so that at least one solution exists for the capital shortage case) have not been investigated in detail.

5. Divergence of Private and Social Product

☞ In section 5 Arrow assigns (the same) lifetime discounted utility function to every agent, allowing him to evaluate the welfare of any consumption stream. It also enables him to deduce the fraction of output agents want to consume every period in competitive equilibrium, rather than to take it as exogenous.

Arrow achieves two goals in this section: He shows how to compute competitive equilibrium, given the utility function he will define and using the wages and profits defined earlier in sections 2 and 3. He also shows how to compute an optimal utility maximizing allocation. His main point is that the optimal allocation has more investment in machines G than the competitive equilibrium. Though his policy conclusion is correct, he makes uncharacteristic errors in his formulation and solution of the two problems. With the benefit of hindsight, the correct analysis is provided in these notes.

As has already been emphasized, the presence of learning means that an act of investment benefits future investors, but this benefit is not paid for by the market. Hence, it is to be expected that the aggregate amount of investment under the competitive model of the last section will fall short of the socially optimum level. This difference will be investigated in detail in the present section under a simple assumption as to the utility function of society. For brevity, I refer to the *competitive solution* of the last section, to be contrasted with the *optimal* solution. Full employment is assumed. It is shown that the socially optimal growth rate is the same as that under competitive conditions, but the socially optimal ratio of gross investment to output is higher than the competitive level.

Utility is taken to be a function of the stream of consumption derived from the productive mechanism. Let

$$c = \text{consumption} = \text{output} - \text{gross investment} = x - g.$$

It is in particular assumed that future consumption is discounted at a constant rate, β, so that utility is

$$U = \int_0^{+\infty} e^{-\beta t} c(t)dt = \int_0^{+\infty} e^{-\beta t} x(t)dt. \\ - \int_0^{+\infty} e^{-\beta t} g(t)dt. \tag{48}$$

Integration by parts yields

$$\int_0^{+\infty} e^{-\beta t} g(t)dt = e^{-\beta t} G(t)\Big|_0^{+\infty} + \beta \int_0^{+\infty} e^{-\beta t} G(t)dt.$$

From (48),

$$U = U_1 - \lim_{t \to +\infty} e^{-\beta t} G(t) + G(0), \tag{49}$$

Arrow begins his demonstration without telling us the two assumptions he must make later to finish his proof. These are $\sigma/(1-n) < \beta < a$.

where

$$U_1 = \int_0^{+\infty} e^{-\beta t} [x(t) - \beta G(t)] dt. \tag{50}$$

The policy problem is the choice of the function $G(t)$, with $G'(t) \geqq 0$, to maximize (49), where $x(t)$ is determined by the production function (9), and

$$L(t) = L_0 e^{\sigma t}. \tag{51}$$

Arrow is very sloppy in his formulation of the problem. He has forgotten to mention that, since consumption $c(t)$ is nonnegative, investment $G'(t)$ cannot exceed output $x(t)$. The problem he says he tackles, with $G(0)$ fixed below, is to maximize utility over all potential paths, including non-stationary paths, beginning at $G(0)$. By ignoring the bounds on G' and also eventually forgetting entirely about $G(0)$, he ends up solving a different problem.

The second term in (49) is necessarily non-negative. It will be shown that, for sufficiently high discount rate, β, the function $G(t)$ which maximizes U_1 also has the property that the second term in (49) is zero; hence, it also maximizes (49), since $G(0)$ is given. Substitute (9) and (51) into (50).

$$U_1 = \int_0^{+\infty} e^{-\beta t} G(t) \left[a - \beta - a \left(1 - \frac{L_0 e^{\sigma t}}{c G^{1-n}} \right)^{1/(1-n)} \right] dt.$$

Let $\bar{G}(t) = G(t) e^{-\sigma t/(1-n)}$.

$$U_1 = \int_0^{+\infty} \overline{e}(\beta - \frac{\sigma}{1-n})t \ \bar{G}(t) \left[a - \beta - a \left(1 - \frac{L_0}{c \bar{G}^{1-n}} \right)^{1/(1-n)} \right] dt.$$

Assume that

$$\beta > \frac{\sigma}{1-n}; \tag{52}$$

This is one of the two assumptions he must make.

otherwise an infinite utility is attainable. Then to maximize U_1 it suffices to choose $\bar{G}(t)$ so as to maximize, for each t,

$$\bar{G} \left[a - \beta - a \left(1 - \frac{L_0}{c \bar{G}^{1-n}} \right)^{1/(1-n)} \right]. \tag{53}$$

Here Arrow forgets about the initial $G(0)$. What he actually does is find a stationary growth path $G^*(t)$ such that there is no better stationary or non-stationary path beginning with the starting point $G^*(0)$ defined in the stationary path. And he provides no intuition for his results.

Before actually determining the maximum, it can be noted that the maximizing value of $\bar{G}$ is independent of t and is therefore a constant. Hence, the optimum policy is

$$G(t) = \bar{G} e^{\sigma t/(1-n)}, \tag{54}$$

so that, from (36), the growth rate is the same as the competitive. From (52), $e^{-\beta t} G(t) \longrightarrow 0$ as $t \longrightarrow +\infty$.

To determine the optimal $\bar{G}$, it will be convenient to make a change of variables. Define

$$v = \left(1 - \frac{L_0}{c\bar{G}^{1-n}}\right)^{n/(1-n)}.$$

so that

$$\bar{G} = \left[\frac{L_0}{(1 - v^{(1-n)/n})}\right]^{1/(1-n)}. \tag{55}$$

The analysis will be carried through primarily for the case where the output per unit capital is sufficiently high, more specifically, where

Here is the second assumption.

$$a > \beta. \tag{56}$$

Let

$$\gamma = 1 - \frac{\beta}{a} > 0. \tag{57}$$

A simpler, more transparent proof of Arrow's stationary optimal growth path goes like this: The crucial idea is that an upgrade at time t upgrades the machines at every $s > t$ by the same amount. Assuming that a path $(G(t), x(t), c(t))$ satisfies $0 < G'(t) < x(t)$ for all t, a necessary and sufficient condition for it to be optimal is that for every t, the true marginal product of G at t $\int_t^\infty e^{-\beta(s-t)} \frac{\partial x(G(t),L(t))}{\partial G} ds$ must be equal to the investment of 1. Investment into G at time t should stop at exactly the point where the extra instantaneous output caused by a better machine at every moment s after t, discounted at rate β, is equal to the investment of 1 at t. Since this equality holds at every t, we must have that at every t, the instantaneous marginal product $dx(t)/dG$ is β. That also implies that the optimal path in which the constraints don't bind is stationary.

The maximizing $\bar{G}$, or v, is unchanged by multiplying (53), the function to be maximized, by the positive quantity, $(c/L_0)^{1/(1-n)}/a$ and then substituting from (55) and (57). Thus, v maximizes

$$(1 - v^{(1-n)/n})^{-1/(1-n)}(\gamma - v^{1/n}).$$

The variable v ranges from 0 to 1. However, the second factor vanishes when $v = \gamma^n < 1$ (since $\gamma < 1$) and becomes negative for larger values of v; since the first factor is always positive, it can be assumed that $v < \gamma^n$ in searching for a maximum, and both factors are positive. Then v also maximizes the logarithm of the above function, which is

$$f(v) = -\frac{\log(1 - v^{(1-n)/n})}{1-n} + \log(\gamma - v^{1/n}),$$

so that

$$f'(v) = \frac{v^{\frac{1}{n}-2}}{n}\left[\frac{\gamma - v}{(1 - v^{(1-n)/n})(\gamma - v^{1/n})}\right].$$

Clearly, with $n < 1$, $f'(v) > 0$ when $0 < v < \gamma$ and $f'(v) < 0$ when $\gamma < v < \gamma^n$, so that the maximum is obtained at

$$v = \gamma. \tag{58}$$

The optimum $\bar{G}$ is determined by substituting γ for v in (55).

From (54), L/G^{1-n} is a constant over time. From the definition of v and (58), then

$$\gamma = \left(1 - \frac{L}{cG^{1-n}}\right)^{n/(1-n)}$$

for all t along the optimal path, and, from the production function (9),

$$\gamma = \left(1 - \frac{x}{aG}\right)^{n} \text{ for all } t \text{ along the optimal path.} \quad (59)$$

This optimal solution will be compared with the competitive solution of steady growth studied in the last section. From (40), we know that

$$W = \left(1 - \frac{x}{aG}\right)^{n} \text{ for all } t \text{ along the competitive path.} \quad (60)$$

It will be demonstrated that $W < \gamma$; from this it follows that *the ratio* G/x *is less along the competitive path than along the optimal path.* Since along both paths,

$$g/x = [\sigma/(1-n)](G/x),$$

it also follows that *the gross savings ratio is smaller along the competitive path than along the optimal path.*

For the particular utility function (48), the supply of capital is infinitely elastic at $\rho = \beta$; i.e., the community will take any investment with a rate of return exceeding β and will take no investment at a rate of return less than β. For an equilibrium in which some, but not all, income is saved, we must have

$$\rho = \beta \quad (61)$$

From (35), $\theta = n\sigma/(1-n)$; hence, by definition (28),

$$\alpha = (1-n)\beta/n\sigma. \quad (62)$$

Since $n < 1$, it follows from (62) and the assumption (52) that (63)

$$\alpha > 1. \quad (63)$$

Equation (33) then becomes the one by which W is determined. The left-hand side will be denoted as $F(W)$.

$$F'(W) = \frac{1 - W^{\alpha-1}}{1-\alpha}.$$

From (63), $F'(W) < 0$ for $0 \geqq W < 1$, the relevant range since the investment will never be profitable if $W > 1$. To demonstrate that $W < \gamma$,

Again Arrow ignores the starting level of machines $G(0)$ in determining the competitive equilibrium. The steady-state competitive equilibrium that he does find can be deduced very quickly from the analysis of section 4. The interest rate or discount factor ρ for the economy must be equal to the common discount rate β assumed for each individual. The zero profit condition determines the rate of growth of wages. As we saw earlier, since in competitive equilibrium the entrepreneurs care only about the profit of the machine they build (ignoring the fact that society benefits in the future from the advance in knowledge), and since profits only go down over time and are finitely lived, the instantaneous profit at any point t where investment occurs must be bigger than the discount rate β. As we saw, the instantaneous profit is equal to the instantaneous marginal product of G, $dx(t)/dG$. Hence in competitive equilibrium the instantaneous marginal product of G is greater than β for every t. From our previous note we deduce that the competitive equilibrium is not optimal, because investment at each moment has stopped at the point where the instantaneous marginal product is too high. In other words, the ratio $G(t)/L(t)$ is lower at each t along the competitive equilibrium path than along the optimal steady state.

it suffices to show that $F(W) > F(\gamma)$ for that value of W which satisfies (33), i.e., to show that

$$F(\gamma) < \theta/a. \tag{64}$$

Finally, to demonstrate (64), note that $\gamma < 1$ and $\alpha > 1$, which imply that $\gamma^\alpha < \gamma$, and therefore

$$(1-\alpha) - \gamma^\alpha + \alpha\gamma > (1-\alpha)(1-\gamma).$$

Since $\alpha > 1$, $\alpha(1-\alpha) < 0$. Dividing both sides by this magnitude yields

$$\frac{1-\gamma^\alpha}{\alpha} + \frac{\gamma - \gamma^\alpha}{1-\alpha} < \frac{1-\gamma}{\alpha} = \frac{\theta}{a}$$

where use is made of (57), (28), and (61); but from (33), the left-hand side is precisely $F(\gamma)$, so that (64) is demonstrated.

The case $a \leqq \beta$, excluded by (56), can be handled similarly; in that case the optimum v is 0. The subsequent reasoning follows in the same way so that the corresponding competitive path would have $W < 0$, which is, however, impossible.

The steady-state competitive equilibrium and the optimal steady state both involve the same growth in output and consumption at rate $\sigma/(1-n)$. The optimal steady state has higher $G(t)$ and $c(t)$ throughout. No other path that starts from the $G(0)$ in the optimal steady state can give higher utility than the optimal steady state. However, the best utility among steady states is the golden rule given by even higher $G(t)$ and starting $G(0)$, namely the $G(t)$ at which $dx(G(t), L(t))/dG = \sigma/(1-n) < \beta$. That steady state maximizes $x(G(t), L(t)) - [\sigma/(1-n)]G(t)$, which is the instantaneous consumption left over after investment is subtracted from output.

Starting from the $G(0)$ of the steady-state competitive equilibrium, the optimal utility maximizing path would eschew all consumption, investing the whole of output x into new machines, until the date t at which $G(t)/L(t)$ rose to the level in the optimal steady state. Similarly, starting from the $G(0)$ of the golden rule steady-state, the optimal utility maximizing path would eschew all investment, consuming the whole of output x, until the date t at which $G(t)/L(t)$ fell to the level in the optimal steady state.

6. Some Comments on the Model

(1) Many writers, such as Theodore Schultz, have stressed the improvement in the quality of the labor force over time as a source of increased productivity. This interpretation can be incorporated in the present model by assuming that σ, the rate of growth of the labor force, incorporates qualitative as well as quantitative increase.

(2) In this model, there is only one efficient capital-labor ratio for new investment at any moment of time. Most other models, on the contrary, have assumed that alternative capital-labor ratios are possible both before the capital good is built and after. A still more plausible model is that of Johansen (1959), according to which alternative capital-labor ratios are open to the entrepreneur's choice at the time of investment but are fixed once the investment is congealed into a capital good.

(3) In this model, as in those of Solow (1960) and Johansen (1959), the learning takes place in effect only in the capital goods industry; no learning takes place in the use of a capital good once built. Lundberg's Horndal effect suggests that this is not realistic. The model should be extended to include this possibility.

(4) It has been assumed here that learning takes place only as a by-product of ordinary production. In fact, society has created institutions, education and research, whose purpose it is to enable learning to take place more rapidly. A fuller model would take account of these as additional variables.

REFERENCES

Abramovitz, M. 1956. "Resource and Output Trends in the United States Since 1870." *American Economic Review* 46 (2): 5–23.

Arrow, K. J., H. B. Chenery, B. S. Minhas, and R. M. Solow. 1961. "Capital-Labor Substitution and Economic Efficiency." *Review of Economics and Statistics* 43 (23): 225–250.

Asher, H. 1956. *Cost–Quantity Relationships in the Airframe Industry, R-291.* Santa Monica, CA.

Haavelmo, T. 1954. *A Study in the Theory of Economic Evolution.* Amsterdam: North Holland.

Hilgard, E. R. 1956. *Theories of Learning.* 2nd. New York, NY: Appleton-Century-Crofts.

Hirsch, W. Z. 1956. "Firm Progress Radios." *Econometrica* 24:136–143.

Johansen, L. 1959. "Substitution vs. Fixed Production Coefficients in the Theory of Economic Growth: A Synthesis." *Econometrica* 27:157–176.

Kaldor, N. 1961. "Capital Accumulation and Economic Growth." In *The Theory of Capital,* edited by F. A. Lutz and D. C. Hague, 177–222. New York, NY: St. Martin's Press.

Lundberg, E. 1961. *Produktivitet och räntabilitet.* Stockholm, Sweden: P. A. Norstedt / Söner.

Muth, J. 1961. "Rational Expectations and the Theory of Price Movements." *Econometrica* 29 (3): 315–335.

Solow, R. M. 1957. "Technical Change and the Aggregate Production Function." *Review of Economics and Statistics* 39:312–320.

———. 1960. "Investment and Technical Progress." In *Mathematical Methods in the Social Sciences, 1959,* edited by K. J. Arrow, S. Karlin, and P. Suppes, 89–104. Stanford, CA: Stanford University Press.

Verdoorn, P. J. 1949. "Fattori che regolano lo sviluppo della produttività del lavoro." *L'Industria* 1:3–11.

———. 1956. "Complementarity and Long-Range Projections." *Econometrica* 24:429–450.

Wright, T. P. 1936. "Factors Affecting the Cost of Airplanes." *Journal of the Aeronautical Sciences* 3:122–128.

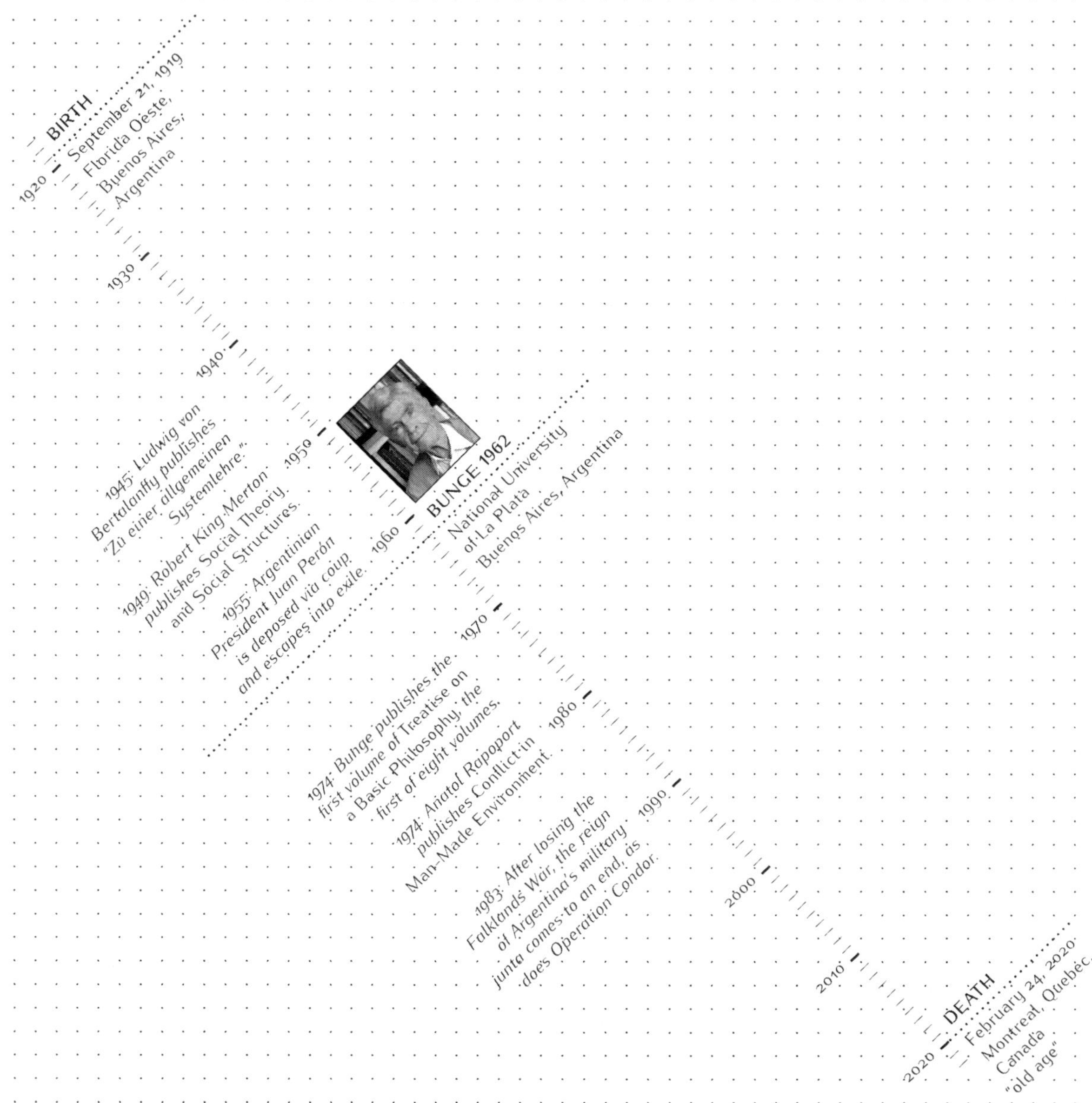

MARIO AUGUSTO BUNGE

[19]

THE UNSTABLE FOUNDATIONS OF SIMPLICITY

David C. Krakauer, Santa Fe Institute

M. Bunge, "The Complexity of Simplicity," *The Journal of Philosophy* 59 (5), 113–135 (1962).

The modern scientific position on simplicity is largely an extension of Newton's argument as presented in his defense of the classical theory of gravity. Namely, in the absence of known mechanism, it is the role of theory to express through laws, all—and only—that which can be deduced from the observation and analysis of phenomena. A value that Newton distilled into an epigram, "Nature does nothing in vain, and more is in vain when less will serve."

The physicist Ernst Mach ([1893]2013) brought the point home when he defined the physical sciences as "sense experience economically arranged." What we now call, somewhat habitually, the principle of parsimony or Ockham's razor is a statement about the compressibility of both compendious observations and rules of inference or generalization. Despite numerous recent efforts to formalize and quantify parsimony, it remains as John Laird described it a century ago, "Every one realizes that it is foolish to be recklessly prodigal of assumptions. That way incapacity lies. But in particular cases it is usually very difficult to prove that any given assumption is necessarily superfluous in every regard" (Laird 1919).

And simplicity is not only a virtue in the natural sciences. Raymond Havens (1953) has reviewed the deliberations of politicians, poets, and authors, from Swift to Walpole, who write of simplicity as, "The best and truest ornaments of most things in life," and that, "Taste . . . can not exist without simplicity." In a recent book on minimalism in art by Kyle Chayka (2020), *The Longing for Less*, the author captures one of the essential tensions experienced in pursuing simplicity as inscribed in a Kyoto garden, "dramatic simplicity side by side with unruly life."

Despite a near-universal cultural admiration for simplicity—its obvious appeal to both intelligibility and the widespread disdain for the unnecessary cost of surplus—there had not been before Mario Bunge's "The Complexity of Simplicity" many systematic efforts to quantify the concept. And Bunge reaches a similar conclusion to the review of Havens, and to Richard Rudner's 1961 summary, "An Introduction to Simplicity" describing the ideas of several philosophers wrestling with the concept, including Bunge's, that while simplicity is almost synonymous with the study of systems, it is upon close inspection uncomfortably various.

Mario Bunge was trained as a physicist in Argentina in the 1950s under Guido Beck and wrote his doctoral thesis on "The Kinematics of the Relativistic Electron." His early work in philosophy from the '60s bears the influence of his physics interests, including papers on chance, phenomenalism, and realism, and interpretations of quantum mechanics. After five years as professor of philosophy in Buenos Aires, Bunge emigrated in 1962 as a political dissident. He ultimately became Professor of Logic and Metaphysics at McGill University, Canada. He authored eighty books, including a *Treatise on Basic Philosophy* published in eight volumes between 1974 and 1989.

Martin Mahner (2021), in his obituary of Bunge in the *Journal for General Philosophy of Science*, describes one of Bunge's core contributions as his work on nomological state spaces. That is, systems whose elements can be covariantly connected (such as energy and mass in physics or metabolic rate and mass in biology) and which provide the basis for constrained state spaces, or "real possibilities," as opposed to the "logical possibilities" favored by philosophers. This is clearly a topic of central concern to complexity science where connections between energy, information, and adaptation impose severe limits on the space of possible phenotypes.

In "The Complexity of Simplicity" Bunge makes several key points: the first is to provide a typology of simplicity: these include syntactic (or logical), semantic (or functional), epistemological (or empirical), and pragmatic (or psychological) varieties. The second and perhaps most valuable insight is that these categories of simplicity are often, and perhaps necessarily, in quantitative disagreement. For example, maximizing the empirical fit of a model to data (= epistemological simplicity because fewer abstractions mediate a description) will frequently lead to complicated expressions (= syntactic complexity required to fit all contingencies). In modern terms, a Large Language Model (LLM) is very close to data (it is to first approximation the data itself coupled to some interesting rule-based constraints) and thereby in Bunge's language, epistemologically simple, but it is horribly complex in terms of the trillions of parameters required to fit the model (= syntactic complexity).

Based on this typology, Bunge defines complexity as "a four-dimensional manifold with an affine rather than a metrical geometry, in so far as the heterogeneity of the various kinds of complexity renders the summation of their various values meaningless." This is his third key insight—that there there is no sensible sum of complexity or simplicity measures that provides a meaningful summary of the final economical state of a theory. The researcher needs to declare which simplicity they are most interested in and ideally explore the implications of their choice on the various other kinds of simplicity that they have disregarded:

> *This should not be taken as a declaration of skepticism, but as a warning against one-sidedness and superficiality. If we wish to advance the subject we must face the complexity of simplicity instead of seeking refuge in model languages that are simple by construction. No oversimplification of simplicity will replace its analysis.*

It is interesting to note how various kinds of syntactic simplicities have come to dominate modern approaches to parsimony (Ding, Tarokh, and Yang 2018). Theories of model selection, including the Minimum Description Length Principle, Bayesian regularization penalties, the Aikaike information criterion, structural risk minimization, Vapnik–Chervonenkis theory, all seek some form of syntactic simplicity in terms of numbers of states or parameters that achieve fit to data or generalize—cross validate—to new data. Bunge makes the important point that most of these approaches do not address the underlying semantic complexity upon which a model is first constructed. As Bunge writes of differential equations applied in physical science,

> *In the first place, the degree of a differential equation is even more important than its order; thus a linear equation of the second order is in every sense simpler than a first-order and*

> *second-degree equation (like Hamilton–Jacobi's). Secondly, the absolute value of the derivation order is not a faithful index of simplicity: a fourth-order equation is not essentially more complex than a second-order equation, whereas fractional-order derivatives are enormously more complex than first-order derivatives.*

Model selection that simply counts and cross validates parameters of models without considering the formal structure of the generative model is in essence sweeping complexity under the carpet in order to arrive at a single, preferred dimension of simplicity. And that may not be the dimension of greatest interest.

There are many gems in this paper and it behooves readers to get past Bunge's rather awkward arithmetic of simplicity in the opening sections of the paper. After all, he only conducts this banal exercise to demonstrate its futility. Much of the paper's depth derives from later sections where distinctions are drawn between basis, operations, algorithms, intelligibility, and utility. If complexity science describes an ontology of adaptive agents, then semantic simplicity is perhaps forever lost, and syntactic simplicity needs to be defined on the basis of defensible and consistent (inter-theoretically) emergent rule systems.

REFERENCES

Chayka, K. 2020. *The Longing for Less: Living with Minimalism.* London, UK: Bloomsbury Publishing.

Ding, J., V. Tarokh, and Y. Yang. 2018. "Model Selection Techniques: An Overview." *IEEE Signal Processing Magazine* 35 (6): 16–34. https://doi.org/10.1109/MSP.2018.2867638.

Havens, R. D. 1953. "Simplicity, a Changing Concept." *Journal of the History of Ideas* 14 (1): 3–32. https://doi.org/10.2307/2707493.

Laird, J. 1919. "The Law of Parsimony." *Monist* 29 (3): 321–344. https://doi.org/10.5840/monist191929317.

Mach, E. [1893]2013. *The Science of Mechanics: A Critical and Historical Exposition of Its Principles.* Translated by T. J. McCormack. Cambridge, UK: University of Cambridge Press.

Mahner, M. 2021. "Mario Bunge (1919–2020): Conjoining Philosophy of Science and Scientific Philosophy." *Journal for General Philosophy of Science* 52 (1): 3–23. https://doi.org/10.1007/s10838-021-09553-7.

Rudner, R. 1961. "An Introduction to Simplicity." *Philosophy of Science* 28 (2): 109–119. https://doi.org/10.1086/287793.

THE COMPLEXITY OF SIMPLICITY

Mario Bunge, University of Pennsylvania

To live is to face complexities, and the foremost rules for approaching the problems they raise are, perhaps, "Try the simplest course first," and "Simplify: polish blurs, and decompose the given problem into simpler tasks." The analysis of the complex into simples—or rather into simpler units—is a desideratum in both practical and theoretical endeavors. But simplicity is easier to preach than to define. Now, if simplicity is a complex concept—as I shall attempt to show—then it is essential to disclose and distinguish its various dimensions before advising the choice of the simplest practical or theoretical course, and even before approaching problems such as those of establishing the relation of simplicity to testability, confirmation, or truth. That is to say, we should analyze simplicity before using the term, unless we want it to remain vague to the point of meaninglessness.

Various kinds of simplicity—or, if preferred, several usages of the word 'simplicity'—will first be distinguished; then some attempts to gauge degree of complexity will be examined, and finally the reasons for seeking certain kinds of simplicity while avoiding others will be investigated. But before analyzing the predicate 'simple' we should know to what it can be attributed. It seems clear that two kinds of object can be ordered in respect to complexity (or simplicity): material and cultural objects (things, events, processes) and their properties, on the one hand, and ideal objects (such as concepts, propositions, and theories) and their properties, on the other hand. Or, stated briefly and rather inaccurately, two kinds of object can be simple: things and signs. We shall speak accordingly of *ontological simplicity* and of *semiotic simplicity.*

The distinction between the simplicity of the world and the simplicity of our theories has more or less been absorbed into a de facto "positivism" in the current literature, the idea being that there are only theories and their verifiable statements.

Ontological simplicity has been postulated—and challenged—from the most ancient times. Semiotic simplicity has been sought—and sometimes avoided—of old as well, by poets and by scientists alike, although the theory of the simplicity of signs is still in its infancy. That nature and man are "basically" simple is an ancient ontological hypothesis-more precisely, a heuristically valuable prejudice-that can adequately be examined on the basis of science alone, since we judge the simplicity (or the complexity) of reality through that of scientific knowledge of it. Consequently, before approaching ontological simplicity, we should examine the simplicity of signs, a semiotic problem that has only recently been faced in a scientific spirit. Let us then proceed to study semiotic simplicity, leaving ontological simplicity for another occasion.

1. Syntactical Simplicity: Economy of Forms

There are basically four types of sign in the field of discourse: terms, propositions, proposals, and theories. Hence, we must begin by studying the formal or structural simplicity of terms (designating concepts), sentences (expressing propositions and proposals), and theories (systems of propositions). In turn, since propositions and proposals are built out of predicates (like 'between'), names of constants (like 'Argentina'), and variables (like 'x'), logical constants (like 'or'), logical prefixes (like 'all'), and modal prefixes (like 'possibly'), a methodical study of logical simplicity should begin by examining the formal complexity of predicates. If we assume that one and the same system of logic underlies all language systems occurring in rational discourse—which is true to a first approximation—we may leave out of account the complexity of logical signs for constituting a constant background common to nearly all discourse. The logical complexity of extralogical predicates will, accordingly, be studied in the first place.[1]

1.1. *L*-SIMPLICITY OF TERMS

Predicates may be formally simple (e.g., 'extended') or complex (e.g., 'extended over a sphere'); they are also called *atomic* and *molecular*. Further, predicates may be one-placed (e.g., 'long') or many-placed (e.g.,

[1] This was the approach of Adolphe Lindenbaum in his pioneering paper "Sur la simplicite formelle des notions," *Actes du congrbs international de philosophie scientifique* (Paris: Hermann, 1936), 7: 28.

'longer'). They may also be of the first order (designating properties of individuals), such as 'hot', or of a higher order (designating properties of properties), such as 'physical property'. From a further point of view, predicates may be dichotomic (presence/absence predicates), like 'curved', or metrical (quantitative properties or numerical functors), like 'curvature'. A rough measure of the complexity C of a dichotomic (nonmetrical) predicate might be the sum of the number A of its atomic constituents (degree of molecularity), the total number P of places of the atomic predicates (i.e., their degree), and their total order number O; i.e.,

$$C = A + P + O \tag{1}$$

Thus, the complexity value of 'black' would be $C = 1 + 1 + 1 = 3$, which suggests normalization to unity by dividing the complexity value by 3. With this natural normalization, the simplicity value of a predicate would be given by

$$S = \frac{1}{C} = \frac{3}{A + P + 0} \tag{2}$$

a function ranging from 0 (infinite complexity) to 1 (minimum complexity).

Notice that the preceding estimation of C presupposes the context of common-sense knowledge. In science we regard 'black' as a derivative, hence molecular, predicate, explicated in some such way as this: 'completely absorbing all electromagnetic wavelengths'. Further analyzed, the definition becomes

$$\text{black } x =_d f(y)[\text{electromagnetic wavelength } y \subset x$$
$$\text{absorbs } y \cdot (\text{absorption coefficient } x = 1)]$$

Three extralogical predicates, all of the first order, occur in the definiens. The first is molecular (degree of molecularity = 2), and the others are atomic ($A = 1$); their degrees or place numbers are 1, 2, and 1, respectively. Hence, the complexity value of 'black' in physics would be 4 (= total molecularity $= 2 + 1 + 1$) +4 (= total number of places $= 1 + 2 + 1$) + 3 (= total order number $= 1 + 1 + 1$) = 11, which, upon normalization, yields 11/3. The corresponding simplicity value is 3/11, approximately one-fourth the value found in the context of ordinary language. The complexity of predicates is, then, contextual; moreover, scientific analysis does not decrease but, on the contrary, is apt to increase the complexity of predicates. To require the simplification of concepts amounts to the stopping of analysis.

In the above elucidation of the predicate 'black' we have introduced two metrical predicates: 'electromagnetic wavelength' and 'absorption coefficient', and we have rated them on a par with the nonmetrical dyadic relation 'absorbs'. However, metrical predicates (numerical functors), the most characteristic of science, are, so to say, "infinitely" more complex than nonmetrical ones. 'Black' is either true or false of a given object, whereas the numerical variable 'wavelength' may take on an infinity of values. The above measure neglects this complexity of metrical predicates and must, therefore, be changed.

In order to account for the above-mentioned complexity of metrical predicates, we might decide to measure this aspect of complexity by some function $f(v)$ of the number of values v that predicates may take on, and such that $f(2) = 1$ (for dichotomic or classificatory predicates), and $f(\infty) = V < \infty$ (for metrical predicates), V being the same for all metrical predicates. Then the total complexity value of a term t in a system L (a given language system or a given scientific theory) would be

This whole exercise strikes us now as at best arbitrary and at worst futile. But perhaps this is Bunge's point. This kind of calculation comes across as veiled numerology.

$$\begin{aligned} C(t, L) &= \frac{1}{3}[\text{total molecularity of } t \text{ in } L \\ &\quad + \text{ total number of places of } t \text{ in } L \\ &\quad + \text{ total order number of } t \text{ in } L + V(t, L)] \\ &= \frac{1}{3}[A(t, L) + P(t, L) + O(t, L) + V(t, L)] \end{aligned} \tag{3}$$

But the following objections suggest themselves immediately.

(1) Some metrical predicates, like 'wavelength' or 'distance', have infinite range, whereas others—such as 'absorption coefficient' or 'velocity'—have finite range (though, of course, an infinity of values). Why should we rate on a par predicates having ranges so widely different as those of 'distance'-which varies between 0 and ∞—and 'absorption coefficient', which is bounded by 0 and 1?

(2) There are infinitely many numbers V and, moreover, infinitely many well-behaved functions $f(v)$ that might fulfill the border conditions

$f(2) = 1$ and $f(\infty) = V < \infty$ required above. Without further postulates any choice among the double infinity of numbers and functions would be much too arbitrary.

(3) Even if these shortcomings could be remedied with the addition of reasonable postulates, there does not seem to be any reason why the four numbers A, P, 0 and V, which are manifestly nonhomogeneous, should be added without further ado, particularly since the predicate order is so much weightier than, say, the number of places.

These difficulties may be some of the reasons why the few logicians who have tried to set up measures of the complexity of predicates have not taken into consideration whether they were metrical or not. As a consequence, their theories are not even applicable to ordinary language, where more and more metrical predicates occur. In fact, the search for an adequate measure of the structural richness of sets of extralogical predicates is under way,[2] but up to now only very modest results have been obtained: they are significant in relation to pauper model languages in which the predicates are both nonmetrical and mutually independent—which is not the case for scientific sign systems.

Take, for instance, Kemeny's proposal to measure the richness of the basic extralogical vocabulary (i.e., of the set of extralogical predicates that are taken as primitives in a theory) by a function of the number of models, or linguistically possible worlds, of the set of predicates. Now, since a single continuous numerical functor, like 'position', yields infinite "models" of that kind, Kemeny is forced to restrict his calculus of simplicity to classificatory and order predicates; and even for these dichotomic concepts he requires a finite domain of individuals—which is consistent with empiricist ontologies but totally at variance with the needs of science. A generalization of Kemeny's measure to cover infinite universes is not possible; so an approach not based on the number

[2] Adolphe Lindenbaum. *op. cit.* Nelson Goodman, *The Structure of Appearance* (Cambridge, Mass.: Harvard University Press, 1951), ch. III; "Axiomatic Measurement of Simplicity," this JOURNAL, 52 (1955): 709, and "The Test of Simplicity," *Science*, 128 (1958): 1064. John G. Kemeny, "Two Measures of Complexity," this JOURNAL, 52 (1955): 722.

of "models" of a logical system should be sought. That the Kemeny approach is inadequate can be seen from a further angle.

The possible states to be found in the universe—or, if preferred, the possible state descriptions-are not determined by a set of independent predicates, and, even less, by a set of nonmetrical independent variables, but rather by the (partially known) set of all the laws of nature and society, which laws consist of definite relations among predicates which are very often metrical. If we want to know, say, what are the "possible worlds" that might be built with such a "simple" object as a single vibrating string (i.e., its infinite possible modes of vibration), we must solve the vibrating-string equation with definite boundary and initial conditions. And this is not a language game; moreover, its results are useless for the approach under consideration, for the number of possible "models" (or state descriptions) of a vibrating string is infinite. State descriptions are of little interest in isolation from law statements.

Here are echoes of Wittgenstein's language games, and later ideas behind Kuhn's paradigms. Both emphasize the larger matrix of meaning required to understand the meaning of a sign. The analysis of simple signs is like the analysis of undetermined systems: there are not enough constraints to reach a solution.

Further complications ought to be accounted for. One of them is the possible mutual dependence of predicates by way of law statements. A second source of complexity not taken into account in the available theories of complexity stems from the interrelations of terms in any system worth this name. Even if isolated symbols could be assigned finite complexity values in an unambiguous way, the fact should be faced that there are no isolated symbols: every symbol belongs to some language system. Now, a symbol immersed in a language may be manifestly simple yet potentially complex, owing to the possible metamorphoses allowed by the transformation rules of the system. Thus the numeral '1' can be written as

$$m/m, \quad -n/-n, \quad i \cdot (-i), \quad (-1)^{2n}, \quad a^0, \quad \cos 0, \quad \lim_{n \to \infty} a^{1/n},$$

Here is a nice demonstration of the limitations of syntactic simplicity. Behind each one of these expressions is a whole area of mathematics. They are more or less as meaningless as scant letters and numbers on a page without the disciplinary (semantic) context.

and so on *ad infinitum*. In this sense every number sign is infinitely complex, despite its manifest simplicity. Consequently, separate measures should be set up for *manifest complexity* and *latent complexity* when dealing with nonprimitive predicates. Also, measuring the complexity of primitives does not exhaust the problem unless the rules of transformation allowed by the system are extremely poor.

We conclude that the theory of the syntactical simplicity of terms is an inspiring yet underdeveloped enterprise, the results of which are so far inapplicable to the estimate of the complexity of propositions actually occurring in ordinary life, let alone in science. If the theory of simplicity is to be of any use, it must cease avoiding complexities in actual sign systems. The rule, "Simplify," should not lead us to circumvent complexities by building ideally simple linguistic toys; theories of simplicity need not be simple themselves.

1.2 *L*-SIMPLICITY OF PROPOSITIONS

Atomic propositions are obviously simpler than molecular ones—at the propositional level at least. It is tempting to measure the formal complexity of a proposition p by assuming that the connectives occurring in it are all equally simple—hence irrelevant to the complexity value—and by counting the number of its logically independent (atomic) constituents, or by adding unity to the number of its connectives.[3] We should then have $C(p) = N(p)$, whence the simplicity of p could be defined as $S(p) = 1/N(p)$. In particular, $C(p \cdot q) = C(p) + C(q)$.

But this is too coarse a measure, for it neglects no less than the inner complexity of the atomic constituents themselves, which may contain complex predicates. A better measure would be the inverse of the sum of the complexity values of the atomic propositions p_i, namely,

$$S = \frac{1}{\Sigma C(p_i)} \tag{4}$$

But here we find again the unsolved difficulties mentioned in the previous section, notably the one regarding the measure of metrical predicates. And even if we decided to measure only propositional complexity, we should meet a grave difficulty posed by universal propositions. In effect, let $(x)Fx$ be such a proposition; if the universe of reference consists of n individuals, named $c_1, c_2, \cdots, c_n$, the complexity value of $(x)Fx$ will be that of $Fc_1 \bullet Fc_2 \bullet \cdots \bullet Fc_n$. Since the complexity of these factors is the same, the complexity of the universal statement is $C[(x)Fx] = nC(Fc_i)$, which tends to infinity

[3] This is, essentially, the proposal of Horst Kiesow, "Anwendung eines Einfaehheitsprinzip auf die Wahrseheinliehkeitstheorie," *Archiv für Mathematische Logik und Grundlagenforschung*, 4 (1958): 27. See also Hans Hermes, "Zum Einfachheitsprinzip in der Warscheinlichkeitsrechnung," *Dialectica*, 12 (1958): 317.

as n approaches infinity. According to (4), then, all strictly universal propositions have the same simplicity value: zero. And this renders the proposed measure entirely irrelevant to science.

Wrinch and Jeffreys[4] have attempted to measure the complexity of a certain class of universal propositions containing metrical variables and only metrical predicates. They did not analyze the predicates themselves, and they restricted their treatment to physical laws expressible as differential equations, on the controvertible assumption that "Every law of physics is expressible as a differential equation of finite order and degree, with rational coefficients."[4] Later on[5] Jeffreys proposed to restrict the coefficients to integers, defining the complexity of an equation of this kind as the sum of the order, the degree, and the absolute value of the coefficients. This proposal is open to the following objections.

☞ The philosophy of science benefits from attention to the specificities of models such as Bunge considers here. For example, if we just considered the simplicity of the Dirac equation (at parametric face value) we would have no sense of the oddity, dare I say latent complexity, of the quantum world.

In the first place, the degree of a differential equation is even more important than its order; thus a linear equation of the second order is in every sense simpler than a first-order and second-degree equation (like Hamilton-Jacobi's). Secondly, the absolute value of the derivation order is not a faithful index of simplicity: a fourth-order equation is not essentially more complex than a second-order equation, whereas fractional-order derivatives are enormously more complex than first-order derivatives; thus, nobody would count a $\frac{1}{2}$-order differential equation as simpler than a first-order equation. Moreover, first-order (partial) differential equations are often more complex than second-order (partial) differential equations, this being the original reason for introducing potentials in Maxwell's theory. Thirdly, the numerical value of the coefficients is irrelevant; at most their ratio (not their sum) could count. Fourthly, the number of variables, on the other hand, is important, but it does not occur in the approach under consideration. Fifthly, a measure of complexity obtained by

---

[4] Dorothy Wrinch and Harold Jeffreys, "On Certain Fundamental Principles of Scientific Inquiry," *Philosophical Magazine*, 42 (1921): 369. Jeffreys abandoned this assumption in the second edition of his *Theory of Probability* (Oxford: Clarendon Press, 1948), p. 100. The pioneer work of Wrinch and Jeffreys does not seem to have been noticed by other workers in the theory of simplicity, with the exception of Popper.

[5] Harold Jeffreys, *Scientific Inference*, 1st ed. (Cambridge: Cambridge University Press, 1931), p. 45.

adding entirely heterogeneous characteristics all equally rated looks suspiciously simple.

Another proposal is to measure the complexity of an equation, whether algebraic or differential, by the number of adjustable parameters in it.[6] But again, this proposal overlooks the degree and the number of variables of an equation, as well as the fact that fractional derivation orders are more complex than integral ones. Compare the equations of light absorption $u_x + \lambda u = 0$ with the wave equation $u_{xx} + u_{yy} + u_{zz} - u_{tt} = 0$. If simplicity is to be equated with paucity of parameters, the second equation must be assigned the complexity value 0 and the first the value 1—but this is surely counterintuitive, even from a typographical point of view. Also, $y = x$ and $y = x^2$ are equally complex by the same token and also "because their consequences will usually differ so much that discrimination between them by means of observation will be easy."[7] But this is certainly an undesirable procedure: the gauge of a logical property like syntactical complexity should not depend on empirical operations—which will be completely out of place if the signs happen to be devoid of empirical meaning.

We need only consider the logistic equation here, the bifurcation series and transition to chaos. A zoo of phenomena hiding within a rather trivial 1–2 parameter model. It is the real value of the parameter where the true complexity hides.

It would seem that complexity has been confused here with generality and with derivativeness, as I shall try to explain by means of an example. The ellipse equation, $(x/a)^2 + (y/b)^2 = 1$, contains two adjustable parameters, a and b; the hyperbola equation, $(x/a)^2 - (y/b)^2 = 1$ has the same complexity value on the criterion under examination, although we would rate the hyperbola, roughly, as twice as complex as the ellipse, for having two branches. Again, the circle equation, $(x/r)^2 + (y/r)^2 = 1$ contains the single parameter r, whence it would be half as complex as the ellipse. However, a simple gauge transformation, namely, $x' = x/a, y' = y/b$, leads from the ellipse to the circle, which shows that the former is more general than the latter ($a = b = r$ being a special case), but not necessarily more complex-or, if it is more complex, then its complexity is not intrinsic but depends on

[6] Jeffreys, *Theory of Probability*, 2nd ed., p. 100. Karl R. Popper, *The Logic of Scientific Discovery* (1935; London: Hutchinson, 1959), secs. 44-46 and *Appendix VIII.

[7] Jeffreys, *loc. cit.*

the choice of the coordinate system, so that C ought to be relativized to the latter.

In the second place, the more freely adjustable parameters a *physical* equation contains, the less *fundamental* it is, insofar as fundamentality can be defined as absence of numerical and material constants (in contrast with universal constants such as c, e, or h);[8] but fundamentality is an epistemological, not a syntactical characteristic. Besides, fundamental equations are *methodologically* simpler than phenomenological relations, in that they are easier to dispose of, whereas the latter protect themselves against refutation by suitable adjustment of parameters. But, again, this could not count for syntactical simplicity.

What is valuable in the proposal that the complexity of an expression of mathematical (more exactly, quantitative) form be measured by the number of its adjustable parameters, is the methodological aspect that has just been mentioned (ease of test), as well as an epistemological consequence of it, namely, that the larger the number of parameters an expression contains, the greater its probability will be—hence the less acceptable on any standpoint except those of phenomenalism and conventionalism, which anyhow are inimical to science. But formal complexity should not be confused with difficulty of test, although the former may entail the latter. Besides, the identification of simplicity with paucity of parameters is powerless to gauge the complexity of propositions that cannot be given the form of equations connecting metrical variables: it assigns the same complexity value (namely, zero) to all expressions in which no parameters occur, no matter how many atomic predicates they contain and what their degree and order are.

We conclude that no adequate measure of the complexity of propositions and propositional functions is available.

---

[8] The occurrence of unexplained numerical constants is typical of empirical generalizations and auxiliary hypotheses. The frequent occurrence of mathematical constants such as π and e, on the other hand, is perfectly admissible in fundamental equations, because they are characteristic of the mathematical tools employed.

1.3. *L*-SIMPLICITY OF THEORIES

The complexity of theories (systems of propositions) must obviously depend on the number and complexity of their postulates. But how? No definite answer to this question can be given before a theory of the simplicity of propositions is available. However, a few remarks can be made. In the first place, it is usually granted that the number of independent analyzed postulates of a theory is relevant to its complexity. Not, of course, the mere number of postulates—which, by conjunction, can always be reduced to one—but the number of statements containing no further independent statements, i.e., no further statements that might be regarded as postulates. But this is insufficient.

In fact, it is easy to see that the simplicity of the basis of a theory does not warrant its over-all formal simplicity: L-simple assumptions may have L-complex consequences, and equally L-simple assumptions may have consequences of unequal L-complexity. Thus, the hypothesis $dy/dx = ax$ leads, by integration, to $y = \frac{1}{2}ax^2 + b$, whereas the equally simple assumption $dy/dx = ay$ yields $y = b\exp(ax)$, which is obviously more complex than the former (being, in fact, a function not reducible to a polynomial). As in the case of terms (compare 1.1), we have to distinguish the manifest from the latent complexity of a theory's postulates; the latent complexity of a theory's basis is given by the number and variety of its transformations, which in turn is determined by the number of its primitives and by the rules of transformation accepted by the theory. Unless we realize this, we shall be tempted to assign all physical theories the same L-complexity value, since they can all be derived from a single variational principle each.

That not only the number and complexity of a theory's postulates but also the number and complexity of its transformation rules are relevant to its L-complexity is plain: after all, we are interested in the complexity of systems that, from a semiotic point of view, can be regarded as languages, and the formal complexity of a language is determined by the complexity of both its vocabulary and its grammar.[9] Thus, mathematical analysis is more complex than arithmetic, because

[9]Willard van Orman Quine, *From a Logical Point of View* (Cambridge, Mass.: Harvard University Press, 1953), p. 26, calls *postulational* this kind of economy, in contrast with paucity of practical expression, which he calls *notational* economy. In the nomenclature

it contains a larger number of operations. But how are we to measure the complexity of the rules of transformation that regulate such operations? Their syntactical complexity, or even their theorematic power (their yield of theorems), will not do: the former does not say much about the complexity of the resulting system, and the latter may be infinite. What matters most is the degree of conceptual unity, or cohesiveness, the rules of transformation lead to, a cohesiveness to which definitions contribute as well. But there is no obvious way of measuring this cohesiveness.

To conclude this section: the ultimate aim of the theory of *L*-simplicity ought to be to gauge the structural simplicity of theories. But we have no satisfactory *L*-complexity measure of theories. We may well counsel the adoption of the rule: "Choose the *L*-simplest among all the theories consistent with the available body of empirical evidence," but, if we do not know how to gauge or at least how to compare degrees of *L*-complexity of theories, our advice will be no more effective than wishing good-night.

Besides, are we sure that we should prefer the *L*-simplest among all the theories consistent with the known empirical data? The simplest of all would be just the conjunction of the given observational propositions, but this would be neither a rich nor a deep system. Besides, we do not want *L*-simple transformation rules either, because the desideratum with regard to complexity is not just paucity but economy, i.e., maximum value of the ratio of yield to means. And this desideratum is inconsistent with un- qualified *L*-simplicity.

Semantical Simplicity: Economy of Presuppositions

2.1. S-SIMPLICITY OF TERMS

This subject seems to be virgin: logicians have been interested in formal simplicity of simple languages, and, as a consequence, they have rated as equivalent all concepts having the same logical form and type, e.g., 'one' and 'Aristotle'. However, complexity of meaning is at least as important as complexity of form, although the latter must certainly be approached

---

of the present paper, notational economy is a subclass of pragmatic simplicity (see sec. 4).

first, both because it is less elusive and because the meanings that can be assigned to a sign depend in part on its form. A few informal remarks will be enough to show the richness of the subject.

There is a trivial sense in which a term 't_1' may be said to be semantically more complex than a second term 't_2', namely, if 't_1' has a larger number of different acceptations than 't_2'. We shall not be concerned with this kind of semantic complexity, because it is a linguistic problem. On the other hand, the body of assumptions and beliefs that stands "behind" any given concept poses a genuine philosophical problem, to which we now turn.

That 'theory' denotes a concept more complex than 'proposition' does, although the two terms are syntactically equally complex, will easily be conceded; the same thing happens with 'living' in relation with 'assimilating' and with 'electron' in relation with 'mass'. The reason seems to be this: 'proposition' occurs in the definiens of 'theory', but normally not vice versa, and the same is true for the other examples. The more complex terms presuppose a larger number of concepts and propositions (e.g., laws of nature) than the simpler do. If preferred, the meaning of the more complex notions is specified (not necessarily in an explicit and unambiguous way) by a larger number of concepts and propositions than is the case with the simpler ones.[10] Thus, 'velocity' presupposes 'distance' and 'duration', which may in turn be regarded as primitives in the context of present-day physics; whence we say that 'velocity' is semantically more complex than either 'distance' or 'duration'. On the other hand, 'force' is as complex as 'mass' in the context of Newtonian mechanics.

The above remarks suggest establishing the following measure of the simplicity of a concept from a semantical point of view. The S-simplicity of a term t (denoting a concept) in the language system L (e.g., that of a theory) is the inverse of the number of its extralogical specifiers of meaning:

$$S(t, L) = 1/M(t, L) \qquad (5)$$

For example, in the arithmetic of natural numbers as formalized by

[10] For an explication of the notion of speeifieation of meaning, see A. Kaplan, "Definition and Specification of Meaning," this JOURNAL, 43 (1946): 281.

Peano, 'one' and 'successor' are primitives, so their S-complexity value is zero; on the other hand, since '2' $=_{df}$ 'successor of 1', the complexity value of '2' is 2. Notice that, on the above stipulation, intensional simplicity is contextual; if we choose 'O' as primitive instead of '1', the S-complexity value of '2' becomes 3.

However, various objections may be raised against the above proposal. Firstly, according to it the S-complexity value of the number $10,000$ would be one thousand times larger than that of 10—and this is counterintuitive. This shortcoming might be met by choosing some smoothing function of the number M of specifiers of meaning, such as $\log M(t, L)$. With this choice we would then have $C(10,000$, Peano arithmetic$) = \log 10^4 = 4$, whereas $C(10$, Peano arithmetic$) = \log 10^1 = 1$, which seems more satisfactory. But then the S-complexity value of 1 would no longer be zero, but $C(1) = \log 0 = -\infty$, which has no obvious interpretation; moreover, the corresponding S-simplicity value would be indeterminate. Of course, a different function might be chosen that would not yield an undesirable value for the beginner of the sequence of integers; for example, $C = \log(1 + M)$, $C = 1 - e^{-M}$, and $C = \tanh M$ would do. But why just these among the infinity of functions—for instance, $\log^n(1 + M)$, $\log(1 + M)^n$, $\tanh M^{1/n}$, $1 - e^{-M^2}$, and so on—that might do the same job? The obvious rejoinder is, of course, that $\log(1 + M)$, $1 - e^{-M}$, and $\tanh M$ are to be preferred because they are simpler and more familiar than most other functions with the same smoothing property. Yet, this appeal to simplicity does not reduce the ambiguity altogether, and it employs an intuitive notion of simplicity. Secondly, the specifiers of meaning cannot be counted in nonformalized languages, which are by far the most numerous. This is partly because the specification of meaning is in them circular or interconceptual rather than linear,[11] and partly because the disclosure of the ideas "behind" a given concept is both a logical and a historical task: terms denote concepts, concepts belong to views, theories, and conceptions, and the latter are parts of intellectual culture, which is anything but

[11]See C. West Churchman, "Concepts without Primitives," *Philosophy of Science*, 20 (1953): 257, for the interdependence of scientific concepts and the accompanying circularity of definitions.

a closed set. An exhaustive conceptual analysis would have to take the whole cultural background into account, particularly as regards critical words like 'cause' or 'idea'. Even so, not all of the presuppositions would be dug out—(*a*) if the analysandum is familiar to us, because we should overlook precisely what was most obvious, (*b*) if it belongs to a past period, because we cannot reconstruct with arbitrary accuracy the ideas, attitudes, and desiderata of the past. Counting presuppositions is an almost hopeless task. Consequently, *S*-simplicity could be measured in formalized languages alone—i.e., where it is hardly an interesting problem. Thirdly, even if the number of specifiers of meaning could be counted in interesting cases, there is no reason why they should all be assigned the same weight unless they all occurred as primitives in the given system.

In short, no reasonable way of measuring the *S*-simplicity of terms is in sight.

This seems to be inevitable. Without a complete genealogy of research fields, and a relational exposition of terms, semantic simplicity would be baseless.

2.2 S-SIMPLICITY OF PROPOSITIONS

Supposing we had a reasonable theory of the semantical simplicity of terms, we still would have to build a theory of the simplicity of propositions with respect to meaning. We might, of course, try something like the reciprocal of the sum of the *S*-complexity values of the extralogical predicates occurring in the proposition. But it does not seem likely that such complexity values are summable, unless they are intensionally independent, which, again, is an unrealistic assumption in the case of science, which may be described as a quest for interconnection. A simpler—and probably more fertile—approach would be to count the number of presuppositions a proposition has in a given context. Take, for instance, the rival cosmological hypotheses "The universe exists from all eternity" and "The universe has had a beginning in time." The former is *S*-simpler than the latter, in that it does not require any of an unlimited number of hypotheses about the alleged creation mechanism. A comparative, not a quantitative, *S*-simplicity concept is thereby suggested, for only in formalized theories is it possible to count ideas.

2.3 S-SIMPLICITY OF THEORIES

There is urgent need for the treatment of this problem, because scientists actually apply (mostly in a tacit way) criteria of evaluation and selection of theories with respect to their semantical complexity. Thus physicists dislike complicated mathematics in so far as it obstructs the interpretation of their formulas in physical terms: they make every effort to simplify the mathematics because syntactically complex expressions are difficult to "read." They do not mind too much if the intermediary computations are lengthy or even clumsy—although they care for expediency and elegance as much as the mathematicians do—on condition that both the postulates and their testable consequences be semantically simple-because, as Fresnel said, nature does not worry about our analytical difficulties and because, as experience teaches, graduate students can always be found to do the boring computations. The comparative semantical simplicity of both the postulates and the testable consequences of general relativity is one of the reasons for its acceptance despite its notorious epistemological and pragmatic (particularly, psychological and algorithmic) complexity.

But how are we to gauge the semantical complexity of theories if we barely know how to estimate that of propositions? The most that we can state at the moment is the truism that the semantical complexity of theories is determined by both the formal and the semantical complexity of their postulates and rules of designation. But this is both too obvious and too vague. It shows that, if the theory of L-simplicity is an infant, that of S-simplicity is not yet born. A good chance for semanticists.

3. Epistemological Simplicity: Economy of Transcendent Terms

Let us call epistemological simplicity degree of closeness to sense experience and, in particular, to observation. Degree of epistemological complexity (or abstractness) is, then, a sort of "distance" from sense experience, a kind of measure of the gap between constructs and percepts. If preferred, epistemological simplicity is degree of ostensiveness, or fewness of transcendent or non-instantial concepts[12]—terms which, like

[12]This is the sense of 'simplicity' dealt with by John O. Wisdom in *Foundations of Inference in Natural Science* (London: Methuen, 1952), ch. vii.

'conductivity', 'choice', and 'love', have no referents that can be pointed to and no sense data that can be reduced to.

Phenomenalistic utterances—like 'I feel a white patch'—are the simplest with regard to epistemological simplicity. (But, as was pointed out in 1.1, phenomenal predicates are, in the context of scientific theory, syntactically more complex than physical ones, hence also semantically more complex, this being one of the reasons for their not being used as primitives.) Then come, in order of epistemological complexity, ordinary physicalist expressions—like 'This is chalk'—and finally, in the same order, scientific physicalist sentences-like 'This is calcium carbonate'. Predominantly phenomenalist languages—to the extent to which they are possible—achieve epistemological simplicity, or triviality, at the cost of both syntactical complexity and epistemological shallowness: it takes longer to say less in phenomenalist languages. The latter are not economical, but just poor.

Syntactical economy and epistemological depth are won by introducing constructs,[13] that is, by increasing the epistemological complexity of theories. One of the most important results of recent epistemology is Craig's demonstration[14] that it is possible to dispense with all the transcendent or diaphenomenal terms of a theory, by building another theory containing observable predicates only—but at the price of giving up a finite postulational basis in favor of an infinite one. In other terms, once a theory has been built, it is possible to destroy its rich superstructure and use only its debris to build a huge one-floor store of observational data: the new theory's postulates will be the infinite conjunctions of the infinite theorems derived from the original theory. In short, epistemological economy is possible, but it must be performed a posteriori (it does not save theoretical work) and must be paid for by a loss of insight and an infinite syntactical complexity.

Margenau's idea of "constructs" is a more sophisticated version of Holland and Gell-Mann's schema. They are semi-autonomous encodings of environmental regularities. A kind of atomic concept.

Predominantly physicalist languages, in which theoretical con- structs occur, allow for greater logical cohesiveness and compactness, deeper insight, and easier refutability (they take more risks than phenomenalist constructions, which ultimately inform about our subjective feelings).

[13]The essential role of constructs has been emphasized by Henry Margenau in *The Nature of Physical Reality* (New York: McGraw-Hill, 1950), chs. 4 and 5.

[14]William Craig, "Replacement of Auxiliary Expressions," *Philosophical Review*, 65 (1956): 38.

Besides, physical-object languages enable us to distinguish subject from object, appearance from reality, the finite known from the infinite unknown, and so on; finally, they render communication possible (a strictly phenomenalist language would be private), and they are interesting.

Modern science did not discourage phenomenalist philosophers, but it definitely involved the downfall of phenomenalism within the field of science: since the seventeenth century scientists have not striven for epistemological simplicity but, on the contrary, have been inventing more and more transcendent (transempirical) concepts, more and more theoretical entities, with the sole restrictions that they be part of theories, scrutable (not necessarily observable), and fertile. Suffice it to recall the atomic analysis of phenomenally continuous bodies, the replacement of gravitational pull by a higher-order unobservable like space-time curvature, and the introduction of the psi-function, that bewildering source of both observables and further unobservables.[15] Physicists do not choose secondary or sensible qualities as fundamental variables; on the contrary, fundamental physical variables—like position and time—and fundamental physical parameters—like mass and charge—are not reducible to sense perception: they are constructs, and it is not possible to reconstruct them as "logical constructions" in the sense of Russell. A further aspect of scientific departure from immediate experience—hence another source of epistemological complication—is the enrichment of the set of relations among constructs, namely, the system of definitions and law statements. In short, the progress of science is accompanied by epistemological complication.

This is where Bunge connects to Popper and Lakatos. These epistemological complications are the anomalies that add up to the eventual refutation of an existing theory or framework.

However, science does not strive for epistemological luxury either: it does not seek epistemological complexity for its own sake; it does not multiply transempirical entities without necessity. A high order of mediacy (a low degree of epistemological simplicity) is not established out of servility toward some system of metaphysics—as was the case with medieval science—but is forced upon us by (*a*) the desiderata of formal, semantical, and pragmatic simplicity, and (*b*) the complexity of reality,

[15] For an examination of physical terms lacking material and/or empirical correlates, see Mario Bunge, *Metascientific Queries* (Springfield: Thomas, 1959), pp. 252 ff.

which is intelligible but not sensible in its entirety. At least, the invention of constructs more and more distant from sense experience would be incomprehensible if reality were simple and if we were not under the pressure of trying to cope with it by organizing our conceptual equipment around a set of highly fertile basic constructs.

The invention of scientific constructs is subject to some rules that prevent their useless multiplication. Firstly, every construct, however high its order of mediacy (remoteness from sense data), must somehow, somewhere, be related, by means of correspondence rules (e.g., coordinative definitions) and law statements, to lower-order concepts. Secondly, constructs must belong to testable theories. (Terms such as 'superego' and 'telepathy' are objected to not because they are transempirical but because they do not occur in testable theories.) Thirdly, conceptual entities should not be multiplied in vain (Occam's razor), but should be welcomed whenever they lead either to deeper understanding of reality or to syntactical simplification of theories. In this respect, science seems to take a middle course between the poverty of phenomenalism and the luxury of transcendentalism.

The above discussion refers primarily to epistemological simplicity of terms; much the same applies to the E-simplicity of propositions and theories. Unnecessarily complicated assumptions and theories should be avoided; that is, hypotheses and theoretical systems employing inscrutable predicates, such as 'Providence' and 'collective unconscious,' should be shaved with Occam's razor. Notice, however, that Occam's razor does not hang in the air, but falls under the more general rule, "Do not propose ungrounded and untestable hypotheses."

Besides, complication of initially simple assumptions may well be unavoidable in the face of increasing incompatibility with the body of evidence. We should encourage such an increase in E-complexity if it does not consist merely in adding arbitrarily adjustable parameters (frequent among phenomenological theories) or untestable *ad hoc* hypotheses, but does involve an enrichment of our picture of the world, e.g., an increase in its accuracy. Thus, for example, to a first approximation, the acceleration of gravity can be regarded as independent of height, but, as soon as we wish to compute the orbit of an artificial satellite, we need a more exact formula, which contains both height and the radius of the earth, scrutable predicates to which we unambiguously assign objective properties—which is never the case with the adjustable parameters of phenomenological theories.

All this, even if it were true, is too sketchy. One should attempt to find at least an ordering relation among concepts of different degrees of epistemological complexity; one should try to explicate such intuitively acceptable orderings as, e.g., 'quantity of heat'—'temperature'—'energy'—'entropy' (an order which is not determined by physical theory). But a point may be reached where epistemological and psychological simplicity become indistinguishable: one cannot help imagining that, if our remote descendants were endowed with a sensory apparatus finer than our own, they would employ scientific constructs that we would regard as entirely unintuitive. This leads us to the last class of simplicity to be distinguished here.

4. Pragmatic Simplicity: Economy of Work

From a pragmatic viewpoint, sign complexes can be arranged in respect to simplicity of at least the following kinds.

This section opens up into possible discussion on the "psychology" of artificial intelligence. Ideas of algorithmic and notational simplicity depend to a certain degree on the capacity of the comprehending agent. This makes real, observer-dependent parsimony, and is an important topic for current machine-learning science.

4.1 *Psychological simplicity:* obviousness, ease of understanding, or familiarity. Thus Ptolemy's postulate is psychologically (and epistemologically) simpler than the heliocentric hypothesis, and psychoanalysis requires far less previous training than psychology does (although it is epistemologically more complex than psychology, on account of the number of inscrutable predicates it contains). The psychological complexity $C(x, s)$ of an object, problem, or procedure x, for a subject (animal or person) s, might be measured by the time needed by s to apprehend the object, to solve the problem, or to master the procedure. In maze problems, where trial and error is at work, C might be measured by the number of unsuccessful trials. An alternative measure, proposed by Birkhoff,[16] is the sum of the efforts or tensions of the various automatic adjustments occurring in the subject during apprehension of the object. This proposal is tempting because it involves a psychological model, but it poses the difficult problem of objectifying the tensions, and it is obviously inadequate in relation to nonautomatic behavior.

[16] George D. Birkhoff, *Aesthetic Measure* (Cambridge, Mass.: Harvard University Press, 1933).

This is the type of simplicity envisaged by the partisans of economy of thought, and particularly by Mach,[17] who described scientific research as a business (*Geschäft*) pursued for the sake of saving thought. Psychological simplicity is desirable for both practical (e.g., didactic) and heuristic purposes: if preliminary theoretical models are to be set up, details must be brushed aside, and easily understandable notions ("intuitive" ideas) must be seized upon. The obvious must be tried first, if only to dispose of it early: this is a well-known rule of intellectual work. However, it should be borne in mind that (1) psychological simplicity is culturally and educationally conditioned; i.e., is not an intrinsic property of sign systems; (2) the deliberate neglect of a given factor should always be justified; (3) we must be prepared to sacrifice psychological economy to depth and accuracy whenever the former becomes insufficient—for, whatever science is, it certainly is not a business whose concern is to save experience with the minimum expenditure of work.

4.2 *Notational simplicity,* or economy and suggestive power of signs. Thus, vector and tensor representations are notationally simpler than the corresponding "analytical" (expanded) mode of writing, and the symbol '$P(h/e)$' for the probability of the hypothesis h on the evidence e is suggestive, hence easy to retain. Needless to say, compactness facilitates interpretation and favors retention, and well-chosen symbols may be heuristically valuable. Notational clumsiness and bulkiness, on the other hand, are obstacles to symbol manipulation and consequently to thought—as must have been felt by the Greek mathematicians in connection with their system of numerical notation. Yet, compactness is of little use if achieved through a long chain of nominal definitions, for then notational economy disguises semantical complication.

4.3 *Algorithmic simplicity,* or ease of computation, is something like the reciprocal of the number of steps in logical or mathematical calculation (this number is actually computed in the programming of machines). Algorithmic simplicity depends on logical, semantical, and psychological simplicity, though not in a simple way. Thus the relation '$y < x$' is S-

[17]Ernst Mach, "Die Gestalten der Fliissigkeiten" (1868), in *Populärwissenschaftliche Vorlesungen*, 4th ed. (Leipzig: Barth, 1910), p. 16.

simpler than '$y = x$' in the sense that the equality relation is defined in terms of the two inequality relations. But the corresponding point set $\hat{x}\hat{y}(y < x)$ is algorithmically more complex than $\hat{x}\hat{y}(y = x)$, because finding the frontier of the former set presupposes the determining of the latter set.

Algorithmic simplicity is sometimes accomplished though complication of the basis; thus certain infinite series can best be computed by approximating them by integrals, and certain integrals of real functions become easier to compute if the real variable is replaced by a complex argument. Conversely, theories that are comparatively simple at the basis may be algorithmically complex; thus Dirac's theory of the electron, which starts with a notationally simple equation, requires such lengthy and difficult calculations that it is hardly used in the computing of electron behavior in accelerators, although there is no doubt about its comparative accuracy and depth.

A desideratum of science is to invent an algorithm—a "mechanical" decision procedure—for every class of problems. However, such "thoughtless" procedures, while pragmatically valuable, may not replace alternative, more complex procedures, but may supplement them, because the simpler procedures save time at the price of giving up some insight.

4.4 *Experimental simplicity:* simplicity in the design, performance, and interpretation of observations or experiments. Hypotheses having the operationally simplest consequents, however complex they may be in all remaining aspects, will be preferred by experimentalists. This may have the undesirable consequence of delaying the empirical test of valuable complex theories.

4.5 *Technical simplicity:* ease of application to practical (non-cognitive) purposes. In applied science, where standards of tolerance are often large and where expediency and low cost may count more than accuracy, rough but simple theories are often preferred to more refined and complex ones.

5. Which Simplicities are Desirable, Why, and to What Extent?

Various kinds of semiotic simplicity have been distinguished in the foregoing: syntactical, semantical, epistemological, and pragmatic. We do not know for certain how to measure every kind of complexity nor, indeed, whether they are all measurable. Even if we knew this, we still would have to find a suitable function of the various measures that might serve as an index for over-all complexity. Metaphorically, we may say that complexity is a four-dimensional manifold with an affine rather than a metrical geometry, in so far as the heterogeneity of the various kinds of complexity renders the summation of their various values meaningless.

Yet, although we lack an exact and complete theory of the simplicity of sign systems, we do possess some half-baked notions on the subject. This semi-intuitive (insufficiently analyzed) conception of simplicity may suffice for approaching the most interesting problems in relation with its use, namely, "What kinds of simplicity do we seek and to what extent?," and "Why do we seek simplicity in some cases and avoid it in others?"

We want *syntactical simplicity* because it favors (*a*) systematicity or cohesiveness (exact formulation and conceptual connectedness being further factors of systematicity); (*b*) easy checking of consistency and/or completeness of the postulate basis; and (*c*) empirical testability (accuracy and scrutability being two other factors). In turn, we require systematicity and testability to the extent to which we want science. Now, it is often asserted that formal simplicity is decisive in the choice among competing hypotheses and theories. But this is not true: other criteria, such as accuracy and depth, which are manifestly incompatible with simplicity, are far weightier. Even formal criteria, such as symmetry and extensibility, may predominate over simplicity. Thus, when working in relativity physics, we would trust an assumption or a theorem of the form A_4B_i less than the formula $A_4B_i - A_iB_4$: the latter is more complex, but is also more symmetrical and, consequently, capable of extension (namely, to $A_\mu B_\nu - A_\nu B_\mu$). In this case, symmetry involves extensibility and is preferable to simplicity.

As to *semantical simplicity* (economy of presuppositions), we value it because, although we want systematicity, we also wish to be able to start research into new subjects without having to know everything else and because economy of presuppositions facilitates interpretation. There must be interconnections among the various fields of research, and these must be compatible with one another, but such ties must allow for some play if the enterprise of knowing is to be possible at all. Yet, it should be realized that semantical simplicity may be incompatible with syntactical economy, as shown by the following examples.

(1) Mathematical transformations are simple in special relativity and in thermodynamics, but the physical interpretation of the formulas is often difficult; suffice it to recall the clock paradox and irreversibility.

(2) Syntactical simplification of physical formulas can be achieved by adopting certain systems of units (e.g., the "natural" units $c = h = 1$) or by making certain changes of variables (e.g., by introducing $u = (4\pi^2 kcm/h^2)^{\frac{1}{4}} x$ in place of the coordinate of the quantum-mechanical linear oscillator). But such procedures, expedient as they are for purposes of computation, becloud the interpretation of the final results, since dimensional considerations are among the main clues to the meaning of complex expressions occurring in physical theory.

With regard to *epistemological simplicity* (economy of transcendent concepts), we definitely do not want poverty of transcendent concepts, but seek rather a middle course between poverty and waste, and this because we wish to attain the richest and most accurate possible picture of reality with the fewest indirectly testable assumptions. The motto of science is not just *Pauca*, but rather *Plurima ex paucissimis*—the most out of the least. In short, we wish economy and not merely parsimony.

Finally, *pragmatic simplicity* is prized on quite different grounds: because we are limited, short-lived, hungry, lazy, stingy, impatient, and ambitious. One should not be ashamed of recognizing that the norm of pragmatic simplicity does not rank much higher than the counsel "Take it easy." While in many cases both are reasonable rules of conduct, there are occasions on which it is inconvenient or even immoral to choose the widest door. Sticking to pragmatic simplicity at all costs is making a virtue out of an imperfection and may lead—has led—to serious damage to knowledge. Granted, useless toil should be avoided; but economy of work should not imperil universality and depth. Scientists do not regard their work as a Minimumaufgabe, but just the opposite; in particular, they know that simplification itself is not always a simple task but may require great labor and ingenuity. Moreover, they are aware that syntactical and pragmatic simplicity are to some extent mutually incompatible desiderata, because paucity in the basis will be overcompensated by lengthy developments (e.g., chains of definitions, and computations).[18] For example, the binary base is both the L-simplest and the S-simplest, to such an extent that human computers find it inconvenient; on the other hand, automatic computers are endowed with it (or, rather, with its material counterpart) because simplicity of the basis is essential for artifacts, whereas the number of operations is comparatively irrelevant in them owing to their high speed. Likewise, scientists and technicians are aware that semantical and epistemological simplicity are consistent with pragmatic complexity: the test or the application of a comparatively simple idea may be exasperatingly difficult, particularly in the sciences and technologies of man; reciprocally, some other things may be more easily done than described and analyzed. One might think that *conservation*, rather than minimization, of mental energy was at work in scientific matters, in that an increase in complexity of a given kind tends to be compensated for by a simplification of a different sort, and vice versa.

Here Bunge discusses the various simplicity trade-offs. My feeling is that this section opens up a fascinating space of future research into the nature of scientific inquiry. The ability to construct alternative synthetic cognitions means that we might now explore the "possible worlds" of theory.

It would be difficult to conclude any definite moral about the value of the various competing kinds of simplicity. One thing at least is certain, however, viz., that the simple injunction "Simplify" is inadequate, if only because it is ambiguous. If some rule has to be proposed, let it be the following: "Simplify as long as simplification does not eliminate interesting problems and does not carry a severe loss of generality, testability, or depth." This maxim is no less ambiguous, but at least it is less harmful than unqualified simplicism.[19]

Conclusion

In short, 'simplicity' is a multivocal term, not all kinds of simplicity are desirable or even compatible with one another, and the theory of simplicity, though still in a very rudimentary stage, threatens to become highly complex. Moreover, the search for a single measure of over-all

[18] Cf. Quine, *op. cit.* That there might be a contradiction between syntactical and practical simplicity was denied, e.g., by Lindenbaum, *op. cit.*

[19] The dangers of simplicism are examined in Mario Bunge, "The Weight of Simplicity in the Construction and Assaying of Scientific Theories," *Philosophy of Science*, 28 (1961): 120.

complexity does not seem to make much sense. This should not be taken as a declaration of skepticism, but as a warning against one-sidedness and superficiality. If we wish to advance the subject we must face the complexity of simplicity instead of seeking refuge in model languages that are simple by construction. No oversimplification of simplicity will replace its analysis.

REFERENCES

Birkhoff, G. D. 1933. *Harvard University Press.* Cambridge, MA.

Bunge, M. 1959. *Metascientific Queries.* 252. Springfield, IL: Thomas.

———. 1961. "The Weight of Simplicity in the Construction and Assaying of Scientific Theories." *Philosophy of Science* 28:120–149.

Churchman, C. W. 1953. "Concepts without Primitives." *Philosophy of Science* 20:257.

Craig, W. 1956. "Replacement of Auxiliary Expressions." *Philosophical Review* 65 (38).

Goodman, N. 1951. *The Structure of Appearance.* Cambridge, MA: Harvard University Press.

———. 1958. "The Test of Simplicity." *Science* 128 (3331): 1064–1069.

Hermes, H. 1958. "Zum Einfachheitsprinzip in der Warscheinlichkeitsrechnung." *Dialectica* 12:317–331.

Jeffreys, H. 1931. *Scientific Inference.* 1st ed. Cambridge, UK: Cambridge University Press.

———. 1948. *Theory of Probability.* 2nd ed. Oxford, UK: Clarendon Press.

Kaplan, A. 1946. "Definition and Specification of Meaning." *The Journal of Philosophy* 43:281–288.

Kemeny, John G. 1955. "Two Measures of Complexity." *The Journal of Philosophy* 52:722–733.

Kiesow, H. 1958. "Anwendung eines Einfachheitsprinzips auf die Wahrscheinlichkeitstheorie." *Archiv für Mathematische Logik und Grundlagenforschung* 4:27–41.

Lindenbaum, A. 1936. "Sur la simplicité formelle des notions." In *Actes du Congres International de Philosophie Scientifique,* edited by Hermann, vol. 7. 28. Paris.

Mach, E. 1910. *Die Gestalten der Flüssigkeiten.* 4th ed. Leipzig: Barth.

Margenau, H. 1950. *The Nature of Physical Reality.* New York, NY: McGraw-Hill.

Popper, K. R. 1959 [1935]. *The Logic of Scientific Discovery.* London: Hutchinson.

Quine, W. V. O. 1953. *From a Logical Point of View.* Cambridge, MA: Harvard University Press.

Wisdom, J. O. 1952. London, UK: Methuen.

Wrinch, D., and H. Jeffreys. 1921. "On Certain Fundamental Principles of Scientific Inquiry." *Philosophical Magazine* 42:369–390.

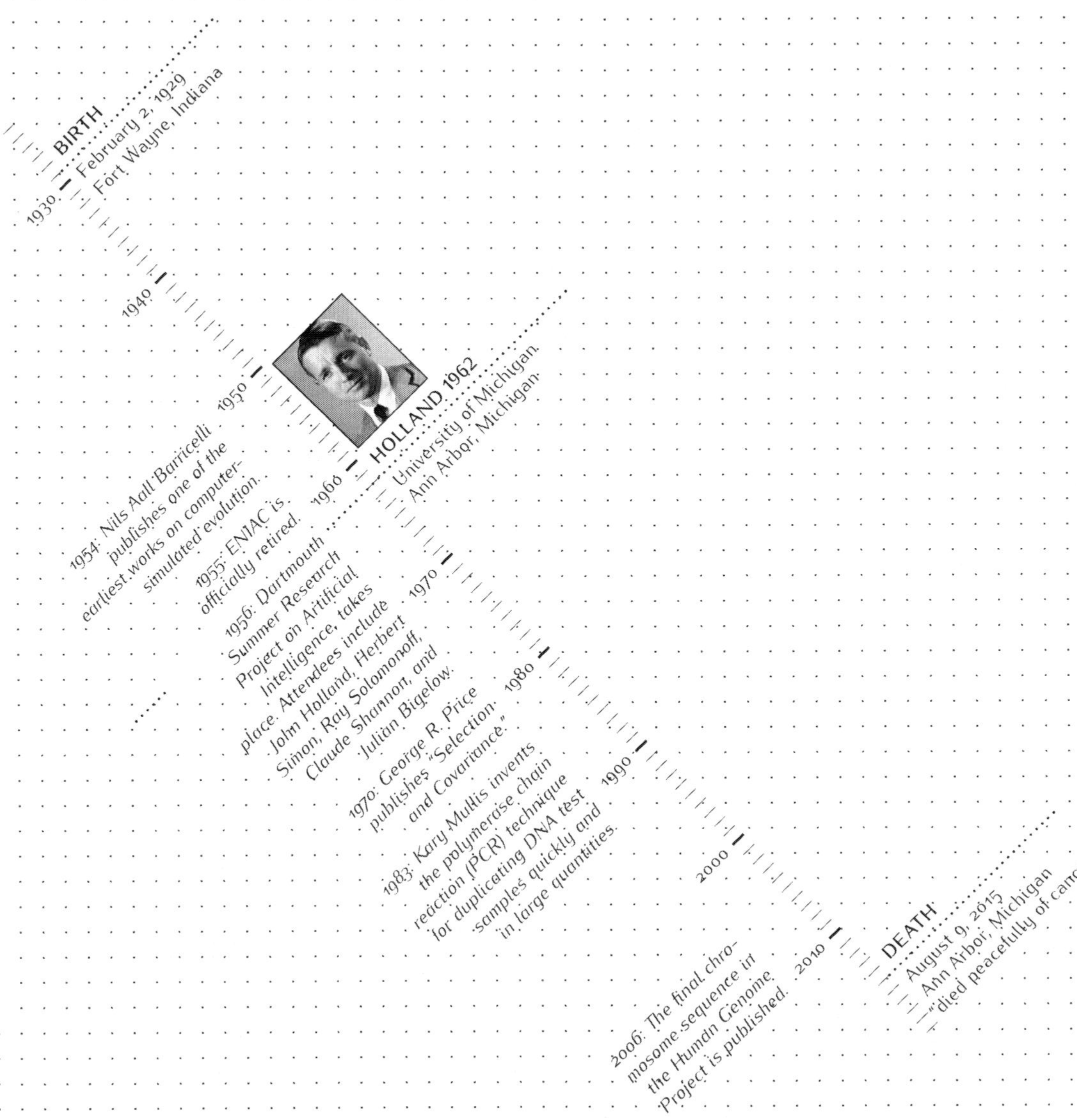

JOHN HENRY HOLLAND

[20]

MASTERING THE GLASS BEAD GAME

John H. Miller, Carnegie Mellon University and Santa Fe Institute

The epigraph of this paper is from Hermann Hesse's novel *The Glass Bead Game*. This novel takes place many centuries hence, and to play the eponymous game requires a deep synthesis of the arts and sciences. The epigraph appears in the third line of the following paragraph (translated using Google Translate):

> *So don't expect a complete history and theory of the Glass Bead Game from us, and authors more dignified and skillful than us would not be able to do that today. This task is reserved for later times if the sources and the intellectual prerequisites for it are not lost beforehand. And this essay of ours is even less intended to be a textbook on the Glass Bead Game, such a one will never be written. One does not learn the rules of this game of games in any other way than in the usual, prescribed way, which requires many years, and none of the initiates could ever have an interest in making these rules easier to learn.*

Thus marks the beginning, over sixty years ago, of John Holland's[1] pursuit in understanding the behavior of what he would eventually come to call "complex adaptive systems."

The goal of the paper is "to suggest how questions concerning adaptation can be treated in a logical context." Such a simple statement

J. H. Holland, "Outline for a Logical Theory of Adaptive Systems," *Journal of the ACM* 9 (3), 297–314 (1962).

[1] In 1961 Holland was part of the *Program in Communication Sciences* at the University of Michigan. This program, founded in 1957, blended computation, language study, and biology, a melange of topics that would easily fit within today's institutes focused on complex systems. Holland and his colleagues Arthur Burks (who helped design the ENIAC and, with John Von Neumann, developed the "Von Neumann architecture" that is the basis of modern-day computing), Robert Axelrod (a political scientist with ground-breaking work on cooperation), and Michael Cohen (a broad-thinking organizational theorist) would go on to form the BACH group at Michigan, a rebel alliance of complex systems thinking in the midst of a large, public university.

belies the breadth and depth of the thinking embodied within this research. As is clear from the citations contained within the paper, it draws upon a broad and key set of ideas, ranging from notions about morphogenesis and universal computation developed by Alan Turing, recent advances in artificial intelligence by Marvin Minsky, Alan Newell, Arthur Samuel, and Herb Simon, insights about self-replicating automata and game theory by John von Neumann, and work on evolution and genetics by Ronald Fisher and Sewall Wright. All of this work, the majority of which had been published within a year or two of this paper, has withstood the scientific test of time. Ultimately, the paper combines nascent ideas about adaptation and computation in a way so prescient that the research agenda it put in motion continues to yield insights, surprises, and new research directions.

Along with developing a theoretical approach for investigating complex adaptive systems, there are other key insights contained within the paper. For example, one finds the beginnings of Holland's ideas about artificial adaptive systems that will eventually be developed in his landmark 1975 book on *Adaptation in Natural and Artificial Systems* (currently, with over 79,000 citations) that introduced key ideas about genetic algorithms and learning classifier systems. He also embraces many ideas that are now central to computer science, such as parallel computers, the application of abstract trees to define programs (the programming language LISP was developed around 1960), and the usefulness of having systems probabilistically wander across various spaces. Foreshadowing other parts of his future work, Holland also develops ideas about using templates (akin to schema in his later work) to direct how the system behaves, new notions about assigning credit to partial solutions and sub-goals, and the use of economic ideas to direct the search for new solutions.

Holland forms his logical theory of adaptation by imagining a machine that is able to generate any possible program that could be run on Turing's universal computer. Adaptation is viewed as a process that modifies such generation process in response to feedback from the environment that arrives in the form of "activation" or "reward." Such feedback gets released as the generation procedures attempt to solve problems inherent in the environment. An interesting element of the model is that it is easily applied to (and, in fact, designed for) populations of both generation processes and environmental problems. The adaptive system is altered using Darwin's notion of selection with modification. The success of a given generator is tied to its ability to obtain net rewards from confronting problems in the environment. The actual implementation of the system is somewhat complicated (a general outline is given in section 3, and a particular implementation using iterative circuit computers is given in section 4).

The theoretical goals of the project, the set of inspirations driving the modeling, and the set of tools used to realize the research, are as relevant today as they were sixty years ago. Early in the paper, Holland notes that the theory he is outlining is "intended as *a* theory and not *the* theory." He returns to this issue at the very end of the paper and notes "There are, however, the larger questions of whether the theorems within reach will be useful and whether one should attempt a theory at all . . ." followed by a quote from von Neumann and Morgenstern's seminal *Theory of Games and Economic Behavior* (1947) about how theories begin by predicting what is easily known and eventually advance to "genuine predictions by theory."

Sixty years on, a *general* theory of complex adaptive systems is still elusive. Various theories about specific complex adaptive systems have certainly been brought near enough to "genuine predictions by theory" to declare success, while a more general theory remains in the midst of the extremes outlined by von Neumann and Morgenstern. The path suggested by Holland in this early paper may well yield the desired general theory. Perhaps over the intervening years we have learned enough to play the Glass Bead Game.

OUTLINE FOR A LOGICAL THEORY OF ADAPTIVE SYSTEMS

John H. Holland, University of Michigan

"Und ein Lehrbuch des Glasperlenspiels soll dieser unser Aufsatz ja noch weniger sein, ein solches wird auch niemals geschrieben werden." —Hesse

1. Introduction

The purpose of this paper is to outline a theory of automata appropriate to the properties, requirements and questions of adaptation. The conditions that such a theory should satisfy come from not one but several fields: It should be possible to formulate, at least in an abstract version, some of the key hypotheses and problems from relevant parts of biology, particularly the areas concerned with molecular control and neurophysiology. The work in theoretical genetics initiated by R. A. Fisher (1958) and Sewall Wright (1960) should find a natural place in the theory. At the same time the rigorous methods of automata theory should be brought to bear (particularly those parts concerned with growing automata: Burks 1960; Burks and Wang 1957; Church 1960; Holland 1959, 1960; Moore 1959; Rabin and Scott 1959; Turing 1936; von Neumann 1962). Finally the theory should include among its models abstract counterparts of artificial adaptive systems currently being studied, systems such as Newell–Shaw–Simon's "General Problem Solver" (1960), Selfridge's "Pandemonium" (1959), von Neumann's self-reproducing automata (1951) and Turing's morphogenetic systems (1950; 1952).

The theory outlined here (which is intended as *a* theory and not *the* theory) is presented in four main parts. Section 2 discusses the study of adaptation via generation procedures and generated populations. Section 3 defines a continuum of generation procedures realizable in a reasonably direct fashion. Section 4 discusses the realization of generation procedures as populations of interacting programs in an iterative circuit computer. Section 5 discusses the process of adaptation in the context of the earlier sections. The paper concludes with a discussion of the nature of the theorems of this theory.

Before entering upon the detailed discussion, one general feature of the theory should be noted. The interpretations or models of the theory divide into two broad categories: "complete" models and "incomplete" models. The "complete" models comprise the artificial systems—systems with properties and specifications completely delimited at the outset (cf. the rules of a game). One set of "complete" models for the theory consists of various programmed parallel computers. The "incomplete" models encompass natural systems. Any natural system involves an unlimited number of factors and, inevitably, the theory can handle only a selected few of these. Because there will always be variables which do not have explicit counterparts in the theory, the derived statements must be approximate relative to natural systems. For this reason it helps greatly that the statements can be verified in terms of specific programming schemes for parallel computers; in this way the "complete" models provide confirmation and corroboration of the theory. At the same time one can discover for each "complete" model a set of analogous "incomplete" models. Thus the results pertaining to a given "complete" model, when suitably interpreted, also apply to the analogous "incomplete" models.

This contains a very early statement of using a known artificial world (a "complete" computational model) to explore an unknown natural one (an "incomplete" model), to provide "confirmation and collaboration of the theory." Holland also notes how computational models can be used to verify theories about natural phenomena and to generate theories about such phenomena. Such an enlightened view of computational modeling took many decades to become widely accepted.

2. General Plan

The study of adaptation involves the study of both the adaptive system and its environment. In general terms, it is a study of how systems can generate procedures enabling them to adjust efficiently to their environments. If adaptability is not to be arbitrarily restricted at the outset, the adapting system must be able to generate any method or procedure capable of an effective definition. The intuitive idea of an effectively defined procedure has several equivalent characterizations, here, following Turing, the set of all effectively defined procedures will be identified with the set of all programs of some suitably specified universal computer. In these terms, unrestricted adaptability (assuming nothing is known of the environment) requires that the adaptive system be able initially to generate any of the programs of some universal computer. The process of adaptation can then be viewed as a modification of the generation process as information about the environment accumulates. This suggests that adaptive systems be studied in terms of associated classes of generation procedures—the associated class in each case being the repertory of the adaptive system.

There is an important change in viewpoint when one speaks of adaptation in terms of generation procedures. It is no longer a question of producing this or that particular program, rather it is a question of the ability to generate some particular population of programs. That is, with each generation procedure we associate the population of programs it generates; successive modifications of the generation procedure by the adaptive system correspond to successive alterations in the population of programs. It quickly becomes apparent that there is no reason to restrict the generation procedure to producing one program at a time. There is in fact a gain in generality if the generation procedure operates in parallel fashion, producing sets or populations of programs at each moment rather than individuals. (All this provided we can find a means of realizing such procedures, a question we shall tackle further on.) In the same vein we can treat the environment as a population of problems and, in fact, it soon becomes apparent that we can replace individuals by populations everywhere in the theory. By so doing we gain both in generality and in facility, opening the way to the application of statistics as well as logic.

A generation procedure which eventually produces any arbitrarily chosen program of some universal computer will henceforth be called a *universal generation procedure*. As noted earlier, it is sometimes essential that the class of generation procedures under study include universal generation procedures. To make subsequent discussion more concrete, a class of procedures containing an infinite subclass of universal generation procedures will be described briefly here (in the next section the same class will be more precisely defined):

Each procedure will be defined in terms of a distinguished finite set of programs, called generators, and a graph, called a generation tree. No restriction is to be placed upon the programs selected for the set of generators; in one case the set of generators may include just the equivalents of individual instructions, in another case it may include highly sophisticated heuristic programs. The set of generators chosen will depend in part upon what phase of adaptation is being investigated—primitive sets of generators being indicated, for instance, when the study concerns the origin of complexity, sophisticated sets when problem-solving is the primary aim. If the procedure is to be a universal generation procedure, the set of generators must be complete with respect to some universal computer. That is, by combining copies of the generators it must be possible to construct any program of the underlying universal computer. The combining processes, whereby the programs are formed from the generators, are specified by the generation tree. The generation tree will be presented here only in terms of a descriptive model (it will be discussed in detail in the next section): The generators can be thought of as embedded in a discrete or cellular space, each type occurring with a given density (the expected number of generators in some fixed number of cells). Each generator undergoes a random walk in the space. Upon coming into contact with another generator it may, with a probability determined by generators involved, connect to it (connected sets of generators correspond to programs). At the same time combinations of generators may, with a probability again determined by the types of generators involved, separate into component combinations. The rate at which generators come into contact (the generation rate) together with the connection and disconnection probabilities determine the density of each type of program as a function of time and the initial densities. That is, these factors determine what programs are generated and in what order.

The core idea here is that generators (that implement various programs, ranging from simple rules to sophisticated heuristics), are combined by a generation tree into a set of behaviors that allows the adaptive system to interact with the environment. A system adapts as it produces new combinations of programs better suited to the environment. A simplified version of this idea becomes a central part of "genetic programming" that is developed around 1985.

It is important to note that a change in the connection probabilities alters the population of programs to be expected at any given time. In particular, assume that the generation rate and disconnection probabilities have been fixed. Then each choice of a set of connection probabilities selects a particular generation procedure from the class of admissible procedures. In short, control of the connection probabilities amounts to control of the generation procedure. We will see (section 5, Models) that special programs for modifying connection probabilities—so-called *templates*—can be added to the models described above. Let the generation process associated with a given template-free model be called a *free generation procedure*, and let the result of adding some set of templates to the given model be called a *modified generation procedure*. If the free generation procedure is taken as a basis, then each modified generation procedure produces a population of programs skewed relative to that of the free procedure. If it is assumed that there is a "cost" involved in producing templates, then we can associate a "skewing cost" with each modified generation procedure.

Adaptation is driven in this system by "activation" (the reward or the utility of a solution) that is received from the environment when a problem is solved by one of the generated programs. These rewards are tempered by "skewing costs" tied to the production of templates that skew the types of programs created by a generation procedure. The adaptive system attempts to discover generation procedures that maximize the net value given estimates of rewards based on recent samples from the environment.

The generated population of programs will act upon a population of problems (the environment) in an attempt to produce solutions. For adaptation to take place the adaptive system must at least be able to compare generation procedures as to their efficiency in producing solutions. In order to provide this comparison it will be assumed that to every problem is assigned a numerical quantity called "activation". ("Activation" is the name used in section 5; the quantity might also have been called "reward".) This quantity is consigned or "released" to the adaptive system whenever the associated generation procedure solves the problem by means of one of its programs. Let $r_G(E)$ be the average rate of activation release (assuming this quantity defined) when the generation procedure G is faced with environment E. Let c_G be the skewing cost per unit time (assuming the template population to be in dynamic equilibrium) associated with G. The net rate of activation release, $r_G(E) - c_G$, can be assigned to G as its rating relative to E.

(Other rating procedures are of course possible; see sections 5 and 6). Two generation procedures confronted by the same environment can be compared in terms of their ratings. Adaptation within this context involves two concurrent processes: (1) sampling of the environment in order to produce estimates of the environment; (2) modification of the generation procedure to obtain the generation procedure with highest rating relative to the current estimate.

It still remains to implement the conditions just set down. In terms of the example so far discussed a natural form of implementation consists of providing each adaptive system with a central control over its associated generation procedure. Central control can be achieved by introducing a process for producing templates while constraining each generation procedure to occupy a finite region in the model. (Boundedness allows definition of local populations and evaluation of their effectiveness.) By providing a special set of generators, the process for producing templates can also be presented as a program, called a *supervisory program* in section 5. The supervisory program serves as a device to gather the threads of control at a single point. It produces a given distribution of templates over time and the templates in turn give rise to a modified generation procedure. In this sense the supervisory program serves as an implicit definition of the distribution of problem-solving programs to be expected at any time t—to any given supervisory program will correspond a specific sequence (in time) of distributions (over type) of problem-solving programs. Changes in the supervisory program will, of course, result in changes in the population of problem-solvers.

Adaptation, then, is based upon differential selection of supervisory programs. That is, the more "successful" a supervisory program, in terms of the ability of its problem-solving programs to produce solutions, the more predominant it is to become (in numbers) in a population of supervisory programs. In order to implement differential selection two provisions are necessary: (1) supervisory programs must be capable of self-duplication in order to provide successive generations as grist for the selection principle; (2) the rate of duplication of a supervisory program must be determined by the effectiveness of the problem-solving programs it controls. The two provisions are satisfied if

Here the importance of selection by fitness is emphasized, where fitness is tied to both the reward from the environment and the associated costs of deploying various supervisory programs (captured by templates) that help determine the final structure of the programs that attempt to solve the problems coming from the environment. The next paragraph suggests how variation can be included in the system by "modification during duplication."

we assume, as with the templates, that there is a cost associated with the duplication of supervisory programs and that this cost is met by released activation. Then the rate of duplication of a supervisory program will be determined by the net rate of activation release of the associated local problem-solving population. Differential selection follows since (assuming spontaneous disconnection and fixed density of generators—see section 5) the program with the higher rate of duplication becomes ever more predominant in the population of supervisory programs. Of course a single supervisory program, even should it deviate widely from the "norm", would affect the rates of connection only locally. Thus useless deviations would initially have little effect upon the overall level of adaptation and (be cause of spontaneous disconnection) would shortly disappear from the population. On the other hand, if the deviation should prove effective, then an ever larger proportion of supervisory programs would incorporate it. In effect, "successful" local variations in the modification of the generation procedure would be propagated throughout the discrete space of the model.

Operation of the selection principle depends upon continued generation of new varieties of supervisory programs. There exist several interesting possibilities for producing this variation. Here we only note that it is possible to subject each position (generator) in the supervisory program to modification during duplication. More precisely, if there are k generator types $(g_1, \cdots, g_k)$, we can associate with the ith position of the supervisory program the vector $(p_{i1}, \cdots, p_{ik})$ where $p_{i1} + \cdots + p_{ik} = 1$ and p_{ij} is the probability that the jth generator will occur at position i. Thus we can associate an ordered set of probability vectors, i.e. a probability matrix, with each supervisory program. Under this condition the successive generations (duplications) of a given supervisory program will give rise to a population made up of a variety of supervisory programs. In the absence of selection, the distribution of types to be expected will be determined by the probability matrix associated with the initially given supervisory program. If a change is made in the set of initially given probability vectors, changes in the distribution of problem-solving programs can be expected. Our next concern will be the nature of these changes.

First, let us examine the successive generations (duplications) of two supervisory programs which differ only slightly in their corresponding probability matrices. It can be shown, in the absence of differential selection, that at first the two populations of supervisory programs arising from the initially given supervisory programs have similar expected distributions. This in turn means that the resulting distributions of problem-solving programs will be similar. In other words supervisory programs which are similar or resemble each other in terms of probability matrices can be expected to produce distributions of problem-solving programs which for a while differ only slightly. But then the level of adaptation, in terms of solutions produced, must be closely related in the two cases. Thus supervisory programs which are similar under the above measure will have related levels of adaptation in any given environment. This observation has several important consequences and the remainder of this section will be devoted to its ramifications.

Before going further note that the criterion of similarity mentioned above—a criterion which is relevant to the effects produced by the supervisory program—can be given a precise formulation. Consider the set of all distinct supervisory programs and the related collection of all probability matrices. In the collection of matrices we can define a distance function or metric as one of the usual measures of difference defined on matrices. Then, given two supervisory programs, the *distance* between them will be taken to be the difference of the associated sets of probability matrices. With the help of this metric, neighborhoods *in the set of supervisory programs* can be defined and discussed. As we consider neighborhoods of smaller and smaller diameter about a given supervisory program the effects of other supervisory programs in the neighborhood will progressively approach that of the given program. This relation can be stated in another way. Let a random sample be drawn from some neighborhood in the collection of supervisory programs. Then, for neighborhoods of progressively smaller radius, the sample will give an increasingly good estimate of the adaptation of the other programs in the neighborhood.

When the factors producing differential selection are added to these models, the idea of sampling takes on additional significance.

Under differential selection, the success of a supervisory program in effect biases the probability of trying further samples in that neighborhood. In greater detail: Duplication with modification, acting on an initially given population of supervisory programs, provides a continual supply of new variants. The degree of similarity to be expected between any given variant and its parent will depend upon the probability matrix, which itself will be subject to selection. New variants may occur as if they were drawn at random from the collection of supervisory programs; or, for example, probability matrices may predominate which make the likelihood of a given variant proportional to its similarity to the parent program. (In section 5 we will discuss conditions under which the latter arrangement would have a definite selectional advantage). If a supervisory program is relatively successful it will duplicate, producing a population which contains not only the original program but modifications of it. If the likelihood of a variant is proportional to its similarity to the original, many of the modified supervisory programs will be similar to the original program. If one of these modified supervisory programs develops a more successful population of problem-solving programs it will begin to predominate in the population of supervisory programs. The result will be that new modifications will increasingly come from the new supervisory program. In other words, under differential selection, the sequence of supervisory programs will "tune in" on the best supervisory program in the neighborhood of similar programs. On the other hand, not only will an unsuccessful supervisory program disappear from the population (because it does not duplicate or does so inefficiently) but, as a result, the chance of similar programs arising from it by modification will be eliminated. In this way not only is the program itself selected against, but in effect so is its entire neighborhood. The combination of the "tuning in" effect with this "avoidance" effect produces the sampling bias described at the beginning of the paragraph.

This completes the discussion of the general plan of the study—the next three sections will consider some of the more difficult points in greater detail.

3. A. Class of Generation Procedures

The preceding section introduced, as an example, a class of generation procedures containing an infinite subclass of universal generation procedures. There the class was discussed informally via a description of one of its models; here the class will be precisely defined in preparation for a more detailed discussion of its models in the next section. This particular class has been chosen because it is relatively easy to define and yet extensive enough to be used to illustrate the main points of a study of adaptation based upon generation procedures.

In this and the next section, Holland provides a more formal definition and a particular instantiation (using iterative circuit computers) of how one could create a space in which to embed the adaptive system. The purpose of such an embedding becomes clear in the last sentence of the first paragraph of section 4: "Because iterative circuit computers have been characterized mathematically, we can use this mathematics to deduce theorems about the embedded systems."

As mentioned earlier each generation procedure is defined in terms of a set of generators and a graph called a generation tree. Let $(g_1, \cdots, g_k)$ be the set of generators. Each permissible combination of generators (each program) is represented by a vertex in the generation tree. The vertices are arranged in levels so that the vertices of the first level correspond to the individual generators (one vertex for each generator), the vertices of the second level correspond to all permissible combinations of two generators, etc. More precisely the generation tree can be defined as follows: Let $V(i, j)$ be the connected set of i generators corresponding to the jth vertex at level i, v_{ij} (where $V(1, j) = g_j$). Let H_k be the set of all pairs $(V(i', j'), V(i'', j''))$ such that $i' + i'' = k$. Let L_k be the collection of all *distinct* connected sets of generators which result from a single connection between the paired elements of H_k. Note that in general there will be several ways of combining a given pair; also different pairs may yield the same program after combination. (To keep things simple programs are constrained to combine only pairwise; three or more programs cannot be combined in a single operation.) To each element of L_k assign a single vertex, v_{kh}, at level k. For each pair in H_k which gave rise to $V(k, h)$ in L_k let there be a directed edge of the graph terminating in v_{kh}; each such edge will be joined, via an auxiliary vertex v^r_{kh}, to the vertices $v_{i'j'}$, and $v_{i''j''}$ corresponding to the pair of elements combined, $(V(i', j'), V(i'', j''))$. Note that the auxiliary vertex in the resulting "Y" configuration does not belong to a level; it will be used to represent the specific connection process which yields $V(k, h)$ from $V(i', j')$ and $V(i'', j'')$. Each auxiliary vertex v^r_{kh} will be labeled with two numbers, p^r_{kh} and q^r_{kh}, $0 \leqq p^r_{kh} \leqq 1$, $0 \leqq q^r_{kh} \leqq 1$, called the *connection* and *disconnection probabilities* respectively. It will be required for each main vertex v_{kh} that $\sum_r p^r_{kh} \leqq 1$ and $\sum_r q^r_{kh} \leqq 1$. Each main vertex v_{ij} will be labeled with a variable $d(i,j,t)$ designated *density*. In addition a single number, c, the *generation rate*, is associated with the graph as a whole.

The connection probabilities, the disconnection probabilities, and the generation rate are constants of the process; under interpretation the densities specify the number of programs of each type in a selected volume of the embedding space (the space in which the generation procedure is embedded). It remains to specify the transition equations giving the densities at time $t+1$, $\{d(i,j,t+1)\}$, as a function of the constants of the process and the densities at time t. The transition equations, for any generation procedure of the class, can be presented in the following form:

A vertex v' will be termed an *input vertex* (*output vertex*) of vertex v if there is a directed edge of the graph from (to) v' to (from) v. In particular, an auxiliary vertex will be termed an input (output) auxiliary vertex of vertex v if there is an edge from (to) the auxiliary vertex to (from) v.

Let $C(t), t = 0, 1, 2, \cdots$, be a sequence of random variables having a common distribution for which the mean is defined and equal to the generation rate, c.

Similarly, let $X_u(t)$ be a random variable taking as values the triples (i,j,h) with probability

$$2p^h_{iy}\frac{d\left(i',j',t-1\right)\cdot d\left(i'',j'',t-1\right)}{\left(d_0(t-1)\right)^2}$$

where $v_{i'j'}$, and $v_{i''j''}$ are input vertices of the auxiliary vertex v^h_{ij} which in turn is an input auxiliary vertex of

$$v_{ij};\quad d_0(t-1) \stackrel{\text{df}}{=} \sum_i\sum_j d(i,j,t-1),$$

the total density of programs at $t-1$.

Let $M_{ij} \stackrel{df.}{=} \{h \mid v^h_{ij}$ belongs to the set of input auxiliary vertices of $v_{ij}\}$.

Define $\delta_{ijh}\left(i',j',h'\right) = \begin{cases} 1 & \text{if } \left(i',j',h'\right) = (i,j,h) \\ 0 & \text{otherwise.} \end{cases}$

$e_1(i,j,t) \stackrel{df.}{=} \sum_{u=1}^{c(t)} \sum_{M_{ij}} \delta_{ijh}(X_u(t)) \stackrel{df}{=}$ production of $V(i,j)$ through connection at time t.

Let $Y_u(i,j,t)$ be a random variable taking as values the triples (i,j,h) with probability q_{ij}^h and the symbol ϕ with probability $1 - \sum_h q_{ij}^h$.

$e_2(i,j,t) \stackrel{df.}{=} -\sum_{u=1}^{d(i,j,t-1)} \sum_{M_{ij}} \delta_{ijh}(Y_u(i,j,t)) \stackrel{df.}{=}$ loss of$V(i,j)$ through disconnection at time t.

Let $N_{ij} \stackrel{df.}{=} \{i', j', h' \mid v_{i'j'}^{h'}$ belongs to the set of output auxiliary vertices of $v_{ij}\}$.

$e_3(i,j,t) \stackrel{df.}{=} \sum_{N_{ij}} \sum_{u=1}^{d(i',j',t-1)} \delta_{i'j'h'}(Y_u(i',j',t)) =$ production of $V(i,j)$ through disconnection of various $V(i',j')$ (where, if there are two edges of the graph from v_{ij} to $v_{i'j'}^{h'}$, $\delta_{i'j'h'}(Y_u(i',j',t))$ is counted twice in the sum).

$e_4(i,j,t) \stackrel{df.}{=} -\sum_{N_{ij}} \sum_{u=1}^{c(t)} \delta_{i'j'h'}(X_u(t)) \stackrel{df.}{=}$ loss of $V(i,j)$ through connection to form various $V(i',j')$ at time t (where, if there are two edges of the graph from v_{ij} to $v_{i'j'}^{h'}$, $\delta_{i'j'h'}(X_u(t))$ is counted twice in the sum).

$d(i,j,t) = d(i,j,t-1) + \sum_{b=1}^{4} e_b(i,j,t)$

The class of generation procedures just defined (assuming an appropriate set of generators has been selected) forms a continuum containing, as a subset, a continuum of universal generation procedures. Given the generation tree and the transition equations of any particular procedure, one can calculate the expected values of the densities as a function of time. From the general form of the transition equations one can determine such things as conditions under which the resulting generation procedures are stationary processes. Derivation of such properties—made possible by the abstract definition—in turn opens the way to results concerning adaptive efficiency. (This subject will be touched upon in section 6). There are many other formulations yielding continua of generation procedures, some simpler and some more intricate; however, few have models as simple as those of the present class. The models will be discussed next.

4. Models

Models of the generation procedures of section 3 can be based upon the class of iterative circuit computers (a mathematical characterization of iterative circuit computers has been given in Holland 1960). It should be emphasized at once that the iterative circuit computer is *not* the adaptive system. Rather, the computer is to be identified with the space in which the adaptive system is embedded. The word "space" is used here with a sense close to that of the words "physical space". With a physical space we associate a geometry and a set of universal laws which any particle at any position in that space must obey. Similarly each iterative circuit computer has an associated geometry and a set of "state-transition" rules holding, without change, for each location in the computer. Because iterative circuit computers have been characterized mathematically, we can use this mathematics to deduce theorems about the embedded systems.

Iterative circuit computers, with appropriate interpretation of the symbols, include representatives structurally and behaviorally equivalent to automata of any of the following types: Turing machines (with 1 or more tapes) (Holland 1960; Rabin and Scott 1959), tessellation automata (von Neumann 1962; Moore 1959), growing logical nets (Burks and Wang 1957; Burks 1960), and potentially-infinite automata (Church 1960). Each computer in the class is constructed of a single basic module (a fixed logical network) iterated to form a regular array of modules. In the mathematical characterization the arrangement of the modules is specified by a finitely-generated abelian group, A, and the scheme of connection of a module to its neighbors is given by a finite set of elements, A^0, selected from A. Any module, being a fixed finite logical net, can assume only a finite number of states. The state of the module at coördinate α at time t will be designated by $S(\alpha, t)$. Each module can control the information that flows through it from its neighbors—each channel to or from a neighbor is in effect gated and at any time, t, any given gate may be either open or closed. Which gates are open and which are closed is specified by a portion, $Y(\alpha, t)$, of the state, $S(\alpha, t)$. Each module also has a fixed storage capacity which may be thought of as a storage register. The other part of the state $S(\alpha, t)$, symbolized by $X(\alpha, t)$, specifies the content of this register. Thus the state at time t of the module at coördinate α is composed of two parts, $S(\alpha, t) = (X(\alpha, t), Y(\alpha, t))$; in the mathematical characterization this means that the set S is the direct product of two finite sets X and Y, $S = X \otimes Y$. Each module also has a fixed number of free inputs which serve as inputs to the computer; the input state of the module at α at time t will be designated by $B(\alpha, t)$. As information flows into a module at time t, through the open channels from its neighbors and through its free inputs, the module processes the information and passes the result along, without delay, through those outgoing channels which are open. The same information is also used to determine what state, $S(\alpha, t + 1)$, the module is to have at the next instant of time. The information used to determine the new storage state, $X(\alpha, t+1)$, is processed or transformed according to a function, f, called a subtransition function. The new pattern of open and closed gates at $\alpha, Y(\alpha, t + 1)$, is determined by a function P. Thus the rules for determining the state, $S(\alpha, t + 1)$, of a module are essentially specified by f and P. These rules are "locally effective". That is, given f and P, and given the states $S(\beta, t)$ and input states $B(\beta, t)$ of selected modules lying within a radius $r(t)$ of α, the state of the module at α at the next instant, $S(\alpha, t+1)$, is unambiguously determined.

The main properties of iterative circuit computers the reader should keep in mind are:

1. Once selections are made for A, A^0, S, f, and P, the structure of the basic module and the arrangement of the modules to form the computer are completely determined. Thus the quintuple (A, A^0, S, f, P,) completely identifies the iterative circuit computer, and with each distinct quintuple is associated a distinct computer.

2. In general, an assignment of states to a connected set of modules can be treated as a sub-program stored in the computer. Each of these sub-programs can be active at the same time; thus from the computing point of view, a given iterative circuit computer can execute arbitrarily many sub-programs simultaneously (again within limits imposed by size).

3. Any given growing automaton (see Burks 1960; Burks and Wang 1957) can be simulated by a connected set of sub-programs in the iterative circuit computer.

4. Important properties which can be given to programs are:

   a) Sub-programs can be written so that, under local control, they shift themselves from one set of modules to another set. Thus the geometry of the underlying iterative circuit computer becomes the geometry of the space in which the embedded program moves.

   b) Sub-programs which are independent initially can move into contact (occupy adjacent sets of modules) and connect so as to form a larger sub-program capable of moving and acting as a unit.

   c) Sub-programs can be written so that they can directly produce a copy of themselves (duplicate) in an adjacent set of modules.

5. Given any iterative circuit computer it is possible to select a finite set of sub-programs (individual instructions in the limiting case) to serve as generators such that any sub-program possible for that computer can be achieved by an appropriate combination of copies of these sub-programs.

6. In an iterative circuit computer the primary technique for locating an operand is relative addressing—instead of referring to an operand by an address, a path (a sequence of opened gates) is opened to the operand. Any module belonging to the path can use the storage registers of other modules along the path as operands. As Newell (1960) describes the process we point to a location rather than addressing it. Relative addressing permits the set of generators to be quite small.

7. By suitably restricting the functions f and P it can be arranged that generators do not interpenetrate or "over-write" when moving (cf. the notion of a "billiard ball" physics).

The free generation procedure of section 2, when modelled in an iterative circuit computer, requires the generators (and combinations of generators) to "shift" and "connect" at random in the computer. The simplest form of random shift occurs under the following conditions: (1) At each moment of time a generator has a fixed probability of shifting to one of its neighboring modules. (2) If a generator attempts to shift to a module already occupied by another generator such a shift is prohibited. A similar prohibition is to apply when two generators attempt to shift to the same (initially "empty") module. Under conditions to be discussed further on, two or more generators occupying adjacent modules ("in contact") may become connected. Such connected sets of generators are to shift as a unit and for such sets the above conditions are still to hold. In particular if a connected set of generators attempts to shift into a set of adjacent modules and any one of these modules is occupied then the shift is to be prohibited.

Conditions (1) and (2) can now be translated into requirements on the iterative circuit computer involved. First of all a portion of the storage register in each module must be used to record the type of generator present (or the absence of any generator). In terms of the characterization this means that the finite set X will be the direct sum of two sets, $X = X_1 \otimes X_r$. If there are k generator types $g_1, \cdots, g_k$ then X_1 will consist of these k elements together with an element g_0 indicating no generator. The complete state of the module at coördinate α and time t is designated by

$$\begin{aligned} S(\alpha, t) &= [X(\alpha, t), Y(\alpha, t)] \\ &= [X_1(\alpha, t), X_r(\alpha, t), Y(\alpha, t)] \,. \end{aligned}$$

If $X_1(\alpha, t) = g_j$, $1 \leqq j \leqq k$, then we will say that generator g_j is present at coördinate α at time t. The transition equations, f and P, must be chosen so that the two conditions on shifting are satisfied. For example, let generator g_j be present at an immediate neighbor, $\beta = a_i(\alpha)$, of α. Let $X_r(\beta, t)$ indicate that this generator is to shift to position α. Then $X_1(\alpha, t+1)$ will equal $X_1(\beta, t)$ only if no other immediate neighbor of α indicates a similar shift. Proceeding in this way we can without too much trouble translate all of the earlier requirements into formal requirements on the transition equations. To provide for random shifts we choose a stochastic function, $B'(\alpha, t)$, which for each module α and time t designates a random variable which assumes the value j with probability p_j, $1 \leqq j \leqq n_i$, where n_i is the number of input states. Then we set the input function $B(\alpha, t)$ equal to the jth input state when $B'(\alpha, t) = j$. The question of whether or not a generator shifts to one of its immediate neighbors depends in a completely deterministic fashion on $B(\alpha, t)$ and the inputs from neighboring modules. By giving the transition equations the appropriate form we can assure that the input sequence is used in effect as a source of random numbers for determining the shifts. Thus the probabilistic aspects of the theory are made to depend solely upon the probabilistic properties of the input. Since the usual theorems about deterministic automata are stated in terms of an arbitrary input function, such theorems can still be used directly in the present theory. It should be emphasized again that the procedures just introduced do not preclude systems in which highly sophisticated sub-programs (connected sets of generators) have been set into the computer initially.

The conditions for connection will be stated in terms of two general quantities which will be associated with the generators as components of X: (1) valence and (2) activation. Valence will be specified by part of the X_1 component of X and activation by part of the X_r component. Thus, letting X_{11} specify the instruction type, X_{12} the valence, and X_2 the activation, X will take the form:

$$\begin{aligned} X &= X_1 \otimes X_r \\ &= X_{11} \otimes X_{12} \otimes X_2 \otimes X_{r'}. \end{aligned}$$

Under interpretation the elements of X_{11} are to correspond to instruction types such as ADD, STORE, etc. The object will be to choose f and P to bring about the following desired relations between instruction type, X_{11}, valence, X_{12}, and activation, X_2: Activations are to be treated as integers and accordingly summed and ordered. Let g_i and g_j (elements of X_1) occupy adjacent modules as the result of a shift. Let E be their combined activation. The valence of g_i and g_j determines an interval, $E^*_{ij} = (a, b)$, such that the pair will connect only if E lies in the interval $a \leqq E \leqq b$. To provide for disconnection (essential if there is to be "competition" and differential selection) E^*_{ij} will also be used to determine the probability that g_i and g_j disconnect at any given time after connection. For many models the probability of disconnection will be specified by $(E^*_{ij})^{-1}$; the "stronger" the connection in terms of the activation required for it to form, the lower the probability of disconnection.

The valence-activation combination provides for the introduction of modified generation procedures as follows: Let A be a particular connected configuration of generators and let n be the number of copies of A in some region, V, of the computer. For simplicity, select a particular generator on the periphery of the configuration A and record the activation associated with the corresponding generator in each copy of A. Let $n(E)$ be the number of copies of A having an activation E at the given position. $n(E)$ can be made exponentially decreasing through the following provision: When two generators come into contact (occupy adjacent modules) the corresponding activations E_1 and E_2 are summed, an activation between zero and the sum, $0 < E_1{}' < E_1 + E_2$, is chosen at random (using $B(\alpha, t)$) to replace the activation E_1, and the activation level $E_2' = E_1 + E_2 - E_1'$ is used to replace activation E_2. (That the resulting distribution of $n(E)$ is exponentially decreasing can be proved as a theorem by applying an argument from statistical mechanics). With an exponential distribution if the activation required for a given type of connection is reduced by half, the number of connections of that type (in region V over time T) will be squared. The problem of modifying the generation procedure thus reduces to one of manipulating the activation required for a given type of connection. However this is not so simple as it might seem. The level of activation required for the connection, E^*_{ij}, is completely specified by the valences of the two generators to be connected. Thus E^*_{ij} cannot be directly manipulated. There is nevertheless an indirect method. This method of control depends upon the introduction of strings of generators—templates—which facilitate connection of other generator strings. (From this point until the end of this section I will assume that we are dealing only with linear strings of generators; the approach can be expanded to arbitrary configurations.) Let us assume that it is desired to increase the rate at which strings of the form $A_1 g_1$ connect to strings of the form $g_2 A_2$ (where g_1 and g_2 are generators,

It is here that templates are given a more concrete definition. In Holland's future work, template-like ideas appear in various guises. For example, the schema theorem from genetic algorithms suggests how key patterns in a population can grow or decline depending on how successful that pattern is across the entire population versus any particular instantiation. In learning classifier systems, input from the environment is recognized by pattern matching rules of various specificity, and based on such matches individual problem-solving programs are deployed.

A_1 and A_2 arbitrary strings). Let E^*_{12} be the activation required for g_1 and g_2 to connect (as determined from their valences). Consider now a string of the form $B_1 g'_1 g'_2 B_2$ and let the level of activation required for g_1 to connect to g'_1 be $\frac{1}{2}E^*_{12}$, similarly for g_2 and g'_2. This string can be used as a template. Note first that B_1 and B_2 are generator strings and hence, in effect, sub-programs. They can control alignment when $A_1 g_1$ and $g_2 A_2$ contact the template string; B_1 and B_2 can also select certain subsets of the possible strings A_1 and A_2 as "admissible" so that only when $A_1 g_1$ (or $g_2 A_2$) involves an admissible A_1 (or A_2) is the process allowed to continue. If the pair (g_1, g'_1) has a total activation $E = \frac{1}{2}E^*_{12}$, then $A_1 g_1$ will connect to the template. A similar statement holds for the pair (g_2, g'_2) and the string $g_2 A_2$. Once the two strings are connected to the template, B_1 and B_2 can use the activation involved, $\frac{1}{2}E^*_{12} + \frac{1}{2}E^*_{12} = E^*_{12}$, to form a direct connection between $A_1 g_1$ and $g_2 A_2$ in place of the connections to the template. That is, B_1 and B_2 will "erase" the connections between g_1 and g'_1, g_2 and g'_2, and "transfer" the total activation involved, E^*_{12}, to g_1 and g_2. Thus, enough activation is provided for a connection between g_1 and g_2. The net result is that $A_1 g_1 g_2 A_2$ is formed and the template is freed to repeat the process (cf. the action of a catalyst or enzyme). Because only $\frac{1}{2}E^*_{12}$ is required, the number of connections of $A_1 g_1$ (similarly $g_2 A_2$) to the template will be the square of the number of connections it would make on direct contact with $g_2 A_2$. The template thus has the desired effect of increasing the rate of connection of the two strings. This of course is only a sketch of one way in which such "templates" can alter the rate of connection—the intention of the sketch is to indicate that such control is possible and that it fits within the framework so far presented.

Templates exert an effect far out of proportion to their numbers because: (1) the exponential distribution of activation amplifies the effect of individual templates, (2) templates modify connection processes without being altered themselves (the catalyst effect), (3) in modifying the generation rate of a given type of program, the template also affects the generation rate of all programs which use the given program as a component. Thus, by introducing programs which construct templates—supervisory programs—one gains extremely flexible control over the generation procedure. In effect the supervisory program modulates the free generation process through the amplification factor of the templates it produces.

The procedure described in section 2 requires that the supervisory program duplicate, with some probability of variation or mutation, upon accumulation of sufficient activation. Briefly, duplication amounts to the supervisory program copying itself into an adjacent set of modules. The likelihood of variation at a given position in the supervisory program can be made a function of the E^*_{ij} and the random inputs $B(\alpha, t)$ (other procedures exist). Operation now proceeds as described in section 2 if the supervisory program, in order to duplicate, is required to collect from the environment an amount of activation equal to the activation involved in its own connections. The nature of the environment and the way in which the supervisory-template-solver complex adapts to it will be discussed in the next section.

5. Environment and Adaptation

In section 2 it was suggested that the environment could be treated as a population of problems. For present purposes let us restrict the problems to well-defined problems—problems which are presented by means of a finite set of initial statements (statement of the problem) and an algorithm for checking whether a purported solution of the problem is in fact a solution. For example, the initial statements could specify an axiom system and a theorem to be proved therein; a tentative solution would be any sequence of statements which purports to be a proof of the theorem; the checking algorithm would be the usual routine which checks that each statement in the sequence follows from previous statements by the allowed rules of inference, with the last statement being the one to be proved. Well-defined problems can easily be embedded in the space defined by the iterative circuit computer. The checking algorithm becomes a program and the initial statements are coded as configurations of generators. The problem is solved when a problem-solving program transforms the coded initial statements into a configuration which the checking program accepts as a solution. In accordance with other suggestions in section 2, each embedded problem will be assigned a quota of activation (the same quantity that was involved in the connection processes discussed in section 4). This activation is released to the controlling supervisory program whenever a problem-solver solves the embedded problem.

When we consider the interaction of an adaptive system with its environment we come very soon to questions of partial solutions, subgoals, etc. Here such questions become questions about the relation of supervisory programs to the checking routines of embedded problems. One of the simplest cases occurs when there is an *a priori* estimate of the nature of a partial solution and, perhaps, a measure of the closeness of its approach to the final solution. (Such *a priori* estimates are quite common in the problem-solving programs constructed to date; see Samuel (1959), Gelernter and Rochester (1958), and particularly Newell, Shaw, and Simon (1960).) To make use of these estimates we simply incorporate them in the checking routines; the checking routine is written so that some of the activation associated with the problem is released whenever a partial solution is presented.

☞ Adaptive systems often confront the credit-assignment problem, that is, how to reward a system for achieving subgoals or partial solutions. For example, while it is important to recognize that there is, say, a triple jump opportunity in a game of checkers, rewarding the sequence of moves that proceeded that opportunity is at least as, if not more, important. Throughout his work Holland confronted the credit-assignment problem in a variety of innovative ways.

A more interesting question of partial solution occurs when the environment involves stochastic sequences, the problem being to predict values of the sequence at a future time. (The stochastic sequence may be produced by an embedded sequence generator or it may be a component of the input sequence $B(\alpha, t)$; the checking program simply matches the prediction for time $t + j$ against the actual value at $t + j$ and releases activation as a function of the "closeness" of the prediction). Note that even a program which uses random number generation as its mode of "prediction" will succeed in releasing some activation. However, programs that improve upon this procedure, by utilizing some of the data to produce better than chance predictions, will gain a selective advantage for the corresponding supervisory programs. The adaptive system can accumulate subroutines which improve its predictions; these can in turn be combined in various ways to generate still more sophisticated prediction procedures. Thus, the adaptive system is not required to leap to "the" solution (if any), but can approximate it by stages.

The prediction problem is one example of a "rich" environment. In the simplest sense an environment is rich when it is composed of graded sequences of problems. By its very nature, such an environment permits the adaptive system to develop subroutines for problems at one level of difficulty which, through recombination, minor changes, insertion, etc., will be useful in solving problems at the next level of difficulty. Operations such as recombination involve only minor modification of the generation procedure; thus, the "tuning in" procedure mentioned at the end of section 2 becomes very effective. Faced with a rich environment, the generation procedure can be adjusted in a continuous fashion to ever more difficult segments of the environment. In effect, the adaptive system can accumulate a repertoire of programs which serve as a new set of generators for programs aimed at problems of the next level of difficulty. Under these conditions the supervisory programs can develop problem-solving programs which are both complex and general in a surprisingly short time. For a rough indication, compare the number of j-step-programs on k generators, k^j, with the number when each level of the hierarchy requires just h-step-programs on k generators, $k^h(\log_h j)$.

Rich environments are not rare. Frequently problem environments which closely approximate some natural situation will prove to be rich in the above sense. The problem of predicting future states of a natural system (e.g. a weather prediction problem) can often be given the format of the prediction problem just discussed. One further example, the general question of stability (survival) of biochemical systems, may give some hint of the profusion and variety of rich environments. In the present context this becomes a question concerning embedded automata. The generators, connections, activation, templates, etc., become the abstract (and highly idealized) representatives of atoms, bonds, activation, enzymes, etc., in the biochemical system. Stability becomes preservation of structure in the face of spontaneous disconnection. Stability is a matter of degree; for an initial random mixture of unconnected generators some early stages of increasing stability are: (i) aggregates of generators formed by randomly occurring connections and disconnections; (ii) restricted aggregates with only selected generators being admitted to the system; (iii) self-modifying, restricted aggregates wherein structural changes in one part of the aggregate are induced by another part; (iv) catalysed aggregates where the rate of aggregation and modification is controlled by templates. Eventually, this sequence should lead to: (n) aggregates with supervisory control and self-reproduction. The more stable system, by its very nature, will have a higher expected density (in space-time) than its less stable companions. Therefore, trial formation of any new, potentially more stable system will incorporate previously formed systems with a probability related to their stability.

Mathematical characterization of classes of "rich" environments relative to a given class of adaptive systems constitutes one of the major questions in the study of adaptive systems.

The advantages pertaining to a rich environment suggest that an adaptive system could enhance its rate of adaptation by somehow enriching the environment. Such enrichment occurs if the adaptive system can generate subproblems or subgoals whose solution will contribute to the solution of the given problems of the environment (cf. Newell, Shaw, and Simon 1960; Minsky 1961). The simplest way to bring this idea into the present context is to allow supervisory programs to cause "auxiliary" checking routines to be formed. Since a checking routine is after all just a program, it is certainly possible for a supervisory program to bring about its formation. Such auxiliary checking routines constitute well-defined problems introduced by the supervisory program. This amounts to the introduction of a "law of effect": Each time a subgoal is achieved and then a given problem subsequently solved, the procedures for forming the relevant subgoal are rewarded by survival. In greater detail: A given supervisory program, under the stated conditions, will implicitly define both a population of problem-solving programs *and* a population of auxiliary checking routines. This mixed population can be expected to release activation from the given well-defined problems at some rate, $r(t)$. Some of this released activation, say $r'(t)$, will be tapped off to supply activation to the auxiliary checking routines. The remainder, $r(t) - r'(t)$, will be channeled to the supervisory program to be used eventually in its duplication. The net rate of accumulation of activation at the supervisory program, $r(t) - r'(t)$, determines its relative advantage under differential selection. For example: Let $r_0(t)$ be the rate of release of activation obtained by the problem-solving population of a supervisory program S_0 which generates no subgoal problems. Let $e_1(t)$

be the change in the rate of release obtained by a supervisory program S_1 which differs from S_0 only in that it generates some subgoal problems too. Let $r_1'(t)$ be the activation tapped off for the subgoal problems of S_1. Then S_1 will have a selective advantage over S_0 only if

$$\left[r_0(t) + e_1(t) - r_1'(t)\right]_{\text{ave}} > [r_0(t)]_{\text{ave}}.$$

In more intuitive language, subgoals confer a selective advantage only if they do not cost too much relative to the additional problem-solving capability they provide.

6. Comment

In the framework outlined, the problem of adaptation becomes one of modifying a free generation procedure in order to maximize activation release from the environment. However, the modification can only be carried out in terms of available information about the environment. Moreover, this information (unless it is given *a priori*) can only be obtained from trials or experiments performed on the environment. Thus the rate of adaptation is limited by the rate of information accrual. In fact, for certain classes of generation procedures, the maximum justifiable modification (skewing) of a free generation procedure has been determined as an explicit function of the accumulated information. (The proof of this and some related theorems will be published in another paper.) For such systems, it can also be shown that two apparently distinct criteria for determining the amount of skewing are identical: The first criterion selects that amount of skewing which, on the basis of available information, maximizes the expected activation release; the second criterion adjusts the skewing, on the basis of available information, according to a minimax estimate of the environment. In other words, for the systems described, the most opportunistic modification of the generation procedure is at one and the same time the most conservative—a somewhat surprising corollary.

In this and the next paragraph, Holland hints at an "efficiency" theorem for adaptive systems that sets a limit on the ability of the system to "fit more perfectly for existence in its environment" (1962, 33) tied to the flow of information it receives from the environment. The details and proof of such a theorem are relegated to a companion paper that was published a year later (Holland, 1962). This latter paper picks up on key themes developed here, and (near its end) introduces the notion of adaptive systems improving their behavior by gaining information from the exploration of new sources of rewards—this idea, now known as the explore–exploit trade-off, is an important area of ongoing research.

The theorem concerning maximum justifiable modification is in effect an "efficiency" theorem for adaptive systems . . . it sets a limit on adaptation in terms of information. In analogy with other "limiting" theorems (such as the channel capacity theorems) one would like to establish a companion "relizability" theorem. That is, one would like to exhibit a sequence of models with efficiencies which approach the efficient limit asymptotically. The iterative circuit computer models (section 4) were developed with this in mind, but there still remains a great deal to be done in the investigation of asymptotically efficient sequences of models.

The central purpose of this paper has been to suggest how questions concerning adaptation can be treated in a logical context. There are, however, the larger questions of whether the theorems within reach will be useful and whether one should attempt a theory at all (rather than contribute, say, to the growing body of heuristics concerned with adaptation). Von Neumann has summed up the situation quite clearly:

> "(The theory's) first applications are necessarily to elementary problems where the result has never been in doubt and no theory is actually required. At this early stage the application serves to corroborate the theory. The next stage develops when the theory is applied to somewhat more complicated situations in which it may already lead to a certain extent beyond the obvious and the familiar. Here theory and application corroborate each other mutually. Beyond this lies the field of real success: genuine prediction by theory." (von Neumann and Morgenstern 1944) ❦

Acknowledgments

I would like to thank Professor Arthur Burks and Dr. Jesse Wright for listening to a reading of the first draft of this paper and for making many helpful suggestions concerning the exposition. I would also like to thank Dr. David Willis and Dr. Richard Tanaka for the opportunity to begin the writing of this paper while at Lockheed Missile and Space Division on leave from the University of Michigan. Finally, I would like to thank students in the Program in Communication Sciences at the University of Michigan for their questions and discussion in a course where many of the ideas in this paper were first presented. This work has been supported at various stages by the National Science Foundation through grants G-4790 and G-11046 and currently by the U. S. Army Signal Corps through contract DA-36-039-sc-87174.

REFERENCES

Burks, A. W. 1960. "Computation, Behavior, and Structure in Fixed and Growing Automata." In *Self-Organizing Systems,* edited by M. C. Yovits and S. Cameron, 282–311. New York, NY: Pergamon Press.

Burks, A. W., and H. Wang. 1957. "The Logic of Automata." *Journal of the ACM* 4:193–218, 279–297.

Church, A. 1960. "Application of Recursive Arithmetic to the Problem of Circuit Synthesis." In *Summaries of Talks Presented at the Summer Institute for Symbolic Logic, Cornell University, 1957,* 2nd, 3–50. Princeton, NJ: Communications Research Division, Institute for Defense Analyses.

Craig, W. 1957. "Linear Reasoning. A New Form of the Herbrand-Gentzen Theorem." *Journal of Symbolic Logic* 28 (3): 250–268.

Fisher, R. A. 1958. *The Genetical Theory of Natural Selection.* Mineola, NY: Dover.

Gelernter, H. L., and N. Rochester. 1958. "Intelligent Behavior in Problem-Solving Machines." *IBM Journal of Research and Development* 2 (4): 336–345.

Holland, J. H. 1959. "A Universal Computer Capable of Executing an Arbitrary Number of Sub-Programs Simultaneously." In *Papers Presented at the December 1-3, 1959, Eastern Joint IRE-AIEE-ACM Computer Conference,* 108–113. New York, NY: Association for Computing Machinery.

———. 1960. "Iterative Circuit Computers." In *Papers Presented at the May 3-5, 1960, Western Joint IRE-AIEE-ACM Computer Conference,* 259–265. New York, NY: Association for Computing Machinery.

Jakowatz, C. V., R. L. Shuey, and G. M. White. 1960. *Adaptive Waveform Recognition.* Technical report. Report 60-RL-2435 E. GE Research Lab.

Lyndon, R. C. 1959. "An Interpolation Theorem in the Predicate Calculus." *Pacific Journal of Mathematics* 9 (1): 129–142.

Minsky, M. 1961. "Steps Toward Artificial Intelligence." In *Proceedings of the IRE,* 49:8–30. 1.

Moore, E. F. 1959. *Machine Models of Self-Reproduction.* Paper presented at meeting of American Mathematical Society, Cambridge, MA, October, 1959.

Newell, A. 1960. "On Programming a Highly Parallel Machine to be an Intelligent Technician." In *Papers Presented at the May 3-5, 1960, Western Joint IRE-AIEE-ACM Computer Conference,* 267–282. New York, NY: Association for Computing Machinery.

Newell, A., J. C. Shaw, and H. A. Simon. 1960. "A Variety of Intelligent Learning in a General Problem Solver." In *Self-Organizing Systems,* edited by M. C. Yovits and S. Cameron. New York, NY: Pergamon Press.

Rabin, M. O., and D. Scott. 1959. "Finite Automata and their Decision Problems." *IBM Journal of Research and Development* 3 (2): 114–125.

Samuel, A. L. 1959. "Some Studies in Machine Learning, Using the Game of Checkers." *IBM Journal of Research and Development* 3 (3): 210–229.

Selfridge, O. G. 1959. "Pandemonium, a Paradigm for Learning." In *Mechanization of Thought Processes: Proceedings of a Symposium held at the National Physical Laboratory No. 10.* London, UK: Her Majesty's Stationery Office.

Turing, A. M. 1936. "On Computable Numbers, with an Application to the Entscheidunsproblem." *Proceedings of the London Mathematical Society* s2-43 (1): 230–265.

———. 1950. "Computing Machinery and Intelligence." *Mind* 59:433–460.

———. 1952. "The Chemical Basis of Morphogenesis." *Philosophical Transactions of the Royal Society B* 237 (641): 37–72.

von Neumann, J. 1951. "The General and Logical Theory of Automata." In *Cerebral Mechanisms in Behavior—the Hixon Symposium,* edited by L. A. Jeffress, 1–41. New York, NY: Wiley.

———. 1962. *The Theory of Automata. Construction, Reproduction, Homogeneity.* Unpublished manuscript.

von Neumann, J., and O. Morgenstern. 1944. *Theory of Games and Economic Behavior.* Princeton, NJ: Princeton University Press.

Wright, S. 1960. "Physiological Genetics, Ecology of Populations, and Natural Selection." In *The Evolution of Life.* Chicago, IL: Chicago.

BIOGRAPHICAL ODDITIES, VOLUME 1

ALFRED JAMES LOTKA | March 2, 1880–Dec. 5, 1949 | *c.f. Chapter 01*

A founding member of the Population Association of America (PAA) and its president 1938-39. Founded initially as a coalition of population scientists, birth control activists, immigration restrictionists, and eugenicists.

In addition to his work in population dynamics, known for "Lotka's law" (a variation on Zipf's law which relates to the productivity of scientists), thereby contributing to the groundwork for the field of scientometrics.

LEO SZILÁRD (BORN SPITZ) | Feb. 11, 1898–May 30, 1964 | *c.f. Chapter 02*

Encouraged everyone he knew to flee Germany after Hitler's appointment as chancellor. He was a cofounder of the Academic Assistance Council to get scholars out of dangerous zones in Europe; by the start of WWII they had gotten over 2,500 people out successfully.

Founded Council for a Livable World in 1962 to deliver "the sweet voice of reason" about nuclear weapons to Congress and the White House.

"If the uranium project could have been run on ideas alone, no one but Leo Szilárd would have been needed."
—EUGENE WIGNER

Do your work for six years; but in the seventh, go into solitude or among strangers so that the memory of your friends does not hinder you from being what you have become.
— from Szilárd's Ten Commandments

SEWALL GREEN WRIGHT | Dec. 21, 1889–March 3, 1988 | *c.f. Chapter 03*

In 1897, at the age of seven, Wright wrote his first "book," entitled "Wonders of Nature;" he published his last paper in 1988 and could therefore be considered the scientist with the longest science-writing career.

Interestingly, in light of Wright's major work on inbreeding: his parents were first cousins.

Wright was known for writing on chalkboards and erasing in a hurried frenzy. Legend has it that, once while lecturing, he supposedly tried to erase the board using as an eraser one of the guinea pigs he had been observing. (Wright always denied this.)

CONRAD HAL WADDINGTON | Nov. 8 1905–Sept. 26, 1975 | *c.f. Chapter 04*

Scientific advisor to the Royal Air Force Commander in Chief of Coastal Command during WWII.

Known as "Wad" to his friends and "Con" to family.

Waddington was a poet; according to his friend Alan Robertson, he thought of Ludwig Wittgenstein as really a poet made to act like a philosopher.

WARREN STURGIS MCCULLOCH | Nov. 16, 1898–Sept. 24, 1969 | *c.f. Chapter 05*

McCulloch published sonnets, enjoyed a good whiskey, and hosted big parties at his farm in Connecticut, where he often encouraged skinny dipping.

McCulloch and Pitts were hospitalized at the same time in April of 1969, in two separate hospitals, across the street from one another. They wrote to each other from their neighboring hospital beds. That same year, both of them passed away, four months apart from one another, as a result of the ailments that originally hospitalized them.

WALTER HARRY PITTS, JR. | April 23, 1923–May 14, 1969 | *c.f. Chapter 05*

At age twelve, Pitts wrote to Bertrand Russell with ideas on how to improve *Principia Mathematica*. Russell, impressed, offered preteen Pitts the opportunity to study with him at Cambridge, which Pitts declined.

Pitts never really attended the University of Chicago; he audited lectures, beginning at age fifteen, when he was an unsheltered runaway. McCulloch brought Pitts and Jerome Lettvin home to live with him and his family when Pitts was eighteen.

Pitts never wanted anyone to know his full name and refused to sign employment contracts for this reason. However, he eventually needed a passport, and for that McCulloch procured his birth certificate for him, so now his full name is public knowledge.

ARTURO ROSENBLUETH | Oct. 2, 1900–Sept. 20, 1970 | *c.f. Chapters 06, 08*

Known for his research and as a pioneer in cybernetics, but was also a physician.

One of eight siblings, all of whom were instructed in some form of art or music, Arturo played the piano. As a student, he supported himself by playing music in restaurants and accompanying silent films in local movie houses.

"The best material model for a cat is another [cat], or preferably the same cat." —Arturo Rosenblueth from *The Role of Models in Science* with Norbert Wiener

NORBERT WIENER | Nov. 26 1894–March 18, 1964 | *c.f. Chapters 06, 08*

A child prodigy, he graduated from high school at age 11 and received his BA in mathematics from Tufts at 14, his MA in philosophy from Cornell at 17, and a PhD from Harvard at 19, with a dissertation in mathematical logic.

When asked in 1952 letter to write an intro to a sci-fi anthology, Wiener complaind that science fiction had lost its way: "I am tired of the space-man cliché. I am tired of the mastery-of-the-robots cliche. I am tired of the working to death of psychoanalytic ideas. And, frankly, I am just plain tired."

In 1952, he submitted to Alfred Hitchcock a "horror and suspense" screenplay which he had cowritten with his daughter, Peggy, and Morris Chafetz.

JULIAN HIMLEY BIGELOW | March 19, 1913–Feb. 17, 2003 | *c.f. Chapter 06*

Worked as chief engineer for John von Neumann to design and build a stored-program computer at the Institute for Advanced Study in Princeton after Norbert Weiner recommended him. Bigelow had collaborated with Wiener during WWII on the creation of fire-control systems for weapons.

George Dyson said "In a way, Julian was the missing link," in that, throughout his career, Bigelow connected the work of theoreticians like Weiner and von Neumann to the real world.

He was a pilot, flying his own plane regularly into his 80s, at which time he restored a plane of his own as a hobby.

FRIEDRICH AUGUST VON HAYEK | May 8, 1899–March 23, 1992 | *c.f. Chapter 07*

Famous mathematical philosopher and logician Ludwig Wittgenstein was Hayek's second cousin.

Though often credited with predicting the 1929 stock market crash, Hayek actually predicted the opposite in a German journal just three days before the crash, saying the market would surely stabilize and that there was nothing to fear.

Hayek was a proponent of environmental protection efforts around the globe, and he authorized his name for use by numerous conservation organizations after he became a Nobel laureate.

CLAUDE ELWOOD SHANNON | April 30, 1916–Feb. 24, 2001 | *c.f. Chapter 09*

Shannon's hobbies included chess, juggling, and unicycling.

In 1950, Shannon created a mechanical mouse-like device, which he named Theseus, designed to solve a maze. On a relay circuit, the robot was capable of finding a target in the maze and could return to the target from memory when removed and placed back in the maze again. It is considered one of the earliest artificial-learning devices ever made.

He invented the world's first wearable computer. Its purpose: to beat a roulette wheel.

WARREN WEAVER | July 17, 1894–Nov. 24, 1978 | *c.f. Chapter 10*

During World War II, Weaver led an initiative funded by the Rockefeller Foundation to collect current copies of every American scholarly journal and store them all for later distribution to libraries at universities in war-torn areas. In all, nearly 5,000 libraries received crates of up-to-date collections of journals when the war was over, to help bolster their ravaged collections.

A devoted Lewis Carroll fan, Weaver published *Alice in Many Tongues: The Translation of Alice in Wonderland* in 1964, a translation history of *Alice's Adventures in Wonderland*, which included 160 versions of the story in 42 different languages.

ALAN MATHISON TURING | June 23, 1912–June 7, 1954 | *c.f. Chapters 11, 13*

In the early 1940s, fearing a German invasion, Turing spent most of his savings on two silver ingots, weighing 90 pounds each. He buried them in two separate places in the woods behind Bletchley Park and made a cipher to help him find them later. When the war ended, Turing tried to find them, but his cipher proved difficult to break, even for him. Once deciphered, the clues were almost useless because all the landmarks had changed in the intervening decade. Turing never found his treasure. As far as anyone knows, it is still out there!

Turing was a world-class distance runner, who missed qualifying for the marathon event in the 1948 Olympics due to an injury at the time of his tryout.

JOHN FORBES NASH, JR. | June 13, 1928–May 23, 2015 | *c.f. Chapter 12*

In 1959, Nash was hospitalized and his schizophrenia was treated with insulin shock therapy, a now-debunked pseudo-scientific practice of injecting patients with insulin to induce comas. Despite modern evidence to the contrary, Nash felt that his condition noticeably improved for a short time after the treatment.

Sylvia Nasar, Nash's biographer, tried to interview him for three years for the biography *A Beautiful Mind*, but Nash did not want to be involved, and told her he would prefer to maintain "Swiss neutrality" towards the project. However, when the book was being made into a feature film, Nash was supportive of the project because it allowed him to set aside something of an estate for his two sons, both named John.

EDWIN THOMPSON JAYNES | July 5, 1922–April 30, 1998 | *c.f. Chapter 14*

During World War II, Jaynes worked on microwave theory and applications at the Navy Research Laboratory in Washington, DC.

Jaynes was a talented pianist. During his time at Washington University, he would host a weekly gathering of students and faculty at his home, often inviting them to bring instruments so that they could gather around his piano and play together.

Often rejected by mainstream physics journals, Jaynes published many of his papers in lab journals and conference proceedings volumes instead.

RUDOLF EMIL KÁLMÁN | May 19, 1930–July 2, 2016 | *c.f. Chapter 15*

Kálmán loved music. He did extensive research on turntable needles and stored his record collection in a special temperature-controlled room.

The Isaac Newton quote *hypotheses non fingo*, "I frame no hypotheses," was a favorite of Kálman's.

In 1977 Kálmán organized a five-day conference on a sailing yacht in the Caribbean. Lectures were given each day while the yacht sailed from island to island, and lecturers often had to contend with their papers blowing away while speaking above-deck.

ROLF WILLIAM LANDAUER | Feb. 4, 1927–April 27, 1999 | *c.f. Chapter 16*

Landauer's family fled Germany for New York in 1938 when Landauer was eleven years old.

Landauer was a critic of quantum computation, and he once devised a disclaimer that he asked colleagues to use in their papers on the subject: "this scheme, like all other schemes for quantum computation, relies on speculative technology, does not in its current form take into account all possible sources of noise, unreliability and manufacturing error, and probably will not work."

He enjoyed rowing and skiing, but did not enjoy spectator sports.

MARVIN LEE MINSKY | Aug. 9, 1927– Jan. 24, 2016 | *c.f. Chapter 17*

Minsky created a "gravity machine" that would ring a bell in the event that the Earth's gravitational constant were ever to change.

He was a member of the scientific advisory board for Alcor Life Extension Foundation, a cryonics research non-profit. Upon his death, Alcor released a statement saying they could neither confirm nor deny that Minsky had been cryogenically preserved.

Minsky was a scientific adviser for Stanley Kubrick's *2001: A Space Odyssey*, and he was also a skilled pianist.

KENNETH JOSEPH ARROW | Aug. 23, 1921–Feb. 21, 2017 | *c.f. Chapter 18*

When Arrow was a child in New York, his parents would often go without buying meat for the family so that they could afford the 10-cent subway fare needed for Arrow to attend high school in Queens.

Arrow was a true polymath. Colleagues of Arrow's, including fellow Nobel laureate Eric Maskin, once set up a test to see if they could find a subject Arrow knew nothing about. They began debating about the breeding and migration patterns of gray whales in front of Arrow, even discussing a migration theory put forth by a particular scholar. Arrow chimed in: not only had he read the paper in question, he knew for a fact that the theory they were discussing had since been debunked

MARIO AUGUSTO BUNGE | Sept. 21, 1919–Feb. 24, 2020 | *c.f. Chapter 19*

Bunge was born into a political family in Buenos Aires. His mother said something off-color about the local government one day in public, and the next morning someone shot at her through the family's kitchen window (and missed). Bunge was arrested twice in his early adulthood, and his career stalled as a result of his anti-Peronist sympathies. He left Argentina in 1963.

Bunge considered psychoanalysis to be a form of pseudoscience and was very vocal about this opinion.

Bunge felt that the key to a successful career and a happy life was to be passionate about one's work. He lived to be a centenarian.

JOHN HENRY HOLLAND | Feb. 2, 1929–Aug. 9, 2015 | *c.f. Chapter 20*

Holland was the first person to earn a PhD in computer science, then dubbed communications science, at the University of Michigan. He would go on to be one of the first professors in what was then Michigan's communications science division.

According to Holland, he came up with most of his best ideas while he was in the shower.

Holland gave the Santa Fe Institute's inaugural Stanisław Ulam lecture in 1994.

EDITOR

DAVID C. KRAKAUER is the President and William H. Miller Professor of Complex Systems at the Santa Fe Institute. His research explores the evolution of intelligence and stupidity on Earth. This includes studying the evolution of genetic, neural, linguistic, social, and cultural mechanisms supporting memory and information processing, and exploring their shared properties. He served as the founding director of the Wisconsin Institutes for Discovery, codirector of the Center for Complexity and Collective Computation, and professor of mathematical genetics, all at the University of Wisconsin, Madison. He has been a visiting fellow at the Genomics Frontiers Institute at the University of Pennsylvania, a Sage Fellow at the Sage Center for the Study of the Mind at the University of California, Santa Barbara, a longterm fellow of the Institute for Advanced Study, and visiting professor of evolution at Princeton University. In 2012, he was included in the *Wired Magazine* Smart List: Fifty People Who Will Change the World. In 2016, he was included in *Entrepreneur Magazine*'s list of visionary leaders advancing global research and business.

Krakauer was previously chair of faculty and a resident professor and external professor at the Santa Fe Institute. A graduate of the University of London where he earned degrees in biology and computer science, he received his D.Phil. in evolutionary theory from Oxford University in 1995 and continued there as a postdoctoral fellow.

THE SANTA FE INSTITUTE PRESS

The SFI Press endeavors to communicate the best of complexity science and to capture a sense of the diversity, range, breadth, excitement, and ambition of research at the Santa Fe Institute; To provide a distillation of work at the frontiers of complex-systems science across a range of influential and nascent topics;

To change the way we think.

SEMINAR SERIES

New findings emerging from the Institute's ongoing working groups and research projects, for an audience of interdisciplinary scholars and practitioners.

ARCHIVE SERIES

Fresh editions of classic texts from the complexity canon, spanning SFI's four decades of advancing the field.

COMPASS SERIES

Provocative, exploratory volumes aiming to build complexity literacy in the humanities, industry, and the curious public.

SCHOLARS SERIES

Texts featuring foundational ideas, systems of knowledge, emerging methodologies, and areas of application in the complex-systems science world.

— Also from SFI Press —

Foundational Papers in Complexity Science, Volumes 2, 3 & 4
David C. Krakauer, ed.

Worlds Hidden in Plain Sight: The Evolving Idea of Complexity at the Santa Fe Institute, 1984–2019
David C. Krakauer, ed.

For additional titles, inquiries, or news about the Press, visit us at

WWW.SFIPRESS.ORG.

ABOUT THE SANTA FE INSTITUTE

The Santa Fe Institute is the world headquarters for complexity science, operated as an independent, nonprofit research and education center located in Santa Fe, New Mexico. Our researchers endeavor to understand and unify the underlying, shared patterns in complex physical, biological, social, cultural, technological, and even possible astrobiological worlds. Our global research network of scholars spans borders, departments, and disciplines, bringing together curious minds steeped in rigorous logical, mathematical, and computational reasoning. As we reveal the unseen mechanisms and processes that shape these evolving worlds, we seek to use this understanding to promote the well-being of humankind and of life on Earth.

COLOPHON

The body copy for this book was set in EB Garamond, a typeface designed by Georg Duffner after the Ebenolff-Berner type specimen of 1592. Headings are in Kurier, created by Janusz M. Nowacki, based on typefaces by the Polish typographer Małgorzata Budyta, and Cochin, a typeface produced in 1912 by Georges Peignot and based on the copperplate engravings of French 17th century artist Nicolas Cochin, for whom the typeface is named. For footnotes and captions, we have used CMU Bright, a sans serif variant of Computer Modern, created by Donald Knuth for use in TeX, the typesetting program he developed in 1978.

The SFI Press complexity glyphs used throughout this book were designed by Brian Crandall Williams.

SANTA FE INSTITUTE

COMPLEXITY GLYPHS

ZERO

ONE

TWO

THREE

FOUR

FIVE

SIX

SEVEN

EIGHT

NINE

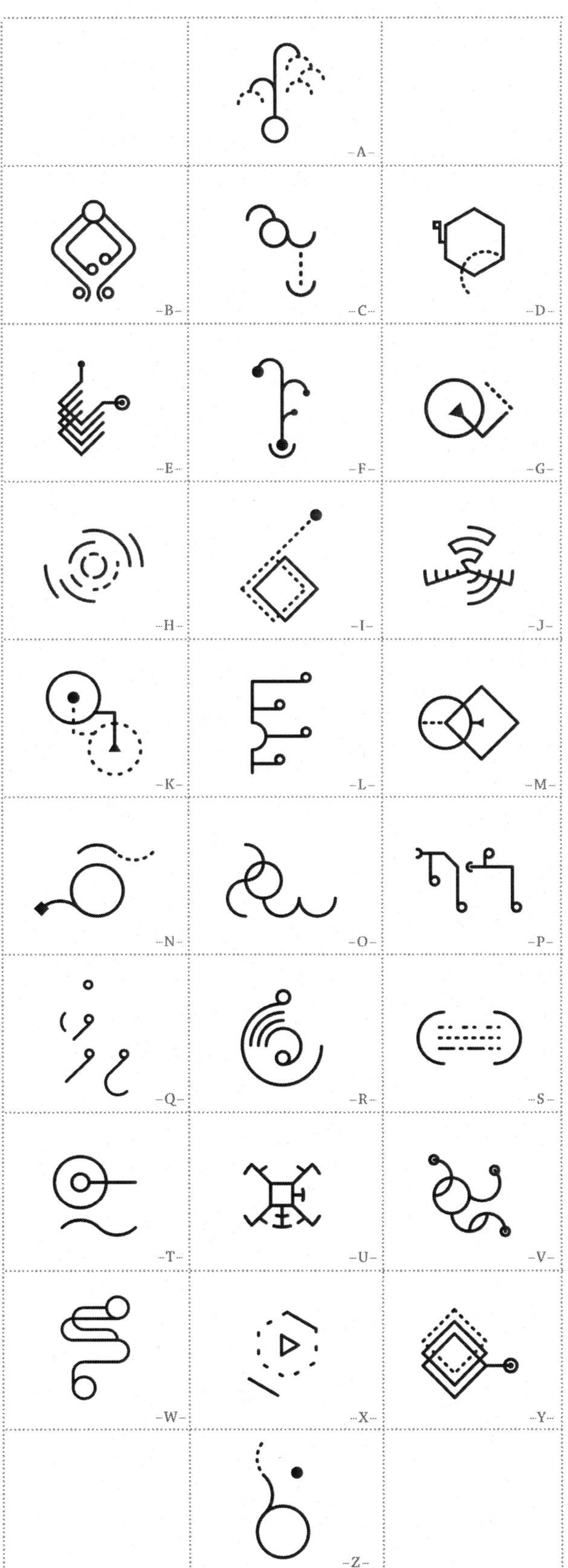

SFI PRESS
SCHOLARS SERIES

Made in United States
North Haven, CT
12 May 2024